DEVELOPMENTAL BIOLOGY OF TELEOST FISHES

Developmental Biology of Teleost Fishes

by

Yvette W. Kunz
Ph.D., D.Sc.
University College Dublin,
National University of Ireland,
Dublin

Springer

A C.I.P. Catalogue record for this book is available from the Library of Congress.

ISBN 1-4020-2996-9 (PB)
ISBN 1-4020-2995-0 (HB)
ISBN 1-4020-2997-7 (e-book)

Published by Springer,
P.O. Box 17, 3300 AA Dordrecht, The Netherlands.

Sold and distributed in North, Central and South America
by Springer,
101 Philip Drive, Norwell, MA 02061, U.S.A.

In all other countries, sold and distributed
by Springer,
P.O. Box 322, 3300 AH Dordrecht, The Netherlands.

Printed on acid-free paper

springeronline.com

All Rights Reserved
© 2004 Springer
No part of this work may be reproduced, stored in a retrieval system, or transmitted
in any form or by any means, electronic, mechanical, photocopying, microfilming, recording
or otherwise, without written permission from the Publisher, with the exception
of any material supplied specifically for the purpose of being entered
and executed on a computer system, for exclusive use by the purchaser of the work.

Printed in the Netherlands.

To William John, Oengus Niall and Ciaran Oisín

> Ag Críost an mhuir;
> ag Críost an t-iasc.
> I líonta Dé
> go gcastar sinn.
>
> (Gaelic)

Scientific illustrator: Paula DiSanto Bensadoun

Contents

Series Editor's Preface   xiii-xiv
David L. G. Noakes

Preface   xv-xvi
Yvette W. Kunz

1	Introduction		1
	Summary		7
	Endnotes		8
2	The egg		9
	2.1	Types of eggs	10
	2.1.1	Pelagic (planktonic) eggs	10
	2.1.2	Demersal eggs	12
	2.1.3	Demersal/pelagic eggs	13
	2.2	Egg care	14
	2.2.1	Guarding	14
	2.2.2	External bearers	15
	2.2.3	Internal bearers (mouthbrooders)	16
	2.2.4	Ovoviviparous and viviparous fish	16
	2.3	Shape of eggs	17
	2.4	Number of eggs	18
	2.5	Size of eggs	19
	2.6	Deposition of eggs	20
	Summary		21
	Endnotes		21

3.	Yolk (vitellus)		23
	3.1	Lipid yolk	24
	3.2	Proteid yolk	26
	3.3	Yolk formation (vitellogenesis)	27
	3.3.1	Genesis of lipid (fatty) yolk	29
	3.3.2	Precursors of proteid yolk	30
	3.4	Relation between yolk and cytoplasm	36
	3.5	Hydration of the oocyte during maturation	38
	Summary		39
	Endnotes		40
4.	Cortex and its alveoli		41
	4.1 Cortex		41
	4.2 Cortical alveoli (CA) and cortical granules		42
	Summary		46
	Endnotes		47
5.	Egg envelope		49
	5.1	Thickness of envelope	49
	5.2	Number of layers in the envelope	50
	5.3	Oolemma	51
	5.4	Zona radiata (ZR)	52
	5.4.1	Structure	52
	5.4.2	Ontogeny of radial canals	58
	5.4.3	Occlusion or disappearance of radial canals	60
	5.4.4	Hardening of the zona radiata	61
	5.4.5	Zona radiata externa and interna	62
	5.4.6	Variation in number of zona radiata layers	67
	5.4.7	Chemical composition of the zona radiata	69
	5.4.8	Origin of the zona radiata	70
	Summary		75
	Endnotes		76
6.	Accessory structures of egg envelope		77
	6.1	Pelagic-nonadhesive	78
	6.2	Demersal-nonadhesive	83

	6.3	Demersal-adhesive	86
	6.4	Special structures for floating, attachment etc.	103
	6.5	Mouthbrooders (oral incubators)	106
	6.6	Ovoviviparous- and viviparous fish (livebearing fish) (dealt with in Chapter 20)	107
	Summary	107

7.	Micropyle		109
	7.1	Types of micropyles	109
	7.2	Discovery of the micropyle	110
	7.3	The micropyle of Gasterosteidae	111
	7.4	The micropyle of Salmonidae	112
	7.5	The micropyle of Esocidae	113
	7.6	The micropyle of Percidae	114
	7.7	The micropyle of other species	114
	7.8	Ultrastructure of the micropyle and its surroundings	114
	7.9	Formation of the micropyle	124
	Summary	129
	Endnote	129

8.	Sperm		131
	8.1	Early investigators	131
	8.2	EM-analyses	134
	Summary	146
	Endnotes	146

9.	Fertilization		147
	9.1	Activation	148
	9.2	Formation of the perivitelline space	148
	9.3	Bipolar differentiation	151
	Summary	154
	9.4	Motility of sperm	155
	9.5	Fertilizing capacity of sperm	157
	9.6	Sperm entry through micropyle	158
	9.7	Fusion of sperm and egg membranes	159
	9.8	Internalization of sperm	161

	9.9	Block of polyspermy	162
	Summary		166
	9.10	Cortical reaction following fertilization	166
	9.11	The perivitelline space following fertilization	172
	9.12	Hardening of the envelope	175
	Summary		176
	9.13	Bipolar distribution (ooplasmic segregation) following fertilization	177
	Summary		180
	9.14	Formation of pronuclei	180
	9.15	Formation of polar bodies	181
	Summary		183
	Endnotes		183
10	Cleavage and formation of periblast		185
	10.1	Cleavage	188
	10.2	Periblast (YSL)	198
	10.2.1	DNA of periblast nuclei	204
	10.2.2	Functional significance of the periblast	204
	Summary		204
	Endnotes		205
11	Gastrulation		207
	11.1	Epiboly	208
	11.2	Involution	214
	11.3	Formation of the yolksac (epiboly and involution)	231
	11.4	Tailbud	232
	11.5	Invagination	233
	Summary		234
	Endnotes		236
12.	Neurulation		239
	12.1	Early investigators	240
	12.1.1	Neural fold	240
	12.1.2	Solid CNS	241
	12.1.3	Closed neural fold	241

	12.2	More recent investigations	244
	12.3	Other midline structures	248
	Summary		249
	Endnotes		250
13.	Fate-maps		251
	13.1	Teleostean fate-maps	251
	Summary		257
	Endnotes		258
14.	Kupffer's vesicle		259
	Summary		265
	Endnotes		266
15.	Ectodermal derivatives		267
	15.1	Neuroectoderm	267
	15.1.1 Brain (encephalon)		268
	Summary		270
	(a) Epiphysis (glandula pinealis) in the adult		271
	(b) Epiphysis in the embryo		272
	Summary		274
	15.1.2 Neuromast organs		274
	Summary		276
	15.1.3 Neural crest		276
	Summary		276
	15.2	Integument	277
	15.2.1 EVL or periderm		277
	15.2.2 Epidermis		278
	(a) Headfold		278
	(b) Finfold		278
	(c) Chloride cells		280
	(d) Pseudobranch		281
	(e) Adhesive organs		285
	Summary		288
	Endnotes		291

Contents

16.	Hatching	293
	16.1 Distribution of unicellular hatching glands (UHG)	294
	16.2 Histological structure of UHG	295
	16.3 Ultrastructure of UHG	296
	16.4 Ultrastructural development of UHG	296
	16.5 Secretions by UHG and their death by apoptosis	298
	16.6 Triggering of UHG secretion	300
	16.7 Hatching enzyme	300
	Summary	302
	Endnotes	302
17.	Development of the eye	303
	17.1 Eye of sexually mature teleost	303
	17.1.1 Bruch's membrane (BM)	311
	17.1.2 Retinal pigment epithelium (RPE)	311
	17.1.3 Photoreceptors (visual cells)	311
	(a) Ultrastructure	312
	(b) Visual pigments	318
	(c) Retinomotor movements (photomechanical changes)	326
	(d) Shedding and renewal of photoreceptor outer segments	326
	17.1.4 Outer nuclear layer and outer plexiform layer	326
	17.1.5 Inner nuclear layer	329
	17.1.6 Inner plexiform layer	330
	17.1.7 Ganglion cell layer	330
	Summary	330
	17.2 Cone mosaics in salmonids	332
	17.3 Developing teleost eye	337
	17.3.1 Eye primordium to optic cup	337
	17.3.2 Development of Bruch's membrane (BM)	339
	17.3.3 Development of the retina	339
	17.3.4 Developmental sequence in the retina of salmonids and the guppy	341
	(a) Retinal pigment epithelium (RPE)	341
	(b) Visual cell layer (VCL) including outer nuclear layer (ONL)	344
	(c) Outer plexiform layer (OPL) of the trout	353
	(d) Inner nuclear layer (INL) of the trout	357

Contents

(e) Müller cell of the guppy	357
(f) Inner plexiform layer (IPL) of the trout	357
Summary	359
17.3.5 Developmental sequence of the zebrafish eye and retina	360
(a) Retinal pigment epithelium (RPE)	361
(b) Visual cell layer (VCL) and outer nuclear layer (ONL)	365
(c) Outer plexiform layer (OPL)	370
(d) Inner nuclear layer (INL)	370
(e) Inner plexiform layer (IPL)	370
(f) Ganglion cell layer (GCL)	371
Summary	371
17.4 Development of the lactate dehydrogenase (LDH) pattern in the teleost eye	372
Summary	374
Endnotes	375
18. Mesodermal derivatives	**377**
18.1 Kidney	378
Summary	378
18.2 Haematopoietic sites (for red blood cells and vascular endothelia)	381
18.2.1 First haematopoietic sites	382
(a) Surface of the yolk	382
(b) Endocardium	384
(c) Intermediary cell mass (ICM)	385
(d) Blood islands	394
(e) Intermediary cell mass and blood islands	398
(f) Extended blood mesoderm	403
Summary	406
18.2.2 Secondary haematopoietic sites in the embryo (endocardium and kidney)	410
Summary	414
18.3 Erythropoietic stages in the developing pike, *Esox lucius*	414
18.4 Haemoglobin	415
18.5 Heart	415
18.5.1 Heart of sexually mature teleosts	415
18.5.2 Development of the teleostean heart	415

	Summary	420
	18.6 Blood circulation	421
	18.6.1 Sexually mature fish	421
	18.6.2 Yolksac circulation	421
	Summary	428
	Endnotes	428
19.	Entoderm and its derivatives	431
	19.1 Formation of the pharynx	435
	19.2 Formation of the thyroid	437
	19.3 Formation of the intestine	438
	19.4 Postanal gut	439
	Summary	441
	19.5 Formation of liver	441
	Summary	442
	19.6 Liver/yolksac contact	442
	Summary	452
	Endnotes	455
20.	Viviparity	457
	20.1 Matrotrophic species	460
	20.2 Lecithotrophic species	493
	Summary	502
	Endnotes	503
21.	Synthesis	505
	Endnote	510
References		511
Species Index		591
Subject Index		603

Series Editor's Preface

This book is a monumental accomplishment. The coverage ranges across 2 millennia, from Aristotle to zebrafish. Yvette Kunz Ramsay is one of a very few scientists who could write such a book. She has the first hand knowledge, including fluency in languages, with the European literature. She has a life-long dedication to detailed studies of developmental biology in a number of fish species. Her volume is remarkable for its scholarly tone, encyclopaedic coverage of the subject matter, and broad synthesis of the material. There is a wealth of information on the ontogeny of a great diversity of species. A large body of that information relates to questions of ontogeny and phylogeny, with implications as broad as systematics and gene regulation.

The study of teleost fishes has long been a significant component of vertebrate developmental biology. Along with chick embryos and fetal pigs, we studied whitefish blastulas as part of our formal course work in embryology laboratories. The book has a special personal resonance for me in this regard. One of the first researchers to publish on the developmental biology of the zebrafish, Danio rerio, was Helen Battle, the professor in my undergraduate embryology course. As a postgraduate student teaching assistant I was assigned to a course in vertebrate embryology taught by Richard Eakin, who had studied in the laboratory of Professor Hans Spemann in Germany. Spemann was awarded the Nobel Prize in 1935 for his research on embryonic induction. The influence of that experience, and of the classical developmental anatomy from Europe, was a continuing influence on Eakin's teaching and research, as it was for many others in North America of that generation. Eakin was justly famous for his impersonations of famous scientists, including Spemann, in his undergraduate lectures.

Routine descriptions of early morphological and anatomical development abound for various freshwater fishes. The descriptive literature on marine and brackish water species is even more extensive. Studies of early development of marine fishes have been typified by a need to identify species of commercial interest, such as the clupeids, gadids and anguillids. Regulations for harvest, or conservation, of designated species often depend upon identification and enumeration of fishes before they reach adult size or age. Another reason for practical applications of developmental morphology of fishes has been the broad area of environmental impact assessment. Many jurisdictions have regulations that require the identification and enumeration of species of concern in this regard. There is a continuing need to identify these species at any time during their ontogeny, from fertilized eggs to adults.

The practical demands for identification of fish species have had a lasting effect on the study of the ontogeny of many species. Handbooks of practical keys for identification of marine and freshwater species have been produced and continue to be used for those purposes. The practicality of this approach has also resulted in the curiously idiosyncratic fixation on defining and naming "stages" of fish ontogeny. This has produced a strange assortment of terms, including eleutheroembryo, fry, protopterygiolarva, preflexion larva, and yolksac larva, among others. The situation is made even more complicated by the apparent inability of those involved with this nomenclature to agree on the terms, or even the application of them to specific examples of a given species.

It is clear that the rationale for these descriptive studies has often been a matter of availability and ease of observations on the species in question. The commercial importance of some species favoured them as study species. The tradition of those detailed studies of developmental morphology and anatomy has all but disappeared. The proximate cause is the absence of journals or monographs prepared to publish the results of simple descriptions of developmental morphology of yet another species. The ultimate cause of course, has to do with changing norms in science, and a focus on experimental studies and testing of hypotheses. Some have viewed the large number of teleost species as a virtually unlimited pool for descriptions of developmental stages. If nothing else, they could be the fodder for generations of uninspired graduate students and technicians. As for any area of science, simple description for its own sake is soon replaced by synthesis and hypothesis testing in focused experimental studies.

Now we are in the midst of a new era in studies of vertebrate developmental biology, with a major emphasis on the zebrafish. Our understanding has been completely revolutionized as a result of this change in our approach to studies of the developmental biology of fishes. The rapidly expanding literature on developmental biology of zebrafish is readily and widely available through electronic access. Specific references, or even entire bibliographies, can be found and downloaded with little effort, or much understanding of the process or the product. We can screen for developmental mutants or verify gene sequences for zebrafish by means of electronic access to online databases.

Two things are now desperately needed to continue to extend our understanding of developmental biology of fishes. One is the transfer of information and knowledge from the zebrafish model to all the other teleost species. The consequences of that will be truly revolutionary for fish biology. The other, equally important need, is to disseminate the comparative information from years of study of the other teleost species to those working on the zebrafish model system. This volume is uniquely qualified to serve both these important needs.

Dr. David L. G. Noakes
Editor, Kluwer Fish & Fisheries Series
Professor of Zoology and Director, Axelrod Institute of Ichthyology
University of Guelph

Preface

'Unlike centenarians who are reaching the end of life, developmental biology is basking in its full-blown prime. Indeed the excitement and promise of the field have never been greater, as researchers close in on the secret of how a single fertilized egg cell goes through the complex and beautifully orchestrated series of changes that create an entire organism '(Baringa, 1994)'.

How I wish I were at the beginning of a career, now that teleost fish finally have entered the modern era of developmental biology! And how much I agree with the sentiments expressed by 'Metscher and Ahlberg (1999)' that 'late 20th century biology [by now early 21st century] requires a reincorporation of developmental, morphological and phylogenetic levels of description if it is to continue to develop our understanding of organismic systems.'

Cellular, subcellular, genetic and molecular aspects have been analyzed for several, largely unrelated, species of teleosts, such as *Heterandria, Fundulus, Oryzias*. In addition, ecology, behaviour, epigenetics and phylogeny have been used to study interdisciplinary relationships. At present much of the emphasis centres on the zebrafish, which is often referred to as 'the new *drosophila*' or 'the *drosophila* with a spine'. The popularity of the zebrafish is probably due to the fact that the embryos are easy to raise, are transparent and readily observed under a dissecting microscope; moreover, they develop rapidly and females lay hundreds of eggs at weekly intervals.

However, for the analyses, by any modern method, of developing teleosts, histological knowledge of the different stages and their organs is a *sine qua non* basis. The science of embryology (now called developmental biology) has a solid basis in the meticulous studies of late nineteenth and early-to-middle twentieth century scholars in Europe at a time when English was not yet the international language. To-date, these early studies have never been translated. This problem of language was brought home to me, when, as a Swiss graduate in Cornell, I was asked by Europeans at an international conference on cheese to translate the summaries of their papers into English and likewise was approached by an American professor to translate a German book on insects into English. Moreover, when I started to lecture in Ireland on developmental biology I realized that the main difficulty for my undergraduates was the lack of a textbook of developing fish in English.

The purpose of this book is to make the early work and the up-to-now continued morphological and histological research results available to modern 'molecular researchers' and to assist graduates in general to compare teleost development with that of *amphioxus*, amphibians, birds and mammals.

In reading the old publications in the original language I found the wealth of information was much greater than anticipated. I also noted how these early researchers were, even violently, disagreeing with each other — and this at a time when research was not dependent on grant-money. Though promotion in rank was not achieved by an accumulation of results, they often quoted results from other authors out of context in order to suit their purpose, or even engaged in misquotation.

The drawings of my book are the work of Paula DiSanto Bensadoun, my close friend and colleague since our postgraduate year at Cornell. Without her help this work could never have been published.

I am very grateful to many colleagues, personally known and unknown, who let me use, or even supplied me with, their original electron micrographs or drawings. And I thank Herma Boyle from our Audio-Visual Department for her expert photography.

Particular thanks are owed to Maurice O'Sullivan and Howard Kealy, who with immense patience and good humour guided me through the word processing of this book.

I am specially indebted to Madeleine O'Dwyer and Avril Patterson of our Main Library, who through the medium of the International Library Loan in England succeeded in providing me with copies of publications as far back as 1819.

Writing this book offers me the opportunity to pay homage to my PhD supervisor, Adolf Portmann (1897-1982), zoologist, anthropologist and philosopher, whose many books, articles and lectures in German are beginning to be translated. Many years ago he instilled in me an interest in fish ontogenesis. This rose from liver-yolksac contact, to development of isozymic patterns of lactate dehydrogenase, haematopoiesis, viviparity and finally development of the eye.

A combined thank-you I offer to Edward F. MacNichol jun., Ferenc I. Hárosi, Jim K. Bowmaker, Ellis R. Loew and John E. Dowling for allowing me to share their laboratory and avail of their specific expertise.

David Noakes, editor of the Fish and Fisheries Series, offered his care and skill in reading the text and making valuable suggestions. To him I am especially grateful.

Kluwer Academic Publishers have been most helpful and supportive with their special expertise.

I acknowledge, with thanks, the grant provided by the National University of Ireland.

My final thanks are to my husband and our two sons for their patience and continued encouragement.

Yvette W. Kunz
Maurice Kennedy Research Centre for Emeritus Staff
University College Dublin
The National University of Ireland - Dublin 4

Chapter one

Introduction

'Now is certainly the time to fish — not to cut bait.'
Detrich, Westerfield and Zon (1999)

The ontogeny of fishes starts with fertilization, or more precisely, with activation[1] of the egg. At the end of the endogenous food supply the major primary organs are usually established. In some fish this stage is attained in a few hours (e.g. zebrafish), in others it takes a few months (salmonids). Since temperature within certain limits enhances or reduces developmental time, so-called degree-days (temperature in °C multiplied by days) have been introduced for comparative analyses. However, this path did not lead to reliable answers. 'Ignatieva (1976a)' proposed a hypothesis that in early embryogenesis 'an interaction between the yolk and active cytoplasm, which determines time of the onset of ribosomal RNA synthesis, is of importance in controlling the temporal relationships'.

'Kunz (1964)' employed a method in which the development of the different fish species is normalized at 100% and the events she was interested in are expressed as percentage values. Using this method it emerged immediately that hatching occurs at different % development (Fig. 1/1). According to the nomenclature introduced by 'Portmann', fish that are helpless at birth are called **nidicolous**[2] and species that hatch at an advanced developmental stage are referred to as **nidifugous**[3] (Figs. 1/2, 1/3). These terms were originally used by 'Portmann (1976)' for bird development, then applied to mammals and finally, by us, for teleosts. Similarly, 'Balon (1985)' in a review of his scientific publications from 1976-1984, and American authors in general, used other terms, derived also from a former division of birds: **precocial** referring to nidifugous fish and **altricial** to nidicolous fish.

Since hatching is not instantaneous for all individuals of a clutch — it even may involve a long span — the zero value in the publications by 'Kunz' refers to a time when 50% of a clutch are hatched. In the following the term **'embryos'** will be used when dealing with fish before hatching and **'postembryos'** after hatching has occurred to the end of yolk absorption. The term 'postembryo' was first used by 'Kryzhanovskii (1948, 1949)' while 'Rass (1946, 1948)' employed the

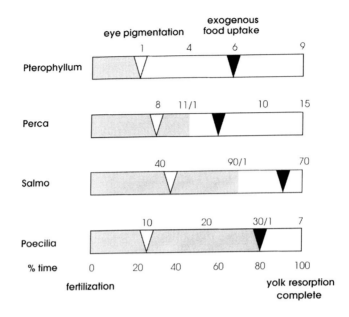

Fig. 1/1 Comparison of developmental pattern from fertilization to complete yolk resorption of *Pterophyllum*, *Perca*, *Salmo* and *Poecilia* (modified from Kunz, 1964, 1990, with kind permission of the director of the Muséum d'Histoire naturelle, Geneva and Pergamon Press, Oxford and New York). Developmental time in days with embryonic time shaded and postembryonic time white. Numbers indicate embryonic days followed by /postembryonic days. White triangle = eyes pigmented and liver primordium visible. Black triangle = exogenous food uptake has started in 50% of the postembryos.

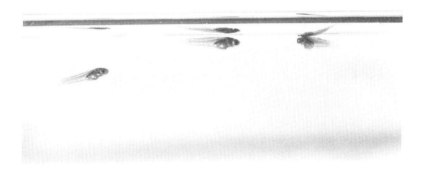

Fig. 1/2 Newborn *Poeciliae reticulatae* in an aquarium. They are typical nestfleers: shortly after birth they swim up to the water surface to fill their swimbladder (Stoeckli, 1957) (magnified newborn see Fig. 20/37d).

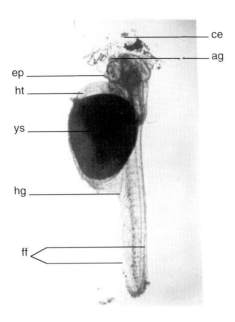

Fig. 1/3 Newborn *Pterophyllum scalare*. This is a typical nestsitter, deposited on substratum and depending on parental care. The newly-hatched fish is attached to plants by the material secreted by its adhesive (cement) glands situated on top and in front of the head. (Detailed drawing of a newly-hatched is displayed in Fig. 18/3a and a model is shown in Fig. 18/12). Similar adhesive glands of other fish are displayed in Figs. 15/15, 15/16, 15/17 and 16/18. The fuzzy material in front of the head represents particles caught by the sticky secretions on removal of the postembryo. ag - adhesive gland; ce - cement; ep - eye *primordium*; ff - finfold; hg - hindgut; ht - heart; ys - yolksac.

term 'prelarva' (cf. 'Soin, 1971'). 'Balon' defined the onset of exogenous feeding, and not hatching, as the end of the embryonic period. In 1975 he referred to fish after hatching as 'eleutheroembryos' and after absorption of the yolk as 'protopterygiolarvae'. Since other authors did not adopt the term 'eleutheroembryos', he replaced it with 'free embryo' and in a most recent publication (2001) on *Latimeria chalumnae* with 'fetus'.

The age will be given in embryonic and postembryonic days and in the fast developing zebrafish embryo in hours post fertilization (hpf).

The term **'larva'** is often used for the period after hatching of fish with altricial development since they bear little resemblance to the sexually mature fish. They may possess adhesive glands (*Herotilapia multispinosa*), large spines (sea basses) and trailing fins (lanternfish *Loweina*) and other appendages that may give them protection from predators. The deep-sea larvae of the black dragonfishes

(Idiacanthidae) are stalkeyed (stylophthalmus) and show trailing guts. The larval eyestalks may measure about one third of the length of the larva. According to 'Balon (1975b)' the larval period begins with exogenous nutrition and extends until the loss of the median fin fold and formation of the axial skeleton.

Larvae will undergo metamorphosis to turn into a juvenile (similar in appearance to a small adult). According to 'Balon (1975b)' the juvenile period begins with fulll differentiation of the fins and the assumption of an adult-like form. It extends until the first maturation of the gametes. Therefore, this development is also referred to as **'indirect'**. There is no larval period for precocial fish: they become juveniles in a development called **'direct'**. Another division of 'teleost developmental programs' refers to 1) indirect development with metamorphosis (abrupt transformation) as found in marine pelagic fish, 2) intermediate development, with a period of transformation, as displayed by the salmonids and 3) direct development as found generally in coastal, marine and virtually all freshwater species '(Evans and Fernald, 1990)'. 'Kendall et al. (1984)' and 'Helfman et al. (1999)' gave a comparison of the terminology used by various authors.

Nidicolous postembryos are often equipped with various types of adhesive organs, which ensure development in surface layers of water with a sufficient concentration of oxygen. (The structure of the different adhesive organs is dealt with in Chapter 15). As an example, the newly hatched *Pterophyllum scalare* is blind, not able to feed or swim, and, therefore, completely dependent on parental care. It becomes attached to plants by the secretions of two pairs of adhesive glands located on the head (Fig. 1/3, 18/30a-c). The parents clean and fan the postembryos; care is predominantly exercised by the male. The postembryos show other temporary organs such as a respiratory vascularization of the ventral finfold and a capillary net furnished by the vena subintestinalis and vena caudalis (Fig. 18/30a,b).

The eggs of *Perca fluviatilis*, as all planktonic eggs, are not cared for by parents. Salmonid eggs are deposited in the cold waters of the rivers and lakes in the autumn. The newly hatched young leave the nest called a redd, which is made up of pebbles, and disperse to gain access to clean oxygen rich water. The newly hatched move from time to time, swim obliquely until they finally reach the water-surface to fill their swimbladder. When exogenous feeding starts, fin primordia have become established.

The nidifugous *Poecilia reticulata*, when it hatches, spends up to 5 days in the ovarian cavity until parturition. The newly born fish immediately swims to the water surface to fill its swimbladder and to start feeding (Figs. 1/2, 20/37d).

A comparison of the development of the above-mentioned fishes (*Pterophyllum scalare, Perca fluviatilis, Poecilia reticulata, Salmo salar*) makes it obvious that the only features that appear at the relatively same time (at around 30% of developmental time) are pigmentation of the eye and the anlage of the liver (which in all cases lies adjacent to the yolk[4]) (Fig.1/1) (see Chapters 17 and 19).

In the middle of the 20th century a new discipline, **ecomorphology** applied to fish embryology, came into being in Soviet Russia. It was founded by 'Kryzhanovskii (1949)'. The essence of this new concept stresses that 'at each moment of life, the organism finds itself in unity with its environment and that its development is both an expression and a result of this unity'. 'Kryzhanovskii' had divided the oviparous fish into ecological groups according to the locality of egg deposition. This ecological trend in embryology has been influential especially on fishery measures such as artificial breeding and the improvement of reproduction. The emphasis of the work of 'Soin (1971)' centred on features of fish development, which represented an adaptation to different conditions. He showed the relationship between colour of the egg (carotenoid pigmentation) and environmental oxygen supply. 'Barancheev and Razumovskii (1937)' had been the first to establish that under equal conditions a deeply pigmented *Coregonus* egg has a higher rate of respiration than a pale one.

Newly hatched postembryos are characterized by the presence of adaptive structures to ensure respiration. For instance, nidicolous young — as the name suggests — cannot move around after hatching. The pelagic postembryos of *Ophiocephalus* (Anabantidae) literally hang from the water surface (with the yolksac uppermost) owing to their low specific gravity resulting from a large oil drop. When the yolksac is gradually depleted the oil drop forms two lateral bulges. These serve as floats to keep the postembryos in the surface layers of the tropical waters until the onset of exogenous food uptake '(Soin, 1971)'. Other newly-hatched marine fish keep to the upper or midwaters by the strong hydration of their yolksac and, in some, due to the presence of small oil-droplets. The sinus cephalicus, termed by 'Kryzhanovskii (1953)' the 'fin cavity' in the dorsal fin fold, is filled with a fluid of low specific gravity. This keeps the postembryo in midwater. The fin cavity disappears once food uptake starts. *Mullus barbatus ponticus* is given as an example of this condition. 'Balon (1975)' described a similar buoancy aid for the newly hatched *Calanx*, which he calls 'dorsal sinus' and in *Loweina rara* he observed both dorsal and ventral finfold sinuses. Some marine fishes, such as *Trachypterus* (lampridiformes) display greatly elongated anterior rays of paired and median fins which keep the postembryos suspended in water '(Soin, 1971)'. Also *Adioryx vexillarius* solves the buoyancy problem with its elongated spines according to 'Balon (1975)'. These types of fish are referred to by 'Soin' as pelagophilic[5]. So-called phytophilic[6] fish hatch in standing or weakly flowing water, i.e. in near-bottom water deficient in oxygen. Examples quoted are crucian carp, roach and bream. They have various adaptations to an existence in a suspended state, such as glandular organs in *Cyprinus carpio*, *Esox lucius*, *Betta splendens*, or adhesive, also called 'cement', organs on the head of *Anoptichthys jordani* and *Pterophyllum scalare* (see Figs. 15/15, 15/16, 18/30). These postembryos attach themselves to floating plants in the surface layers with improved concentration of oxygen. *Misgurnus fossilis* possesses barbels with glandular formations, in addition to the frontal cement organs. So-called lithophilic[7], psammophilic[8] and ostracophilic[9] fish spawn at the bottom of flowing waters. Such benthic postembryos have the following adaptations of structure and behaviour to guarantee respiration: Newly-hatched lithophilic cyprinids of the group *Barbina*, which develop in swiftly flowing mountain rivers and lakes of the Caucasus and Central Asia, display 'jumping up movements'. Representatives of the psammophilic cyprinids and loaches, such as *Gobionina, Gobio gobio* and *Nemachilini*, have unusually long pectoral fins which are kept spread widely to the side and rest their margins against the sandy bottom. The pectoral fins of *Pseudogobio rivularis* are especially large with long digitate processes. These modified pectoral fins prevent the postembryos from keeling over on their side and allow them to float. An example for the psammophilic fish is *Acanthorhodeus asmussi*, which develops in the mantle cavity of bivalves, hatches at an early stage of development, covered with epidermal scales of the tegulate type, with their apex directed backward. After hatching, the postembryo moves with the help of these scales out of the egg mass and spreads along the gill of the mollusc thereby taking advantage of an excellent supply of oxygen '(Soin, 1971 and Balon, 1975)'. *Macrozoarces americanus* seems to be unique and has long been the subject of controversy '(rev. Breder and Rosen, 1966)'. It is oviparous (with egg diameters of 8.5 mm) and the postembryos are nidifugous in contrast to their viviparous European representatives (see 'Zoarcidae' in Chapter 20). It has been reported that in the aquarium the female wraps herself around the egg mass. At hatching *M. americanus* is adult-like in appearance, has teeth and possesses the adult complement of fin rays '(Methven and Brown, 1991)'. 'Balon (1990)' summarized his views on the classification of reproductive styles.

According to 'Fuiman (1997)' 'metamorphosis in flatfish, i.e. morphological transformation from a symmetrical pelagic larva to the ultimate bottom mimic is an unparalleled feat of ontogeny'. As a result of preliminary investigations he concluded that 1) different flatfish species undergo settlement at a common ontogenetic state; 2) they have a common set of skills at settlement that differ from those of pelagic species of the same ontogenetic state; 3) skills or features that appear earlier in the ontogenetic program of flatfishes than in pelagic fishes suggest attributes that are important to survival in a benthic habitat. 'Osse and Boogaart (1997)' discussed the requirements for a successful change from a symmetric pelagic larva to a typical asymmetric juvenile benthic flatfish. — Metamorphosis resulting in migration of the eyes will be specifically referred to in Chapter 17 (Development of the Eye).

A review based on the development of different **behaviour patterns**, with a concentration on social behaviour of Salmonidae and Cichlidae, was given by 'Noakes (1978)'. This was followed by a detailed analysis of teleostean ontogeny of behaviour and concurrent developmental changes in sensory systems (visual, chemosensory, lateral line and inner ear) '(Noakes and Godin, 1988)'.

So-called **annual fish** represent a modification of the typical teleost development caused by the adaptation to harsh living conditions. These fish undergo a diapause in their early cleavage stages, followed by a phase in which embryoforming DC blastomeres completely disperse during epiboly and after its completion reaggregate. The embryos are capable of survival for many months in dry mud. As a consequence, hatching may be delayed for up to six months in nature and possibly longer than a year under extreme conditions. Annual fishes include the genus *Cynolebias, Austrofundulus* and *Nothobranchus* (subfamily Rivulinae of the family Cyprinodontidae) '(Wourms, 1972a,b,c; Kowalska-Dyrcz, 1979; Van Haarlem, 1981; Able, 1984; Carter and Wourms, 1991, 1993)'.

Among vertebrates, the zebrafish [*Brachydanio (Danio) rerio*] is particularly amenable to both embryologic and **genetic studies** of development (Kimmel, 1989; Kimmel *et al.*, 1995). In the recent past, developmental processes of fish embryos have been studied by using various species as experimental material. Transparency of the embryo and speed of development have been important considerations. For example, transparent *Fundulus heteroclitus* embryos have been used by 'Trinkaus' and others to analyze cell migration during gastrulation '(Trinkaus and Erickson, 1983; Trinkaus *et al.*, 1992; Trinkaus, 1993, 1998)'. Also, the medaka, *Oryzias latipes*, has been a useful species because it is easily obtained and maintained and because of the availability of many mutant strains '(Kageyama, 1977; Iwamatsu, 1994; Ozato and Wakamatsu, 1994)'. However, a relatively thick envelope and large oil droplets in the yolk are sometimes hindering clear observation. The transparent eggs and embryos of the ice goby (Shirouo), *Leucopsarion petersii*, are readily available commercially from fishermen, and their mature adults can be easily kept and bred in the laboratory. The embryos develop swiftly and may be kept under simple culture conditions. Cleavage is finished in 17 hours at $20°$ C and most organs are established in 3 days. The main stages of embryonic development and methods of handling have been described by 'Nakatsuji *et al.*, 1997)'.

Summary

Development of teleosts starts with fertilization, or activation, and postembryonic development ends with the complete resorption of the yolk. In order to compare fish that develop at different speeds — some literally in days, others in many months — a method is shown in which the development of the different fish species is normalized at 100% and the events dealt with in detail in later chapters are expressed as percentage values. Such a comparison reveals that the only features appearing at the relatively same time are pigmentation of the eye and the appearance of the liver primordium. This method also shows that hatching occurs at different % development. According to 'Portmann' fish such as *Pterophyllum scalare*, which are helpless at hatching, are called nidicolous and species, which are independent at that stage, such as the viviparous *Poecilia reticulata*, are referred to as nidifugous. Similarly, 'Balon' uses the term 'precocial' for nidifugous fish and 'altricial' for nidicolous fish. There are a great many different species taking up position between the extreme nidicolous and extreme nidifugous species.

The term embryo is used to describe fish before hatching and postembryo for those after hatching until complete absorption of the yolk. Nidicolous postembryos are equipped with various types of adhesive organs to ensure development in oxygen rich surface layers of water. Nidifugous postembryos swim up to feed and fill their swim bladder immediately after hatching.

A new discipline, ecomorphology, was applied to fish embryology by 'Kryzhanovskii'. This concept stresses the dependence of an organism on its environment. It divides fishes into ecological groups according to the locality of egg deposition: pelagophilic (free-floating), phytophilic (hatching in standing water, deficient in oxygen, and subsequent attachment to floating plants). Benthic specimens are lithophilic, lying quietly at the bottom and 'jumping up' to facilitate respiration. Psammophilic postembryos have very long pectoral fins to prevent keeling over. Ostracophilic fish spawn into the mantle cavity of bivalves, which guarantees a continual supply of oxygen.

Another division of 'teleost developmental programs' differentiates among 1) indirect development with metamorphosis, 2) intermediate development and 3) direct development. — The most obvious signs of metamorphosis in flatfish involve eye migration which coincides with changes of the natural habitat from pelagic to demersal.

Study of the ontogeny of behaviour coupled with an emphasis on the development of different sensory systems is another approach '(Noakes, 1981)'.

Various fish have recently become very popular for genetic and molecular biological analyses. Transparency of the embryo and speed of development have been important considerations for choosing them. The types of fish currently most studied by these methods are inter alia the zebrafish (*Brachydanio rerio*/*Danio rerio*), *Fundulus heteroclitus* and *Oryzias latipes*.

Endnotes

1. activation takes place when an egg comes into contact with water. It is, therefore, independent of the presence of sperm. It is characterized by the cortical reaction and the formation of the perivitelline space (see chapters 9 and 11).
2. nidicolous = nestsitters, G. Nesthocker
3. nidifugous = nestfleers, G. Nestflüchter
4. see Figs. 3/1-3/5, 18/9, 18/27, 18/30, 19/8, 19/9, 19/12, 19/13.
5. pelagophilic = laying eggs in open waters. From L. pelagicus, from Gk. pelagikos = level surface of the open sea. And from Gk. phile = loving.
6. phytophilic = laying eggs on plants. From Gk. phyto (plant) and phile (loving).
7. lithophilic = depositing eggs in stony places. From Gk. lithos (stone) and phile (loving).
8. psammophilic = laying eggs on sand. From Gk. psammos (sand) and phile (loving).
9. ostracophilic = laying eggs into mussle shells (lamellibranchs/bivalves). From Gk. ostracon (hard shell) and phile (loving)

In all endnotes the following abbreviations have been used:
F. French
G. German
Gk. Greek
I. Italian
L. Latin

Chapter two

The egg

'The panoramic display to be witnessed in small transparent eggs, like those of the starfish and sea urchin, reveals the order and relations of consecutive scenes, and thus supplies the needed vantage-ground for comparative study.'

Agassiz (1885)

The eggs of oviparous fish, having undergone a growth period in which yolk[1] has been formed, are ovulated from the ovarian follicles into the ovarian lumen or peritoneal cavity, usually after completion of the first meiotic division. The large nucleus[2], which originally has taken up a central (or slightly eccentric) position, has now migrated to the periphery. There the nuclear envelope breaks down. However, in the oviparous *Poecilia reticulata* and the ovoviviparous *Zoarces viviparus* the nucleus is already in an eccentric position before onset of the growth phase of the oocyte, an indication of early 'bipolar differentiation' (see Chapter 9). When the mature egg of oviparous fish has attained the metaphase of the second maturation (meiotic) division, it is deposited into the water (referred to as spawning or shedding) through the urogenital orifice or genital pore and is ready for fertilization. The mature egg contains a large mass of yolk (with oil-droplets of various number and size or no oil-droplets) occupying the central portion of the cytoplasm[3]. The whole egg is protected by an envelope, which plays a role in gas exchange and provides physical protection and selective transport of materials into the egg. The teleostean ovary is one of the following types: 1) Synchronous — all oocytes grow and ovulate in unison, 2) Group-synchronous — at least two populations are encountered: a synchronous population of large oocytes (clutch) and a heterogenous group of smaller oocytes which supply the clutch; 3) Asynchronous — oocytes of all stages are present in the ovary. Pattern 2) is the most common. However, pattern 3) is experimentally the most ideal in that it allows the study of the sequence of events of oogenesis in one and the same ovary (e.g. in *Fundulus*). In the ovary of the pipefish *Syngnathus scovelli* the oocytes are even sequentially arranged according to developmental age '(Begovac and Wallace, 1988)'.

Chemical analysis of a teleost egg (*Cyprinus carpio*) showed 64.9-75% water, 17.6-27.7% protein, 2.2-7.3% lipid and 1.4-2.2% ash. Glycogen as an energy source amounted to 0.2-3 milligram per gram of eggs (the amount is dependent on the quality of food taken) '(Linhart *et al.*, 1995)'.

A detailed classification of the reproductive styles of fishes is given by 'Balon, 1990'.

2.1 TYPES OF EGGS

Pelagic eggs are transparent whereas demersal ones are not '(Hoffmann, 1881)'. Due to their transparency pelagic eggs are not commonly found in the stomachs of fishes, while ova deposited on the bottom are (e.g. eggs of *Cyclopterus*, *Cottus* and *Clupea harengus*).

2.1.1 Pelagic (planktonic) eggs

These eggs float on or near the surface of the water (or at any depth of the water column). Pelagic eggs are spawned mainly by marine species but also by some freshwater fish. Eggs of freshwater (Anguillidae) are pelagic and non-adhesive and are deposited in the open sea '(Breder and Rosen, 1966)'. According to 'Soin (1971)' the functional significance of eggs floating at the surface or in midwater is to guarantee the presence of a high oxygen concentration.

Eggs of pelagic fish are colourless without any trace of pigment (in contrast to demersal eggs, see below). Pelagic eggs are produced in great numbers; e.g. *Mola mola*, the Atlantic sunfish, is reported to produce 300 million eggs '(Potts, 1984)'. The pelagic eggs of most marine spawners are usually small and develop in the upper layers while large eggs develop at considerable depths (bathypelagic) '(Ginzburg, 1968)'. Eggs take up oxygen from the surrounding water and carotenoids in the egg (which give its colour) are of respiratory importance. Demersal eggs are highly coloured due to the low concentration of oxygen in their surrounding, while pelagic eggs are colourless. Therefore, there seems to be a clear connection between carotenoid pigmentation of the egg and the oxygen regime of the environment.

The first to discover that the eggs of the cod float at the surface was 'G.O. Sars (1864)' at the Lofoten Island [cf. 'Agassiz and Whitman (1884)'; 'Henshall (1888)'; 'Raffaele (1888)']. 'Haeckel (1875)' called attention to the pelagic eggs of some Gadidae [cf. 'Agassiz and Whitman (1884)']. Soon afterwards a debate arose as to the reason why certain fish ova float. This topic was extensively deliberated by 'Ryder (1885/87)'. He divided buoyant ova into several types: 1) those in which the specific gravity of the yolk is diminished, as in the egg of the cod; 2) those in which large oil-drops in an eccentric position aid in causing the eggs to float; 3) those in which a very large oil-drop causes the ovum to float even in fresh water. However, he stressed that these three categories are connected by intermediate types. The other conditions for buoancy were a) that the egg be free (not adhesive), with a thin membrane, and b) that it be immersed in water having a greater density than 1.014.

According to 'Soin (1971)', 'Craik and Harvey (1984)' and 'Blaxter (1988)' ovulated pelagic eggs of marine species (Atlantic cod, *Gadus morrhua*, whiting, *Merlangius merlangus*, haddock, *Melanogrammus aeglefinus* and plaice, *Pleuronectes platessa*) contain markedly high levels of water, i.e. in the range of 83-92%. The influx of water into the eggs during maturation is suggested to be due to the osmotic effect of free amino acids '(Matsubara and Sawano, 1995)'. The buoyancy of the eggs of some herrings and flounders and other fish is due to the inclusion of an oil drop.

Eggs without oil-drops, too, are found floating: The eggs of *Gadus aeglefinus* are an example as are those of *Tautoga onitis* (Labridae). The eggs of *Alosa sapidissima*, deposited in fresh or brackish water, remain in a state of suspension near the bottom '(Cunningham, 1888)'.

Some freshwater species spawn eggs, which float in rivers but sink to the bottom in calm waters (e.g. *Ctenopharyngodon idellus*, *Hypophthalmichthys molitrix*, *Erythroculter erythropterus*, *Elopichthys bambusa*, *Hemiculter cultratus* and also the anadromous Caspian shad. It has been suggested that in most of these species buoyancy is achieved by reducing specific gravity as a result of a large perivitelline space. Buoyancy may also be due to a hydrated egg membrane as observed in *Paraleucogobio soldatovi* and *Gobiobotia pappenheimi*, or due to the presence of a large oil drop as in the Chinese perch *Siniperca chua-tsi*. Freshwater fish that spawn floating eggs in standing water are found mainly within tropical species (anabantoidei) [rev. by 'Soin (1971)'].

The freshwater *Clupeonella delicatula* which spawns in Russian lakes lays pelagic eggs with a large oil-drop and a voluminous perivitelline space '(Soin, 1971)'.

As a rule, ova, which are buoyant, float loosely connected or singly near the surface. However, there are several exceptions to this: The eggs of *Lophius piscatorius*, about 50 000, or several millions, are enclosed in a pelagic ribbon, up to 1 metre broad and from 8-9 metres long. They were first mentioned by 'Agassiz (1882b)' [cf. 'Agassiz and Whitman (1884)']. The male paradisefish constructs a floating raft by expelling adhesive bubbles from his mouth. He subsequently carries the fertilized eggs from the bottom and ejects them into the raft where they develop '(Henshall, 1888)'. The eggs of *Fierasfer acus* and *F. dentatus* (Ophidiidae) are united in their thousands in a thick gelatinous envelope and float at the surface of the sea '(Cunningham, 1888)'. The case of *Perca fluviatilis* was already described by 'Aristotle' ('the perch deposit their spawn in one continuous string') and later in detail by 'von Baer (1835)' and 'Lereboullet (1854)'. The eggs of *Perca* (about 5 000 of them) cohere so as to form a long flat band of eggs folded in an accordion-like fashion. This collapsed tube is made up of a network of eggs with irregular meshes, likened to a netted bead purse and is usually attached to freshwater plants '(Breder and Rosen, 1966)' (Fig. 6/19). The ova of *Antennarius* are deposited into a pelagic nest, which is guarded by the male ['Cunningham

(1888)'; Henshall (1888); Breder and Rosen (1966)']. Scorpaenids spawn gelatinous egg-masses '(Erickson and Pikitch, 1993)'. The eggs of *Betta splendens* have no oil-drop. The male constructs a nest of mucus-covered air-bubbles and guards the nest and the developing eggs '(Breder and Rosen, 1966)'.

Most pelagic fishes are of the non-guarding type. Egg-scattering pelagic spawners include altricial (see Chapter 1) fish with a long larval period terminated by metamorphosis (e.g. *Clupeonella delicatula*). An example of a rock and gravel spawner which lays unattended eggs is represented by *Stizostedion vitreum*.

Eggs are held together in considerable number by thin threads as observed in some flying fishes (Exocoetidae). They are found entangled in floating seaweeds and, if detached, the eggs will sink. The eggs of *Belone longirostris* are fastened together by means of filaments, and the clusters of eggs usually become attached to foreign objects in the water (Fig. 6/12b,c). The male four-spined stickleback builds a basket-like nest resting on water plants ['Cunningham (1888)'; 'Henshall (1888)'; Bell and Foster, 1994)'].

Pelagic eggs have their animal pole directed downward. 'Soin (1971)' suggests that this arrangement protects the blastodisc from harmful influence of direct sunlight.

2.1.2 Demersal eggs

Demersal eggs sink to the bottom where they rest. They are found mainly in freshwater fishes and in marine fishes spawning in the littoral zone. The eggs are called benthonic (or benthic) when they are found at the bottom of the ocean or lake. If they rest on inshore, rocky sites or plants (often exposed to strong tidal currents) they are called littoral. On ecological grounds, 'Kryzhanovskii (1949)' and 'Soin (1971)' differentiated the demersal fish of Soviet waters, according to the oviposition sites, into lithophilic (eggs deposited on sandy-stony bottom), phytophilic (eggs attached to plants), psammophilic (eggs developing on open sandy bottom after spawning) and ostracophilic (eggs spawned into mantle cavities of bivalve molluscs) (see endnotes [3-7] in Chapter 1). Benthic eggs have a relatively high specific gravity, due to their low water content (60-70%) and a small perivitelline space. Demersal eggs are less delicate in appearance than the floating ones. They, too, can contain oil droplets: *Cottus* with scattered oil globules; *Liparis* with three or more oil globules of various sizes; *Cyclopterus lumpus* (lump sucker) with numerous oil globules of various sizes, arranged in a cluster at the ventral pole) '(Cunningham, 1888)'.

The colour of the eggs varies from yellow to whitish orange (many Cyprinidae) to crimson (Salmonidae) and orange-violet (Cyclopteridae) and depends on the amount of carotenoid pigment contained in the yolk and the oil-droplets. However, in some fish, which live in favourable respiratory conditions, such as herring of the genus *Clupea* (Atlantic and Pacific) and *Betta splendens*, the carotenoids are concentrated in the blastodisc while there is hardly any in the yolk. The egg of *Cyclopterus lumpus* displays abundant carotenoids in both yolk and plasma. Carotenoids are not synthesized in the maternal fish but are taken up with the food. During oogenesis the ova take up carotenoids, which are supplied by various organs, but mainly by the muscles. There is clearly a link between the intensity of carotenoid pigmentation of the egg and the oxygen conditions in which its development takes place. The first investigators to suggest that there is such a connection were 'Barancheev and Razumovskii (1937)' who worked with the eggs of *Coregonus lavaretus*. The strongest pigmentation was reported for *Oncorhynchus masu*, which spawns in slowly-moving shallow streams with very poor oxygen conditions [rev. 'Soin (1971)'].

Eggs deposited in masses on the sea-bottom strongly adhere to each other. Adhesive eggs are also deposited in great numbers in shallow nests [black bass, sunfishes (Centrarchidae), catfishes (Ameiuridae, now Ictaluridae]. '(Helfman *et al.*, 1997)'. The littoral eggs of *Cottus scorpius* and *Blennius pholis* become attached to the rocks between the tide marks. The eggs of *Gobius niger* and *G. minutus* adhere to the substratum by a thick 'rope' '(Kupffer, 1868)'. Osmeridae and some species of Exocoetidae attach their eggs to weeds, sticks and rocks '(Potts, 1984)'.

The non-adhesive ova of some species, such as the salmonids, *Alosa* and turbot (*Scophthalmus maximus*, also *Rhombus maximus*), sink to the bottom and remain separate. However, in most cases demersal eggs adhere to each other or to stones, sand, plant or other substrata (e.g. cyprinids, pike, *Clupea* and loaches) '(Ginzburg, 1968)'. The eggs of the European smelt *Osmerus eperlanus* swing in the water attached by a flexible suspensory membrane formed by the ruptured adhesive outer layer '(Buchholz, 1863)'. The eggs of the pike, *Esox lucius*, are attached to aquatic plants in freshwater creeks. Eggs of the goldfish (and most carps) adhere to the substratum, in most cases aquatic plants ['Cunningham (1888)'; 'Henshall (1888)'; 'M'Intosh and Prince (1890)'; Breder and Rosen (1966)'.].

Demersal eggs are oriented with their animal pole facing upwards due to the fact that the yolksac has a relatively higher specific gravity than the blastodisc and that most of the oil drops (not present in all) are concentrated below the disc. This orientation secures better condition for respiration of the developing blastodisc.

The ostraphilic reproductive habit is displayed by the bitterlings. The eggs of *Rhodeus amarus* develop in the gill cavity of mussels (Unionidae). The female fish develops a long ovipositor, which she introduces into the inhalent syphon of a freshwater mussle for deposition of the eggs. Then the nearby male ejects the sperm which is taken up by the inhalent current of the mussle '(Olt, 1893)'.

2.1.3 Demersal/pelagic eggs

There are differences within families of fishes also. The ova of the freshwater *Lota vulgaris* are found separate and loose at the bottom of water, while the rest of the Gadidae are marine and produce pelagic ova '(Cunningham, 1888)'. According to 'Agassiz (1878, 1882a,b)' the eggs of *Cottus groenlandicus* are found floating on the surface, while other Cottidae lay their eggs in bunches attached to the bottom or singly between stones in shallow water.

2.2 EGG CARE

Apart from pelagic eggs, all egg types are generally associated with some type of egg care. This has been characterized into the following principal types: guarding, external bearer and internal bearer ['Balon (1975, 1985)'; 'Potts (1984)'] wheras 'Blumer (1979)' discussed 15 types of parental care. Overall parental care has been described by 'Gross and Shine (1981)' in 45% of 182 teleostean families examined. These authors interpreted the prevalence of male parental care with external fertilization as resulting from male territoriality, which in turn results from female discrimination of oviposition sites. Internal fertilization preadapts females to selection for embryo retention, leading to live-bearing.

2.2.1 Guarding

Most fishes with complex patterns of parental care are observed in freshwater fishes. Sixty percent of families reproducing in freshwater show some degree of parental behaviour, as opposed to 16% of marine families. When guarding of larvae was investigated, only 2 marine species of the same family were found. Of the 52 families listed as embryo guarders, 39 lack species with internal fertilization '(Baylis, 1981)'. Guarding refers to more than aggressive defense by the parents. It also includes oxygenation, removal of debris and diseased or infertile eggs (all achieved by fanning), splashing the eggs with water, etc. '(Keenleyside, 1979; Baylis, 1981)'. 'Blumer (1979)' carried out a survey of all families which exhibit parental guarding and concluded that male care occurs in 61 and female care in 44 families. Similarly, 'Gross and Shine (1981)' reported that in the 182 families tested paternal care is significantly more common (61%) than female care (39%). Solitary parental care is dominant and biparental care occurs in less than 25% of families tested. Another analysis investigating 22 families revealed 50% male carers, 18% male and female carers and 32% female carers '(Sargent and Gross, 1986)'.'Kawase and Nakazono (1995)' reported that of the 86 families tested 62% showed male care, 24% families have male and female care, and only 14% families female care.

A certain lability of these conditions should be stressed. For instance, variation in parental care in one species, *Ictalurus nebulosus*, a freshwater catfish, revealed 56.2% biparental, 39.3% uniparental paternal and 4.5% uniparental maternal care. An example of predominantly maternal egg care is shown by the Japanese filefish, *Rudarius ercodes* (Monacanthidae). It involves blowing water on eggs, fanning them with their pectoral fins and driving away fish passing nearby. Under laboratory conditions male egg-guarding can occur when an egg-tending female is removed. After reintroduction of the female, the male is driven away '(Kawase and Nakazono, 1995)'. The same authors described biparental care (watching and blowing of the egg mass) in the filefish *Paramonacanthus japonicus* '(Nakazono and Kawase, 1993)'. The males of *Antennarius* and of *Callichthys* build a nest and guard the eggs deposited into it. The adhesive ova of *Gobius ruthensparri* are deposited into the shell of a barnacle and watched over and fanned by the male.

The eggs of Gasterosteidae (*Gasterosteus* and *Spinachia*) are deposited into nests, held together by plant matter and guarded by the male ['Cunningham (1888)'; 'Henshall (1888)', 'Gudger (1918)'; 'Wooton (1976)'; 'Giles (1984)'; 'Potts (1984)'; 'Bell and Foster (1994)']. The plant material of the nest is glued together by a substance secreted by the kidneys of the male '(Keenleyside, 1979)'. After spawning with one female, the male will spawn with further females during the next two to three days. He then looks after the embryos for 2-3 weeks. Each clutch contains about 50-150 eggs '(Wootton, 1976)'. It has been shown by 'Ridley and Rechten (1981)' that female sticklebacks prefer to spawn in nests that already contain eggs. Nests of the four-spined stickleback, *Apeltes quadracus*, are typically built one on top of the other, with holes left open for aeration at each level '(Wootton, 1976)'. In this species, rival male interruptions are often resonsible for prolonged courtship, reduced spawning success and nest attention '(Willmott and Foster, 1995)'.

The fact that *Gasterosteus* does not allow the young to leave the nest and takes any 'disobedient young' into his mouth to bring him back to the nest was first noted by 'Coste (1848)', followed by 'Hancock (1854)' and 'Warington (1856)'. 'McKenzie, (1979)' observed that the male brook stickleback, *Culaea inconstans*, spits the retrieved stray young into a 'nursery', formed previously by pulling apart the upper part of the nest. However, this behaviour should not be confused with mouthbreeding, though several attempts have been made to regard the above mentioned behaviour as the beginning of oral gestation '(Gudger, 1918)'. This holds also true for species of Anabantidae (labyrinthfish) described by 'Carbonnier (1874)', 'Hall and Miller (1968)' and 'Kramer (1973)'. In *Pomatoschistus minutus* it is the male which fans the developing embryos '(Riehl, 1978a)'. The same holds true for *Ameiurus albidus* '(Cunningham, 1888)'. The adhesive eggs of *Lepadogaster decandolii*, attached to stones or shells, are guarded by both parents while the ova of *L. bimaculatus* adhere to the inner surface of the lamellibranch *Pecten opercula-tus* and are watched over by at least one of the parents ('Cunningham, 1888'). The adhesive eggs of *Centronotus gunnellus* form a spherical mass the size of a walnut. This ball is free and both parents lie coiled around it ['Cunningham (1888)' and 'Gudger (1927)'].

2.2.2 External bearers

These carry the ova in various portions of the body until they hatch. The female of *Oryzias latipes* (medaka) carries a cluster of fertilized eggs attached to the vent for some hours before they are deposited '(Keenleyside, 1979)'. The male catfish *Aspredo* is provided during the breeding season with numerous little stalks on the under-surface of his abdomen, which receive the eggs and retain them until hatching '(Henshall, 1888)'. However, according to 'Cunningham (1888)', and summarized by 'Breder and Rosen (1966)', it is the female *Aspredo* which attaches the eggs to the skin of her own ventral surface, where they remain until they are hatched. The female of Solenostomidae carries the eggs attached to filaments developed on the ventral fins. The male sea-horse *Hippocampus* possesses a pouch, under the tail, into which the ova are placed and develop until hatching. Twenty to 1 000 embryos can be found and development continues for 20-28 days '(Kornienko and Drozdov, 1998)'. The male of *Siphonostoma typhle*, and similarly the male *Syngnathus*, carry the ova till the time of hatching in a pouch formed by longitudinal folds of the skin behind the anus. In *Nerophis* the ova are attached to the abdomen of the male by a viscoid secretion in front of the anus ['Cunningham (1888)'; 'Henshall (1888)'; 'Korschelt (1936)'; 'Keenleyside (1979)'; 'Baylis (1981); 'Wooton (1984)' and 'Bell and Foster (1994)']. The eggs carried by the male *Syngnathus nigrolineatus* in a closed pouch are pigmented a deep yellow-orange. This again has been linked with a dearth in oxygen. In contrast, the eggs adhering to the abdomen of the male of *Nerophis ophidion* are, therefore, in direct contact with water and are very weakly pigmented '(Soin, 1971)'.

2.2.3 Internal bearers (mouthbrooders)

'Bloch (1794)' was the first to describe oral gestation in siluroid fishes. In the next century a report of a male catfish as buccal incubator was given by 'Hensel (1870)' on *Arius commersonii*, a sea catfish found in the brackish waters of southern Brazil. Subsequently, the mouthbrooding habit of the male of the gaff-topsail catfish, *Felichthys felis* (*Aelurichthys*) was described by 'Clarke (1883) and again by 'Ryder (1883a, 1887)' and 'Holder (1904)'. The fish keep the developing embryos in their oral cavity until the young are hatched '(Gudger, 1918)'. According to 'Henshall (1888)' the male marine catfish *Galeichthys felis* takes the fertilized eggs into his mouth, where they undergo further development between the leaves of his gills. It is always and only the male that carries the eggs in catfish. 'Henshall (1888)' further reported that the male mouthbreeders such as *Tilapia* (later called *Sarotherodon* and now *Oreochromis*) and *Arius* carry the young in the buccal cavity well beyond their hatching time. The same has been reported for *Tilapia nilotica* and *T. galilaea*. However, in *Tilapia macrocephala* from Nigeria the female is the carrier '(Kraft and Peters, 1963)'. In *Tropheus morii* the female carries the eggs '(Ziegler, 1902)'. The males of *Luciocephalus* are mouthbrooders '(Riehl and Kokoschka, 1993)', whereas the cichlids *Gnathochromis pfefferi* and *Ctenochromis horei* are maternal mouthbrooders '(Ochi, 1993a,b)'. The yolk of mouthbrooders contains a high amount of carotenoids which have been shown to serve as an endogenous source of oxygen '(Balon, 1981, 1990)'.

2.2.4 Ovoviviparous and viviparous fish
(dealt with in Chapter 20)

2.3 SHAPE OF EGGS

Teleost eggs are usually not egg-shaped (i.e. oval, and elliptical in section) but round (spherical). Aberrant forms also occur (Fig. 2.1).

Oval shaped eggs are produced by many cichlids, e.g. *Gadus morrhua*, *Motella*, *Rhodeus amarus*, the annual fish *Austrofundulus myersi*, *Scorpaena porcus* and *S. rufa* and the anchovy *Engraulis japonica* as well as *Firasfer acus*, *Luciocephalus* sp.. The eggs of most engraulids are ovoid, which is interpreted as an adaptive value to reduce cannibalism by the filter-feeding adults '(Blaxter, 1988)'.

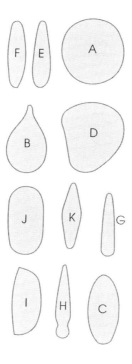

Fig. 2.1 Shape of teleostean eggs (after Breder, 1943). **A** - *Anchoviella*; **B** - *Stolephorus*; **C** - *Engraulis*; **D,E,F** - *Gobius*; **G** - *Glossogobius*; **H** - *Bathygobius*; **I** - *Lepadogaster*; **J** - *Pomacentrus*; **K** - Indetermined species.

The ovum of *Gobius minutus* has a peculiar elongated pyriform (pear) shape and the egg of the pipefish, *Syngnathus scovelli*, is oval-to-pear-shaped. *Gobius niger* and *Luciocephalus pyriform* lay also pyriform eggs. The eggs of *Bathygobius soporator*, when entering the water, assume the shape of a coffin. An unidentified 'star-shaped' egg has been encounterd in the Alaska region; however its occurrence is infrequent '(Matarese and Sandknop, 1984)'.

2.4 NUMBER OF EGGS

The number of eggs produced varies according to the age and weight of the female. Therefore, the number of eggs is calculated per body weight. The number is greatly reduced in ovoviviparous teleosts (in which yolk furnishes all the food) or viviparous teleosts (in which a greater amount of food is furnished by the ovary).

Estimates of 100,000-300,000 eggs per kilogram body mass have recently been reported for the carp, *Cyprinus carpio* '(Linhart *et al.*, 1995)'. *Mola mola* is said to produce 300 million eggs '(Potts, 1984)'. In the viviparous *Anableps dowi* brood size is a function of body size, i.e. large females have larger broods '(Knight *et al.*, 1985)'. As an example the average number of eggs per kilogram body mass of the female, in a descending order, is given, according to 'Cunningham (1888)', 'Fulton (1898)' and 'Hofer (1909)'in Table 1. However, according to 'Wooton (1979)' size of the female alone often accounts for only some 50-60% of the variation. In many fish such as *Gasterosteus*, females of a given size can vary in reproductive output by a factor of 2 or more; possible reasons are discussed by 'Wooton (1976, 1979)', 'Baker (1994)' and 'Baker *et al.* (1995)'.

Table 1

Species	Number of eggs per kilogram body mass of the female
Scophthalmus maximus	1 200 000
Molva molva	1 000 000
Lota lota	1 000 000
Scomber scrombus	800 000
Tinca tinca	600 000
Gadus aeglefinus	300 000
Gadus morrhua	500 000
Lucioperca lucioperca	200 000
Pleuronectes platessa	150 000
Perca fluviatilis	100 000
Cyprinus carpio	100 000
Clupea harengus	100 000
Osmerus eperlanus	50 000
Abramis brama	50 000
Esox lucius	30 000
Acipenser sturio	25 000
Thymallus thymallus	8 000
Barbus barbus	6 000
Phoxinus phoxinus	5 000
Salmo salar	2 000
Salmo fario	2 000

2.5 SIZE OF EGGS

On average, demersal are larger than pelagic eggs. Egg size is also related to external conditions: the farther north the distribution area of the species, the larger its eggs '(Ginzburg, 1968)'.

Most pelagic eggs are very small measuring not more than 0.3 millimetre in diameter '(Eigenmann, 1890)'. Examples are *Fundulus heteroclitus*: 0.16, *F. diaphanus*: 0.25 millimetre, *Brachydanio rerio*: 0.1 millimetre to 0.3 millimetre, *Crenilabrus*: 0.24 millimetre, *Cymatogaster aggregatus*: 0.3 millimetre, *Sardinops* 0.3 millimetre. According to 'Ahlstrom and Moser (1980)' pelagic eggs range from 0.5 millimetre (*Vinciguerria*) to about 5.5 millimetres (Muraenidae).

Demersal eggs range from small to large; mouthbrooders have some of the largest eggs. Eggs with a large diameter are observed in the salmonids, e.g. trout (*Salmo trutta*) 4-6 millimetres, *Salmo salar* 6 millimetres, *Oncorhynchus keta* 8-9.5 millimetres. Moreover, salmonid eggs taken from different rivers vary in size. The mouthbrooder, *Tropheus moorii*, from Lake Tanganyika is reported to have eggs of 7 millimetres diameter, while marine mouthbrooders possess eggs of 15-22 millimetres diameter (examples: Ariidae and *Bagre marinus*). The marine *Anarrhichas lupus* possesses eggs of 5.5-6 millimetres in diameter. Large eggs were reported for South African fish, such as *Gymnarchus* (Mormyridae), having 10 millimetres in diameter, or for *Arius commersonii* with 18 millimetres ['Clapp (1891)'; 'Henshall (1888)'; 'M'Intosh and Prince (1890)'; 'Assheton (1907)'; 'Gugder (1918)'; 'Wallace, 1981)'; 'Blaxter (1988)']. The largest eggs have been reported by 'Merriman (1940)' for the Ariidae *Bagre marinus* and *Galeichthys felis* (14-21 millimetres). Accordingly, the number of eggs of these species is small (20-40).

Seasonal patterns in egg size within, as well as among, marine species have been documented. Egg size declines from winter to spring and increases again during autumn. Geographical differences were also reported '(Rijnsdorp and Vingerhoed, 1994)'.

Egg size may vary between populations and stocks. Moreover, great differences in size occur not only amongst the eggs of different females but sometimes amongst the eggs of the same female. As already mentioned it can vary also over the life time of a female or seasonally. There can occur a variation in the average size of eggs from one spawning district to that of another. Seasonal and geographical patterns in egg size of the sole, *Solea solea*, have been reported. Geographical difference in egg size can amount to as much as 44% in volume. The eggs of the mouthbrooding *Tilapia macrocephala*, *mossambica* and *galilea* are considerably larger than those of the substrate-spawning *Tilapia tholloni* and *guinensis* ['Kraft and Peters (1963)'; 'Blaxter (1988)'; 'Rijnsdorp and Vingerhoed (1994)'; 'Miller et al. (1995)'; 'Chambers and Leggett (1996)'].

A trend for deeper water species of New Zealand marine fish to produce larger eggs than those spawning in shallow waters has been observed. It was suggested that the larger eggs ascend more rapidly than the smaller eggs '(Robertson, 1981)'. In marine fishes, generally, spawning time is said to be adapted to the availability of suitable food for larval development. It has been stated that marine fishes typically have more and smaller eggs than do freshwater fishes. A special study was recently undertaken to assess the maternal influences on variation in egg sizes within populations of marine fishes. At a species-level egg sizes of 309 North Atlantic fishes were shown to range from 0.3 to 18.00 millimetres in diameter. In addition, significant variation in egg sizes of species that spawn multiple batches per year have been reported. This variation usually shows a decrease in egg size as spawning proceeds '(Chambers and Legett, 1996)'.

When comparing the size of eggs one has to consider that some authors measure live eggs, others eggs after fixation (which causes shrinkage) and others again eggs that have already been in contact with water (which causes swelling) (see Chapters 5 and 9 under 'hardening').

2.6 DEPOSITION OF EGGS

In most teleosts the lumen of the ovary is continuous with the oviduct, a condition termed cystovarian (as opposed to gymnovarian of the more primitive osteichthyes, which shed their eggs into the body cavity). In the Salmonidae and *Anguilla* the oviducts have been secondarily lost in whole or in part. As a result their ova, when ripe, pass directly into the body-cavity from the ovaries and reach the outside by an external pore. In the other teleosts, the eggs pass to the posterior end of the ovary as they become mature, are ovulated into the ovarian cavity and shed via an oviduct to the exterior. The latter is the more prevalent mode, and it presents two types: the act of deposition is very rapid, as in the cottoids, *Discoboli*, and certain blennies or it is slower, as in the Gadidae and may be even prolonged, as appears to be the case with *Trigla*. If the eggs become mature simultaneously, as in *Cottus scorpius*, oviposition occupies only a few seconds. If the eggs reach maturity by successive strata, as in *Trigla gurnardus*, extrusion extends over a prolonged period. '(M'Intosh and Prince, 1890; Blaxter, 1959; Hempel, 1979; Wooton, 1990)'.

When placed into water, teleostean eggs become significantly larger. This is due to the uptake of water and is not dependent on fertilization. This was first noticed by 'von Baer (1835)' in *Cyprinus blicca*, subsequently by 'Vogt (1842)' in *Coregonus palea*, by 'Aubert (1854)' and 'Truman (1869)' in *Esox reticulatus* and by 'Lereboullet (1854)' in the perch (*Perca fluviatilis*) and the pike (*Esox*), also by 'Reichert (1856, 1857)'. It was further described by 'Ransom (1867)' in the stickleback (*Gasterosteus*), the trout (*Salmo fario*), the salmon (*Salmo salar*), the ruffe (*Acerina vulgaris*) and, again, *Perca fluviatilis*. 'Oellacher (1872)' made the same observations in the trout. 'Miescher (1878)', the discoverer of nucleic acids in the salmon, observed that the egg of *Salmo salar* weighs 127 milligrams when leaving the abdominal cavity and reaches 133 milligrams after resting in water. 'Blanchard (1878)' described the water uptake of the trout egg only seconds after immersion. 'Brook (1884)' observed in the freshly extruded eggs of *Trachinus vipera* and in 1885b in the eggs of the herring, *Clupea pallasii*, a 'breathing- cavity', which becomes filled with water. 'Raffaele (1888)' observed the uptake of very little water by *Labrax lupus*. The early investigators talked about 'imbibing or absorption or inception of water', forming a 'water chamber' or 'breathing chamber'). 'Groot and Alderdice (1985)' again stressed that the flaccid ovulated salmonid eggs become rigid when shed into the water. More recently, the pelagic egg of the wrasse *Tautogolabrus adspersus* was shown to increase in volume 4.3 times over 20 hours '(Wallace and Selman, 1981)'.' Hurley and Fisher (1966)' reported that the nature of force responsible for the uptake of water has not yet been elucidated and that it has been suggested for many fish that substances released from the eggs immediately after deposition cause the inflow.

Summary

Egg types are either pelagic or demersal. The latter category comprises both adhesive, and non-adhesive eggs. Often care by the male, female or both is observed. Male caring usually predominates. Developing embryos may also be carried by a parent either externally (e.g. a pouch) or internally (mouthbrooders and ovoviviparous and viviparous fish). The shape of the eggs is usually spherical though pyriform (pearshaped) and other forms also occur. The number of eggs depends on the age of the female. The size of the eggs (diameter) varies from 0.2-25 millimetres. Variation depends on the age of the female but also on a range of other conditions. When eggs are deposited water uptake occurs.

Endnotes

[1] see endnote[1] of Chapter 3.
[2] see endnote[2] of chapter 10.
[3] cytoplasm or ooplasm was originally called formative yolk, germinative protoplasm, germ plasm, germ protoplasm or simply germ. 'Cavolini (1787)' used the term 'cicatrix', which was later taken over by the French embryologists (see below). The term 'cytoplasm' was created by 'Strasburger'. Note: the often-used term 'protoplasm', used mostly, but not exclusively, by early investigators, strictly speaking encompasses cytoplasm + nucleus.
G. Keim, Hauptkeim of 'His', Bildungsdotter, Nebendotter (which includes the cortex) of 'van Beneden, 1878'
F. germe, cicatricule, cicatrice, cicatrile, vitellus plastique, substance plastique.
I. citoplasma

Chapter three

Yolk (vitellus)

> 'M'Intosh and Prince cite various authors in support of the idea that the protoplasm during development is nourished and grows at the expense of the yolk; but surely that fact is sufficiently obvious — what else is the yolk for?'
>
> Cunningham (1891)

Yolk[1] contains all the nutrients necessary for the developing embryo of oviparous fishes. 'Haeckel (1855)' considered the yolk as an appendage, a caenogenetic[2] addition or adaption 'not directly contributing to the building up of the tissues, but mainly serving to furnish pabulum food to the delicate and rudimentary embryo on emerging from the egg'. 'Prince (1887)' held a similar view '(Thomas, 1968a)'. Yolk, besides providing soluble nutrients to the blastomeres, contributes maternal membraneous and ribosome-like material to the embryonic cells. Yolk includes complex carbohydrates, lipids and proteins. Conjugated lipoproteins (including vitellogenin) predominate. They may comprise more than 80% of the dry weight of the egg '(Callen et al., 1980)'. The lipids isolated from teleostean eggs are characterized by a high iodine number indicating a high degree of unsaturated fats. The fatty acids of eggs of the common carp contain 11% cholesterol and 59.2% diolein lecithin '(Ginzburg, 1968; Kim, 1981)'. Various carotenoids, responsible for the colour of the eggs, have been investigated in salmonids, both freshwater and marine species. Astaxanthin ester made up the greatest part, i.e. 67.8% of all carotenoids. Their content is greatly influenced by the type of food the maternal fish eats '(Czeczuga, 1979)'. According to 'Ginzburg (1968)' proteins generally predominate over fats. The ratio may reach 20:1 or even higher e.g.in pike, Baltic cod and pikeperch. The amount of carbohydrates in teleostean eggs is insignificant.

'Henneguy (1888)' has reported finding a particular protein in the yolk he called 'ichtuline', which is similar to albumin and contains sulphur and phosphorus. 'Grodinski (1954)' notes that the yolk of the bitterling Rhodeus amarus consists of a small amount of fluid ichtulin. Most egg yolk proteins are derived from the high-molecular-weight precursor vitellogenin. Vitellogenin of the trout contains 0.5% calcium and 0.6% phosphoprotein phosphorus '(Babier and Vernier, 1989)'. It has been suggested that a whole series of components other than vitellogenin are vital for the production of viable eggs and are selectively sequestered by the developing oocyte '(Tyler, 1993)'.

Morphologists divide yolk into so-called proteid yolk and lipid (fatty) yolk. Fatty yolk is formed before the proteid yolk '(Cunningham, 1898; Kraft and Peters, 1963; Linhart *et al.*, 1995)', but is absorbed later during embryonic development '(Smith, 1958; Kunz, 1964)'. Lipid droplets (of different size and number) are mainly made up of triglyceride fat. They may be heavily pigmented and moveable. The so-called yolk droplets consist mainly of protein in a relatively well-hydrated phase, which is less dense than the main bulk of yolk. So-called 'intravesicular yolk' is not yolk at all. It is the name given by many authors to the cortical alveoli (see Chapter 4). Their content is not nutritive and is secreted after activation or fertilization of the egg (see Chapter 9). Other authors [as e.g. 'Yamamoto and Oota (1967b)] use the term 'carbohydrate yolk' when they refer to cortical alveoli.

According to 'Brook (1885a)' at the 4-cell stage of the herring 'the animal pole at this stage consists of four segments or cells, while the whole of the vegetative pole has the value of one cell. The whole of the vegetative area may be compared to a gigantic fat-cell in which the fat is replaced by food yolk.' Similarly, 'Oellacher (1872)' and 'van Beneden (1878)' regarded the deutoplasmic globe in a pelagic teleostean ovum as a large endodermic cell, with a constitution analogous to a fat-cell. This might explain why several present-day molecular biologists use the term 'yolk cell' when they refer to the yolk.

As the yolk mass enlarges during vitellogenesis, the yolk vesicles (future cortical alveoli) and the nucleus (germinal vesicle) shift towards the periphery. The nucleus becomes finally located in the cortical cytoplasm[3], i.e. the future animal pole.

3.1 LIPID YOLK

Lipid yolk is present either as a number of oil droplets, varying much in size (e.g. in *Osmerus, Poecilia reticulata, Pterophyllum*) (Figs. 3/1, 3/2), or just as one large oil globule (as in Coregonus) (Fig. 3/3) '(Ziegler, 1902; Retzius, 1912; Bertin, 1958; Kunz, 1964)'. Of a total of 515 pelagic species, 60% had one oil globule, 15% multiple oil globules and 25% none '(Ahlstrom and Moser, 1980)'. An example of eggs lacking oil droplets are those of the mouthbrooder *Luciocephalus* and the herring *Clupea harengus* '(Cunningham, 1888; Riehl and Kokoschka, 1993)' and *Callionymus* sp. '(Raffaele, 1888)'. It has been suggested that the oil droplets have a hydrostatic function '(Kryzhanovskii, 1960)'. However, there are pelagic eggs without and demersal eggs with oil droplets. It has been suggested, that the droplets may serve as a reserve of vitamin A to be used for the onset of vision '(Kunz, 1964)'.

Yolk (vitellus) 25

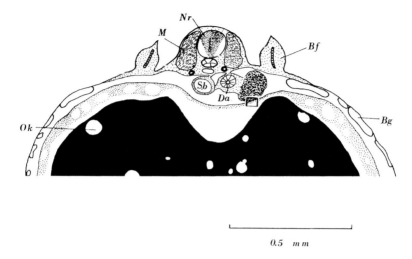

Fig. 3/1 Transverse section through *Poecilia reticulata* (*Lebistes reticulatus*), 10th embryonic day, to show dense yolk with small oil globules. (Kunz, 1964, with kind permission of the director of the Muséum d'Histoire naturelle, Geneva). Bf - pectoral fin; M - somite; Bg - blood vessel; Nr - neural tube; Da - intestine; Ok - small oil globules; Sb - swim bladder. The square denotes the liver/yolksac contact (dealt with in Chapter 19).

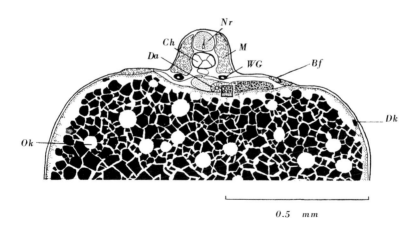

Fig. 3/2 Transverse section through *Pterophyllum scalare*, immediately after hatching (which occurs on the 2nd embryonic day). Yolk is present as platelets with interspersed oil globules. (Kunz, 1964, with kind permission of the director of the Muséum d'Histoire naturelle, Geneva). Bf - pectoral fin; M - somite; Ch - notochord; Nr - neural tube; Da - intestine; Ok - small oil globules oil sphere; Dk - periblast *nucleus*; WG - Wolffian duct. The square denotes the liver/yolksac contact (dealt with in Chapter 19).

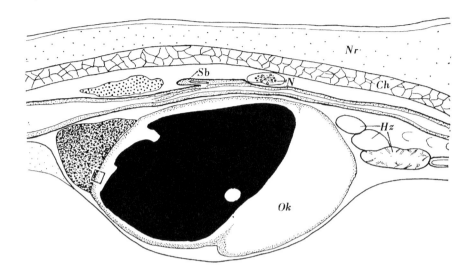

Fig. 3/3 Longitudinal (median) section through *Coregonus alpinus* on the day of hatching (60th embryonic day), characterized by dense yolk with a few small oil globules and an anterior large oil sphere. (Kunz, 1964, with kind permission of the director of the Muséum d'Histoire naturelle, Geneva). N - kidney; Ch - notochord; Nr - neural tube; Ok - large oil sphere; Sb - swim bladder; Hz - heart. The square denotes the liver/yolksac contact (dealt with in Chapter 19).

3.2 PROTEID YOLK

This represents the large yolk mass, so characteristic of the teleost egg. In the preserved egg it occurs either as distinct spherules (unmassed) traversed by cytoplasm[3], as in the herring, salmon and *Pterophyllum scalare* (Fig. 3/2), or as a single large spherule (massed) as in *Coregonus alpinus* (Fig. 3/3) and in pelagic eggs generally (see 'Cunningham, 1891'). During the later stages of vitellogenesis, the yolk globules of some teleosts fuse with each other to form a single mass of yolk; the transparency of live eggs of pelagic species may be due to this fusion, as first suggested by 'Fulton (1898)'. In *Poecilia reticulata* the massed yolk disintegrates into small particles during further development (Figs. 3/4, 3/5).

EM analysis of the yolk platelets in *Noemacheilus barbatulus, Gobio gobio* and *Barbatula barbatula*) revealed that they are zonated with an electron-dense central stripe and two less electron-dense lateral parts. Ultrahistochemical tests showed that the median part is made up of proteins and lipids, while the lateral portions contain mostly proteins and acid mucopolysaccharide '(Riehl, 1977a; Lange et al., 1983)' (Fig. 3/6).

The yolk globules of the mature egg of the zebrafish *Brachydanio rerio* are scattered throughout the cytoplasm. EM studies have shown that each is made

Yolk (vitellus) 27

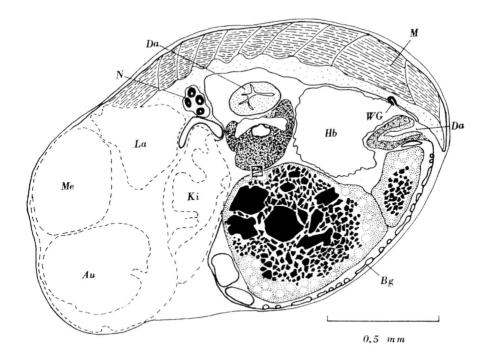

Fig. 3/4 *Poecilia reticulata* (*Lebistes reticulatus*) in egg envelope, longitudinal (sagittal) section. Massed yolk shown in Fig. 3/1 has changed into small pieces. (Kunz, 1964, with kind permission of the director of the Muséum d'Histoire naturelle, Geneva). Au - eye; M - skeletal muscle; Me - mesencephalon; Bg - blood capillaries; N - kidney; Da - intestine; Hb - urinary bladder; WG - Wolffian duct; Ki - gill; La - labyrinth. Note area of liver-yolksac contact (indicated by square), dealt with in Chapter 19.

up of three components, i.e. an enclosing membrane, a superficial layer of fine, dense granules and one to several main bodies which display a crystalline pattern. Under high magnification, the main body may appear as a rectangular network with meshes of approx. 100 x 50 Å, or as parallel electrondense bands depending on the plane of the section '(Yamamoto and Oota, 1967b)' (Fig. 3/7).

3.3 YOLK FORMATION (VITELLOGENESIS)

Previtellogenesis immediately precedes vitellogenesis. It denotes the stage at which the different cytoplasmic organelles are differentiated (so-called yolk nuclei, ribosomes, mitochondria, Golgi apparatus, endoplasmatic reticulum,

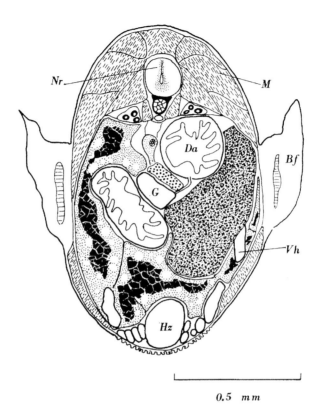

Fig. 3/5 Transverse section through *Poecilia reticulata* (*Lebistes reticulatus*) after hatching (30th embryonic day) showing that yolk (black platelets surrounded by periblast) has not yet been fully resorbed. (Kunz, 1964, with kind permission of the director of the Muséum d'Histoire naturelle, Geneva). M - skeletal muscle; Bf - pectoral fin; Da - intestine; Nr - neural tube; G - gall bladder; Vh - *vena hepatica*; Hz - heart Speckled area = liver.

lamellary bodies and lipid bodies). It is towards the end of this period that the microvilli start to lay down the zona pellucida '(Ulrich, 1969; Riehl, 1991)' (see Chapter 5).

Vitellogenesis is the last stage in oocyte development, prior to ovulation. (rev. 'Mommsen and Walsh, 1988'). Formation of lipid yolk starts prior to that of proteid yolk.

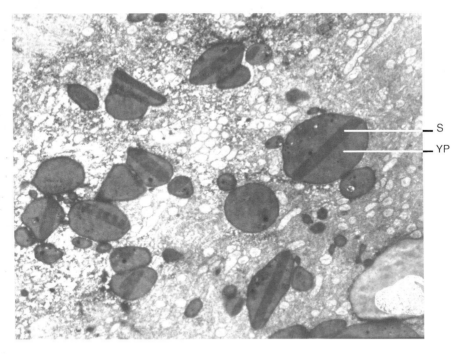

Fig. 3/6 TEM of yolk of *Barbatula barbatula*. × 5 700 (Courtesy of Rüdiger Riehl). Yolk platelets (YP) are zonated with electron dense central stripe (S) and two lateral less electron dense parts.

3.3.1 Genesis of lipid (fatty) yolk

This can take place in two different ways: Lipid droplets in the oocytes of the guppy *Lebistes reticulatus* (*Poecilia reticulata*) form in the periphery. Fusing vesicles of the Golgi body have been reported to be engaged in formation of the lipid droplets in the guppy while in *Blennius pholis* no Golgi complexes are associated with the development of the droplets '(Droller and Roth, 1963; Shackley and King, 1977)'. The lipid droplets of *Blennius pholis* increase in size during oogenesis and the mature droplets do not possess a limiting membrane as was observed in the eggs of the guppy. In the latter lipid yolk forms in the periphery of the oocyte and it has been suggested that the vesicles of the Golgi complex are thought to fuse to form the lipoid yolk. The Golgi body is not associated with the development of lipid droplets in *Blennius pholis* and it has not been established where its lipid precursors are synthesized. As shown for *Noemacheilus barbatulus* and *Gobio gobio* there are two ways of lipid yolk formation in this species: 1) small vesicles originate in the cytoplasm; they increase in size and eventually fuse 2) another type of lipid is

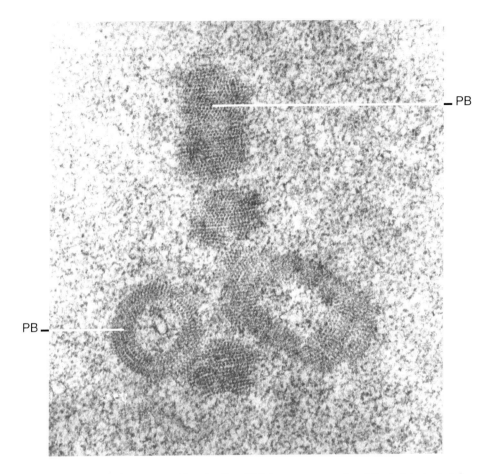

Fig. 3/7 TEM of paracrystalline bodies (PB) in young oocyte of *Barbatula barbatula*. × 50 300 (Courtesy of Rüdiger Riehl). The tubular bodies, some of them orientated at right angles to each other, are protein precursors at early developmental stages.

formed by the Golgi body. According to histochemical analyses the lipid yolk consists mainly of neutral fats (triglycerides) '(Riehl, 1977a)'. The precise origin of the lipid yolk has not yet been settled satisfactorily (rev. 'Guraya, 1986').

3.3.2 Precursors of proteid yolk

The precursors are produced exogenously (from an organ other than the ovary) or endogenously (from organelles of the oocyte) or by both sites. Exogenous formation is also called heterosynthetic and endogenous formation intraovarian or autosynthetic.

Exogenous origin

Vitellogenesis is characterized by a rapid growth of the oocyte. The young oocyte of the trout, which measures about 20 microns in diameter, increases to about 5 millimetres at maturity '(Selman and Wallace, 1989)'. In *Brachydanio rerio* the increase is from 120 to 600 microns '(Ulrich, 1969)'. This growth is mainly due to the incorporation of the yolk protein precursor vitellogenin (a lipo-glycophosphoprotein of 250-600 kDa) into the oocyte '(Norberg and Haux, 1985)'. The term vitellogenin had been coined by 'Pan *et al.* (1969)'. The uptake of vitellogenin is the result of receptor-mediated endocytosis from the blood by membrane-bound receptors '(Tyler *et al.*, 1988a, 1988b)'. The degree of growth of intraovarian eggs is also expressed in the gonadosomatic index GSI (weight of the gonads as a percentage of the body weight of the maternal fish). It was recently shown that there is another lipoprotein, formerly known as 'peak protein A' and now as VHDLII (very high density lipoprotein II) incorporated in vivo by the ovary of mature female winter flounders, *Pleuronectes americanus* '(Nagler and Idler, 1989)'.

It has been suggested that early production of yolk of the eggs of *Fundulus heteroclitus* and the rainbow trout is of endogenous nature. Their early follicles contain a whole series of yolk proteins of unknown origin '(Selman *et al.*, 1986; Inoue and Inoue, 1986)'. 'Tyler (1993)' showed with sensitive radioimmunoassays that in the rainbow trout even small oocytes (diameter 0.3 mm) are able to sequester vitellogenin (which is present in the circulating blood throughout the life of the fish). In the small oocyte (follicle diameter 0.3 mm) vitellogenin makes up only 0.5 % and in the follicle of 0.5-0.8 mm 3%, while in 1.0 mm follicles it rises to 50% of total protein. The rise is probably due to the increased activation of the vitellogenin receptors. Therefore, the endogenous contribution of yolk in this species is doubtful.

As mentioned above, vitellogenin in the plasma of teleosts is a VHDL (very high density lipid). Its lipid content has been reported as 21.5% and 18% in rainbow trout, 19% in sea trout, 21% and 20% in goldfish. Phospholipids predominate and represent, by weight, about two-thirds of these lipids '(Babier and Vernier, 1989)'.

In vitellogenic oocytes vitellogenin is proteolytically cleaved into smaller proteins, lipovitellin, phosvitin and possibly others. It has been suggested that in salmonids, the flounder *Pseudopleuronectes americanus* and *Verasper moseri* other derivatives of vitellogenin, so-called beta-components (which, in contrast to lipovitellin and phosvitin contain neither lipid nor phosphorus) may also enter the oocyte '(Nagler and Idler, 1989; Matsubara and Savano, 1995)'. In contrast, several authors claim that vitellogenin is not processed into smaller peptides after uptake into the oocyte '(cf. Tyler, 1993)'. It has been maintained that further degradation of these proteins takes place only later in embryonic life. However, it emerges that lipovitellin, phosvitin and beta components are further cleaved into smaller polypeptides during egg maturation (so-called secondary proteolysis) as shown in several marine species including the halibut, *Hippoglossus hippoglossus*, the cod, *Gadus morrhua*, the plaice, *Pleuronectes platessa*, the killifish, *Fundulus heteroclitus* and the gilthead sea bream, *Sparus aurata* '(Wallace and Selman, 1985; Wallace and Begovac, 1985; Greeley et al., 1986)'. In the oocytes of the barfin flounder, *Verasper moseri*, the beta-component and phosvitin disappear during maturation and lipovitellin is also partially degraded '(Matsubara and Sawano, 1995)'.

Microscopic evidence for exogenous supply of yolk precursors has been supplied by several authors: Membrane vesiculations observed at the base of the microvilli over the whole surface of the oocyte are indicative of pinocytosis of material into the egg '(Flügel, 1964c; Wartenberg, 1964; Götting, 1966)'. Ultrastructural evidence for ovarian incorporation of exogenous vitellogenin by micropinocytosis has been provided by 'Droller and Roth (1963)' for the guppy *Poecilia reticulata*, by 'Anderson (1968)' and 'Begovac and Wallace (1988)' for the pipefish *Syngnathus fuscus*, by 'Wegmann and Götting (1971)' for the swordtail *Xiphophorus helleri*, by 'Burzawa et al. (1994)' for the silver eel stage of *Anguilla anguilla* and by 'Kobayashi (1985)' for *Oncorhynchus keta*.

In order to establish the origin of the yolk in the zebrafish oocytes, injections of tritiated amino acids into the female *Brachydanio rerio* were followed by autoradio-graphy. This revealed after 3 hours a maximum uptake of the tracer by the liver, which was followed by the circulating blood and the oocytes. Tracer was incorporated into the cortical alveoli (Chapter 4) within three hours. However, tracer accumulated only later — after 24 hours and more — in the yolk globules. This suggests that the main mass of the yolk needs the participation of an exogenous protein component for its formation '(Korfsmeier, 1966)'. Subsequently, incorporation of exogenous vitellogenin into oocytes has been followed by autoradiography in the minnow, *Cyprinodon variegatus*, the killifish, *Fundulus heteroclitus*, and the rainbow trout, *Salmo gairdneri* '(Selman and Wallace, 1982a,b, 1983)'. In the rainbow trout, *Oncorhynchus mykiss*, vitellogenin uptake of follices takes place throughout the ovarian development, rather than being confined to certain stages of oocyte growth. This was shown by following with radioimmunoassay the in vivo uptake and processing of labelled $^{32}P^3H$-vitellogenin. It was concluded that vitellogenin gives rise to more than 60% of the total protein in mature oocytes '(Tyler et al., 1988a, 1991; Tyler, 1993)'.

Plasma vitellogenin in teleosts circulates generally as a dimer '(Babin and Vernier, 1989)'. In the rainbow trout, (*Salmo gairdneri*) (*Oncorhynchus mykiss*), plasma levels of vitellogenin begin to increase a year before ovulation '(Copeland et al., 1986)'. Immunoreactivity of vitellogenin has been shown in oocytes 9 to 10 months before ovulation (oocyte diameter 0.6 mm). At this time, membrane-bound vitellogenin receptors, which mediate endocytosis of vitellogenin, are expressed over the whole cell surface '(Tyler et al., 1991)'. Uptake of vitellogenin continues but decreases in oocytes close to ovulation (4-4.5 mm diameter). In *Oncorhynchus mykiss* endosomes containing vitellogenin are thought to fuse with Golgi derived vesicles and are subsequently called multivesicular bodies (MVB). They reside in the cortex[4] and contain acid phosphatase and the proteolytic enzyme cathepsin D. Yolk proteins (lipovitellin and phosvitin) derived from vitellogenin, together with cathepsin D, will eventually be included in the central coalescent yolk mass '(Sire *et al.*, 1994)'.

'Nagler *et al.*, (1994)' have developed a method for culturing ovarian follicles from the rainbow trout. They succeeded in keeping the follicles over a 6-day period and noted sequestration of vitellogenin from the medium and subsequent proteolytical processing of the incorporated vitellogenin. This has opened a new way of studying the mechanisms — hormonal and intraovarian — involved in oocyte growth.

Immunohistochemical techniques using antiserum against rainbow trout vitellogenin confirmed that vitellogenin is taken up from the plasma into the oocyte '(Hyllner *et al.*, 1994)'. It has been suggested that in *Heterandria formosa* Na^+,K^+ ATP-ase participates in the micropinocytosis of vitellogenin '(Riehl, 1980a, 1991)'.

Triggering of vitellogenin

As mentioned above, pulse-chasing of tritiated amino acids in *Brachydanio rerio* had revealed an uptake into liver, which was followed by uptake into blood plasma and finally into yolk globules of the oocytes '(Korfsmeier, 1966)'. 'Shackley and King (1977, 1978, 1979a,b)' have shown that in *Blennius pholis* vitellogenin is formed in the liver and transported to the ovaries by the blood, where it is taken up rapidly at the oocyte periphery by micropinocytosis. Coated vesicles (bristle coat) fusing to form tubules and finally yolk granules have been observed. The peripheral region of the oocyte at this stage showed many mitochondria, ribosomes and polyribosomes.

It was subsequently revealed that injection of **oestrogen** resulted in induction of vitellogenic proteins in the plasma of stickleback, sea bass, sea horse, pipe-fish, rainbow trout and roach. These proteins are subsequently present only in the yolk granules and near the theca layer of the follicle '(Covens *et al.*, 1987)'.

Already 'Zahnd and Porte (1966)' had observed in the **liver** an increase in ribonucleic acid (RNA) during the breeding period of various teleosts. EM analyses pointed to an increased activity of the liver during vitellogenesis, such as development of the rough endoplasmic reticulum (RER), enlarged Golgi bodies and mitochondria with 'densely packed and concentrically arranged membrane configurations' '(Ng and Idler, 1983)'. The synthesis and secretion of vitellogenin by the liver under the influence of oestrogen was subsequently reviewed by 'Guraya (1986)'. Synthesis of vitellogenin in the maternal liver is dependent on a high oestrogen titer. The hormonal rise is referred to as 'second increase of oestrogen'. The 'first increase in oestrogen' takes place when the SF (spawning female-specific) proteins of the egg envelope are induced (Chapter 5). It was finally established that oestrogen triggers in the hepatic cells of *Oryzias latipes* the expression of the genes for both the SF substances and vitellogenin '(Murata *et al.*, 1994)' and that the control is effected by a **nuclear oestrogen receptor** '(Burzawa-Gerard *et al.*, 1994)'.

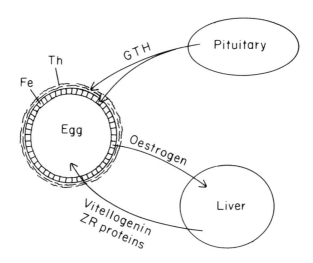

Fig. 3/8 Diagram of vitellogenesis: interaction of pituitary, ovary and liver. Gonadotrophic hormone (GTH) released by the pituitary stimulates the formation of oestrogen in the follicle cells of the ovary. Oestrogen induces synthesis of vitellogenin and ZR proteins in the liver. These hepatic proteins are carried to the ovary. Fe - follicular *epithelium*; GTH - gonadotrophic hormone; Th - follicular *theca*.

Additionally, recent studies have revealed that vitellogenin is also inducible in immature and male teleosts by oestrogen treatment '(Murata *et al.*, 1994)'. 'Buerano *et al.*, (1995)' identified two immunochemically different forms of vitellogenin in the plasma of female and oestrogen-induced males of *Oreochromis niloticus*. Oestrogen is produced by the follicle cells of the ovary in response to the secretion of **gonadotropin(s)** by the pituitary.

In the eel, *Anguilla anguilla*, vitellogenesis starts at the stage of silver eel. (The 'previtellogenic' stage is called 'yellow eel' during which the fish take up food and grow. This is followed by the 'silver eel' stage during which the fish fast and start their migration for reproduction into the sea). It has been recently shown that treatment with growth hormones increases the sensitivity of the response of the liver to oestrogen. The action of growth hormone was confirmed with in vitro studies indicating that the hormones act directly on the liver cells '(Burzawa-Gerard *et al.*, 1994)'. Similar results had been obtained by 'Carnevali *et al.* (1992)' on hepatic cell cultures of *Anguilla japonica*. There are also indications of a hormonal induction of vitellogenesis by androgens (at pharmacological doses) and by androgens applied to cultures of hepatocytes. Such experimental approaches may in future help to elucidate the mechanisms of genome stimulation '(Burzawa-Gerard *et al.*, 1994)'.

'Srivastava and Srivastava (1994)' showed a positive correlation between liver and serum proteins, serum calcium, inorganic phosphate and magnesium levels during the annual ovarian cycle of *Heteropneustes fossilis* which seems to be related to vitellogenin synthesis.

It should be stressed that the sequence of events reported above — synthesis of vitellogenin in the liver triggered by oestrogen produced by the ovary — was already put forward as a hypothesis for vitellogenesis in the goldfish by 'Bailey (1957)'.

Endogenous (autosynthetic) intraovarian origin

According to 'Ludwig (1874)' 'yolk of fish is formed endogenously by the oocyte'. In contrast, 'Gegenbauer (1861)', 'Waldeyer (1883)' and 'van Beneden (1877)' were of the opinion that the follicular epithelium forms the yolk.

It has to be stressed that under the term 'endogenous yolk formation' many authors understand the formation of the content of cortical alveoli. They call the alveoli 'yolk vesicles' and their contents are referred to as 'intravesicular yolk' (G. intravesikuläre Dotterschollen). In our treatise, cortical alveoli are treated separately (Chapter 4) and are not considered part of the yolk.

The above-mentioned results are based on light-microscopical analyses. From 1960 onwards it emerged, largely on the basis of EM analysis, that proteid yolk is formed both endo- and exogenously.

Dual (exogenous + endogenous) origin of protein yolk

'Droller and Roth (1966)' concluded on EM-evidence that in *Lebistes reticulatus* (now *Poecilia reticulata*) the proteid yolk is formed by the egg's own Golgi bodies and endoplasmic reticulum as well as by the occurrence of highly specific and selective micropinocytotic processes. 'Yamamoto and Oota (1967a,b)' stressed that in *Brachydanio rerio* the yolk globules of mature oocytes were formed by modified mitochondria, and not in the Golgi body and rough endoplasmic reticulum (RER), which agrees with results obtained with light microscopy by 'Malone and Hisaoka (1963)' on the same species. The changes observed by the former authors in the mitochondria of the oocyte show the gradual disappearance of cristae and intramitochondrial granules, followed by the appearance of dense round particles of about 160 Å in diameter. At the same time the double membrane of the mitochondria changes into a single limiting membrane. Through the action of large pinocytotic vesicles (300 microns in diameter) unto the modified mitochondrion the primary yolk globule arises, filled with dense minute particles, and finally the mature yolk globule with a crystalloid body appears. 'Ulrich (1969)' suggested that in *Brachydanio rerio* the crystalline part of the protein globules are of exogenous origin whereas the amorphic mass surrounding them is of endogenous origin. 'Anderson (1968)' reported both endo- and exogenous production of yolk in the pipefish, *Syngnathus fuscus*, and the killifish, *Fundulus heteroclitus*.

He concluded that the protein portion of the yolk is produced by the endoplasmic reticulum and is subsequently transferred to the Golgi body via vesicles while the Golgi body, in addition, produces the polysaccharide component of the yolk (cortical alveoli).

'Riehl (1991)' reports for *Noemacheilus barbatulus* and *Gobio gobio*, in addition to exogenous sources, various ways of endogenous proteid-yolk formation: 1) via transformation by mitochondria; 2) from multivesicular bodies; 3) in annulate lamellae; 4) from lamellary bodies; 5) from paracrystalline bodies (observed only in young oocytes) (Fig. 3/7, 3/9). Paracrystalline bodies at the EM-level consist of many osmiophilic particles (diameter 150-300 Å) arranged in tubes, sometimes at right angles to each other. Ultrahistochemically the bodies have been identified as yolk precursors '(Riehl, 1977b)'. Multivesicular bodies commonly occur in pre-vitellogenic and vitellogenic oocytes and are absent in later stages of vitellogenesis (Fig. 3/9). According to 'Anderson (1968)' and 'Shackley and King (1977)' it is not clear whether these bodies fuse with micropinocytotic vesicles and whether they contain digestive enzymes.

As the yolk mass enlarges, the yolk vesicles (future cortical alveoli) and the nucleus (germinal vesicle) shift towards the periphery. The nucleus becomes finally located in the cortical cytoplasm, i.e. the future animal pole.

3.4 RELATION BETWEEN YOLK AND CYTOPLASM

The yolk of all teleostean eggs is surrounded by a thin layer of cytoplasm (ooplasm, protoplasm[3]) which, in addition to cortical alveoli and yolk components, contains many mitochondria, numerous Golgi complexes, large quantities of rough endoplasmic reticulum, ribosomes and particulate glycogen (see Chapter 4).

Due to the large amount of yolk, teleostean eggs have been generally classified as macrolecithal (polylecithal). As regards the distribution of the yolk, they are referred to as telolecithal, with the cytoplasm confined to the periphery and thickened at the animal pole.

However, from the middle of the 19th century onwards many reports have been published stressing the intermingling of cytoplasm and yolk in the ovulated teleostean oocyte. This condition is referred to as 'plasmolecithal' by 'Soin (1981)'. Already 'Coste (1850)' and 'Lereboullet (1861)', followed by 'Gerbe (1875)' and 'Henneguy (1888)' thought that the plasmic and nutritive elements (F. éléments plastiques et nutritifs) in salmonids, perch and pike are mixed in the egg until deposition. 'Brock (1878)' noted that in thin sections liquid droplets had often fallen out leaving behind a protoplasmic net. 'Balfour (1881)' quoted 'van Bambeke's' findings (1875) of fine protoplasmic strands, which radiate from the blastodisc and 'plunge' into the vitelline sphere. The cytoplasm is also mixed with the yolk in *Alosa*, as reported by 'Ryder (1881)'. 'Hoffmann (1881)'mentions that in the herring and *Heliasis* yolk and cytoplasm are mixed. 'Waldeyer (1883)' described how in meroblastic unfertilized eggs numerous fine cytoplasmic strands emanat-

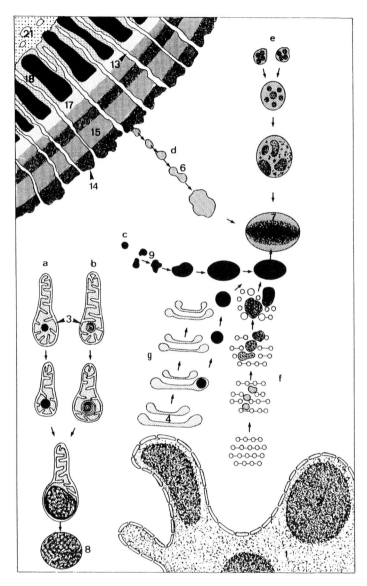

Fig. 3/9   Different modes of yolk formation in *Noemacheilus barbatulus* and *Gobio gobio*. (Riehl, 1991. Illustration by Alois Bleichner; with kind permission of Acta Biologica Benrodis). a, b - by transformation of *mitochondria*; c - from paracrystalline bodies; d - from exogenous material; e - from multivesicular bodies; f - in annulate *lamellae*; g - in Golgi bodies; 1 = *nucleus*; 2 = *nucleolus*; 3 = *mitochondria*; 4 = Golgi body; 6 = exogenous material; 7 = yolk platelet, 8 = yolk precursor, 9 = paracrystalline body, 13 = microvilli, 14 = Zona radiata externa, 15 = Zona radiata externa, 17 = Zona radiata interna, 18 = Zona radiata interna, 21 = ooplasm.

ing from the superficial cytoplasm enter the yolk and enclose it in a network. As illustrated by a woodcut, the cytoplasmic processes become finer — and the net larger — the deeper they dip into the yolk. Many other authors described in different terms how in teleost eggs there is no sharp delimitation between the cytoplasm and the yolk before activation, e.g. 'Kupffer (1878)', 'Agassiz and Whitman (1884)', 'Brook (1885a, 1887a)', 'von Kowaleski (1886a)', 'Stuhlmann (1887)', '(Henneguy, 1888)', 'McIntosh and Price (1890)', 'Cunningham (1891)', 'Fulton (1898)', '(Gudger, 1905)', '(Assheton, 1907)'. This is followed by later publications: For *Brachydanio* '(Rosen-Runge, 1938; Yamamoto I. and Ohta, 1967a,b; Kane, 1999)' and for *Clupea* '('Kanoh (1949; Yanagamachi and Kanoh, 1953)'.

While in the above-mentioned eggs before fertilization the protoplasm is evenly distributed in the yolk, on activation it becomes concentrated at one pole '(Henneguy, 1888)'. This process is known as 'ooplasmic segregation' or 'bipolar segregation', dealt with in Chapter 9 (Fertilization).

It should be stressed that already the early investigators noted that 'interfusion of the yolk with protoplasm' is largely restricted to demersal eggs. In contrast, pelagic eggs are translucent and the cytoplasm forms only a superficial layer around the yolk-sphere '(Brook, 1887a; Fulton, 1898). 'Hoffmann (1881) in his treatise on teleosts describes the transparent eggs of *Julis vulgaris, Scorpaena scrofa, Serranus* and *Fierasfer*. Also a 'read-through' of the monography 'Fauna e flora del golfo di Napoli' by '*Salvatore lo Bianco* (1931)' shows that pelagic eggs in general are translucent.

The demersal opaque ova of the Salmonidae and *Amiurus nebulosus* take up an intermediate position. At oviposition the cytoplasm and yolk are clearly separate and the cytoplasm surrounding the egg is already thickened into a blastodisc at the animal pole '(Reis, 1910)' (see Chapter 9).

3.5 HYDRATION OF THE OOCYTE DURING MATURATION

'Fulton (1898)' concluded that the 'clear, glassy appearance and larger size' of pelagic eggs 'seemed to be due to sudden accession of fluid from the ovarian follicles, which increases the bulk of the ovum and renders the opaque contents clear by dilution. In demersal eggs the augmentation of the size is much less.' According to 'Milroy' (cf. 'Fulton') the percentage of water is 91.86% in pelagic and 65.5% in demersal eggs. Hydration of oocytes following maturation (germinal vesicle breakdown) was achieved in vitro by 'Hirose (1976)', 'Iwamatsu and Katoh (1978)' and 'Wallace and Selman (1985)'. They concluded that hydration is especially pronounced in marine fish with pelagic eggs. 'Kjesbu and Kryvi (1993)' reported that in the cod the crystalline yolk granules disintegrate, swell and coalesce prior to maturation. The water content of these granules increases sharply during the last stage of maturation. Concomitantly, a steady increase in water was also noted in vitellogenic oocytes: the eggs of *Fundulus heteroclitus* swell two to three times their original volume by the time of ovulation. A loss of phosphorus from yolk protein has been suggested as a supply of energy for water uptake. On the other hand, a rise in oocyte potassium during maturation has been associated with hydration. Yolk proteolysis has also been thought to assist hydration '(Greeley et al., 1986)'.

Summary

Yolk is divided into lipid yolk and protein yolk. Vitellogenesis is the main event responsible for oocyte growth. Yolk precursors are supplied from outside the oocyte (exogenous) and yolk is produced within the oocyte by its own organelles (endogenous). Vitellogenins are the major precursor for protein yolk. The synthesis of vitellogenin takes place in the liver of females and is dependent on high oestrogen titers. In males and immature females, in which vitellogenins are normally absent, their synthesis in the liver can be induced by injection of oestrogen. In mature females oestrogen is produced by the follicle cells and triggered by gonadotrophic hormones secreted by the pituitary. The vitellogenins are transported by the blood to the ovaries where they are taken up by micropinocytosis. Their incorporation into the follicles is brought about by membrane-bound receptors with specific binding sites for vitellogenin. In the oocytes, vitellogenins are processed into lipovitellin and phosvitin, which are the major yolk proteins. It has been shown for several teleosts that phosvitin and lipovitelllin are further cleaved into smaller polypeptides, a process which is referred to as secondary proteolysis. These results are based on studies with radioactive isotopes, immunological and ultrastructural techniques.

EM analysis of several teleost oocytes revealed that proteid yolk platelets are zonated with an electrondense central stripe (made up of proteins and lipids) and less electrondense lateral parts (made up mostly of proteins and mucopolysaccharides), while the electrondense bodies in the mature oocyte of the zebrafish displayed a crystalline pattern. Influx of water occurs during maturation of the oocytes. The high water content of ovulated pelagic eggs, compared with demersal eggs, is thought to be a cause of their buoyancy.

The precise origin of lipid yolk has not yet been resolved satisfactorily. The term 'carbohydrate yolk' refers to the cortical alveoli dealt with in Chapter 4.

Teleost are commonly known as being telolecithal (yolk concentrated at one end), with the cytoplasm confined to the periphery and surrounding the yolk. However, many analyses from the early investigators to the recent ones, point to the fact that many mature teleostean eggs are centrolecithal to begin with. In other words, in the ovulated egg the cytoplasm permeates, to a greater or lesser degree, the yolk. Changes in the arrangement of the egg contents occur after activation or later in the process of fertilization. These changes are brought about by cytoplasmic streaming towards the animal pole leaving the rest of the yolk surrounded by a very thin cortex[4]. At this stage the egg has become truly telolecithal.

Endnotes

[1] yelk, foodyolk, nutritive yolk, nutritive plasm. Yolk was originally known as deut(er)oplasm (a term coined by 'van Beneden')
G. Dotter, Nahrungsdotter, Nebenkeim of 'His', Nebendotter of 'His', Nahrungsdotter (as opposed to Bildungsdotter) of 'Reichert', Nahrungsdotter + Rindenschicht = Dotterhaut of 'Oellacher'
F. substance plastique, substance nutritive, vitellus nutritif, globe vitellin
I. vitello nutritivo of 'Raffaele', tuorlo

[2] caenogenetic (from Gr. kaino = recent): Originally, the development of features in ontogenesis which are the result of environmental adaptation. Now, specifically the development in embryonic or larval forms of functional adaptations not present in the adult.

[3] see endnote [3] in Chapter 2.

[4] see Chapter 4

Chapter four

Cortex and its alveoli

'Cortex, in Latin bark or rind, is not only of botanical usage, but is also the name given to e.g. the 'rind' of the human brain and adrenal, and also of animal eggs generally.'

4.1 CORTEX

Cortex[1] refers to the peripheral layer of the egg, subjacent to the oolemma[2]; it has been known since the publications of 'Lereboullet (1862)' and 'Kupffer (1868)'. 'Van Bambeke (1872)' called it 'couche intermédiaire' (intermediary layer). 'His (1873)' described the cortex as a granular (G. 'trüber') layer, which surrounds the liquid yolk[3] and contains many stainable drops. 'Klein (1876)' studied in detail the cortex of the trout. 'Owsiannikow (1885)' described the cortex of the egg of *Gasterosteus*. 'Eigenmann (1890)' maintained that the 'membrane within the ZR', described by 'Vogt (1842)', 'Ransom (1856)' and 'Oellacher (1872)', is not to be considered as a vitelline membrane but represents 'the superficial part of the protoplasm of the egg', i.e. the cortical layer or cortex. Many authors described a zonoid layer as outermost layer (G. 'Zonoidschicht' of His). However, it was soon realized that this layer disappears in the maturing egg and that in its place appears the cortex.

Fig. 4/1 *Cortex* of the trout egg (Henneguy, 1888). cc - *cortex* with small cortical granules; ch - envelope (*chorion*); h - cortical *alveoli*.

The cortex of teleost eggs is distinct and measures, e.g. in the zebrafish, 15-20 microns in thickness '(Hart, 1990)'. Ultrastructurally it can be subdivided into two layers. The outer layer, subjacent to the oolemma and continuing into the microvilli[4] or microplicae[5] is electron-dense. It has been shown to contain a meshwork of filaments, which continue in the form of bundles into the microvilli. (It is most probably the presence of microvilli which give this layer a radiate appearance as e.g. described by 'Brock (1878)' for *Alburnus lucidus* and also by the same author for *Salmo fario* and *Perca fluviatilis*). The deeper layer, subjacent to the yolk, is less dense and contains organelles, such as ribosomes, mitochondria, smooth endoplasmic reticulum, Golgi apparatus, cortical granules, lipid droplets and, occasionally, small yolk bodies '(Kobayashi, 1985; Hart, 1990)'. Filaments with a diameter of about 8 nanometres, confined to the peripheral layer of the cortex, have been described for *Brachydanio* and *Oncorhynchus*. They are sensitive to treatment with cytochalasin B, which would suggest that the cortical cytoplasm includes an actin-containing cytoskeleton made up of microfilaments. Such a meshwork would be necessary to maintain the shape of the egg and its surface specializations. Staining with rhodamine phalloidin indicated a localization of F-actin beneath the oolemma '(Hart and Wolenski, 1988; Hart, 1990)'. 'Farias et al. (1995)' analyzed the main features of tubulins and microtubule-associated proteins in the cortex of the unfertilized versus fertilized egg of *Oncorhynchus mykiss*. 'Becker and Hart (1996)', on the basis of SDS-PAGE, immunoblotting and fluorescence microscopy, applied to the zebrafish egg, report the presence and distribution of filamentous and non-filamentous actin and myosin II, which are thought to modulate the restructuring of the egg cortex upon fertilization (Fig. 4/2). 'Jesuthasan and Strähle (1997)' noted in the zebrafish after fertilization an ordered arrangement of parallel layers of microtubules at the vegetal pole. Activities occurring at the egg surface, following sperm-egg union (formation of fertilization cone, exocytosis of cortical granules, retrieval of membrane by endocytosis, migration of pronuclei, polar body formation, cytoplasmic aggregation in some) have been dealt with in the appropriate chapters.

4.2 CORTICAL ALVEOLI (CA) AND CORTICAL GRANULES

Rapid growth of the oocyte is due to the deposition of yolk and the accumulation of CA. While the formation of CA and yolk may proceed simultaneously, the CA are the first to develop. The alveoli are often arranged in one layer (chum salmon, lake trout) or may be arranged into two or three rows (goldfish). CA are generally absent in the region of the micropyle '(Ginzburg, 1968)' (Chapter 7).

'Ransom (1854)', in his description of the ovum of *Gasterosteus*, mentioned 'The layer of the yelk immediately internal to the inner membrane... is formed by yellowish highly refractive drops which disappear in water, undergoing some remarkable changes, and by a fluid substance which water precipitates in a finely granular form'. Subsequently, 'Boeck (1871)' described a layer of 'round strongly-refractive globules of an oily nature' immediately under the egg-membrane. Still, 'Sars (1876)' is usually quoted as the first to call attention to CA in teleosts.

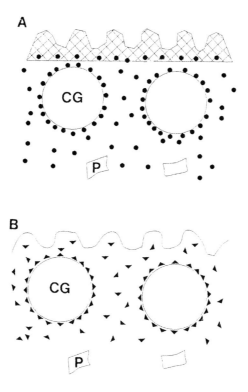

Fig. 4/2 Schematic representation of the spatial distribution of organelles in the *cortex* of the unfertilized zebrafish egg. (Becker and Hart, 1996, with kind permission of Wiley-Liss, Inc., a subsidiary of John Wiley & Sons, Inc.). **A** - displays distribution of filamentous actin (cross-hatch) and non-filamentous actin (black dots). **B** - shows localization of myosin-II (black triangles). While filamentous actin is restricted to the *microplicae* and the *cytoplasmic* face of the plasma membrane, in the *cortex* proper myosin-II colocalizes with the filamentous and non-filamentous actin. Both non-filamentous actin and myosin-II seem to be concentrated over the surface of the cortical granules. CG - cortical granule; P - yolk platelet.

He called them 'oil bodies' '(cf. T. Yamamoto, 1961)'. 'Ryder (1884)' made the observation that CA of the mature eggs of the cod disappear after fertilization. 'Ziegler (1882)' agreed with 'Balbiani' that the bodies observed in the cortex of *Salmo salar* are vesicles of an albuminoid nature and not nuclei, as several other authors suggested. 'Owsiannikov (1885)' referred to 'yolk bodies' (G. Dotterkörper) contained in the cortex of mature eggs of *Gasterosteus*. A series of studies performed around the middle of the 20th century indicated that CA arise from yolk vesicles, a term which is inappropriate since these vesicles are not involved in yolk production; they are rather homologous to the cortical granules of other vertebrates. The cortical granules or vesicles first appear at the periphery. They continue to increase in number as well as in size until they almost fill the oocyte. This is followed by movement to the periphery at which time, as already mentioned, the yolk production in the centre starts '(T.S. Yamamoto, 1955; Osanai, 1956; K. Yamamoto, 1956a,b; Malone and Hisaoka, 1963; Tesoriero, 1980; Wallace and Selman, 1981; Selman et al., 1986; 1988; Bazzoli and Godinho, 1994)'.

The mature CA are arranged in one to several peripheral layers '(Kudo, 1991; Bazzoli and Godinho, 1994; Linhart et al., 1995)'. In the mature oocytes of the zebrafish the alveoli form several irregular rows '(Hart and Donovan, 1983)'. A single oocyte of 500 microns diameter of *Brachydanio rerio* contained approximately 3 300 CA of an average diameter of 13 microns '(Hart et al., 1977)'. CA are spherical or oval and are of variable size. In chum salmon they measure 2-25 microns in diameter, in *Oryzias* 10-40 microns, in *Syngnathus fuscus* 10-12 microns '(Anderson, 1968)' and in *Fundulus* up to 50 microns '(Hart, 1990)'. There exists also considerable variation in the morphology of the contents of these granules, even within the same oocyte. For instance, the mature CA in *Syngnathus fuscus* and *S. scovelli* contain one or two eccentrically situated filamentous bodies; these are absent in the alveoli of *Fundulus heteroclitus*. The alveoli generally appear at the periphery and continue to expand centripetally. In the mature oocyte they have moved again centrifugally to come to lie in the cortex '(Riehl, 1991)'. However, in the oocytes of the marine *Lutjanus synagris* the alveoli are formed above the nuclear membrane and spread centrifugally, and in *Phoxinus* and other freshwater teleosts the alveoli appear simultaneously in the whole oocyte '(Arndt, 1956; Riehl and Schulte, 1977a)'. Another variation was found in *Melanotaenia nigricans*, where the alveoli are observed first in the cortex of the animal pole and further formation proceeds towards the vegetal pole '(Sterba, 1957)'. The functional significance of these variations is still unknown.

Subsequently, transmission-EM of differentiating oocytes of *Syngnathus fuscus*, *Fundulus heteroclitus* and of *Oryzias latipes* eggs indicated that the Golgi body and endoplasmic reticulum are involved in the formation of the alveoli and their contents '(Anderson, 1968a)'. 'Ulrich (1969)' using EM-cytochemical techniques confirmed that in *Brachydanio rerio* the Golgi plays a role in the formation of the CA. EM-autoradiography, using ^{3}H-glucose and ^{3}H-galactose, showed that endoplasmic reticulum and Golgi body participate in the formation and transport of precursor material of the CA of *Oryzias latipes* '(Tesorerio, 1980)'.

SEM of CA of the zebrafish indicated that they consist of an electron-dense core surrounded by particulate, electron-lucent material '(Hart, 1990)'. Similar results had been reported for *Phoxinus phoxinus* by 'Riehl and Schulte (1977a)' (Fig. 4/3).

Cytochemical experiments revealed that the cortical granules are PAS-positive, often display metachromasia with toluidine blue and stain with alcian blue at low pH as well as with colloidin iron and bromophenol blue. These results would sug-

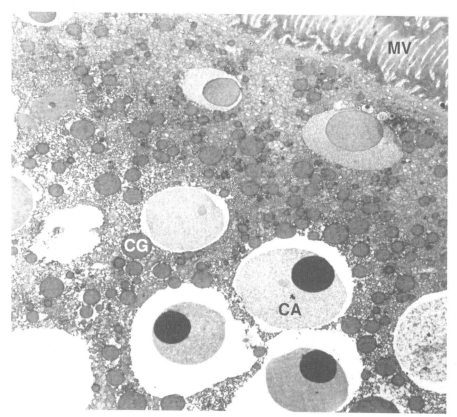

Fig. 4/3 TEM of egg *cortex* of *Phoxinus phoxinus* (Courtesy of Rüdiger Riehl). × 3 600. CA - cortical *alveolus*; CG - cortical granule; MV - *microvilli*. Note that in the CA an electron-dense '*nucleus*' is surrounded by less electron-dense material.

gest that the contents of the granules contain acid mucopolysaccharide and/or mucoprotein '(Kudo, 1976; Hart and Yu, 1980)'. Autoradiography of the ovaries of fish injected with tritiated amino acids or sugars revealed the presence of glycoprotein '(Aketa, 1954; T.S. Yamamoto, 1955; Korfsmeier, 1966; Riehl, 1977a; Bazzoli and Godinho, 1994)'. Cytochemistry, autoradiography, gel electrophoresis and indirect immunolabelling of oocytes of *Fundulus heteroclitus* documented that yolk vesicles contain an acidic glycoconjugate and confirmed that this is synthesized endogenously and that synthesis continues throughout oogenesis '(Selman *et al.*, 1986, 1988)'. These results would, therefore, confirm that the cortical vesicles of small oocytes give rise to the CA of later stages. When mature, they fuse, as CA, with the oolemma and discharge their contents as a result of activation or fertilization of the egg. This process is known as cortical reaction (see Chapter 9 on fertilization).

Other constituents of the cortical granules include lectins '(Nosek, 1984a)' and enzymes '(Guraya, 1982)'. Hexokinase has been shown within CA of the oocytes of *Misgurnus fossilis*. It may be interesting to note that the granules of the carp and goldfish display acid phosphatase activity, which is absent in the granules of the zebrafish '(Hart, 1990)'.

The alveoli of 102 species studied were shown to contain polysaccharides, which formed glycoprotein and glycoconjugates: neutral glycoproteins (35% of the species), carboxylated glycoconjugates (28%), neutral glycoproteins plus carboxylated glycoconjugates (18%), sulfated glycoconjugates (12%), neutral glycoproteins plus sialic acid-rich glycoproteins (6%), and neutral glycoproteins plus sulfated glycoconjugates (1%). The alveoli proved to be sudanophobic and did not contain any glycogen '(Bazzoli and Godinho, 1994)'. In contrast, 'Eggert (1931)' and 'Gilch (1957)' had reported the presence of fat in the alveoli of *Salarias flavoumbrinus* and labyrinthfish respectively '(cf. Kraft and Peters, 1963)'.

Comparison of mucopolysaccharides within alveoli of freshwater species revealed a relation of their type (acid or neutral) to habitat: marine fish contained neutral only, freshwater: acid only, and anadromous: neutral plus acid '(K.Yamamoto, 1956; Bazzoli and Godinho, 1994)'. The CA of mature common carp oocytes (*Cyprinus carpio*) contain A-granules (about 0.4-2 microns in diameter) which are enzyme-cytochemically distinct from the rest of the CA since they have no acid phosphatase activity. This suggests species-specific enzyme-cytochemical differences ('Kudo, 1978, 1991'). B-granules also occur in groups in the cortical cytoplasm after initiation of the cortical reaction and are also formed by the Golgi apparatus '(Linhart et al., 1995)'.

The usage of different names, apart from cortical granules CA, for the structures in the cortex and the inconsequential use of most of them, has lead to great confusion. The most frequent alternate terms are: yolk vesicles, cortical vesicles, intravesicular yolk, vacuolar yolk, vacuomes, carbohydrate yolk (F. gouttes claires).

Summary

Cortex refers to the cytoplasmic layer surrounding the oocyte. It is made up of two layers. The peripheral layer, bonded by the oolemma (with microplicae or microvilli), contains microfilaments and microtubules whereas the deeper layer contains ribosomes, mitochondria, smooth endoplasmic reticulum (SER), Golgi bodies, cortical granules, cortical alveoli, lipid droplets and also elements of a cytoskeleton.

Cortical alveoli (CA) are mostly spherical or ovoid structures embedded in the cortex. They arise from cortical granules and are often erroneously referred to as intravesicular yolk. Their content consists mainly of mucoproteins and is discharged as a result of activation or fertilization. The process is known as cortical reaction (see Chapter 9 on fertilization). The alveoli occur in one or more rows and are absent near the micropyle.

The Golgi body and the endoplasmic reticulum are involved in the formation of the CA and their contents. A significant variation in the morphology and contents of the CA, even within the same oocyte, has been observed.

Endnotes

[1] cortical layer of 'Ransom'
   G. Rindenschicht of 'His'
   F. Manteau protoplasmique of 'van Bambeke'
   I. Strato corticale of 'Raffaele'
[2] see endnote [2] of Chapter 5
[3] see endnote [1] of Chapter 3
[4] from Gk. mikros = small and L. villus = tuft of hair
[5] from Gk. mikros = small and L. plica = fold

Chapter five

Egg envelope

'Der interessanteste Theil des Fisheies sind unstreitig seine Hüllen, deren complicirter Bau von jeher die Aufmerksamkeit der Forscher auf sich gelenkt hat.'

Brock (1878)

Mature eggs are surrounded by a non-cellular envelope which serves to attract sperms but prevents polyspermy, protects the developing embryo, provides gas exchange and excretion, and, in viviparous fish, selects the transport of nutrients to the developing embryo. The envelope of some fish has proved to have an antibacterial function '(Riehl, 1991)' (see Chapter 6). Most teleostean egg envelopes consist of more than one layer. They are each of complex architecture and structure, which are of taxonomic importance and are also considered to be of possible phylogenetic value.

5.1 THICKNESS OF ENVELOPE

According to 'Kendall *et al.* (1984)' 'Most marinae fishes, regardless of systematic affinities, demersal or pelagic habits, coastal or oceanic distribution, tropical or bolear ranges, spawn pelagic eggs that are fertilized externally and float individually near the surface of the sea.' It is generally assumed that the envelope of pelagic eggs is much thinner than that of demersal eggs '(Stehr and Hawkes, 1979; Loenning *et al.*, 1988; Riehl, 1991; Hirai, 1993)'. There are exceptions to this: For instance the envelope of the demersal, non-adhesive eggs of *Brachydanio rerio* is thin, approximately 1.5-2.0 microns in thickness, while the envelope of the pelagic egg of *Pleuronectes platessa* is exceptionally thick, measuring about 15 microns. In the pelagic eggs capsule thickness (t) as a proportion of egg diameter (t/2r) is low, largely less than 0.005, and the capsule is thin, mostly less than 10 microns. In the demersal eggs, t/2r is greater, above 0.006, and the capsule is thicker, mostly greater than 10 microns. Although there is substantial variability, in general smaller eggs appear to have thinner capsules '(Stehr and Hawkes, 1979)'.

The demersal eggs which are exposed to currents have an especially thick envelope: Salmonidae deposit their eggs in a 'redd' (nest of pebbles) and their egg envelope measures 30-60 micron in thickness '(His, 1873; Becher, 1928; Hurley and Fisher, 1966; Riehl and Goetting, 1975; Riehl, 1980; Groot and Alderdice, 1985; Riehl, 1991)'. In *Agonus cataphractus*, which spawns in the intertidal zone, the envelope is 70-80 microns thick '(Goetting, 1964, 1965)'. In the lumpfish, *Cyclopterus lumpus*, it measures 60 microns '(Loenning et al., 1984, 1988)'. However, the herring spawns at sea in large schools and deposits adhesive eggs. 'Kupffer (1878)' measured the envelope of the herring as 6-8 microns in thickness.

Viviparous fishes, such as *Zoarces viviparus* '(Götting, 1976)' or *Dermogenys pusillus*, a representative of the order Beloniformes '(Flegler, 1977)', display a very thin envelope, of only 2-3 microns. Eight species of the viviparous Goodeidae (with intraluminal or intrafollicular gestation) possess an envelope between 0.5 and 1.5 microns in thickness '(Riehl, 1991; Riehl and Greven, 1993)'. In the ovoviviparous Poeciliidae the egg envelope is even thinner '(Erhardt and Götting, 1970; Flegler, 1977)'.

There may be differences in one and the same family such as in the Syngnathidae: The thickest envelope has been observed in the eggs of *Nerorphis* and *Entelerus* which, attached to the belly of the male, are unprotected. However, when the eggs are sheltered by a bellyfold of the male, such as in *Syngnathus*, the envelopes are much thinner. The envelope of the eggs of the seahorse (*Hippocampus*) is exceedingly thin since the eggs undergo their development in the marsupium of the male. Analysis of four species of *Pleuronectinae* revealed that the three pelagic species displayed a thin egg membrane (2.7-3.3 microns) while the demersal species had an envelope thickness of 12.4 microns '(Hirai, 1993)'.

On the other hand, marked differences in the thickness of the ZR among the pelagic non-adhesive eggs from the same species but different geographic localities have been reported for *Platichthys flesus* and *Limanda limanda*. It was suggested that these variations might be correlated with differences in the salinity, temperature and viscosity of the sea water, which necessitate adjustments in the structure to secure adequate buoancy. An apparently homogeneous population of a single species, *Fundulus heteroclitus*, which spawns demersal adhesive eggs, was shown to display variability in thickness of the ZR, which may reflect phenotypic variation in the population or variablity in developmental maturity of extruded ova '(Kuchnow and Scott, 1977)'.

5.2 NUMBER OF LAYERS IN THE ENVELOPE

While we deal here essentially with the mature egg, it is nevertheless important to refer again and again to certain stages of oogenesis which reflect best the increasing specializations of the oocyte and its parts.

'Forchhammer (1819)' stated for *Blennius viviparus*: 'membrana vitellina duplex est'. 'Rathke (1833)' mentions that the above fish possesses three layers, i.e. a chorion, a liquid layer, and a very thin transparent layer surrounding the yolk (see also 'Hoffmann, 1881'). 'Von Baer (1838)' described for *Cyprinus blicca* a soft outer layer, followed by a more rigid albuminous layer and innermost the very thin yolk membrane. According to 'Boeck (1871)' the teleost envelope is 'tough, elastic and when torn appears to be composed of several layers'. According to 'Kupffer (1878)' there are two tightly connected layers in the envelope of the herring: the inner one is radially and the outer one concentrically striated. Superimposed is an adhesive layer. 'Owsiannikov (1885)' summarized that many researchers on teleost eggs insist that there is only one membrane, others opt for two and some even for three.

To this day the fact that the eggs of teleosts are covered by more than one envelope has given rise to a plethora of names for these[1]. To quote 'Dumont and Brummet (1980)': 'Perhaps as impressive as the structure of the developing envelope itself is the variety of terms applied to it'.

Each of these terms is based on justifiable characteristics, but it is the inconsistency in their usage that has led to considerable confusion: homologous structures may be given different names whereas different structures may be given the same name. In an attempt to overcome the difficulties it was suggested to classify the various covers according to their site of origin. This was first put forward by 'Ludwig (1874)', then reemphasized by 'H.V Wilson (1889/1891)' and again by 'E.Z. Wilson (1927)', 'Anderson (1967, 1974)' and 'Dumont and Brummet (1980)'.

The classification suggested was the following:
1) primary membranes are those formed by the oocytes within the ovary
2) secondary membranes are those formed by the follicle cells (granulosa cells)
3) tertiary membranes are those formed by the oviduct or other structures not immediately connected with the egg.

However, to this day different views have been held as to the origin of the various membranes, which has rendered such a classification system invalid.

The terms which we shall use here are: Oolemma[2] for the innermost layer enclosing the oocyte, Zona radiata (ZR) for the middle and Viscous Layer for the outermost layer. (The last layer is dealt with in Chapter 6, 'Accessory structures of egg envelope').

5.3 OOLEMMA

'Von Baer (1835)' mentioned that the dissection of the egg of cyprinids reveals that the yolk is surrounded by a very thin membrane, which is not discernible under the microscope. The same observations were reported for the Salmonidae by 'Agassiz and Vogt (1845)'. This membrane is the plasmamembrane of the egg cell. It should be stressed that it is visible only by EM. This technique also reveals surface specializations such as microvilli or microplicae (as e.g. in the zebrafish)

'(Hart and Donovan, 1983)'. The surface of the egg of the pelagic *Mugil capito* and *M. cephalus* was shown to be covered by innumerate microvilli following the retraction of the oocyte either by mechanical force or as a result of fixation '(Stahl and Leray 1961)'. It is interesting to note that the number of microvilli does not increase during the growth of the oocyte [as shown for *Novodon modestus* by 'Hosokawa (1985)'].

5.4 ZONA RADIATA (ZR)

5.4.1 Structure

'Owsyannikov (1885)' called the layer 'zona perforata'. The term zona radiata was first proposed by 'Waldeyer (1883)' for the layer exterior to the oolemma. Some authors stressed that the term 'zona' is incorrect because it means a girdle or belt. Others feared this term was to be confused with the Zona radiata in the folliculus cavus of placental mammals and suggested the term 'cortex'. But 'cortex' had already been in use to designate the cortical cytoplasm of the egg (Chapter 4). At present, the term 'chorion' is widely used in English publications. However, since it is generally employed to refer to the membrane surrounding the embryo of amniotes and insects, we do not favour its application and we will use the term Zona radiata (ZR).

The ZR shows many variations in its development and ultrastructure in oviparous, ovoviviparous and viviparous fishes. As mentioned earlier, it gets its name from the presence of radial canals. These canals are the means by which substances reach or leave the developing oocyte.

'Von Baer (1835)' mentioned two membranes in the egg of *Perca fluviatilis* with the outer one traversed by long narrow dark spots ('dunklere Flecken'). On the basis of this observation 'von Baer' was erroneously hailed as the first to note the striations in the ZR. However, his description clearly refers to the outer fibrillar layer (see Chapter 6, Fig. 6/19), and not to the ZR layer of the *Perca* egg, which is also, but more finely, striated [see 'Lereboullet (1854)' below].

'Vogt (1842)' was the first to describe and illustrate the ZR. He observed opaque dots on the egg envelope of salmonids. These dots were arranged in a 'chagrin' (shagreen) pattern. He mentioned with illustrations 'the punctated or dotted marking produced by the external apertures of the fine canals which run through the membrane'. 'Ransom (1854/1855)' illustrated his observations on trout and salmon eggs, which clearly display a finely striated ZR (Fig. 5/1). 'Aubert (1854)' noted that the envelope of the mature egg of the pike (*Esox reticulatus*) is surrounded by a transparent membrane with fine dots, regularly arranged into intersecting circular lines. On contact with water the envelope separates into two layers; the outer one is very thin while the inner, somewhat thicker one, shows fine radial stripes. However, 'Aubert' did not mention any perforations. 'Ransom (1854)' in his description of the ovum of *Gasterosteus pungitius* and *G. leirus* men-

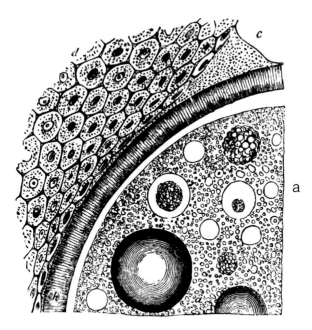

Fig. 5/1 Semi-diagrammatic view of the section of a portion of the yolk, porous membrane and external layer of cells in an ovarian *ovum* of the salmon (Ransom, 1854/1855). a - yolk. ch - section of the porous or dotted external membrane (now called ZR). c - portion of the outer surface of the same turned towards the observer so as to show the punctuated or dotted marking produced by external apertures of the fine canals which run through the membrane. d - flat surface of the hexagonal follicular cells

tioned that 'the inner membrane' shows a fine and regularly dotted structure which 'is seen to cease suddenly at the margin of the clear spot' (i.e. the aperture of the micropyle).

'Müller (1854)' is generally, but wrongly, hailed as the 'discoverer of the radial canals'. He reported that the points he observed on the envelope of *Perca fluviatilis* and *Acerina vulgaris* represent the funnel-like openings of radial canals, positioned in the middle of each hexagonal mesh (called facet) of a reticulum. (Fig. 5/2). The canals open also at the inner surface of the layer; they are much shorter in *Acerina* than in *Perca*. In this context it should be stressed that 'Remak (1854)' also mentioned fine canals which traverse the thick envelope of the egg of *Gobio fluviatilis*. He spoke about radially arranged 'thin cylinders' alternating with canaliculi. He engaged in disputes with 'Müller (1854)' since the facets of *Gobio* measured 1/1 000 lines in diameter (1 line = 2.26 millimeters) and the observed openings covered each 5x5 or 5x6 facets. In contrast, 'Müller' had reported facets of a diameter of 1/120 to 1/80 lines and an opening for each facet. To settle the dispute, 'Remak (1854)' tested the envelope of eggs of a young *Perca* and found values of 1/3 000 lines for 'radial granules' (nearing the values for *Gobio*). However, 'Müller' responded that the fine radial canals were an optical illusion. It was left to 'Lereboullet (1854, 1862)' to discover that the envelope of *Perca* is traversed by two independent systems of canaliculi. The outer system (assumed to be the ZR by 'Müller') is represented by hollow piliform appendices which hold the different eggs together (Chapter 6) while the inner system possesses more and infinitely finer (microscopic) tubules which run perpendicularly from the outer to the inner surface of what we now call ZR. This publication was followed later by a study from 'Leuckart (1855)' who also clearly differentiated between the two radial systems of the *Perca* envelope. 'Reichert (1856)' doubted as to whether the two canaliculi systems could be continuous since the number of the canaliculi of the outer systems was so much lower and since these tubes were provided with funnel-shaped openings at the contact zone with the inner layer of fine canaliculi. 'Kolliker (1858)' further discussed the origin of the 'tubules' mentioned by 'Müller' in the viscous envelope of *Perca* and considered them outgrowths from the follicular cells. 'Brock (1878)' and 'Eigenmann (1890)' clearly differentiated between the two radial systems of the *Perca* envelope (Figs. 5/3, 5/4). 'Ransom (1855)' stressed, too, that 'in some fishes, as the perch', the egg is covered externally with villous, reticular or other appendages, which serve to connect the ova in masses or strings.' 'Ransom (1867)' described for *Perca* canals passing through the outer portion of the egg as having a double contour for each wall, and as filled with material; they did not seem to convey anything, either fluid or solid, into or out of the egg. This description relates again to the outer viscous envelope. Other investigators such as 'His (1873)', 'Owsiannikow (1885)' and 'Retzius (1912)' confirmed that the *Perca* envelope consists of two layers and that both are traversed by canals (Fig. 5/5A,B,C)

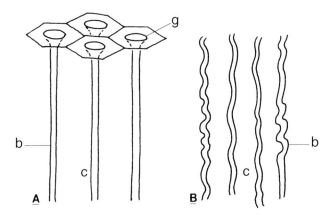

Fig. 5/2 Section through 'fibrous coat' (ZR) of perch egg (redrawn from Müller, 1854). **A** - 'canaliculi' (b) originate as funnels from the hexagonal follicular *epithelium* (g). **B** - boiling causes spiralling of 'canaliculi' (b). c = mucus.

Egg envelope

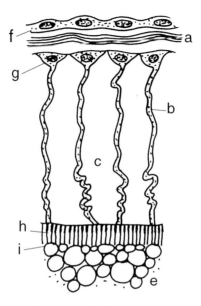

Fig. 5/3 Section through 'fibrous coat' of perch egg (redrawn from Brock, 1878). a - follicular *theca* f capillary; b - follicular cell processes; g - follicular *epithelium*; c - mucus, jelly (*zona pellucida*); h - ZR *externa*; e - yolk; j - ZR *interna*.

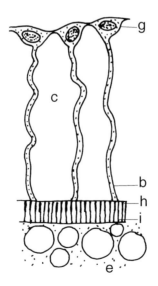

Fig. 5/4 Section through envelope of perch (redrawn from Eigenmann, 1890). b - follicular cell processes; g - follicular *epithelium*; c - mucus, jelly (*zona pellucida*); h - ZR *externa*; e - yolk; j - ZR *interna*.

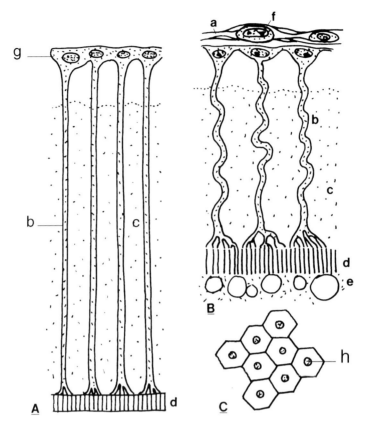

Fig. 5/5 A,B,C **AB** - Section through envelope of perch. **C** - Surface view of follicular *epithelium*. Hexagonal cells with central *nucleus* (redrawn from Retzius, 1912). a - follicular *theca*; f - capillary; b - follicular cell processes; g - follicular *epithelium*; c - mucus, jelly (*zona pellucida*); d - *zona radiata* (ZR); n - nucleus; e - yolk.

'Bruch (1856)', as before him 'Vogt (1842)', reported a 'chagrin-like dotted' envelope in the trout. 'Reichert (1854)' noted the absence of radial canals in the salmon while 'André (1875)' described for the trout 'so-called radial pores and canals' ('des canaux dits poreux, très-fins') and concluded that they are not canals but 'des linéaments serrés les uns contre les autres, linéaments rectilignes se dirigeant sans interruption, et dans le sens des rayons du globule, de la superficie à la face interne'. 'Chaudry (1956)' concluded that in the trout the pore canals are the openings by which, in the ovary, projections of the egg emerge to bring nutrients to the egg from the ovarian tissue surrounding it. But again, their existence was doubted by some as late as 1965 and 1977. In some cases, perforations were reported and then denied even for the same species. 'Fisher (1963)' reported that in the trout the capsule has many holes through it — perhaps as many as a million in one egg. Their diameter is at the limit of resolution of the LM and hence there was no real agreement about their significance when they were first observed by 'Mark (1890)' and later by 'Chaudry (1956)'. EM showed that the canals are of the order of a tenth of a micron in diameter.

'Meckel von Helmsbach (1852)' '(cf. Mark, 1890)' observed a 'radial structure of the zona pellucida' in *Cyprinus auratus*. However, he did not express any opinion as to the nature of these striations. 'Reichert (1856)' described dots on the outer and inner surface of the envelope of *Leuciscus erythrophthalmus*, which were more pronounced in *Esox reticulatus* and most obvious in *Cyprinus carpio*. He mentioned that this would suggest that this layer is traversed by small canals and that the dots are 'an optical expression' of the openings of the funnels. 'Kölliker (1858)' analyzed the envelope of 27 freshwater teleosts. He differentiated between two envelopes: the inner one, which we now call ZR, and the outer viscous layer. He noted a dotted appearance in the latter and was in no doubt that — on the basis of light microscopical observations — in all fish tested the inner envelope was traversed by fine pore canals. 'Buchholz (1863)' reported that the envelope of *Osmerus eperlanus* displays a fine 'chagrin-like dotting' as observed on many eggs up to that date. He concluded that this dotting results from very fine pore canals. 'Ransom (1867)' noted on the egg surface of *Gasterosteus leiurus* and *G. pungitius* 'a tolerably regular fine dotting, the dots being arranged in lines which cross each other, so that lozenge-shaped spaces are left between them. These spaces are but obscurely hexagonal'. He further stated that 'the dots have a similar aspect on both inner and outer surfaces of the sac'. 'His (1873)' stressed that the envelope of the pike, *Esox lucius*, shows parallel striations besides the radial striations. 'Brock (1878)' mentioned the occurrence of the ZR in all teleosts studied while he doubted the universal occurrence of an outer homogeneous lamella as put forward by 'Reichert (1856)' and 'Kölliker (1858)'.

'Kupffer (1878)', 'Brock (1878)', 'Hoffmann (1881)' and 'Brook (1885a)' described pore-canals in the ZR of the herring *Clupea harengus*. In the outer part they give way to lamellae in the mature egg (see ZR externa and interna).

Radiate striations were subsequently confirmed by many authors in various fish until the end of the century. However, 'List (1887)' again denied that the egg membrane of *Crenilabrus* was perforated by many pore canals, as reported by 'Hoffmann (1881)'. 'Ryder (1882)' found no striations in the envelope of *Belone longirostris* or the cod, whereas 'Eigenmann (1890)' observed them in fresh eggs. The latter author described pore canals also in *Morone americana*, *Pygosteus pungitius* and *Fundulus heteroclitus*. 'Owsiannikow (1885)' confirmed that radial canals traverse the envelope of the trout *Salmo trutta*, *Lota vulgaris*, *Osmerus eperlanus* and *Gasterosteus*. 'Raffaele (1888)' found the canaliculi absent in the clupeids and the trachinid *Uranoscopus scaber* but described 'pori- canali' in *Mullus surmuletus*. 'Holt (1890)' noted that the ZR of *Gobius minutus* showed 'the usual closely-set minute dots or punctures'. 'Fusari (1892)' described that the envelope of *Cristiceps argentatus* seemed to be constituted of several concentric layers which were penetrated at right angles by the pore canals.

From the beginning of the 20th century onwards numerous observations of radial canals in the ZR were again reported, e.g. by 'Retzius (1912)' (Fig. 5/5A,B). The fact that the striations were filled by dyes, air or paraffin, was taken as proof that they consisted of open canaliculi '(Becher, 1928)'.

The reason for all these discrepancies lies in the fact that the appearance, distribution and disappearance of the radial canals is an ontogenetic phenomenon, as already put forward by 'Reichert (1856)' and subsequently again by 'Flügel (1967a)'. The absence of radial canals is due to the fact that originally open canals become blocked, partially fuse or disappear altogether. Closure of the canals allows for increased water uptake from the perivitelline space. The functional significance of these phenomena is to protect the embryo against exposure to the natural environment.

5.4.2 Ontogeny of radial canals

The single layer of the follicular epithelium (granulosa) is surrounded by a vascularized theca containing connective tissue. A basal membrane lies between the follicle epithelium and the connective tissue. The oolemma of the young oocyte is in close association with the plasmamembrane of the follicle cells. In contrast to observations on eggs of marine fish, those of the guppy display desmosomes between oocytes and follicle cells '(Götting, 1964)'. During the growth of the oocyte its oolemma develops microvilli, which eventually cover the whole surface of the oocyte. As early as 1890 'Eigenmann' observed in the pike and *Notemigonus chrysoleucus* that 'fine threads, which have the appearance of being prolongations of the substance of the yolk continued into pore-canals of the zona'. Chaudry (1956) described on the basis of light microscopical studies that there are plasmatic processes in the fully developed oocytes of *Trichiurus savala* and *Triacanthus brevirostris*. Subsequent EM analyses revealed that the microvilli of the mature egg are slender and have an interior core of microfilaments, which extends into the peripheral ooplasm. They also contain endoplasmic reticulum and the club shaped ones many mitochondria. The microvilli of *Oncorhynchus keta* were shown to contain additionally free ribosomes '(Kobayashi, 1985)'.

While the microvilli are being formed, a homogeneous substance begins to be deposited between their bases. This interstitial layer increases in thickness and becomes the zona pellucida. While the microvilli increase in length and decrease in width they become enclosed within the presumptive pore canals '(Kessel *et al.*, 1985)'. In other words, the microvilli offer passive resistance to the deposition of material for the envelope and thereby participate in the formation of the radial canals. We prefer to call the membrane from now on ZR. During the growth phase the microvilli increase considerably in length and project radially and run parallel to the surface of the oocyte. In some species the long microvilli bend and run parallel to the membranes of the follicular epithelium, which brings them into direct contact with the follicle cells '(Flügel, 1967a,b; Riehl and Schulte, 1977a; Riehl, 1991)'. In other species that present long microvilli (e.g. in *Phoxinus*, also to some extent in *Perca fluviatilis*) these are also in direct contact with the follicular cells, which guarantees the exchange of substances. 'Flügel (1967a)' reported tight junctions between microvilli and granulosa cells in *Perca*. Gap junctions between microvilli and follicle cells were reported for *Plecoglossus altivelis* by 'Toshimori and Yasuzumi (1979)' as a result of EM studies including freeze-fracture. The existence of gap junctions between microvilli and granulosa cells was suggested for *Oncorhynchus keta* on the basis of an interspace width between the two membranes of less than 5 nm. While some of the microvilli establish such contact with the granulosa cells, others project only into the subfollicular space '(Kobayashi, 1985)'. Gap junctions were also reported for *Plecoglossis altivelis* '(Toshimoro and Yasuzumi (1979)' and in *Brachydanio rerio* by 'Kessel et al. (1985)'. Freeze-fracturing of the periphery of the oocyte of *Heterandria formosa* revealed gap-junctions between oocyte and follicular epithelium '(Riehl, 1991)'.

While microvilli are developing from the oocyte the follicular cells may begin to send out processes (macrovilli) which, too, contain longitudinally-arranged microfilaments '(Kessel et al., 1985)'. The macrovilli lie parallel to or interdigitate with the microvilli in the radiate pore canals. Follicular cell extensions (macrovilli) in the radial canals were first observed by 'Waldeyer (1870)' and 'Brock (1878)' (Fig. 5/3) and subsequently by 'Owsiannikow (1885)' in *Coregonus* and other fish. Their occurrence was again questioned by 'Becher (1928)'. Macrovilli are not frequently encountered in the ZR of *Phoxinus phoxinus*, where some microvilli are intertwined and the long microvilli are in direct contact with the follicular epithelium. In contrast, there are about seven macrovilli per microvillus in *Noemacheilus barbatulus* '(Riehl and Schulte, 1977b)'. In *Brachydanio* oocytes macrovilli are less numerous than microvilli '(Kessel et al., 1985)'. EM-analyses revealed that macrovilli are absent in the silver eel stage of *Anguilla anguilla* after stimulation with gonadotrophic hormone. However, these fish had only reached the stage of a gonadosomatic index of 6 '(Burzawa-Gérard et al., 1994)'. The close contact between follicle cell processes and oocyte microvilli ensures the transfer of yolk proteins and metabolites. The observations of micropinocytotic vesicles in the follicle cells and the oocytes are a morphological expression of this. In *Pleuronichthys coenosus* only the microvilli of the oocyte traverse the egg envelope and establish contact with the follicle cells or their macrovilli in the subfollicular space '(Stehr and Hawkes, 1983)'. The same situation has been described by 'M.Yamamoto (1963)' and 'Jollie and Jollie (1964)'. (cf. Stehr and Hawkes, 1983). In the chum salmon *Oncorhynchus keta* macrovilli pass into the radial canals. However, some display expanded distal ends, which seem to come into contact with depressions of the oolemma. Coated invaginations in the ooplasm facing follicular processes are indicative of exocytotic activity '(Kobayashi, 1985)'. These observations, too, would indicate that the functional significance of the pore canals is to bring nutrients to the oocyte from the surrounding ovarian tissue. $Na^+K^+ATPase$ activity observed on the microvilli are indicative of transport (im- or export) function '(Riehl, 1991)'.

In the young oocyte the radial pore canals are highly coiled, an observation which was already made by 'Kölliker (1858)', 'His (1878)', 'Owsiannikow (1885)' and 'Eigenmann (1890)'. According to the last author and 'Flügel (1967a)' the canals straighten radially in older oocytes. A ribbing arrangement spirally around the pore canals observed by EM confirms the earlier light microscopical findings in the brook trout *Salvelinus fontinalis* '(Hurley and Fisher, 1966)', in *Salmo salar* '(Riehl, 1991)' and *Pleuronyichthys coenosus* '(Stehr and Hawkes, 1983)' A helical pattern was observed in *Oncorhynchus keta* and *O. gorbuscah* (chum and pink salmon) but was absent in the *O. nerka* (sockeye), *Salmo gairdneri* (steelhead) '(Groot and Alderdice, 1985)' and in *Phoxinus phoxinus* and *Pomatoschistus minutus* '(Riehl, 1991)'. The cross-sectioned radial canals in *Salmo irideus* are stellar shaped (an indication of the presence of spiralled ribs); this is in contrast to the observations in *Salvelinus, Salmo trutta trutta, Coregonus albula* and *C. lavaretus* reported by 'Flügel (1967b)'. In some salmonids, e.g. pink (*Oncorhynchus gorbuscah*), chum (*O. keta*) and steelhead trout (*Salmo gairdneri*) a thickened annulus surrounds the external pore opening '(Groot and Alderdice, 1985)' which may correspond to the ring-shaped reinforcements described by 'Flügel (1964a,b)'.

As was mentioned earlier the number of microvilli does not increase during the growth of the eggs of *Novodon modestus*; at the same time the number of radial canals per oocyte remains unchanged while the egg surface increases 11-14 times '(Hosokawa (1985)'. However, in *Salvelinus fontinalis* the diameter of the microvilli was observed to increase by about 10 times during development of the oocyte '(Hurley and Fisher, 1966)'.

5.4.3 Occlusion or disappearance of radial canals

As development proceeds, the microvilli as well as the processes of the follicular epithelium (macrovilli) are withdrawn. This usually coincides with the completion of vitellogenesis (see below). In *Perca fluviatilis* the radial canals become occluded by a diaphragm and in the mature eggs of *Salvelinus* and other salmonids they are closed by 'plugs' immediately prior to ovulation (Fig.5/7a). However, the pore canals themselves remain visible even at the LM level ('Becher, 1928'; 'Flügel, 1967b'; 'Groot and Alderdice, 1985'; 'Kügel *et al.*, 1990' and 'Riehl, 1991'). According to observations by 'Hurley and Fisher (1966)' in *Salvelinus fontinalis* some of the pore canals originally present are filled in as the oocyte matures. 'Flügel (1970)' suggested that the plugs closing the pore canals of *Coregonus albula* are formed by the Golgi apparatus of the follicular cells. Radial sections of the ovulated eggs of *Carassius auratus* reveal that in these fish, too, the pore canals are closed up by plug-like material '(Cotelli *et al.*, 1988)'. In *Brachydanio rerio* most of the canals remain open, while their outer openings become plugged '(Hart and Donovan, 1983)' (Fig. 5/8). The radial canals of *Fundulus* and other cyprinodont fish as well as of cichlids are completely obliterated before ovulation and a compact egg envelope is formed. Marine eggs, in general, seem to suffer the same fate '(Flügel, 1967a,b; Wourms, 1976; Shackley and King, 1977; Dumont and

Brummet, 1980)'. This difference has lead to the terms 'cyprinodont type ZR' (in which radiations disappear) and 'salmonid type ZR' (in which radiations persist).

5.4.4 Hardening of the zona radiata

On contact with water the egg 'hardens'. This was already noticed by the early investigators: 'Vogt (1842)', 'Aubert (1854)', 'Lereboullet (1854)', 'Ransom (1854)'; 'Reichert (1856)' and 'Fulton (1898)'. Hardening is associated with so-called 'activation' of the egg and is independent of the process of fertilization. It is the result of the discharge of colloidal material from the cortical region (see 'cortical alveoli', Chapter 9) and results in the formation of the perivitelline space. It should be stressed that closing or obliteration of the pore canals does not prevent the uptake of water and minerals by the ZR. By contrast, the ZR is a semipermeable leakage or sieve membrane and it is the closure or disappearance of the canals which allows the colloid-osmotic uptake of water (whether it be freshwater or seawater) into the perivitelline space ['Yamagami et al. (1992)'; 'Scapigliati et al. (1994)']. Hardening can also be elicited by temperature shock, changes in the ionic balance of water and other chemical influences ['Kanoh (1953)'; 'Devillers et al. (1954)'; 'Zotin (1958)'; 'Detlaf (1959)'; 'Luther (1966)'; 'Lönning et al., (1984)'].

It has been stressed that the hardening of the egg envelope is an increase in toughness rather than in hardness. When oviduct eggs of the trout are placed in fresh water they increase in weight by as much as 10-25 % ['Fisher (1963)'; 'Wallace and Selman (1979)'; 'Wallace (1981)']. The egg of the salmon increases by 10.83% according to 'Miescher (1860)', (cf. 'His, 1873'). This confers on the egg a remarkable change in its resistance to mechanical damage.

The weight required to destroy an oviduct trout egg by crushing is approximately 40 grams '(Fisher, 1963)'. After the formation of the perivitelline space by the uptake of water, the egg can be broken by less than 100 grams. But by 10 hours later, up to 5 kilograms or more are required to destroy the egg by crushing ['Gray, (1932)'; 'Hayes (1942)'; 'Hayes and Armstrong (1941)'; 'Zotin (1953, 1958)']. By contrast, the envelope of the unfertilized egg of the lake trout has been said to resist a load of 120-160 grams '(Ignatieva, 1991)'. The eggs of *Agonus cataphractus* burst after application of a pressure of 3-3.5 kilograms '(Götting, 1964)'.

It has been suggested that both in freshwater and marine teleosts the hardening process depends on the presence of calcium ions which are necessary for the proper function of the enzymes involved '(Zotin, 1958; Hurley and Fisher, 1966; Lönning et al., 1984)'. Hardening was considered to be related to disulfide cross-linking by 'Ohtsuka (1960)' (cf. 'Yamagami et al., 1992' and cf. 'Fisher, 1963'). 'Hagenmaier et al. (1976)' demonstrated glutamyl-lysine isopeptide cross-linking in proteins of the envelope of fertilized eggs. ' Lönning et al., (1984)' suggested that different habitats may be the reason for the difference in hardness, e.g. teleost eggs may float near the surface (cod, plaice, flounder), be suspended in mid water (halibut) or be laid upon the substratum (herring, capeline, lumpsucker and salmon).

While it is known that the demersal eggs of the lumpsucker and salmon have a thicker envelope than the pelagic and mesopelagic eggs so far studied, there are not enough data for firm conclusions to be drawn.

There is a mechanism in the fine structure of the egg membrane of the coho salmon, *Oncorhynchus kisutch*, that relieves some of the tension caused by increased hydrostatic (internal turgor) pressure: the central plugs in the pore canals of the membrane can move in response to increased pressure. It was shown that e.g. the central plugs in coho eggs incubated at 8°C are embedded in their respective pore canals, but at 13°C the central plugs protrude from their pore canals. This suggests that the plugs are not rigidly set in the ZR, which confers on this membrane a certain amount of flexibility '(Cousins and Jensen, 1994)'. For more information, the reader is referred to Chapter 9 on fertilization).

5.4.5 Zona radiata externa and interna

As observed already by 'Boeck (1871)', 'Kupffer (1878)', 'Brock (1878)', 'Flügel (1967a)', 'Brook (1887a)' and by 'Eigenmann (1890)' in *Clupea, Gadus morrhua, Morone americana, Esox reticulatus, Cyclogaster lienatus* and *Amiurus catus* the ZR is made up of two layers, the ZR externa and interna (referred to from now on as 'externa' and 'interna'). 'Becher (1928)' described two ZR layers in *Perca, Scomber* and *Raniceps*. EM revealed that the two layers differ in ultrastructure and stainability and confirmed that, as a rule, the externa is much thinner than the interna. A ratio of 1:9 or 1:10 has been reported for *Phoxinus*. In some other cyprinids, however, the ratio is 1:1. In most teleostean families the interna is considerably thicker than the externa, as e.g. in *Salvelinus fontinalis, Agonus cataphractus, Perca fluviatilis, Salmo gairdneri, Lutjanus synagris* and *Zoarces viviparus* '(Götting, 1964; Riehl and Schulte, 1977a,b)'. Also in the buoyant eggs of the Russian freshwater cyprinids, the silver carp *Hypophthalmichthys molitrix*, the grass carp, *Ctenopharyngodon idella*, the black carp, *Mylopharyngodon piceus* and the bighead, *Aristichthys nobilis*, the interna is much thicker (silver carp 6.7-9.1 microns, grass carp 11 microns) than the externa, which is made up of two layers, the outer homogeneous with high electron density and the adjacent one finely grained and porous (in both silver and grass carp both layers of the externa measure in toto ~0.8 microns in thickness). During water uptake the diameter of these eggs increases 3-5 times, and their surface area 6-14 times '(Mikodina and Makeyeva, 1980)'.

Externa

There is a lack of uniformity in the description of the ZR externa. EM studies showed that in most of the species studied it is represented by three electrondense layers [*Fundulus heteroclitus* by 'Flügel (1967a)' (Fig. 5/7b), and 'Dumont and Brummet (1980)'; *Platichthys stellatus* by 'Stehr and Hawkes (1979)'; *Gadus morrhua* by 'Lönning (1972)'; *Lutjanus synagris* by 'Erhardt (1976)'; *Phoxinus phoxi-*

nus by 'Riehl and Schulte (1977a)'; *Oryzias latipes* by 'Yamamoto and Yamagami (1975)'; *Oncorhynchus gorbusha* by 'Stehr and Hawkes (1979)'; chum salmon by 'Kobayashi and Yamamoto (1981)'. In contrast to the results reported by 'Flügel (1967b)', the externa of *Fundulus heteroclitus* was described by 'Kemp and Allen (1956)' as consisting of tangentially orientated filaments in an adhesive matrix and by 'Kuchnow and Scott (1977)' as a thin homogeneous zone. Salmonidae show an externa with osmiophilic 'condensations' which before ovulation fuse to form a compact dense band sandwiched between two less electron-dense layers ['Hurley and Fisher (1966)' and 'Flügel (1967b)' (Fig. 5/7a)]. The surface view of the externa of four *Oncorhynchus* species showed a hexagonal pattern of pore openings which were open in *O. nerka* but closed as thickened annuli in *O. gorbuscha, O. keta, O. kisutch* and *O. tshawytscha* '(Groot and Alderdice, 1985)'. These annuli probably correspond to the ring-shaped reinforcements described for *Salvelinus fontinalis* by 'Flügel (1964a,b)'. *Perca* displays an electron-dense, highly organized crystalloid externa. The externa of the cyprinodont *Cynolebias* is said to be represented by a homogeneous layer '(Sterba and Müller, 1962; Müller and Sterba, 1963)'. *Cichlasoma* also contains just one thin and dense homogeneous layer '(Busson-Mabillot, 1973)' (Fig. 5/7c). The viviparous *Dermogenys pusillus* (Hemirhamphidae) shows an irregular very thin layer '(Flegler, 1977)'. *Brachydanio* and *Anguilla anguilla* have just one homogeneous electron-dense layer '(Hart and Donovan, 1983; Burzawa-Gerard et al., 1994)'. At ovulation the externa of *Brachydanio* displays plugged openings of the pore canals (see 5.4.3 and Fig. 5/8).

Interna

'List (1887)' reported that the 'lower part' of the zona of *Crenilabrus tinca* displays parallel layers. And 'Fusari (1892)' noted that the thick capsule of the mature egg of *Cristiceps* (Blenniidae) is made up of many concentric layers, which are at right angles to the radial pore canals. In other (non-identified) eggs he counted as many as 24 layers.

With the advent of EM, it became obvious that the ultrastructure of the interna changes during development. 'Flügel (1967b)' suggested that the interna displays two main types: The Salmonid type, in which the number of lamellae increases more and more up to ovulation, and the Cyprinodont type, in which the interna first displays a reticular framework (bundles of filaments), which is only later changed into lamellae.

Salmonid type

According to 'Hurley and Fisher (1966)' in *Salmo fontinalis* the matrix of the interna of the small oocyte is homogeneous in structure. At the beginning of vitellogenesis striations appear in the ZR which are orientated tangentially to the egg surface. Since they are inclined in one direction and then in another a so-called herringbone pattern emerges. The authors suggest that this alternating arrangement is responsible for the formation of 'ribs' spirally arranged around the pore canals. Microvilli formed by the oolemma are observed in the pore canals as are projections from the follicle epithelium, which are called macrovilli. Micro- and macrovilli never fuse. In *Salmo fontinalis* the concentric lamellae thicken more and more in the last phase before ovulation. Immediately before ovulation the microvilli and macrovilli are withdrawn from the pore canals. As mentioned earlier, the external openings of the pore canals at the level of the externa are closed by 'plugs' and the pore canals remain (Fig. 5/7a). The radial canals persist also in *S. irideus*, *S. trutta trutta* and *Coregonus albula*. The interna of the thick ZR of the demersal salmonid *Oncorhynchus gorbuscha* is composed of numerous short and discontinuous lamellae '(Stehr and Hawkes, 1979)'. SEM performed by 'Groot and Alderdice (1985)' revealed the same picture of discontinuous lamellae in several species of the Pacific salmon. However, the authors observed this only in cross-sections that had fractured perpendicularly to the membrane surfaces (i.e. along the pore canals). In differently fractured specimens the concentric lamellae appeared to be almost entirely separate from each other. The inner surface of the interna in this species displayed a fibrous texture and a regular arrangements of the inner opening of the ZR pore canals (Fig. 5/6)

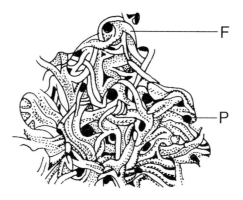

Fig. 5/6   Inner surface of egg envelope of pink salmon *Oncorhynchus gorbuscha* (Drawing based on SEM of Groot and Alderdice, 1985, with kind permission of NRC Research Press, Ottawa). × 3 500. Note fibrous texture (F) and regularly arranged pores (P).

In some salmonids the interna consists of 'logs' arranged in a herring-bone fashion or concentrically. In some salmonids such as *Coregonus albula* up to 22 concentric layers occur '(Flügel, 1967b)'. 'Riehl (1980, 1991)' found that the microvilli are spirally arranged in the pore canals of *Phoxinus phoxinus* and *Salmo salar*.

The interna of the prehatching pike, *Esox lucius*, is made up of osmiophilic lamellae alternating with interlamellae of about the same thickness. Irregularities of the parallel arrangement are often observed. The interna of the annual fish *Nothobranchius korthausae* is of a lamellar construction with the thickness of individual lamellae gradually decreasing from the innermost toward the more external lamellae. Radial canals are said to be absent '(Schoots *et al.*, 1982c)'. The radial canals remain in *Salvelinus*, *Salmo irideus*, *S. salvelinus*, *S. trutta trutta* and *Coregonus albula* (Fig. 5/7a).

Cyprinodont type

While vitellogenesis proceeds, the oocyte of *Fundulus heteroclitus* displays an interna composed of fibrils arranged in a herring-bone pattern ['Shanklin and Armstrong (1952)' and 'Kemp and Allen (1956)'], while 'Flügel (1967)' reported both for *Fundulus heteroclitus* and *Pleuronectes platessa* the presence of a network which will eventually 'fuse' and yield 6-7 electron-dense layers (Fig. 5/7b). *Cynolebias belotti* was described as having a meshwork type interna made up of 'bundles' '(Sterba and Müller (1962)'. According to 'Anderson (1967)' the interna (called by him Z3) of *Hippocampus erectus* and *Syngnathus fuscus* displays a reticular-like network which becomes multilamellate before ovulation. *Cichlasoma* shows lamellae, with a pattern of folds in its interna '(Busson-Mabillot (1973)' (Fig. 5/7c). The interna of the Russian cyprinids is made up of fibrous bundles '(Mikodina and Makeyeva, 1980)'.

In Cottidae and Agonidae the interna displays tangentially orientated bundles of interwoven filaments which eventually fuse into a dense horizontally striated layer '(Götting, 1964; Erhardt and Götting, 1970; Wourms, 1976)'. Analysis of the ZR of the annual fishes *Austrofundulus transilis*, *Cenolebias melanotaenia* and *C. ladigesi* revealed the ontogenesis of the interna as thin filaments which consolidate into highly organised tangentially oriented bundles of interwoven filaments, which eventually fuse to form a compact layer. The appearance of the transitory bundles of filaments was described as 'art-nouveau arabesques' by 'Wourms (1976)'. Transitory bundles were also observed in the interna of *Pomatoschistus minutus* '(Riehl, 1978)'. The interna of *Pleuronichthys coenosus* changes from a reticular network of fibrils, with spirally arranged lamellae, into a thin layer composed of lamellae lying parallel to the surface of the oocyte '(Stehr and Hawkes (1983)'. Horizontal lamellae have been described in mature pelagic eggs of several other species of marine fish '(Hagström and Lönning, 1968; Lönning, 1972; Ivankov and Kurdyayeva, 1973; Lönning and Hagstrom, 1975; Stehr and Hawkes, 1979)'.

'Schoots *et al.*, (1982b)' did not detect a laminated appearance of the interna in the prehatching zebrafish egg. In contrast, a 'superficial' interna with criss-crossing fibre bundles and a 'deep' zone of 16 horizontal electron-dense lamellae were reported for the mature eggs of *Brachydanio rerio* by 'Hart and Donovan (1983)' (Fig. 5/8). Since a developmental study of the zebrafish by 'Ulrich (1969)' did not mention any lamellae, it has been suggested that their formation takes place only late in development.

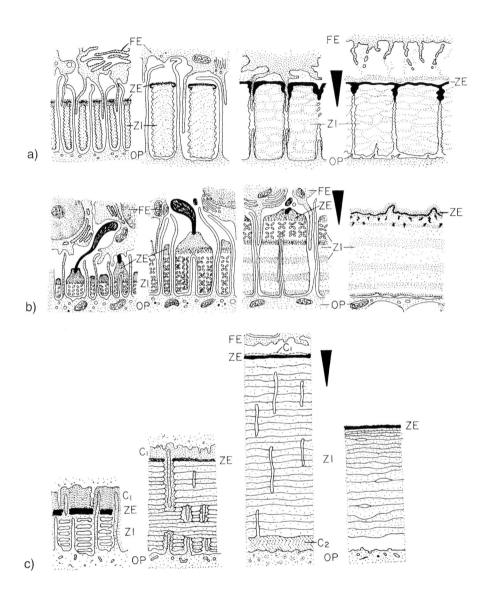

Fig. 5/7 Successive stages in the development, differentiation and ultrastructure of the envelope of *Salmo fontinalis*, *Fundulus heteroclitus* and *Cichlasoma*. Micro- and macrovilli meet in the pore canals of the ZR and are withdrawn before ovulation. [redrawn from Flügel (1967b) and Busson-Mabillot (1977). With kind permission of Springer-Verlag, Heidelberg and Biology of the Cell, Paris]. a) *Salmo fontinalis*; b) *Fundulus heteroclitus*; c) *Cichlasoma*; C_1 - apical cell coat; C_2 filamentous coat; FE - follicular *epithelium*; OP - ooplasm; ZE - *zona radiata externa*; ZI - *zona radiata interna*; Black triangle ▼ = ovulation.

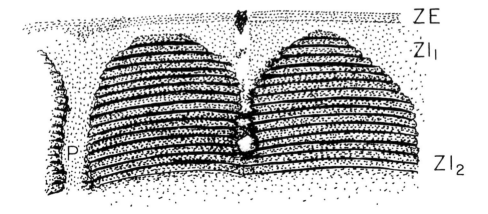

Fig. 5/8 Transverse section through envelope of unfertilized egg of *Brachydanio rerio* (Drawing based on a TEM by Hart and Donovan (1983), with kind permission of Wiley-Liss, Inc., a subsidiary of John Wiley & Sons, Inc.). × 17 000. P - pore canal; ZE - electron-dense *zona radiata externa*; ZI_1 - fibrillar zone of *zona radiata interna*; ZI_2 - lamellar zone of *zona radiata interna*. Note the plugging of one of the pore canals.

The occurrence of bundles in the interna of *Phoxinus phoxinus* was described by 'Riehl and Schulte (1977a)'. Subsequent subtangential ultrathin sections of this region revealed that the bundles are arranged 'as the sprays of a fountain' '(Götting, 1965, 1966, 1967; Riehl, 1991)'. The interna of *Blennius pholis* displays a fibrillar, highly organized network '(Shackley and King, 1977)'. The ultrastructure of the interna in the cod *Gadus morrhua* revealed five lamellae with a featherlike appearance and capped with a very thin (0.3 microns thick) homogeneous layer. In the labrids the demersal eggs of *Centrolabrus exoletus* displayed an interna with 17-18 featherlike lamellae, whereas in the pelagic eggs of *Ctenolabrus rupestris* the number of featherlike lamellae was usually five or six '(Lönning, 1972)'.

5.4.6 Variation in number of zona radiata layers

The ZR of viviparous fish lacks the complexity of zonation so obvious in the oviparous forms. However, lectin-binding procedures using colloidal gold have shown a zonation '(Schindler and Vries, 1989)'. The ZR of the Poeciliidae has one or two very thin layers. The species described are *Molliensia sphenops* and *Xiphophorus helleri* '(Zahnd and Porte, 1966)', *Poecilia reticulata* '(Jollie and Jollie, 1964; Dépêche, 1973)', *Gambusia* sp. '(Chambolle, 1962)'. In *Platypoecilius maculatus* bundles are lacking altogether since, as it has been suggested, a protective envelope is not needed in ovoviviparous development. The overall thickness of the ZR is 0.7-0.8 microns '(Erhardt and Götting, 1970)'. In another ovoviviparous fish, *Zoarces viviparus*, bundles are present in the interna which is overlain by a thin homogeneous layer '(Götting, 1976)'. A single layered ZR was reported for four species of viviparous goodeids '(Riehl and Greven, 1993)'. Only a single layer was observed also in the livebearing *Hubbsina turneri, Xenophorus captivus* and *Xenotoca eiseni*, while five other goodeids showed a thin externa and interna '(Riehl and Greven, 1993)'. The viviparous *Dermogenys pusillus* displayed an interna characterized by periodically cross-banded fibrillae and lamellae, which undulate and form spirals and loops suggestive of tubules. This layer subsequently decreased and its structure decomposed into rather thick parallel lamellae '(Flegler, 1977)'.

According to some authors the ZR is divided into three layers, named from the outside to the inside, Z1, Z2, Z3. [These terms should be avoided since they have been used to name electrophoretically separable subunits of vitellogenin (see Chapter 3) and subunits of SF (see below) with different molecular weight.] Z1 is thin, electron-lucent and homogeneous, while Z2 is thin, opaque and granular. Z3 is large and reticular, changing into multiple lamellae before maturation. In the pipefish *Syngnathus scovelli* Z2 disappears during development '(Begovac and Wallace, 1989)' while in *Syngnathus fuscus* and *Hippocampus erectus* Z1 degenerates '(Anderson, 1967)'. In *Brachydanio rerio* the so-called middle layer is a superficial portion of the ZR interna '(Hart and Donovan, 1983)'. In the annual fish *Cynolebias melanotaenia* and *C. ladigesi* Z2, before ovulation, is characterized as a thin layer of residual material '(Wourms, 1976)'. It would seem then, that the envelope of the above fish essentially represents a two-layered externa/interna ZR.

A three-layered ZR has been reported for *Xenoophorus captivus* by 'Schindler and Vries (1989)' (as opposed to 'Riehl and Greven, 1992' mentioned above) and for *Dicentrarchus labrax* by 'Fausto *et al.* (1994)'. The ZR of the sea bass *Dicentrarchus* consists of a 200 nanometres thick outer layer of fine granular appearance, a middle layer of about 600nanometres thickness and of paracrystalline structure with fibres arranged in parabolic arcs. The inner layer is 5 microns thick and has 12 electron-dense fibrous lamellae which are mutually interconnected. For *Fundulus heteroclitus* a third zone, internal to the interna, is mentioned. It is homogeneous, crystalline and varies from 1 to 13 microns in thickness. However, it is not always present '(Kuchnow and Scott, 1977)'. The sockeye, pink and chum salmon and the steelhead (*Salmo gairdneri*) display an additional layer called subinterna. This layer is not present in the coho and chinook salmon '(Groot and Alderdice, 1985)'. The authors suggested that the subinterna may be simply an inner region of the interna characterized by more loosely packed fibrils (Fig. 5/6). Three layers termed ch1, ch2, ch3 have been described for *Carassius auratus*. During maturation ch3 (corresponding to the interna) thickens continuously. Ch1 and ch2 remain almost unchanged but ch3 stretches considerably at ovulation '(Cotelli *et al.*, 1988)'.

Four layers have been reported for the ZR of *Cichlasoma*. However, two of these are only transitory: The innermost layer is called 'filamentous coat' (F. couche filamenteuse). It is formed before ovulation and is no longer visible in the shed egg.

There is another layer called 'cell coat', which is formed first and, therefore, is external to the externa. It is no longer present once the interna has greatly increased in thickness (at the end of vitellogenesis) '(Busson-Mabillot, 1973)' (Fig. 5/7c).

The cyprinodont *Tribolodon hakonensis* has five layers, the outermost two showing peroxidase activity '(Kudo et al., 1988)'.

5.4.7 Chemical composition of the zona radiata

Glycoproteins are key components of the teleost envelope '(Anderson, 1967)'. 'Kaighn (1964)' determined quantitatively the amino acid composition of the ZR of *Fundulus heteroclitus* and compared it with other structural proteins. He detected three types of carbohydrate components, i.e. non-amino sugar, hexosamine and sialic acid. On the basis of studies of solubility and chemical composition he concluded that the ZR-protein is not stabilized by disulfide bonds. 'Riehl (1991)' summarized the heretofore results of histo- and ultrahistochemical experiments which showed the presence of mucopolysaccharides, glycoproteins or a combination of proteins and polysaccharides.

SDS-Page electrophoresis of the mature ZR of *Carassius auratus* yielded 20 bands of polypeptides [(30 to 250 kiloDaltons (kDal)]. Of these those of 40-60kDal were most prominent and represented about 70% of the total of ZR components. A PAS modified SDS-Page technique yielded four bands of glycoproteins (46,84,110 kDal) ['Cotelli et al. (1988)'; 'Brivio et al. (1991)']. Glycoproteins of a molecular weight of 47 kDal and 170 kDal were detected in the envelope of the sea bass *Dicentrarchus labrax*. These values are roughly similar to the main glycosylated polypeptides present in the trout envelope, i.e. 130 kDal and 47 kDal. '(Scapigliati et al., 1994)'.

The biochemical analysis of the whole envelope of the mature egg of *Oncorhynchus mykiss* (rainbow trout) lead to the identification of four major proteins (129, 62, 54 and 47 kDal) which represent about 80% of total envelope-proteins. After application of chemical and enzymatic deglycosylation treatments only the 129 and 47 kDal proteins proved to be glycoproteins '(Brivio et al., 1991)'.

The ZR externa is rich in polysaccharides whereas the interna is rich in protein ['Hagenmaier (1973)', 'Riehl (1991)']. In contrast, histo- and cytochemical analyses of the ZR of *Cichlasoma nigrofasciata* revealed that the externa is made up of proteins (highly resistant to proteases) and devoid of polysaccharides while the interna is made up strictly of protein (ichtulokeratine). The same author observed two additional, though transitory, coats: 1) an apical 'cell coat', i.e. exterior to the externa, observed only during vitellogenesis, and made up of glycoproteins and 2) a fibrillary layer of glycoprotein and polysaccharides, which lies between the oolemma and the interna, after the end of vitellogenesis and unto the advent of ovulation (F. 'revêtement de pré-ovulation', pre-ovulation coat) '(Busson-Mabillot, 1973)'. Histo- and ultrachemical investigations on *Noemacheilus barbatulus* and *Gobio gobio* revealed in the externa mainly polysaccharides with some protein; acid mucopolysaccharides prevailed in *Gobio*. The externa of both species contained some protein. Proteins and neutral fats occurred in the interna of both species '(Riehl, 1977a,b)'. In *Pleuronichthys coenosus* both interna and externa contain mucopolysaccharides '(Stehr, 1982; Guraya, 1986)'.

In the zebrafish the first major synthetic activity of the oocyte is said to involve the synthesis of polysaccharides. This would confirm that the externa is the first layer to be formed. Judging from the available literature, it seems that in all teleosts so far investigated structurally and ultrastructurally, the ZR externa develops before the interna and that the externa remains relatively thin throughout oogenesis.

The eggs of *Noemacheilus barbatulus* and *Blennius pholis* are glued unto the substratum by mucopolysaccharides secreted by the externa '(Shackley and King, 1977; Riehl, 1991)'.

5.4.8 Origin of the zona radiata

The question as to whether the components of the envelope are synthesized by the follicle cells or the oocyte, or the two working in concert — a problem already discussed by 'Kollesnikov (1878)', 'Owsiannikow (1885)' and 'Eigenmann (1890)' — has not yet been satisfactorily answered.

For instance, 'Ransom (1868)' argued that the ZR cannot grow by apposition of layers from within or without, and that it must grow by interstitial deposition of material. However, he did not wonder where this material comes from: 'There are no facts known to me to point out whether the pabulum for the growth of this membrane (ZR) is derived directly from the currents passing inwards, or from the material elaborated in the egg and passing out of it, or from both sources indifferently'. 'Chaudry (1956)' analysed the oocytes of *Trichiurus savala* and *Triacanthus brevirostris* and considered the externa of follicular origin and the interna 'partly follicular and partly ooplasmic in nature'.

Synthesis of ZR by oocyte

The notion of the early investigators that the ZR of teleosts is a synthetic product of the oocyte was based upon morphological observations. The first to suggest this origin was 'Reichert (1856)' who called it the 'problem of the day'. He was followed by 'Kölliker (1858)', 'Gegenbaur (1861)', 'Lereboullet (1861)', 'His (1873)' and 'Ludwig (1874)', 'van Beneden (1880)' (rev. by 'Stahl and Leray, 1961'). 'Hoffmann (1881)' while admitting that the origin of the ZR presented many problems, proposed that the oocyte produces the ZR. 'Henneguy (1885)' agreed but stressed that his opinion was not based on personal investigations. Then followed 'Balfour (1885)', 'Mark (1890)', 'Eigenmann (1890)', 'Retzius (1912)', 'Becher (1928)', 'Spek (1933)', 'Korschelt and Heider (1902)', 'Sterba (1958)', 'Arndt (1960)', 'Stahl and Leray (1961)', 'Hurley and Fisher (1966)', 'Erhardt and Götting (1970)', 'Flegler (1977)', 'Laale (1980)' who all stipulated for various species that the ZR is of oocyte origin.

Ultrastructural investigations into the origin of the ZR revealed that subsequent to the formation of microvilli by the oolemma, an electron-dense substance is deposited between the microvilli. This intermicrovillous non-cellular layer is the Zona pellucida (originally so named because it does not stain with dyes commonly used in histology). This layer subsequently increases until it reaches the tip of the microvilli, when due to its striated appearance it becomes the ZR. It has been suggested for *Hippocampus erectus* '(T.S.Yamamoto, 1963)' and for the annual fish *Cynolebias melanotaenia* and *C. ladigesi* that the material which gives rise to the Zona pellucida and subsequently to the ZR has been transported to this site by intraoocytic vesicles '(Wourms, 1976)'. Similar observations had been provided by 'Anderson (1967)'. He described exocytotic vesicles in the cytoplasm of *Syngnathus fuscus* and *Hippocampus erectus* oocytes, which he interpreted as an argument in favour of formation of the ZR by the oocyte. Moreover, endoplasmic reticulum (ER) and Golgi complexes, respectively, were rather numerous in the peripheral cytoplasm of the oocyte just before deposition of the ZR layers.

Again, on the basis of EM studies, it has been proposed for *Oryzias latipes* that precursor material for the protein-polysaccharide envelope is transferred from the Golgi body of the oocyte to the ZR by means of a population of smooth-surfaced dense-cored vesicles '(Tesoriero, 1977a)'. This hypothesis was subsequently supported by cytochemical techniques for the visualization of polysaccharides at the ultrastructural level, which yielded a positive reaction in the material of the ZR and in that of the dense-cored vesicles '(Tesoriero, 1977b)'. As a further approach autoradiography with an isotopically labelled amino acid (^{3}H protein) was used. Lightmicroscopy revealed a transfer of up to 60% incorporated label from the cytoplasm of the oocyte to the ZR at the end of 48 hours. At the EM level the label was followed from the Golgi body to the dense cored vesicles and the ZR '(Tesoriero, 1978)'. Similarly, the thickening of the ZR interna of *Novodon modestus* was thought to be caused by material from the oocyte. This, too, was based on the observation of dense-cored vesicles ['Hosakawa (1985)' in 'Yamagami et al. (1992)']. Extensive rough endoplasmic reticulum (RER), Golgi body and smooth surfaced vesicles were noted in the peripheral cytoplasm of the oocytes of the annual fish *Cynolebias melanotaenia* and *C. ladigesi*. This suggested that dense-cored vesicles fuse with the oolemma and thereby deposit envelope material on the surface of the oocyte '(Wourms, 1976)'. Similarly, observations on the oocyte of the annual fish *Austrofundulus transilis* showed that amorphous material consolidating into the ZR interna had been secreted by the oocyte. The presence of dense-cored vesicles and rough endoplasmic reticulum (RER) in the ooplasm of *Pleuronichthys coenosus* suggested at least some involvement of the oocyte in the formation of the ZR '(Stehr and Hawkes, 1983)'. Subsequently, autoradiography with ^{3}H-prolin confirmed that the zona material of *Pomatoschistus minutus* is formed by the ooplasm. The material is concentrated by the Golgi body and transported in dense-cored vesicles. Pinocytotic vesicles observed below the oolemma are thought to carry proteins to the cell surface to be incorporated into the developing envelope ['Tesoriero (1978)'; 'Riehl (1978b, 1991)'].

The following circumstantial evidence supports the view that the oocyte furnishes the zona: morphological observations showed that in goodeids the surface of the oocyte is active in exocytotic and/or endocytotic processes while the follicular surface is not. Moreover, polarized distribution of binding sites for concanavalin A and wheat–germ agglutinin in the ZR was observed '(Schindler and deVries, 1989)'.

Synthesis of ZR by follicle

'Thomson (1855)' and 'Waldeyer (1870)' (cf. Eigenmann, 1890) and later 'Hurley and Fisher (1966)' and 'Laale (1980)' maintained that the ZR is formed from the follicular epithelium, with which 'Eigenmann (1890)' agreed. 'Kollesnikov (1878)' mentioned that the ZR is a secretory product of the follicle in *Perca* and *Gobio* and stressed that most authors consider the ZR to be a follicular product. Others named were 'Ryder (1881)' and 'Owsiannikow (1885)'. The latter mentioned for *Acerina vulgaris* that the granulosa cells secrete a substance that surrounds the oocyte, layer by layer.

Subsequently, the ZR of *Salmo salar* and *Salvelinus fontinalis* was presumed to be formed by the follicle cells according to 'Battle (1944)'and 'Hurley and Fisher (1966)'. 'Chambolle (1962)' considered the ZR of *Lebistes reticulatus* (*Poecilia reticulata*) of follicular origin. 'Flügel (1967)' discussed a possible origin of the ZR from the follicular epithelium since the latter is abundant in rough endoplasmic reticulum (RER) at the time when the ZR is being formed. The species analyzed were several salmonids (*Salvelinus fontinalis*, *Salmo irideus*, *S. trutta trutta*, *Coregonus albula*, *C. lavaretus*) as well as *Fundulus heteroclitus*, *Perca fluviatilis* and *Pleuronectes platessa*. Similarly, the presence of RER in the follicle was observed in *Dermogenys pusillus* by 'Flegler (1977)'. However, he hesitated to draw a conclusion as to the origin of the ZR. Immunocytochemical analysis and in vitro tracer experiments of the ZR interna proteins of the eggs of the pipefish, *Syngnathus scovelli*, suggest a synthetic source 'within the follicle' '(Begovac and Wallace, 1989)'. 'Riehl (1977b)' concluded from histochemical and EM results that the interna of *Noemacheilus barbatulus* and *Gobio gobio* may be formed by the follicle epithelium. 'Shackley and King (1977)' noted that in the eggs of *Blennius pholis* the simple, squamous cell layer of the follicle changes into a cuboidal and columnar epithelium at the commencement of vitellogenesis. Additionally, ultrastructural findings suggest that the granulosa cells are possibly involved in the formation of the ZR of *Blennius pholis*. This is based on observations of concentric whorls of rough endoplasmic reticulum, large mitochondria and large nuclei with peripherally situated chromatin in these cells at the beginning of vitellogenesis.

Source for ZR outside ovary

The above-mentioned studies lacked the identification of the molecules associated with the ZR. Therefore, an entirely new aspect of ZR ontogenesis had been opened up when 'Hamazaki et al. (1984, 1985)' found, as a result of immunological experiments, that a glycoprotein synthesized in the liver is a precursor of the major component of the ZR interna of the medaka, Oryzias latipes. The relative absence of previous information on teleost ZR proteins was probably due to the apparently insoluble nature of teleost egg envelopes. However, partial digestion of the medaka envelope had been achieved by applying to it a hatching enzyme. This pioneering study had been performed by 'Iuchi and Yamagami' as far back as 1976/1979 and eventually led in 1984/85 to the above-mentioned results by 'Hamazaki et al.'. Shortly afterwards intact macromolecules associated with the ZR of goldfish and Salmo gairdneri were described by 'Cotelli et al. (1986)'.

Further immunological investigations followed. Reactivity with anti-ZR glycoprotein antibody, detected in the ovary, liver and blood plasma of the spawning female medaka, Oryzias latipes, disappeared on cessation of oviposition, firstly from the liver and finally from the ovary. The reactive substance was found neither in the non-spawning female nor in the male. However, it was synthesized in the liver of oestrogen-treated males and oestrogen-treated non-spawning female fish. This suggested that the reactive substance is normally produced in the liver of the female under the influence of an oestrogenic hormone (E2), and is closely related to oogenesis. The name given to it is 'Spawning Female-Specific Substance" (SF) '(Hamazaki, Iuchi and Yamagami, 1987a)'. When the proteins of the ZR were examined by electrophoresis and subjected to SDS-PAGE, three major bands were obtained, named ZI-1, ZI-2, ZI-3, in order of decreasing molecular weight. ZI-3, the most abundant of the major components, had a molecular weight of about 49 kDal similar to the SF substance. In other words, the SF-substance consists of a glycoprotein closely related to ZI-3 and is probably its precursor. When the ^{125}I labelled SF-substance was injected into spawning female fish, radioactivity was shown to be incorporated into the ovary. The labels were first found in intercellular spaces; they then moved into the envelopes of growing oocytes and finally were localized exclusively in the ZR interna. It was further found that ZI-1 and ZI-2 (MW 76 kDal and 74 kDal) of the ZR interna are also immunoreactive with anti-egg envelope glycoprotein antibody and are found only in the liver, blood plasma and ovary of spawning female fish under natural conditions. In order to avoid confusion, they are named 'High Molecular Weight Spawning Female-specific Substances' (H-SF substances) while the former is now called 'Low Molecular Weight Spawning Female-specific Substance' (L-SF substance). Therefore, the general term 'SF substance' applies to ALL proteins, which are immunoreactive with anti-egg envelope glycoprotein antibody. Recently, cDNA clones for the L-SF substance, the precursor of ZI-3, were isolated. It was revealed that expression of the L-SF gene occurs exclusively in the liver of spawning female and oestrogen-treated male fish and that there is no mRNA encoding L-SF in the ovary of the spawning female fish. It remains possible that the minor components of the interna, i.e. H-SF substance (ZI-1 and ZI-2), are synthesized in the oocyte and released as additional constructive material of the ZR ['Hamazaki et al. (1987a,b,c, 1989)'; 'Murata et al. (1991, 1994, 1995)'; 'Yamagami et al. (1992)']

Subsequently, experiments with polyclonal antibodies against ZR proteins revealed that oestrogen induces the three major ZR proteins (60 kDal, 55 kDal and 50 kDal) in the rainbow trout (*Oncorhynchus mykiss*), brown trout (*Salmo trutta*) and turbot (*Scophthalmus maximus S.*) '(Hyllner et al., 1991)'. 'Oppen-Berntsen et al. (1992a)' concluded that ZR proteins of the cod, *Gadus morrhua*, were synthesized in an extra-ovarian tissue and transported in the blood for deposition in the ovaries. 'Oppen-Bernsten et al. (1992b)' further reported that the three major ZR proteins (60, 55 and 50kDal) are synthesized by isolated in vitro hepatocytes of rainbow trout *Oncorhynchus mykiss* injected with oestradiol-17 beta. This suggests that the three ZR proteins are co-ordinately regulated by the hormone and are synthesized in the liver.

'Hyllner et al. (1994)' used an antiserum directed against ZR proteins to investigate the formation of the ZR and the uptake of vitellogenin in *Oncorhynchus mykiss*. Although envelope-proteins were present in the plasma, they did not form an envelope until the oocyte was 450 microns in diameter. This coincided with an increase in plasma oestrogen and took place about one year before ovulation. When the oocytes reached a diameter of 600 microns they showed an immunoreactive envelope with a thickness of 3 microns. The same authors showed that the formation of two major ZR proteins of the halibut *Hippoglossus hippoglossus* are controlled by oestrogen.

'Koya et al. (1995)' noted that during the formation of the ZR in the masked greenling *Hexagrammos octogrammus* the RER was not well developed in the cortex of the egg, which suggested a low activity of protein synthesis in the oocyte. They suggested, therefore, that a precursor for Z2 and Z3 (externa and interna) may be synthesized in some extra-ovarian tissue.

In contrast, 'Begovac and Wallace (1989)' found that major ZR interna proteins (109 and 98 kDal) (first reported by the same authors in 1986) of the oocytes of the pipefish *Syngnathus scovelli* make up approximately 72% of the ZR dry mass. A monoclonal antibody specific for ZR proteins recognized only the ZR proteins and did not react with liver or plasma proteins. The antigen made its first appearance when the interna was being formed. This would indicate that the ZR proteins are synthesized within the follicle. This is, according to the authors, consistent with previous suggestions by others that teleost ZR proteins are synthesized locally by the oocyte rather than in the liver. These results were obtained on the basis of immunocytochemistry (immunogold staining of ultrathin sections), autoradiographic and electrophoretic methods.

Thus the results based on biochemical and immunological criteria differ. There may indeed be multiple mechanisms for the formation of the envelope among different species of fish. On the other hand, the glycoprotein from the liver (called SF, see above) is only a precursor, which, having arrived in the ovary, still has to be converted to a ZR protein, by the oocyte or the follicle. We are thus faced with a new question as to where this conversion takes place. Already 'Hamazaki et al. (1987b)' had speculated that the hepatically produced plasma protein may be taken up by the oocyte, processed, and subsequently resecreted.

Summary

Demersal eggs possess a thick envelope while the envelope enclosing pelagic eggs is thin. The thinnest envelope is encountered in the eggs of viviparous fish. However, there are exceptions, as e.g. shown for the thin envelope of the demersal eggs of the zebrafish. The teleostean envelope is made up of three layers: the innermost 'oolemma', the middle one 'zona radiata (ZR)' and the outermost 'viscous layer'. (5.1-5.2)

Different names have been given to the teleostean egg envelope, which is made up of concentric layers. The names used here are the following: oolemma for the innermost layers, ZR for the adjacent layer and viscous layer for the outermost layer. The oolemma corresponds to the plasmamembrane of the egg cell; it forms microvilli, which eventually cover the whole surface. The ZR is so named due to the presence of the radial canals. It was first described in the middle of the 19th century. The canals provide the means by which material reaches or leaves the oocyte. There has been controversy about the presence of radial canals. This is due to the fact that they, dependent on the species, become totally or partially closed. The 'outer' striations of the *Perca* envelope reported by the early investigators, and wrongly referred to as ZR, are dealt with in Chapter 6. (5.3-5.4.1)

While during ontogeny microvilli are formed at the surface of the oocyte, material of the future zona pellucida is laid down between them. The channels harbouring the microvilli are called pore-canals or radial canals. The canals often contain also membraneous processes (macrovilli) of the follicle cells. Once the microvilli and radial canals are formed, their number does not increase with the growth of the egg. Macrovilli are less numerous and wider than the microvilli. Once the zona pellucida is traversed by the pore canals the zona pellucida becomes the ZR. (5.4.2)

Radial canals disappear in some teleosts after vitellogenesis is complete. In others they are plugged at both ends or partially fused. This explains the ongoing controversy as to the presence or absence of radial canals in various types of teleosts (5.4.3)

Hardening of the ZR occurs as a result of intake of water into the perivitelline space. Calcium is involved in the process. In eggs with plugged radial canals the plugs are raised upward in response to raised temperature, which in turn increases hydrostatic pressure of the egg. When the temperature is decreased the plugs retract into the canals. (5.4.4)

The ZR is made up of two layers, the externa and the interna, which are ultrastructurally different. The externa of many teleosts is made up of an osmiophilic layer sandwiched between two less densely staining layers. In other teleosts the externa is composed of only one layer. The interna of the salmonid type eggs is made up of concentric lamellae while the interna of the cyprinodont type eggs consists of a fibrillar network, which later changes into lamellae. Some teleostean ZR contain only one, others 2 or more layers. (5.4.5-5.4.6)

Biochemical analyses revealed that the teleostean ZR is made up of mucopolysaccharides and glycoproteins. The externa is rich in polysaccharides while the interna is rich in protein. (5.4.7)

Structural and ultrastructural analyses did not yield a consensus as to the origin of the ZR, from the oocyte or the follicle cells or both. Recently, it has been established that the precursor, Spawning Female-specific Substance (SF) of the ZR proteins is synthesized in the liver under the influence of oestrogen. When released from the hepatic cells it enters the blood circulation and is incorporated into the developing oocytes. It still remains to be shown if the conversion into ZR proteins takes place locally within the oocyte or within the follicle epithelium. (5.4.8)

Endnotes

[1] Egg envelope, egg capsule, extracellular matrix
G. Eihülle, Eikapsel, Schalenhaut, Dotterhaut, Eimembran
F. membrane coquillière, capsule de l'oeuf, capsule ovulaire, membrane vitelline, lame de revêtement
I. capsula
L. zona pellucida, zona radiata, corona radiata
Gk. chorion

[2] plasma membrane, plasmalemma, vitelline membrane, egg membrane, oolemma.
F. membrane perivitelline
I. sottile membrana anista of 'Raffaele'

Chapter six

Accessory structures of egg envelope

> 'Si dans le domaine de l'embryologie, ils sont le prix d'efforts plus pénibles, les fruits n'en sont aussi que plus doux de cueillir.'
> Henneguy (1888) quoting 'Kölliker, 1884'

The surface patterns of teleostean eggs show species-specific differences and present one of the main tools for identification. For instance the egg-surface of *Coregonus nasus* shows a honeycombed relief pattern, while such a relief is missing in *C. lavaretus*. The different patterns also shed light on spawning ecology and type of egg development. According to 'Kraft and Peters, (1963)' the genus *Tilapia* includes substrate brooders (*tholloni, guineensis and zillii*), female mouthbrooders (*mossambica* and *nilotica*) and male mouthbrooders (*macrocephala*). The eggs of the substrate spawners are smaller and their number at each spawning is less than found in the mouthbrooders. Species-specific differences in Antarctic icefishes are even more pronounced '(Riehl, 1993)'.

Pelagic eggs are said to have generally smooth surfaces while demersal eggs are often endowed with elaborate structures. However, eggs that appear smooth under LM may show distinct sculptures such as buttons, papulae, warts and cones when viewed with SEM. Therefore, we have to resort to some other method of distinguishing between the types of eggs and we shall use the following categories:

1) pelagic-non-adhesive, 2) demersal non-adhesive, 3) demersal-adhesive, 4) eggs with special structures for flotation or attachment 5) eggs of mouthbrooders and 6) eggs of annual fish. (Eggs of oviparous and viviparous fish are 'non-chorionated' and are dealt with in Chapter 20).

It should be noted that the area around the micropyle is generally endowed with different structures. A report on these will be found in Chapter 7 on the micropyle.

6.1 PELAGIC-NONADHESIVE

'Sars (1864)' was the first to discover that the eggs of the cod float at the surface (cf. 'Henshall, 1888)'. The egg of the dragonet *Callionymus lyra* presents exteriorly a reticulum of slightly elevated ridges, the meshes of the reticulum being hexagonal '(Cunningham, 1887; M'Intosh and Prince, 1890)'. 'Padoa (1956)' has since reported similar observations on the eggs of other Callionymidae. A honeycomb structured envelope was described for the labrid *Crenilabrus tinca* by 'List (1887)' and 'M'Intosh and Prince (1890)'. While other Labridae, such as *Centrolabrus rupestris* and *Coris julis* also produce pelagic eggs (rev. 'Breder and Rosen, 1966'), the majority seem to lay demersal eggs (see 6.3.20). The regular net of hexagonal meshes on the eggsurface of the stargazer *Uranoscopus saber* (Uranoscopidae) and of *Saurus lacerta* were discovered by 'Raffaele (1888)'. He concluded that after ovulation a hexagonal ridge pattern, corresponding to the intercellular spaces between adjacent follicle cells, remains on the egg surface and produces the honeycomb effect.

On the basis of LM observations the eggs of the sole *Parophrys vetulus* appeared as a 'ball of yarn' due to the presence of surface ridges '(Orsi, 1968)'. EM analyses of the mature egg of the pleuronectine fish *Pleuronichthys coenosus*, the C-O sole, revealed hexagonal walls which, as observed earlier by 'Raffaele', correspond to the lateral margins of the adjacent follicle cell. Within each hexagonal chamber there is a subpattern of polygonal areas circumscribing a ZR pore-opening in their centre (Figs. 6/1, 6/2). The paucity of organelles in the cortex of the oocyte, in contrast to the presence of Golgi bodies, RER and vesicles in the adjacent follicular cells, suggests that the latter contribute the hexagonal ridges (no follicular cell processes are seen to extend into the canals of the ZR) '(Stehr and Hawkes, 1983; Boehlert, 1984)' (Fig. 6/3). Analyses by SEM of the surface of the egg of *Pleuronichthys cornutus* revealed the same ornamentation '(Hirai, 1993)'.

The hexagonal pattern of *Mauroclinus mülleri* (Gonostomidae) overlies a highly porous surface structure '(Robertson, 1981; Boehlert, 1984)'(Fig. 6/4). On comparing its surface with that of *Pleuronichthys coenosus*, it is striking that in the latter the facets are relatively very small and much more regularly hexagonal. Hexagonal patterns have been observed also in the ovarian eggs near ovulation of the Pacific Coryphaenoides '(Boehlert, 1984)'. The hexagonal packing of follicle cells is also thought to determine the pattern of the outer egg surface of *Cynolebias melanotaenia* without involvement of the Golgi body '(Wourms and Sheldon, 1976)'. The role of the micropyle as coordinator for the hexagonal pattern formation will be discussed in Chapter 7 on the micropyle. A hexagonal network suspended on stilts on the egg surface was observed by 'Raffaele (1888)' on an unnamed macrurid and on *Macrurus coelorhynchus* by 'Sanzo (1933)' and on *Caenorhynchus australis* (Macruridae) by 'Robertson (1981)' (Fig. 6/5a-c). Different ecological functions have been put forward for the hexagonal surface structures such as protection, resiliency and buoyancy.

Accessory structures of egg envelope 79

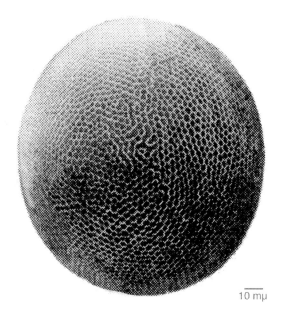

Fig. 6/1 SEM of an egg of *Pleuronichthys coenosus*. [Stehr and Hawkes (1983) with kind permission of Wiley-Liss, Inc., a subsidiary of John Wiley & Sons, Inc.)].

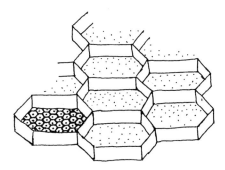

Fig. 6/2 Drawing of SEM of egg envelope of *Pleuronichthys*. Surface view showing hexagonal walls enclosing a polygonal subpattern. [modified after Stehr and Hawkes (1983) with kind permission of Wiley-Liss, Inc., a subsidiary of John Wiley & Sons, Inc.]. × 750. Dots = openings of the radial canals.

80 Chapter six

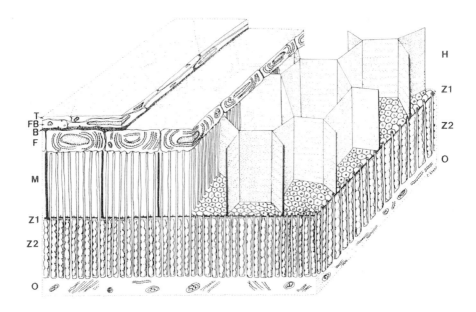

Fig. 6/3   Schematic drawing of the relationship between the theca cell layer (T), fibroblasts (FB), basal lamina (B), follicle cells (F), hexagonal walls (H), microvilli (M), polygonal subpattern (arrow), ZR externa (Z1), ZR interna (Z2) and ooplasm (O) of stage 7 oocyte of the C-O sole, *Pleuronichthys coenusus*. [modified from Stehr and Hawkes (1983) with kind permission of Wiley-Liss, Inc., a subsidiary of John Wiley & Sons, Inc.)].

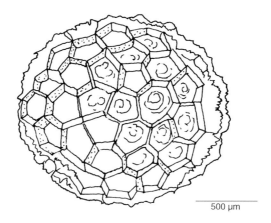

Fig. 6/4   Surface view of *Mauroclinus mülleri* [drawing of SEM by G.W. Boehlert (1984) with kind permission of Allen Press, Lawrence, KS).

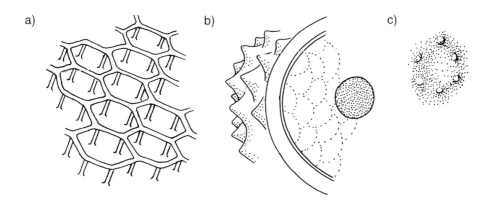

Fig. 6/5 Surface of *Macrurus* egg. **a** - Hexagonal pattern 'on stilts' (after Sanzo, 1933). Length of hexagon 20 microns; **b** - Surface view of envelope of *Macrurus* egg; **c** - Surface view of single hexagon [b) and c) redrawn from 'Raffaele, 1888'].

Other ornamentations observed are the following: The envelope surface of *Ctenopharyngodon idella, Mylopharyngodon piceus, Hypophthalmichthys molitrix, Aristichthys nobilis*, as observed by SEM, is described as rough. The cyprinids belong to the ecological group of pelagophilous freshwater fish according to '(Kryzanovski, 1949)'. Their eggs are deposited in flowing water and develop in pelagic water. However, they are characterized by their slight stickiness (due to acid mucopolysaccharides on the surface of the envelope) observed only in the first 2-3 minutes in water. Subsequently, the eggs swell and become buoyant '(Mikodina and Makeyava, 1980)'. (However, other cyprinids, as e.g. *Brachydanio rerio*, deposit demersal-non adhesive eggs, see 6.2.2).

Random ridges are observed in *Paracallionymus costatus* and *Mugil cephalus*. The eggs of *Oxyporhamphus micropterus* display spines and those of *Lactoria diaphana* pits and pores '(Boehlert, 1984)'.

The surface of the egg of *Engraulis japonicus* revealed an envelope, which possesses evenly distributed small knobs '(Hirai and Yamamoto, 1986)'. It has been possible to identify eggs of the antarctic fishes Channichthyidae and Nototheniidae on the basis of their surface structures. The surfaces of the unfertilized and fertilized (incubated) eggs of *Trematomus eulepidus* are practically identical.

The filaments of the eggs of Scomberesocidae were first described by 'Haeckel (1855)'. According to him the fibrils are as long as the diameter of the ovum, and are uniformly distributed over the surface of the egg. However, he misinterpreted his results and mistook the position of the filaments, describing them as lying inside the ZR. It was 'Kölliker (1858)' who corrected the mistake, and this was subsequently accepted by 'Haeckel'. According to 'Retzius (1912)' *Esox bellone* displays irregularly spaced appendages reminiscent of those of *Gobius* (Fig. 6/6a,b).

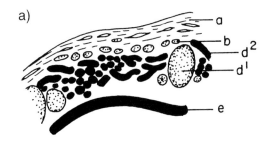

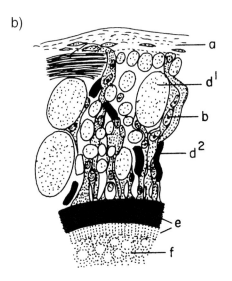

Fig. 6/6 Accessory surface *appendices* of the envelope of *Esox bellone* (*scomberesocidae*) (redrawn from Retzius, 1912). **a** - *Nuclei* of follicular *epithelium* in one layer subjacent to follicular *theca*. Two types of *appendices* present between follicular *epithelium* and ZR. **b** - *Nuclei* of follicular layer at different levels; processes project as far as ZR. a - follicular *theca*, b - follicular *epithelium*, d^1, d^2 - two differently staining types of filaments, d^1 - 'wormlike logs', staining black with haematoxylin. d^2 - logs' stained red with eosin, surrounded by a thin layer stained with haematoxylin. e - ZR [in b) ZR*externa* and ZR*interna*]; f - yolk.

The eggs of most Scomberesocidae are pelagic and without long filaments (*Scomberesox, Nanichthys, Elassichthys*). However, *Scomberesox saurus* displays on its egg-surface short bristles which may represent the remnants of chorionic filaments. 'Collette *et al.* (1984)' in a table described the filaments as uniform, or in groups, numerous, short and rigid. The eggs of *Scomberesox saurus* display 'stumps' or 'tufts' 14 microns apart according to 'Robertson (1981)'. However, the structures on the egg-surface of this species depend on the method of their fixation. They may resemble small bundles of hair, or simply coalesce to tufts 'with a relatively complex basal morphology'.

Cololabis saira, the North Pacific saury, is a scomberesocid which undisputedly possesses filaments. Relatively long filaments are gathered in a polar cluster and one single long filament is laterally attached. The egg adheres to floating objects such as kelp '(Collette *et al.*, 1984)' (Fig 6/7). On the other hand, 'Robertson (1981)' maintained that the filaments in this species serve to hold the eggs together, which, as egg clumps, become free-floating. Finally, according to 'Blaxter (1969)' tendrils found on *Scomberesox* eggs 'are as likely for attachment as for buoancy'.

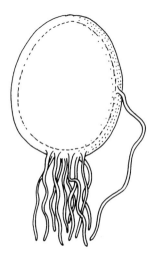

Fig. 6/7 Oval egg of *Cololabis saira* displaying a polar cluster of filaments and a long 'lateral' filament. [redrawn after Collette *et al.* (1984) with kind permission of Allen Press, Lawrence, KS.]. Diameter of egg 1.7 × 1.9 mm.

6.2 DEMERSAL-NONADHESIVE

Salmonidae

The eggs of Salmonidae are buried in unguarded nests called 'redds'. 'Vogt (1842)' observed that the egg of the salmonid *Coregonus palea* was showing a 'very elegant' granulated netlike surface, which seems to form an integral part of the external envelope. 'His (1873)' noted a generally smooth envelope, though, after careful inspection, he observed small streaks (German: 'Striemen') and flat erosions. The eggs receive a thick layer (2-10 microns) of jelly-like material only after ovulation. The jelly of *Salmo trutta* is made up of mucin '(Owsiannikov, 1885)'. He pointed out that since the jelly is not able to glue the deposited eggs together, its functional significance is problematical. 'Retzius (1912)' in his survey mentioned that the surface of the eggs of Salmonidae is smooth, i.e. devoid of 'cones' (which are common in demersal-adhesive species). This was reemphasized by 'Becher (1928)'.

However, SEM of the eggs of the pink salmon *Oncorhynchus gorbuscha* and the chum salmon *O. keta* revealed elongated blebs overlying the mostly closed pores. The pores have been closed by plugs probably formed by material from the follicular epithelium. EM analyses showed puffball-like plugs on the unfertilized egg surface of the coho salmon *Oncorhynchuys kisutch*. The area between the plugs is covered with small nodules and small ridges, some of them probably composed of several coalesced nodules ['Groot and Alderdice (1985)'; 'Johnson and Werner (1986)']. Previously, 'Flügel (1967b)' and 'Loenning (1981)' had also observed plug-filled pores on the egg-surface of different Salmonidae.

The presence of 'Zöttelchen' (small cones) covering the egg surface of *Coregonus* has been reported by 'Owsiannikov (1885)'. *Coregonus macrophthalmus* and *C. wartmanni* have a smooth envelope according to 'Riehl and Götting (1975)' while the surface of the eggs of *Coregonus fera, C. macrophthalmus* and *C. nasus* display a honeycomb relief with the pore canals opening in the centre of each hexagon ['Riehl and Schulte (1977b)'; 'Riehl (1993)']. A honeycombed relief is missing in *Coregonus lavaretus* and *C. wartmanni*, the egg surface of which is covered with minute knobs '(Riehl, 1993)'. The egg surface of *Salvelinus fontinalis* is covered with amorphous material (small knobs), which are absent in *S. alpinus* '(Riehl and Schulte, 1977b)'.

Cyprinidae

SEM analysis of the egg surface of the zebrafish *Brachydanio rerio* showed two types of papillae: small, homogeneous, densely distributed ones, with a height of 0.10 to 0.15 microns, and large, heterogeneous, sparsely distributed papillae of unequal height (up to 3-4 microns) '(Schoots *et al.*, 1982a)'. According to 'Hart and Donovan (1983)' surface views by SEM revealed that the eggs were covered by granular material and scattered dome-shaped projections, anchored to the surface by threadlike processes or short vertical spikes. Neighbouring projections were seen to be connected by ridges (microplicae). The dome-like projections are interpreted as remnants of follicle cells (Fig. 6/8a,b).

Gasteroidae (sticklebacks)

The nest-building of *Gasterosteus pungitius* was first reported by 'Coste (1848)', of *Gasterosteus aculeatus* and *G. spinachia* by 'Hancock (1854)' and of *Gasterosteus leiurus* by 'Warrington (1855)' (rev. Gudger, 1918). 'Ransom (1856)' mentioned

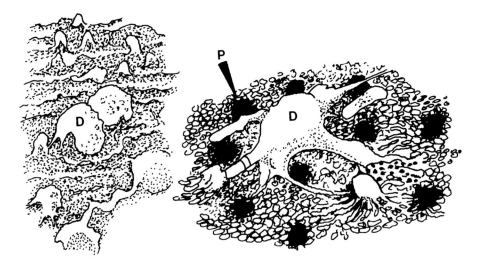

Fig. 6/8 Outer surface of egg of *Brachydanio rerio*. [drawn from SEMs of Hart and Donovan (1983) with kind permission of Wiley-Liss, Inc., a subsidiary of John Wiley & Sons, Inc.]. Left - glutaraldehyde fixed showing hemispherical domelike projections (D) × 4 700. Right - Fixation with formaldehyde containing calcium chloride. This failed to preserve the ZR *externa* and exposed the plugged pore canals (P).

that one portion of the investing membrane of the egg of *Gasterosteus leiurus* and *G. pungitius* 'presents a number of cup-shaped pediculated bodies or buttons scattered over its surface'. 'Kölliker (1858)' described special 'mushroom-type appendages' covering half of the egg-surface of *Gasterosteus*. The eggs of *Gasterosteus pungitius* are covered by spherical appendages with a small stalk (G. 'Wärzchen') according to 'Kölliker (1858)' (Fig. 6/9). 'Ransom (1867)' further observed that in the oviduct the eggs become surrounded by a viscid layer. The freshly deposited eggs cohere firmly together by this 'colourless, transparent, viscid, mucoid matrix', which resists for some time the action of water. 'Brock (1878)' confirmed the findings of 'Ransom'. 'Eigenmann (1890)' described for *Pygosteus pungitius* an outer membrane with short appendages (Fig. 6/10). The eggs of *Gasterosteus aculeatus* and *Apeltes quadracus* (four-spined stickleback) adhere to each other firmly in small clumps '(Kuntz and Radcliffe, 1917)'.

The observations repeat themselves. According to 'Swarup (1958)' *Gasterosteus aculeatus* deposits simultaneously 100-150 eggs, which stay together surrounded by mucus. Soon after oviposition the eggs harden and become firmly attached to each other.

As observed already in the middle of the 19th century the male stickleback builds a muff-like nest. This consists of plant material held together by a mucoid thread which he produces; hypertrophy of the kidney has been associated with thread production in nest builders generally. According to 'T.S Yamamoto

(1963)' the envelope of *Pungitius thymensis* has a reticular pattern and button-like processes are restricted to the animal hemisphere. It should be mentioned here that 'Ransom (1854)' had already described and drawn bleb-like processes in the micropylar region of the bitterling The eggs are surrounded by a mucus liberated from the oviduct. Parental care in *Gasterosteus* is undertaken exclusively by the male according to 'Giles (1984)'.

Fig. 6/9   Egg surface of *Gasterosteus pungitius* with mushroom-like appendages (redrawn from 'Kölliker, 1858'). b - appendages zre - *zona radiata externa*; zri - *zona radiata interna*.

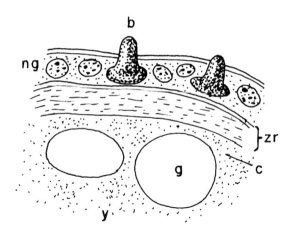

Fig. 6/10   Egg surface of *Pygosteus pungitius* with appendages (modified from 'Eigenmann, 1890'). The appendages are rivet shaped and appear to be in intimate contact with the *granulosa* cells. This becomes obvious when the *granulosa* is lifted from the egg and the appendages remain attached to it and not to the ZR. b - rivet-shaped process; c - cortex; g - oil gobule; n - *granulosa*; y - yolk; zr - torn ZR.

6.3 DEMERSAL-ADHESIVE

These types of eggs are covered with a gelatinous or adhesive substance. Attachment of the egg to the substratum or each other is due to such an adhesive surface layer or adhesive appendages. The latter structures are known as fibrils, filaments, spikes and, by the early authors, in German, as 'Zöttelchen', 'Zotten', 'Zotteln' (referred to in English as cones). The very short appendages are called little warts (G. Wärzchen). The terms fibrils and filaments are interchangeably used by the early investigators. 'Guitel (1891)' mentioned that the fibres are secreted by the follicular cells, a fact which has been accepted to this day.

The basic structure of fibrils at the EM level is an electrondense matrix surrounding a paracrystalline array of tubule-like elements '(Dumont and Brummet, 1980)', or according to 'Riehl (1976)' an osmiophilic inner rod surrounded by lighter finely granulated material into which short processes of the follicle cells project.

Clupeidae (herrings)

The Atlantic herring, *Clupaea harengus*, spawns twice a year in deep water on the sea floor while the Pacific herring, *Clupea pallasi*, deposits its eggs during a single spawning season on seaweed along the shoreline. 'McGowan and Berry (1984)' stressed that the egg-envelope of the herring is devoid of ornamentation and sculptures.

'Kupffer (1878)' was the first to observe a homogeneous gluey layer covering the two-layered envelope of the herring egg. This layer was slightly facetted due to the imprints of the follicle cells; cones were not observed. Subsequently, 'Hoffmann (1881)' and 'Brook (1885a, 1887a)' agreed that substrate-spawning fish, such as herring, glue their eggs unto the substratum. 'Eigenmann (1890)' mentioned that the egg of *Clupea vernalis* has a thin homogeneous outer layer without appendages while *Clupea sprattus* shows appendages, in some regularly, in others irregularly arranged '(Retzius, 1912)'. [In contrast, 'McGowan and Berry (1984)' maintain that *Sprattus sprattus* lays pelagic eggs.] The eggs of the Japanese herring *Clupea pallasi* possess a mucus layer which covers the cone-like structures, and the whole surface displays shallow netlike indentations ['Kanoh (1949)'; 'Ohta (1984)']. According to 'Shelton (1978)' the egg surface of *Dorosoma petenense* is covered by a thick coat of glycoprotein, which is composed of 'transformed ovarian follicle cells, an unusual feature among teleosts'. A thin electrondense sticky layer was observed to cover the whole ovulated egg, except for the micropylar region, of *Clupea pallasi* '(Ohta, 1984)'. The eggs of *Alosa* are non-adhesive according to 'Fuiman (1984)'.

SEM of the surface of the eggs of *Alosa pseudoharengus* revealed 'undecorated' pores and interporal granular areae with small nodules '(Johnson and Werner, 1986)'.

Osmeridae (smelts)

A suspensory ligament (also called a stalk), situated opposite the micropyle, has been observed on the eggs of the rainbow smelt *Osmerus mordax* and *O. eperla-*

nus ['Buchholz (1863)'; 'Cunningham (1887)'; 'Brook (1887a)'; 'Mark (1890)'; '(Riehl and Schulte, 1978b)']. This anchoring stalk results from the rupturing and inversion of the ZR during spawning. The randomly-distributed pores appear 'decorated' ['(Kanoh, 1952'; 'Yamada, 1963'; 'Scott and Crossman (1973)'; 'Cooper, 1978'; 'Johnson and Werner, 1986']. The development of such an anchoring stalk was found in all 10 species of osmerids '(Hearne, 1984)'.

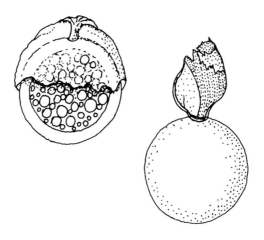

Fig. 6/11 Anchoring stalk of *Osmerus eperlanus* resulting from rupture and inversion of the ZR. [modified from Brook, 1887 and Riehl and Schulte, 1978 (with kind permission of Schweizerbart'sche Verlagsbuchhandlung, Stuttgart)].

Esocidae (pikes)

'Reichert (1858)' mentioned that the outer envelope of the pike egg is transparent and homogeneous. The fact that under the influence of water-uptake a facetted appearance could be observed would indicate that this hexagonal pattern of the covering mucus is of a follicular cell origin. Similarly 'Aubert (1854)' had observed on eggs kept in water that dots of the chagrin type envelope fuse to form irregular squares. The eggs of the pike were found to cohere to each other very strongly after 5 minutes in water. It was suggested that they would adhere to rough stones or gravel in the same way as to each other '(Truman, 1869)'. According to 'Becher (1928)' the envelope of the pike is devoid of cones and this was later confirmed by 'Riehl and Götting (1975)', Riehl and Schulte (1977)' and 'Riehl and Patzner (1992)'. According to SEM analysis of the eggs of the pike, *Esox lucius*, the outermost surface shows a 'rope-nettinglike appearance' '(Schoots *et al.*, 1982a,b)'. SEM of the egg surface of the muskellunge, *Esox masquinongi*, revealed that it is dotted with nodules of varying sizes and overlain with rounded plug structures. Fertilized eggs displayed a regular hexagonal pattern with six plugs surrounding one central plug. It had been suggested that the nodules are simply remnants of the outer mucus layer '(Johnson and Werner, 1986)'.

Characidae (piranhas)

The eggs of *Serrasalmus nattereri* exhibit a honeycomb-like relief and irregularly scattered plug-like processes. Fibrils are restricted to the animal pole where they form an adhesive disc (G. Haftfeld) '(Wirz-Hlavacek and Riehl, 1990)'.

Cyprinidae (carps and minnows)

'Von Baer (1835)' described the outer membrane of *Cyprinus blicca* as having many small cylindrical processes ('Zöttelchen') responsible for a velvet-like appearance of the egg surface. Observations of cones/appendages/papillae were reported by 'Müller (1854)' for *Cyprinus erythrophthalmus*, and for several cyprinids by 'Reichert (1856)',' Brock (1878)' and 'Hoffmann (1881)'. ' Kölliker (1858)' described external appendages on the eggs of *Abramis brama, Cyprinus rufus, Leuciscus erythrophthalmus* and *Chondrostoma nasus. Leuciscus rutilus* and *Alburnus lucidus* eggs were reported to be covered by clubshaped cones by 'His (1873)' and 'Brock (1878)'. The surface of the egg of *Rutilus rutilus* is papillary ('Hoffmann, 1881; Riehl, 1978c'). According to 'Riehl (1978d)' *Leuciscus cephalus* shows only few papillae while in *Alburnus alburnus* the papillae are short and distant from each other.

'Riehl and Götting (1975)' reported a layer of cones for *Phoxinus phoxinus*. However, a later publication stressed that EM revealed a soft surface '(Riehl and Schulte, 1977a)'. The eggs of *Phoxinus* are deposited into 'nests' and the surface of the envelope displays a hexagonal arrangement of pores closed by plugs according to 'Groot and Alderdice (1985)' and 'Hirai, (1986)'. SEM showed that the surface of the egg of the fathead minnow *Pimephales promelas* conveys a honeycomb-like appearance due to the presence of 'regular depressions' '(Manner et al., 1977)'.

A facetted surface of the egg of *Gobio fluviatilis* was described by 'Remak (1854)', while 'Kölliker (1858)' noted the presence of cones on the egg surface. Long clubshaped cones cover the whole envelope of *Gobio* according to 'Becher (1928)'. 'Riehl (1978d)' confirmed the presence of long papillae by LM. At the EM level an elaborate adhesive mechanism of the envelope of *Gobio gobio* eggs is displayed. Short projections (microvilli) of the follicle cells fit into excavations in the apical surface of the cones. It has been suggested that, in a similar way, the cones attach the deposited eggs to the substratum '(Riehl, 1976, 1991)'

The surface layer of *Gobio gobio* was analyzed histochemically and revealed the presence of polysaccharides and acid mucopolysaccharides and reacted weakly for aromatic proteins. The surface of the envelope becomes adhesive and attached to the substratum only on contact with water '(Riehl, 1991)'. Peroxidase activity with a possible antibacterial function has been found in the most superficial layer of *Tribolodon hakonensis* ['Kudo and Inoue (1986, 1989)'; 'Riehl (1991)'].

Cobitidae (loaches)

'Kölliker (1858)' was the first to describe the cones of the egg-surface of *Cobitis barbatula*. They were also observed in *Noemacheilus barbatulus*. Ultrahistochemical analyses of the eggs of this species confirmed the presence of polysaccharides in the cones '(Riehl, 1991)'.

Ictaluridae

The mass of eggs deposited by the female *Ictalurus albidus* measures about 20 centimetres in length and nearly 10 centimetres in width containing about 2 000 eggs. These are covered not with a gelatinous envelope but with an adhesive coat '(Ryder, 1887)'. An adhesion was observed in *Italurus nebulosus* by 'Armstrong and Child (1962)'.

Belonidae (needlefishes)

'Von Baer (1835)' reported that the surface of the eggs of *Belone* is covered with cones. 'Haeckel (1855)' described for *Belone* and other fish a system of peculiar fibrils, which he thought were positioned under the envelope (Fig. 6/12a,b). (According to 'Ryder (1881)' 'Haeckel unfortunately observed only unripe ova, contrary to what he supposed, as is clearly shown by his figures'.) Subsequently, 'Kölliker (1858)' and 'Hoffmann (1880, 1881)' corrected these findings and added that the fibrils are restricted to the area of the micropyle and serve to attach the eggs. In contrast, a uniform distribution of tentacular filaments of equal length over the egg surface was reported by 'Ryder (1882)' for *Belone longirostris*. After oviposition they twist together into strands and become entangled with the filaments from neighbouring eggs so as to form large clusters (Fig. 6/12c). According to 'M'Intosh and Prince (1890)' and 'Becher (1928)' the cones of *Belone* are restricted to the zone of the micropyle. In contrast, 'D'Ancona (1931)' and 'Russel (1976)' again stressed that the fibrils of Belonidae are uniformly distributed over the whole egg surface. According to 'Collette *et al.* (1984)' the same holds true for the various species of *Strongylura* (Fig. 6/12e). The fibrils of *Tylosurus* are organized into uniformly-spaced tufts (Fig. 6/12d). 'Tsukamoto and Kimura (1993)' stress that the eggs of beloniformes in Japan are identified on the basis of number and arrangement of fibres.

Hemirhamphidae (halfbeaks)

'Müller (1854)' described the presence of long fibrils in *Hemirhamphus* of the Red Sea while 'Haeckel (1855)' reported fibrils covering the whole egg-surface but erroneously believed by him to be positioned between envelope and yolksac.

Various structures and distribution of appendages are reviewed by 'Collette *et al.* (1984)': *Rhynchorhamphus marginatus* with long filaments on one pole and

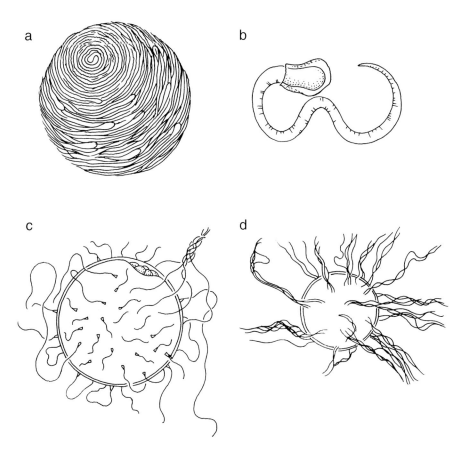

Fig. 6/12   a) Egg surface of *Belone* (redrawn from Haeckel, 1855); b) Filament of *Belone*, magnified [redrawn from Haeckel (1855)]; c) *Belone longinostris* with fibrils uniformly distributed (redrawn from Ryder (1882)); d) *Tylosurus acus* with groups of filaments uniformly distributed (redrawn from Mito, 1958).

a single thicker filament on the opposite pole were described by 'Kovalevskaya (1965)' and *Hemiramphus marginatus* with 8-12 tufts, each made up of 4-6 fibrils, by 'Talwar (1968)'. *Hyporhampus ihi*, the New Zealand garfish, possesses long 'tendrils' for attachment to sea grass '(Robertson, 1981)'.

Exocoetidae (flying fishes)

'Haeckel (1855)' noted that in *Exocoetus exiliens* the fibres are concentrically arranged. Fifteen different demersal species display a uniform arrangement of filaments and 9 species a bipolar distribution [rev. by 'Collette *et al.* (1984)'; 'Tsukamoto and Kimura (1993)']. *Exocoetus rondoletti* has the fibrils regularly distributed while in *Exocoetus heterurus* they are restricted to one pole '(D'Ancona, 1931).

Cyprinodontidae (killifish)

'Ryder (1886) and 'Eigenmann (1890)' found filaments on the eggs of *Fundulus heteroclitus*. In the same species a jelly coat with filiform appendages was subsequently mentioned by 'M'Intosh and Prince (1890)'. The filaments vary in length, also in *F. diaphanus*, but most of them are several times longer than the diameter of the egg (Fig. 6/13a,b).

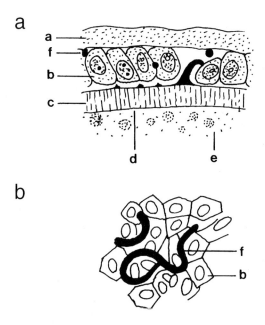

Fig. 6/13 *Fundulus heteroclitus* (redrawn after Eigenmann, 1890). × 750. **a** - Radial section of an egg, 0.4 mm in diameter. a - *theca folliculi*; b - follicular *granulosa*; c - *zona radiata*; d - *cortex*; e - yolk f filament. **b** - Tangential section of an egg 0.25 mm in diameter. b - follicular *epithelium*; f - filament.

Adhesive threads for *F. diaphanus* were subsequently also described by 'Richardson (1939)' and for *F. notatus* by 'Carranza and Winn (1954)'. According to 'Wickler (1959)' the eggs of *F. diaphanus* and *F. notatus* remain attached to the female for a short time. 'Kemp and Allen (1956)' noted extracellular strands of filaments among *Fundulus* follicle cells and concluded [as had earlier 'Eigenmann

(1890)'] that they were derived from the follicle cells but did not remain in contact with them. However, there has been to this day controversy about the coating of the eggs of *Fundulus heteroclitus*; e.g. 'Flügel (1967b)' does not note any evidence of a jelly layer or fibrils although numerous investigators from 1938 to 1966 had reported their existence. Subsequently, EM investigations by 'Anderson (1974)' described hyaline fibrils which, cross-sectioned, revealed tightly packed tubular units, each of approximately 200 Å in diameter. 'Kuchnow and Scott (1977)' noted that the outer surface of the *Fundulus heteroclitus* ovum is covered with a dense pile of short fibrils (diameter 0.3-0.6 microns). Peripheral to the micropyle the fibrils are thicker (0.8-1.0 microns) and longer, while a fibre-free zone of 50-100 microns surrounds the micropyle. According to 'Dumont and Brummet (1980)' the ovulated eggs possess a thin layer of jelly and fibrils composed of bundles of small 15-20 nm filaments. Structural differences between the egg-surface of *F. ocellaris* and *F. heteroclitus*, both collected in Woods Hole, have been observed while both species taken in South Carolina show a similar structure. Apart from attaching the eggs to the substratum, the fibrils may also aid in reducing water loss of the eggs at low tide '(Brummet and Dumont, 1985)'.

The egg of *Cyprinodon variegatus* is characterized by 'adhesive threads', evenly distributed over the whole egg surface '(Kuntz, 1915)'. They are thicker and probably longer around the micropyle. 'Wickler (1959)' provided a list of cyprinodont species (including *Cyprinodon variegatus*), the eggs of which are equipped with adhesive threads. *Cynolebias belotti* (a member of the genus *Cynolebias* sensu strictu) as opposed to the annual fish *Cynolebias melanotaenia* and *C. ladigesi*, which are members of the subgenus *Cynopoecilus*, is reported to be covered by hairlike appendages. The development of annual Cyprinodontidae, occurring in South America and Africa, deviates from the typical pattern because the eggs of these fish, buried in mudholes during the dry season, undergo developmental arrest (also referred to as diapause), while juveniles and adults die. The eggs are characterized by a tough envelope and a highly developed surface ornamentation, which is thought to act as a respiratory system during the dry season. *Austrofundulus myersi* was the first species to be studied. The unfertilized egg is surrounded by a thick envelope bearing many small and short adhesive fibrils '(Wourms, 1972)'. There followed an extensive analysis of the mature oocyte of *Cynolebias melanotaenia* and *C. ladigesi* by 'Wourms and Sheldon (1976)'. They reported that the ZR is covered with uniformly-spaced hollow conical projections (about 120 microns in length), which terminate as a crown of recurved spikes arranged in equilateral triangles. Ribs at the base form a system of interconnecting pentagons and hexagons which are thought to correspond to the intercellular spaces between follicle cells. Judging by the size (250 Å in diameter) of the tubular structures of the spikes they represent cytoplasmic microtubules. While the macroscopic surface pattern is highly ordered, the microtubules do not seem to be arranged in any order. Putative tubular electron-dense components are formed and transported by the rough endoplasmic reticulum (RER), bypassing the Golgi body and reach the exterior of the cell by exocytosis.

The egg-surface of seven species of the genus *Epiplatys* was analyzed by 'Thiaw and Mattei (1996)'. While transverse sections by TEM were similar in all species, species-specific differences were observed by SEM. In *E. bifasciatus* small granules and in *E. lamottei* small and large granules are scattered over the whole surface. The small granules aggregate into walls of polygons in *E. spilargyreius, E. fasciolatus, E. hildegardae, E. ansorgei* and *E. chaperi*. The width of the polygons and the height of the granules vary from species to species. The polygons of *E. hildegardae* have thick walls and contain a few small granules while the polygons of *E. ansorgei* are incomplete and numerous small granules are disseminated within them. In short, the pattern of ornamentation varies between the seven species analyzed.

Adrianichthyidae

The ovary of the medaka, *Oryzias latipes*, secretes a liquid at the time of egg maturation, which facilitates the extrusion of eggs at spawning '(K. Yamamoto, 1963)'. While the number of eggs per cluster may range up to 67 it usually is between 12 and 19 '(Egami, 1959; Breder and Rosen, 1966)'. The eggs of *O. latipes* display a small number (on average 200) evenly-distributed (interfilamental distance 65-70 microns) non-attaching filaments (called villi by some authors) over the whole surface, whereas a tuft of long adhesive filaments (about 30 in number) arises from the vegetal pole ['T.Yamamoto (1975)'; 'Hart *et al.* (1984)'; 'Collette *et al.* (1984)'; 'Iwamatsu (1994)'] (Fig. 6/14). The non-attaching filaments serve to unite with those of neighbouring eggs while the attaching filaments anchor the egg cluster to the gonoduct of the female where it remains attached from 2-10 hours as already noticed by 'Wickler (1959)'. When shed, the eggs adhere to vegetation. Both non-adhesive and adhesive filaments are divided into basal and distal segments and internally consist of packed unbranched tubular units with an outside diameter of 19.5 nanometres (in non-attaching fibrils) and of 18.8 nanometres (in attaching fibrils) (Fig. 6/15). As a result of optical rotational analysis it was determined that the wall of each tubule is a cylinder composed of 14 globular subunits. SEM analysis by 'Hart *et al.* (1984)' revealed that the surface of the basal segments of non-adhesive filaments is covered with fine microprotrusions and is not smooth as reported by 'Yamamoto and Yamagami (1975)'. Two ultrastructural types of attaching filaments were distinguished: Type I was similar in internal organization to the non-attaching filament in that it consisted of only tubules while type II showed a highly osmiophilic electrondense bar surrounded by packed tubules '(Tsukahara, 1971; Hart *et al.*, 1984)'. 'K. Yamamoto (1963)' and 'Tesoriero (1977a,b, 1978)' suggested that the filaments of *Oryzias* are formed by the oocyte while 'Anderson (1974)' and 'Tsukahara (1971)' maintained that they are secretory products of follicle cells. 'Tesoriero (1978)' showed by EM-autoradiography that labelled proline moves from the cytoplasm to the Golgi apparatus and then by way of dense cored vesicles to the developing envelope. He deemed it highly probable that the filaments are formed in the same way. The cluster of long filaments, which arises from the vegetal pole area of the egg of *Oryzias*, displays both a right- and a left-handed spiral pattern of attachment. It has been suggested that this is caused by the rotation of the oocyte as a result of movement of the follicular cells during oogenesis (see also micropyle) '(Iwamatsu et al., 1993b).

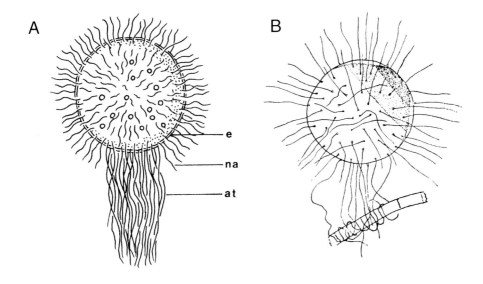

Fig. 6/14 A. Egg-surface of *Oryzias latipes* uniformly covered by short non-attaching filaments. A tuft of about 25 long adhesive filaments arises from the vegetal pole. [redrawn from Yamamoto, T. (1975) with kind permission of Kaigako Publishing Co., Tokyo. at - attaching adhesive filaments; e - envelope; na - non-adhesive filaments. B. Appendages of *Strongylura strongylura* (after Job and Jones (1938).

Goodeidae (Mexican topminnows)

The envelopes of the eggs of Goodeidae have their fibres restricted to the animal pole ['Eggert (1931)'; 'Riehl (1978c, 1984)'].

Atherinidae (silversides)

'Ryder (1882)' described that *Chirostoma notata* (*Menidia menidia*) has only four filaments attached close together (tuft) at one pole of the egg. They are wound

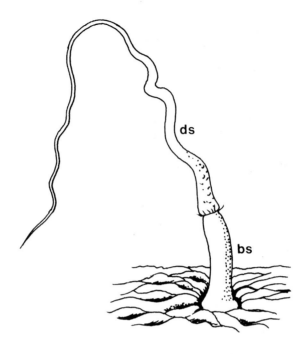

Fig. 6/15 Non-attaching filament of *Oryzias latipes* is made up of basal (bs) and distal (ds) segments. [redrawn from Hart *et al.* (1984) with kind permission of Wiley-Liss, Inc., a subsidiary of John Wiley & Sons, Inc.].

round the equator of the egg and unwind after oviposition. 'Henshall (1888)' observed the same in the gudgeon *Menidia notata*. 'Kuntz and Radcliffe (1917)' described that the eggs of *Menidia menidia notata* and *M. beryllina cerea* were 'held together in ropy clumps by a tangle of adhesive threadlike processes, a tuft of which arises from the membrane of each egg'. 'Hildebrand and Schroeder (1928)' confirmed that *Maenidia maenidia* and *M. beryllina* have adhesive threads. 'Ryder (1882)' and 'Eigenmann (1890b)' found also tightly coiled filaments on the egg of *Atherinopsis californiensis*. While in this species the filaments are single and terminate in loose ends, the filaments of *Artheriopsis affinis* are looped, without free ends '(Boehlert, 1984)' (Fig. 6/16). 'Crabtree cf. White *et al.* (1984)' described the same species with only 6 filaments and without loops. 'Tsukamoto and Kimura (1993)' reported that the distribution of filaments in *Atherion elymus* is bipolar while the eggs of *Hypoatherina bleekeri*, *H. tsurugae* and *Iso* sp. showed numerous filaments uniformly distributed over the whole egg surface.

Accessory structures of egg envelope

Fig. 6/16 *Atherinopsis. affinis* with looped filaments without free ends. (redrawn from Curless, 1939) Bar 1 000 microns.

Pseudomugilidae

These were placed with the Melatonaeiidae in 1984, but referred to as a separate family by 'Howe (1987)'. This author described four species, which displayed significant differences in the length and position of the filaments: in *Pseudomugil signifer* and *P. mellis* short fibrils are distributed regularly over the whole egg surface. Both *P. tenellus* and *P. gertrudae* possess long (5-20 millimetres) filaments. In the former they are restricted to a tuft at the vegetal pole and in the latter to two tufts, one at the vegetal and one at the animal pole.

Melanotaeniidae

The ovarian egg of *Eurystole eriarcha* is unusual 'with a brownish band swirling over its surface'. Numerous small anchor-shaped unpigmented pedicels are distributed over the surface. Either one major filament arises from the side of one of these or there are a small number of finer filaments attached to some of the other pedicels. It has been reported that each filament can become entangled in the pedicels of its own and neighbouring eggs '(White *et al.*, 1984)' (Fig. 6/17).

Pseudochromidae

This family is abundant on Indo-Pacific coral reefs. The eggs occur in egg-balls, which are guarded by the male parent. Eggs of all subfamilies possess filaments to

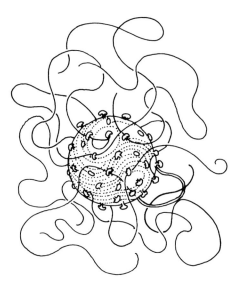

Fig. 6/17 Egg of *Eurystole eriarcha* with numerous small anchor-shaped pedicels. Each egg has one major filament arising from the side of one of these unusually shaped pedicles. Egg diameter on average 1.7 mm [redrawn from SEMs from White *et al.* (1984) with kind permission of Allen Press, Lawrence, KS.].

attach the eggs to the substratum or to other eggs to form an egg-mass. Some display modified filament attachments but only the congrogadinae possess cruciform hooks. The number of arms per hook varies from three to six. Only a few hooks bear filaments, which are of considerable length (at least 750 microns). The eggs entangle with each other by means of the filaments and the hooks. The multi-armed hooks of mature oocytes, which have developed from buttonlike knobs, are distributed equidistantly over the surface of the oocytes. The filaments often form intricate interlacing patterns '(Mooi *et al.*, 1990)'.

Cichlidae (cichlids)

The analysis of the oval eggs of fourteen cichlid species showed that they adhere to the substratum either horizontally, i.e. with their longitudinal side (l-eggs) or vertically, i.e. with their long fibrils concentrated at the vegetal pole (p-eggs). A mucous coat covering the ZR, together with the viscous fibrils, hold the eggs together. The adhesive substance is formed the moment the eggs touch water '(Wickler, 1956)'. In *Cichlasoma nigrofasciata* the whole adhesive apparatus (filaments and gelatinous covering) is formed during vitellogenesis. EM analysis showed that each filament is composed of bundles of protein tubules measuring about 200 Å in diameter. The tubular proteins are synthesized by the RER and secreted directly — not via the Golgi body — into the extracellular space where they polymerize into contiguous tubules. This implies that, for a short time, the membrane of the rough endoplasmic reticulum fuses with the cell membrane '(Busson-Mabillot, 1977)'. A bypassing of the Golgi body has also been observed in the follicular cells of the annual fish *Cenolebias melanotaenia* and *C. ladigesi*, as mentioned above on Cyprinodontidae (6.3.11)'(Wourms and Sheldon, 1976)'. However, in that case the secretory product had been transported by vesicles of the rough endoplasmic reticulum to be discharged to the cell surface. In *Cichlasoma nigrofasciata* the mucous coat (F. gangue muqueuse) made up of glycoproteins is also accumulated by the rough endoplasmic reticulum and secreted during ovulation, bypassing the Golgi body. The mucous coat is formed not as a result of fusion of membranes but in the manner of an apocrine secretion, however, with loss of a small amount of cytoplasm '(Busson-Mabillot, 1977)'. Similar results have been reported for another cichlid, *Tilapia*, by 'Kraft and Peters (1963)'. The layer of fibrils enclosing the eggs of *Cichlasoma meeki* measures 12 microns '(Riehl and Götting, 1975)'.

Pomacentridae (damselfishes)

Their eggs are attached in patches to submerged objects, especially coral reefs. *Heliasis* (Heliastes) has been described as having an adhesive pedestal (peculiar long fibres) in the region of the micropyle ['Hoffmann (1880, 1881)'; 'M'Intosh and Prince (1890)']. However, according to 'Richard and Leis (1984)' demersal labrid eggs are adhesive, but do not have an adhesive pedestal. 'Shaw (1955)' mentions the presence of adhesive filaments in the sergeant-major *Abdudefduf saxatalis*. 'Cones' are reported to cover the surface of the egg envelope of *Helias chromis*. *Amphiprion clarki*, *Paraglyphidodon nigroris* and *Chromis weberi* have their adhesive threads attached to one pole, the micropylar end. The filaments are present in two layers. The outer layer consists of an anastomosing network of filaments and the inner layer of separate filaments. In the last species the inner filaments are attached in pairs via button-like knobs. About 30 knobs, with 60 filaments, are present in *Amphiprion clarki*, and *Paraglyphidodon nigroris* have their filaments individually attached by only slightly expanded bases. Each species possesses 2 000 filaments. The eggs of the last three species peel off their egg surface to form an adhesive disc or cup which connects the eggs to the substrate '(Mooi, 1990)'.

Labridae (wrasses)

Labrus and *Crenilabrus* produce demersal adhesive eggs according to 'Raffaele (1888)'. *Labrus berggylta, L. turdus, L. festinus* and *Crenilabrus quinquemaculatus* and various other *Crenilabrus* species lay demersal eggs into a nest guarded by the male '(Breder and Rosen, 1966)'. The eggs of *Crenilabrus* form a glue the moment they touch water '(Hoffmann (1880)'. Other labrids produce pelagic, non-adhesive eggs (see 6.1).

Chapter six

Gobiidae (gobies)

According to 'List (1887)' the outer part of their egg consists of regular six-sided areas. 'Hoffmann (1880, 1881)' described the presence of 'cones' on the surface of the eggs of *Gobius minutus* and *G. niger*. They are not distributed over the whole egg and there are long filaments in the region of the micropyle '(Guitel, 1891, 1892)'. 'Cunningham (1887)' reported that the ovum of *Gobius ruthensparri* has adhesive eggs watched over by the male and fanned by him. The eggs of *Gobius minutus* were subsequently analyzed by 'Holt (1890)'. The male chooses a shell and lines it with sand and mucus. The author described the 'apparatus for attachment of the eggs' as follows: 'From the facet or pedicle of attachment springs a hyaline structure, which spreads outward in the form of an umbrella. Under a high power this structure is seen to be pierced by alternate concentric rows of diamond-shaped or ovoid apertures, which increase in size the further they lie from the pedicle, whilst, on the contrary, the proximal interstitial hyaline matter is more massive than that surrounding the more remote rows of apertures. Three or four such rows of apertures can be made out, beyond which the structure is continued in the form of a fringe of long and tapering threads, which adhere to the shell and to the threads of the adjacent ova'. The attachment area stained very deeply with carmine (Fig. 6/18).

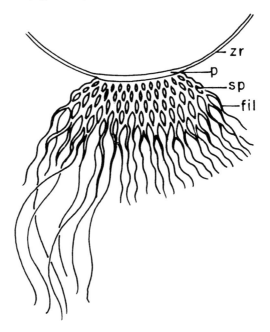

Fig. 6/18 Process of attachment of ovum of *Gobius minutus*; the filaments are mostly curtailed. (redrawn from Holt, 1892). fil - filament; sp - apertures in process of attachment; zr - *zona radiata*; p - pedicle.

'Guitel (1891)' investigated in detail the construction of the nest in a shell by the male *Gobius minutus*. He also noticed that the eggs are glued onto the shell by adhesive filaments which occur at one pole and which were secreted by the follicular cells. After some hours in water the filaments harden. Subsequently, 'Guitel (1892)' described the pattern of the filaments as a 'crinoline' which is very similar to the description given by 'Holt, 1890'. He added that, apart from the filaments, the eggs are also attached by a disc. 'Kuntz (1915)' described bundles of filaments at one pole of the egg in *Ctenogobius stigmaticus* (= *Gobionellus boleosoma*), and in *Gobiosoma bosci* he called them peduncles

According to 'Retzius (1912)'external appendages cover the whole egg surface of *Gobius fluviatilis, G. (Pomatoschistus) minutus, G. niger* and *G. flavescens* (*G. ruthensparri*). He called the threads in German 'Balken' (logs). Ramifications of the threads were observed, especially in the polar regions. The eggsurface of *Gobius flavescens* (*Gobius ruthensparri*) showed similar results to that of *G. niger*. It was suggested that the adhesive fibres of *G. niger* were formed from the follicle cells. This was reiterated by 'Eggert (1929, 1931)' and 'Wickler (1956b)', and again confirmed by 'Riehl (1978c)'. The last author analyzed *Pomatoschistus minutus* by LM and EM and observed that adhesive threads in this species may branch dichotomically. They are restricted to the animal pole and wind around the egg. The number of threads varies; there may be up to 220 fibrils with a length of 100 to 200 microns. EM revealed that the fibrils consist of a homogeneous, deeply osmiophilic substance embedded in a less electrondense matrix. Autoradiography with tritiated proline showed that the fibres are derivatives of the follicular cells '(Riehl, 1991)'.

Hexagrammidae (greenlings)

The eggs of the masked greenling, *Hexagrammos octogrammus* are congregated into an egg mass with a central cavity. It has been reported that the ovulated eggs develop adhesiveness after they are immersed in sea-water '(Munehara and Shimazi, 1989; Koya *et al.*, 1995)'. Before ovulation an outermost transparent layer (2-4 microns) is formed from diffused portions of the down-like ZR externa components. After the eggs come in contact with sea-water, the transparent layer disappears and the eggs become attached to each other with the down-like layer functioning as adhesive material. It has been concluded that the transparent layer prevents the down-like covering of the eggs from making contact with each other while still in the ovary. Chloride ions are responsible for the disappearance of the transparent layer and calcium and magnesium ions are necessary for egg adhesiveness. The adhesive material is formed early in oogenesis. At least part of the precursor substance is synthesized in the RER of the granulosa cells, then transported by vesicles to the Golgi body, which buds off vesicles that fuse with the outer cell membrane.

Cottidae (sculpins)

'Kölliker (1858)' mentioned that spherical warts on small stalks cover the envelope of *Cottus gobio*. Similarly, 'Becher (1928)' observed widely dispersed long club-shaped cones covering the eggs of the same species.

Agonidae (poachers)

The eggs occur in clumps attached to seaweeds. The surface of the egg of *Agonus cataphractus* is covered with well-marked areolae, also called reticula and covered with G. 'Mikro-Zotten' (small cones) '(McIntosh and Prince, 1890)'. Before deposition the eggs become surrounded by mucus of the oviduct '(Götting, 1964)'.

Cyclopteridae (lumpfish)

Liparis monagni eggs have a 'minutely areolate appearance'. When freshly deposited, a regular hexagonal pattern becomes obvious. However, after exposure to water this pattern seems to make room for a series of elevations. *Cyclopterus lumpus* ova are fixed together in sponge-like masses so as to permit free aeration. '(M'Intosh and Prince, 1890)'. SEM analyses confirmed sculpturing of the envelope of *Liparis liparis* '(Able et al., 1984)'.

Blenniidae (blennies)

'Cones' are reported to cover the surface of the egg in *Blennius pholis*. The attached eggs are guarded by the male '(Hoffmann, 1881)'. The fibrils of the Blenniidae are reported to be restricted to the animal pole ['Hoffmann (1880, 1881)'; 'M'Intosh and Prince (1890)'; 'Eggert (1931)'; 'Patzner (1984)']. The fibrils are of follicular-cell origin as shown for *Blennius sphinx* by 'Guitel (1893)' and for *Salarius flavoumbricus* by 'Eggert (1929)'. According to 'Wickler (1957)' the eggs of *Blennius fluviatilis* are attached at the animal pole by an adhesive disc. This disc or foot can reach almost to the equator of the egg and is made up of closely attached adhesive threads. A central round opening in the disc is probably the site of the micropyle. The adhesive eggs of *Centronotus gunnellus* form a spherical mass about the size of a walnut. The parents lie coiled around it ['Cunningham (1887)'; 'Potts (1984)'].

Gobioesocidae (clingfishes)

'Holt (1893)' described the eggs of *Lepadogaster lepadogaster* as follows: 'The whole lower surface of the shell and even the circumference of the convex surface is equipped with a number of forked filaments, all of which are oriented to follow a ray. Those on the edge of the shell are bigger and longer than the others, the terminal filaments are very long, extending much around it. The eggs are laid under stones where both parents are found'. 'Stahl and Leray (1961)' suggested that the F. 'filaments de fixation' on the eggs of *Lepadogaster gouani* are formed by the follicle cells.

6.4 SPECIAL STRUCTURES FOR FLOATING, ATTACHMENT ETC.

Already 'Aristotle' noted that the 'glanis or sheat-fish and the perch deposit their spawn in one continuous string'.

Siluridae (sheatfish)

The jelly of the European *Silurus glanis* is produced by a specialized mucus producing follicle layer ['Abraham *et al.*, (1993b)']. EM analyses of the Japanese species have been performed by 'Kobayakawa (1985)'. Secretory vesicles called mucosomes or acorn bodies have been shown to release their contents, which results in adhesion of the eggs to the substrate.

Percidae (perches)

As already mentioned above, 'Aristotle' was the first to observe the string of eggs of the perch. Interest in this egg mass became very intense from the middle of the 19th century onwards. 'Von Baer (1835)' noted that the eggs of *Perca* were attached to each other in a netlike fashion. The eggs were described as transparent, semi-buoyant and included in accordion-folded ribbons by 'Müller (1854)' and 'Retzius (1912)'. The American congener perch, *Perca flavescens*, deposits its eggs as well in long zig-zagging ribbons '(Breder and Rosen, 1966)'. The string of the perch may be made up of thousands of eggs (other authors quote 200 000-300 000). It contains an interior passage throughout its length, which is closed at both ends. However, there are apertures in the walls of the string. They are thought to be responsible for circulation of the water essential for respiration. The whole egg mass, arranged like a spring, is said to be vibratory in that the least agitation of the water puts its string into motion '(Breder and Rosen, 1966)'. All the micropyles (see Chapter 7) face towards the lumen of the ribbon. The eggs adhering to each other exhibit a flattened area at the contact zones '(Mansueti, 1964)'. The formation of ribbons is due to the presence of a jelly, which surrounds each egg and glues the eggs together. The thickness of the viscous layer is about 40 microns, i.e. it is about five times as thick as the ZR. The string or band appears as a collapsed tube. The eggs adhere to each other before extrusion and cease to be adhesive after they are expelled (Fig. 6/19).

At the beginning of 1854 'Lereboullet' was the first to report at the Meeting of the Academy in Paris that he observed in the envelope of *Perca fluviatilis* 'appendices filiformes creux, qui traversent toute l'épaisseur de la coque' (filamentous hollow appendages which traverse the whole thickness of the envelope'). These

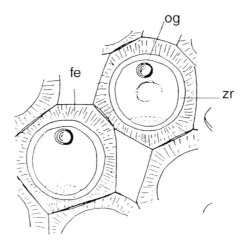

Fig. 6/19 Netlike string of eggs of *Perca fluviatilis* or *Perca flavescens* containing 200'-300'000 eggs. Eggs are flattened at contact area. The string is hollow and closed at both ends (Chevey, 1925). og - oil globule; zr - *zona radiata; fe - fibrous coat.*

interlace, and with the mucous envelope, hold the eggs together in 'élégants réseaux' (elegant nets). He continues 'Outre ces espèces de poils creux, la coque est traversée par des tubes beaucoup plus petits qui sont les véritables organes d'absorption de l'oeuf' (Apart from these hollow fibres, the envelope is traversed by tubes which are much smaller and which are the real organs of yolk absorption). In other words 'Lereboullet' also identified the striations of the ZR. Somewhat later in the same year 'Müller' misinterpreted the jelly layer as being the ZR, which was traversed by funnel-shaped long fine 'processes' or 'canaliculi', often dichotomous and with corkscrew windings and funnel-shaped outer and inner endings. In spite of this, the discovery of the ZR has been wrongly attributed to this author (Fig. 5/2 A,B). 'Müller' discussed the question as to whether the canaliculi originate each from a cell each or if they are intercellular. 'Reichert (1856)' suggested that the canaliculi are a derivative of the granulosa and stressed that the viscous layer was much thicker than the ZR. It was 'Kölliker (1858)' who had no doubt that the canaliculi are protrusions of the follicular cells and that the jelly is nothing else but a secretion of these cells. He called the canaliculi G. 'Saftröhrchen' (literally translated: juice tubules). 'Lereboullet (1862)' published a more extensive work which, compared with the publication of 1854, did not contain any new views on the above-mentioned surface of the perch egg. 'His (1873)' agreed with the conclusions of 'Kölliker' and so did 'Brock (1878)'. The latter noted that the striations within the jelly are much coarser than those of the ZR. His drawing of the section of the egg of *Perca* was taken up by the textbooks of the time (Fig. 5/3). However, 'Hoffmann (1881)' disagreed with the above conclusions and considered the canaliculi as part of the ZR. Detailed observations by 'Owsiannikow (1885)' followed. He observed in the jelly 'canals and tubes' with funnels at both ends and stated further that the canaliculi are in contact with those of the ZR but doubted that the former were formed by the granulosa cells. 'Eigenmann (1890)' quoted his own observations of the eggs of *Perca americana*. 'The radially arranged spiral structures traversing this layer (jelly capsule) arise as funnel-shaped tubules, one beneath each cell of the granulosa. In the early stages of their development the tubules have a more or less spiral course, while in the later stages they become more nearly straight. In 'February-eggs' the inner ends are slightly expanded, and terminate in a thin structureless film overlying the zona. In radial sections of eggs taken in May, the tubules often appear triangular at the base, and their contents divide into branches which enter the pores of the zona.' He stressed that the thick jelly capsule is produced by a secretion from (and metamorphosis of) the granulosa cells (Fig.5/4). 'M'intosh and Prince (1890)' suggested that the canaliculi of the jelly do not serve, as stated by 'Lereboullet (1854)', for absorption like the minute canals of the ZR, though both structures penetrate the capsule. 'Waldeyer (1883)' still stressed the various opinions on the jelly layer. 'Retzius (1912)' reviewed the then up-to-date literature and added his own observations. He confirmed the hexagonal netting (facetted hexagons) of the outer surface of the jelly, which coincided with the shape of the overlying granulosa cells. Their protoplasmatic processes fill the funnel-like outer endings of the canaliculi and follow through the canaliculi and their inner dichotomic ramifications. He stressed that the spiral winding of the canaliculi is due to shrinkage of the eggs during histological processing since they remain radially straight in the fresh condition. The thin follicular cell processes proceed into the radial canals of the ZR (Fig. 5/5A,B). 'Chevey (1925)' stated that the canaliculi do not ramify as stated by Müller, 1854 (Fig. 6/19).

'Flügel (1967a)' analyzed the envelopes of *Perca fluviatilis* with EM techniques. The viscous coat appears 'rather homogeneous' also under the EM. He found embedded in the matrix some electrondense bodies 'of a complicated organisation', and of unknown function. He confirmed that the canaliculi in the jelly coat contain processes of the follicular epithelium, which enter the canaliculi of the ZR. They enclose some ER and many mitochondria. Treatment with Ruthenium-red yielded a positive reaction of the jelly, especially in the region near the follicular cells. This staining reaction indicates the presence of acid mucopolysaccharides. The walls of the follicular processes traversing the layer become thicker in the mature egg, which makes them visible by light microscopy. 'Flügel' suggested — as did the early investigators before him — that the follicular cells produce the jelly. At ovulation the egg becomes surrounded by the jelly.

'Müller (1854)' described *Acerina cernua* as having eggs with a velvet-like covering due to the presence of 'Zöttelchen' (small cones). He contended that *Acerina* does not produce long strings of eggs but lays separate eggs, though 'Seeley (1886)' wrote that 'the eggs of *A. cernua* are deposited on the roots of plants in connected strings like those of the yellow perch' (cf. 'Breder and Rosen, 1966)'.

Serranidae (sea basses)

A similar situation to that seen in *Perca* has been described for *Serranus heptatus* by 'Brock (1878)'. The viscous coat is also traversed by fine processes emanating from the follicle cells which are 'loosened', i.e. do no longer form a close epithelial layer. Several thin canals emanating from each cell form ramifications and eventually can be followed as far as the ZR.

Scorpaenidae (scorpion fishes)

Pelagic gelatinous egg-masses have been described for *Scorpaena porcus, S. scrofo* '(Raffaele, 1888)'. The eggs of *Dendrochirus brachypterus* and *Sebastolobus alascanus* are released at ovulation into a gelatinous mass which has been secreted by specialized cells lining the ovary. The oocytes develop on ovarian protrusions called peduncles (also pedicles, stems, branches, delle, stalks) ['Fishelson (1978)'; 'Riehl (1991)'; 'Erickson and Pikitch (1993)'].

Lophiidae (goosefishes)

The floating eggs of *Lophius piscatorius* adhere together in long 'ribbon-shaped veils', which may be as long as 10 metres and up to 1 metre in width near the surface. But the eggs may also become free ['Agassiz (1882)'; 'Henneguy (1885)'; 'M'Intosh and Prince (1890)'.

Antennariidae (frogfishes)

Antennarius builds a nest among floating seaweed '(Henshall, 1888)'. According to the review of 'Breder and Rosen (1966)' Antennariidae produce rafts or veils of eggs and according to 'Mosher (1954)' the rafts of the sargassumfish *Histrio histrio*, in contrast to the egg strands of *Perca*, do not show spaces which would allow for the flow of fresh water.

6.5 MOUTHBROODERS (ORAL INCUBATORS)

'Bloch (1794)' was the first to describe mouthbrooding in siluroid fish. 'Clarke (1883)' reported the mouthbrooding habit of *Felichthys feus* and 'Ryder (1883a, 1887)' followed with a brief mention of *Aelurichthys (Felichthys)* '(cf. Gudger, 1918)'.

The eggs of the mouthbrooder *Tilapia* secrete, before ovulation, a mucous layer directly by the ER, i.e. bypassing the Golgi apparatus '(Kraft and Peters, 1963; Busson-Mabillot, 1977)'. Both male and female carry the developing eggs: in *Tilapia mossambica (Oreochromis mossambicus), Tilapia nilotica* and *T. galilaea* the female, in *Tilapia macrocephala* the male only. In contrast to the filament-bearing eggs of the substrate-spawning tilapias, the eggs of the mouthbrooders are not adhesive, with the exception of those of *Tilapia galilaea*. The latter take up a middle-position between the substrate-depositing and mouthbrooding species. Immediately following deposition the eggs adhere to one another and to the substratum. However, the eggs become loose once they are taken up into the mouth '(Kraft and Peters, 1963)'. 'Wickler (1956)' had already noted that the adhesive apparatus has vastly atrophied in the mouthbrooders and that their eggs adhere scarcely or not at all. Also the mouthbrooding cichlid *Geophagus jurupari* first glues the eggs onto the substratum and only after a day takes them into his mouth '(Kraft and Peters, 1963)'. A unique surface pattern, vegetal different from animal pole, has been observed by SEM in the eggs of the mouthbrooder *Luciocephalus* sp. Spiral ridges run from the animal pole (the micropyle) to the vegetal pole. This arrangement has been interpreted as a sperm guidance system '(Riehl and Kokoschka, 1993)' (see Chapter 7).

6.6 OVOVIVI- AND VIVIPAROUS FISH (LIVEBEARING FISH)
(dealt with in Chapter 20)

Summary

1) There is almost unanimous agreement that all accessory surface-structures — be they filaments or additional layers for egg adhesion — are formed by the follicular epithelium (granulosa). This was already suggested by some of the investigators of the 19th century. More recently, EM investigations support the same site of origin, based on the presence of Golgi bodies, abundant rough endoplasmic reticulum (RER), ribosomes, microtubules, mitochondria, electron-lucent vesicles and on experiments with tracers (autoradiography).
2) Pelagic-nonadesive eggs. Most display a honey-comb structured envelope produced most probably by the hexagonically-packed follicle cells. However, filaments are reported to be present on the egg surface of all Scomberesocidae.
3) Demersal-non-adhesive eggs, surrounded by a thick layer of gelatinous material added only after ovulation, are found buried in unguarded nests. Jelly plugs are seen filling the openings of the pore canals. In some fish, nodules and blebs have been observed. Zebrafish eggs are different in that they are dropped singly and their surface displays hemispherical domes and vertical spikes interconnected by ridges. Sticklebacks build nests from mucoid threads and mucus keeps the eggs together.
4) Demersal-adhesive eggs adhere to the substrate by appendages given different names such as cones, fibrils, filaments, papillae, pedicles or hooks, also suspensory ligaments and adhesive discs. The total number and size of fibrils can vary and so also their distribution (regularly distributed, in clusters, bipolar, at one pole only). Apart from the adhesive appendages, non-adhesive ones

may be present. The covering jelly is secreted by the Golgi, or by the RER bypassing the Golgi body, of the follicular granulosa cells. Fish, the development of which is arrested in the dry season, are called annual fish. Their eggs display a tough envelope adorned with highly developed surface structures thought to act as respiratory systems.

5) Special structures for floating and attachment occur as long zig-zagging collapsed bands embedding the eggs. The jelly of the eggs displays a radial striation, which is coarser than that of the ZR. Follicular processes extend through the often ramified canaliculi of the jelly to reach the canaliculi of the ZR. The eggs are attached to one another but the ensuing band does not adhere to the substratum.
6) Mouthbrooders. The surface structures differ in the few species studied.
7) Ovoviviparous and viviparous fish are dealt with in Chapter 20.

Chapter seven

Micropyle[1]

'Pour bien savoir les choses il faut en savoir le détail.'
LaRochefoucauld

The tough egg envelope of teleost fishes is an effective barrier against penetration of sperms. However, there is a special structure, the micropyle, situated at the animal pole of the egg, which allows the passage of a sperm to the oocyte membrane. The inner diameter of the micropyle is about 1.5 microns. The fine structure of the micropyle and the surface pattern around its opening are species-specific and are, therefore, used for identification processes. It is of phylogenetic interest that *Acipenser*, which belongs to the chondrostei, has 7 micropyles. Some authors quote up to 15 for chondrostei. Holostei have only one micropyle '(Ginzburg, 1968, 1972; Cherr and Clark, 1982; Riehl, 1991)'. The sperm-activating factor in the micropyle area will be dealt with under fertilization (Chapter 9).

7.1 TYPES OF MICROPYLES

Micropyles in freshwater fish have been differentiated into four types: Type I with a deep pit or funnel-like depression (vestibulum) on the surface of the oocyte, leading into a short canal which opens on the inner surface of the envelope. Examples: *Pungitius, Leuciscus cephalus, Gobio gobio, Noemacheilus barbatulus, Phoxinus phoxinus, Tinca tinca, Cichlasoma meeki, C. nigrofasciatum*. Type II with a shallow saucer-like depression (flat pit) leading into a long canal. Examples: *Oncorhynchus keta, Brachydanio rerio, Salvelinus alpinus, Salmo salar. Chionobathiscus dewitti, Cyprinus carpio, C. carassius*. Type III has only a canal. Examples: *Coregonus lavaretus, C. nasus, C. oxyrhynchus, C. pidschian, C. fera, C. macrophthalmus, C. wartmanni, Esox lucius, Salmo gairdneri, S. trutta, S. trutta morpha fario. S. trutta morpha lacustris, Salvelinus fontinalis, Engraulis japonica, Pagrus major, Chionodraco myersi* and *Anguilla japonica*. Type IV with two vestibula (outer and inner) at the egg surface and a short canal. Examples: some antarctic fish '(Riehl and Götting, 1974; Riehl, 1980b; Riehl and Kock, 1989; Riehl, 1991, 1993; Riehl and Kokoscha, 1993)'.

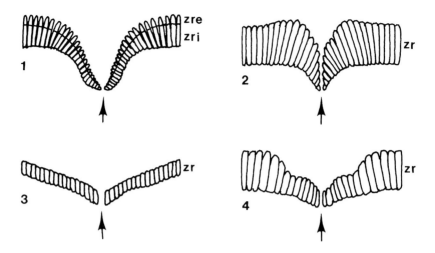

Fig. 7/1 Schematic drawing of the different types of micropyles. (modified from Riehl, 1991, with kind permission of Acta Biologica Benrodis). **1** - Type I; **2** - Type II; **3** - Type III; **4** - Type IV, zr - *Zona radiata*; zre - *Zona radiata externa*; zri - *Zona radiata interna*; arrow = micropylar canal.

The inner opening of the canal is consistently smaller than the outer opening and often ends on a papilla as first observed by 'Hoffmann (1881)'. The outer opening is found in a depression and the striations of the ZR present an inclination towards the micropyle, which is increased as the aperture is approached, and still more so down the walls of the crater. The area around the external opening of the micropyle is usually different from the remainder of the egg surface (Figs. 7/2-7/10).

7.2 DISCOVERY OF THE MICROPYLE

Because of the protracted controversy as to the priority of the detection of the micropyle in fish, full consideration is given to the various contributions.

'Von Baer (1835)' was the first to describe the micropyle in the egg of *Cyprinus blicca* (*Blicca bjoerkna*) as a circle above the 'Keim' (germ, i.e. blastodisc or discus proligerus), which he rightly interpreted as the entrance into a funnel. By inserting a needle, he observed that the tip of it almost reached the yolk. After the intake of water, the funnel gradually disappeared. However, 'von Baer' did not appreciate the significance of this structure. 'Rusconi (1836 and 1840) had artificially inseminated eggs of *Cyprinus tinca, C. alburnus* and the pike by pouring seminal fluid over the ova but he did not mention the process of fertilization. 'Doyère' was the first to rightly interpret the micropyle in a teleostean egg (in a communication to the Société Philomathique de Paris on December 15th 1849, which was printed in 1850)'. He observed in the egg of *Syngnathus ophidium* (*Neophis ophidion*) an aperture for sperm penetration though he did not substantiate his conclusion by experimental data. 'Leuckhart (1855)' found the micropyle in the related species *Syngnathus acus*. Observations by early investigators on other species were the following: 'Leuckhart (1855)' noted the micropyle in *Silurus glanis* but had only one egg at his disposal. 'Bruch' announced his discovery of the micropyle in the eggs of trout (*Salmo fario*) and salmon (*S. salar*) in a letter to Professor Siebold, dated December 28, which appeared in print in 1855. He claimed to have first established on a firm basis the existence of a micropyle in vertebrata. However, he was unsuccessful in proving that it serves for the entrance of sperms. There followed observations of trout and salmon, *Salmo salar*, by 'Ransom' in December 1854 and January 1855, which were communicated in the latter month in a letter to Dr. Allan Thompson. The Report of the 25th Meeting of the British Association for the Advancement of Science in 1855 revealed that 'Dr. Ransom announced the existence of an aperture through the yolksac of the ovum, in several freshwater fishes; pointed out its relation to the formative yolk[1], and its importance as permitting the entrance of the spermatozoa'. The micropyle in the stickleback *Gasterosteus leiurus* and *G. pungitius*, was subsequently also described by 'Thomson (1855)' (Fig. 7/2). 'Ransom' as well as 'Bruch' had been unaware of the earlier discoveries by 'von Baer' and 'Doyère'. 'Reichert (1856)' observed the micropyle in the pike, in all cyprinoids available and also in *Acerina* and others. He was unable to see it in *Perca fluviatilis* [while, apart from 'Leuckhart (1855)', also 'Kölliker (1858)' and 'Ransom (1867)' had observed it in this fish.] Subsequently, the general occurrence of the micropyle in teleosts was suggested by 'Leydig (1857)' as follows: 'Les poissons sont les vertébrés qui présentent d'une manière non douteuse un canal infundibuliforme traversant les enveloppes de l'oeuf' (The fish are the vertebrates which without any doubt present an infundibuliform canal traversing the envelopes of the egg). 'Kölliker (1858)' observed the micropyle in several species, apart from *Perca fluviatilis* and *Syngnathus* mentioned above, i.e. in the pike, *Esox lucius*, *Cobitis barbatula*, *Gobio fluviatilis*, *Cobitis fossilis*, *Acerina cernua*, *Rhodeus amarus*, *Phoxinus laevis* and *Aspius alburnus*. In both species of *Cobitis*, in *Aspius alburnus* and *Phoxinus* the micropyle was funnel-shaped. In the other fishes it was represented by a canal of similar width throughout. Finally, 'Ransom (1868)' provided experimental proof that the sperm penetrates the micropyle of the egg of the stickleback, *Gasterosteus pungitius*. In other words, he followed microscopically how the spermatozoon penetrates the egg via the micropyle.

7.3 THE MICROPYLE OF GASTEROSTEIDAE

'Ransom (1854/55)' depicted the type 1 micropyle of *Gasterosteus*. A ring of flask-like processes surrounds the funneltype opening (Fig. 7/2A,B). A detailed account of the micropyle was given by 'Ransom (1867)' and by 'Owsiannikov (1885)'. 'Eigenmann (1890)' illustrated the regions of the micropyle in *Pygosteus pungitius*. The micropyle of *Pungitius tymensis* was described by 'T.S.Yamamoto (1963)'. It is in a funnel-shaped indentation with a small inner opening and a thickened margin at the inner and outer opening. The canal opens eccentrically at the outer margin. 'Thomopoulos (1953b)' and 'Wickler (1959)' suggested that the mushroom-like outgrowths around the micropyle of *Gasterosteus* secure a space for the sperms to reach the micropyle. The micropyle of the above fish belongs to type I.

7.4 THE MICROPYLE OF SALMONIDAE

As mentioned above, 'Bruch' discovered the micropyle in the mature eggs of *Salmo salar* and *S. fario* in 1854/55. The micropyle was always found in an area above the 'Keimfleck' (spot of the germ = animal pole). He described it as a shallow depression surrounding a crater, which in *Salmo fario* terminated in a funnel. However, he did not observe the penetration of the sperm, neither through the micropylar canal (the diameter of which corresponded to the size of the 'sperm body') nor through any other site. He agreed with 'Remak (1854)' that the pore canals of the ZR were far too thin for the entrance of a spermatozoon. He disagreed with 'von Baer' (on *Cyprinus blicca*), who found that 'the funnel' collapsed after the eggs came into contact with water. He did not observe any collapse of the micropyle, neither in the unfertilized nor in the fertilized salmonid ovum kept in water. As mentioned under 7.2 'Ransom' observed the micropyle in salmon, *Salmo salar*, and the trout, *Salmo fario*, at the same time as 'Bruch', and again in 1868 referred to these results but in a more extensive publication dealing with the ovum of various osseous fishes (Fig. 7/2C,D). 'His (1873)' compared and contrasted the micropyle of the trout and salmon. In the latter there is a large but shallow vestibulum from which emanates a straight canal while in the former the vestibulum is narrow and deep. 'His' was the first to maintain, with the help of diagrams, that in the salmon never more than one spermatozoon can pass the micropylar canal. 'André (1875)' confirmed that the micropyle is destined for the intake of sperms. He described for the trout, *Salmo fario*, an outer rim or annulus of a crater leading into a canal, which is not perfectly cylindrical, but is sinusoid, i.e. midway along its course enlarges and narrows again, and opens on an internal small papilla. In contrast, 'Blanchard (1878)' reported that the micropyle is used for water uptake. 'Ziegler (1882)' described for *Salmo salar* a funnel-shaped micropyle situated in an 'indentation of the envelope'. 'Henneguy (1888)' compared and contrasted the micropyle of salmon and trout.

The micropyle of salmonids belongs to Type II or Type III. It should be stressed at this point that the presence of a rim around the outer opening of the micropyle, in the trout, *Salmo fario*, is a morphological component of the micropyle and should be incorporated into its definition. (Figs. 7/2C,D,7/3,7/4).

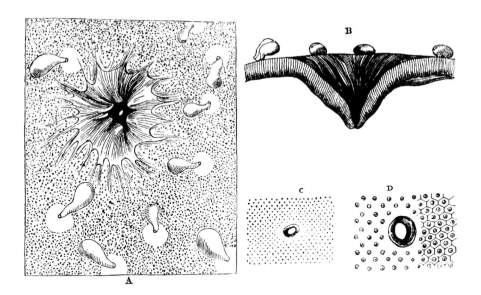

Fig. 7/2 Surface view of a type I micropyle in the mature ovarian egg of *Gasterosteus* (A,B) and of the trout (C,D) (Ransom, 1854/55). **A** - Flask-like processes in the vicinity of the micropyle are seen. The radiated shading represents the funnel-shaped depression leading to the opening of the micropyle. **B** - Semidiagrammatic view of longitudinally-sectioned micropyle depicted in A. The fine canals of the ZR are represented fewer and wider than they are in nature. **C** - Small flattened portion around the *apex* of the funnel containing the aperture. 500 ×. **D** - Same as C but at a magnification of 1 000 ×. Hexagonal divisions between the apertures of the ZR canals (representing surface views of follicle cells) are indicated on the right hand side.

7.5 THE MICROPYLE OF ESOCIDAE

As described in 7.2 'Rusconi (1840)' was the first to attempt artificial insemination. He reported that the sperms of the pike are so large that it is impossible for them to penetrate the envelope via the radial canals. He found it impossible to say how the sperms act on the egg. (The micropyle was not known at that time.) He noted that if the eggs are laid in water containing sperm after about half an hour the envelope detracts from the yolk, 'possibly by imbibing water'. The discovery of the micropyle in the pike, *Esox lucius*, was reported by four investigators around the same time '(Bruch, 1855; Leuckart, 1855; Reichert, 1856; Kölliker, 1858)'. 'Ransom (1867)' described the micropyle in the pike as trumpet shaped. 'His (1873)' studied only surface views of the micropyle of the pike. He concluded that there is a shallow pit which leads into a funnel and then into the canal. The shape of the shallow surface pit is indicated by the oblique pore canals surrounding the area. The micropyle of the pike, *Esox reticulatus*, belongs to Type III as is evident from Fig. 7/17 by 'Eigenmann (1890)'.

7.6   THE MICROPYLE OF PERCIDAE

'Leuckhart (1855)' described the micropyle of *Perca fluviatilis* and observed rosettes around the internal opening of the canal. The eggs of *Perca fluviatilis* are attached to each other in a net-like tube (see Chapter 6) and 'Ransom (1868)' observed that the eggs are all so placed that the micropyles face the inside of the collapsed tube. He suggested that one effect of this arrangement must be to prevent the occlusion of the micropyles by oocyte contact.

7.7   THE MICROPYLE OF OTHER SPECIES

'Ransom (1868)' described the micropyle in the following species: *Acerina vulgaris, Cottus gobio, Cyprinus gobio, Leuciscus phoxinus, L. cephalus*. 'Kölliker (1858)' found the micropyle in the eggs of *Cobitis barbatula, Gobio fluviatilis, Cobitis fossilis, Acerina cernua, Rhodeus amarus, Phoxinus laevis* and *Aspius alburnus*. 'Buchholz (1863)' described the micropyle of *Osmerus eperlanus* which seemed to be very different from the micropyles of other fishes described to that date. 'Kupffer (1878)' observed a type I micropyle in the egg of the herring. 'Hoffmann (1881)' analyzed the herring, *Leuciscus rutilus, Tinca vulgaris, Scorpaena scrofa, Julis turcia, J. vulgaris*, various species of *Crenilabrus*, and *Heliasis chromis*. 'Ryder (1882a)' described the micropyle of the cod, *Gadus morrhua*, and the white perch, *Roccus americanus*, and 'Brook (1884)' that of *Trachinus vipera*. 'Owsianniskow (1885)' dealt with the micropyle of *Osmerus eperlanus* and 'List (1887)' with that of *Crenilabrus tinca*. 'Holt (1890)' described the micropyle in *Gobius minutus* and 'Agassiz and Whitman' in *Ctenolabrus*.' M'Intosh and Prince (1890)' concluded, as had 'Leydig (1857)', that the micropyle seems to be present in all teleostean eggs, and 'Mark (1890)' reviewed the publications on 'egg membranes' and associated structures of all teleosts.

'Riehl and Götting (1974)' reported that up to their time the micropyle of about 70 teleostean species had been reported and they added another 17 species. A type III/IV micropyle was reported for *Coregonus macrophthalmus* and *C. wartmanni* by 'Riehl and Goetting (1975)' and 'Riehl and Schulte (1977b)'. Subsequently, a review of the literature on the light microscopical analyses of the micropyle was given by 'Laale (1980)'.

7.8   ULTRASTRUCTURE OF THE MICROPYLE AND ITS SURROUNDINGS

SEM analyses of the rim of the micropyle in the rainbow trout *Oncorhynchus mykiss* revealed a rough-surfaced ring around the edge of the funnel. This ring is composed of more or less regularly arranged small knobs, which are not to be found on the other regions of the egg surface. The diameter of the internal protuberance of the canal is nearly the same as that of the funnel '(Szöllösi and Billard, 1974; Riehl, 1980b)' (Fig. 7/3). However, a ring around the border of the vestibule was absent in the egg of the chum salmon, Oncorhynchus keta, and small knobs were distributed over the whole egg surface '(Osanai, 1977; Riehl, 1980b)'. Generally speaking, in Salmonidae the surface of the region around the micropyle shows species-specific differences. In Oncorhynchus keta the absence of the ZR externa from the vestibule was confirmed by TEM; in other words, the envelope in the micropylar region consists only of the ZR interna. The interna is not perforated by pore canals in the protruded region around the inner opening of the micropylar canal but is made up of fibrous material. Because of the absence of pore canals in this region, numerous free but shorter compressed microvilli emanating from the plasma membrane of the oocyte are in evidence '(Kobayashi and Yamamoto, 1981)'.

SEM revealed the presence of spirally arranged ribs within the walls of the micropylar canal of S. trutta morpha fario, Salvelinus fontinalis, Salmo gairdneri and Oncorhynchus gorbuscha '(Szöllösi and Billard, 1974)'. These probably correspond to the edges of the horizontal lamellae of the ZR interna. In Salmo trutta the opening of the canal is not situated in the centre of the funnel '(Riehl and Schulte, 1977b; Riehl, 1980b; Stehr and Hawkes, 1979)' (Fig. 7/3, 7/4).

'Groot and Alderdice (1985)' analyzed five Pacific salmon, Oncorhynchus nerka, O. gorbuscha, O. keta, O. kisutch, O. tsawytscha, and the steelhead trout Salmo gairdneri. They concluded that the micropyles of these six species of Salmonidae are basically similar. They observed that larger plugs fill the pores in the vicinity of the exterior opening of the micropyle even in O. nerka, which otherwise possesses no obvious plugs. Within a circular area around the micropylar canal the ZR externa was ostensively thinner than in the rest of the egg surface, or absent altogether.

The micropyle of two limnic species Coregonus nasus and C. lavaretus showed species-specific differences. The micropyle of the former is devoid of a pit and the canal opens outward like a funnel. The honeycombed surface pattern of the egg is absent in the region of the funnel. The micropyle of the latter is larger than that of the former and also lacks a pit. It widens outward like a funnel and the surface pattern consisting of minute knobs is absent in the funnel area. The micropyles of both species belong to type III '(Riehl, 1993)'. In the closely related antarctic ice-fishes Chanodraco myersi and Chionobathyscus dewitti (Channichthyidae) the species-specific differences are even more pronounced. The micropyle of Chanodraco myersi shows a pit with a surface pattern resembling that of the remaining egg-surface except in the very direct neighbourhood of the opening of the micropyle canal, where pores are found. The micropyle of Chionobathyscus dewitti displays a small pit and a narrow canal. The region surrounding the exit of the micropyle canal is not different from the remaining egg-surface.

The micropyle of the egg of Brachydanio rerio shows a cone-shaped vestibule bordered by folds descending towards the canal in a right-handed spiral. The ves-

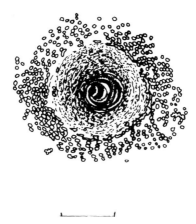

Fig. 7/3   Micropyle of *Salmo gairdneri* (drawn from a SEM of Riehl, 1980 and a SEM from Szöllösi and Billard, 1974, with kind permission of Kluwer Academic Publishers and Biology of the Cell, Paris). Bar = 5 microns. Note rib-like thickenings to reinforce the external opening of the micropyle, which is surrounded by a wreath of small knobs.

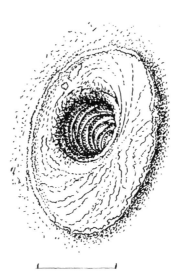

Fig. 7/4   Micropyle of *Salmo trutta* (combined drawing of two SEMs of Riehl, 1980, with kind permission of Kluwer Academic Publishers). The opening of the oblique micropylar canal is not situated in the centre of the micropyle funnel. Bar = 5 microns.

tibule is also lined by dome-like projections which are characteristic for the whole ZR surface. The walls of the canal are formed by circular, rib-like thickenings descending in a right-handed spiral '(Hart and Donovan, 1983)'. The ribs were interpreted as the edges of the lamellae of the ZR interna. The vestibule and canal with its internal opening are seen on Figs. 7/5-7/7. SEM of the micropyle of the walleye, *Schizostedion vitreum*, revealed a long canal situated in a slight depression which housed accessory openings. The depression was surrounded by a circular elevated portion of the egg envelope. The canal was described as being composed of concentrically arranged rings of diminishing size '(Johnson and Werner, 1986)'. The micropyle of both *Brachydanio rerio* and *Schizostedion vitreum* belong to type II. The adhesive eggs of the rosy barb, *Barbus conchonius*, and *Sturisoma aureum* display a micropyle with a funnel-shaped vestibule. Several micropylar grooves and ridges are thought to function as a sperm-guidance system '(Amanze and Iyengar, 1990; Riehl and Patzner, 1991' (see Chapters 8 and 9) (Fig. 7/11a-c).

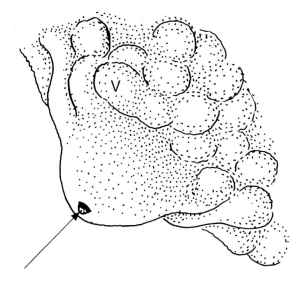

Fig. 7/5 Drawing of regionalization of the micropyle of *Brachydanio rerio* seen from the inside of the egg [drawn after a SEM by Hart (1990) with kind permission of Academic Press, London, subsidiary of Harcourt Brace & Company Ltd.]. × 1 700. MC micropylar canal V vestibule. Arrow points to the inner aperture of the canal.

A micropyle with the canal opening on the summit of a conical elevation of the ZR is displayed by *Oryzias latipes* according to 'Hart et al. (1984)' (Fig. 7/8), but not recognized for the same species by 'Iwamatsu and Ohta (1981)', 'Iwamatsu et al. (1993a,b)' and 'Iwamatsu (1994)'. According to 'Hart et al. (1984)' the fuzzy

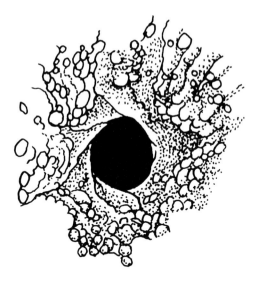

Fig. 7/6 Topography of micropylar apparatus of *Brachydanio rerio* egg. [drawn after a SEM by Hart and Donovan (1983) with kind permission of Wiley Liss Inc., a subsidiary of John Wiley & Sons, Inc.]. 2 500 ×. Surface view into funnel. Note the diaphragm (clockwise spiral folds).

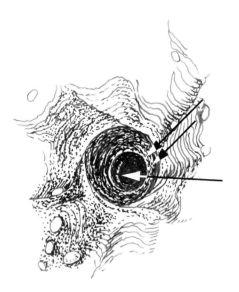

Fig. 7/7 Overhead view of the micropylar canal of *Brachydanio rerio*. [drawn after a SEM from Hart and Donovan (1983) with kind permission of Wiley Liss Inc., a subsidiary of John Wiley & Sons, Inc.]. 2 700 ×. Note inner opening of the canal (long arrow); outer opening shorter arrows.

layer of the ZR is 'organized into tightly packed, bulblike protrusions' around the opening of the micropyle. The walls of the micropyle (vestibule and canal) of *Oryzias latipes* were shown to form a spiral pattern. Of 85 oocytes examined, 85% displayed a right-handed spiral and the remainder did not show any spiral folds. In fully-grown oocytes, the ZR externa lined two-thirds of the wall of the micropyle '(Hosokawa, 1979; Iwasmatsu and Ohta, 1981; Iwamatsu *et al.*, 1993a,b)'.

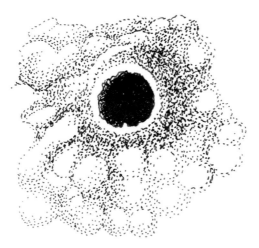

Fig. 7/8 External micropylar opening of egg of *Oryzias latipes*. (drawing after a SEM from Hart *et al.*, 1984 with kind permission of Wiley Liss Inc., a subsidiary of John Wiley & Sons, Inc.). 1 600 ×. Opening is at the apex of a cone-like elevation of the envelope.

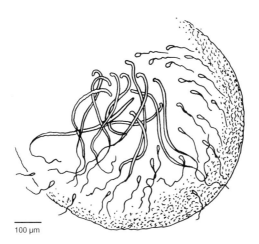

Fig. 7/9 Vegetal pole of the egg of *Oryzias latipes* showing left-handed spiral arrangement of the filaments. (modified drawing of a SEM of Iwamatsu *et al.*, 1993a with kind permission of Wiley Liss Inc., a subsidiary of John Wiley & Sons, Inc.).

When the oocytes of this fish were treated with cysteine or thioglycolate, the ZR interna was shown to be composed of many layers '(Nakano, 1956)'. The rib-like appearance of the canal wall might therefore be due to the presence of layers in the interna. The attaching filaments on the vegetal pole showed a corresponding spiral arrangement. The micropyle of *Oryzias* belongs to type I (Figs. 7/9, 7/10).

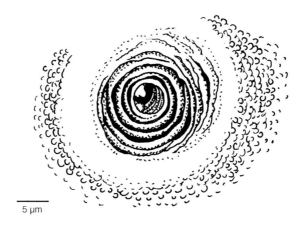

Fig. 7/10 Left-handed spiral in the wall of the micropyle of *Oryzias latipes*. (modified drawing from a SEM of 'Iwamatsu *et al.*, 1993b' with kind permission of Wiley Liss Inc., a subsidiary of John Wiley & Sons, Inc.).

SEM of the ovum of *Fundulus heteroclitus* revealed that it is covered by a thin jelly coat comprised of fibres except for a fibril-free region around the micropyle. The absence of fibrils would obviously help in collecting sperm and facilitate their access to the micropyle. '(Kuchnow and Scott, 1977)'. 'Dumont and Brummet (1980)' reported essentially the same findings. However, they added that small tapered projections adorn the walls of the depression. They displayed an ultra-structure unlike that of the fibrils, i.e. particulate rather than tubular. In stark contrast to the above is the situation reported by 'Wirz-Hlavacek and Riehl (1990)' who analyzed with TEM the egg surface of the piranha, *Serrasalmys nattereri*. The external opening is surrounded by abundant short adhesive filaments which form a disc. When the adhesive disc attaches the eggs onto the substrate, the micropyle becomes inaccessible. To guarantee fertilization before attachment of the eggs the parents are engaged in a complex reproductive behaviour lasting about 1 hour. The resulting copulation (during which the male encloses the female genital opening with his anal fin) takes place 20-30 times, each lasting for up to 9 seconds, while eggs are discharged for only 3-5 seconds. The micropyle is represented by a simple canal and, therefore, belongs to type III.

Pelagic eggs, with the exception of a few groups, do not exhibit, at light microscopical level, a specialized micropylar apparatus '(Ahlstrom and Moser, 1980; Hirai and Yamomoto, 1986)'. However, at the EM-level species-specific differences were observed. The micropyles of the anchovy, *Engraulis japonica* '(Hirai and Yamomoto, 1986)', the porgy, *Pagrus major* '(Hosokawa et al., 1981)' and the eel, *Anguilla japonica* '(Ohta et al., 1983)' are all composed of a simple canal without a depression of the vestibule and belong to type III. However, the micropyle of the anchovy and eel eggs is smaller than that of the porgy egg and the outer opening of the micropyle is smaller in the other two. It was concluded that the pelagic eggs, too, display species-specific ultrastructural differences in the micropylar region.

Pelagic eggs of nine marine species were studied with SEM, with regard to their micropyle and surrounding structures, by 'Hirai (1988)'. The diameter of the micropylar canal varied between the different species; it ranged from 2.6 microns in *Kareius bicoloratus* to 12.0 microns in *Inimicus japonicus*. A distinct convoluted structure was observed in the canal of *Lateolabrax japonicus* and *Hypodytes rubripinnis*. The micropylar area was characterized by a circular elevation in *Sardinops melanostictus, Sillago japonica, Parapristipoma trilineatum, Oplegnathus fasciatus, Inimicus japonicus* and *Hypodytes rubripinnis* with larger pores than those observed over the remaining egg surface. However, *Kareius bicoloratus* displayed neither an elevation nor distinct pores. In *I. japonicus* and *H. rubripinnis* some pores were found inside the micropylar canal. A funnel opening outwards was described for *Sillago japonica, Parapristipoma trilineatum, Hypodytes rubripinnis* and *Kareius bicoloratus*. In *Sardinops melanostictus* a circular elevation measuring about 20 microns in diameter was observed, in which distinct cavities of various sizes were obvious. The micropylar canal opened outward as a funnel of about 4.2 microns in diameter. *Saurida elongata*, too, revealed a funnel like opening (8.3 microns in diameter) of the micropylar canal. The funnel was surrounded by about 70 shallow cavities. There was no circular elevation in the micropylar region. Annular lamellae were observed on the inside of the canal. *Lateolabrax japonicus* displayed a flat micropylar region of 18 microns diameter, which contained about 60 pores. The canal did not open with a funnel and the opening measured 4.8 microns in diameter. A convoluted structure was observed inside the canal. Unique micropyles were observed in the pelagic ostraciid fish *Ostracion meleagris* and *Lactoria fornasini*, where the micropyle is surrounded by 'hollow bumps' '(Leis and Moyer, 1985)'. The regions around the outer opening of the micropyle of *Stolephourus* and of some paralichthyid fishes (*Escheneida* sp. and *Pleuronectida* sp.) displayed warts '(Delsman, 1931; Mito, 1963 and Hirai, 1987, 1988)'.

While species-specific differences have been stressed, the micropyle and its surrounding area can display distinct similarities among distantly related species. The micropyle may also vary in different geographic localities '(Hirai, 1988)'.

The micropyle of the mouthbrooder *Luciocephalus* sp. consists of a pit and a canal and, therefore, belongs to type I. The egg surface displays a spiralling pattern of ridges partially terminating in the micropylar region '(Riehl and Kokoschka,

1993)' (Figs. 7/11a-d). Grooves and ridges directed towards the micropylar canal have been described also for other species, e.g. for the cyprinid *Barbus conchonius* by 'Amanze and Iyengar (1990)' and for the catfish *Sturisoma aureum* by 'Riehl and Patzner (1991)'. However, the pattern of the surface furrows of these species is radial, running from the animal to the vegetal pole (Figs. 7/12, 7/13).

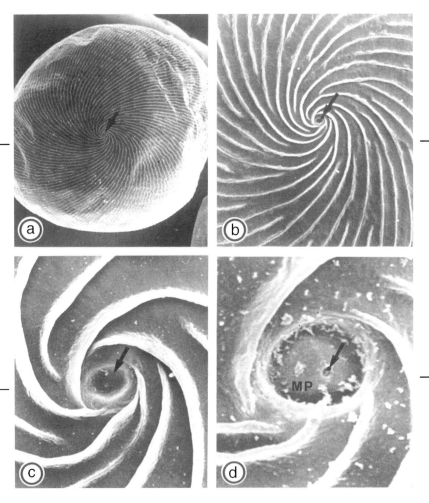

Fig. 7/11 Micropyle at animal pole of *Luciocephalus* sp. *egg* (Riehl and Kokoschka, 1993, with kind permission of Academic Press, London). **a** - Low power micrograph with arrow pointing to the micropyle. Bar = 500 microns. **b** - High magnification of micropyle and spiral ridges, some of which enter micropylar pit (arrow). Bar = 100 microns. **c** - High magnification displaying micropylar pit (arrow) and central canal. Bar = 25 microns. **d** - High magnification to show central canal (arrow). MP pit. Bar = 10 microns.

Micropyle 123

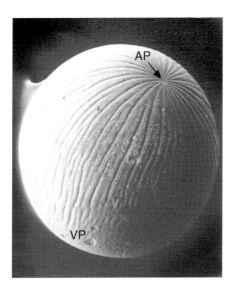

Fig. 7/12 Egg of the catfish *Sturisoma aureum* displaying radial furrows running from the vegetal to the animal pole (Riehl and Patzner, 1991 with kind permission of Farrell Kuhns Publications). Egg diameter = 2.1 mm. AP - animal pole VP - vegetal pole. Arrow points to location of micropyle.

Fig. 7/13 Micropyle of *Sturisoma aureum* surrounded by a smooth region (Riehl and Patzner, 1991 with kind permission of Farrell Kuhns Publications). The micropyle opened outwards like a funnel (C). There was no pattern around the entrance of the micropyle: the furrows terminated near the micropyle canal and left around it a naked smooth region (SM) ranging from 35 to 55 microns in diameter.

SEM and TEM of the micropyle of mature unfertilized eggs of the sea bass *Dicentratus labrax* (Percichthydae) showed the absence of a vestibule or funnel which is indicative of type III according to 'Riehl (1991)'. The proximal part of the canal wall is made up of the borders of the layers of the ZR interna: lamellae alternate with interlamellar spaces, which are crossed by fibrous bridges. This conveys a ribbing effect as already observed in the Salmonidae at a light microscopical level '(Fausto et al., 1994)'.

7.9 FORMATION OF THE MICROPYLE

'M'Intosh and Prince (1890)' stated that they were left in doubt as to the way in which the micropyle arises. 'Mark (1890)' described for *Lepidosteus* the 'micropylar cell' which 'may serve to form the canal by resorption or to prevent the occlusion of the canal by less penetrable matter at the time of oviposition' (Fig. 7/14). 'Eigenmann (1890)' stated for *Pygosteus pungitius* that in the 0.4 millimetre egg the granulosa cells, which cover all other parts of the oocyte, are composed of one layer and of two or three layers in the vicinity of the micropyle. At this site 'a single cell larger than the others is always to be found directly above the canal. It usually sends a prolongation into the canal itself'. He, too, named this modified follicular cell, the 'micropylar cell' (Fig. 7/15). 'Eigenmann (1890)' also observed micropylar cells in *Perca fluviatilis* (Fig. 7/16), *Fundulus heteroclitus, F. diaphanus, Esox reticulatus* (Fig. 7/17) and *Cyclogaster lineatus* (*Liparis lineatus*). The drawings of the micropylar cell by 'Eigenmann (1890)' have been incorporated in the textbook by 'Nelsen (1953)'. Subsequently, 'Padmanabhan (1955)' and 'Laale (1980)' have described an almost identical development of the micropyle of *Macropodus cupanus*. In order to avoid any misunderstanding it should be stressed that the micropylar cell does not enter a preformed canal but that, on the contrary, it is the elongating micropylar cell which forms the canal in the developing ZR. This is brought about by mechanical pressure on the growing ZR rather than by digestion or dissolution of the egg envelope '(Kobayashi and Yamamoto, 1985)'. The formation of the micropyle was also described by 'Eggert (1929)' for *Salaras flavo-umbricus* and by 'Sterba (1958)' for *Noemacheilus barbatula*. Further results were reported by 'TS Yamamoto (1963)', 'Riehl and Goetting (1974, 1975)' 'Riehl (1977a, 1980)','Laale, (1980)', 'Dumont and Brummett (1980)', 'Takano and Ota (1982)' and 'Nakashima and Iwamatsu (1994)'. The micropylar cell is easily observed, because of its cytoplasmic process and because it does not show any affinity for basic dyes while the granulosa cells take up the stain.

EM analysis of the formation of the micropylar cell in the oocytes of *Noemacheilus barbatulus* and *Gobio gobio* was described by 'Riehl (1977a)'. Both fish belong to type II. In *Noemacheilus* the first indication of a micropylar cell within the follicular epithelium is noticed by an enormously increased nucleus (5 x 2.1 microns) which belongs to a cell of 10x3 microns (measured in radial sections). The swelling of the nucleus is responsible for the formation of a flat pit on the surface of

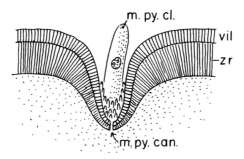

Fig. 7/14 Formation of micropyle in *Lepidosteus* (redrawn after Mark, 1890). m.py.can - micropylar canal m.py.cl - micropylar cell. vil - villous layer; zr - *zona radiata*.

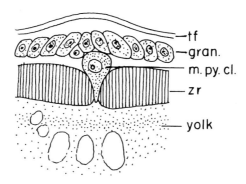

Fig. 7/15 Radial section through micropyle of egg of *Pygosteus pungitius* (about 0.4 mm in diameter) (redrawn from Eigenmann, 1890). gran - *granulosa*; m.py.cl - micropylar cell; tf - *theca folliculi*; zr - *zona radiata*.

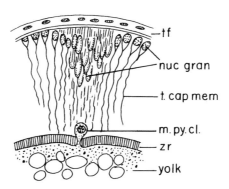

Fig. 7/16 Radial section through micropyle of an egg of the perch (redrawn from Eigenmann, 1890). m.py.cl - micropylar cell; nuc.gran - *nuclei* of *granulosa* cells; t.cap.mem - tubules of capsular membrane; tf - *theca folliculi*; zr - *zona radiata*.

the oocyte. Organelles are rare; some spherical mitochondria and some vesicles are observed. An absence of desmosomes between the micropylar cell and the adjoining follicle cells has been reported. The micropylar cell has not yet formed a protrusion. The anlage of this cell precedes the formation of the ZR interna. Once the protrusion of the micropylar cell has formed, the nucleus appears smaller and contains a nucleolus. The protrusion sinks into the ZR surrounded by a thick ridge (German: 'Wulst') of the interna, which prevents the collapse of the newly established micropylar canal. The micropylar cell sends microvilli into the radial canals of the surrounding ZR externa and interna. The micropyle of *Gobio gobio* is about twice as large as that of *Noemacheilus*. The micropyle is formed almost entirely within the interna. The micropylar cell of this species is smaller and with it several follicle cells are engaged in filling the vestibular space. The number of microvilli is smaller too. 'Kobayashi and Yamamoto (1985)' studied the micropylar cell and its changes during oocyte maturation in the chum salmon, *Oncorhynchus keta*. In this species the vestibule consists of a shallow depression (type II), which is not filled with follicular cells. The authors report clear morphological evidence indicative of the release of 'some components' at the distal end of the micropylar process. The long process contains microtubules and thin filaments whereas in the loach, *Misgurnus anquillicaudatus*, which possesses only a short cytoplasmic process, these organelles are distributed throughout the whole cytoplasm. It is worth mentioning that in this species the micropylar cell showed activity of 3ß-hydroxysteroid dehydrogenase, known to be essential for the biosynthesis of steroid hormones '(Ohta and Terenishi, 1982)'. When ovulation draws near, the process of the micropylar cell shows degenerative changes and shortens. During this process, an 'ooplasmic outgrowth' appears in the mature oocyte. So-called sperm motility-initiating or -enhancing factors are believed to be present in the vicinity of the micropyle of the chum salmon and other fish.

SEM revealed that the inner wall of the micropylar vestibule and the canal of the medaka *Oryzias latipes* display several folds. These form a spiral pattern, which correlates with the surface formation of the micropylar cell. 'Iwamatsu *et al.*, (1993a,b)' analyzed by EM the differentiation of a micropylar cell among the granulosa cells in this species. The cell, with a flat or irregularly shaped nucleus, projects its large cytoplasmic extension into the thin envelope. While desmosomes between adjacent granulosa cells are observed, junctions are not formed between the micropylar cell and its adjacent granulosa cells. A portion of the electrondense ZR externa was compressed by the elongating cytoplasmic extension of the mushroom shaped micropylar cell towards the cell surface. This caused the adjacent ZR interna to be distorted. The centre of the twisted cytoplasmic extension of the micropylar cell contained meandering bundles of microtubules and microfilaments. The spiralling wall of the forming micropyle is a replica of the surface of the micropylar cell. Tight junctions are observed between the apical surface of the cytoplasmic extension and the oolemma. The observed spirals are a result of the capability of the micropylar cell to rotate freely but for the above-mentioned tight connection with the oocyte (Fig. 7/18).

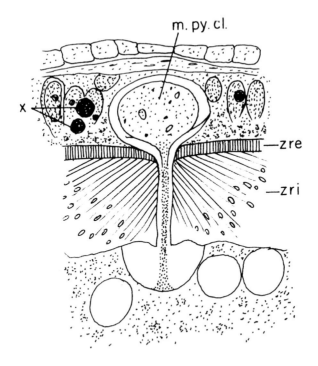

Fig. 7/17 Radial section through a micropyle (in formation) of an egg of *Esox lucius* (redrawn from Eigenmann, 1890). m.py.cl - micropylar cell; x - highly refractive homogeneous body; zre - *zona radiata externa*; zri - *zona radiata interna*.

Withdrawal of the micropylar cell from the newly formed micropylar vestibule and canal was studied on oocytes of *Oryzias latipes* by 'Nakashima and Iwamatsu (1994)'. They incubated isolated pre-ovulatory follicles at the stage of germinal vesicle breakdown of the oocyte and observed the in vitro changes in the micropylar and granulosa cells. The micropylar cell began to shrink about 3 hours after germinal vesicle breakdown and gradually withdrew from the micropylar canal as a result of the breakdown of the tight junctions between the protrusion and the oocyte. Concomitantly, the micro- and macrovilli in the pore canals of the ZR were withdrawn. While the fully developed micropylar cell protrusion contained large twisted bundles of microtubules and tonofilaments (considered the main architectural agents for cell shape), as well as mitochondria, Golgi bodies and rough endoplasmic reticulum (RER), neither spiralling bundles of microtubules and tonofilaments nor mitochondria arranged in parallel with the spiral bundles were observed in the shortening protrusion. However, tonofilaments were still present in the main body of the cell after ovulation. At the time of ovulation, the degrading micropylar cells can be distinguished from adjoining granulosa cells by the presence of lysosomal vesicles and dilated RER. However, in the Pacific

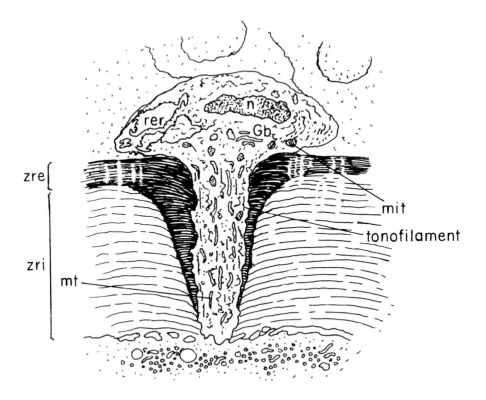

Fig. 7/18 Modified drawing of SEMs of *Oryzias latipes* displaying the formation of the micropyle [after Iwamatsu *et al.* (1993) and Nakashima and Iwamatsu (1994) with kind permission of Wiley Liss Inc., a subsidiary of John Wiley & Sons, Inc.]. Gb - Golgi body; mit - *mitochondrion*; mt - microtubule; n - *nucleus*; rer - rough endoplasmatic *reticulum*; zre - *zona radiata externa*.

herring *Clupea pallasi*, the pond smelt, *Hypomesus transpacificus nipponendis*, and the chum salmon, *Oncorhynchus keta*, the RER was much more dilated than in the medaka '(Ohta and Takano, 1982; Kobayashi and Yamamoto, 1985)'. The presence of microtubules and tonofilaments in the micropylar cell had also been reported for other teleost species '(Ohta and Tetranishi, 1982)' (loach), 'Takano and Ohta, 1982 (pond smelt).

The micropylar cell of the chum salmon contains the above observed organelles only in the cytoplasmic protrusion '(Kobayashi and Yamamoto, 1985)'. Microtubules and tonofilaments are thought to be responsible for the cytoskeleton of a cell and it is likely, therefore, that they play an important role in the formation of the micropylar cell protrusion.

Similar changes in granulosa cells have been reported in several fish species '(Hirose, 1972; Nagahama *et al.*, 1978; Kagawa and Takano, 1979; Kobayashi and Yamamoto, 1981, 1985)'. The changes in these cytoplasmic inclusions seem to be associated with the changes in the cytoskeleton and the shape of the micropylar cell.

'Eigenmann (1890)' observed that in *Esox reticulatus* the micropylar canal is partially filled with a plug of substance which appears to be continuous with the yolk. 'Riehl and Götting (1975)' found that in 80% of the eggs stripped off from *Esox lucius* females the micropylar canal was plugged. This plug was the remnant conical process of the micropylar cell. However, when they added sperms to these eggs, fertilization occurred almost without exception. They, therefore, concluded that the egg envelopes of the pike may be traversed by sperms also in sites other than the micropyle.

Summary

The micropyle in the teleost egg envelope is an opening, which allows the passage of a sperm to the enclosed egg. Teleosts possess only one micropyle per egg, which is located at the animal pole. The absence of an acrosome in the sperm head (see Chapter 8) corresponds to the presence of the micropyle in the egg envelope. The type of micropyle is also considered to be a criterion for the identification of teleost eggs.

There are four morphological types of micropyles. Type I has a deep pit or funnel-like depression in the egg surface leading into a short cone. Type II has a shallow saucer-like pit and a long canal. Type III consists only of a canal and type IV has an inner and outer vestibulum and a short canal. The inner wall of the micropylar canal is often ribbed and folded. The area surrounding the external opening of the micropyle is usually different from the rest of the envelope.

The micropyle is formed from a modified follicular cell, which sends out a thin process that leaves the micropylar canal open. It can also be formed by several coalescing cells. The canal is lined partly by the ZR externa and/or the ZR interna.

Endnote

[1] small opening, from Gk. micro = small, pute = gate

Chapter eight

Sperm

> 'Ramon J. Cajal would envisage the sperm-cells as activated
> by a sort of passionate urge in their rivalry for penetration
> into the ovum cell...'
>
> Sherrington (1949)

The sperm has a dual function: activation of the oocyte and transmittance of paternal hereditary material. Sperms of teleosts have the simplest structure among fishes. The head is on average 2-3 microns long and the total length of the spermatozoon 40-60 microns '(Ginzburg, 1968)'. Sperm length of internally-fertilizing species is generally greater than that of externally-fertilizing species. In the latter, sperm length is positively related to number of ova but not to ovum size '(Stockley *et al.*, 1996)'. Internally-fertilizing sperms are known as 'intra-sperms' while the term 'aquasperm' is used for sperm with an aquatic, free-swimming phase.

In the most recent review of spermatozoon ultrastructure, 'Mattei (1991)' concluded that 'there is a great diversity in the spermatic structure of fishes. It is not possible to construct a spermatic model for the 'fishes', as is the case, for example, for the snakes and mammals. Although this may be explained by the fact that the fishes do not constitute a monophyletic group, it should be noted that it is also impossible to construct a spermatic model for the monophyletic Actinopterygii or even the Teleostei'. The great value of spermatology in fish is based on taxonomy and phylogeny.

8.1 EARLY INVESTIGATORS

Due to the small size of the sperms the descriptions of the early investigators are very meagre. The sperms are referred to as pin-shaped with a small spherical head and a very fine threadlike tail '(Remak, 1854)'. A thickening between the head and tail of the sperm of *Cobitis* was observed by 'Wagner (1854)'.

'Miescher (1872, 1873, 1874, 1878)', the discoverer of the nucleic acid, reported in two lectures that on analysis of the sperm of the salmon, *Salmo salar*, the perch, *Perca fluviatilis* and the pike, *Esox lucius*, he found, besides head and tail, a middle piece adjacent to the head. He described that the head of the salmon sperm, viewed laterally, has the shape of a cross-sectioned bean with rounded edges and in the middle an umbilicus yielding the heart-shape of the head. The tail of the salmon sperm is very thin and about 45 micron long while the head measures 3.3 microns in length, 2.5 microns in breadth and about 1.3 microns in thickness. Microscopical analysis revealed a thick envelope of the head and a dye-resistant inner part. In the latter he observed a flat, straight rod which continues as a fine canal (he called microporus) into the middle piece. The findings on the perch and pike were similar. He concluded that the head of the sperm is essentially made up of nuclear material. He forecast the tremendous repercussions that would ensue were it possible to prove that the entrance of the sperm into the egg is essentially of similar value to the entrance of a physiologically 'vollgiltigen' (of full value) cellular nucleus. How right he was and what a prophet!

'Kupffer (1878)' described the sperm for the herring, *Clupea harengus*, as having a head, tail and middle piece as reported by 'His (1873)' for the salmon, *Salmo salar*. However, the head of the herring sperm is round while that of the salmon sperm is flat. The measurements given by Kupffer for the salmon sperm are: length of head 0.0025 millimetres, breadth of head 0.0020 millimetres and length of tail 0.062-0.075 millimetres. The middle portion, which he called 'tail globule', when stained with aniline (gentian violet, rosaniline) revealed 4-5 intensely stained granules (which later ultrastructural analyses, described below, revealed to be mitochondria.) '(Retzius, 1905, 1910; Ballowitz, 1915, 1916)'.

'Leydig (1883)' analyzed the sperms of *Tinca chrysitis* and *Gasterosteus aculeatus*. The pear-shaped head of the former is made up of a dark cortex and a light interior. He further observed a 'shining dot' (the centriole of the electron microscopists) in the region where the thread is inserted. The sperm of *Gasterosteus* is similar, with the exception of the head which is not round but polygonal; additionally a part containing four little buttons (mitochondria?) is separated posteriorly from the 'shining dot'.

'Ballowitz (1890)' supplied drawings of the sperms of *Clupea harengus, Esox lucius, Leuciscus rutilus* and *Zoarces viviparus*. He observed a middle piece in all the species tested and mentioned that the head is taken up by the nucleus while the middle piece seemed to be composed of dense cytoplasm and the axial thread, which passes obliquely through it (the middlepiece). In some species the tail is composed of a main part (often supplied with a thin seam) and a terminal part (often wavelike). The tail thread of *Zoarces viviparus* is made up of two parallel threads and in the pike, *Esox lucius*, each of these can be split again into two threads.

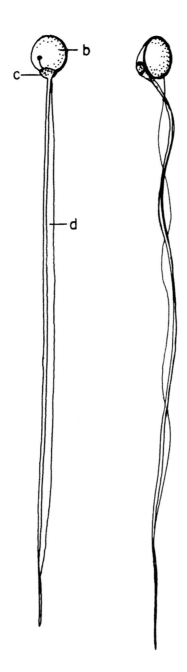

Fig. 8/1 Spermatozoon of *Esox* (left) and *Perca* (right) (Retzius, 1905). b - head; c - middle piece d tail.

8.2 EM-ANALYSES

Sperm morphology has been reviewed by 'Mattei (1970, 1991)', 'Billard (1970)' and 'Grier (1981)'. If not stated otherwise, the following results are based on the extensive treatise edited by 'Baccetti (1970)' and 'Jamieson and Leung (1991)'. The latter authors used differences in sperm morphology extensively for phylogenetic analyses, which, however, are not part of this survey.

The spermatozoon is divided into a head housing the nucleus, a midpiece containing the mitochondria and the tail made up of a flagellum. Teleost sperm are of two kinds: the term 'aquasperm' refers to sperm with an aquatic, free-swimming phase and 'introsperm' to internally fertilizing sperm. These terms were first proposed by 'Rouse and Jamieson (1987)'. Introsperms are packed into spermatophores[1], which, in *Cymatogaster aggregata* (Pomacentridae) each contains approximately 600 spermatozoa lying parallel. The sperms are released from the capsule within 1 hour of insemination '(Gardiner, 1978)'. While the term spermatophore is used to describe encapsulated sperm bundles, the term spermatozeugmata[2] refers to the sperm bundles which are not encapsulated '(Ginzburg, 1968; Grier, 1981)'.

Head

The head is composed mainly of the nucleus. It is spherical as e.g. in the carp, *Cyprinus carpio*, the pike, *Esox lucius*, the loach, *Misgurnus anguillicaudatus*, the goldfish, *Carassius auratus* and the flat bitterling, *Acheilognatus rhombeus*. It is elongated cylindrical in the trout, *Salmo fario* and *S. gairdneri* and the butterflyfish, *Pantodon bucholzi*, elongated flattened parallel the long axis of the sperm in *Cephalacanthis volitans*, the guppy, *Poecilia reticulata*, *Tilapia nilotica*, and *Mimagoniates barberi* '(Mattei, 1970; Billard, 1970; Ohta et al., 1993; Pecio and Rafinski, 1999)' (shapes of heads of other sperms are shown in Figs. 8/3, 8/4, 8/5).

Freeze fracture techniques revealed intramembranous particles (IMP) on the head of teleostean spermatozoa. In the zebrafish, *Brachydanio rerio*, the IMP display a characteristic array (parallellogram or hexagon) and occur in a specialized region of the sperm head, frequently localized in an equatorial position '(Kessel et al., 1983)'. Similar observations on the rose bitterling, Rhodeus ocellatus, were reported. The IMP were usually present at a site anterior to the centrioles. These characteristic arrays were also encountered in two other cyprinids, the goldfish, *Carassius auratus*, and in the flat bitterling, *Acheilognathus rhombeus*. The aggregates were concentrated on the side of the sperm head while irregular scattering of IMP over the whole area was also observed. In the loach, *Misgurnus anguillicaudatus*, and the sweetfish, *Plecoglossus altivelis*, the IMP were scattered over the sperm head and the midpiece (Fig. 8/2). In all species the number of IMP is greater on the protoplasmic face (PF) than on the extracellular face (EF) '(Ohta et al., 1993)'. A possible functional significance of these particles in sperm/egg plasma fusion and/or egg recognition will be discussed in Chapter 9 (fertilization).

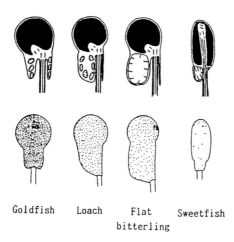

Goldfish   Loach   Flat bitterling   Sweetfish

Fig. 8/2   Diagrammatic representation of the sperm heads of four species (Ohta et al., 1993, with kind permission of Churchill Livingstone, Edinburgh). Upper row shows the structure, lower row the distribution of IMP (intramembraneous particles). From left to right - goldfish, Carassius auratus; loach, Misgurnus anguillacaudatus; flat bitterling, Achailognathus rhombeus; sweetfish, Plecoglossus altivelis.

Acrosome[3] (earlier called apical body or periorator)

The light-microscopists had debated if the teleost sperm possesses an acrosome. After the introduction of EM it was agreed that the teleostean sperm is generally anacrosomal (lacking an acrosome). The presence of a micropyle in the ovum has been related to the absence of an acrosome in the sperm. However, a putative acrosome makes a transitory appearance in the spermatid of *Salmo gairdneri* and the lophiiform *Neoceratias* sp. A possible acrosomal vestige has been described for the mature sperm of *Lepadogaster lepadogaster* (gobioesociform). An acrosome-like structure is also encountered in the salmoniform *Lepidogalaxias salamandroides* which may be related to the fact that this species is lacking a micropyle in the egg envelope '(Ginzburg (1972)'. In two species of *Melanotaenia* (atherinoform) there is a large vesicle (acrosomoid) anterior to the nucleus. An acrosome-like structure may be represented by a number of vesicles in the sperm head of the gymnotoid *Stenarchus albifrons* and of *Lates calcarifer*. 'Mattei (1991)' summarized the presence of so-called pseudoacrosomes in the Argentinoidei, Salmonidae, Photichthyidae, Gobiesocidae, Poeciliidae and Pomacentridae. He continued that 'these formations, which we interpreted as vestiges of a primitive acrosome '(Mattei and Mattei, 1978)', are considered neoformations by 'Baccetti (1985)'. (Since the formation of an acrosome is an important factor of Golgi activity during spermatogenesis, it has been suggested that, in teleosts, the Golgi apparatus is engaged mainly in the production of the flagellar membrane '(T. Yamamoto, 1963; Sakai, 1976)'.

It is interesting to note that the spermatozoa of the holostean fish *Lepisosteus ossseus* also lack an acrosome '(Afzelius, 1978)'.

Nucleus

The *Muraenidae* display a spherical nucleus. The nucleus is elongate e.g. it measures 7 microns in the osteoglossoid *Pantodon buchholzi* and 8 microns in *Lepadogaster lepadogaster* (Gobioesocidae). Another example is the sperm of *Ophidion* sp. (Hexagrammidae), which possesses an apically helical nucleus, also of a length of 8 microns. The nucleus is even longer, 11 microns, and spiral in *Neoceratias spiner* (lophiiform) and 20 microns in salmoniform *Lepidogalaxias salamandroides*. The nucleus is elongate and blade-like in Poeciliidae, Zenarchopteridae and Hemirhamphidae. The shape of the nucleus can be slightly curved with a hook-shaped end, directed anteriorly (*Conger muraena*) or posteriorly (*Albula vulpes*) '(Mattei, 1970)'. 'Grier (1981)' mentions that an elongation of the nucleus is characteristic of most species with internal fertilization. *Mimagoniates barberi*, although an internally fertilizing fish, is oviparous. The nucleus of its sperms elongates to become a very long flat plate, at right angles to the plane of the axoneme '(Pecio and Rafinski, 1999)' (Fig. 8/5).

A very deep fossa in the ventral part of the nucleus has been observed in clupeids (*Sardinella aurita*, *Ethmalosa fimbriata*, *Anchoa guineensis*) (Fig. 8/4 nos. 22, 25, 27), the soleid *Pegusa triophthalmus* (no. 13), the European John Dory, *Zeus faber* (no. 18), *Dactylopterus* (*Cephalocanthus*) *volitans* (Fig. 8/3 no. 8) (Dactyolopteridae), *Balistes forcipatus* (no. 6), *Pseudobalistes fuscus* (Balistidae), *Aluterus punctatus* (Monocanthidae), *Chilomycturus antennatus* (Diodontidae) (no. 4), *Scorpaena angolensis* (no. 3), *Oligocottus maculosus* (Scorpaenidae); *Upeneus prayensis* (Mullidae) (no. 7), *Fundulus heteroclitus* (Cyprinodontidae), *Trachinocephalus myops* (Synodontidae) (no. 12) as well as in the zebrafish *Brachydanio rerio* '(Wolenski and Hart, 1987)'.

In the pointed, or blade-like nucleus a fossa was observed in *Oligocottus maculosus* (Cottidae); *Jenynsia lineata* (Jenynsiidae), *Poecilia reticulata*, *Xiphophorus helleri*, *Gambusia affinis*, *P. latipinna* (Poeciliidae) and *Cymatogaster aggregata* (Embioticidae). Additionally, members of the exocoetoids *Dermogenys pusillus*, *Hemirhamphodon pogonognathus*, *Nomorhamphus celebensis* and *Zenarchopterus dispar* revealed sperms with nuclear flattening and an indented nucleus '(Jamieson and Grier, 1993)'. What these diverse fishes have in common is internal fertilization. 'Grier (1975)' in his study of spermiogenesis in *Gambusia affinis* described the pattern of nuclear fossa-associated microtubules in this species and concluded that it is significantly different from that observed in other poeciliids.

A crescentic nucleus has been described for the sperms of elopomorphs (with the exception of Muraeinidae, Megalopidae, Albulidae (belonging to the clupeiformes) and Notacanthidae and Halosauridae (belonging to the anguilliformes). The sperm of salmonids is characterized by having an osmiophilic body at the base of the nucleus.

Midpiece

The midpiece, posterior to the nucleus, is a collar-like structure housing the mitochondria and the centrioles. Teleosts with external fertilization have sperms with a low mitochondrial collar while a high collar with many mitochondria is typical of viviparous species. The mitochondria form a single ring- or C-shaped body in the salmonoidei, argentinoidei and galaxioidea; further in the engraulid clupeiforms, and the percoid *Macquaria ambigua*. A single unilateral mitochondrion occurs in *Jenynsia* (Cyprinodontidae), *Trachinocephalus* (Synodontidae) (Fig. 8/3 no. 12); *Maccullochella* (occasionally two in *M. Peeli*) (Percichthydae) and in the gobioid *Hypseleotris galii* (Eleotridae). The mitochondrion of the sperms of Elopidae, Albulidae and Anguillidae is embedded in an anterior or lateral nuclear concavity '(Mattei, 1970, Jamieson and Grier, 1993)'.

The midpiece is long in the anglerfish *Neoceratias spinifer* (Neoceratidae). At 45 microns it reaches perhaps its greatest elongation in the osteoglossoid *Pantodon buchholzi*, the only species of the family Pantodonidae. It consists of nine dense helical fibres alternating with mitochondrial derivatives (modified cristae and end to end fusion of the original mitochondria), which describe 20-25 turns, with an inclination of 60-70 degrees, around the central axoneme.

A usually single mitochondrion is sideplaced along the nucleus as far as its tip in *Blennius pholis*. This is unlike the situation in other blennioids such as *B. cristatus* (Fig. 8/4 no. 20), *B. vandervekeni* (no. 26), *Clinus nuchipinnis* (Fig. 8/3 no. 2), *Ophioblennius atlanticus* (Fig. 8/4 no. 16), in which the mitochondria are in a postnuclear position. In the clupeid *Ethmalosa fimbriata* the large fused mitochondrion in the head partially surrounds the flagellum (no. 25) '(Favard and André, 1970)'.

The arrangement of the mitochondria in elongate midpieces varies taxonomically as well. In the viviparous cyprinodont *Jenynsia lineata* (Jenynsiidae) the mitochondria fuse and form a so-called submitochondrial sheath, which at maturity contains only one mitochondrion. The sperms of the live-bearing Poeciliidae display a long mitochondrial sheath with the mitochondria arranged bilaterally into two to five columns. The poeciliids and *Jenynsia* show a so-called submitochondrial net. It is formed by a densification of the membrane lining the inside of the mitochondria, close to the axoneme, and appears as a regular diamond lattice. A comparison of the internally-fertilizing exocoetoid fish *Hemirhamphodon pogonognathus*, *Dermogenys pusillus*, *Zenarchopterus dispar*, *Nomorhamphus celebensis* with the externally-fertilizing *Arrhamphus sclerolepis* revealed that the sperms of the former four species have a much longer mitochondrial sleeve. Their mitochondria are grouped bilaterally (whereas in the internally-fertilizing poeciliids they are circumferentially arranged). In the externally- fertilizing *Arrhamphus* the

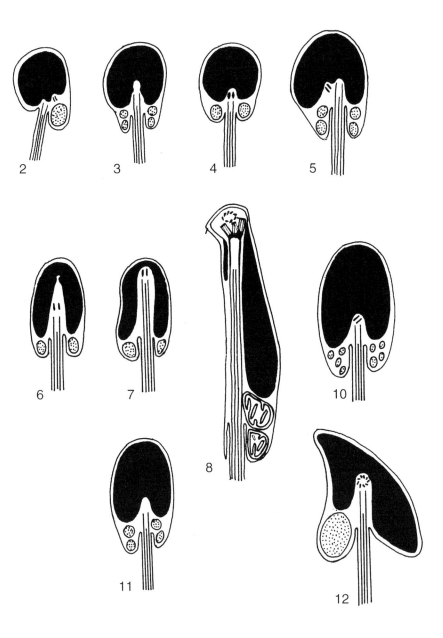

Fig 8/3 Schematic representation of a number of *spermatozoa* (Mattei, 1970, with kind permission of the Academia Nazionale dei Lincei, Roma). **2** - *Clinus nuchipinnis*; **3** - *Scorpaena angolensis*; **4** - *Chilomycterurus antennatus*; **5** - *Fistularia tabacaria*; **6** - *Balistes forcipatus*; **7** - *Upeneus prayensis*; **8** - *Cephalacanthus volitans*; **10** - *Clarias senegalensis*; **11** - *Hemichromis fasciatus*; **12** - *Trachinocephalus myops*.

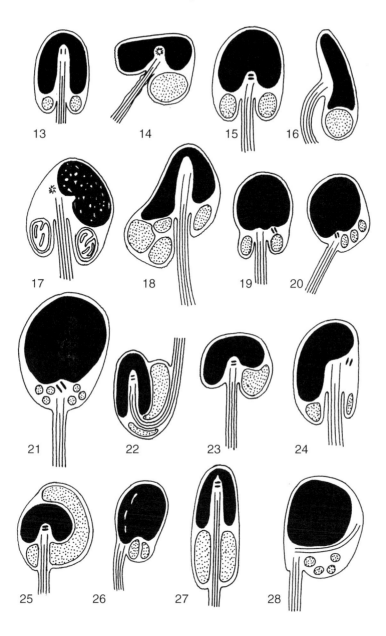

Fig. 8/4 Continuation of Figure 8/3. **13** - *Pegusa triophthalmus*; **14** - *Boops boops*; **15** - *Antennarius senegalensis*; **16** - *Ophioblennius atlanticus*; **17** - *Parapristipoma octolineatum*; **18** - *Zeus faber*; **19** - *Fodiator acutus*; **20** - *Blennius cristatus*; **21** - *Papyocranus afer*; **22** - *Sardinella aurita*; **23** - *Periophthalmus papirio*; **24** - *Galeoides decadactylus*; **25** - *Ethamalosa fimbriata*; **26** - *Blennius vandervekeni*; **27** - *Anchoa* (= *Engranulis?*) *guineensis*; **28** - *Lycodontis afer*.

mitochondria occur in two longitudinal tiers. The inner wall of the sleeve of the sperms of the five exocoetids shows some densification called submitochondrial dense layer which is comparable to the above-mentioned submitochondrial net of the poeciliids '(Jamieson and Leung, 1991; Jamieson and Grier, 1993)'.

A long posterior sheathlike extension of the cytoplasmic collar, devoid of mitochondria, occurs in *Stenarchus albifrons* (Apteronotidae). In the osteoglossoid *Pantodon* there is a 6 microns long fenestrated sheath behind, but not forming an extension of, the midpiece.

The midpiece of the aflagellate sperm of Gymnarchidae and Mormyridae contains few globular mitochondria, which in the latter lie between the centrioles and one pole of the cell as in other teleosts, while in the former they are irregularly arranged surrounded by abundant microtubules and lipid bodies. '(Favard and André (1970)' (Fig. 8/7).

In *Mimagoniates barberi* the mitochondria dislocated to the posterior tip of the head and not in a midpiece next to the diplosome beneath the nucleus '(Pegio and Rafinski, 1999)' (Fig. 8/5).

The fact that 1) the mitochondria in the sperm are isolated from the motile flagellum and, therefore, present problems of energy transfer, 2) that the sperms of coccid insects are motile in spite of the fact that they do not possess mitochondria and 3) that DNA is present in mitochondria, lead Robison to consider 'that it is entirely possible that mitochondria are involved in the transmission of hereditary information in addition to or in lieu of a role in supplying energy for motility '(Billard, 1970)'.

Centriolar apparatus

Vertebrate spermatozoa generally have two centrioles instead of the single basal body found in cilia (Fig. 8/6). In the osteoglossomorpha with a nuclear fossa the centrioles lie within it (type I). In contrast, location of the centrioles external to the fossa or eccentric to the nucleus (type II) is encountered in the percoids [haemulid *Upeneus prayensis* (Fig. 8/3 no. 7), centropomid *Lates calcarifer* and mugiloid *Parapercis*]. However, an intermediate condition occurs in the mugilid *Liz*, the polynemid *Galeodes decadactylus* (Fig. 8/4 no. 24), and the blennoids *Clinus nuchipinnis*, *Blennius cristatus* (no. 20) and *Ophioblennius atlanticus* (no. 16).

The above-mentioned classification is based on 'Mattei (1970)': Spermiogenesis according to type I is made up of four steps: 1) young spermatid 2) migration of centrioles 3) rotation of nucleus 4) migration of mitochondria. The example given is *Upeneus prayensis* (Fig. 8/3 no. 7). In type II step 3, rotation of the nucleus, does not take place. The example stated is *Parapristipoma octalineatum* (Fig. 8/4 no. 17), belonging to the *Pomasidae*. It is characterized by the flagellum remaining tangential to the base of the nucleus and the centrioles outside the nuclear fossa. Out of 29 perciformes tested 25 displayed mode II. *Mimagoniates barberi* represents an extreme case of type II ' (Pecio and Rafinski, 1999)' (Fig. 8/5).

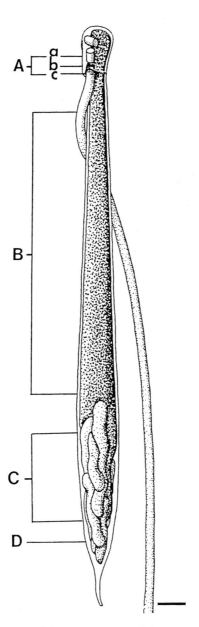

Fig. 8/5 Schematic diagram of the *spermatozoon* of the oviparous, internally fertilizing *Mimagoniates barberi* [Pecio and Rafinski (1999) with kind permission of Blackwell Science Ltd., Oxford). Bar 1 micron. Long *flagellum* runs parallel to the long axis of the *nucleus*. **A** - Tip with strongly compressed *nucleus*, distal centriole (a), *cytoplasmic* canal (b) and (c) cross-section below the opening of the *cytoplasmic* canal. **B** - Middle portion of the head. **C** - Distal portion of the head showing the *mitochondria*. **D** - Posterior tip of the head.

The centrioles are arranged at a right angle to each other e.g. in the Cyprinidae *Alburnus alburnus* and *Barbus barbus*, in *Trachinocephalus myops* (Aulopidae) and the cyprinodont *Poecilia reticulata* (Fig. 8/3 no 12, 8/6), in the lophiiform Neoceratias spinifer (Neoceratidae) and *Oligocottus maculosus* (Cottidae), in the percoids *Plectropomus lepidorus* and *Vomer setapinnis* and in the cichlids *Hemichromis fasciatus* (Fig. 8/3 no 11), *Tilapia nilotica* and *Oreochromis niloticus*. The proximal centriole is also perpendicular to the basal body in the following species: the tetradontiform *Gastrophysus hamiltoni*, the cyprinodont *Fundulus heteroclitus*, *Pseudomugil* (Atherinidae) and *Mimagoniates barberi* (Characidae) (Fig. 8/5). The proximal centriole is positioned at 90° or more obliquely to the distal basal body in the goldfish, *Carassius auratus*, the loach *Misgurnus anguillicaudatus* (Fig. 8/4 no 21), the notopterid *Papyrocranus afer*, the cyprinid *Leuciscus leuciscus*, the African catfish *Clarius senegalensis* (Fig. 8/3 no 10) and the neon tetra *Paracheirodon innesi*. Only rarely do the proximal and distal centrioles lie parallel, as seen in *Pantodon buchholzi* and in the mormyriform *Gymnarchus* (Fig. 8/7). Sometimes the proximal and distal centrioles occur in the same line (coaxial), a condition often accompanied by deep penetration of the nucleus by its basal fossa as encountered in *Scorpaena angolensis* (Scorpaenidae) (Fig. 8/3 no 3), in *Pegusa triophthalmus* (Soleidae) (Fig. 8/4 no 13), in the tetradontiform *Balistes forcipatus* (Balistidae) (Fig. 8/3 no 6), *Pseudobalistes fuscus* and *Chilomycturus antennatus* (Tetradontidae) (Fig. 8/3 no 4). A similar condition was observed in *Scorpaena angolensis* (Scorpaenidae) and in the percoid *Upeneus prayensis* (Fig. 8/3 no 7).

The proximal centriole is small and sometimes modified. The centriole is surrounded by dense material, especially between it and the distal one. This material may show cross-striations or segmentation. Part of the distal centriole displays tubular doublets instead of triplets '(Nicander (1970)'. Division of the proximal centriole into two elongate bundles, of 4 and 5 triplets as a pseudo-flagellum, is diagnostic of the elopomorpha (Elopiformes, Anguilliformes and Notacanthiformes), as is a cross-striated centriolar rootlet (though absent from the Muraenidae).

Of the two centrioles, the proximal one may be reduced in the mature sperm, as seen for example in *Melanotaenia duboulayi*, the poeciliids, and the internally-fertilizing exocoetoids *Hemirhamphodon pogonognathus* and *Dermogenys pusillus*, *Nomorhamphus celebens*, *Zenarchopterus dispar*, or even lost, as in the hagfish *Eptatretus stoutii* '(Jamieson and Grier, 1993)'.

Nine satellite rays, radiating from the distal centriole, are rarely seen in teleost sperm. However, they are well developed in the atheriniformes *Craterocephalus stercusmuscarum*, *C. helenae*, *C. marjoriae*, *Querichthys stramineus*, *Cairnsichthys rhombosmoides* and *Melanotaenia maccullochi* '(Jamieson, 1991)'.

Partial occlusion (septation) of the microtubules of the doublets occurs in the salmonoids and engraulids involving doublets 123567, and in the galaxoids and the argentoids doublets 12567.

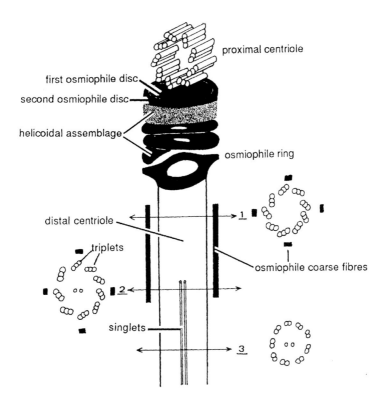

Fig. 8/6 Schematic reconstruction of the centriolar complex of the sperm of *Poecilia reticulata* (*Lebistes reticulatus*) (Mattei and Boisson, 1966, with kind permission of Comptes Rendus Hébdomadaires des Séances, Paris). **1** - Transverse section (TS) of the distal centriole in the midregion. **2** - TS of the base of the distal centriole. **3** - TS of the flagellum. Note: axes of proximal and distal centriole lie at right angles.

Tail filament or flagellum

The filament begins at the centriolar apparatus in the posterior part of the head and penetrates the midpiece before it enters the tail. In *Mimagoniates barberi* the flagellar axis runs parallel to the elongated head (Fig. 8/5). The teleost sperm has one flagellum or, rarely, two flagella, or none at all. The microtubular formula of the axoneme is generally of the 9+2 type, i.e. nine doublets, each with two dynein arms, and two inner singlets. However, the axoneme is of the 9+0 pattern in elopomorphs and the myctophiform *Lampanyctus*. The Mormyridae and Gymnarchidae are devoid of a flagellum with *Gymnarchus* displaying cytoplasmic microtubules in its cytoplasm which enable it to move by amoeboid motion (Fig. 8/7).

Biflagellarity has developed in some siluriforms, such as *Ictalurus punctatus* and in the myctophiform *Lampanyctus*, the batrachoids *Opsanus tau* and *Porichthys notatus*, the gobiesocid *Lepadogaster lepadogaster*, and in a percoid apogonid. In many species the membrane forms folds, commonly called 'fins', which are seen as ridges in a plane through the two central microtubules (Fig. 8/8b,c) '(Jamieson, 1991)'.

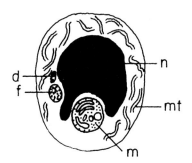

Fig. 8/7 Section of *Gymnarchus niloticus*. Note absence of *flagellum* and the abundance of microtubules in the *cytoplasm* (modified after Mattei *et al.*, 1967, with kind permission of Comptes Rendus Hébdomadaires des Séances, Paris). d - diplosome; n - *nucleus*; f - lipid body; m - *mitochondrion*; mt - microtubule.

A single fin is present on the flagellum of the sperm of *Esox lucius* and the Australian *Maccullochella macquariensis* (Serranidae) and in some regions of the tail in some species with two fins (Fig. 8/8b,c). Two is the usual number, which are in the plane of the central singlets (*Dermogenys, Nomorhamphus* and *Hypseleotris*) (Fig. 8/8c,d). In the suborder Aplocheiloidei one, two or even three fins are present according to the species. Fins are reduced in the Poeciliidae. Within the Exocoetidae fins are absent in *Hemirhamphodon* (internal fertilization, live young), reduced in *Zenarchopterus* (internal fertilization, lays fertilized eggs) and *Arrhamphus* (external fertilization) has two short lateral fins '(Jamieson and Grier, 1993)'. The ostariophysi, which include the cyprinoformes, siluriformes and characiformes, lack fins, which has been interpreted as a tentative endorsement of their taxonomic relationship (Fig. 8/8a).

Most recently, the relationship between reproductive behaviour, length of sperm, size of ova and number of ova has been analyzed by 'Stockley *et al.*, (1996)'. Sperm length is greater in internally fertilizing fish than in externally fertilizing species. This might be related to the fact that longer sperms may have a greater flagellar swimming force, which would facilitate swimming in a medium which is viscous. Externally fertilizing fish spawn in close proximity. However, many would have to cope with tidal and other currents. This prompted the

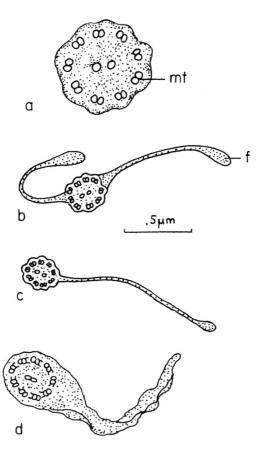

Fig. 8/8 Transverse sections of flagellum; a after Fribourgh *et al.* (1970) with kind permission of Allen Press, Lawrence, Kansas; b,c after Jamieson (1991) with kind permission of Cambridge University Press. d after Nicander (1970) with kind permission of Academic Press New York, London. **a** - Absence of fin. Axoneme 9×2+2 microtubular configuration (example: *Carassius auratus*). **b** - Two lateral fins; central microtubules are in line with fin (example: *Hypseleotris galli*). **c** - One lateral fin. Central microtubules are in line with fin (example: posterior part of *flagellum* in *Aphyosemion gardneri*). **d** - One lateral fin in *Esox lucius* with high ridge in the plane of the central microtubules.

authors to state that competition for external fertilizers 'may resemble a raffle, but may be more like a race for internal fertilizers'. Among external fertilizers the length of sperm is also positively correlated with the number of ova shed but not with their size. In these fishes a positive trend between sperm longevity and ovum diameter and a negative association between sperm longevity and number of ova was noted.

The sperm-activating factor (motility of sperms) will be dealt with in Chapter 9 on fertilization.

Summary

The spermatozoon is made up of a head housing the nucleus, a midpiece containing the mitochondria and a tail made up of one flagellum, rarely two flagella, the axoneme of which is mainly of the 9+2 microtubular pattern. There is no acrosome present. In other taxa the acrosome with its enzymes is needed to aid the sperm in penetrating the egg envelopes to establish connection with the egg cytoplasm, whereas in teleosts the micropyle acts as the entrance door for the sperm (Chapter 7).

The centrioles lie within a fossa of the nucleus or eccentric to it. The centrioles are perpendicularly arranged, or at an angle relative to each other exceeding 90°, or, rarely, lie parallel.

Sperm of internally fertilizing fish are packed into spermatophores[1] or spermatozeugmata/ spermatozeugma[2] (without a capsule), containing up to 600 sperms. Internal fertilization is reflected to a large degree in alterations of sperm morphology. This involves elongation of the nucleus coupled with possible elongation of the midpiece, as encountered e.g. in the Poeciliidae.

The plasma membrane of the tail of the sperm may be drawn out into one or two fins. Some fish with two fins possess only one fin in some regions. However, there are also species in which fins are absent altogether.

Endnotes

[1] spermatophores are aggregates of sperm with a dense outer membrane.
[2] spermatozeugmata or spermatozeugma are aggregates, lacking an enclosing membrane. This term was introduced by 'Ballowitz' for similar aggregates in insects.
[3] from Gr.akron = tip and soma = body

Chapter nine

Fertilization

> 'No matter with what equipment and by what means an organism manages to survive, it must be prepared eventually to unite its genital products, or their equivalent, with another in a successful manner.'
> Breder and Rosen (1936)

Fertilization in its broadest sense refers to the various steps leading to and resulting in sperm-egg fusion. Fertilization is a process, which can last 20 or more hours (as in salmonids) although it is much shorter in most other fish. Fertilization in its narrow sense refers to the fusion of female and male pronucleus[1]. Synonyms of fertilization are fecundation and impregnation whereas the term insemination denotes the bringing together of seminal fluid with eggs.

The year was 1875 when fusion of pronuclei was seen for the first time, namely in the sea urchin by 'Oskar Hertwig'. This major breakthrough in developmental biology prompted 'Agassiz and Whitman (1889)' to comment: "The panoramic display to be witnessed in small transparent eggs, like those of the starfish and sea urchin, reveals the order and relations of consecutive scenes, and thus supplies the needed vantage-ground for a comparative study". And they continued to protest against "the shiftless tendency to neglect the less inviting material", i.e. fish. Their eggs were "less inviting" because those of commercial fish, as e.g. the herring, are deposited on the sea floor and those of Salmonidae are opaque. Therefore, pelagic eggs — which are transparent — such as those of *Oryzias* and *Fundulus* became subsequently especially advantageous for the observation of structural changes at the time of fertilization '(Hoffmann, 1881; Nakano, 1969)'. During maturation of the fish oocyte the centrally positioned nucleus moves towards the periphery. This nucleus is also called vesicula germinativa or germinal vesicle[2]. The site the egg nucleus will finally occupy is in the neighbourhood of the micropyle (Chapter 7), and is known as animal pole (the antipode is called vegetal pole). Meiosis[3] of the oocyte is initiated before oviposition and yields two cells of different size. The smaller cell is the first polar body, which is given off at the animal pole at the stage of metaphase I. After ovulation the egg remains arrested in metaphase of the second meiotic division (metaphase II) until it becomes activated.

When teleost eggs are fertilized before ovulation, polyploidy[4] is found in the embryos that develop. Although there is no definite explanation for such polyploidy, one possiblity is that the sperm nucleus unites with the diploid egg nucleus before the onset of maturation division '(Nakano, 1969)'.

While all modern textbooks quote the teleosts as having telolecithal[5] eggs, it has to be stressed that most teleostean eggs are centrolecithal at oviposition and become telolecithal only after activation or even later in the process of fertilization (Chapter 3).

9.1 ACTIVATION

Development does not start with fertilization as commonly stated but with activation, which -- to quote 'Balon (1985)' -- is the process that releases the developmental block of the egg and triggers irreversible changes. In many eggs mere contact with water, independent of the presence of sperm, results in activation. Activation can be brought about also by temperature shock or chemical influences '(Devillers, 1961; Luther, 1966)'. 'T. Yamamoto (1961, 1975)' further describes stimulating by electrical, photodynamic and ultrasonic means as activating stimuli. Primary activation changes are revealed experimentally in the trout (*Salmo trutta* L., *morpha lacustris* Linné) 20-30 seconds after application of a stimulus '(Ginsburg, 1963)'.

Activation begun without fertilization continues with similar changes as described for fertilized eggs, i.e. emission of the second polar body, cortical reaction, formation of the perivitelline[6] space, hardening of the envelope and bipolar differentiation (as described below). However, the movements are slower and eventually come to a standstill. The most visible processes are formation of the perivitelline space and bipolar differentiation (Fig. 9.1) '(Ransom, 1876; van Bambeke, 1875; Hoffmann, 1880)'. These, and all subsequent processes triggered by activation and followed by fertilization will be dealt with later in detail.

Activation of the teleostean egg increases internal free calcium, possibly modified by membrane potential-dependent changes in Ca^{2+} influx '(Nuccitelli, 1980)'. This is followed by the question: how does the Ca^{2+} wave initiate the cell cycle and protein synthesis. A general overview of the question 'how do sperm activate eggs?' is given by 'Nuccitelli (1991)'.

9.2 FORMATION OF THE PERIVITELLINE SPACE[6]

This fluid-filled space between egg and envelope, without the presence of sperm, was observed by the early investigators. The space was often referred to as 'water chamber' or 'respiratory chamber'. It was realized by various authors that during the formation of this chamber the cortical alveoli disappear. This process is referred to as the cortical reaction (see 9.8). 'Von Baer (1835)' described that in most fish, e.g. the cyprinid *Cyprinus blicca* and *C. erythrophthalmus*, the eggs swell

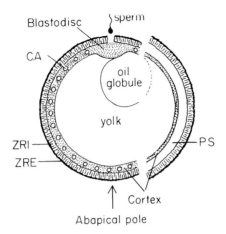

Fig. 9/1 Diagram of an unfertilized and fertilized egg. Left side: unfertilized Right side: fertilized. Unfertilized side: sperm is about to enter the micropyle. The envelope (ZRE and ZRI) is tightly attached to the *cortex*, which contains the cortical *alveoli* (CA). Note that the CA are absent at the apical pole which is occupied by the blastodisc. Fertilized side: The *cortex* has undergone the cortical reaction, i.e. the CA have shed their contents, thereby creating the perivitelline space (PS).

after deposition into water. Formation of a 'transparent space' in *Coregonus palea* was reported by 'Vogt (1842)' and for *Perca fluviatilis* by 'Lereboullet (1854)'. 'Reichert (1856)' and 'His (1873)' noted that in the unfertilized eggs of the pike water enters, which separates the envelope from the yolk. Observations by 'Ransom (1867/68)' on a great number of teleosts followed (*Salmo fario, Cyprinus gobio, Acerina vulgaris, Leuciscus phoxinus, L. cephalus* and *Perca fluviatilis*). 'Truman (1869)', 'Oellacher (1872)' and 'His (1873)' reported that the unfertilized eggs of *Salmo salar* and the trout, *Salmo fario*, 'imbibe water' forming a 'water chamber' (Figs. 9/1, 9/4). Similar results were reported by 'Owsiannikow (1873)' for *Coregonus lavaretus*, by 'van Bambeke (1876)' for *Tinca vulgaris* and by 'List (1887)' for *Crenilabris tinca* (Fig. 9/5). 'Hoffmann (1881)' stressed that *Scorpaena, Julis* and *Crenilabrus* produced only a small perivitelline space. 'Olt (1893)' maintained that the formation of a 'water chamber' in the egg of the bitterling *Rhodeus amarus* is the result of yolk contraction upon contact with water. According to 'Ransom (1867)' in *Gasterosteus* and according to 'Kupffer (1878)' in the herring, *Clupea harengus*, the contact with water (devoid of sperm) does not bring about any of the above-mentioned changes.

Almost a century later, activation merely as a result of contact with water was again reported for the unfertilized eggs, namely of the brook charr, *Salvelinus fontinalis*, the goldfish *Carassius auratus*, the pond smelt *Hypomesus olidus*, the chub *Gnathopogon elongatus*, the dace *Tribolodon hakuensis*, the ayu *Plecoglossus altivelis* and *Salmo irideus* '(Devillers, 1956, 1961; T. Yamamoto, 1958, 1961; review by Laale, 1980)'. 'Armstrong and Child (1965)' described that mature ova of *Fundulus* can be activated merely by stripping them from the female into sea water. Also, induction of the cortical reaction and elevation of the envelope in the unfertilized egg of *Oryzias latipes* were achieved by pricking the egg with a fine glass needle or electrical stimulation. Puncture at the animal pole resulted in a wave of cortical breakdown towards the vegetal pole. A reverse course was achieved by stimulating the egg at the vegetal pole. Stimulation at the equator seemed to indicate that the breakdown of the cortical alveoli is more rapid near the animal pole than near the vegetal pole. The reaction time is also shortest at the animal pole and longest at the vegetal pole '(T. Yamamoto, 1939, 1944)'. The early investigators had reasoned that the uptake of water closes the micropyle and thereby prevents the entrance of sperm '(Henneguy, 1888)'. Later investigators, e.g. 'Ishida (1948)' and 'Iwamatsu *et al.* (1993a,b)' came to the same conclusion. Therefore, in order to retain fertilizability of the eggs, the so-called 'dry method' (méthode russe) has been used since the 19th century for artificial fertilization in fish culture. With this method, sperms and eggs are brought together without the addition of water and activation is achieved after subsequent addition of water. However, if withholding of water is continued before the perivitelline space is formed, the fertilized eggs can be shipped in this state. On subsequent contact with water the perivitelline space can be formed as late as 24 hours afterwards.

Artificial insemination of fish was first performed in 1758 by 'Jacobi' '(cf. 'Henneguy, 1888)' although 'Rusconi (1835, 1840)', followed by 'Vogt (1842)' and 'Aubert (1854)', are usually credited as being the first. Usage of this method was continued by 'Kupffer (1878)' for the herring, *Clupea harengus*, by 'Hoffmann (1881)' with the eggs of *Scorpaena scrofa, S. porcus, Julis vulgaris, Crenilabrus pavo* and *Heliasis chromis*, by 'Ryder (1885/1887)' with the eggs of the common shad, *Clupea sapidissima* and cod, *Gadus morrhua*, and by 'List (1887)' with the eggs of *Crenilabrus tinca, C. pavo, C. quinquemaculatus, C. rostratus and C. ocellatus*.

Later again, i.e. from the middle of the 20th century onwards, it was stressed that as a result of activation unfertilized eggs may lose their fertilizability. 'T. Yamamoto (1944)' noted that in the eggs of *Oryzias latipes* this occurs within 6 mins of exposure to freshwater. In the trout fertilizability rapidly decreases after 60 seconds in water and is complete after 120 seconds '(Rothschild, 1958)'. The *Fundulus* eggs remain fertilizable until 15-20 mins after stripping '(T.Yamamoto, 1944, 1958)'. Loss of fertilizability was reported also for *Salvelinus fontinalis, Carassius auratus, Hypomesus olidus, Gnathopogon elongatus, Tribolodon hakuensis* and *Plecoglossus altivelis* '(T. Yamamoto, 1958, 1961; Iwamatsu *et al.*, 1993)'. 'Fisher (1963)' reported that oviduct eggs of the trout placed in freshwater immediately begin to gain weight and within minutes become unfertilizable. The uptake of water ends in 20-30 minutes, by which time the eggs may have increased their weight by as much as 20-25%. In some Russian freshwater cyprinids the cortical reaction occurs 5-10 seconds after placing the eggs in water and the membrane has become detached from the entire egg after 45-55 seconds. The eggs keep their capacity for fertilization only for 20 seconds at a temperature of 27°C. During swelling of the eggs the diameter of the envelope increases 3-5 times and their surface area 6-14 times '(Mikodina and Makeyeva, 1980)'.

9.3 BIPOLAR DIFFERENTIATION

This term refers to the cytoplasm, which hitherto had enclosed, and in some also permeated, the yolk and on contact with water starts streaming towards the animal pole forming the blastodisc[7]. It leaves behind a thin layer of cytoplasm surrounding the yolk. Bipolar differentiation is not causally connected with the formation of the perivitelline space.

It was 'Haeckel (1855)' who coined the name 'blastodisc' to designate 'the discoidal thickening of the animal pole of the egg, to which the subsequent cleavage will be restricted' (Figs. 9/1-9/6).

'Vogt (1842)' described that the 'Keim' (germ) of *Coregonus palaea* appears on the yolk of the unfertilized egg as soon as it is put into water. 'Ransom (1854/55, 1867/68)' observed in eggs of the following species concentration of the 'germ' regardless if the eggs were put into fresh or inseminated water: *Salmo salar, S. fario, Thymallus vulgaris, Esox lucius, Cyprinus lobio, Leuciscus phoxinus, L. cephalus, Perca fluviatilis, Acerina cernua, Cottus gobio* and *Gobio fluviatilis*. He even observed the beginning of cleavage in the egg of the pike. 'Oellacher (1872)' agrees with the results of *Salmo fario* by 'Ransom (1868)' and mentions a protoplasmic 'network' below the blastodisc. However, while a breathing chamber is formed in the unfertilized eggs of *Coregonus lavaretus* on contact with water, bipolar separation does not seem to occur. 'Lereboullet (1862)' reported that in the trout the 'membrane sous-ajencte au germe' retains its 'aspect fenêtre' and does not take part in cleavage. 'Lereboullet (1854, 1862)' described cytoplasmic streaming in the eggs of *Perca* and the pike on contact with water, while 'Truman (1869)' observed rotation and contraction of yolk of the egg of the pike on contact with water.

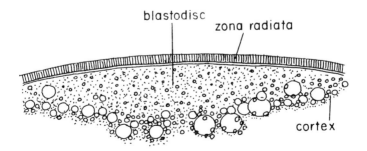

Figs. 9/2 Blastodisc of unfertilized eggs of salmonids (redrawn after His, 1873). Blastodisc of unfertilized egg of the salmon *Salmo salar*.

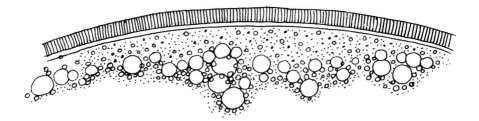

Figs. 9/3 Blastodisc of unfertilized eggs of salmonids (redrawn after His, 1873). Blastodisc of unfertilized egg of the trout *Salmo fario*.

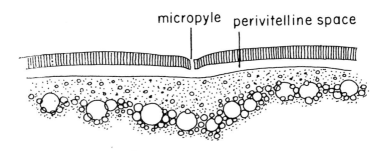

Figs. 9/4 Blastodisc of unfertilized eggs of salmonids (redrawn after His, 1873). Egg of *Salmo salar* ovulated into freshwater. Note that perivitelline space has been formed although the egg is still unfertilized.

'Van 'Hoffmann (1881)' reported that in the eggs of *Scorpaena scrifa, Julis vulgaris, Crenilabrus parvo* and *Heliasis chromis* contraction of the germ is observed in fresh water though it takes several hours in the last-mentioned species. He assumes that in most unfertilized eggs the 'germ contracts' after some time in contact with water and a second polar body is emitted. 'Olt (1893)' reports that concentration of plasma takes place without fertilization in *Rhodeus amarus*.

Many authors assert that the germ (blastodisc) is present before fertilization without mentioning any cytoplasmic streaming. The germ is described as an 'accumulation of cytoplasm' at the animal pole while a thin cytoplasmic layer surrounds the yolk. Salmonidae and the pike are the main examples. 'Von Baer (1835)' stressed that the 'Keim' (germ) is already present in the unfertilized egg of the pike, *Cyprinus blicca* and *C. erythrophthalmus*, even before ovulation. 'His (1873)' maintained that in the salmon and trout the germ is already visible before oviposition (Figs. 9/2-9/4, 9/6) whereas 'Lereboullet (1854)' stressed that in both of these species the blastodisc is formed only after fertilization. 'Van

Bambeke (1875)' observed bipolar separation before fertilization in *Tinca vulgaris* and *Lota vulgaris*. 'Ryder (1884)' mentioned that 'the disk appears in some cases at least to have been differentiated before it leaves the intraovarian cavity, as in some Salmonidae, for example'. 'Gerbe (1875)' maintained that the germ of teleosts manifests itself only after oviposition regardless if it is fertilized or not. He seems to speak essentially of Salmonidae. 'Hoffmann (1881)' described for *Gobius minutus* an accumulation of cytoplasm at the micropylar pole on eggs taken from a gravid female. He mentioned that the unfertilized eggs of *Crenilabrus pavo* display a blastodisc, and emanating from it, a thin cytoplasmic layer surrounding the yolk. 'List (1887)' observed the same in *C. tinca* and *C. pavo* (Fig. 9/5). However, 'Janosik (1885)' maintained that in these species streaming of cytoplasm was observed only after fertilization. Mature eggs of *Gobius minutus* and *G. niger* show, already in the ovary, an accumulation of cytoplasm at the micropylar pole according to 'Hoffmann (1881)'. 'Ryder (1884)' mentions 'that the germinal disk is formed independently of the influence of the spermatozoon in many species there cannot, however, be the slightest doubt. I have witnessed this phenomenon in the eggs of *Chirostoma, Morone, Parephippus* and *Creratacanthus*, while it is known to take place in many other species investigated by European authors'. 'Agassiz and Whitman (1889)' reiterated that the blastodisc of the fish egg forms in many cases entirely independently of fecundation. Some time later, 'Devilliers (1956, 1961)' and 'Arndt, 1956)' reported that in *Leuciscus* and *Salmo irideus* bipolar segregation had ended already at ovulation (Fig. 9/6).

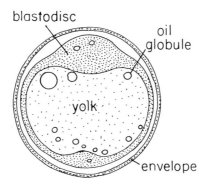

Fig. 9/5 Mature, unfertilized egg of *Ctenolabrus tinca* (redrawn after List, 1887). Note the early presence of the blastodisc.

The reports on the herring vary. 'Kupffer (1878)' observed that the unfertilized eggs of the herring, *Clupea harengus*, when put into water do not undergo any bipolar differentiation. 'Hoffmann (1881)' maintained that in the 'Strömling' (*Clupea harengus membras*), of the Baltic Sea the 'germinal layer' exists before fertil-

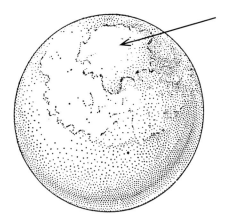

Fig. 9/6 Mature, unfertilized, denuded egg of the rainbow trout *Salmo irideus* (Devillers, 1956, with kind permission of C.A. Reitzel Forlag, Copenhagen). Blastodisc is visible in the upper zone (arrow).

ization. 'Brook (1885a)' has called attention to the fact that in the Pacific herring, *Clupea pallasii*, as in *Tinca vulgaris*, the influence of a sperm is necessary before a true 'germ disc' is formed. [*C. pallasii* and *C. harengus* have somewhat different spawning habits. The former spawns from near the high-tide line to only about a metre below low tide and has only one spawning season while the latter spawns in deeper water during spring as well as autumn '(Breder and Rosen, 1966)']. In *Gasterosteus* the formation of a 'breathing chamber' and the concentration of the 'formative yolk' into a blastodisc depended on the presence of sperm '(Ransom, 1854/55, 1867/68)'.

Summary

After completion of oogenesis the egg enters an inert phase (also called developmental arrest). Activation brings about the release from this state. Even in unfertilizated eggs activation is brought about by contact with water and also by thermal, chemical and other stimuli.

Already the early investigators suggested that the contact with water of unfertilized eggs closes the micropyle and prevents fertilization from taking place. As a result the 'dry method' was used in fish hatcheries for artificial fertilization; i.e. sperms and eggs are mixed before the addition of water.

The following reactions can also be induced by early contact with water or other stimuli: Induction of the cortical reaction which results in a lifting of the envelope as a result of the cortical alveoli releasing their contents. Another result is bipolar differentiation, which refers to the cytoplasm permeating the yolk streaming towards the animal pole to form the blastodisc. However, several species of teleosts seem to display a 'thickened area' at the animal pole already before activation, in some even before ovulation. The above processes, as a result of sperm entry, will be dealt with in detail in 9.10-9.13. (Figs.9.1-9.3)

9.4 MOTILITY OF SPERM

'Ransom (1855/56)' noted in *Gasterosteus leiurus* and *G. pungitius* that the sperms continue to move for a considerable period in the viscous secretion which envelopes the ova, 'but they very quickly become still in water'. The eggs of the bitterling *Rhodeus amarus* are deposited by the long ovipositor of the female into the mantle cavity of a mussel. The males shed their sperms near the mussel. Adhesion of the sperms to the cilia of the interlamellary spaces and gill ducts of the mussel guarantees the sperms a long stay within the gills to facilitate fertilization in loco '(Olt, 1893)'. The cell membrane of the aflagellate *Gymnarchus niloticus* is lined internally with a network of microtubules. It has been suggsted that this network enables the sperm to move by amoeboid motion '(Mattei, 1991)'.

'Scheuring (1924)' described that spermatozoa of *Salmo irideus* are motionless when suspended in the seminal plasma of another trout. However, a dilution of 0.001 with water induces the same degree of sperm activity as observed in fresh water. 'Schlenk and Kahmann (1938)' observed in the same species that sperms suspended in water at first swim vigorously. Then, the rate of their movement gradually decreases and after 30-40 seconds most of them show only vibrating movements. After 125-203 seconds all movement had ceased. Movements of activated spermatozoa are only rarely rectilinear but often curved and spiral '(Schlenk and Kahlmann, 1935)'. 'Lindroth (1946)' mentions that the motility of sperm of *Esox lucius* reported by earlier authors lasts from 20-30 seconds up to 3-4 minutes. He found that there is a linear relationship between the logarithm of the swimming time and the temperature. In simple terms: sperm of pike swim in water 2 minutes at 5°C, 1.5 minutes at 10°C and 1 minute at 15°C. 'Schlenk and Kahmann (1938)' reported for *Salmo trutta m. fario* that the inhibitory substance in *Salmo trutta m. fario* seminal plasma is potassium. This was also suggested in a review by 'Rothschild (1958)'. It seems that only in salmonids is potassium known to suppress sperm motility while in cyprinids the presence of potassium is even somewhat favourable for sperm movement '(Morisawa, 1985; Utsugi, 1993)'. The spermatozoa of the Pacific herring, *Clupea pallasi*, are equally active in isotonic potassium and sodium chloride solutions and are almost motionless in sea water. However, when they drift near the micropyle, they become very active and enter the micropyle '(Yanagimachi, 1957; Yanagimachi et al., 1992)'.

Organic sperm motility suppressors, as found e.g. in the epididimys of the rat, have so far not been detected in teleosts while motility enhancing factors are thought to reside in the micropylar region. A substance increasing the activity of spermatozoa released from eggs of the rainbow trout, *Salmo irideus*, was studied by 'Hartmann (1944)', Hartmann et al., 1947a,b)', from herring, *Clupea pallasii* eggs, by 'Koshi and Ogawa, (1931)' and from three Japanese bitterlings, *Acheilognathus lanceolata, A. tabira* and *Rhodeus ocellatus*, by 'Suzuki (1958)'. Two types of hormone-like substances (gynogamones) responsible for the attraction of sperm and also agglutination of supernumerary sperm were described by 'Hartmann (1944)', 'Hartmann et al. (1947a,b)', 'T. Yamamoto (1961)'and 'Suzuki (1961)'. The possible actions of gamones, i.e. sperm-egg interacting substances, were reviewed by 'Rothschild (1958)'.

Removal of the envelope demonstrated that the motility-initiating factor resides in the envelope of the micropylar region rather than in the egg proper '(Gilkey, 1981)'.

Surface structures unique to the micropylar region of *Platichthys stellatus* (small pores) and *Oncorhynchus gorbuscha* (protrusions and/or secondary openings) were discovered by SEM and TEM '(Stehr and Hawkes, 1979)'. 'Ohta (1984)' showed that the micropylar area of the eggs of the Pacific herring, *Clupea pallasii*, was covered with material thought to be favourable for efficient fertilization.

Calcium ions seem to be necessary for the sperm to reach the micropyle according to 'Yanagimachi et al., (1992)'. This had been established for *Oryzias latipes, Oncorhynchus keta, O. kisutch and O. mykiss* and *Clupea palllasii* by 'Yanagamachi and Kanoh (1953)' and 'Yanagimachi (1953, 1956, 1957, 1988)'. A transient rise in free calcium in the oocyte started at the base of the micropyle ' as observed by 'Ridgway et al. (1977)' and 'Gilkey et al. (1978)'. Activation of the teleost egg increases internal free calcium, possibly modified by membrane potential-dependent changes in Ca^{2+} influx '(Nucitelli, 1980)'. Subsequently, 'Pillai et al. (1993)' have isolated from the micropylar region of the egg of the Pacific herring, *Clupea pallasii*, a factor (glycoproteid) responsible for the initiation of sperm motility. They later showed that cryopreserved sperms of this herring were able to activate and fertilize fresh eggs resulting in embryonic development, while cryopreserved eggs were not fertilizable '(Ward and Kopf, 1993; Pillai et al., 1994)'.

Vigorous swimming movements of sperm in the neighbourhood of the micropyle were observed also in four species of the bitterling, *Rhodeus ocellatus, Acheilognathus lanceolata, A. tabira* and *Sarcocheilichthys variegatus* '(Suzuki, 1958)'. No sperm aggregation could be observed on the surface of denuded eggs. Later, the same author separated the chemical factor responsible for motility enhancement of sperm and suggested that it is produced by the micropylar cell (which is responsible for the formation of the micropyle). The factor is not species-specific within the family of Cyprinidae '(Suzuki, 1961)'. The factor is also found to a lesser degree in the vegetal pole area. This is explained by the fact that in the elongated ovipositor of the mature female the eggs are arranged in a single row. In this way the region of the micropyle of one egg gets into contact with the vegetal pole of the adjoining one. Eggs of the bitterling are deposited by the long ovipositor into the gill cavity of the fresh water mussel where fertilization takes place with sperms that have entered through the inhalent siphon. In these fish the presence of a sperm-stimulating factor on the micropyle seems necessary to facilitate the encounter between sperms and eggs.

The possiblity that there is an intimate relationship between the behaviour of sperm and the structure of the micropyle has been stressed more recently by 'Amanze and Iyengar (1990)'. They showed with SEM that the micropylar region of *Brachydanio rerio* eggs consists of 7-10 grooves and ridges, which directly drain into the micropylar vestibule. The authors suggested that this arrangement provides a sperm guidance role. With time-lapse video microscopic study and computer-aided analysis of sperm motility pattern in the micropylar region they concluded that, once the sperms have reached the micropylar region, the egg penetration by the sperms is increased by as much as 99.7%. After entrance of the first sperm, other sperms are agglutinated preferentially along the grooves. The first sperm reaches the fertilization cone within 2-3 minutes '(Hart et al., 1992)'. Radial grooves running towards the micropyle at the animal pole were also observed in *Barbus conchonius* and *Sturisoma aureum* '[Amanze and Yyengar (1990) and Riehl (1991)]' (Figs. 7/11-7/13). 'Iwamatsu et al. (1993b)' recorded with a video camera the sperm movements in both intact and isolated egg-envelopes of the medaka *Oryzias latipes*. The spermatozoa of the medaka were neither activated nor attracted toward the micropyle, in contrast to the results with three species of the Japanese bitterlings, *Acheilognathus lanceolata*, *A. tabira* and *Rhodeus ocellatus* and the mud loach, *Misgurnus anguillicaudatus* '(Suzuki, 1961)', the Pacific herring, *Clupea pallasii*, '(Yanagimachi, 1957 and Yanagimachi et al., 1992)', starry flounder *Platichthys stellatus* and pink salmon *Oncorhynchus gorbuscha* '(Stehr and Hawkes, 1979)'. However, 'Iwamatsu et al., (1997)' showed with antibodies against glycoproteins of the envelope that these antibodies have an affinity for spermatozoa and, therefore, play an important role in sperm guidance into the micropyle.

Generally, spermatozoa of fishes spawning in brackish and sea water swim much longer than those of most freshwater species '(Ginzburg, 1963)'. 'Morisawa (1985)' studied the initiation mechanism of sperm motility in some freshwater and marine fish. He concluded that it is the cAMP-dependent phosphorylation of a protein which triggers the initiation of motility.

9.5 FERTILIZING CAPACITY OF SPERM

Mobility of sperm as such does not always guarantee fertilizing capacity. The sperms have to be capable of forward motion and penetration of the micropyle (see below). When salmonid semen was diluted in equal proportion with water or coelomic fluid or Ringer's solution, sperms in water had completely lost their fertilizing capacity in 40 seconds, while those in Ringer fertilized 15.4% of eggs after 2 minutes and those in coelomic fluid 78.8% after 5 minutes, and 12.5% after 20 minutes. It was concluded that salmonid sperms are not only activated by coelomic fluid or Ringer but that in these fluids they retain their activity and fertilizing capacity much longer than in water '(Ginzburg, 1963)'.

More recently, it was shown that cryopreserved sperms of *Clupea pallasi* remain viable and capable of fertilizing fresh eggs for over 7 weeks '(Pillai *et al.*, 1994)'.

9.6 SPERM ENTRY THROUGH MICROPYLE

Four sperms were seen to pass into the micropyle of the ovum of *Gasterosteus leiurus* and *G. pungitius* by 'Ransom (1854/1856)'. However, 'Aubert (1854)' could not establish how the large sperms of the pike are able to enter the egg. 'Bruch (1856)', who described the micropyle in *Salmo fario* and *S. salar*, failed to see any sperm entering. However, 'His (1873)' observed in the egg of *Salmo salar* that never more than one sperm enters the micropyle. 'Oellacher (1872)' reported that in the egg of the trout spermatozoa are 'drawn in' (G. hineingerissen) through the micropyle. In contrast, 'Kupffer (1878)' saw a great number of sperms entering the breathing chamber and subsequently the yolk. He also observed sperms in the egg of the herring, *Clupea harengus*, but he could not determine how they entered. 'Hoffmann (1881)' reported the penetration of the sperm, through the micropyle, in the eggs of *Scorpaena, Julis* and *Crenilabrus*. 'Ryder (1884)' never observed the entrance of the sperm into the egg of the cod, *Gadus morhua*. 'List (1887)' described the sperms swarming vigorously around the micropylar opening, which was followed by the entrance of one sperm. Penetration of further sperms was prevented by a plug formed in the inner part of the micropyle. (Figs. 9/7AB1, AB2)

The outer diameter of the micropyle funnel is larger than the inner which is only as wide as the head of a sperm, i.e. it offers access of one sperm only to the egg '(Sakai, 1961; Ginsburg, 1963; Brummet and Dumont, 1979; Yanagimachi, 1988)'. In the eggs of the common carp, *Cyprinus carpio*, however, the internal aperture of the micropylar canal is wide enough (5 microns) to admit two sperm cells at once '(Kudo, 1980, 1991; Linhart *et al.*, 1995) (Fig. 9/7B1). The sperm head averages 2.4 microns in width according to 'Kudo (1991)', 3.8 microns according to 'Stein (1981)' and 1.8-1.9 microns according to 'Emeliyanova and Makeyava (1985)', and the diameter of the micropyle amounts to 3-3.5 microns according to 'Kiselev (1980)'.

In the medaka a significantly larger number of sperms enter the micropyles of unfertilized than of fertilized eggs. This is also true for unfertilized eggs with closed micropyles. The authors concluded that there must be a factor or factors in the unfertilized egg that guide the sperms towards the micropyle. Chemical attraction does not seem to play a role '(Takano and Onitake, 1989)'.

As shown for the herring, *Clupea pallasii*, spermatozoa enter the micropyle one by one. Only the first one fuses with the egg plasmamembrane. The others are expelled '(Yanagimachi *et al.*, 1992)'. Similarly, more than 20 spermatozoa of the trout, *Oncorhynchus mykiss*, and the silver salmon, *Oncorhynchus kisutch* enter the micropyle and only the first one fuses with the egg membrane (Fig. 9/7A1). Under experimental conditions salmon spermatozoa entered trout micropyles and vice versa. However, they entered only occasionally the micropyle of the Pacific herring,

Clupea pallasii '(Yanagimachi et al., 1992)'. It was shown for the herring that a sperm-guidance factor is critical for the entrance of the sperms and fertilization, and that it facilitates the entrance of homologous sperms. It was concluded that the sperm guidance factor of the herring has both sperm motility-initiating and sperm-guiding activities, whereas the salmonid guidance factor possesses only sperm guidance activity. Salmonid sperms are already actively motile in water.

9.7 FUSION OF SPERM AND EGG MEMBRANES (Fig. 9/7AB2)

In the medaka, *Oryzias latipes*, attachment of the sperm to the egg occurs 1-3 seconds after insemination '(Iwamatsu and Ohta, 1981)'. Since teleost sperms lack an acrosome, fusion between the plasma membranes of the gametes is based on a direct interaction which occurs at the so-called sperm entry site (usually referred to as SES) which lies at the bottom of the micropyle. (Some authors call the micropyle itself the sperm entry site.).

The SES of *Cyprinus carpio* is characterized by the presence of a tuft of 19-27 microvilli (Fig. 9/7B1). The extended surface provided by the microvilli is thought to expedite the binding and fusion between the sperm and egg membanes. The eggs respond to the penetration of one sperm with the formation of a fertilization cone in two steps '(Kudo, 1980; Kudo and Sato, 1985; Kudo, 1991; Linhart et al., 1995). Twenty seconds after immersion of the egg in fresh water the so-called 'fusion cone' is observed. It may be as long as 10-21 microns. Its ball-like tip reaches the micropylar vestibule and creates a transient plugging of the micropylar canal. Several sperms lying on the apex of the cone can be seen. During incorporation of a sperm and with initiation of the cortical reaction the 'fusion cone' reduces in size '(Kudo, 1980; Kudo and Sato, 1985; Kudo, 1991)'. At this stage, 105 seconds after insemination, a secondary conformation, called the 'later fertilization cone' below the top of the much shortened fusion cone is formed and keeps growing (Figs. 9/7B2,3). A similar 'two-storied fertilization cone' was observed in *Plecoglossus altivelis* '(Kudo, 1983b)'. It is clear that the response of the formation of two cones (one above the other) was induced by sperm penetration since no fertilization cone of either type was observed in artificially activated eggs.

The fertilisation cone ranges in height from 1-3 microns in *Fundulus* '(Brummet and Dumont, 1979)' and from 2-10 microns in the chum salmon *Oncorhynchus keta* '(Kobayashi and Yamamoto, 1981, 1987)'. The fusion occurs between the sperm head and the short-lived fertilization cone. No involvement of microvilli was observed (Fig. 9/7A,2).

The first spermatozoon to reach the micropyle of the egg of the rose bitterling, *Rhodeus ocellatus*, becomes surrounded by microvilli, and a mass of egg cytoplasm, corresponding to the fertilization cone of other species, becomes visible. This mass subsequently plugs the micropyle '(Ohta and Iwamatsu, 1983)'. Rapid fusion between egg and sperm membrane following sperm contact has been observed by 'Ohta, (1985)'. The extended surface provided by the microvilli of the SES at the base of the micropyle is thought to expedite the binding and fusion between the two membranes. It has been suggested that these microvilli possess receptors for the recognition and/or binding with the sperm plasma membrane. However, it was shown by 'Iwamatsu and Ohta (1978)' and 'Ohta (1985)' that spermatozoa of the rose bitterling, *Rhodeus ocellatus*, can enter denuded eggs in all areas and attach to different sites.

'Ohta and Nashirozawa (1996)' showed that even water activation without sperm did transform the sperm entry site from a tuft of microvilli into a swollen mass, although it took a longer time and the mass was smaller than in the case of inseminated eggs. They further established that cytochalasin B, which acts on actin microfilaments, did not prevent formation of the swollen mass, irrespective of insemination or activation.

The sperm-binding site in the egg of *Brachydanio* is also specialized as a circular tuft of microvilli (Fig. 9/7B1). The diameter of the tufts closely correlates with the inner micropylar diameter and the head of the spermatozoon '(Hart and Donovan, 1983)'. The sperm attaches to the SES within 5 seconds of insemination. Subsequent to the fusion of the two membranes and 45-60 seconds post-insemination a nipple-shaped fertilization cone is observed '(Wolenski and Hart, 1987)'.

SEM of the eggs of *Oryzias latipes* revealed that the successful sperm attaches itself to the egg surface within 1-3 seconds after insemination. The sperm head is rapidly engulfed by a folded egg surface which is followed by a slow engulfment of the whole sperm by the fertilization cone. Fusion of egg and sperm plasmamembranes occurs at least 60 seconds after insemination. In this species no highly differentiated cytoplasmic structure for binding with the fertilizing sperm at the base of the micropyle has been observed '(Iwamatsu and Ohta, 1978.1981)'.

'Kessel *et al.* (1983)' have shown on freeze-fractured replicas the presence of intramembraneous particle-arrays in the equatorial position of the zebrafish sperm head. These may represent receptors specialized for binding to complementary receptors at the microvillar SES. Fusion of the sperm and egg membranes appears restricted to the equatorial or lateral region of the sperm head in *Brachydanio rerio* '(Wolenski and Hart, 1987)' (Fig. 8/2). Circumstantial evidence pointing to the presence of receptors had been presented earlier. For instance, 'Hartmann *et al.* (1947a,b)' had shown that a substance in trout sperm supernatant and in ovarian fluid agglutinates sperm. 'Gilkey (1981)' reviewed the possibility of the presence of a receptor involved in sperm-egg contact.

SEM and TEM, in conjunction with RhPh (rhodamine phalloidin) staining, have demonstrated that in *Brachydanio* polymerized filamentous actin is present in the cortex of ovulated unfertilized eggs. At the SES actin is present in the surface microplicae of the plasma membrane, the microvilli and the peripheral cytoplasm of the cortex as a tightly knit meshwork. It has been suggested that the formation of the fertilization cone requires reorganization of this meshwork which will thicken in order to surround the sperm head. However, eggs pretreated with either cytochalasin B or D, which selectively inhibit the formation of actin, still form a fertilization cone but they fail to incorporate sperm. Under normal conditions, one consequence of fusion between the gametes is to bring the nucleus of the sperm head into direct contact with the actin-containing cortex. Once entered, the nucleus becomes surrounded by the actin meshwork. On the basis of results with mammalian eggs, it has been suggested that the actin filaments become structurally linked to the sperm nuclear membrane '(Hart and Donovan, 1983; Wolenski and Hart, 1987, 1988; Hart and Wolenski, 1988a,b; Hart et al., 1992)'.

'Farias et al. (1995)' analyzed tubulin and microtubule-associated protein pools in the unfertilized and fertilized eggs of Oncorhynchus mykiss and suggested that a redistribution of tubulin pools occurs during the first 20 minutes after fertilization.

9.8 INTERNALIZATION OF SPERM (Fig. 9/7AB3,4)

'Agassiz and Whitman (1885)' stressed that it is not yet known if the cytoplasm of the engulfed sperm plays a part or even if it has any function at all. 'Blanc (1894)' reported that in Trutta lacustris also the tail enters the egg and is later resorbed. 'Behrens (1898)' mentions that the tail of the sperm of Salmo iridea and/or S. trutta fario enters the egg; polyspermy was never observed.

'Brummet et al. (1985)' report that in Fundulus heteroclitus the internalized nucleus and midpiece of the sperm become surrounded by the fertilization cone. The 50 micron long flagellum eventually (i.e. within 10-15 minutes) enters the egg (not observed in any other fish) while the nucleus and the midpiece penetrate more deeply. Twenty seconds post insemination the nucleus begins to fragment followed by chromatin decondensation. Nuclear envelope fragmentation seems to occur much more rapidly than in the egg of Oryzias '(Iwamatsu and Ohta, 1978)'.

The membrane of the sperm of Fundulus heterclitus and Oncorhychus becomes incorporated into the egg plasma membrane to form a so-called mosaic '(Brummet et al., 1985)'. Fusion takes place between the membranes of the sperm head and the tuft of microvilli in Brachydanio rerio '(Wolenski and Hart, 1987)', in Rhodeus ocellatus '(Ohta and Iwamatsu, 1983)' and Cyprinus carpio' (Kudo, 1980; Kudo and Sato, 1985)'. Fusion of the plasma membrane of the egg and of the sperm is usually limited to the area below the micropyle. However, in the Pacific herring, Clupea pallasi, the plasma membrane in other areas of the egg has the potential to fuse with a sperm as e.g. exemplified in denuded eggs '(Yanagimachi, 1988)'. SEM of the Oryzias eggs revealed that the successful sperm attaches itself to the egg surface within 1-3 seconds after insemination '(Iwamatsu and Ohta, 1981)'.

Attachment of Brachydanio rerio spermatozoa to the SES can be achieved within 5 seconds of mixing a gamete suspension. By 60 seconds the spermhead is partially engulfed by the fertilization cone and incorporation of the whole sperm is achieved 2-3 minutes after insemination '(Wolenski and Hart, 1987)' (Fig. 9/7B).

According to TEM and SEM analyses of Oryzias eggs, the penetration of the sperm is divided into: 1) a rapid engulfment of the head or its greatest portion 10 seconds after insemination and 2) slow engulfment of the tail portion by minute 'wave-like folds' on the surface of the fertilization cone, 30-90 seconds after insemination. The production of the fertilization cone concurs approximately with the beginning of the cortical reaction. By 3 minutes after insemination the whole spermatozoon, deprived of its plasma membrane, is completely enclosed by the egg cytoplasm. '(Iwamatsu and Ohta, 1978; 1981)'. 'Kobayashi and Yamamoto (1987)' were unable to observe the above-mentioned sperm engulfment in the eggs of *Oryzias* though they describe this process for *Oncorhynchus keta* (Fig. 9/7A3,4). As mentioned above, the sperm nucleus after entering the egg becomes surrounded by a network of actin which may become structurally linked to the sperm nucleus membrane '(Hart *et al.*, 1992)'.

9.9 BLOCK OF POLYSPERMY

Since the prevention of polyspermy in the echinoderms and mammals is diphasic, two or more preventative processes were suggested for teleosts also: 1) invisible changes caused by a so-called fertilization impulse or activation impulse, which occurs within the course of a few seconds and makes the surface of the egg envelope unreceptive to further spermatozoa. 2) These changes catalyse the cortical reaction (Fig. 9/8) '(Rothschild (1958); Ginsburg (1961)'. 'T. Yamamoto (1944a,b, 1975)' produced experimental evidence that in *Oryzias* eggs the fertilization-wave or impulse propagates itself with decreasing intensity and velocity. Although 'Nuccitelli (1980)' reported that in the medaka electrical changes 'accompany' fertilization, there is no electrical block to polyspermy [see also 'Gilkey (1981); rev. Hart *et al.* (1993)']. 3) It is the micropyle — the single entry for sperm which avoids the sperm having to penetrate the envelope — which seems to constitute an inherent block to polyspermy. 'Hoffmann (1881)' reported for *Scorpaena* that 25 minutes after fertilization the polar body exits via the micropyle and prevents further sperm from entering. It was noted in *Crenilabrus pavo* that after the entrance of the first spermatozoon into the micropyle, the inner part of it gets plugged. This prevents the entrance of further sperms '(List, 1887)'. A 'dome' formed over the entrance of the micropyle was reported for *Sfizostedion vitreum* by 'Reighard (1890)'. When several sperms have entered, they advance in a single file due to the narrowness of the canal. In the flounder, *Limanda shrenki*, a sperm takes about 2 minutes to reach the base of the canal. However, within a few minutes the remaining spermatozoa are expelled through the mouth of the micropyle '(K. Yamamoto, 1952, T. Yamamoto, 1961)'. Extrusion of supernumerary sperms from the micropyle has been reported also for the herring, *Clupea pallasii*, and the salmon, *Oncorhynchus kisutch*, by 'Yanagimachi and Kanoh (1953)', 'Ginsberg (1961, 1963)', 'Yanagimachi *et al.* (1992)', for *Oncorhynchus keta* by 'Kobayashi and Yamamoto (1987)', for the medaka, *Oryzias latipes*, by 'Sakai (1961)' and for *Fundulus heteroclitus* by 'Brummet and Dumont (1981)'. A colloidal substance (released from the egg surface) traps and evicts the supernumerary sperms. This same substance may also subsequently seal the outer open-

ing of the micropyle in the form of a ball-like body or plug '(Sakai, 1961; Ohta and Iwamatsu, 1983)'. In *Oncorhychus keta* 300 seconds after insemination, the micropylar canal decreases in diameter near the lower end, which prevents supernumerary sperms from passing through '(Kobayashi and Yamamoto, 1987)'. The micropylar canal becomes narrow due to the swelling of the ZR interna '(K. Yamamoto, 1952)'. Five minutes after sperm attachment to the medaka egg the inner one third of the micropyle becomes completely occluded due to the contraction of the ZR interna around the micropyle '(Takano and Onitake, 1989)'.
4) In all the above fish the micropylar canal is long whereas in *Rhodeus ocellatus* it is short, leaving no space for supernumerary sperms '(Ohta and Iwamatsu, 1983)'. Therefore, shortness of the canal, which does not allow more than one spermatozoon through, seems to present another mechanism to prevent polyspermy.
5) The fertilization cone may also be instrumental in preventing polyspermy. In the egg of the chum salmon '(Kobayashi and Yamamoto, 1981)' an extension of the fertilization cone extends into the micropylar canal and plugs it (Fig. 9/7A2,3). Within 1 minute of fertilization the cone of the egg of *Cyprinus* changes dramatically in appearance from its earliest stage to its regression; as mentioned above, the 'fusion cone' (the upper part of the two-storied cone) prevents penetration of subsequent sperms (Fig. 9/7B3) '(Kudo, 1980; Kudo and Sato, 1985)'. But, as 'Kudo (1983b)' already stresses, 'it is not yet known whether block of polyspermy by the cone is caused in part by the sperm-head plasma-membrane inserted during cone formation, by redistribution of surface carbohydrate moieties of the egg plasma membrane, by a combination of these factors, or by something else'.
6) Other possible candidates are the microvilli projecting into the micropylar canal. Sperm-binding sites on their tips may have been taken up by the first entering sperm, which prevents subsequently entering sperms from fusing. The microvilli of the SES are usually converted into a fertilization cone.
7) The 'classical inhibitor' for polyspermy is thought to be provided by the perivitelline space filled with secretions from the cortical alveoli discharged during the cortical reaction (see 9.8) '(Bazzoli and Godhino (1994)'. [The perivitelline fluid of the Atlantic salmon, *Salmo salar*, consists of 58% water, 25% protein, 12% lipid and 1.7% carbohydrate '(Eddy, 1974)'.]

As mentioned above, substances expelled by the cortical reaction may be available to agglutinate supernumerary sperm in the perivitelline space. The sperms are presumably there because the retracting fertilization cone has allowed them access to this space. 'Nosek (1984)' suggested that lectins, emanating from cortical alveoli, block polyspermy. The cortex itself does not seem to constitute a blockage since removal of the envelope permits polyspermy, i.e. sperms seem to be able to penetrate any part of the cortex '(Kanoh and Yamamoto, 1957; Yanagimachi, 1957; T. Yamamoto, 1961; Sakai, 1961; Iwamatsu 1983)'. Since penetration of numerous spermatozoa becomes possible after removal of the perivitelline fluid, it has been suggested by 'Ginsburg (1961, 1963)' that in denuded (already fertilized or activated) eggs it is the absence of substances contained in the cortical alveoli which renders them accessible to spermatozoa. 'Ohta (1985)' analyzed by EM

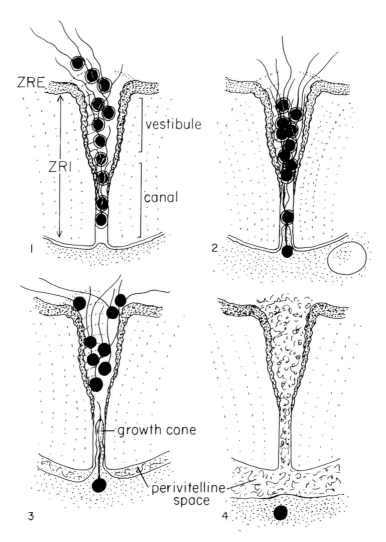

Fig. 9/7A Sperm entry through micropyle and sperm-egg fusion (modified from Kobayashi and Yamamoto, 1981, 1987, with kind permission of Wiley-Liss, Inc., a subsidiary of John Wiley and Sons, New York). Examples: Salmonids, *Fundulus*: **1** - Narrow vestibule lined by ZRE and micropylar canal. Indication of growth cone at base of canal. Width of canal accommodates one sperm head. **2** - Growth cone formed. Head of admitted sperm without membrane. Supernumerary sperms remain in vestibule. **3** - Perivitelline space formed as a result of cortical reaction. (Note that cortical *alveoli* do not reach the region of the micropyle.). **4** - Growth cone receded. Micropyle filled by secretions of the shed cortical *alveoli*, which agglutinate and eject additional sperms and prevent further entrance of supernumerary sperm. ZRE = *Zona radiata externa*; ZRI = *Zone radiata interna*.

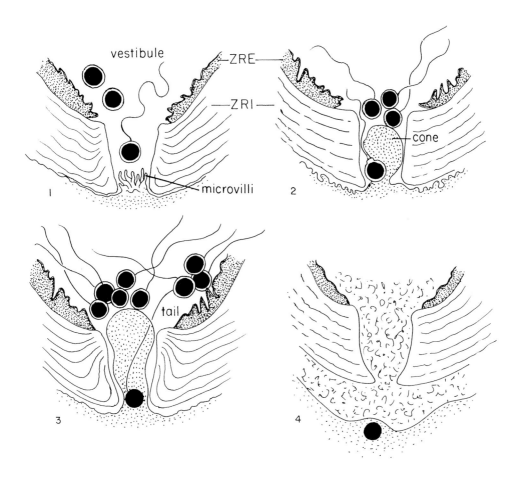

Fig. 9/7B Sperm entry through micropyle and sperm-egg fusion. (modified from Kudo, 1980, with kind permission of Blackwell Science Asia P/L, Melbourne). Examples: Carp, *Brachydanio rerio:* 1. Large vestibule and canal. The sperm entry site is characterized by a tuft of *microvilli*. 2. One sperm admitted and surrounded by growth cone, which fills the relatively wide micropylar canal. 3. Fusion of the egg and sperm membranes has taken place. Enlarged cone prevents entry of further sperms. (Carp possesses a two-storied fertilization cone, not shown in the diagram.). 4. Fertilization cone receded. As a result of the cortical reaction micropylar canal and vestibule are plugged by colloid. ZRE = *Zona radiata externa*; ZRI = *Zone radiata interna*.

the sperm entry into unfertilized eggs of denuded (called by him 'dechorionated') eggs of the rose bitterling, *Rhodeus ocellatus*. Following insemination, sperms were found on the surface of the eggs. The egg membrane at the sperm attachment sites swelled as the microvilli and microplicae disappeared and small fertilization cones were formed. These cones retracted about 10 minutes after insemination. Many sperms were observed to enter the cytoplasm where their nuclear envelope vesiculated and disappeared, the nucleus decondensed and development into a male pronucleus occurred. In the medaka the nuclear membrane of the sperm disintegrates within 5 mins after sperm entry, and a new one is formed by 15 minutes resulting in a male pronucleus '(Iwamatsu and Ohta, 1978)'.

The inner structure of the fertilization cones is similar in monospermic and polyspermic eggs. Fusion between egg plasma membrane and sperm plasma membrane in dechorionated (polyspermic) eggs of the medaka *Oryzias latipes* has been described by 'Iwamatsu and Ohta (1978)'.

Summary

Sperms swim vigorously after contact with water. They reach the micropyle due to a 'motility enhancing factor' present in the micropylar region. In some eggs, surface grooves seem to guide the sperm towards the micropyle. After entering the micropyle only the first spermatozoon will fertilize the egg. Since teleostean sperms do not possess an acrosome, the first step is fusion between the head membrane of the sperm and the egg membrane resulting in a 'mosaic membrane'. This takes place in some fishes at the site of a fertilization cone, in others the sperm binding site is supplied with a tuft of microvilli. An actin meshwork of the egg cortex is thought to be responsible for the formation of the fertilization cone. In some fishes the sperm head and midpiece enter the egg, in others the tail enters as well. Supernumerary sperms are blocked from entering the egg by the fertilization cone. They are agglutinated by a substance released by the cortical alveoli and then ejected. The 'classical' inhibitor of polyspermy is thought to be provided by the perivitelline space, though nowadays it is thought to play only a secondary role in this event. (9.4-9.9)

9.10 CORTICAL REACTION FOLLOWING FERTILIZATION

This reaction, starting from the area of the micropyle, involves the rupture of the cortical alveoli[8] (CA) and the discharge of their contents into the extracellular space, called perivitelline space (Fig. 9/8). Therefore, the creation of this space is caused by the discharge of the CA and/or a decrease in volume of the vitellus (yolk) and the lifting of the envelope. The cortical reaction can be initiated by immersion of fertilized (or artificially activated) eggs in fresh water (see 9.1-9.7).

'Ransom (1855)' noted 'yellow drops of the superficial layer of the yolk grow pale and disappear'. The change began near the micropyle. 'Ryder (1884)' suggested that in the cod, *Gadus morhua*, the disappearance of the CA is connected with fertilization. More than 100 years later the morphological details of the dis-

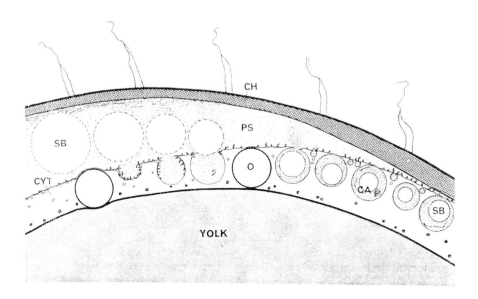

Fig. 9/8 Diagram of the cortical reaction in the *Oryzias* egg (Iwamatsu and Ohta, 1976, with kind permission of Springer Verlag, Heidelberg). Breakdown of cortical *alveoli* (CA) and elevation of the envelope or chorion (CH) are accompanied by a reduction in the thickness of both the envelope and cortex (CYT). The process begins with the release of colloidal material and spherical bodies (SB), following the formation of a large aperture into the perivitelline space (PS), and ends with shrinkage of the envelope which transforms into many *microvilli*. O - oil-droplet.

appearance of the CA and its significance in fertilization were still incomplete, according to 'Brummet and Dumont, 1981'. SEM data of these authors on *Fundulus heteroclitus* suggested that the membrane of the CA fuses with the oolemma, ruptures at the points of fusion and releases its contents into the perivitelline space. The cortical reaction is also referred to simply as exocytosis and the breakdown of the CA as CA-dehiscence. Part of the alveolar membrane fragments are lost to the perivitelline space. 'T. Yamamoto (1962)' reported that in *Fundulus* the remaining alveolar membrane becomes an integral part of the microvillous oolemma while in *Oryzias* the alveolar membrane dissolves. In contrast, 'Iwamatsu and Ohta (1976)' in their study of the breakdown of the CA in *Oryzias* suggest that the membrane of the shrunken alveoli is transformed into many microvilli. A marked drop in the capacitance of the egg of *Oryzias* before completion of the cortical reaction would be consistent with removal of membrane from the surface '(Nucciteli, 1980)'.

According to 'Wessels and Swartz (1953)', 'Armstrong and Child (1965)' and 'Brummet and Dumont (1981)' the cortical reaction in *Fundulus* both begins earlier (at 1-3 seconds after insemination) and ends earlier (at 60 seconds) than in

Oryzias latipes, in which the average initiation time is 28.6 seconds and the average completion time 152.8 seconds '(Iwamatsu and Ohta, 1981)'. Water-temperature for both species is given as 26°C. In the area of the micropyle, dehiscence of the CA does not begin until 60 seconds post-activation and is generally complete within 1-7 minutes. This delay in alveolar discharge may facilitate incorporation of the sperm '(Hart, 1990)'. In the common carp, *Cyprinus carpio*, the cortical reaction is observable in 87.8% of the eggs at the 30 seconds stage (20-22°C); after 60 seconds the reaction has spread over the whole surface. Subsequently, it takes another 3-4 minutes for all the CA to complete their discharge '(Kudo and Sato, 1985)'.

The CA are evenly embedded in the cortex of the egg, except for a small area adjacent to the micropyle at the animal pole (Fig. 9/1). Their absence in the vicinity of the sperm entry site was descibed e.g. for *Fundulus* '(Brummet and Dumont, 1979)', *Cyprinus carpio* '(Kudo, 1980)', *Oncorhynchus* '(Kobayashi and Yamamoto, 1981, 1987)' and *Brachydanio rerio* '(Hart and Donovan, 1983)'. Following activation they disappear in a wave spreading out from the micropyle and ending at the vegetal pole '(Kagan, 1935; Tschou and Chen, 1936; T. Yamamoto, 1939, 1964; Nakano, 1956; Ohtsuka, 1957, 1960; Fisher, 1963; Iwamatsu 1968, 1969; Hart et al., 1977; Hart, 1990)'. The disappearance of the CA is relatively slow (5-7 minutes). 'Donovan and Hart (1982, 1986)' described that in *Brachydanio* the mosaic egg surface (membrane of CA fused with oolemma) is only temporary and they propose that the CA membrane is retrieved by the egg cytoplasm by endocytosis. Labelling with ferritin suggested internalization of this membrane by coated vesicles. This would be followed by a recycling of the membrane for use in membrane synthesis of cleaving blastomeres. The Golgi body is involved to some degree in handling the retrieved membrane '(Hart *et al.*, 1977; Hart and Yu, 1980; Donovan and Hart, 1982, 1986; Hart and Collins, 1987; Becker and Hart, 1999)'. 'Hart *et al.* (1987)' used the localization of lysosomal enzymes by cytochemical methods to follow the fate of the surface membrane during the cortical reaction of zebrafish eggs. They showed that membrane retrieved from the egg surface is shuttled into organelles with lysosomal properties (i.e. secondary lysosomes and multivesicular bodies) and thereby undergoes degradation. However, the authors do not exclude other fates of the internalized membrane, such as storage and recycling to the plasma membrane. 'Kudo (1983b)' describes ultracytochemical modifications in the distribution of surface carbohydrates that take place in the fertilized egg of *Cyprinus*, which may be an expression of the mosaic arrangement of the limiting membrane of alveoli and original plasma membrane.

EM revealed that the contents of the CA consist of an electrondense spherical body (which often contains membraneous structures) embedded in a colloid. This was reported by 'T. Yamamoto (1962)', 'Iwamatso and Ohta (1976)', 'Iwamatsu and Keino (1978)' and Guraya (1982)' for the medaka and by 'Kudo (1976)' for the goldfish. (Fig. 9/9). Biochemical and histochemical tests revealed that the alveolar colloid is a polysaccharide '(Kusa, 1954; Aketa, 1954; Nakano, 1956; Arndt, 1960a)'. 'Sakai (1961)' had noted earlier that spherical bodies are

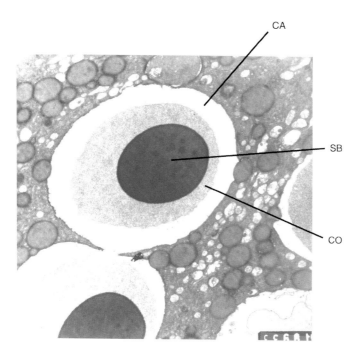

Fig. 9/9 EM of the *cortex* of a *Phoxinus phoxinus* egg revealing that the cortical *alveoli* consist of an electron dense spherical body embedded in a colloid. 5 000 × (Courtesy of Rüdiger Riehl). CA - cortical alveolus; CO - colloid; SB - spherical body.

released from the cortical alveoli of the naked medaka egg activated by spermatozoa but she did not follow the destiny of the granules. Subsequently, 'Iwamatsu and Ohta (1976)' observed that the spherical bodies of cortical alveolar origin persist in the perivitelline space during development of the egg. While the emptied CA shrink and disappear, the spherical bodies of the cortical alveoli remain and are thought to play a role in maintaining the distance between the cell membrane and the vitellus. In other words they are the cause of the elevation and hardening of the envelope. On the basis of enzymatic digestion 'Iwamatsu and Ohta (1976)' reported that the spherical bodies consist mainly of protein and lipids. However, 'Guraya (1965, 1982)' did not observe sudanophilic lipids in the CA. As regards the functional significance of the spherical bodies, he writes that it still remains uncertain as to whether the spherical bodies are involved in the hardening of the envelope.

'T. Yamamoto (1961)' described for *Oryzias* the presence of granules (of a diameter of 0.3-0.5 microns) surrounding the CA (diameter 2-28 microns). He named these bodies a-granules while in an earlier account ('T. Yamamoto, 1961) he had referred to them as 'cortical granules A'. He maintained that the a-granules start to dissolve before the breakdown of the CA, whereas b-granules (which stain differently, are fainter in colour and much smaller in size) remain intact. 'Aketa (1954)' demonstrated that RNA-rich granules are embedded in the cortical cytoplasm of *Oryzias* eggs and that they disappear after fertilization. He assumed these granules to be the a-granules. There is also ultrastructural evidence of a- and b-granules in the egg of *Cyprinus carpio* and *Carassius auratus*. The CA, a-granules and b-granules are discharged at different times, i.e. the latter two earlier than the former. Therefore, the ensuing fertilization membrane appears to contain the plasmalemma of the ovum and the limiting membranes of the granules '(Kudo, 1976)'.

Ultrastructurally, there are differences among the three types of granules. B-granules appear in clusters and are associated with tubular elements from which they originate. It was suggested that the b-granules may be elaborated by the Golgi body. Subsequently, investigations into enzymo-cytochemical changes between the unfertilized and fertilized egg of the common carp and goldfish were performed. The results for acid phosphatase and choline-esterase activities were interpreted as lending further support to the conclusion that there are two types of granules besides the CA: a-granules lack an enzyme (acid phosphatase) present in the CA and b-granules can be distinguished from a-granules by the presence of non-specific cholinesterase activity in their membrane '(Kudo, 1976, 1978, 1983b)'.

Around the beginning of the 20th century it was realized that if unfertilized eggs are pricked in the absence of calcium in the water, the cortical reaction does not take place. 'Dalcq', in the 1920s, postulated that the signal which activates eggs is an increase in the concentration of free calcium in the egg cytoplasm (cf. Gilkey, 1981). Two decades later it was confirmed that fertilization is not possible in calcium-free Ringer solution though the sperms are active '(T. Yamamoto, 1939, 1944a; Rothschild, 1958)'. 'Ridgway *et al.* (1977)' and 'Gilkey *et al.* (1978)' were the first to demonstrate that a transient rise in free calcium triggers activation of *Oryzias* eggs.

The luminescence of the photoprotein aequorin micro-injected into eggs made it possible to record transient increases of cytoplasmic Ca^{2+} upon fertilization. 'Gilkey *et al.* (1978)' observed a free calcium wave, initiated at the site of sperm entry, traversing the activating egg of the medaka, *Oryzias latipes*. They further showed that propagation of the wave is brought about by a process of 'calcium-stimulated calcium release'. In other words, a chain reaction is precipitated by the sperm. It was reported that the peak of free calcium does not diminish as the wave traverses the egg and that a calcium wave may be initiated anywhere in the peripheral cytoplasm by ionophore treatment. The wave of Ca^{2+}-release precedes the wave of the breakdown of CA and it was suggested that it is the increase in free calcium in the cortex which triggers the cortical reaction, i.e. disintegration of CA. In the words of 'T. Yamamoto (1961, 1962)' 'the presence of Ca ions is a sine qua non for fertilization'. He quoted 'Hamano (1949)' who demonstrated that also in the salmon *Oncorhynchus keta* Ca ions are indispensable for fertilization and he showed that that Ca ions dissolved from the cortex of the ovum are sufficient to ensure fertilization. He further quoted 'Yanagimachi and Kanoh (1953)' who showed in the herring, *Clupea pallasii*, that no fertilization occurs in Ca-Mg-free solution and that the activity of the sperms is inhibited around the micropyle.

'Gilkey *et al.* (1978)' suggested that leakage of external calcium into the medaka egg triggers the release of internally sequestered calcium and according to 'Lessman and Huver (1981)' the resulting high free calcium level in the egg triggers activation. 'Yoshimoto *et al.* (1986)' confirmed the results of 'Gilkey *et al.* (1978)' on the medaka. They stressed that 'all the characteristics of the Ca^{2+} wave strongly suggest that its propagation is due to an active process involving Ca^{2+}-stimulated Ca^{2+}-release rather than to passive diffusion of Ca^{2+}'. They showed that in the medaka the leading edge of the calcium transient just precedes CA fusion with the cell membrane. No fusion occurs in regions which the calcium does not enter. Upon fusion, the alveoli discharge their contents into the perivitelline space.

According to 'Nuccitelli (1987)' calcium in the medaka is released through the action of the membrane factor IP3 (inositol 1,4,5-triphosphate). Subsequently, 'Iwamatsu *et al.* (1988a,b)' reported that microinjection of Ca^{2+}, Sr^{2+}, Br^{2+}, and cGMP (cyclic guanosine 5'-monophosphate) into aequorin-loaded oocytes of *Oryzias* brought about a delay in the initiation of the calcium wave, while microinjection of IP3 (inositol 1,4,5-triphosphate) or the calcium ionophore A23187 induced calcium release without a time lag. They concluded that calcium may not be the only regulatory signal of calcium release and that cytoplasmic Ca^{2+} induces Ca^{2+} release from the cytoplasmic stores in an indirect way, probably via IP3.

Another study was undertaken to examine with the ionophore A23187 (which exhibits a high specificity for calcium over other divalent ions) if the cortical granule exocytosis is self-propagating around the egg of *Brachydanio rerio* '(Schalkoff and Hart, 1986)'. The results revealed that the CA breakdown is partial and restricted to less than 50% of the egg surface. The complete exocytosis appears to require the stimulation and release of calcium from multiple sites over the cortex. This agrees with earlier results on *Brachydanio rerio*, which showed that calcium-induced exocytosis of cortical granules does not start at one point to spread in a wave, but rather is initiated randomly over the egg surface more or less simultaneously '(Hart and Yu, 1980; Schalkoff and Hart, 1986; Hart, 1990; Becker and Hart, 1999)'. It was concluded that some factor(s) other than calcium is/are needed to coordinate exocytosis.

Data from SDS-PAGE, immunoblotting and fluorescence microscopy revealed the presence of filamentous and non-filamentous actin and non-muscle myosin (myosin II) in the cortex of the unfertilized zebrafish egg. A meshwork of filamentous actin was restricted to the microplicae and the cytoplasmic face of the egg membrane. In the deeper region of the cortex, which contains the cortical granules, non-filamentous actin was sequestered (see Chapter 3). Myosin II co-localized with both types of actin. Possible functions of these proteins before and after fertilization were discussed by 'Becker and Hart (1996)'. Subsequently, the same authors showed that in *Brachydanio* eggs the cortical granules are released non-synchronously over the whole surface. They tested the hypothesis that filamentous actin presents a physical barrier to exocytosis. They showed that cytochalasin treatment, which depolymerizes actin filaments, accelerates the cortical reaction. In contrast, phalloidin, which stabilizes actin filaments, inhibits the cortical reaction. In other words: actin blocks exocytosis. The authors further suggested that the remodelling of the cell surface (i.e. closure of cortical crypts and restoration of the plasma membrane) following the cortical reaction appears to be driven by an actomyosin contractile ring '(Becker and Hart, 1999)'.

'Iwamatsu (1998)' followed the movements of cortex, inclusions and pronuclei in the newly fertilized *Oryzias latipes*. The cortical reaction was followed by a cortical contraction towards the animal pole. The oil-droplets were moved towards the animal pole and their movement was twisted in correlation with the unilateral bending of the non-attaching filaments (Chapter 6).

9.11 THE PERIVITELLINE SPACE FOLLOWING FERTILIZATION

As mentioned above, the release of the cortico-alveolar contents is followed by elevation of the envelope from the underlying egg surface and the creation of the so-called perivitelline space. These processes start at the animal pole and progress wavelike over the whole egg (Figs. 9/8, 9/10A,B). Already the early investigators noted that the space is filled with a fluid medium containing yolk or protein in solution and is formed as a result of the elevation of the egg envelope and maybe 'contraction of the yolk'. It develops first in a ring around the micropyle as the centre and gradually extends to the opposite pole of the egg '(Aubert, 1854; Ransom, 1854; Reichert, 1856; Lereboullet, 1861; Truman, 1869; His, 1873; Hoffmann, 1881; Ryder, 1884 and Cunningham, 1887)'. 'Ransom (1854)' had already noticed that, while 'the respiratory chamber' [perivitelline space] is in the process of formation, 'the yellow drops of the superficial layer [= CA] grow pale and disappear'. Many of the early investigators spoke of a 'contraction of the yolk' as a reason for the formation of the space. Some suggested that the fluid is due to water intake but others mentioned also the presence of a 'protein-containing substance'. 'Kupffer (1878) observed that in the egg of the herring, *Clupea harengus*, the CA disappear during the formation of the perivitelline space. 'Impregnation seems to be necessary in some species before any water can be absorbed; this is especially the case with the ovary of the shad [*Clupea sapidissima*] but in a lesser degree necessary in the case of the cod [*Gadus morhua*]' '(Ryder, 1884)'. 'List (1887)' observed that 45 minutes after fertilization, contraction of the 'contents of the egg' are observed creating the perivitelline space. 'His (1873)' observed changes in the shape of the blastodisc in the unfertilized egg of the pike, *Esox*. His drawing of the egg immediately after fertilization shows a perivitelline space, which he mistakenly thought was caused by the intake of water (Fig. 9/12). One hundred years later it was argued that the formation of the perivitelline space in teleosts is partly due to a decrease in the volume of the ooplasm and partly due to

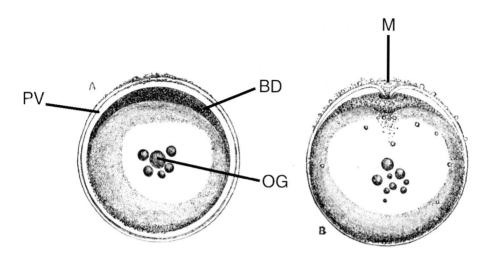

Fig. 9/10 Formation of the perivitelline space after fertilization in the egg of *Gasterosteus* (Ransom, 1854). A. 8-10 min. after fertilization 'a clear respiratory space' (perivitelline space) surrounds the whole egg. The blastodisc indicates the site of the animal pole. B. 3 min. after fertilization. The lifting of the envelope has taken place arounjd the whole egg. BD - blastodisc; M - micropyle; OG - oil globules; PV - perivitelline space.

an expansion of the envelope caused by increased osmotic pressure in the space, the result of some colloidal substance released by the erupting CA ['T. Yamamoto (1939, 1944a), Thomopoulos (1953), Nakano (1956), Ohtsuka (1957, 1960), 'Rothschild (1958), Armstrong and Child (1965)], Luther (1966), Iwamatsu (1969), Iwamatsu and Ohta (1976)']. 'Lessman and Huver (1981)' concluded that the decrease of the egg diameter in *Catastomus* which they observed, as well as the data for *Brachydanio* reported by 'Roosen-Runge (1938)' and 'Hart et al. (1977)', are due to the cortical reaction.

The thickness of the perivitelline space has been used for the quantification of the cortical reaction in the carp, *Cyprinus carpio*, by 'Renard et al. (1990)'. The space elevates progressively from 0.05 millimetre within 12 minutes after immersion in freshwater and after 18 minutes in saline solution (150 milliosmoles per liter).

While the unactivated salmonid oocyte is flaccid (internal hydrostatic pressure at or near 0 mm Hg), the release of CA colloids upon activation causes a net flow of water into the egg, i.e. into the perivitelline space. A hydrostatic pressure for salmonids between 30 and 50 mm Hg is brought about by a pressure difference across the membrane of 50-60 mm Hg '(Groot and Alderdice, 1985)'.

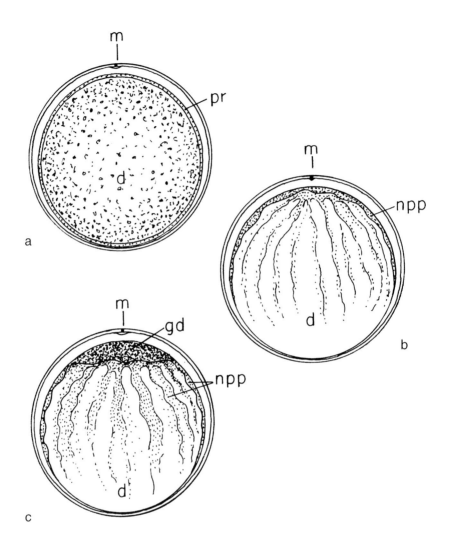

Fig. 9/11 Formation of blastodisc and perivitelline space following fertilization of the egg of *Gadus morrhua*, the cod (Ryder, 1884). The eggs float with the apical pole facing downwards. However, the diagrams are shown with the apical pole upwards for easier comparison with other eggs. × 55 a) mature unfertilized egg. b) 1½ h. after fertilization. The *cytoplasm* travels in beaded streams towards the apical pole. c - 3 h. 40 min. after fertilization. The blastodisc is defined and the *cytoplasm* continues to stream apically. d - yolk; gd - blastodisc; m - micropyle; npp - nodose protoplasmic processes; pr - *cytoplasm*; pr' - cortex.

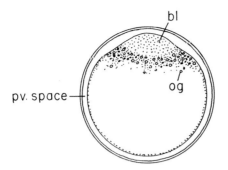

Fig. 9/12 Egg of pike immediately after fertilization (His, 1873) × 17. bl - blastodisc; og - oil globules; pv. space - perivitelline space.

9.12 HARDENING OF THE ENVELOPE

Envelopes of teleostean eggs are known not only to be elevated but also to harden when transferred to water '(T. Yamamoto, 1961; Blaxter, 1969; Ginzburg, 1972 and Lönning et al., 1984)'. As a result of the cortical reaction (exocytosis), the ejected contents of the alveoli are deposited on the inner surface of the envelope. Therefore, the soft, and often fragile, egg envelope of most fishes becomes changed into a strong and tough structure. This process is referred to as 'hardening'. This term is not exact in the strict sense of the word, since the changed envelope displays toughness rather than hardness (It is mechanically non-elastic and chemically resistant.). The hardening of the envelope is due to the strengthening of the ZR interna '(Nakano, 1956; Ohtsuka, 1957, 1960; Yamamoto, 1961 and Cotelli et al., 1988)'. 'Zotin (1958)' reasoned that a special enzyme responsible for the hardening is released from the egg surface. The hardening enzyme is unstable and its effect is only necessary during the first 6-7 minutes following fertilization. From then onwards hardening progresses equally well in the absence of the enzyme. However, calcium ions are necessary for the accomplishment of the enzyme action. 'Iwamatsu et al. (1995)' propose that in Oryzias latipes hardening enzyme(s) is/are released from the cortex into the perivitelline space during the cortical reaction. 'Zotin (1958)' measured the increase in envelope toughness of salmonids and 'Pommeranz (1974)' that of the plaice Pleuronectes platessa a short time after fertilization. The thickness of the membrane first decreases and then increases. The forces endured by the envelope of the plaice increased from 1.5 grams to 700 grams, that of the whitefish Coregonus lavaretus from 40-50 grams to 1 800-2 200 grams and that of the trout and salmon from 160-280 grams to 2 500-3 500 grams. It has further been suggested that the process underlying envelope hardening involves an oxidation process. A phospholipid originating from the cortical alveoli was thought to be of prime importance '(Ohtsuka, 1957)'. However, 'Nakano (1956)' had observed that non-oxidative substances such as gum and gelatin are capable of hardening the envelope. 'Ohtsuka (1960)' suggested that hardening is due to a combination of SH groups with the aldehydes produced by oxidation of alpha-glycol groups. Calcium ions have been shown to be of great importance in envelope hardening in *Salvelinus fontinalis* and *Oncorhynchus keta* '(Ohtsuka, 1960; Oppen-Berntsen et al., 1990)'. Hardening of the envelope is followed by a reduction in its thickness '(Nakano, 1956)'. 'Lönning et al. (1984)' and 'Davenport et al. (1986)' reported that in the plaice, *Pleuronectes platessa*, (which lays pelagic eggs) and lumpsucker, *Cyclopterus lumpus*, (with demersal eggs brooded by the male parent) the resistance of the envelope to tearing increases greatly after egg activation. By 24 hours after activation the envelopes of the latter required about 3 times as much force to tear them as those of plaice (107.5 grams versus 35.2 grams). 'Hagenmaier et al. (1976)' (cf. Davenport et al., 1986) found that Glu-Lys peptide appeared in the envelopes of *Salmo gairdneri* eggs after fertilization which indicates the formation of cross linkages by isopeptide bonds in the proteins of the envelope.

Extracts from envelopes of artificially activated eggs of *Salmo gairdneri* were found to exert an action on the bacteria *Acromonas hydrophila* or *Vibrio anguillarum*. The main source of the extracts was the outermost layer, as confirmed by EM and agglutination tests of fish sperm or human B-type erythrocytes. It has also been suggested that the envelope of the fertilized egg has antifungal or fungicidal activity against the fungus that is the major cause of mycosis in fish and fish eggs '(Tunru and Yokoyama, 1992)'.

A review of the hardening of the envelope (and of other processes regarding the envelope) has been presented by 'Yamagami et al. (1992)'.

Summary

The formation of the perivitelline space, due to the cortical reaction, and bipolar differentiation are the most conspicuous results of fertilization. During the cortical reaction the breakdown of the cortical alveoli (CA) is closely accompanied by the release of their colloid content and spherical granules. The envelope of the egg becomes elevated and encloses the perivitelline space, which is the result of the colloid osmotic pressure of the substance discharged by the CA. The role of calcium in this process has been stressed and its origin debated. Hardening of the envelope is observed parallel with this process. The hardening is calcium-dependent and the presence of a so-called 'hardening enzyme' has been put forward as the cause. Formation of the perivitelline space and hardening of the envelope are also involved in the prevention of polyspermy though this action is deemed to be secondary. (9.10-9.12)

9.13 BIPOLAR DISTRIBUTION (OOPLASMIC SEGREGATION) FOLLOWING FERTILIZATION

As mentioned earlier (9.3) in some teleosts the cytoplasm permeating the yolk streams to the animal pole before encountering the sperm. However, in others, fertilization seems to be a prerequisite for bipolar differentiation.

'Coste (1850)' described that in the adhesive eggs of the herring, *Clupea harengus*, the 'plastic elements' (cytoplasm) and the 'nutritive elements' (yolk) are mixed to the moment of oviposition and will separate only at fertilization. This was confirmed by 'Brook (1887a)'.

The distinction between the distribution of cytoplasm and yolk is also reflected in the type of developmental site of the ova, according to 'van Bambeke (1876)', 'Kupffer (1878)' and 'Ryder (1884)'. In the adhesive ova on the sea floor e.g. the herring, *Clupea harengus*) the yolk consists of a large number of spherules and the germinal cytoplasm distributed as a network between them, whereas in the pelagic group of ova (for example, *Trachinus* and Gadidae) the cytoplasm usually forms a comparatively-even, superficial layer around the large yolk sphere. The Salmonidae take up a somewhat intermediate position between the adhesive and pelagic types '(Brook, 1887)'. The thickened cytoplasm at the animal pole is referred to generally as blastodisc, a term introduced by 'Haeckel (1855)'.

'Ryder (1884)' observed in the pelagic eggs of the cod, *Gadus morhua*, waves of contraction and his illustration shows 'one and a half hour after impregnation the protoplasmic layer travelling in beaded streams towards the lower pole in an amoeboid manner' (cod eggs float with the animal pole downwards.) (Figs. 9/11a-c). The germinal disc is defined 3 hours and 40 minutes after fertilization. Streaming of superficial cytoplasm after fertilization has been described also for the pelagic eggs of *Serranus atrarius* by 'Wilson (1889/1891)'.

'Ransom (1855, 1867)' observed in the eggs of *Gasterosteus leiurus* and *G. pungitius* that the 'germinal disk' has visibly increased in bulk 3 minutes after fecundation but probably begins to concentrate much sooner. However, he failed to see any distinct movement of the granules of the cortex towards the animal pole. Contractions of the yolk began 15-20 minutes after fertilization. 'Swarup (1958)' observed in the same species waves of contraction to start about 30 minutes after fertilization. These waves begin at the vegetal pole, pass through the equator and end at the animal pole where the accumulation of cytoplasm yields the blastodisc.

'Owsiannikow (1873)' described the contraction of the cytoplasm and the fusion of oil-droplets after fertilization in *Coregonus lavaretus*. 'Mahon and Hoar (1956)' showed that the blastodisc in the egg of *Oncorhynchus keta* grows progressively thicker after fertilization and within a few hours presents a button-like appearance as described by 'Wilson (1927)' for Coregonus. In the Salmonidae, both before segmentation commences and throughout this process, a comparatively thick cortical layer of cytoplasm is found, which is intimately connected with the yolk.

'Hoffmann (1881)' and 'Behrens (1898)' reported in detail the bipolar distribution that takes place in the trout, *Trutta fario* and *Trutta iridea*, after fertilization. 'Lereboullet (1861)' and 'Devillers (1956, 1961)' stress that fertilization is a prerequisite for blastodisc formation in the trout (*Salmo fario* and *Salmo irideus*). [In contrast, 'His (1873)' maintained that in *Salmo salar* the blastodisc is already present before fertilization]. In the trout, *Salmo irideus*, the formation of the blastodisc has been shown to be transitorily inhibited by KCN '(Devillers, Domurat and Colas, 1959)'.

'Von Kowalewski (1886a)' noticed that hand in hand with the cytoplasmic streaming towards the animal pole the yolk gradually becomes transparent. He suggested that the observation by 'Kupffer (1878)' of the lightening of the egg of the herring, and similar observations of other fish eggs, are due to the same process, i.e. cytoplasmic streaming. 'Fusari (1892)' observed that the eggs of *Cristiceps argentatus* become transparent after being in water for a certain time. He gives as a reason the separation of ooplasm and yolk, which heretofore were intermingled. 'Von Kowalewski (1886a,b)' observed in the goldfish that, in contrast to all other analyzed fish, concentration of the cytoplasm towards the animal pole continues to the end of cleavage.

'Roosen-Runge (1938)' described that in the demersal eggs of the zebrafish *Brachydanio rerio*, as in *Carassius* and *Alosa*, the cytoplasm is mixed with the yolk. Subsequent studies with time-lapse cinemicrography of *Brachydanio rerio* revealed cytoplasmic streaming after fertilization '[Lewis and Roosen-Runge (1942), Huver (1962, 1964) and Beams et al. (1985)]'. The same method was also used to observe the movements of cytoplasm following activation in the egg of *Salmo* by 'Wülker (1953)', 'Thomopoulos (1954)' and 'Abraham et al. (1993)'.

'Oppenheimer (1937)' observed in *Fundulus reticulatus* that at fertilization the yolk platelets flow together and the yolk becomes crystal clear. She concluded that in the unfertilized egg the cytoplasm is probably mingled to some extent with the whole yolk, though concentrated at its surface. 'Nelsen (1952)', in his textbook, stated that the egg of *Fundulus heteroclitus* is centrolecithal before and strongly telolecithal after fertilization. 'Brummett and Dumont (1981)' report that in the unfertilized egg the cytoplasm is a surface layer only and that the blastodisc increases after fertilization, though they did not observe any streaming: 'The flow of the cortically distributed cytoplasm to form a typical blastodisc at the animal pole seems to occur only after normal cortical vesicle breakdown. But whether a causal relationship exists between cortical vesicle breakdown and blastodisc formation in this species has not been established experimentally'.

Bipolar differentiation in the demersal eggs of the medaka, *Oryzias latipes*, takes place immediately after the breakdown of the cortical alveoli. '(T. Yamamoto, 1944, 1958)'. The cytoplasm accumulates at the animal pole, takes up a lens shape and ultimately bulges out as the blastodisc. If the sperm enters at a site other than the animal pole, the real blastodisc is still formed at the animal pole. This means that the sperm has to travel a longer distance than normally '(Sakai, 1961, 1964a,b)'. Wavelike contractions towards the animal pole (running 1 minute behind the calcium wave) were observed in the egg of the medaka before the onset of the first cleavage ['Iwamatsu (1973)'; 'Gilkey *et al.* (1978)'; 'Gilkey 1981)']. While the cytoplasm moves towards the animal pole, the oil droplets accumulate at the vegetal pole ['Sakai (1961, 1964a,b)'; 'Yamamoto (1961)'; 'Iwamatsu (1965,)']. It has been stressed that the movement of oil droplets in a given region is dependent on the streaming of the cortical cytoplasm in that region '(Sakai, 1961, 1965)'. Carbon particles injected into the cytoplasm of medaka eggs move to the animal pole parallel with the plasma membrane while the innermost particles move toward the vegetal pole. No stratification of the cytoplasm was evident with light microscopy '(Gilkey, 1981)'. The same author reported the observations of saltatory movements of natural cytoplasmic inclusions superimposed on a bulk flow.

'Abraham *et al.* (1993)' observed cytoplasmic streaming and blastodisc formation in the eggs of the medaka by time-lapse cinemicrography. They observed two movements: The first consisted of the movement, called streaming, of many inclusions (diameter 1.5 to 11 microns) towards the animal pole at a speed of 2.2 microns/minute^{-1}. The second type of movement was saltatory and was faster (about 44 microns/minute^{-1}. The particles moved towards the vegetal pole; their size was given as 1.0 microns in diameter. A third type of motion towards the vegetal pole involved the oil droplets. All three movements began only after a strong contraction of the cytoplasm towards the animal pole, which, at 25°C, began 10-12 minutes after fertilization and <3 minutes after formation of the second polar body. Microtubular poisons such as colchicine, colcemid, or nocodazole, inhibited the movement of oil droplets and saltatory motion towards the vegetal pole and slowed the growth of the blastodisc.

'Webb *et al.*, (1995)' observed cytoplasmic streaming in the *Oryzias latipes* egg with time-lapse video microscopy. Cytochalasin D (which has an influence on microfilaments) inhibited both cytoplasmic streaming towards the animal pole and the formation of the blastodisc. Formation of the blastodisc in *Brachydanio rerio* and *Misgurnus fossilis* was equally affected by cytochalasin '(Ivanenkov *et al.*, 1987)'. It has been suggested that in the zebrafish an array of microtubules in the yolk transports substances from the vegetal region to the blastoderm since cold treatment, UV irradiation and microtubule poison destroy this array '(Strahle and Jesuthasan, 1993; Jesuthasan and Strahle, 1997)'.

Chlorides reversibly inhibited the formation of the blastodisc in the egg of *Salmo*, which would be indicative of an osmotic response '(Devillers, 1961)'. However, addition of salts did not result in inhibition in the eggs of various freshwater fish ['T.Yamamoto (1939, 1954)'; 'Thomopoulos (1953)'; 'Luther (1966)'].

Cytochalasins, which are actin inhibitors, prevented the formation of the blastodisc in eggs of the medaka, *Oryzias latipes*.Therefore, microtubules as well as microfilaments may be involved in the ooplasmic segregation. Cytosolic Ca^{2+} is elevated at the animal and vegetal poles of the medaka egg during ooplasmic segregation and weak calcium buffer prevented the formation of the blastodisc ['Fluck *et al.*, 1991)'; 'Abraham *et al.* (1993)'].

Summary

Bipolar differentiation, also called 'ooplasmic segregation', refers to the movement of cytoplasm towards the animal pole in response to activation. The movement has been described as streaming or bulk flow.

In demersal ova the cytoplasm is intermingled with the yolk and upon fertilization streams towards the apical pole leaving behind the yolk surrounded by a thin cytoplasmic cortex. In pelagic ova the cytoplasm is located as a relatively thick superficial layer on the large yolk sphere and upon fertilization moves in bulk towards the animal pole. The eggs of Salmonidae take up an intermediate position.

More recent cinemicrographical observations on the demersal egg of medaka revealed a streaming of inclusions towards the animal pole while a second type of movement of innermost particles was saltatory, faster and was directed towards the vegetal pole. A third type of movement towards the vegetal pole involved oil droplets. All these movements were preceded by a cytoplasmic contraction towards the animal pole begun after fertilization. Microtubular poisons inhibited movements of oil droplets and saltatory movements and slowed the increase of the blastodisc, while microfilament-inhibitors prevented a blastodisc from forming. (9.13)

9.14 FORMATION OF PRONUCLEI[1]

Some of the early embryologists called the female pronucleus the 'egg nucleus' (G. Eikern) and the male pronucleus the 'sperm nucleus' (G. Spermakern).'Kupffer (1868)' observed the 'conjugation' of two pronuclei but could not distinguish the male from the female pronucleus. 'His' observations were begun in summer of 1883 and completed in winter of 1885/86, essentially on *Ctenolabrus*. He compared his observations with those of at least a dozen different species, which allowed him to 'affirm with the utmost positiveness that the story of one is the story of all'.

'Hoffmann (1881)' observed in *Crenilabrus pavo*, that 7 minutes after the sperm has entered, the polar body moved through the micropyle towards the outside. Ten minutes later, the author noticed two small nuclei: one lying near the opening of the micropyle as the sperm pronucleus, the other more deeply situated as the female pronucleus. In *Scorpaena* and *Julis*, the two pronuclei moved towards each other and met 35-40 minutes later. Fusion occurred after 45 minutes. 'Behrens (1898)' described the movements and fusion of the two pronuclei in the trout, *Trutta fario* and *Trutta iridea*.

In *Oryzias latipes* the encounter of the two pronuclei follows immediately after sperm penetration, with the egg pronucleus located already under the micropyle '(Sakai, 1961; Iwamatsu, 1998)'. The sperm nucleus begins to decondense while vesiculation of its nuclear envelope occurs. The decondensed chromatin is soon surrounded by smooth-surfaced vesicles which fuse and form the double membrane of the male pronucleus. The male pronucleus is larger than the sperm nucleus and the membranes necessary to surround the male pronucleus have to be supplied by the egg. The most likely source is the endoplasmic reticulum, which is known to connect with the nuclear envelope in other cells ['Iwamatsu and Ohta (1978)'; 'Lessman and Huver (1981)'; 'Ohta and Iwamatsu (1983)'; 'Ohta (1985)'; 'Brummet et al. (1985)'; 'Hart (1990)'].

Several observations indicate that the pattern of nuclear envelope breakdown is highly regulated. For example, disintegration of the sperm nuclear envelope in the medaka begins in the apical portion of the sperm head and progresses towards the basal end '(Iwamatsu and Ohta, 1978)'. By contrast, in *Fundulus* the fragmentation of the nuclear envelope begins in the indented, posterior portion of the sperm head (nuclear fossa) in the vicinity of the centrioles '(Brummet et al., 1985)'. Fragmentation starts within 3-5 minutes of insemination in *Oryzias* and *Rhodeus*, and even as early as 20-60 seconds after sperm entry in some other species ['Iwamatsu and Ohta (1978)'; 'Gilkey (1981)'; 'Ohta and Iwamatsu (1983)'; 'Ohta, (1985)'; 'Brummet et al. (1985)']. It proceeds more rapidly in *Fundulus* than in *Oryzias*. In the case of the latter, the pattern of chromatin dispersion closely follows the pattern of nuclear envelope fragmentation, i.e. it begins at the apical end of the sperm nucleus and includes swelling and increase in the volume of chromatin material. 'Hart (1990)' proposes that the agents responsible for chromatin dispersion probably originate in the egg cytoplasm. Transformation of the sperm nucleus into the male pronucleus is completed by 15 minutes in *Rhodeus* '(Ohta, 1985)' and by 25 minutes in *Oryzias* '(Iwamatsu and Ohta, 1981)'.

Details of the morphogenesis of the female pronucleus in teleosts and the processes leading to the fusion of the male and female pronucleus to form the zygote nucleus are less well known '(Hart, 1990)'. In the medaka egg the second meiotic division is completed and the polar body expelled within the first few minutes after fertilization '(Iwamatsu, 1965)'.

9.15 FORMATION OF POLAR BODIES [9]

The polar bodies are the result of unequal divisions of the oocyte. The first polar body is formed after the first meiotic division of the immature oocyte. The egg remains arrested in metaphase of the second maturation until it is activated and will give off the second polar body.

'Blanc (1894)' and 'Behrens (1898)' observed the formation of the second polar body in the trout, *Trutta fario* (Figs. 9/13a,b). 'Huver (1960)' described the formation of the polar bodies in *Fundulus heteroclitus*, based on time-lapse microcinematography and, unaware of the publication by 'Behrens (1898)', he maintained that, for the first time, both polar bodies were observed. 'They project into the perivitelline space from the rim of the funnel-shaped mouth of the micropyle'. Both polar bodies were of equal size (0.04 millimetres in diameter).

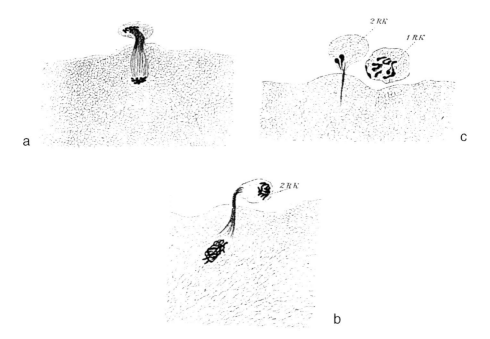

Fig. 9/13 Polar bodies of the trout (*Trutta fario*) egg (Behrens, 1898) × 1 000. **a** - Formation of second polar body. **b** - Second polar body is about to be given off; pro*nucleus* in formation. **c** - First and second polar body. 1 RK first polar body 2 RK second polar body.

SEM analysis of the egg of the medaka, *Oryzias latipes*, revealed the first polar body on the surface of the egg at the time of fertilization. It was located ~50-100 microns from the centre of the sperm entry site ['Hart and Donovan (1983)'; 'Brummet *et al.* (1985)'].

'Ohta (1986)' observed by SEM and TEM the formation of the scond polar body both in intact and 'dechorionated' eggs of *Rhodeus ocellatus*. At the metaphase II stage the site of the extrusion of the second polar body is morphologically specialized and is identified as an aggregation of thick leaf-like projections of the egg cytoplasm. The site is near the sperm entry site characterized by microvilli and by the absence of cortical alveoli. The formation of the second polar body is not sensitive to cytochalasin B. Its nucleus is surrounded by a nuclear envelope. Complete extrusion of the second polar body may take as long as 30 minutes in the rose bitterling. Similarly, the second polar body in the egg of *Fundulus heteroclitus* emerged 4-5 minutes after insemination at a distance of 70 microns from the SES; it was discoidal in shape and had a diameter of 8 microns '(Brummet *et al.*, 1985)'. The second polar body in the zebrafish egg began to form within 4-10 minutes of insemination at a distance of 10-15 microns from the SES. Water-activated unfertilized eggs also produced a fully formed second polar body at their animal pole. This would indicate that sperm binding and/or fusion is not necessary for its formation '(Wolenski and Hart, 1987)'.

Summary

The male pronuclear development involves the breakdown and fragmentation of the sperm nuclear envelope, dispersion of the chromatin and the development of the pronuclear envelope. The pronucleus is larger than the sperm nucleus and additional membrane for the former has to be supplied.

The egg becomes arrested at metaphase II of meiosis at which time the first polar body appears on the egg surface. The second polar body is formed later, i.e. after fertilization.

Endnotes

[1] G. Vorkern
[2] see endnote[2] chapter 10.
[3] Division resulting in two cells with only one half of the chromosome number. From Gk. meiosis = lessening.
[4] Gk. poly = many, ploos = equivalent. Denotes more than normal number of chromosomes.
[5] yolk accumulated at vegetal pole. From Gk. telos = end and lekithos = yolk.
[6] Other synonyms of perivitelline space: sub-zonal space by 'Reighard (1890)'; sub-chorionic space by 'Armstrong and Child (1962)'; water-space by 'Ryder (1883, 1887)'; breathing cavity by 'Ransom (1868)'; 'Kingsley and Conn, (1883)'; 'Brook (1885)'.
F. chambre claire of Henneguy (1888).
I. spazio perivitellino of 'Raffaele (1888)'
[7] see endnote[1] of Chapter 10.
[8] cortical vacuoles, vacuolar yolk, intravacuolar yolk, intravesicular yolk, vacuomes, yolk vesicles, carbohydrate yolk
[9] G. Richtungskörper, Polzelle, Polocyte
F. globule polaire

Chapter ten

Cleavage and formation of periblast

'Omnis cellula e cellula et omnis nucleus e nucleo.'
Strasburger (1881)

In the fertilized egg of teleosts cytoplasm is accumulated at the animal pole as the blastodisc[1], which houses the nucleus[2]. This was first described by 'Coste (1850)'. In some teleostean species the streaming of cytoplasm towards the blastodisc is said to continue after fertilization, e.g. in *Gasterosteus pungitius* '(Ransom, 1867)', the pike '(Truman, 1869)', *Salmo fario* '(Oellacher, 1872)', *Gadus morrhua*, salmonoids, clupeoids and cyprinodonts '(Ryder, 1884)', *Crenilabrus rostratus, C. pavo* and *Tinca vulgaris* '(Janosik, 1885)', *Salmo salar, Cyprinus carpio, Gadus morrhua* and *Rhodeus amarus* '(Reis, 1910)', *Brachydanio rerio* '(Thomas and Waterman, 1978)'. This 'late arriving cytoplasm' contributes also to the thickened edge, the so-called annulus[3], as has been shown by the movements of injected chalk particles '(Long, 1980)'. The blastodisc is facing downwards in most pelagic and upwards in demersal eggs. At the edge the blastodisc diminishes into a very thin layer of cytoplasm (YCL) which surrounds the whole of the yolk[4].

The cell divisions (mitoses) of the zygote are referred to as cleavage[5] until the blastula stage has been reached and the resulting cells are known as blastomeres. Since in teleosts cleavage is restricted to the blastodisc and the yolk remains unsegmented, the cleavage is called partial or meroblastic[6] (first described by 'Rusconi, 1836' and 'Vogt, 1842'). (Cleavage is called total or holoblastic[7] when there is a complete segmentation of the entire egg, as observed e.g. in the amphibia.)

A change from the holoblastic to the meroblastic condition is demonstrated in comparing chondrostei (*Acipenser*) and holostei (*Amia and Lepidosteus*) with the teleostei (Fig. 10/1). *Acipenser* is characterized by a holoblastic unequal condition (similar to the amphibia) while *Amia* displays a transition from holo- to meroblastic cleavage and, interestingly also the development of a periblast. In *Lepidosteus* the cleavage pattern is partial with only superficial clefts in the vegetal hemisphere.

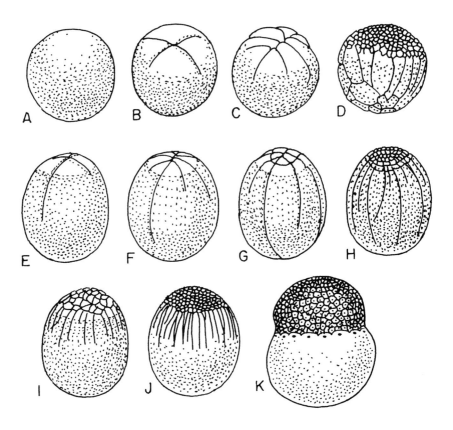

Fig. 10/1 Cleavage from holoblastic to meroblastic type. **A-D** - *Acipenser* (redrawn after Dean) — ***chondrostei***; **E-H** - *Amia calva* (redrawn after Lutman and Eycleshymer) — ***holostei***; **I,J** - *Lepidosteus osseus* (redrawn after Eycleshymer) — ***holostei***; **K** - *Crenilabrus pavo* (redrawn after Kopsch) — ***teleostei***.

In contrast to other meroblastic eggs, such as those of birds and reptiles, the cleavage pattern in teleosts displays great regularity. The mitotic spindle of the dividing nucleus becomes manifest first. This is followed by a furrow (groove)[8] — at right angles to the mitotic spindle — first superficial at the animal side of the cell surface, then penetrating almost to the base (cytotomy) but not perfectly bisecting the disc, as already 'Aubert (1854)' and 'Lereboullet (1862)' had noticed. This process is known as cytokinesis. 'Rusconi (1836)' was the first to describe the formation of cleavage furrows at right angles to each other up to the 32 cell stage (*Cyprinus tinca*). In all teleosts investigated the first few cleavages are meridional, yielding a one-layered blastodisc. This is followed by a horizontal cleavage resulting in a bilayered disc. The upper blastomeres (in demersal eggs) are set off completely from the yolk while the lower ones keep sitting on it. It has been argued that the time of the appearance of the horizontal cleavage is dictated by the time of the total withdrawal of the cytoplasm into the blastodisc. For instance, in *Trutta fario* the horizontal cleavage occurs after the fourth division '(Kopsch, 1911)', or sometimes the third '(Hoffmann, 1888)', in *Belone acus* after the seventh '(Kopsch, 1911)' and in *Cristiceps* after the ninth '(Fusari, 1892)'. The egg of the ice goby *Leucopsarion petersii* seems to be unusual in that already the third cleavage is horizontal '(Nakatsjui et al., 1997)'. In the late blastodisc two cell-types can be distinguished: 1) a cellular envelope called EVL[9], covering 2) the deep cells (DC)[10] destined to form the embryo.

A bilateral arrangement of the blastomeres during the early stages was observed in *Ctenoblabrus* and *Serranus* by 'Agassiz and Whitman (1884)' and 'Wilson (1889/1891)', respectively, which lead to the supposition that there exists some causal relation between this bilaterality and that of the adult body. 'Von Kowalewski (1886a)' observed bilaterality already at the 4-cell stage of *Carassius auratus*. However, 'Clapp (1891)' showed in the egg of the toad-fish, *Batrachus tau*, that no such relation exists. Subsequently, 'Morgan (1893)' tested 22 marked eggs of *Ctenolabrus* and *Serranus* and found no relation.

Irregularities in the cleavage pattern have also been observed, e.g. in *Perca fluviatilis*, from the 16 cell stage onwards '(Chevey, 1925)'. It should be stressed that also minor differences, even between the eggs of the same batch (intraspecific differences), occur as pointed out already by 'Kingsley and Conn (1883)'. A relationship between cell diversification and intercleavage time was not observed in the annual[11] fish *Nothobranchius guentheri*' '(van Haarlem et al., 1983)'.

The fact that even the unfertilized eggs can undergo cleavage has been reported by 'Burnett and Agassiz (1857)' and by 'Boeck (1871)' for the herring (cf. 'Kupffer, 1878) and by 'Ransom (1867)' for the stickleback, *Gasterosteus pungitius*. A segmentation cavity[12] was described by many early investigators and its existence denied by others. According to 'Wilson (1889/1891)' it later changes into the subgerminal cavity [13], which is situated between the periblast and the overlying blastomeres, in an eccentric position. His observations on *Serranus atrarius* were confirmed for the common trout, *Salmo fario*, by 'Oellacher (1873)' and 'Klein (1876)'. The subgerminal cavity was also described in later publications e.g. by 'Price (1934a,b)' for *Coregonus clupeaformis*, by 'Orsi (1968)' for the sole *Parophrys vetulus*, by 'Verma (1970)' for *Cyprinus carpio* and by 'Parihar (1979)' for the Indian carp *Cirrhinus mrigala*.

While at the beginning of cleavage all blastomeres are basally continuous with the thin cytoplasmic layer surrounding the yolk (YCL), later on in many species only the marginal cells, in others only cells on the blastodisc floor, and still in others cells from both sites, remain continuous with the YCL. When these cells divide, one nucleus stays with the new blastodisc cell and the other nucleus is pushed into the YCL. In this way the YCL receives more and more nuclei without its cytoplasm ever cleaving. In other words, the nuclei form the yolk syncytial layer (YSL), which we call the periblast [14]. In contrast, other authors proposed that the periblast nuclei give rise to blastodisc cells.

According to its location, the syncytial layer is called central or peripheral periblast (G. zentrales or Randsyncytium of 'Virchov'). The periblast nuclei continue to divide in a remarkable synchronistic way (this is also referred to as 'parasynchronous mitosis'). Nuclear division ceases only before the onset of gastrulation (Chapter 11). Later in development the periblast nuclei increase in size (up to 125 microns in length in the pike), become vacuolated and divide amitotically, also referred to as gemmation or budding '(Klein, 1876; Kowalewski, 1886a; Wenckebach, 1886; Wilson, 1891; Fusari, 1892; His, 1898; Kunz, 1964; Bachop, 1965; Bachop and Price, 1971; Twelves and Bachop 1979)' (see Chapter 18).

Two regions of the periblast can be distinguished on the basis of their ultrastructure '(Shimizu and Yamada, 1980; Heming and Buddington, 1988)'. One region, which extends throughout the syncytium, is characterized by smooth endoplasmic reticulum, numerous mitochondria and glycogen granules. It is proposed to be responsible for carbohydrate and/or lipid metabolism. A second region, which forms a stratified structure extending in portions across the syncytium, is characterized by the presence of rough endoplasmic reticulum (RER) and Golgi bodies. This region is thought to be involved in the synthesis and transport of proteinaceous substances.

10.1 CLEAVAGE

In the following, as an example of interspecific differences, the cleavage of the sea bass, *Serranus atrarius*, and the trout, *Salmo fario*, will be described in detail.

Serranus atrarius [a) after Ziegler (1902); b)-q) after 'Wilson (1889/91)']

The first two cleavage furrows, yielding the two and four cell stages, are meridional and at right angles to each other. During their formation the grooves run from the outer to the inner surface of the blastodisc. Subsequently the blastomeres round off but at the periphery stay continuous with the YCL. (Fig. 10/2a,b,f)

The third cleavage plane yielding 8 blastomeres is again meridional and parallel to the first cleavage plane. The so-called segmentation cavity (sc) becomes visible (Fig. 10/2 c,h). The fourth cleavage furrows are also meridional but parallel to the long axes of the 8-cell stage, i.e. parallel to the second plane of division. They yield the 16 cell stage consisting of 4 rows of four cells (Fig. 10/2d). As a result from the cleavage 16-32 cells (Fig. 10/2 e) the four central cells lie above the segmentation cavity, whereas the marginal cells remain in connection with the central and peripheral periblast (Fig. 10/2 i,j). As mentioned above, the first cleavage furrows cut almost through the whole blastodisc and leave a mere strand of uncleaved cytoplasm on the yolk which represents the future central periblast (cp).

During the progression from the 32 to the 64 cell stage the four central blastomeres undergo a so-called latitudinal or horizontal cleavage (plane parallel to the surface of blastodisc); as a result the centre of the blastodisc becomes two cells deep (Fig. 10/2k). The cleavage of the four peripheral corner cells (m_2) is meridional while the plane of division of the remaining intermediate peripheral cells (m_1) is equatorial. The marginal cells remain in contact with the central and peripheral periblast. From now onwards the divisions of the blastomeres become irregular.

Salmo fario

The egg at oviposition a) is shown after 'Henneguy (1888)', during cleavage b)-g) after 'Kopsch (1911)' and the embryo at the end of cleavage h) after 'Lereboullet (1861)' (Figs. 10/3b-g, 10/4a,h).

The surface view of the animal pole of the egg immediately after oviposition is shown in Fig. 10/4a. The early furrows are incomplete so that the blastomeres are syncytially arranged. The first 4 blastomeres are of different size (Fig. 10/3b). At the 8 cell stage (third cleavage) all blastomeres still form a syncytium with each other and the periblast (Fig. 10/3c). Two layers of blastomeres become distinct at the 16 cell stage (fourth cleavage) (Fig. 10/3d). At the end of the 32 cell stage (fifth cleavage) the upper layer contains 13 fully cleaved blastomeres and the lower layer 19 syncytially connected with each other and the yolk cellular layer (Fig. 10/3e). The tenth cleavage results in six layers of blastomeres and only the lowest layer remains in the syncytial condition (future periblast) (Fig. 10/3f). Note that in this species the periblast arises directly from the blastoderm cytoplasm. The periblast below the blastodisc is known as central periblast, while the one at the sides of the blastodisc is referred to as peripheral periblast. The extension of the shape of the blastodisc and the increased number of periblast nuclei are shown in Fig. 10/3g. A surface view of the animal pole at this stage is given by 'Lereboullet (1861)' and shown in Fig. 10/4h.

Different aspects of the cleavage in salmonids had been described earlier by 'Vogt (1842)'; 'Coste (1850)'; 'Lereboullet (1861)'; 'Stricker (1865)'; 'Kupffer (1868)'; 'Rieneck (1869)'; 'Goette (1869)'; 'Gerbe (1872)'; 'Oellacher (1872, 1873)'; 'Klein (1876)'; 'Romiti (1873)'; 'His (1873/1875)'; 'Ryder (1884)'; 'Ziegler (1882, 1902)'; 'Henneguy (1888)'; 'Hoffmann (1888)'; 'Samassa (1896)'; 'Sobotta (1896)'.

A special mention should be made of the investigations by 'Weil (1872)' on the trout. She observed great activity of the blastomeres at the rim of the blastodisc. The blastomeres sent out in various directions cell protrusions of varying shape and size, which continuously changed form, were sent out and again withdrawn. Most of these protrusions were cylindrical and on their formation were observed to fill with 'granulated substance' (cytoplasm); sometimes segmentation into two cells was evident.

EM-investigations of the cleavage of the trout egg revealed that the peripheral cells (forming the future EVL) are linked by short zonulae adherentes on the second day after fertilization. These cells display short and widely spaced microvilli projecting from their outer surface. Mitosis of an enveloping cell, with the spindle oriented at right angles to the surface, gives rise to an envelope and a deep cell '(Bouvet, 1976)'. 'Hagenmaier (1969)' analyzed bio- and cytochemically the nucleic acid and ribonucleo-protein status of the early developmental stages of *Salmo irideus* and *S. trutta fario*. The constant low levels observed during the cleavage period indicate the presence of reserves of maternal origin. Both methods show an increase of RNA and basic protein after the end of cleavage, reaching a maximum in the blastula. A gradient of intensity of RNA and basic protein from the yolk through the periblast into the germ has been revealed by histochemical methods.

Fig. 10/2 Development of *Serranus atrarius* [redrawn a) after Ziegler (1902), b) to q) after Wilson (1889/1891)]. **a** - First cleavage, section along animal-vegetal axis 10 ×. bl - blastodisc; c - cortex; e - envelope; og - oil globule; pe - perivitelline space; y - yolk. **b** - Surface view of a segmenting egg revealing two blastomeres in which nuclear division has already taken place. epr - early periblastic ridge. **c** - 8 cell stage. View from below. sc - segmentation cavity; a - line of sectioning for figure g; b line of sectioning for figure h. **d** - Blastodisc showing the beginning of the fourth furrow, i.e. change from 8 into 16 blastomere stage. **e** - Section showing normal cleavage from 16 into 32 cells. a-b and c-d denote levels of section for figures g-j. The division of the corner cells is meridional but in the remaining peripheral cells the plane is equatorial. The four central cells (shaded), on the other hand, suffer horizontal cleavage, i.e. parallel to the surface of the blastodisc. This becomes evident when viewing section i. (This symmetrical cleavage is ideal, i.e. it is not observed in all eggs.) **f** - section through 4 blastomere stage; it does not differ from section through 2 blastomere stage. The two cells are connected by a thin bridge (cp) of *cytoplasm*. At the periphery the blastomeres are in continuity with the *cytoplasm* clothing the whole yolk. The edge thus formed is the 'early periblastic ridge' (epr) shown e.g. in Fig. h. This ridge is observed until the periblastic nuclei are established. It should not be confounded with peripheral periblastic wall (pw) shown in Figs. l and m. **g** - Eight cell stage. Section along line a of Fig. c. **h** - Eight cell stage. Section along line b of Fig. c. **i** - Stage of 16 into 32 cells; section is along the line a to b of Fig. e. **j** - Stage of 16 into 32 cells; section is along the line c to d of Fig. e. **k** - Stage of 32 into 64 cells. The peripheral cells m^2 undergo meridional and the intermediate cells m^1 horizontal division. The four upper central cells c^1 divide in planes parallel to y-axis, the four lower central cells c^2 parallel to x-axis. (c1 and c2 are labelled in Fig. j). **l** - Section through blastodisc of similar age to p. Marginal cells have changed into the periblast wall (pw), which as yet contains but one circle of *nuclei*. *Cytoplasm* from the wall flows towards the centre of the blastodisc, which results in a thickening of the central periblast (cp). **m** - This section through the periblast wall corresponds to stage q and is characterized by an active multiplication of *nuclei* in the pw. **n** - Surface view of edge of blastodisc of 4.40 hrs; marginal cells still distinct. **o** - Surface view of edge of blastodisc of 7.30 hrs; marginal cells not marked off from periblast. **p** - Surface view of edge of blastodisc of 8.30 hrs; outlines of marginal cells entirely lost. **q** - Surface view of edge of blastodisc of 9.30 hrs; multiplication of periblast *nuclei* (shown only in reduced number). bl - blastodisc; c - cortex; corp - cortical periblast; cp *cytoplasmic* bridge; cp - central periblast; e - envelope; epr - early periblastic ridge; og - oil globule; pe - perivitelline space; pw - periblast wall; sc - subgerminal cavity; y - yolk.

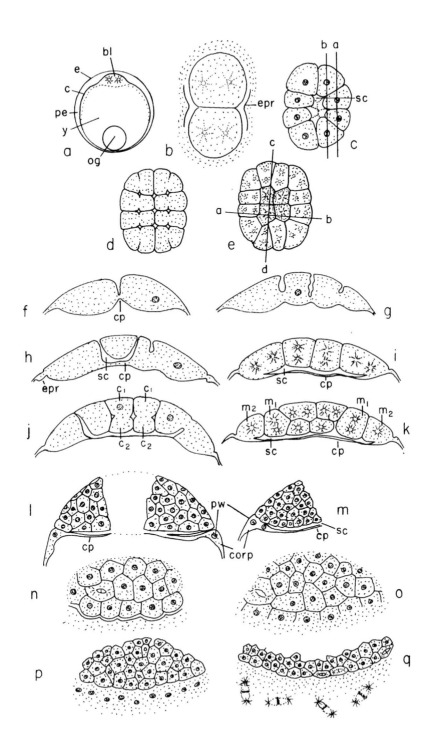

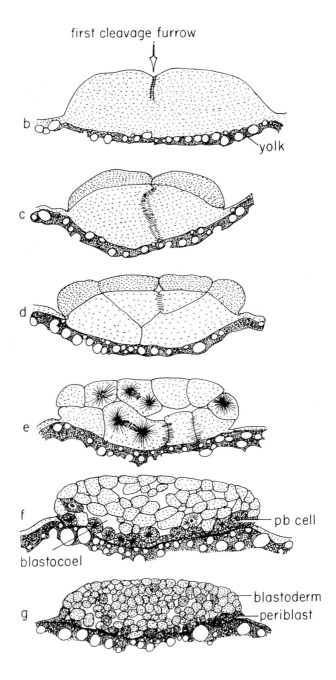

Fig. 10/3 Development of the trout displaying the separation of periblast directly from the cells of the base of the blastodisc [redrawn after Kopsch (1911)]. **b** - Transverse section through blastodisc at the end of the 2nd cleavage. Sectioned at right angles to the first furrow, the incision of which extends only to the upper third of the blastodisc. **c** - Cross-section through blastodisc in the middle of the 3rd cleavage (yielding 8 blastomeres). Sectioned at right angles to the first furrow. At the 8 cell stage all blastomeres are syncytially connected with each other and with the cortex surrounding the yolk. **d** - Cross-section through blastodisc at the beginning of the 4th cleavage. Sectioned at right angles to the first furrow. The furrow of the 3rd cleavage runs parallel with the furrow of the first one. **e** - Transverse section of blastodisc in the middle of the 6th cleavage. At the beginning of this stage the upper 13 blastomeres separated from the lower 19, which remained syncytially connected. **f** - Transverse section of blastodisc at the beginning of the 10th cleavage. A thickening of the peripheral periblast is observed As a result of the 9th cleavage the blastodisc consists of six layers. Only the most basic layer displays a syncytium. **g** - Transverse section of blastodisc at the end of the 11th cleavage. The blastodisc starts to extend and becomes lentil-shaped. The number of periblast *nuclei* has greatly increased.

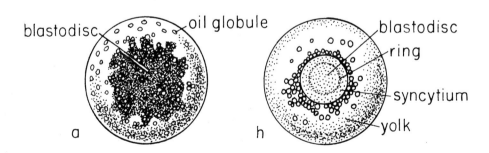

Fig. 10/4 Egg of the trout **a** - after Henneguy (1887). Surface view of animal pole after oviposition **h** - after Lerebouillet (1861). Surface view of animal pole before onset of gastrulation.

EM analyses of the surface structures of the envelope cells of *Salmo irideus* during cleavage revealed the presence of 'cristae', which at gastrulation are transformed into a meshwork of microvilli. The deep cells of the blastodisc presented lamellipodia and filipodia indicative of cell movement [as indicated already by 'Weil (1872)', see above]. The 'cristae' disappeared during mitosis with the exception of the zone of the division furrow '(Boulebache and Devillers, 1977)'. The molecular basis of the motility of lamellipodia, filipodia, cristae and ruffles was recently reviewed by 'Small *et al.* (2002)'.

Brachydanio (Danio) rerio

The development of the zebrafish is very fast (at 27°C fertilization to hatching takes 60 hours). With time-lapse filming 'Roosen-Runge (1938)' and 'Lewis and Roosen-Runge (1942, 1943)' studied the early stages of *Brachydanio rerio*. While the newly fertilized egg is nearly spherical, the animal pole soon flattens and the blastodisc appears as a result of concentration of the cytoplasm. Lengthening of the egg is observed as the first cleavage spindle appears. At the resulting two-cell stage no protoplasmic connection at the base of the blastomeres is observed in contrast to reports for other teleosts. At the beginning of the first cleavage streaming of cytoplasm through the yolk towards the blastodisc is evident. Additionally, a counterstream towards the vegetal pole is observed. The same processes of streaming and concentration of protoplasm in the blastodisc begin again at the prophase of all subsequent mitoses. It is a result of the contraction of the 'superficial gel layer' that the yolk granules are held back as by a sieve. The second cleavage plane is perpendicular to the first, and the third parallel to the first, the fourth parallel to the second and the fifth parallel to the first. The sixth cleavage plane is horizontal (while e.g. in *Serranus atrarius* the first horizontal cleavage takes place at the fifth division and in *S. trutta* at the third). Time lapse studies of the zebrafish development by 'Hisaoko and Battle (1958)' showed streaming of material from the yolk towards the blastomeres until the early blastula stage is reached. 'Thomas and Waterman (1978)' reported that the blastomeres in the zebrafish have become motile by the late blastula stage.

Twenty-five developmental stages were described and illustrated for the zebrafish by 'Hisaoko and Battle (1958)'. They observed that in the zygote initially 'cytoplasm is admixed with the yolk' and that 5 minutes after fertilization the animal pole is differentiated. Phase-contrast optics by 'Thomas (1968a)' revealed round opaque particles of yolk at the periphery of all blastomeres of the so-called high blastula stage (stage 9). At the late high blastula yolk particles are found peripherally to the nucleus of the mitotic apparatus. All cells have a complement of yolk; however, the number of yolk granules is highest in cells near the periblast and from this stage onwards yolk particles are absent from the blastomeres furthest from the periblast. A sequential pattern of yolk utilization in blastomeres of the high blastula stage has been described. Initially, yolk particles possess only amorphous yolk enclosed by a membrane. Later, membraneous elements and ribosomes appear in the centre of the particles and finally a highly complex particle displaying cytoplasm surrounded by little yolk and a unit membrane is encountered. The author concludes that yolk, besides providing nutrients to the blastomeres, contributes membranes and ribosome-like material of maternal origin to the blastomeres.

TEM investigations by 'Thomas (1968b)' on early developmental stages of *Brachydanio* revealed both furrow and vesicle cleavage in the same blastula. Furrow cleavage is characterized by a deposit of fine dense fibrillar material along the inner edge of the furrow. The feature of vesicle cleavage is the presence of a row of vesicles across the plane of separation between the daugther cells. Both types of cleavage have been observed in the same embryo. A so-called midbody has been described in the latter type of cleavage. Both types of cleavage are observed until the one-third epiboly stage is reached, after which only furrow cleavage is present.

While up to cleavage cycle 7 cell divisions are highly synchronous, slight metachrony displaying waves of mitoses emanating from the animal pole was observed at cleavage 8. At the so-called midblastula transition (cycle 10) the cell cycles begin to lengthen. However, this happens not in all the blastomeres at the same time, i.e. asynchrony of cell division starts and cell motility begins, characterized by the presence of pseudopodia and blastomere movement observed by time-lapse recordings. Experiments with partially enucleated blastulae revealed that it is the nucleo-cytoplasmic ratio which appears to regulate the onset of cell cycle lengthening. '(Kane and Kimmel, 1993; Kane, 1999)'.

SEM analysis of the pregastrula of *Brachydanio* showed that the external surface is covered by 'cristae' (which disappear at gastrulation). The furrows of the first cleavages display smooth patches with short 'cristae' and 'microballs' ['microboules' of 'Boulekbache and Devillers (1977)']. Up to the 16 cell stage, small circular depressions are visible which may represent the end of cortical vacuole expulsion. 'Dasgupta and Singh (1981)' noted two types of blastomeres in the zebrafish: cells staining either densely (D) or lightly (L) with a mixture of stains used for light microscopy (toluidene and methylene blue). They ascribed the dense staining to a high density of mitochondria, rough endoplasmic reticulum and polyribosomes. D cells are the first to appear (late cleavage) and when the L cells make their first appearance the D cells move peripherally and surround the L cells. Partially dense blastomeres have been found also and the authors suggest that the YSL plays some rôle in the transformation from L to D cells.

'Laale (1984)' observed naturally occurring diblastodermic (mixed with monoblastodermic) eggs obtained from one pair of zebrafish. He concluded that the duplicity of the blastodisc is due to a separation of the first two blastomeres. In some cases, conjoined twins were formed by subsequent fusion of the adjacent blastodiscs. Though twinning occurs relatively frequently in teleosts, pregastrula twinning is extremely rare. A detailed staging of the developing zebrafish embryo has been provided by 'Kimmel *et al.*, 1995'.

Fundulus heroclitus

This is another species which has been thoroughly studied. 'Oppenheimer (1937)' divided its development (from before fertilization until complete yolk absorption) into 34 stages. She observed that cycles 4-7 are stereotyped and highly synchronous. 'Lentz and Trinkaus (1967)' showed by TEM analyses that the cleaving blastomeres lack cytoplasmic specializations: there is a paucity of ribosomes, mitochondria are small and lipid droplets are few while smooth endoplasmic vesicles are numerous. The most conspicuous organelles are membrane-bounded granules. Blunt microvilli extend into the perivitelline space. Surface projections and vesicles are more common in the uncleaved blastodisc through to the 8-cell stage than later on. The blastomere surface adjacent to the yolk is highly specialized displaying long, branching and anastomosing processes extending into the yolk material. Phagocytosis was not observed although yolk was present at the bases of the microvilli and between the anastomosing arms, which are filled with dense granules presumed to represent glycogen. Once the blastodisc contains a few hundred cells, signs of differentiation become evident: endoplasmic reticulum is developed further and Golgi apparatus, nucleolus as well as polyribosomes, appear. The formation of a segmentation cavity has been observed.

Timelapse studies of *Fundulus* revealed that the early cleavage blastomeres show no surface activity despite the fact that they display surface projections. However, in late cleavage (129 cell stage) 'gentle undulations' of the cell surface are observed, which increase gradually in amplitude and frequency. By the time nuclei appear in the periblast, blunt rounded protuberances (blebs, lobopodia), in size from one-quarter to two-fifths of the diameter of the blastomere, are evident. At first these blebs form slowly and at any one time involve only a small number of blastomeres. At the height of their activity blebs form every 10-30 seconds and are retracted immediately at the same rate giving the picture of 'a constantly jostling mass of surface-active cells'. Since eventually the whole segmentation cavity is densely packed with blastomeres (up to several thousand) the surface activity can be observed with accuracy only near the margin of the blastodisc. Blebbing is confined to the DC, i.e. it is absent in the EVL. It has been suggested that once the motile system of a blastomere is no longer tied up with cytokinesis, it is immediately available for motile activity (blebbing). During mid-blastula and late-blastula locomotion is observed in some blastomeres with the bleb leading the way. During late blastula the lobopodia begin to adhere to the surfaces of other cells, heralding the onset of epiboly (see epiboly in chapter 11) '(Trinkaus, 1973a,b, 1980; Trinkaus and Erickson, 1983; Trinkaus, 1992)'.

Electrotonic (measured by impalement of microelectrodes) and dye coupling have been demonstrated for *Fundulus* blastomeres. It was shown that the structures associated with dye coupling are not cytoplasmic bridges but gap junctions (which allow molecules of up to 1.2 microns to pass between cells). Gap junctions between isolated pairs of blastomeres and between blastomeres in the intact *Fundulus* embryo have been shown to allow passage of Lucifer yellow until the stage of 'late-blastula'. However, junctional electrical coupling was often present when dye coupling was no longer detected '(Bennett et al., 1978, 1981, 1984)'. Passage of dye via gap junctions from blastomeres to the YSL and also from the YSL to blastomeres was also documented up to gastrulation by 'Kimmel et al. (1984)'.

Misgurnus fossilis

In the loach, Misgurnus fossilis, the blastomeres divide at a constant rate in the first hours after fertilization; due to prolongation of the interphase the cleavage rate slows down in the period preceding gastrulation '(Rott and Sheveleva, 1968)'. The duration of the first four cleavage divisions of the loach Misgurnus fossilis and seven other teleosts was investigated by 'Ignatieva (1976a,b)' and the relationship of mitotic cycle with cytokinetic processes during early division has been studied by 'Kostomarowa and Ignatieva (1968)'.

Before furrows of the first cleavage division appear in Misgurnus fossilis, the eggs are non-adhesive. Experiments with chimaerous embryos revealed that before furrows of the first cleavage appear, the eggs are completely non-adhesive. SEM of cleavages II-IV showed that the adhesive part occurs in the depth of the furrow and that it is smooth, with a number of spherical bodies (of one micron diameter) attached. The outer surface of the blastomeres displays numerous random folds and the yolk surface a few crest-shaped folds up to the middle blastula stage. Connecting the rough upper zones with the smooth lower zone of the furrow was a smooth zone of a number of lamellae with microvilli at the end (ruffles). This is the site of intercellular contact. The term 'ruffles' denotes repetitive upfoldings of the cell margin. Colchicine treatment induced reorganization of the smooth surface of the cleavage furrows and at the final stage the smoothing of ruffles which renders the surface non-adhesive '(Bozhkova et al., 1983)'. Gap junctions have been shown to connect blastomeres during cleavage and blastula stages '(Bozhokova et al., 1980)'. The efficiency of dye-transfer between superficial blastomeres increases by the late blastula stage '(Bozhokova and Voronov, 1997)'.

Putative regulators of cleavage

The hypothesis that acetylcholine regulates cell movement early in development has been put forward. Acetylcholine-esterase was shown to be present in homogenates of cleaving eggs of Oryzias latipes. Cytochemical analyses revealed acetylcholine esterase activity in deep blastomeres. It was present nearly throughout the cytoplasm, i.e. only a well-defined peripheral area of the cells where blebbing activity takes place, was devoid of reaction product '(Fluck, 1978)'. The occurrence of low esterase activity during the cleavage of Fundulus heteroclitus had been previously reported by 'Sawyer' (cf. 'Fluck, 1978'). The presence of contractile proteins in cleavage furrows suggests a role for calcium in cytokinesis. Measurements with calcium-selective microeletrodes showed that a significant rise in free calcium-ion concentration accompanies cytokinesis in *Oryzias latipes* '(Schantz, 1985)'.

10.2 PERIBLAST (YSL)[14]

The syncytial periblast was discovered by 'Lereboullet (1854)'. The term periblast, or parablast, was used initially to distinguish this tissue from the 'germ' which was called archiblast '(Klein, 1876)'. 'Hoffmann (1888)' elaborating on the history and meaning of periblast nuclei called it 'a standing puzzle'. Since even much later — in the words of 'Bachop and Price (1971)' — a multilingual and unstable terminology describing this layer had arisen, an account of the multitudes of terms is given in endnote[14]. In the recent treatise of the stages of embryonic development of the zebrafish the term 'yolk syncytial layer' has been used throughout by 'Kimmel *et al.* (1995)'.

Two contrasting views were put forward as to the destiny of the periblast by the early investigators. The first group maintained that the periblast will furnish the 'intestinal sheet' (G. 'Darmblatt'). These included apart from 'Lereboullet (1854)', 'van Bambeke (1872, 1875)', 'Kupffer (1868)', 'Klein (1876)', 'Owsiannikow (1885)' and 'Lwoff (1894)'. Further, 'van Beneden (1877)' and 'Brook (1885a,b)' maintained that the periblast gives rise to ento- and mesoderm. The second group was represented by 'Goronowitsch (1885)', 'von Kowalewski (1886)', 'Wenckebach (1886)', 'Ziegler (1894)' and 'Berent (1896a)'. They put forward that the periblast has only a nutritional function. For this reason, 'Hoffmann (1880)' referred to it as 'provisional blood'.

Nuclei

The periblast nuclei — called 'yolk nuclei'- were first described by 'Lereboullet (1854, 1862)' but their origin was elucidated only much later. It had been held by 'Kupffer (1868)', 'Rieneck (1869)', 'van Bambeke (1872)', 'Klein (1876)', 'Kupffer, (1878)', 'van Beneden (1878)' and 'Brook (1884)' that the periblast nuclei, partly surrounded by very fine cell contours, originate in the yolk and are arranged in concentric rings. These nuclei were said to be formed 'freely', referred to also as de novo, endogenously or in situ: in other words, they were thought to arise independently of the blastodisc '(Wilson, 1881; Veit, 1923)'. On the other hand, 'Weil (1872)' observed in the trout cells separating from the blastodisc and coming to lie in the yolk. Similarly, 'Oellacher (1873)' maintained that cells separate from the blastodisc and fall unto the floor of the segmentation cavity to become periblast cells. According to him, this had already been observed by 'Rieneck (1869)' and 'Stricker (1865)', his supervisor. 'Haeckel (1875)', working with *Gadus lota*, opposed in no uncertain terms the 'false parablast theory of His' and he insisted that apart from the blastodisc, the rest of the ovum is homogeneous. At the same time, 'Balfour (1875)' maintained that there is no evidence that the periblast nuclei derive from preexisting nuclei of the blastodisc; but he left their origin as an open question. As a third possibility of the origin of periblast cells, 'Vogt (1842)' and 'Rauber (1883)' had intimated that the marginal cells were involved in producing periblast cells. Nevertheless, 'Agassiz and Whitman (1884)', are usually credited with proving this origin. They described that in *Ctenolabrus* the marginal cells form the anlage (primordium) of the periblast 'cells', the membranes of which are faint before they disappear altogether. 'Reinhard (1888)' mentioned that the periblast is formed 1) by entering amoeboid blastodisc cells, which reunite (fuse) as the periblast and later 2) by the occurrence of free nuclei. In contrast, 'Henneguy (1888)' concluded that in the stickleback the cells forming a 'wreath' around the blastodisc leave the periblast to join the blastodisc ('pour s'ajouter au germe'). Similarly, several authors between 1869 and 1919 (e.g. 'Rienek, 1869, Weil, 1872') suggested that cells of the periblast take part in the formation of the embryo. 'Klein (1876)', in particular, stressed that 'it is certain, as we have shown, that the deeper layers of cells of the archiblast are in a considerable measure genetically derivatives from the paraplast (compare with 'secondary cleavage' described by later authors under 10.2.2).

At the end of cleavage in *Serranus atrarius* the marginal cells of the blastodisc retain their outlines along the 'early periblastic ridge' (Fig. 10/2n). A few minutes later they lose their basal demarcation (Fig. 10/2o). After 1 hour the lateral cell membranes are no longer in existence, i.e. the marginal cells have fused with one another, which results in a common syncytial tissue (Fig. 10/2p). A cross-section through the blastodisc of this stage is depicted in Fig. 10/2 l. Subsequently, multiplication of the nuclei around the margin of the blastodisc is observed (Fig. 10/2q). A section through the periblast wall of the same stage shows two nuclei, one in the periblast wall and the other moving towards the central periblast (Fig. 10/2m). Division of the nuclei is followed by a gradual migration of more and more nuclei towards the centre of the blastodisc (central periblast). After this stage the periblast nuclei cease to divide mitotically '(Wilson, 1889/1891)'.

The sections of the trout, *Salmo fario*, reveal that in this species the periblast tissue is directly derived from the overlying basal cells of the blastodisc (Fig. 10/3 f,g).

Leaving aside the 'de novo formation' of nuclei which is no longer considered to be an option, there remain three possible geneses of the periblast nuclei, 1) from the marginal (peripheral) blastomeres, 2) from the cells of the blastodisc floor and 3) from a combination of both sites 1) and 2).

Origin of type 1 (genesis from blastodisc margin)

Species	Authors
various teleosts	'Agassiz and Whitman (1887)', 'M'Intosh and Prince (1890)', 'Sobotta (1896)', 'Berent (1896)'.
Anchovia argyrophana	'Kuntz and Radcliffe (1917)'.
Belone	'Sobotta (1896)'.
Belone acus	'Kopsch (1911)', 'Wenckebach (1886)'.
Blennius	'Wenckebach (1886)'.
Brevoortia tyrannus	'Kuntz and Radcliffe (1917)'.
Carassius auratus	'Brook (1884, 1885b)', 'von Kowalewski (1886a)'.
Coregonus palea	'Vogt (1842)'.
Crenilabrus pavo	'List (1887)', 'Ziegler (1902)', 'Wenckebach (1886)', 'Kopsch (1911)'.
Crenilabrus tinca	'List (1887)'.
Ctenolabrus	'Agassiz and Whitman (1884, 1885)', 'Wilson (1891)'.
Cristiceps argentatus	'Fusari (1892)'.
Ctenolabrus coeruleus	'Kopsch (1911)', 'Ziegler (1902)'.
Cyprinidae	'Rusconi (1836)'.
Esox lucius	'Lereboullet (1854)'.
Gadus morhua	'Cunningham (1885)', 'Wenckebach (1886)', 'Reis (1910)'.
Gasteroidea	'Coste (1850)'.
Gobius	'Lwoff (1894)', 'Rauber (1883)'.
Labrax lupus	'Ziegler (1894)', 'Kopsch (1911)'.
Lebias	'Reis (1910).
Menidia menidia notata	'van Harleem et al. (1983)'.
Merluccius	'Kingsley and Con (1883)'. 'Wenckebach (1886)'.
Merluccius bilinearis	'Kuntz and Radcliffe (1917)'.
Nothobranchius guentheri	'van Harleem et al. (1983)'.
Oncorhynchus keta	'Mahon and Hoar (1956)'.
Perca fluviatilis	'Wenckebach (1886)', 'Lereboullet (1854)'.
Polycanthus viridiauratus	'von Kowalewski (1886b)'.
Prionotus carolinus	'Kuntz and Radcliffe (1917)'.
Salmo fario	'Romiti (1873)', 'Weil (1872)', 'Henneguy (1888)', 'Wilson (1891)'.
Serranus atrarius	
Stenotomus chrysops	'Kuntz and Radcliffe (1917)'.
Tautoga onitis	'Kuntz and Radcliffe (1917)'.
Tautogolabrus adspersus	'Kuntz and Radcliffe (1917)'.

Origin of type 2 (genesis from floor of blastodisc)

Species	Authors
various pelagic eggs	'Agassiz and Whitman (1884)', 'Wenckebach (1886)'.
Carassius auratus	'von Kowalewski (1886a,b)', 'Reis (1910)'.
Coregonus lavaretus	'Owsiannikow (1873)', 'Wenckebach (1886)'.
Cyprinus carpio	'Reis (1910)'.
Julis	'Lwoff (1894)'.
Lebias (Cyprinodon) colaritanus	'Reis (1910)'.
Leuciscus erythrophthalmus	'Reinhard (1888)', 'Reis (1910)'.
Polycanthus viridiauratus	'von Kowalewski (1886a)'.
Rhodeus amarus	'Reis (1910)'.
Salmonidae	'Henneguy (1888)', 'Reis (1910)'.
Salmo fario	'Berent (1896)', 'Fusari (1892)', 'His (1878)'. 'Klein (1876)', 'Kopsch (1911)', 'Oellacher (1873)', 'Rieneck (1869)', 'Wenckebach (1886)', 'von Baer (1835)', 'Stricker (1865)'.
Salmo salar	'Hoffmann (1888)'.

Origin of type 3 [combination of 1) and 2)]

Species	Authors
pelagic eggs unidentified	'Wenckebach (1886)'.
Belone acus	'Wenckebach (1886)', 'Sobotta (1896)'.
Carassius auratus	'Wenckebach (1886)', 'Sobotta (1896)', 'von Kowalewski (1886a,b)'.
Coregonus	'Eyclesheimer (1895)'.
Gadus	'Ryder (1887)', 'Wenckebach (1886)'.
Gadus (morhua)	'Ziegler (1882)', 'Wenckebach (1886)'.
Perca fluviatilis	'Kopsch (1911)'.
Salmo fario	'Oellacher (1872, 1873)', 'Wenckebach (1886)'. Kopsch (1911)', 'von Kowalewski (1886a)', 'His (1898)'.
Siphonostoma floridae	'Gudger (1905)', 'Veit (1923)', 'Auerbach (1904)'.
Spinachia	'Kupffer (1868)'.

Recent investigations of periblast and its nuclei

'Bachop' and co-workers, in various publications from 1965-1980 described the periblast nuclei of *Esox masquinongy ohioensis* as a result of light microscopical observations and histocemistry (Feulgen reaction). On the basis of EM analyses two regions of the yolk syncytium of the embryonic rockfish, *Sebastes schlegeli*, could be distinguished '(Shimizu and Yamada, 1980)'. One region, characterized by smooth endoplasmic reticulum, numerous mitochondria and glycogen granules, is proposed to be responsible for carbohydrate and/or lipid metabolism. This region extends throughout the syncytium. The second region is characterized by rough endoplasmic reticulum (RER) and Golgi complexes, and extends in portions across the syncytium forming a stratified structure. This latter region is thought to be involved in the synthesis and transport of proteinaceous substances '(Heming and Buddington, 1988)'.

After the periblast nuclei have ceased to divide mitotically, they grow extremely large and divide amitotically. They lie with their long axis horizontally in the periblast. However, towards the end of yolk resorption they assume a radial orientation and eventually attain a very irregular contour (see Chapter 19).

Brachydanio (Danio) rerio

The periblast of *Brachydanio rerio* contains large closely packed membrane-bound yolk particles at the early high blastula stage. Some yolk particles are observed within others but all are separated by their own membrane. The yolk particles appear uniform under phase-contrast optics but EM-analysis shows membraneous vesicles within some. At high-blastula stage the periblast syncytium is separated from the blastomeres by rows of vesicles rather than by membrane to membrane contact '(Thomas, 1968a,b)'. SEM analysis of the surface of the periblast cells of the zebrafish revealed that it was smooth '(Boulekbache and Devillers, 1977)'.

'Thomas and Waterman (1978)' suggested on the basis of SEM observations that there is no free margin of the EVL and that during epiboly portions of the periblast cytoplasm may be added continuously by cytokinesis onto the EVL margin. This so-called 'secondary cleavage' was also described by 'Roubaud and Pairault (1980)'. 'Kimmel and Law (1985)' proved by intracellular dye injection in *Brachydanio rerio* that the periblast nuclei arise from daughter cells of the marginal blastomeres. These form plasma membranes at their marginal borders, thereby cutting off cytoplasmic confluences between themselves and the periblast. The membranes of the periblast cells are first indistinct (open blastomeres) and ultimately disappear so that their cytoplasm becomes confluent to form a syncytium. These same observations were reported for *Fundulus* by 'Kimmel et al. (1984)' and for *Barbus conchionus* by 'Gevers and Timmermans (1991'. From this point onwards until the end of gastrulation the blastodisc and the periblast syncytium develop separately. However, the marginal cells of the blastodisc remain attached to the periblast by extensive tight junctions as shown for *Fundulus* by 'Betachaku and Trinkaus (1978)'. 'Bozhkova and Voronov (1997)' analyzed the intercellular communication via gap junctions from early blastula to late gastrula in zebrafish and loach (*Misgurnus fossilis*). Observations on dye transfer (Lucifer yellow and Fluorescein) at the sixth to the seventh cleavage showed that communications via gap junctions between EVL cells and underlying DCs was virtually absent. However, dye transfer was present and increased both in zebrafish and loach embryos from early to midblastula. The authors stressed that the absence of dye transfer at the early stages does not mean absence of gap junctions since experiments on electrical conduction and EM analyses showed the presence, though at a low level, of gap junctions.

Fundulus heteroclitus

The development of the YSL of *Fundulus* was further investigated in detail with time-lapse filming of light microscopical pictures obtained with Nomarski differential interference optics. The YSL, which separates the yolk from the overlying blastodisc, is of great importance because, in order to reach the blastodisc and later the embryo, all nutrients from the yolk must pass through it. As already pointed out by the early investigators, during early cleavage [from cleavage 8-11 according to 'Lentz and Trinkaus (1967)'; 'Bennet and Trinkaus (1970)'] certain marginal blastomeres are seen to collapse and merge with the YCL and supply it with nuclei and cytoplasm. The latter authors referred to them as 'open blastomeres'. Nuclei enter the YCL, which now has become the YSL, from the 'open blastomeres' most frequently at the 8th and 9th cleavage but also at the 10th and sometimes the 11th cleavage. Videophotographic records revealed that during that time the margins of the marginal cells change their contour almost constantly. The nuclei that entered the YSL first, divide five times and later nuclei divide with them; in other words YSL mitoses are shown to occur in metachronous waves '(Trinkaus, 1993)'. [Metachrony has also been reported for *Oryzias* by 'Kageyama (1986)']. All nuclear divisions in the YSL cease completely and together before the onset of gastrulation (Chapter 11). It should be stressed that no YSL nuclei were seen to arise from basal submarginal blastomeres although blastomeres during the earliest cleavages are confluent with the cytoplasm of the YCL. After each nuclear division, the syncytium increases in width and the nuclei are evenly spaced. The interphases lengthen between successive mitoses and, as stated above, all nuclear divisions cease together. The giant nuclei in *Pterophyllum* take up a radial position and disappear after absorption of the yolk (see Fig. 19/14a-d in Chapter 19).

Judging from the above, the YSL seems to be programmed for only 5 signals for nuclear division after which cessation is definitive. It has been suggested that an increase in the nucleo-cytoplasmic ratio in the periblast with each succeeding mitosis is responsible for the deceleration and final cessation of the mitoses. In other words, it has been explained as a 'watering down' of a factor in the maternal cytoplasm essential for division '(Trinkaus, 1993)'. EM of the periblast of *Fundulus heteroclitus*, which forms the floor of the segmentation cavity, reveals short apical microvilli. Although the periblast is instrumental in transferring yolk to the overlying blastomeres, it does not form any contact with them. In contrast, 'cytoplasmic arms' project from the basal surface of the periblast into the yolk '(Trinkaus and Lentz, 1967)'.

10.2.1 DNA of periblast nuclei

The giant periblast nuclei in *Esox masquinongy* have been found to contain amounts of DNA far in excess of the amount found in the diploid cells of the embryo. The DNA content of a single chromosome of the periblast of this species and of *Brachydanio* is not increased but there are extra sets of DNA containing chromosomes. In other words, the giant nuclei are polyploid '(Bachop and Price, 1971; 'Bachop and Recinos, 1973'; Bachop and Schwartz, 1974)'. In *Oryzias latipes* the DNA content ranged from diploid, tetraploid, to octaploid at the end of the blastula stage '(Kageyama, 1996)'.

10.2.2 Functional significance of the periblast

The involvement of the periblast in yolk transfer was suggested from the earliest time onwards. Already 'Rieneck (1869)' noticed the presence of yolk granules in the blastomeres of the trout. Since the lowest cells in the blastodisc always contain more yolk than the upper ones, 'Oellacher (1872)' talked about a 'mouth to mouth feeding' of the blastomeres. The food is supplied to the lowest cells through the action of the periblast cells and 'Hoffmann (1888)' put forward that the periblast 'assumes the role of provisional blood'. 'Ziegler (1882)' 'Hoffmann (1883)', 'von Kowalewski (1886a), and 'Brook (1887a)' advanced the suggestion that it is the function of the periblast nuclei to work the yolk into some shape which is easily assimilated by the blastodisc cells. Later, 'van der Ghinst (1935)', 'Cordier (1941)' and 'Trinkaus and Drake (1956)' [cf. 'Lentz and Trinkaus (1967)'] reported the enzymatic hydrolysis of the yolk by the periblast and the transfer of material to the overlying blastomeres. Finally, 'Devillers (1961)' and 'Kageyama (1996)' — as before them 'Kopsch (1911)' — referred to the periblast as an embryonic organ since it mobilizes the yolk reserves and transmits nutrients to the embryo and also plays a role in gastrulation, as will be discussed later.

According to 'Ober and Schulte-Merker (1999)' the external periblast induces mesoderm in the zebrafish.

Summary

Cleavage in teleosts is meroblastic, i.e. restricted to the blastodisc. In contrast to other meroblastic eggs (e.g. birds and reptiles) the teleostean cleavage pattern is regular.

The thin cytoplasmic layer surrounding the yolk (YCL) becomes invaded by nuclei from dividing peripheral blastomeres, from basal cells of the blastodisc or from both sites. These nuclei divide several times, converting the layer into a syncytium known as YSL or periblast. Around the time the nuclei become postmitotic they increase to a gigantic size and divide amitotically, i.e. by nuclear fission. The large amount of DNA in the periblast nuclei of *Esox masquinongy*, *Brachydanio rerio* and *Oryzias latipes* has been shown to be due to polyploidy.

The periblast is interfaced with yolk on one side and the yolksac blood circulation on the other (as regards periblast/liver contact consult Chapter 19). Therefore, substances derived from the yolk must pass through the periblast to enter the embryonic blood stream thereby bypassing the embryonic gut.

Endnotes

[1] Blastodisc (discus blastodermicus) of 'Ransom'. Other synonyms employed especially by the early investigators are: germinal disc (discus germinativus), discus proligerus of 'Haeckel', germ, polar disc, archiblast of 'His', 'Ryder' and 'Klein', lenticular enlargement of the cortex of 'Oellacher'
G: Keim, Blastos, Hauptkeim of 'Häckel', Anschwellung of 'Rusconi', Bildungsdotter of 'Haeckel' and 'Reichert', Furchungshügel, Keimhügel, Keimscheibe, Archiblast of 'His',
F. disque germinatif, disque germinateur, germe, ampoule du germe, vitellus formateur of 'Lereboullet', calotte jaunâtre of 'Lereboullet', ampoule du germe, cicatricule of 'Coste'
I. blastoderma, disco germinale

[2] germinal vesicle
G. Keimbläschen, Kern, Hauptkern
F. vésicule germinatif, vésicule Purkinje (named after its discoverer 'Purkinje, 1825')
I. nucleo

[3] peripheral welt, embryonic rim, embryonic ring, germ ring, thick rim, limiting border of the blastodisc;
G. Keimwall of 'His', Randwulst of 'Goette' and 'van Bambeke', Keimwulst of 'Kupffer', 'Oellacher' and (Agassiz-Whitman).
F. bourrelet marginal, bourrelet périphérique of 'van Bambeke', bourrelet blastodermique of 'Lereboullet'.
I. anello embrionale of 'Raffaele'

[4] see endnote[1] of chapter 3

[5] segmentation

[6] from Gk. mero (part), blastos (germ)

[7] from Gk holo (total), blastos (germ)

[8] G. Furche
F. sillon

[9] Flattened pavement; epidermoid layer, periderm
G. Umhüllungshaut of 'Reichert', Deckschicht of 'Goette', Pflasterzellen of 'Vogt' and 'Rienek'.
F. cellules épidermoidales of 'Lereboullet', couche cellulaire de revêtement of 'Pasteels',

couche enveloppante, couche épidermidale of 'Vogt', épithelium en pavé of 'Lereboullet', lame de revêtement of 'Fusari', lame enveloppante.

I. strato superficiale dell'epidermide, foglietto epidermico superficiale di 'Raffaele'.

[10] F. cellules profondes

[11] Annual fish eggs undergo multiphasic diapause while buried in the mud. Diapause during the cleavage involves complete dispersion and subsequent reaggregation of the blastomeres.

[12] cavity of 'von Baer'
G. Furchungshöhle or Baer'sche Höhle
F. cavité de segmentation

[13] Blastocoeloma
G. Keimhöhle
F. Cavité germinative, cavité au-dessous du germe
I. Cavità germinale

[14] Synonyms for periblast of 'Agassiz and Whitman, cortical layer; embryonic entoderm of 'Agassiz and Whitman', entoderm, free nuclei of 'Kupffer', glandular layer of 'Remak', hypoblast (part of the entoderm) of 'Klein', membrana vitellina, intermediary layer of 'van Bambeke' and of American authors generally, intermediate layer of 'van Beneden', lamina mycogastralis of 'Haeckel', nucleated periblast of 'Agassiz and Whitman' and 'Cunningham', nuclear zone of 'Kupffer', 'Brook', 'M'Intosh and Prince', secondary entoderm, subgerminal plate of 'Kupffer', trophic or glandular layer of 'Remak', 'M'Intosh', vitelline membrane of 'Oellacher', 'Klein', yelk envelope of 'Ryder', yelk hypoblast of 'Ryder', yolk hypoblast of 'Ryder', yolk sac entoblast, yolk syncytial layer of 'Trinkaus (YSL), yolk syncytium, fat cell of 'Oellacher', 'Klein', 'His' and 'Weil'.
G. Dotterhaut of 'Owsiannikov', 'Ryder',' Oellacher', Dotterrinde of 'Waldeyer'. Dottersackentoblast of 'Kopsch', Dottersyncytium of 'Virchow', 'Kupffer', Intermediäre Schicht of 'von Kowalewski', 'Reinhard', Keimwall of 'His', Kernzone of 'Kupffer', Körnerzone of 'Kupffer', Merocyten of 'Rückert', Nebendotter, Nebenkeim of 'His', 'Owsiannikow ', Paraplast of 'His', 'Waldeyer', 'Hoffmann', 'Gasser', 'Kupffer', 'Reis', Paraplast (as opposed to Archiblast) of 'Klein ', Plasmodium of 'Rauber', Primäres Entoderm, Rindenschicht of 'His', Sekundäres Entoderm, Vielkerniges Plasmodium of 'von Kowalewski', 'Reinhard'.
F. couche corticale of 'Hennequy', couche hématogène of 'Vogt', couche intermédiaire of 'van Bambeke' 'Hennequy', couche sous-blastodermique of 'Lereboullet', couche sousajante au germe of 'Lereboullet', 'Hennequy', feuillet muqueux of 'Lereboullet', 'Hennequy', feuillet organique of 'Lereboullet', feuillet végétatif of 'Lereboullet', hypoblaste jaune (Hypoblaste du jaune), hypoblaste vitellin of 'Hennequy', membrane intermédiare of 'van Bambeke', membrane interne of 'Lereboullet', membrane sousajacente au germe of 'Lereboullet', membrane vitellaire of 'Vogt', membrane vitelline of 'Hennequy', parablaste of 'Klein', pellicle of 'Ryder', pellicule vitelline of 'Oellacher', 'Brook', zone nucléaire of 'Kupffer', 'Hennequy', zone nucléea of 'Hennequy', 'van Beneden '.
I. periblasto of 'Raffaele'.

Chapter eleven

Gastrulation

'It is not birth, marriage, or death, but gastrulation, which is truly the most important time of your life.'
Louis Wolpert (1983)

The features of teleost gastrulation are uniquely different from those of other vertebrates. For this reason some authors consider it inappropriate to use such terms as gastrula and blastoporus. However, the conceptual bias derived from comparing teleostean with e.g. amphibian gastrulation led to differing interpretations and a plethora of terminology. Therefore, although still mindful of the differences, we will use the terms gastrula and blastoporus. As already 'Kupffer (1884)' said: 'Es kommt darauf an, ob man das Trennende oder das Verbindende betonen will' (It depends if one wants to stress the differences or focus on the similarities).

During gastrulation the so-called three-layered embryo stage is being reached. This process involves, in general, harmonized morphogenetic movements which establish and juxtapose the three primary germ layers, ectoderm (outer layer), entoderm or endoderm (inner layer) and mesoderm (also referred to as chorda-mesoderm) (middle layer). In teleosts, the epiblast will give rise to ectoderm, and the hypoblast is the rudiment of both meso- and entoderm. Gastrulation is orchestrated by the dorsal organizer situated in the embryonic shield (ES)[1]. Similar to the amphibia, the zebrafish organizer forms where the maternal protein ß-catenin enters embryonic nuclei. Additionally, the dorsal periblast constitutes a domain of the organizer that is active in early teleost development '[Kodjabachian *et al.* (1999)]'.

In vertebrates generally, the prospective entoderm is moved interiorly as a result of either of two processes. In eggs with little yolk, invagination (emboly, buckling-in) takes place, as displayed e.g. in the echinodermata (Fig.11/1). In eggs with a large amount of yolk epiboly (overgrowth) is observed, which denotes a growth of the future ectoderm over the future entoderm (as observed in amphibia, Fig. 11/2). The prospective cells of the chorda-mesoderm move inwardly, over the blastoporus rim, as a result of involution (rolling-under) (Fig. 11/3). [Some authors consider epiboly as being different from gastrulation;

for them gastrulation is synonymous with involution. For instance in the zebrafish gastrulation begins when epiboly has reached 50% according to 'Holder and Xu (1997)'.] Gastrulation in teleosts consists of epiboly and involution only, without the process of invagination. A review of gastrulation in different vertebrates is given by 'Schoenwolf and Alvarez (1992)'. 'Collazo et al. (1994)' provide a phylogenetic perspective of teleostean gastrulation.

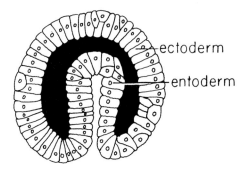

Fig. 11/1 The result of the process of invagination (buckling in) of the *entoderm* (as observed in the *echinodermata*).

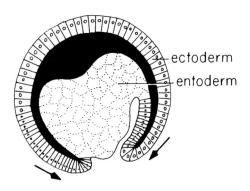

Fig. 11/2 The result of the process of epiboly (overgrowth) of the *ectoderm* (as observed in the *amphibia*).

11.1 EPIBOLY

In teleosts gastrulation begins when the blastoderm, up to now called blastodisc[2], has flattened out into a thin layer of cells and consequently covers now a larger surface of the yolk sphere. It is now referred to as blastoderm. Epiboly is initiated

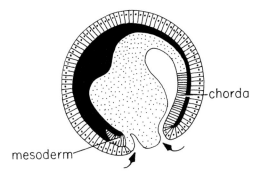

Fig. 11/3 The result of the process of involution (rolling under) of the *chorda-mesoderm* as observed in the *amphibia*.

by the spreading of the outermost flattened layer (EVL)[3] and of the periblast[4] over the yolk sphere. The two layers are continuous at their margins with the non-nucleated yolk cytoplasmic layer (YCL) or cortex and appear on surface view as a thickened ring called germ ring (GR)[5] once involution (11.3) starts. The margin of the EVL is atttached firmly to the surface of the periblast and divides the latter into external and internal periblast. The periblast spreads widely in advance of the EVL; it 'leads the way' and apparently supplies the active force. The two layers spread until they almost entirely cover the yolk. Subsequently the margin of the periblast once more moves ahead of the EVL margin. It closes its 'blastopore'[6] first; followed by the closure of the EVL 'blastopore'.

The GR greatly accentuates the epibolic movement until overgrowth of the yolk is complete. This process was first described by 'Rathke (1833)', followed by 'von Baer (1835)' and 'Rusconi (1835, 1840)'. As a result of convergence, the GR becomes somewhat broader at the point where the future embryo will develop. This triangular area consisting of a massing of cells is referred to as the embryonic shield (ES)[1] (Figs. 11/8A,B). It was first recognized by 'Forchhammer (1819)'. The ES will subsequently elongate to form the antero-posterior long axis of the early embryo, with the future head-portion pointing towards the centre of the blastodisc (Fig. 11/7B). At 80% epiboly of the zebrafish egg the deep cells (DC), from lateral to medial positions, converge towards the dorsal midline. This movement known as convergent extension causes the lengthening of the embryonic axis. It was originally held that the embryo was formed by the coalescence and fusion of the blastoporal lips along the median (dorsal) line of the future embryo. In other words, it was assumed that the GR on each side of the embryonic axis contains the potencies to form a lateral half of the embryo. Both halves would later on coalesce to form the embryo. However, this so-called concrescence theory of 'His and Rabl (1876)', modified by 'Kopsch (1904)', became subsequently untenable '(Schmitt, 1902; Sumner, 1904)'. It has been replaced by the processes of convergence (narrowing) and extension (lengthening). It has been repeatedly suggested that there appears at this stage a cavity between the blastodisc and the periblast called subgerminal cavity[7], which may more often than not be only a virtual cavity. It was never found in the trout, *Salmo fario* ['Stricker (1865)'; 'Rieneck (1869)'; 'Oellacher (1872/1873)' and 'Klein (1876)'].

'Cunningham (1885b)' and 'Prince (1887)' have pointed out that the entire GR represents the rim of the blastopore[7] (Figs. 11/4, 11/7A,B, 11/8-11, 11/13, 11/14). The dorsal lip marks the posterior pole of the developing embryo, which lengthens by growing posteriorwards. The head is considered to be practically a fixed point [punctum fixum of 'His (1876)' and 'Eyclesheimer (1895)'] and remains in the centre of the expanding blastoderm. The part directly opposite to the posterior end of the embryo moves along the same meridian as the developing body of the embryo and, therefore, has to move the largest distance. The trajectory of other parts of the GR is shorter the nearer they are to the posterior end of the embryo (Fig. 11/10). It was 'Trinkaus (1951)' who stressed that epiboly is probably not to be thought of as a problem of growth, but rather as a problem of mass movements.

The degree of epiboly provides a staging index and is expressed as percentepiboly. Therefore, when the margin has reached the equator of the yolk sphere, this is called 50% epiboly and when 3/4 of the sphere are covered, 75% epiboly, 100% is reached at the closure of the blastopore. The degree of epiboly is independent of the organization of the embryo: *Salmo salar* possesses 38-39 somites at closure of the blastopore while in *Serranus atrarius* there are not yet any somites developed at that stage (Figs. 11/8, 11/10). In the zebrafish the yolk is 'doming' (bulging) towards the animal pole just before epiboly begins. Therefore, this stage is referred to as 'dome stage' '(Kimmel *et al.*, 1995)' (Fig. 11/5B,B').

The superficial single layer of squamous polygonal cells of the blastodisc is called the enveloping layer (EVL) (Periderm)[3] (Figs. 11/4, 11/5, 11/6). The terms epidermic stratum, or 'cellules épidermoidales cohérentes' used at times by 'Lereboullet', have subsequently led to great confusion. Some believed this layer to form part of the epidermis, others recognized it as what it is, namely an enveloping layer, which does not take part in embryo formation and remains a simple epithelium. As regards the functional significance of the EVL, it has been suggested that it assumes a protective function over the embryo. Additionally, it maintains a high electrical resistance and provides an effective barrier to movement of even small ions '(Wood and Timmermans, 1988)'. Moreover, it was thought likely that EVL cells covering the yolksac secrete hatching enzymes. EVL cells are shed late in embryogenesis or at hatching, or else they remain for an unspecified time as a top layer of the epidermis. The definitive fate of the EVL has yet to be established '(Fleig, 1993)'.

The inner cells of the blastodisc are known as deep cells (DC). They are loosely packed and irregularly shaped. They intercalate radially beginning at the midblastula stage (Fig. 11/5A) and through the onset of gastrulation (Fig. 11/5B). Underlying the DC — separating the blastoderm from the yolk proper — is the yolk cytoplasmic layer (YCL) [first discovered by 'Lereboullet (1854)']. It encloses the whole yolk sphere, which is called 'yolk cell' by several modern authors. This term may be based on 'Oellacher (1873)' and 'van Beneden (1878)', who regarded the 'deutoplasmic globe' in a pelagic teleostean ovum as a large endodermic cell, with a constitution analogous to a fat-cell, a view shared subsequently by 'Hoffmann (1881)'. Once the YCL receives nuclei it becomes the yolk syncytial layer or periblast[4] (Figs. 10/2,10/3).

As previously stated, the EVL does not take part in the formation of any organs and remains a monolayer called periderm. Mitotic activity is said to be not necessary for epiboly of the EVL, which increases its surface manyfold during the process. The number of cells remains constant or almost constant in the EVL of *Oryzias* and *Fundulus* (while continuous cell division has been reported for *Ctenolabrus* by 'Morgan, 1895'). On the other hand, it was suggested that in *Fundulus* motile EVL cells are 'taken out of action' for a short time to undergo mitosis and subsequently reenter the mass of cell movement. Therefore, it has to be accepted that cell division and mass cell movements are not incompatible. It has been shown that in *Fundulus* only the marginal cells of the EVL are in contact with the underlying periblast by tight and close junctions apically and by wider appositions more proximally. Prior to epiboly the junctional complex measures 2.1-2.4 microns and becomes very restricted (0.5-0.8 microns) during early epiboly up to mid-gastrula. Subsequently, the contact area increases again up to 2.0-2.3 microns towards the end of epiboly. The EVL moves passively over the yolk by virtue of the attachment of its marginal cells to the independently expanding periblast '(Betchaku and Trinkaus (1978)'. 'Thomas and Waterman (1978)' suggested that in *Brachydanio rerio* there is no free margin of the EVL and that during epiboly portions of the periblast cytoplasm are added by cytokinesis on to the EVL margin. In *Fundulus* the EVL cells become elongated towards the vegetal pole '(Keller and Trinkaus, 1987; Weliky and Oster, 1991)'. The EVL layer expands uniformly during 3/4 of epiboly. After that, the marginal cells and their neighbours resume their regular hexagonal shape until the closure of the blastopore, which takes place in a draw-string fashion '(Keller and Trinkaus, 1987)' (Fig. 11/4). The results of immunolocalization of fibronectin carried out on the rainbow trout suggest that this substance is implicated in morphogenetic movements such as the purse-string like closure of the blastopore '(Boulekbache et al., 1984)'. 'Gevers et al., (1993)' studied the involvement of fibronectin during gastrulation in *Cyprinus carpio*. From 60% epiboly onwards extracellular fibronectin reactivity was observed at the basis of the EVL cells facing the periblast. At 100% epiboly staining reactivity of these cells had increased but some contact areas still lacked fibronectin staining. Blocking fibronectin receptors in *Cyprinus carpio* resulted in a retardation of epiboly, which did not exceed 60% '(Ho and Kimmel, 1993)'. The decrease in size of the EVL (starting after 3/4 epiboly) is brought about not only by a change in cellular shape (tapering and narrowing) (Fig. 11/4) but also by the fact that about half of the original marginal cells leave the margin at a rate of about six per hour. Morphometric analyses showed that these cells have adjusted by rearrangement within the EVL-monolayer. '(Keller and Trinkaus, 1987)'. The EVL can generate DC up to the midblastula stage as observed in the *Fundulus* by 'Kimmel *et al.* (1990)'.

The cells of the EVL are slightly adhesive and are bound together by close and tight junctions, as mentioned before. Furthermore, gap junctions, desmosomes, 200Å contacts and interdigitations of their adjoining plasma membranes are observed. Despite these elaborate junctions and an increase in tension during the epibolic expansion, individual EVL cells keep rearranging themselves within the monolayer (as mentioned above). This has been recorded by cinematographical analysis following marking with vital dyes and carbon particles. It has to be borne in mind that immense geometrical problems have to be overcome when an enlarging sheet, such as the EVL, has to enclose progressively a spherical body. It has recently been stated that this meticulous harmonization of cell arrangements presents one of the most precise of all morphogenetic movements known. A mechanical model to gain insight into the pattern of cellular force generation driving the cell rearrangements and cell shape changes has recently been put forward by 'Weliky and Oster (1990, 1991)' (Fig. 11/4).

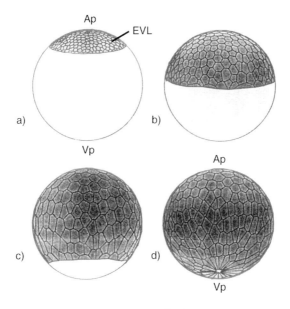

Fig. 11/4 Changes of the shape of the EVL cells during epiboly shown by four frames of a computer movie based on epiboly of *Fundulus* (Weliky and Oster, 1994, with kind permission of Plenum Press, New York). **a**- blastodisc covered by EVL. **b** - blastodisc has spread vegetally towards the equator. Marginal cells elongated parallel to the EVL margin. **c** - and d) once passed the equator, length of margin decreases and marginal cells elongate orthogonally to the boundary. AP - animal pole; EVL - periderm (enveloping layer); VP - vegetal pole.

The ultrastructural analyses showing that the cells of the EVL are joined by a circumferential apical complex of tight and close junctions and appositions '(Lum et al., 1983)' tie in with the early histological observations of 'Boeke (1903)' that 'Kittleisten' (terminal bars) are responsible for the tough adherence of the EVL cells. EM also revealed that the EVL cells do not show any protrusions during epiboly. This explains why the EVL does not crawl over the surface of the yolk. On the contrary, as pointed out before, it is passively carried ahead by a pull exerted from the independently expanding periblast. Ultrastructural analyses of the margin have shown that in this region EVL and periblast are firmly attached to each other by tight and close junctions. If this marginal contact is severed, the EVL recedes while the periblast spreads over the yolk at an increased rate. In other words, the EVL depends for its epiboly on contact with the periblast whereas the latter is able to complete epiboly in the absence of the EVL. However, according to 'Devillers (1951b)' the periblast of the trout, *Salmo irideus*, does not undergo epiboly without the presence of the EVL, in contrast to the results obtained with *Fundulus*. Prior to the onset of epiboly the periblast extends well beyond the margin of the EVL. This band of periblast, therefore, is known as external periblast or external yolk syncytial layer (E-YSL) to differentiate it from the periblast below the blastodisc, the internal or I-YSL (also called central periblast). The E-YSL and I-YSL exhibit numerous microvilli whereas the surface of the YCL is smooth. The long microvilli of the I-YSL become replaced by short ones as epiboly progresses. The belt of E-YSL, which is thicker in the dorsal lip region than in the rest of the margin, is generously endowed with contractile microfilaments which probably contain actin. Soon after its formation, the belt of the E-YSL narrows, brought about by contraction of its cytoplasm. As a consequence, the I-YSL expands and is pulled vegetalwards. It adjusts to the traction of the E-YSL by endocytosing its own microvilli, which, therefore, are considered a source of reserve membrane. Fusiform vesicles, indicative of endocytosis, have been observed in this region '(Betchaku and Trinkaus, 1978, 1982; Trinkaus, 1984)'.

Because the E-YSL has been shown to contract immediately after the blastoderm covered by the EVL is surgically removed, the contractile E-YSL is said to be continually 'straining on the leash' '(Trinkaus, 1984)'. 'Devillers (1961)', on the contrary, had insisted that the EVL possesses an intrinsic spreading force since, when cultivated alone, it shows a capacity for epiboly.

'Trinkaus (1993)' studied the process of periblast formation and development and its relation to the beginning of gastrulation in *Fundulus*. After the fifth and last mitosis of the periblast (beginning of 'permanent interphase') the E-YSL contracts in its animal-vegetal axis and effects the epibolic expansion of the I-YSL and the EVL.

Epiboly in zebrafish embryos is selectively impaired as a result of UV-radiation of the zygote. It has been shown that UV irreversibly damages the microtubules and that application of microtubule-disrupting agents has the same effect. Epiboly is also affected when irradiation is applied to early stages following the zygote, which led to the conclusion that the UV-sensitive targets may be maternally encoded components of the machinery driving epiboly '(Straehle and Jesuthasan, 1993)'. It has been suggested that microtubules in the YCL are instrumental in initiating and driving epiboly. It has been maintained that the earlier mentioned contraction of the E-YSL in itself cannot drive the periblast vegetalwards. Therefore, the periblast may be propelled over the surface of the yolk by microtubule motors that move along microtubule tracks in the YCL '(Solnika-Krezel and Driever, 1994; Topczewski and Solnica-Krezel, 1999)'.

'Long (1983a,b)' applied particles of coloured chalk into the cytoplasm of the germ ring of zebrafish and *Catastomus* embryos and concluded that the periblast of early embryonic stages does not reach the vegetal pole, i.e. 'it lagged behind as epiboly progressed'. However, chalk implanted at a later stage (stage 10 of *Catastomus*) into the germ ring reached the yolk plug.

11.2 INVOLUTION

When epiboly has advanced a certain distance, e.g. when the blastoderm covers about one third of the yolk mass in *Brachydanio*, involution (inflection, immigration) sets in. This is indicated by a thickening of the blastoderm margin, which will be called germ ring (GR)[5] from now on. It greatly accentuates the epibolic movement and will be obliterated once overgrowth of the yolk is complete.

Involution takes place along the whole GR of the expanding blastodisc. (Some authors refer to involution as gastrulation, in contrast to epiboly.) The marginal deep cells (DC) of the outer layer, now called epiblast, which is subjacent to the EVL, involute and form the inner layer, now called hypoblast. This process is exclusive of the EVL, as already observed by 'Goette (1873, 1878)', 'Oellacher (1873)', 'von Kowalewski (1886a)', 'Wilson (1889/1891)' and 'Kopsch (1898)' (Fig. 11/6).

Involution was first described by 'Goette (1869 and 1873)' in the trout. He called it 'Umschlag'. This was followed by reports from 'Haeckel (1875)' for pelagic teleosts, 'His (1878)' for the salmon, *Salmo salar*, 'Henneguy (1880 and 1888)' for the trout, perch and pike and subsequently by 'Ziegler (1882)' for the trout, salmon and *Rhodeus amarus*, 'Agassiz and Whitman (1884)' for pelagic fish, 'Janosik (1884)' for *Ctenolabrus* and *Tinca*, 'Goronowitch (1885)' for the salmonids, 'Cunningham (1885a)' for 3 species of *Trigla*, 'von Kowalewski (1886a,b)' for the goldfish and *Gobius*, 'Wilson (1889/1891)' for *Serranus atrarius* and 'Morgan (1895)' for *Ctenolabrus* and *Fundulus* (Figs. 11/6, 11/9, 11/13, 11/14). 'Kingsley and Conn (1883)' investigated the development of *Ctenolabrus* and noted that involution 'processes much more rapidly from the posterior or embryonic portion of the blastoderm than from any other portion of its margin, and at the anterior portion more rapidly than at the sides'.

Other early embryologists were at variance with this view and maintained that the hypoblast arises from the epiblast by the process of delamination '(von Baer, 1835; Stricker, 1865; Rieneck, 1869; Weil, 1872; Oellacher, 1873; Hoffmann,

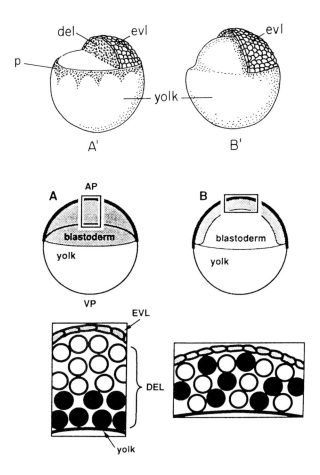

Fig. 1l/5 Organization of the blastodisc/blastoderm in the zebrafish at the mid-blastula stage. Figs. A' and B' modified from 'Solnica-Krezel and Driever (1994)'; Figs. A and B from 'Warga and Kimmel (1990)' (with kind permission from The Company of Biologists, Ltd., Cambridge, England). **A'** - Three-dimensional representation with blastodisc cut open at right angles to the animal-vegetal axis. Note blastodisc filled with deep cells. **A** - Diagram of whole mid-blastula at 30% of epiboly of the blastoderm. Below: cut-away diagram (area indicated by vertical rectangle in A) displaying two types of deep-cells (white and black). **B'** - Three-dimensional representation of specimen at 50% epiboly with animal half cut open at right angles to the animal-vegetal axis. Blastoderm has greatly thinned and yolk bulges towards the animal pole. **B** - Epiboly at 50% showing thinned deep-cell layer (grey). Below: cut-away diagram (area indicated by horizontal rectangle in B shows that deep cells have moved outward (radially) intercalating with superficial deep-cells. This has caused thinning of the blastoderm. AP - animal pole; DEL, - del deep-cells; EVL, - evl envelope cells (*periderm*); P - periblast; VP - ventral pole.

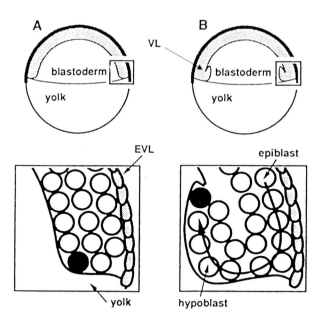

Fig. 11/6 Involution in zebrafish egg (Warga and Kimmel, 1990, with kind permission of The Company of Biologists Ltd. Cambridge, England). **A** - Diagram of onset of involution. Below: cut-away diagram (area indicated by square in A). The black deep-cell is at the front of the wave of involuting cells. **B** - Involution has progressed and a ventral lip (VL) has been formed. Below: cut away diagram (area indicated by square in B). Involution has increased and black deep cell has moved forward. The hypoblast has been generated. EVL, - envelope cells (*periderm*); VL - ventral lip.

1881, 1882a,b and Ryder, 1884, 1887)'. Still other early investigators suggested that delamination as well as involution were involved. 'Brook (1885b)' studied pelagic ova and above all those of *Trachinus vipera* and discussed the results of various previous authors. He concluded 'I take it that invagination in the true sense means an ingrowth or an infolding of a layer already existing, the archiblast. If this be so, there is no true invagination in such pelagic ova as those here described, and the hypoblast is not derived from the archiblast at all, but from the periblast and the yolk by a process of segregation'.

Involution is more extensive in the region of the future embryonic axis (dorsal lip of the blastopore) than elsewhere along the GR (Figs. 11/6, 11/8, 11/13, 11/14). Histologically, involution is displayed by mitotically active DC, which go to form the hypoblast '(Sumner, 1904)'. Already the early embryologists e.g. '(Kupffer, 1884)' talked about 'Wanderung' (migration) and 'Verschiebung' (shifting) of cells. Some had observed amoeboid movements in the trout, *Salmo fario* ['Rieneck (1869)'; 'Romiti (1873)'; 'Weil (1872)'; 'Morgan (1895)'].

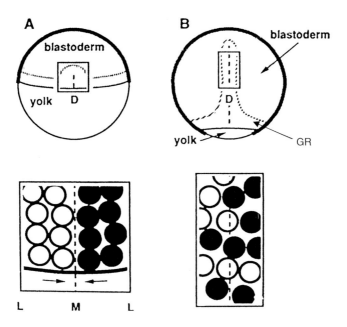

Fig. 11/7 Germ ring established and embryonic shield develops. (Warga and Kimmel, 1990. with kind permission of The Company of Biologists Ltd. Cambridge, England). **A** - Embryonic shield present. Below: Cut-away region (indicated by square in A) shows that the deep-cells converge (arrows) towards the midline. **B** - Embryo lengthened at 80% epiboly. Below: Cut-away region (indicated by rectangle in B) shows intercalation of deep-cells in the mid-line. This so-called convergent extension results in a lengthening of the embryonic axis (dashed line). D - dorsal; GR - germ ring; L - lateral; M - midline.

The hypoblast of the ES is more closely apposed to the undersurface of the epiblast (future embryonic ectoderm) than to the underlying periblast. In *Salmo* and *Serranus* the presumptive entodermal cells (also called primary hypoblast) lie at the caudal edge of the ES but are not exposed to the surface, while in *Fundulus* the primary hypoblast is exposed (Fig. 13/4). It has been suggested by 'Sumner (1904)' that the entoderm arises from a collection of cells called 'peristomial thickening', which arises from the posterior margin of the developing embryo. According to 'Perihar (1979)' they are the first cells to involute. When involution begins, the primary hypoblast moves forward, i.e.in a centripetal direction, below the epiblast (future embryonic ectoderm), and thus becomes the entoderm (or secondary hypoblast) (Fig. 11/13, 11/14). (Disputed origin of entoderm see also

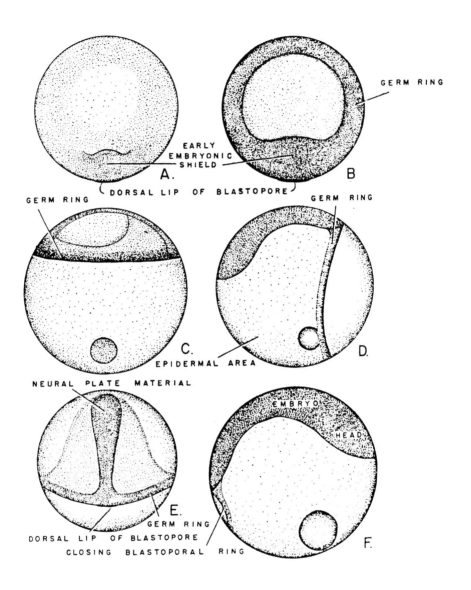

Fig. 11/8 Gastrulation as observed in *Serranus atrarius* (Wilson, 1889/1891). **A.** Surface view of the blastoderm, at 16 hours. Dorsal lip at the posterior pole. **B.** Surface view at 20 hours. Germ ring and blastodisc with embryonic shield evident. **C.** Side view of embryo at 20 hours. Anterior pole of blastodisc at the left and posterior pole to the right. Oil-globule in yolk. **D.** Side view of embryo at 25 hours. Embryo delineated and surrounded by extraembryonic epidermal area. Epiboly about 80%. **E.** Embryo of 25 hours, viewed from above. Neural plate discernible. **F.** Lateral view of embryo of 31 hours. Blastopore almost closed.

Gastrulation

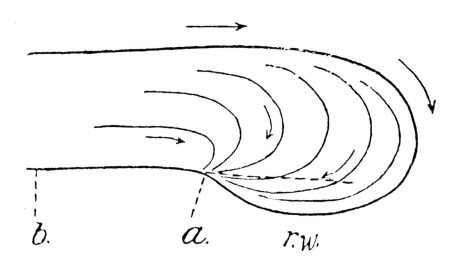

Fig. 11/9 Diagrammatic section to illustrate formation of the germ ring (Randwulst). (Wilson, 1889). **a** - apical line. **b** - blastodisc r.w. germ ring (G. Randwulst). Apical line designates the site where 'the concentric lines of growth by which the 'randwulst' is established, meet'. It marks the inner edge of the 'randwulst'.

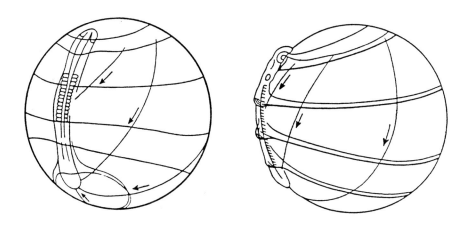

Fig 11/10 Schematic representation of the epiboly in the salmon (His, 1873). Left: as seen obliquely from the side. Right: as seen in profile. Four stages projected on top of each other reveal that the germ ring moves over the yolk sphere in an eccentric manner. The arrows indicate the route from germ ring to embryo; it is evident that the point opposite the embryonic *primordium* has the longest way to overcome to reach the embryo.

under Chapters 13 and 15). 'Ho and Kimmel (1993)' report a complete blocking of involution by fibronectin. In the zebrafish reactivity was not present at the site where involution occurs. Marginal cells remain pluripotent throughout the late blastula and early gastrula. A commitment to a hypoblast-derived fate (mesoderm and entoderm) arises at mid-gastrula.

At the end of epiboly and involution the epiblast, consisting of remaining, non-involuted cells subjacent to the EVL, represents the future embryonic and extra-embryonic ectoderm. The ectoderm will give rise to the epidermis, CNS and the anlagen of the main sensory organs. The inner hypoblast represents the future mesentoderm. The newly-formed notochord, the result of mesoderm involuted over the dorsal blastoporal lip, will induce in the ectoderm above it the neural plate which designates the future dorsal midline of the embryo. This will form the solid neural rod and the brain. The putative non-chordal mesodermal cells involute into the embryo over the dorso-lateral lips of the blastopore. They will give rise to the somites (which lack a cavity, the myocoele) and lateral plate (somatopleura and splanchnopleura) as well as blood cells and vessels. The involuted entoderm is seen as an elongated plate of cells located in different positions (see Chapter 13). It will give rise to the intestine and its accessory structures.

From 1940-1965, apart from histological studies, the introduction of time-lapse motion pictures had allowed the observations of morphogenetic movements in live specimens of the zebrafish and had as a result again confirmed the occurrence of involution ['Lewis and Roosen-Runge, (1942, 1943)'; 'Hisako and Battle (1957)'; 'Hamano, 1964)']. Therefore, after the initial controversies it became generally accepted that the hypoblast arises as a result of involution. This remained unchallenged until 'Ballard (1966-1982)', on the basis of a series of experiments on various fishes, maintained that no involution whatsoever takes place (see Chapter 13 on fate maps). He insisted [as 'Rieneck (1869)' had done] that the hypoblast arises by delamination and that this is followed by outward migration of cells from the centre of the blastoderm. More recently, i.e. from 1988 onwards, scientific evidence based on experiments with an ever increasing degree of sophistication, may now have settled once and for all that involution of the DC is taking place: It was shown that the DC of the rosy barb, *Barbus conchonius*, possess prominent nuclei, which allow their monitoring during epiboly by Nomarski differential-interference-contrast microscopy. This technique provides direct evidence of extensive involution of DC (at least 70% of superficial DC were shown to be involved) during the early stages of epiboly. Subsequently, cell lineage analyses of the zebrafish, another teleost in the same family as the rosy barb, confirmed, too, that involution takes place. Single blastomeres were injected with tracer dyes and followed '(Gevers and Timmermans, 1991)'.

Gap junctions are known to play an important role in controlling embryonic development. They allow the direct transfer of ions and small regulatory molecules from cell to cell without their release into the extracellular space. Gap junctions connect blastomeres with each other and with the periblast during cleavage and blastula as demonstrated with injecting fluorescein dyes into early embryos (see Chapter 10). Gap junctional communication between periblast and blastomeres was studied in the cyprinid *Barbus conchionius* by 'Gevers and Timmermans (1991)'. At the onset of epiboly Lucifer yellow transferred from the periblast to all blastomeres. Between 10% and 40% epiboly dye-coupling became restricted to the marginal region. Between 40% and 60% epiboly a ring-shaped group of labelled cells — probably involuted during early gastrulation and representing the leading edge of the involuting hypoblast — was observed. At 60% epiboly and later, the blastoderm cells are no longer dye-coupled with the periblast. The ringhaped hypoblast cells migrate towards the dorsally located embryonic axis. A similar dye-transfer between periblast and blastoderm was observed in *Cyprinus carpio* '(Meijer and Kronnie, 1996)'. The same authors reported pilot experiments into the developmental function of gap junctional communication using antibodies against connexin. Blastomeres and periblast of *Fundulus* exhibit dye and electrical coupling, as shown by microelectrode impalements, while uncoupling is observed at a well-defined developmental stage, i.e. at the beginning of epiboly when the marginal blastomeres are no longer in cytoplasmic continuity with the periblast '(Kimmel et al., 1984)'. 'Bozhkova and Voronov (1997)' showed that in both *Misgurnus fossilis* and *Brachydanio rerio* the basal layer of DC remained connected with the periblast by gap junctions. Until the late blastula Lucifer yellow and fluorescein transferred from the periblast into the basal cell layer in a uniform way. However, at the 10-20% epiboly stage the intensity of fluorescein increased in the periphery of the blastodisc and at the early gastrula stage the germ ring intensively stained, including besides the marginal DC cells also marginal EVL cells. Subsequently with the development of the embryonic disc the ring-like staining pattern changed to a bilateral one, with increased staining intensity in the dorsal part of the embryo. Accordingly, gap junctions, which initially had been ubiquitous, became concentrated at the blastoderm edge (EVL cells and precursors of ectoderm cells). At the early gastrula stage, involuting prospective mesoderm cells display more effective gap junction connections with the periblast than with the overlying putative ectoderm. Dye transfer was observed until the stage of 60-70% epiboly when the gap junctions finally uncouple blastoderm cells from the periblast.

The above-mentioned contraction of the E-YSL (11.2) causes a crowding of its nuclei, which, as a consequence, are moved to nucleate the I-YSL below the blastodisc. Once epiboly has commenced, nuclear divisions gradually cease in the periblast. Therefore, it has been suggested that once the machinery of cell division in the periblast has come to a halt, it becomes available for E-YSL contraction '(Betschaku and Trinkaus, 1978; Trinkaus, 1992)'. At a later stage, the periblast nuclei will become very large, as already reported by 'Kowalewski (1886b)', and will begin to divide amitotically (see Chapter 10). These observations conform to the descriptions of 'globules parablastiques' described by 'Henneguy (1888)'.

As mentioned above, the multinucleate E-YSL 'epibolizes' vegetalwards, together with, but ahead of, the EVL. This has already been stressed by the early investigators around the end of the 19th and beginning of the 20th century.

However, only recent evidence established the periblast as the motive force and that contraction of the narrow belt of E-YSL signals the commencement of epiboly. It has been suggested that the contraction of the E-YSL continues as a wave moving further and further towards the vegetal pole. In *Fundulus* the surface of the periblast increases 12-fold during epiboly and the surface of the YCL, with which it is confluent, diminishes in tandem and ultimately disappears. This means that the vegetally advancing E-YSL is fed by cytoplasm from the YCL, or, in other words, the nuclei and organelles flow from the E-YSL into the YCL. In *Fundulus* and *Oryzias*, epiboly is divided into three phases (as already mentioned above). During phase I the E-YSL spreads widely in advance of the EVL. In phase II the two layers move together and in phase III the E-YSL again moves ahead. SEM-analysis revealed a bouquet of 150-300 microvilli sprouting from the E-YSL at the site of the closing blastopore (diameter 10 microns). (Microvilli are not observed in the E-YSL of phase I and II.) '(Fluck *et al.*, 1981, 1982)'.

'Clapp (1891)' observed in the toad fish, *Batrachus tau*, that the blastopore closes quite a distance away from the posterior end of the embryo (Fig. 11/11). It should be mentioned at this point, too, that in *Ameiurus nebulosus* the caudal end of the embryo can withdraw from the GR at some stage and reunite again before the blastopore closes '(Reis, 1910)' (Fig. 11/12).

Mention should also be made of the special case of the annual fishes, e.g. *Austrofundulus myersi*. These oviparous cyprinodonts are found in bodies of fresh water that dry up seasonally. In order to survive, these fish undergo multiphasic diapauses. During their annual development a process of complete dispersion and subsequent reaggregation of amoeboid DC is interposed between cleavage and embryogenesis '(Worms, 1972)'. Dispersion is completed prior to the completion of epiboly and the DC remain randomly dispersed for several days. This is followed by a slow process of reaggregation during which embryogenesis begins. The development of the annual *Nothobranchius* is similar to the above-mentioned species. However, fusion of EVL cells is observed just before commencement of epiboly and again during DC aggregation '(van Harleem *et al.*, 1983)'.

Fig. 11/11 Transient separation of embryo and yolk plug in the toad-fish *Batrachus tau*. (Clapp, 1891). **a** - blastodisc with first appearance of axial thickening. **b** - germ ring with well defined embryo. This is shortly after the notch first makes it appearance. **c** - The two sides of the germ ring approach each other forming an acute angle behind the embryo, and giving rise to a very conspicuous notch. **d** - The notch is seen at a little distance behind the embryo; a shadowy connection may be traced between the germ ring and the embryo. Unusual position of KV. **e, f, g** - The notch is seen to retract farther behind the embryo, as the thickened margins of the blastodisc unite, and it disappears shortly before the completion of the closure of the blastopore. The egg is very large, 5 mm in diameter. It is adhesive and may be found attached to stones. Adhesive disc is shown in Figs. c, e, f, g situated at the bottom and in Fig. f in full view.

Gastrulation

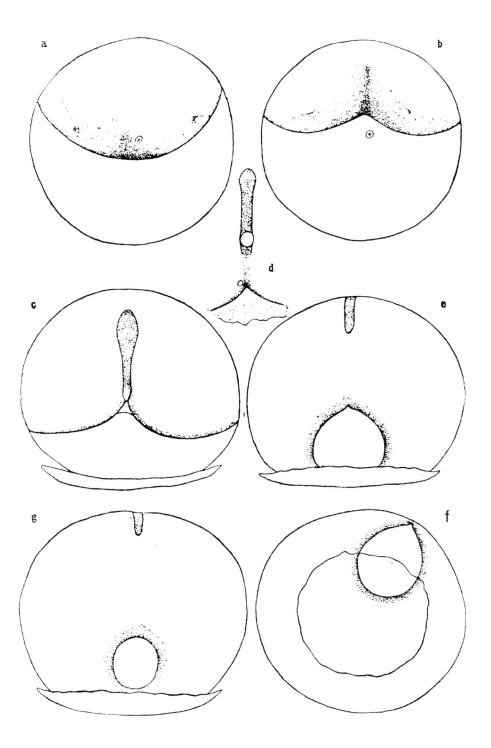

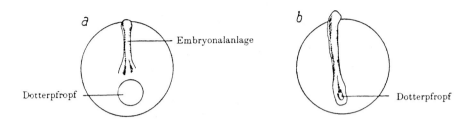

Fig. 11/12 Separation of developing embryo from germ ring in *Ameiurus nebulosus* (Reis, 1910). **a** - The caudal end of the embryo has completely lost the connection with the germ ring. **b** - Secondarily, the caudal end of the embryo connects again with the germ ring and surrounds the yolk plug as observed in teleosts generally. Dotterpfropf = yolk plug Embryonalanlage = *primordium* of embryo.

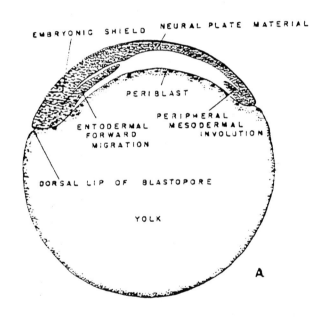

Fig. 11/13 Diagram of early teleost gastrula (Wilson, 1889, with modified labelling).

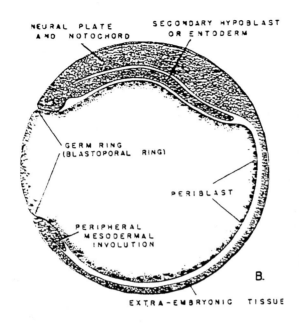

Fig. 11/14 (Wilson, 1889, with modified labelling. Diagram of teleost gastrula not long before blastopore closure.

Notochord and prechordal plate

The prechordal plate (G. praechordale Platte) is represented by a transient embryonic mesodermal tissue, which replaces the notochord anteriorly. It is involved in induction of cephalic structures. While early investigators, and later even 'Nelsen (1953)' and 'Mahon and Hoar (1956)', thought of the pre-chordal plate belonging to the entoderm, it was later shown that it is a derivative of the chordamesoderm. The pre-chordal plate (presumptive head-mesoderm) and presumptive notochord[8] material, which lay anterior to it, move posteriad, involute over the dorsal lip and migrate anteriad between the epiblast and secondary hypoblast. The chordamesodermal material converges posteriad. While the prospective chorda- material involutes over the dorsal lip, the other presumptive mesoderm cells do so over the lateral lips of the blastoporus. As involution progresses the presumptive notochordal cells form a distinct rod (chorda dorsalis) which occupies a median position. Recent cell lineage analyses of the rosy barb and the zebrafish showed that cells that involute early during gastrulation usually form entoderm, and those that involute later, chorda-mesoderm. It was also demonstrated that the order in which cells of a clone involute corresponds to their subsequent position along the anterio-posterior axis of the embryo. The movements of convergent extension (see 11.2) were also followed both in the epiblast and hypoblast and it was shown that mixing among cells occurs within, but not between, these layers '(Wood and Timmermans, 1988; Kimmel et al., 1990; Warga and Kimmel, 1990)'. In the zebrafish *Brachydanio rerio* involution starts at 50% epiboly '(Takeda and Miyagawa, 1994)'. It begins more-or-less simultaneously around the circumference of the blastodisc. Involuting cells move anteriorwards in the hypoblast behind the leading edge of the blastodisc. The DC undergo medio-lateral intercalations and accumulate dorsally to form the embryonic shield which as a result of convergence and extension produces the embryonic axis '(Warga and Kimmel, 1990)' (Fig. 11/7).

In its earliest stages the notochord is made up of two layers of polygonal cells. Subsequently, each cell extends the whole width of the notochord, which is often referred to as a 'stack of coins' (Figs. 11/15, 11/16). In transverse histological sections of the notochord the cells are seen to be vacuolated i.e. to be crossed by cytoplasmic strands, some of them encompassing nuclei (Figs. 11/17, 11/18). Some cells have moved out from the stack and form now an epithelial sheath around the notochord (Fig. 11/17).

The notochord of the early teleostean embryo serves as the major skeletal element for locomotion. It signals the formation of the floor plate and the patterning of the neural tube (Fig. 12/5). Independently, it can also signal the formation of motoneurons. It also plays several roles in the patterning of the somites, the fashioning of the dorsal aorta but seems to be less important for the formation of the axial vein, which may be formed by signals from the entoderm.

Hypochord[9]

Immediately ventral to the notochord, and symmetrically opposite to the position of the ectodermal floor plate, is another transient embryonic tissue, the hypochord[9], also called subnotochordal rod, which is a single cell wide (Fig. 12/4). It has been described by 'Vogt (1842)', 'Balfour (1881)', 'Oellacher (1873)', 'Ryder (1885b)' and others. 'Ryder' mentioned it in *Alosa* and *Salmo* as a well-marked strand of cells and 'Oellacher' was of the opinion that it shares in the development of the aorta along the under-surface of the notochord. 'Balfour'(1885)', 'Wilson (1889/1891)', 'Franz (1897)', 'Pasteels (1934)', 'Brachet (1935)' and 'Mathews (1982)' suggested that it is a derivative of the entoderm. However, 'Oellacher (1873)' came to the conclusion that the small group of cells between the notochord and the entoderm represents the anlage of the dorsal aorta. 'Ziegler (1887)' in a treatise of the origin of blood in teleosts, has drawn the hypochord in many of his figures. 'Henneguy (1888)' observed that in the trout the hypochord (called by him 'tigo subnotochordale') is formed from a mass of 3-4 cells in cross-section which have detached from the underlying entoderm. They are differentiated along an anterio-posterior axis, never form a lumen and never communicate with the subjacent solid primordium of the intestine. In the cephalic region it develops only during the last stages of the closure of the blastopore (induced by the prechordal plate?). According to 'Orsi (1968)' the hypochord in the sole, *Parophrys vetulus*, is a single-cell string of flattened cells formed by the concentration of endoderm cells towards the midline and becomes sandwiched between the gut and the notochord. The hypochord disappears as the aorta develops. However, it persists after hatching in the posterior part of the tail. 'Hatta and Kimmel (1993)' also suggested that it is a derivative of the entoderm. The hypochord is very distinctive with

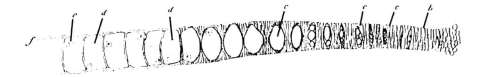

Fig. 11/15 Posterior end of developing notochord in the herring (Kupffer, 1868). b - posterior end; c - *nucleus*; d - secondary notochord cells with their *nucleus* e; f - notochordal sheath.

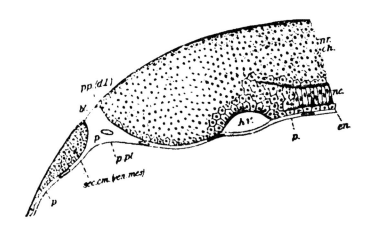

Fig. 11/16 Median longitudinal section showing posterior end of developing notochord in the sea bass (Wilson, 1889/91). en - *entoderm*; kv. - Kupffer's vesicle; nr.ch - neurochord; p - periblast; pp.(dl) - posterior pole of blastoderm (dorsal lip); p.pl. - periblast plug (G. Dotterpropf); sec.cm. - Ventrally involuting mersoderm (ventral lip).

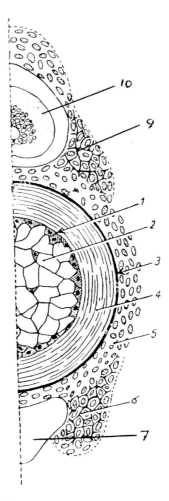

Fig. 11/17 Notochord of teleost in transverse section. Note: *primordia* of neural (9) and haemal arch (6) (Portmann, 1976, with kind permission of Benno Schwabe, Basel). 1 - *epithelium* of notochord; 2 - vesiculated cells; 3,4 - primary and secondary notochordal sheath; 5 - autocentric mesoblast cell material; 6 - *primordium* of ventral *arcualia (basiventralia)*; 7 - blood vessel; 9 - primordium of neural arch (basidorsalia); 10 - neural tube.

Nomarski optics in the body trunk and particularly in the tail of normal embryos. The only tissue besides the notochord that associates closely with the hypochord in the 42 hours old zebrafish embryo is the endothelium of the dorsal aorta. All these specific midline structures do not extend into the anterior head '(Hatta and Kimmel, 1993)' (Fig. 12/4).

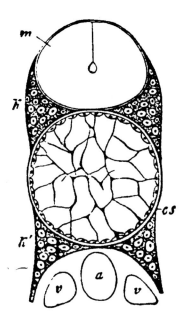

Fig. 11/18 Transverse section through spinal column of a young salmon (Gegenbauer, 1881). a - dorsal *aorta*; cs - sheath of notochord; k - neural arch; k' - haemal arch; m - spinal chord; v - cardinal veins.

'Kodjabachian *et al.* (1999)' put forward a new model for the specification of midline fates. They refer to an earlier view where chordal and prechordal fates are established within the organizer during gastrulation. Subsequently, the notochord sends signals to adjacent tissues to induce additional cell types such as the floor plate in the neuroectoderm and the hypochord in gut entoderm. In contrast, in the new model, based on recent results obtained from the zebrafish (and *Gallus domesticus*), the progenitors of notochord, floor plate and hypochord arise from a common region within the organizer (based in the embryonic shield).

Formation of the secondary hypoblast

The presumptive entodermal cells lying at the caudal end of the embryonic shield (see fatemap, Chapter 13) migrate forward below the epiblast. Thus they produce the entodermal layer or secondary hypoblast. In the trout the entoderm of each side grows under the notochord to form one layer ['Weil (1872)'; 'Goette (1878)'; 'Sumner (1904)'], while in the sea bass it is discontiunuous across the midline '(Wilson, 1889)' (Figs. 19/1, 19/2). In both cases the entodermal cavity is formed only at a later stage.

Cell movements

EM analyses of involution revealed that in the GR of *Fundulus* there were observed the same types of locomotion as in the blastula, namely by cellular projections containing actin fibres. Involution and convergence are not affected by UV while epiboly is. This would indicate that these movements are timed by a clock, which is independent of the degree of epiboly. EM has shown that convergence and involution are the result of active migration of individual DC, which move between EVL and the periblast with the help of filopodia, lobopodia, blebs and occcasionally lamellipodia. The cells move in the direction of the protrusions. Their locomotion is not limited by contact inhibition, which is in contrast to the behaviour of the EVL cells. Most DC move in clusters and it is only the cells at the free surface that show protrusions and appear to be contact-inhibiting. Nevertheless, the clusters are moved by constant forces which impose a directional bias on what otherwise would be random cell movements. This is analogous to the movement of birds in a flock. It is not yet known how these movements are initiated and propagated. Converging DC in the GR of *Fundulus* move by filopodia and lamellipodia and, much less frequently, by protuberances called blebs. Blebs appear hyaline in vivo, most probably due to the lack of organelles. Some of the filopodia extend up to 50 microns or more '(Trinkaus, 1966; Kageyama, 1977)'. To use Trinkaus' words 'teleost gastrulation thus offers a classic example of convergent extension, a hallmark of vertebrate gastrulation' '(Trinkaus, 1992)' (Fig. 11/7A,B). The recent review by 'Small *et al.* (2002)' on cell motility should be mentioned in this context.

Cells that translocate by blebbing (pulsatory activity) are not contact-inhibiting and, therefore, do not need free space to move into. Both involuted and non-involuted DC converge towards the ES while cell division continues at about 12%. Time-lapse filming of involution in *Fundulus* gastrulae has shown that, while there is random, often meandering, cell movement, it is with a directional bias towards the ES. It is the involuted cells that are responsible for the antero-posterior extension of the embryonic axis '(Hamano, 1964)'.

When *Fundulus* gastrulae are treated with colchicine, which disrupts microtubules, locomotion by means of filipodia and lamellipodia is eliminated while lobopodia, which appear to be formed by cytoplasmic flow into blebs, remain intact.

'Trinkaus *et al.* (1992)' studied the gradient in convergent cell movement in the GR ring of *Fundulus*. They observed that the cells in the dorsal GR nearest to the embryonic shield move towards it at a net faster rate than those further away. They suggested that this was due to increased meandering of the more distant cells. This led to the hypothesis that there was a gradient of cues directing cells to the embryonic shield. 'Trinkaus (1998)' tested this hypothesis by pursuing the cells with Nomarski differential interference contrast optics and time-lapse video. He concluded that although both dorsal and ventral cells meander, the latter do so much more and, as a result, show little or no directional movement towards the embryonic shield. This would suggest that the embryonic shield attracts cells, perhaps chemokinetically, chemotactically, galvanotactically or haptotactically (by adhesiveness).

EM revealed the absence of junctional complexes and interdigitations between the adjoining plasmamembranes of motile DC of *Fundulus* '(Hogan and Trinkaus, 1977)'. Instead, appositions of plasmamembranes separated by a distance of 26-28 microns are observed and, therefore, appear to be slightly adhesive. The fast locomotion of *Fundulus* DC is thought to be due to this light adhesion, strong enough to give traction but also weak enough to break readily. On the basis of freeze-cleave electron- microscopy it has been suggested that the observed intramembraneous particles are indicative of the presence of glycoprotein molecules responsible for adhesivity. (EVL cells, which are non-adhesive, have been shown to lack intramembraneous particles.) Gap junctions, though few in numbers, are also seen in DC which would explain results showing electrical coupling between DC '(Hogan and Trinkaus, 1977)'. It seems, therefore, that the motiliy of DC in *Fundulus* depends partly on their not being too firmly joined together.

Tracing of vitally stained areas from the fate map (Chapter 13) to gastrulating stages

'Pasteels, 1936' showed with the help of diagrams the territorial displacements of the various areas of the fate map (see Chapter 13). The prospective prechordal plate and notochord involute first from the surface over the dorsal lip of the blastopore. Most of the mesoderm cells involute over the lateral lips and move caudad within the germ ring to be incorporated in the posterior region of the embryo (Fig. 11/20). The delineation of the future entoderm and mesoderm was followed by time-lapse cinemography in other species without giving clear results '(Lewis and Roosen-Runge, 1942; Hamano, 1964)'. It must be stressed again that the limits of the areas in teleostean fate-maps are less precisely demarcated than those of the amphibia. It should also be emphasized that 'Pasteels' and 'Oppenheimer' seemingly did not take into consideration that the whole developing organism is covered with the EVL. It would appear, therefore, that the dye they applied had penetrated the EVL and marked the subjacent layer of DC. 'Brummet (1955)', on the contrary, inserted a needle loaded with carbon particles into the germ ring of *Fundulus*. The carbon particles adhered either to the EVL or to underlying DC. However, the goal of 'Brummet (1955)' was to elucidate the formation of the tail bud.

11.3 FORMATION OF THE YOLKSAC (EPIBOLY AND INVOLUTION)

While embryogenesis is taking place, the rest of the GR 'epibolizes' over the yolk sphere and the EVL and periblast eventually surround it completely. It was already noted by 'Morgan (1895)' that the yolksac receives new DC as a result of cell divisions. In the mouth-brooding cichlid *Haplochromis* only very few DC are present in the yolksac at early epiboly. After 50% epiboly no DC, and therefore no

GR, are seen extraembryonically in this species, while in all other teleost tested extra-embryonic presumptive ectoderm expands and presumptive mesodermal cells involute '(Fleig, 1993)'. It is obvious that the space between extraembryonic EVL and periblast does not contain observable sheets of lateral line mesoderm as would be the case in the amniotes. Instead, a whole mixture of motile cells is encountered (see Chapter 12). The entoderm does not participate in this extraembryonic overgrowth.

The mesodermal cells disperse and move over long distances by circus movements (at random) by means of filolamellipodia and blebbing. The average speed for these cells is given as 0.5 microns/minute in *Oryzias* '(Trinkaus, 1973a)'. Some authors refrain from talking about 'extraembryonic areas' and call them instead 'the belly region of the future embryo'.

In the yolksac of the *Fundulus* embryo several types of motile cells were distinguished and followed by time-cinemicrography with phase contrast optics '(Armstrong, 1980)': 1) A layer of highly flattened stellate cells underlies the EVL. Their margins extend into processes, which contact neighbouring cells. The borders are in constant motion and thereby continuously change the shape of the cells. During epiboly the stellate cells can be seen trailing immediately behind the margin of the GR. Mitoses are observed and the daughter cells stay in the stellate layer, which remains a monolayer adherent to the EVL. The ultimate fate of these cells has not been determined. 2) The flattened elongated epithelioid cells, below the stellate layer, occur in clusters. Many migrate into the ES. Those that remain in the yolksac may contribute to endothelia 3) The compact rounded amoebocytes, which do not contact each other, wander rapidly over the yolksac in an undirected fashion. They move by the extension of lamellipodia, sometimes lobopodia, with scalloped margins. These cells also invade the embryo proper. 4) The highly mobile melanocytes possess extendable processes. Some investigators suggest that these cells originate from the GR. Others think they are of late embryonic (neural crest) origin. Their movements are irregular, often in zig-zag fashion, but generally away from the embryo proper.

The presumptive endothelial cells and the small spherical erythroblasts which occur in clusters will eventually form the yolksac circulation (see Chapter 18).

11.4 TAIL BUD [10]

At the end of the period of gastrulation a tail bud has formed. It consists of a mass of undifferentiated cells, which will give rise to the posterior trunk and the tail (Fig. 11/19). 'Peter (1947)' and others refer to the gastrulation culminating in the tail bud as a 'primary way'. In contrast, the subsequent development of the tail and some parts of the posterior trunk is thought to be effected by bypassing gastrulation and neurulation, i.e. as a direct development from an undifferentiated cell mass, often called blasteme. This process is referred to as 'secondary way'.

'Ryder (1882b)' in his publication on *Belone longirostris* reports 'The conversion of this caudal plate [tail bud] into the mesoblastic, epiblastic, and hypoblastic structures of the tail end of the embryo accordingly appears to me to be beyond question'. Others, like 'Goronowitsch (1885)' and 'Henneguy (1888)' likewise considered the tail bud as a growth centre. 'List (1889)' had observed that in *Salmo* the 'embryonic tail bud of Oellacher', made up of undifferentiated cells, lies above Kupffer's vesicle (KV). After the disappearance of KV, 'Oellacher's tail bud' and cells of the GR form the definitive tail bud consisting of undifferentiated cells '(Schwarz, 1889)'. 'Sumner (1904)' called the tail bud a 'zone of growth' or a 'noeud vital'. According to him, at its anterior end cells are continually differentiating into the neural axis, chorda and somites, and the 'gut'hypoblast' is continually completing itself in the middle line beneath the newly-formed portions. At its posterior end, it is continually receiving new material from the laterally-situated portions of the blastoderm margin. Its growth is, however, to a large degree intrinsic and not dependent upon external sources of supply.' Also 'Kopsch (1904)' characterized the tail bud as an 'undifferentiated caudal mass' or site of intense cell proliferation. This latter notion was disproved for the trout by 'Pasteels (1936)'. He had made mitotic counts of this region and had concluded that the number of mitotic figures at that locus was not higher, in fact it was lower, than in any other regions. Subsequently, he characterized the tail bud as 'un carrefour de mouvements cellulaires où les courants d'invagination, de convergence et d'extension se rencontrent' (a traffic jam of cellular movements where the streams of invagination, convergence and extension meet) (Fig. 11/19). While the vital staining experiments of 'Oppenheimer' and 'Pasteels' had suggested that the cells of the germ ring 180° from the posterior embryonic axis form the bulk of the tail bud, 'Brummet (1955)' showed that the entire tail bud was formed by axial convergence of cells situated less than 90° on each side of the posterior embryonic axis (posterior embryonic shield). These cells were incorporated into the tail bud at the time of the closure of the blastopore or shortly afterwards. Cells of the germ ring marked between 90° and 180° appear on the yolksac epithelium posteriorly and slightly laterally to the formed tail bud. A mark placed 180° from the embryonic axis at the 'yolk plug stage' or stages before that, was not incorporated into the tail bud but was observed posterior to it.

Since the vital staining and carbon techniques did not furnish an unequivocal answer as to the origin of the tail bud, 'Brummet (1968)'devised another experimental approach, employing deletion-transplantation techniques. These confirmed that the posterior embryonic shield of *Fundulus* supplies most of the material for the formation of trunk and tail.

11.5 INVAGINATION

There is a total absence of any experimental evidence of invagination or ingression from the surface of the teleost blastodisc. It should be stressed again that

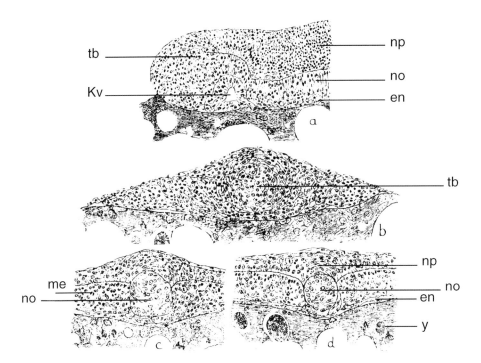

Fig. 11/19 Tail bud of *Salmo irideus* (Pasteels, 1936). **a** - Median longitudinal section through the tail bud region. The cavity is the lumen of Kupffer's vesicle. **b** - A transverse section through the same region slightly posterior to Kupffer's vesicle. **c** - A more anterior section, which displays the differentiation into the different germ layers. **d** - A still more anterior section. en - *entoderm*; kv - Kupffer's vesicle; me - *mesoderm*; np - neural plate; no - notochord; tb - tailbud; y - yolk.

several authors, over the years, mistakenly used the word 'invagination' to describe involution. The word invagination should be used only when the end product is a vaginal cavity such as the gastrocoel or archenteron as e.g. present in the amphibia.

Summary

Gastrulation involves the transformation of the blastula into a layered embryo, the gastrula. The process of **epiboly** consists of the spreading of the EVL (a cellular tissue) and the periblast or YSL (a syncytial layer) over and across the globe of yolk. Since the margin of the blastodisc/blastoderm (GR) is regarded as the blastoporal lip, epiboly ends with the closure of the blastopore. This event is not

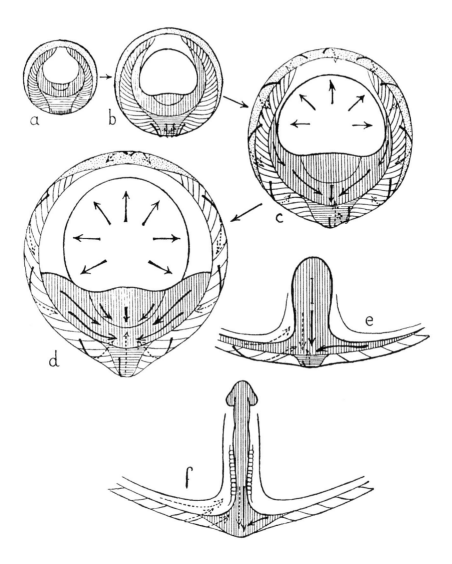

Fig. 11/20 Representation of the presumptive (prospective) areas and their displacement during gastrulation of the trout (Pasteels, 1936). a-d shows a dorsal surface view of the blastodisc. e-f depicts only the embryonic part and the adjacent germ ring. a blastula b-f successive stages of gastrulation. The solid arrows indicate the movements at the surface; the dashed arrows signify movements below the disc. White area: *ectoderm* (in the centre the embryonic *ectoderm* and in the periphery the extraembryonic *ectoderm*). Vertical striations: *neuroectoderm*. Anteriorly, the cerebral *primordium* is separated from the medullary part by a curved line. Horizontal striations: zone of the future notochord. Oblique striations: zone of the future somites. Dotted area: lateral plate *mesoderm*. Dense vertical striations, present in diagram a) only: *entoderm*.

connected with the age of the embryo proper, which arises as a thickening in the GR, called the embryonic shield (ES).

Involution: At various stages of epiboly blastomeres involute over the whole expanse of the GR of the expanding blastodisc. Involution is most pronounced at the dorsal lip. Involuting cells give rise to the hypoblast (entoderm and mesoderm), in an anterior-posterior sequence, while the epiblast does not involute and represents the future ectoderm (epidermis and neuroectoderm). EVL cells, too, do not undergo involution. Shortly after gastrulation has begun, the embryonic axis appears and lengthens along the dorsal side of the embryo. During convergence and extension of the embryo the cells in the epi- and hypoblast mix. It should be stressed that mixing occurs only within, and not between, these layers. The cells move with the help of filo-, lobo- and lamellipodia during convergence and involution. Involuted mesoderm gives rise to so-called midline structures of the embryo. The mesodermal notochord, which anteriorly does not extend to the tip of the head, is replaced in this region by the pre-chordal plate. Immediately ventral to the notochord stretches the one cell wide hypochord. Both prechordal plate and notochord are transitory structures, which play a key role in inductive events. At the end of involution the non-involuted cells are the future ectoderm which will form the embryonic and extraembryonic integument and the nervous system of the embryo. The origin of the entoderm has not been definitely settled (see Chapters 13 and 15).

The yolk sphere becomes overgrown by the EVL and the yolk syncytial layer (YSL), which form the walls of the **yolksac**. Cells representing the future extraembryonic ectoderm and mesoderm move into the space between the two layers.

Application of dye was used with the aim to establish the final destination of certain areas of the **fate map** (see Chapter 13).

The **tail bud** is a mass of cells, which has to this day intrigued the researchers into fish development. It forms the tail, which does not arise as a result of gastrulation and neurulation processes. The tail develops as a solid outgrowth and contains the same mass of cells which give rise to the notochord. 'New matter' seems to be added by apposition and intossusception. Some investigators suggested that amoeboid cell movement may be involved. The tail seems to build up all its structures long before a caudal vascular system has been established.

Endnotes

1. G. Embryonalschild of 'Kupffer (1868)' and 'Oellacher (1873)'
   F. Bande primitive of 'von Baer' Bandelette primitive of 'Vogt' Bandelette embryonnaire of 'Lereboullet' Ecussion embryonnaire of Henneguy
   I. disco germinale of 'Romiti' scudo embrionale of 'Raffaele'
2. endnote[1] of chapter 10
3. see endnote[9] of chapter 10
4. see endnote[14] of chapter 10[5]; see endnote[3] of chapter 10

[6] blastoporus
G. Dotterloch
F. trou vitellaire
I. blastoporo

[7] see endnote [13] of Chapter 10

[8] G. Rückensaite
F. corde dorsale
I. chorda dorsalis

[9] G. subchordaler Stab
F. tige subnotocordale, tige sous-notocordale

[10] G. Schwanzknospe of 'Oellacher' Indifferente Zellmasse of 'Schwarz' Terminale Zellmasse of Schwarz
F. bourgeon caudal of 'Henneguy', écusson embryonnaire, nodosité terminale, noeud postérieur noeud terminal
I. eminenza codale

Chapter twelve

Neurulation

'Ce qui caractérise le développement de l'embryon des Téléostéens, c'est que j'appellerai le développement massif.'

P. Henneguy (1891)

During gastrulation the neural chord (neurocord, neurochord) is formed, induced by the underlying notochord, as mentioned in the previous chapter. The process is called neurulation and the ensueing embryonic stage the neurula.

The thickening first observed in the embryonic shield is the neural plate[1]. In amphibians, birds and mammals this plate invaginates to form a neural groove, which subsequently closes to form the neural tube. This is often referred to as the traditional, or normal, or primary neurulation. Textbooks of vertebrate morphology, when referring to teleostean development, usually stress that neurulation in bony fish differs markedly from that in the other vertebrates though the end result is the same. In contrast to other vertebrates the neural plate of teleosts thickens but does not fold. It separates from the rest of the ectoderm and begins to sink below the surface and the overlying putative epidermis fuses. The solid neural rod[2], often referred to as neural keel, ridge or carina, which projects ventrally towards the yolk, later becomes a hollow neural tube by cavitation. This process is referred to as 'secondary neurulation'. In higher vertebrates secondary neurulation is observed in the tailbud region. In addition to the neural cord the teleostean eyecup and ear arise as solid (massif) thickenings and only subsequently acquire a lumen. All authors who mention cavitation in the teleostean CNS agree that the lumen is first observed in the primordium of the eye. The separation of the solid neural rod from the epiblast (which additionally gives rise to the epidermic strata) takes place relatively very late. Before it has been completed the first traces of the olfactory, auditory and optic primordia are already visible. Clusters of neural crest cells, wedged between the cord and the overlying epidermis, are observed.

The majority of the material for the nervous system (NS) of the teleostean embryo lies within the early embryonic shield. The presumptive NS undergoes considerable antero-posterior stretching and very little convergence. However, the cells lying immediately laterally to it in the shield (Fig. 11/8E) do converge toward the midline, contributing to the formation of the mid-brain, hind-brain, neural chord, and the NS in the tailbud-blastema. The embryonic axis coincides most frequently with the second cleavage plane. The ectoderm forming the dorsal section of the embryonic shield is thicker than the extra-embryonic ectoderm and increases in thickness as the ectoderm cells move toward the midline of the embryonic shield, creating a deep keel that is especially well developed at the anterior end of the shield — the future brain. The early brain appears in living embryos as a pronounced ventral swelling.

In contrast to the evidence of the so-called secondary neurulation in teleostei, recent investigations with dye labelling suggest that the neural cord in *Brachydanio rerio* is formed as a result of infolding similar to primary neurulation in the other vertebrates '(Papan and Campos-Ortega, 1994)'. However, close analysis of the early literature reveals that more than 100 years earlier, and intermittently during the meantime, it was shown already that the neural cord, usually referred to as keel, in certain teleosts represents a 'closed fold' '(Goette, 1873, 1878)' (Fig. 12/1ABC-12/3ABC). But these results were overlooked for a long time and have not been included in textbooks on comparative embryology.

12.1 EARLY INVESTIGATORS

12.1.1 Neural fold

The following early investigators took it for granted that the cerebro-spinal axis in teleosts is formed as a tube as found in amphibia and amniotes: 'Rathke (1833)' for *Blennius*, 'von Baer (1835)' for *Cyprinus blicca*, 'Vogt (1842)' for *Coregonus palea*, 'Lereboullet (1861, 1862)' for the pike, perch and trout, 'Stricker (1865)' for the trout, *Salmo fario*, 'Truman (1869)' for the pike, 'Hoffmann (1881)' for teleosts generally. They noted that a thin skin (epidermic stratum) covers the open furrow. Already 'Vogt (1842)' had noted that the 'cellules épidermoidales' (epidermic stratum) do not take part in neurulation. According to 'Calberla (1877)' the keel (ventral ridge) in *Syngnathus* and *Salmo* is formed as a result of the folding together of the two sides of the primitively uniform epiblastic layer. The epidermic stratum is carried down into the keel as a double layer just as if it, too, had been 'folded in'. He suggested that the 'epidermic stratum' will form the ependyma while the 'nervous layer' will give rise to the true nervous tissue. He was unaware of the results by 'Romiti (1873)' (see 12.1.2). 'His (1878)' described the appearance of a shamrock-shaped anlage of the brain with cross-shaped lumina leading to the neural cord with its canal. He, too, concluded that brain and neural cord are formed as a result of folding. 'Kowalewski (1886b)' mentioned briefly the presence in teleosts of a connection between the alimentary canal and the 'neural groove', in other word a to a 'neurenteric canal'. Equally, 'Kingsley and Conn (1883)' and 'Raffaele (1888)' observed medullary folds but were unable to observe their fusion.

12.1.2 Solid CNS

'Kupffer (1868, 1884)', analyzing the embryos of *Gasterosteus* and *Gobius*, was the first to specify that the teleostean nervous system is laid down as a solid convex keel and suggested that the central canal (canalis centralis) results from a dehiscence of the cells in the centre of the keel. He also stressed that the anlage of the brain is solid. 'Romiti (1873)' accepts the solid formation of the neural cord but observed that the 'epidermic stratum' evaginates dorsally into it forming the future ependyma. According to him, the central canal arises as a result of the separation of the ependymal cells. Solid formation of the neural system was upheld also by 'Schapringer (1871)', 'Weil (1872)', 'van Bambeke (1876)', 'Ryder (1884, 1885a)', 'Henneguy (1885)' and 'Reinhard (1888)'. The neural keel forms in an anterior to posterior progression along the axis as does the neural fold in higher vertebrates. According to the first two authors, the central canal is formed as a result of the simple separation (dehiscence) of the medially situated and radially arranged cells whereas 'Oellacher (1873)' in his treatise on the development of the trout stated that the ependymal canal in the solid neural cord is formed by the destruction and following resorption of the median cells. An internal lumen is first observed ventrally and then proceeds dorsally. 'Balfour (1881)' stressed that the separation of the solid nervous system from the epiblast takes place relatively very late; and before it has been completed, the first traces of the auditory, optic and olfactory anlage are evident. 'Hoffmann (1882a)', on the basis of different staining methods, suggested that the central canal in the nervous system of the trout is partly formed by a dissolution of median cells. 'Cunningham (1885a)' in a treatise on Kupffer's vesicle mentioned that 'the neurocord is present as a thick cord of cells derived from the epiblast, and containing no canal'. 'Henneguy (1888)' described that in the solid CNS the lumina both in the optic lobes and the neural cord are formed by the separation of cells before the ventricles in the brain proper appear. The neural keel forms in an anterior to posterior progression along the axis as does the neural fold in higher vertebrates.

12.1.3 Closed neural fold

'Goette (1873, 1878)' was the first to stress that in the trout the open neural (medullary) furrow observed in amphibia is present as a 'closed fold'. The presence of a fold becomes especially clear when the lateral sheets of the neural cord separate after the epidermis has been cut off (Fig. 12/1A,B-12/3A,B).

'Goette' suggested that the formation of this 'closed fold' takes place as follows: The broad shieldlike ectodermal medullary plate[1] is medially thinned by the inpushing ventral entoderm and more posteriorly by the notochord. This results in an incomplete division of the CNS into two halves and the clear presence of a closed fold. 'His (1878)', 'Hoffmann (1881-1883)', 'Ziegler (1888)', Henneguy (1888)', 'Wilson (1891)' and 'Kopsch (1898)' agreed essentially with Goette's interpretations. 'Wilson (1891)' gave detailed illustrations of the development of

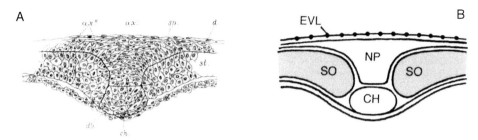

Fig. 12/1 (Goette, 1878) Transverse sections through the trunk regions of trout embryo. (A) Histological section. (B) Diagrams near the head. Notochord (ch), neural keel (ax^I) and side plates (ax^{II}) can be distinguished. *Entoderm* (db) is continuous under the notochord (ch). The EVL (d) covers the embryo.

Fig. 12/2 (Goette, 1878) Transverse sections through the trunk regions of trout embryo older than Fig. 12/1 (A). (B) Diagrams. The side plates of the neural keel have 'shrunk' and the neural keel presents a pseudostratified structure. A 'dorsal dip' (f) is visible and putative epidermal cells (g) have emerged from the axial plate.

the neural chord in the sea bass *Serranus atrarius*. 'Ziegler (1902)' in his textbook on the comparative development of lower vertebrates confirmed that the ectoderm forms medially the solid medullary plate which only later acquires a lumen as previously shown in *Lepidosteus*. The median part of the plate sinks ventrally thereby forming a closed fold. The epidermic stratum does not participate in this fold. 'Jablonowski (1899)' described a closed fold in the region of the fore- and middle-brain of the pike.

All teleosts tested showed a dorsal 'dip' situated above the keel. This dorsal groove, by some authors called neural furrow[4], subsequently straightens and closes. However, this furrow is nothing but a transient longitudinal indentation and is not homologous with the neural groove of the higher vertebrates, which later folds to form the neural tube ['Kupffer (1868, 1878, 1884)'; 'Goette (1869)';

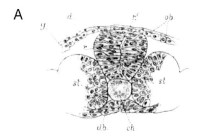 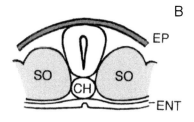

Fig. 12/3 (Goette, 1878) Transverse sections through the trunk regions of trout embryo older than 12/2. (A) Historical section. (B) Diagram. The previously solid primordium of the neurochord exhibits a crevasse (G. Spalt) which eventually changes the neural chord into a neural tube with a lumen. The neurochord has become separated from the *epidermis* (ob). The *entoderm* (db) has thickened, which forecasts the formation of the intestine.

Abbreviations figs 12/1-3, for A: ax^I - median keel of neural plate (axial plate); ax^{II} - lateral parts of neural plate; ch - notochord; d - EVL/*periderm*; db - *entoderm*; f - dorsal 'fold' (dip); g - future *epidermis*; ob - *epidermis*; r - neurochord; r^r - lumen of neurochord; sp - the continuation of the neural plate (sensory plate) into the trunk region; st - somite. 'Goette' refers to the early neural plate as 'axial plate' because it still contains the cells of the future *epidermis*. Once deprived of these cells the neural plate is made up of a 'neural keel', which represents the brain *anlage*, and the 'wings', which represent the 'sensory plates' which give rise to the main sensory organs.

Abbreviations figs 12/1-3, for B: CH - notochord; ENT - entoderm; EP - epidermis; EVL - enveloping layer (periderm): F - 'dip' in dorsal neurochord; NC - neurochord; NP - neural plate; SO - somite.

'Hoffmann (1881)'; 'M'Intosh and Prince (1890)'; 'Eycleshymer (1895)';]. According to 'Goette (1878)' this transitory 'dip' is due to the abovementioned cellular movements resulting in a neural chord (see 'f' in Fig. 12/2). M'Intosh and Prince (1890)' mention that the dorsal groove in teleosts 'wholly passes away'.

'Goette (1878)' stressed that Carlbera's statements (12.1.2) were not to be trusted and goes on to say 'if they could be accepted, the difference in the formation of the medullary canal in teleostei and in other vertebrates would become altogether unimportant and consist simply in the fact that the ordinary open medullary groove is in teleostei obliterated in its inner part by the two sides of the groove coming together. Both layers of epiblast would thus have a share in the formation of the CNS: the epidermic layer giving rise to the lining epithelial cells of the central canal, and the nervous layer to the true nervous tissues.' He stressed, too, that if Calberla's interpretation could be accepted, the difference between the neurulation in teleosts and other vertebrates would become altogether unimportant. 'Balfour (1881)' put forward that 'the explanations of

Goette and Calberla appear to me to contain between them the truth in this matter. The groove above in part represents the medullary groove; but the closure of the groove is represented by the folding together of the lateral parts of the epiblast plate to form the medullary keel.'

'Hertwig (1906)' stressed that there is no special difference between the primary and secondary formation of the neural tube. The former displays an externally protruding fold with an open and broad furrow versus the latter with an internally oriented fold with the folding sheets tightly pressed together. 'Assheton (1907)' refers to the neurulation in teleosts as a secondary modification of the usual method of folding and mentions that a trace of such a folding is seen in the head region of the pike and in the middle trunk region of *Gymnarchus*.

12.2 MORE RECENT INVESTIGATIONS

'Holmdahl (1932)' stressed that although the way of neural tube formation is different in primary and secondary neurulation, the difference is not very pronounced. He referres to the observations by 'Goronowitsch (1885)' and 'Jablonowski (1899)' on the pike, which seemed to represent both types of neural tube formation: the cranial part was said to be formed as a result of primary neurulation while the rest of the body followed the secondary type. According to 'Holmdahl (1932)' secondary neurulation is restricted to the caudal region of the trunk in many vertebrates that otherwise exhibit primary neurulation, from *Petromyzon*, through *Gallus domesticus* to *Homo sapiens*. 'Oppenheimer (1936b)' showed with vital staining of *Fundulus* embryos that most of the material for the nervous system lies within the early embryonic shield, while a smaller part is found in the extraembryonic region anterior to the shield (see Fig. 13/3). The cells in the shield form the solid keel and the hind- and midbrain while regions of the anterior shield and the neighbouring extraembryonic region will form the solid eye and forebrain primordia. 'Price (1934a,b)' described the primordia of the CNS in the whitefish *Coregonus clupeaformis* as a solid neural keel and three solid primary brain lobes and the optic outgrowths. The indifferently placed cells in the neural keel become later arranged into two parallel rows. Their subsequent separation will form the cavity of the neural canal. The rearrangement of the brain cells starts in the prosencephalon and continues posteriorly.

In a biological study on vertebrate development 'Peter (1947)' reported that ectodermal organs of teleosts develop from a 'compact mass'. In the neural keel only later a lumen is produced by separation of cells. Alternatively, a 'fold without a lumen' is found. Its walls adhere to each other and later separate. In the textbook of 'Comparative Embryology of Vertebrates' by 'Nelsen (1953)' the teleostean neural keel has been described as an axial cord, the cells of which are organized according to an irregular pattern.'Gihr (1957)' in her thesis on the development of the pike, *Esox lucius*, mentions that at the 4-somite stage the 'neural material' is shifted from a transverse into a dorso-ventrally directed arrangement.

The embryology of the English sole, *Parophrys vetulus*, was investigated by 'Orsi (1968)'. The fusiform cells of the solid early neural keel run ventro laterally from an apex in the dorsal midline, while in the brain region the deeper cells 'run more or less straight across it'. This author, too, noticed a temporary dorsal furrow prior to the end of epiboly. He stressed two basic changes in the neural system: 1) the main areas of the brain are defined by constrictions into fore-, mid- and hindbrain, 2) the cells simultaneously rearrange themselves to permit the opening of the brain ventricles and the neural canal. The formation of the latter begins in the medulla and then moves down the neural cord. This is followed by the formation of the third ventricle and the dorsal part of the fourth ventricle. Apart from changing position, the neural cells become smaller, more numerous and densely packed. However, all other authors who mention cavitation agree that the lumen is first observed in the primordium of the eye. It appears as a slit and later becomes a ventricle (see Chapter 17).

Organogenesis of the viviparous embiotocid *Hyperprosopon argenteum* was described by 'Engen (1968)'. He points out that the CNS development is of the typical teleostean type. 'The neural plate tissue proliferates inward to form a solid cord of cells along the midline called the neural keel'. The lumen develops secondarily and is first noted in association with the formation of the optic vesicle.

'T. Yamamoto (1975)' in his extensive treatise of the medaka *Oryzias latipes* assumed a solid primordium for the whole CNS with the neurocoele appearing at a later stage. This was followed by 'Miyayama and Fujimoto (1977)' who reported light microscopical studies of the neural tube formation in the medaka. These authors were the first to perform additional analyses by EM. They observed, too, a transient neural groove (dip), which was very shallow. The outermost cells of the solid neural chord were continuous with the deep cells of the epidermis. The central cells of the neural chord were randomly arranged and desmosomes were observed between the apical part of neighbouring cells. During development the cells moved apart into groups at each side of the midline. In both groups, starting from the ventral and proceeding to the dorsal side, cells arranged themselves into a pseudostratified epithelium. As development proceeded, a distinct zig-zag midline was observed separating the left from the right pseudostratified epithelium. [No separating membrane was observed, as was suggested on the basis of light microscopical investigations by 'Goette (1873, 1878)' and 'Carlbera (1877)'.] The primitive neurocoele started to open in the ventral part of the chord, then widened and extended to the dorsal part. Transitory cytoplasmic processes, originating near the desmosomal junctions, as well as short cilia, began to protrude into the lumen. At the neck of these cells microfilaments and microtubules were observed; the latter were also present in the cell body. When the neural chord changed from its wedge-shape into an elliptical form, the appearance of cluster of neural crest cells, wedged between the chord and the overlying epidermis, became evident. These cells later separated and moved ventrad. 'Raible *et al.* (1992)' described the segregation and early dispersal of trunk neural crest cells in *Brachydanio (Danio) rerio*, which is similar to the development of these cells in other species. 'Reichenbach *et al.* (1990)', using SEM, observed a neural groove and a narrow neurocoele in the neural keel of *Cichlasoma nigrofasciatum*.

The zebrafish, *Brachydanio (Danio) rerio*, has been the subject of the most recent investigations. A paper by 'Hisaoka and Battle (1958)' contained the first complete description of the normal stages of the zebrafish. They mention that a solid nerve chord becomes visible when the blastopore is closed and two somites have developed and the neurocoele makes its appearance at the five somite stage. 'Schmitz *et al.* (1993)' studied the process of neurulation by iontophoretically labelling cells either with a dye (DiI) or fluorescine dextran amine. They observed that the neural plate developed between 90% epiboly and the 10-somite stage. They mention that a solid nerve chord becomes visible when the blastopore is closed and two somites have developed. At this stage the neural tube anlage consisted of a medial thickening above the notochord with a smaller one on each side. This was followed by a fusion of the two bilateral ectodermal thickenings with the median thickening which gave rise to the neural keel. Labelling of cells and observations at the late gastrula stage revealed that the medial region contains exclusively neural progenitor cells (from the 90-100% epiboly stage onwards), while the lateral regions contain a mixture of neural precursor, prospective epidermal and neural crest cells. On the basis of these results the authors concluded that the lateral thickenings are homologous to the neural folds in higher vertebrates. The prospective neural and epidermis cells separated within 1-2 h, during which time some cells of the lateral thickenings would have had to move a distance of up to 400 microns to reach their destination. EM analyses revealed the presence of intercellular junctions along the dorsal-ventral extent of the neural cord. From the 14-somite stage onwards clefts were observed along the midline, first ventrally then dorsally, until they eventually fused to form the neural cavity. In the meantime epidermal cells had overgrown the neural tube. Within the neural tube of higher vertebrates most cell divisions take place at the ventricular side of the epithelium and, as mentioned above, in teleosts, due to the absence of a neurocoele in early stages of their development, cell divisions are observed along the midline of the neural keel and the nervechord. 'Papan and Campos-Ortega (1994)' further studied the morphogenetic movements that transform the neural plate into the neural tube by injecting a dye marker into single neural plate cells. They established that the latero-medial position of injected cells in the neural plate was correlated with the dorso-ventral position of their progeny in the neural tube. Based on these results, which confirm those of 'Schmitz *et al.* (1993)', the authors suggested that the neural plate 'folds in' at the dorsal midline to form the neural keel and, therefore, that neurulation in zebrafish does not differ essentially from that of higher vertebrates. Moreover, they observed mitoses of labelled cells along the medial divide of the neural chord preceding neurocoele formation.

'Kimmel *et al.* (1994)' reported cell lineage analyses by video time-lapse recording in vivo in the CNS of zebrafish embryos. When cells of a single clone become arranged as a bilateral pair of discontinuous lines, these are referred to as 'strings'. These strings are oriented along the neural axis as described already by 'Kimmel and Warga (1986)' and 'Kimmel (1993)'. Cell intercalations during interphases, alternating with mitotic divisions, generate these strings. In the more recent publication the relationship between cell cycles and cellular morphogenetic behaviour of single strings was investigated. Cells in single clones were shown to divide together and to maintain this synchrony up to the end of gastrulation (cycle 16). As a result of convergence, cells drift away from the EVL and are moved, as anteroposteriad oriented strings, towards the dorsal midline. After cycle 16 an extension of the strings, oriented preferentially in the mediolateral direction, follows due to cell intercalations. On reaching the dorsal midline, the cells often divide and during the following interphase the sister cells move, one to the left, the other to the right, producing a bilaterally symmetrical pattern. By division 18 the solid neural keel hollows into a neural tube and the neurocoele probably blocks cell passage across the midline. The authors suggested that this divergent lateral movement of sister cells takes place underneath, not into, the stream of cells still converging medially, and that such inward movement is important for the formation of the neural tube. It has been proposed that the progressive thickening of the neural anlage, as well as its dorso-ventral organization, is the result of the following cell movements: cells near the dorsal midline will go to form the bottom of the stack, i.e. the ventral part of the neural tube, and later arriving cells will divide and 'dive' later, and consequently will form the more dorsal parts of the neural tube.

In vertebrates with primary neurulation, i.e. folding of the neuroectoderm, the dorsoventral position of a cell in the neural tube can be predicted from its mediolateral position on the neural plate. Medial cells of the plate will form ventral structures and lateral cells will form dorsal structures of the neural tube. The same holds true for the zebrafish neural keel despite its peculiar mechanism of formation. Dye-labeling of single cells in the zebrafish neural plate showed that the mediolateral positon of a cell in it is projected faithfully into the dorso-ventral axis of the neural keel. However, in contrast to other vertebrates, the left-right separation of cells in the neural plate is not preserved in the zebrafish neural keel. Cells can cross the midline to colonize the contralateral side of the neural keel.

It has been proposed again and again that the mechanism of teleost neurulation is similar to the secondary neurulation seen in the tailbud of higher vertebrates. In many respects, however, zebrafish neural tube formation resembles that of other vertebrates and may thus be just a variant of primary neurulation. This idea is supported by the observation that evolutionarily more primitive fish, like the sturgeon, generate the neural tube by folding of the neural plate. It is tempting to speculate that the teleost type of neurulation is an adaptation to the altered egg architecture characteristic of teleosts.

As mentioned above, due to this infolding, cells from both halves of the neural plate are juxtaposed at the midline. As in the case of other epithelia, cells of the neural keel and nerve rod divide at their apical side; thus, sister cells can be distributed on either side of the nerve rod and this leads to a bilateral distribution of cell clones. This is a peculiarity that clearly distinguishes zebrafish neurulation from that in other vertebrates.

The neural keel like the neural plate is organized as a two-layered pseudostratified columnar epithelium which would suggest that the neural keel forms by infolding of the neural plate, probably caused by the continued convergence of ectodermal cells towards the prospective dorso-medial regions of the embryo. These results endorse the conclusion of 'Reichenbach et al. (1990)' regarding the 'epithelial genesis' of the neural keel of teleosts.

Most cell divisions within the neural tube of vertebrates take place at the ventricular side of the epithelium, i.e. at the side of the neurocoele. Although there is no neurocoele in early stages of neurulation of teleosts, cell divisons can be observed in the middle of the neural keel and neural rod, where the neurocoele will form '(Reichenbacha et al., 1990; Schmitz et al., 1993)'.

'Strähle and Blader (1994)' and 'Takeda and Miyagawa (1994)' have produced a review of the neurogenesis and cell fate in the zebrafish embryo.

12.3 OTHER MIDLINE STRUCTURES

The hypochord was thought to be an endodermal derivative induced by the notochord though doubts were recently expressed (see Chapter 13). Its role during embryogenesis is still unknown. The ectodermal floor plate, another median structure, is formed due to a secondary embryonic induction by the notochord (Fig. 12/4). It is present along the ventral neural tube from the forebrain/midbrain junction to the tail. As shown in the zebrafish, the floor plate forms anteriorly a two cell-wide strip, which narrows to a single row in the spinal cord. Along most of the length of the spinal cord the floor plate is a single linear cell row but changes at about the fourth spinal segment to a staggered row, which continues into the hindbrain. The floor plate can readily be distinguished from the neural tube since its small cells have a very distinctive cuboidal epithelial shape. Moreover, antibodies are available that label selectively the floor plate '(Hatta and Kimmel (1993)'. 'Schmitz et al. (1993)' observed ciliary structures and basal bodies in the ectodermal floor plate. 'Kimmel (1993)' reported that 'ectoderm, like the mesodermal prechordal plate, also seems to make a singular midline derivative anterior to the floor plate: Along the floor of the diencephalon is a midline tissue that, argued from genetic and transplantation studies, is somewhat equivalent to the floor plate, both developmentally and in its signaling function'. The floor plate serves as a guidepost for the growth of commissural axons '(Hatta et al., 1991)'. The medial longitudinal fascicles (MLF), which flank the floor plate, consist of axons the nuclei of which are located in the midbrain ['Hatta et al., (1991)'; 'Kimmel (1993)'; 'Strähle and Blader (1994)']. According to 'Hatta et al. (1994)' the floor plate may participate in a signaling cascade that establishes dorsoventral patterning of the neural primordium. As mentioned above, the formation of the floor plate is the result of a so-called secondary embryonic induction because its genesis depends on inductive signals from the underlying notochord. It has been suggested that in the early neurogenesis of the zebrafish Reissner's substance, expressed in a temporal-spatial pattern, may play a role in axonal decussation '(Lichtenfeld et al., 1999)'. Reissner's fibres were first described by 'Ollsson (1964)'. 'Portmann, 1976)' mentions the presence of so-called Rohon-Beard cells which are expressed in the embryos of selachians, 'yolksac stages' of teleosts and in the larvae of amphibia. These cells are formed at the same time as those of the neural crest and occur along the dorsal neural tube. They send long processes to the periphery and in teleosts reach as far as the ventral part of the yolksac. Their function is unknown.

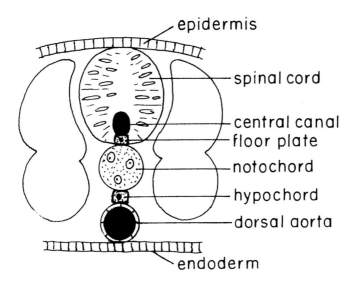

Fig. 12/4 Schematic representation of midline structures: spinal cord (neural tube), floor plate, notochord, hypochord (subnotochordal rod), dorsal *aorta*.

Summary

Although the mechanisms of neurulation in teleosts differ from those observed in amphibia and higher vertebrates, the endresult is the same in all embryos examined. Primary neurulation, as observed in amphibia, birds and mammals, proceeds by invagination of the neural plate to form a neural groove, which subsequently closes to form a neural tube. In secondary neurulation, as displayed by the teleosts, cells initially form a dense nerve rod, in which a lumen originates secondarily by cavitation. However, the observation of a 'closed neural fold' by some early investigators in the trout and other fish, and recently observed in the medaka and zebrafish, reveal that teleostean neurulation resembles primary neurulation of other vertebrates. It differs only from primary neurulation in that neurocoele formation occurs secondarily (12.1, 12.2).

At the ventral aspect of the neural tube, and in close contact with the notochord, the ectodermal floor plate extends anteriorly through the midbrain. The floor plate has been shown to be implicated in inductive interactions that pattern the neural tube and the somites (12.3).

Endnotes

[1] G. Axenplatte of Goette, Medullarplatte of 'Ziegler'.
[2] G. Axenstrang of 'His'
F. cordon axial of 'Henneguy'
[3] Salar Ausonii. Val., *Salmo fario* L.Bt.
[4] G. Rückenfurche of von Baer, Primitivrinne, Primitivstreif, Achsenstreif
F. Sillon dorsal of 'Vogt' and 'Lereboullet'; gouttière primitive, sillon primitif, sillon médullaire de 'Henneguy', ligne primitive

Chapter thirteen

Fate-maps[1]

> *'To understand is to perceive patterns.'*
> Isaiah Berlin

A knowledge of the origin as well as destiny of migrating cells within the embryo and outside it, is paramount for our understanding of the processes of morphogenesis. So-called fate-maps indicate the positions of the prospective (presumptive) areas at the onset of gastrulation, or even earlier. In other words, they indicate the arrangement of the future germ layers. In order to generate such a pattern, cells must have been provided with positional information, supposedly by a gradient-like distribution of a putative morphogen. Morphogenetic gradients appear to be mediated by gap junctions '(Wolpert, 1969, 1978; Bozhkova and Voronov, 1997)'. It should be mentioned that a cell predisposed to differentiate into a certain layer or organ does not necessarily reflect the state of the commitment of the cell. Therefore, cell fate can be divided into prospective (presumptive) fate i.e. what the cell normally develops into, or prospective (presumptive) potency i.e. what the cell can be made to develop into. Only prospective fate will be dealt with in this chapter.

13.1 TELEOSTEAN FATE-MAPS

The first to attempt a fate-map of a teleost was 'Sumner (1904)' who compared and compiled both longitudinal and cross sections in the brook trout, *Salvelinus*. 'Pasteels (1934)' was the first to study the movement of cells by means of vital staining, which formed the basis for the establishment of a detailed fate-map for the trout, *Salmo irideus* (Fig. 13/1). Since he realized that the dye often coloured more than a certain primordial area, he analyzed a great number of embryos. But he stressed that the localization still was not as exact as obtained for the frog gastrula. Two years later he supplied a much more detailed fate-map (Fig. 13/2). It is noticeable that in this figure the entoderm is no longer evident. This is due to the fact that he could not observe this area in the live blastula. However, in the sectioned embryo a very thin cellular entodermal sheet was evident. He showed a thin section of the dorsal lip revealing just one cell of the putative future entoderm in the process of involution, followed by the establishment of an entodermal layer '(Pasteels, 1936)'(Fig. 13/5). 'Nelsen (1953)' in his textbook displays a section through the blastodisc of the trout showing that presumptive entoderm is not exposed to the surface (Fig. 13/4). 'Pasteels (1936, 1940)' stressed that he could not establish on the live blastula the areae which will give rise to the 'intermediary cell mass', the pronephros and the Wolffian duct (see Chapter 18).

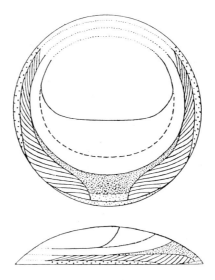

Fig. 13/1 Fatemap of the trout, viewed from above and from its side (Pasteels, 1934). Densely dotted, notochord; loosely dotted, prechordal plate and *entoderm*; small crosses, lateral plate; blank: *ectoderm*. An interrupted line separates cephalic *ectoderm* from *ectoderm* of the trunk.

'Oppenheimer (1936a)' followed by describing a fate-map of *Fundulus*. In a later publication she contrasts her fate map of *Fundulus* with that of *Salmo*, simplified after 'Pasteels (1936)', which highlights distinctly the similarities and differences (Fig. 13/3). Schematized sections through the blastodisc just previous to gastrulation reveal that the putative entoderm in *Fundulus* lies clearly exposed at the caudal end of the blastodisc while in *Salmo* it is not exposed to the surface (Fig. 13/4).

In this context it should be mentioned that already 'Berent (1896a)' and before him 'von Kowalewski (1886a)' have recognized the entoderm cells and followed their involution in the trout, *Salvelinus* sp., and in *Gobius* and *Carassius* respectively (Fig. 13/6).

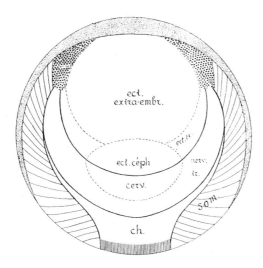

Fig. 13/2 Fatemap of the trout, viewed from above. More elaborate than Fig. 13/1 (Pasteels, 1936). Note absence of *entoderm*. Vertical lines = prechordal plate; finely dotted = lateral line and ventral *mesoderm*; small crosses = caudal regions (not identifed); cerv. - brain; ch. - notochord; ect.céph. - cephalic *ectoderm*; ect.extra-embr. - extra-embryonic *ectoderm*.; ect.tr. - *ectoderm* of trunk; nerv.tr. - nervous system of trunk; som - somites (first 23).

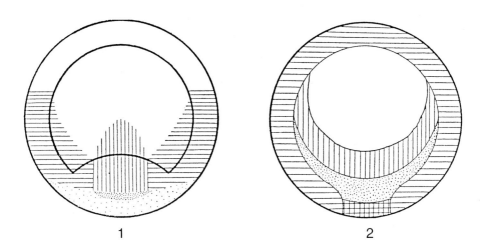

Fig. 13/3 Comparison of *Fundulus* and *Salmo* fate maps. **1** - *Fundulus* fate map after Oppenheimer (1936). **2** - *Salmo* fate map, simplified after Pasteels (1936). *Entoderm* = light stipple; notochord = heavy stipple; prechordal plate = cross-hatching; nervous system = vertical lines; *mesoderm* = horizontal lines; *ectoderm* = white areas.

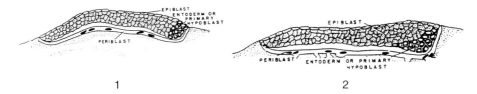

Fig. 13/4 Blastodisc section at the site of the dorsal blastopore lip to show location of *entoderm* (Oppenheimer, 1936; Pasteels (1938)). **1** - *Fundulus* **2** - *Salmo*. In *Fundulus* the *entoderm* is exposed while in *Salmo* the *primary hypoblast* (putative *entoderm* and *mesoderm*) lies below the surface.

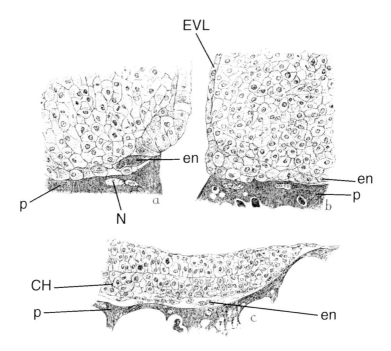

Fig. 13/5 Detail of *quasi* involution of *entoderm* in *Salmo irideus* (Pasteels, 1936). **a** - detail at the site of the blastoporal lip. Darkened cell is a putative entodermal cell ready to be involuted. **b** - detail of an older stage than a. **c** - parasagittal section of an older embryo with a layer of *entoderm* cells. CH - notochord; en - *entoderm* cell; EVL - enveloping layer; N - *nucleus* of *periblast*; p - *periblast*.

After establishing the fate-map of the trout, *Salmo irideus*, 'Pasteels (1936)' described in diagrammatic form the presumptive (prospective) fate of the various regions during gastrulation (Fig. 11/20).

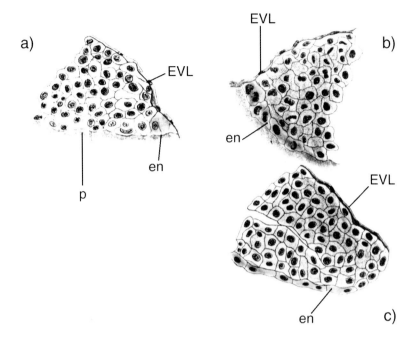

Fig. 13/6 Involution of *entoderm* in the trout (Berent, 1896). **a** - median longitudinal section through blastodisc on day 15. One entodermal cell about to involute. **b** - day 16 displaying layer of involuted *entoderm*. **c** - day 17 dorsal lip with involuted *mesoderm* and *entoderm*. en - *entoderm* cell; EVL - enveloping layer; p - *periblast*.

'Ballard, from 1966 to 1982' challenged the fate-maps put forward by 'Pasteels' and 'Oppenheimer'. On the basis of injecting fine grains of coloured chalk and following their movements, he insisted that there is a total absence of any experimental evidence of involution (in his words 'inturning' or 'inwheeling movements') of pre-chordal plate and chordamesoderm material at the rim followed by centralward movement of cells. On the contrary, he stressed that the hypoblast arises by delamination and outward migration of centrally situated DC.

According to 'Ballard' the areas of cells destined for different fates overlap each other within the depths of the blastodisc and move around at random before the morphogenetic movements begin. Therefore, his fate-map is three-dimensional. The embryonic axis is formed after the hypoblast cells have arrived at their destinations. The fish he studied were *Salmo gairdneri*, *Salvelinus fontinalis*, *Catostomus commersoni* and *Gobius niger*. 'Ballard' argued that the 'fate-mapping' techniques of 'Oppenheimer' and of 'Pasteels' followed solely particle and dye displacements, which convey only a static picture (indirect evidence) of what is essentially a dynamic process. He, therefore, called the fate-maps of 'Oppenheimer' and 'Pasteels' 'diagrams of tentative localization'. As opposed to these techniques, his method involved direct observations using Nomarski DIC (differential interference contrast) microscopy '(Wood and Timmermans, 1988)'. Ballard's view became generally accepted in the United States [textbook by 'Nelsen (1953)' excepted] while the publications and embryology textbooks in Europe continued referring to the fate-maps proposed by 'Pasteels' and 'Oppenheimer' '(Portmann, 1976; Siewing, 1969, Starck, 1975; Fioroni, 1987)'. A comparison of the fate-maps of the frog, a sauropsid and teleosts was put forward by 'Portmann (1976)' (Fig. 13/7).

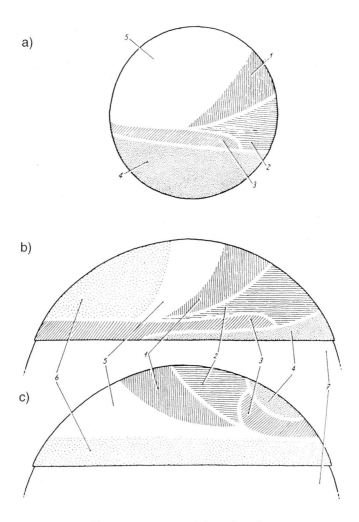

Fig. 13/7 Comparison of fate maps — **a** - amphibian, **b** - teleostean **c** - avian (Portmann, 1976, with kind permission of Benno Schwabe, Basel). 1 - nervous sytem; 2 - notochord; 3 - *mesoderm*; 4 - *entoderm*; 5 - embryonic *ectoderm*; 6 - extraembryonic *ectoderm*; 7 - yolk.

'Strehlow and Gilbert (1993)' established with the help of fluorescent dyes on very high-molecular mass carriers that the first cleavage defines dorsal-vegetal, the second cleavage left-right, and the third, anterior-posterior adult body axes of the zebrafish. The authors further summarized the contribution of each blastomere at the eight cell stage. The fate-mapping analyses in zebrafish were hampered by the scattering of early blastomeres in late blastula stages. However, by the beginning of the gastrula stage nearly all lineages had become tissue-restricted and the ensuing summary fate-map looks very like the ones for *Salmo* and *Fundulus*. However, no delineation of presumptive areas can be supplied since considerable overlaps are observed '(Kimmel et al., 1990, 1995; Ho, 1992; Ho and Kimmel, 1993)'. This is reminiscent of Ballard's emphasis on the overlapping of areas in the teleosts he studied. He, therefore, had suggested a three-dimensional organization to account for the overlap, whereas indeterminancy rather than three-dimensional organization was suggested to account for the recently observed overlaps in the zebrafish fate-map '(Kimmel et al., 1990; Wilson et al., 1993)'. 'Ho (1992)' and 'Ho and Kimmel (1993)' reported that marginal cells of the zebrafish blastodisc are as yet uncommitted to a specific fate, i.e. they are still pluripotent throughout the late blastula and early gastrula stages. These results were based on experiments involving transplantation of single cells at various stages of development into different regions. Finally, 'Warga and Nüsslein-Volhard (1999)' succeeded in establishing the origin of the zebrafish entoderm and its separation from the mesoderm by labeling single marginal cells with lineage tracer dye in mid-blastula stage. Structures of entodermal origin are derived predominantly from the more dorsal and lateral cells of the germ ring while cells located further than 4 cell diameters from the margin give rise exclusively to mesoderm. They stressed that before the segregation of the entoderm into a germ layer often both ento- and mesoderm share a common precursor.

We can appreciate the sentiments expressed by 'Wilson' as far back as 1889: 'The teleostean gastrula is such a complicated embryonic form that it has given rise to many interpretations, and the disagreement as to the proper one still continues'. Sometime later, ' Veit (1923)' when dealing with teleost gastrulation quoted the old saying 'Soviele Köpfe, soviele Sinne' (so many heads, so many senses = interpretations). He adds that, in fact, there has hardly been a scientist who has not once changed his/her opinion on the early development of teleosts.

We seem to have come full cycle since it should now be accepted that the results introduced more than 60 years ago, present, with minor species-dependent adaptations, the classical teleost fate-map for our times. However, the differentiation of the entoderm seems to be species-specific.

Summary

A fate-map of an early embryo (before gastrulation) illustrates what the blastomeres will eventually develop into. This is made possible by marking specific areas and observing where they reside in later stages. The fate-map established recently for the zebrafish is very like the one put forward for the salmon and *Fundulus* in 1934-1936. It is now agreed that mesoderm involutes over the germ ring (GR); however, the different presumptive areas cannot be exactly delineated. The origin and significance of the prominent mass of apparently undifferentiated cells which consitutes the anlage of the tail are still being debated (see Chapter 11).

Endnotes

[1] G. Anlagemuster
F. plan des ébauches

Chapter fourteen

Kupffer's vesicle

'Qu'il n'y ait pas de réponse n'excuse pas l'absence de questions.'
Claude Roy

Kupffer's vesicle[1] (KV) is a transitory organ peculiar to early teleost embryos. It lies at the end of the solid postanal gut and disappears as the intestine is formed. The ultimate fate of the regressing KV is as yet unknown. The fact of the development of a lumen in an early teleostean embryo, which otherwise shows only solid primordia, has been intriguing to this day.

'Vogt (1842)' observed in *Coregonus palea* above the anus a vesicle, which later disappeared. He interpreted it as a rudiment of the allantois. 'Coste (1847-59)' and 'Lereboullet (1854)' showed Kupffer's vesicle (KV) in their illustrations without, however, giving any interpretation of its importance. As its name suggests, the vesicle was thought to have been first described by 'Kupffer (1866, 1868)' (Fig.14/1, 14/2). He observed it in the transparent embryos of the herring, *Gasterosteus aculeatus, Gobius minutus* and *G. niger* and some time later in *Osmerus eperlanus, Leuciscus rutilus, Abramis brama* and the pike '(Kupffer, 1884)'. He believed that the vesicle arises as an ectodermal invagination from the dorsal surface and opens onto it, as did 'Boeke (1903)'. 'Kupffer' (as previously 'Vogt, 1842') regarded KV as a rudimentary allantois of the higher vertebrates. He suggested that it finally develops into the urinary bladder. 'Van Bambeke (1875)' observed in the living *Leuciscus rutilus* a 'vésicule allantoide de Kupffer'. 'Kupffer, 1878' in a treatise of teleost development mentioned that, in addition to the species quoted above, KV is also readily visible in *Spinachia vulgaris, Clupea harengus, Perca fluviatilis, Acerina cernua, Cyprinus brama* (*Scardinius erythrophthalmus*), but less so in the different species of *Gobius*. He mentioned that he could observe concretions in KV of *Gasterosteus* and *Esox lucius*, but never in that of the herring. He admitted that to ascertain the functional significance of this structure is very difficult and that he had not yet clarified the eventual fate of this structure. Subsequently, 'Henneguy (1880, 1888)' described KV in the trout as a small vesicle of temporary existence and surrounded by cylindrical cells. He thought it was identical with KV in *Gasterosteus*. Again, in the perch, *Perca fluviatilis*, KV is of similar shape and location to the bitterling but appears only after closure of the blastopore.'Henneguy', too, considered KV as an allantois. Most of the subsequent investigators of the early fish embryo departed widely from 'Kupffer's' account and thought that KV was part of the gut, or even a vestige of the archenteron or that it represents the enteric part of the neurenteric canal (which, in teleosts, would have no lumen), or even that it has an opening to the exterior. Some thought that it may be formed as a result of the coalescence of granules or spaces. Some authors referred to it as 'anal or postanal vesicle' '(Balfour, 1881; Hoffmann, 1881; Agassiz and Whitman, 1884; Ryder, 1884; Ziegler, 1887 and List, 1887)'. Similarly, 'Cunningham (1885a)' in his treatise of KV concludes 'I think there is no room for doubt that the significance of Kupffer's vesicle is completely elucidated by the facts and comparisons I have (thus) given; it is the last rudiment of the invagination cavity in the teleosteans'. 'Von Kowalewski (1886a,b)', analyzed histological sections of *Gobius* and *Carassius*. He believed that the entoderm, involuted at the posterior site of the blastodisc of *Gobius*, represents the anlage of KV (which he also calls allantois). He followed the development in the goldfish, *Carassius auratus*, where it attains a large size and remains surrounded by a cellular coat. The ventral wall of KV continues into the still solid intestine. 'Wilson (1889/1891)' agrees with the interpretation that KV represents the archenteron. KV of *Cristiceps* is covered dorsally by the entoderm and laterally and ventrally by the plasma membrane according to 'Fusari (1892)'. 'Eigenmann (1892)' observed various vesicles at the posterior end of the embryo in *Clinocottus analis* and concluded that the larger vesicles represent KV. However, 'Budd (1940)' maintains that the larger vesicles represent oil globules.

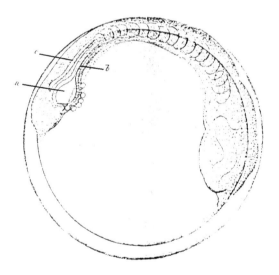

Fig. 14/1 *Gasterosteus* embryo, showing Kupffer's vesicle (KV) (Kupffer, 1866). a - KV; b - intestine; c - notochord.

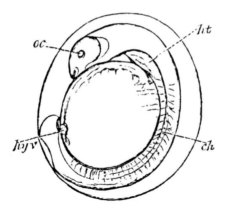

Fig. 14/2 Advanced herring embryo (Kupffer, 1878). ch - notochord; ht - heart; hyv - Kupffer's vesicle; oc - eye.

'Berent (1896a,b)' reviewed the results of the various authors. According to him 'Cunningham (1885b)' mistook a yolk vacuole for KV. 'M'Intosh and Prince (1890)' reported that, in some species, the advent of KV is preceded by vesicles or elongated spaces. 'Kingsley and Conn (1883)', too, noted a group of vesicles which coalesced to form KV. Secondary vesicles, which show the same features as the normal KV, seem to be frequent. '(Brook, 1885a,b; M'Intosh and Prince, 1890; Eycleshymer, 1895; Sumner, 1904)'. 'Berent (1896a,b)' also disagreed with 'Kupffer' and 'Henneguy (1888)' who suggested that KV represents an allantois, especially since KV is not surrounded by any splanchnic cell cover. After reviewing, and disagreeing with, the various authors 'Berent' added his own results on the trout. He suggested that KV is comparable with the postanal intestine (G. Schwanzdarm) of the selachians, and this was put forward also by 'Balfour (1881)' and 'Schwarz (1889)'. 'Ziegler (1902)' depicted KV of *Salmo salvelinus* situated at the posterior end of the intestinal primordium, i.e. at the end of the solid postanal intestine.

'Sumner (1900)' in his review came to the conclusion that 'the generally accepted view that KV represents a certain part of the archenteron seems to me to be true beyond doubt'. It is needless to point out that KV cannot be connected with the archenteron since there is no archenteron developed in the teleostei.

'Sumner (1903)' further elaborated that KV is a 'precociously dilated post-anal gut, formed in connection with a neurenteric canal, in some cases, and opening through it to the exterior'. He further remarked that in most pelagic eggs he had described, the lower wall is at first completely lacking. Later, in some cases at least, the lateral walls draw together and unite on the ventral side, thus forming a cellular floor.

'Kupffer'considered the walls of KV as a continuation of the entoderm. However, as regards the lining of KV, different observations have been reported. In *Belone acus* and the herring the roof only is lined by an epithelium while the periblast (Chapter 10) provides the sides and the floor '(Sobotta, 1898; Cunningham, 1885a)' (Figs. 14/3, 14/4). Roof and sides of the vesicle are cellular in *Coregonus, Perca, Serranus atrarius* and in pelagic eggs of other species '(Agassiz and Whitman, 1884; Wilson, 1891; Sobotta, 1898 and Chevey, 1925)' (Figs. 14/5, 14/6). The vesicle is completely surrounded by epithelium in the Salmonidae. 'Berent (1896a,b)' and 'Sumner (1900a)' observed it in *Salvelinus* as did 'Ziegler (1902)' in the salmon (Figs. 14/7, 14/8). The dorsal wall is formed by cylindrical entoderm cells while the lateral and the ventral walls are formed by squamous cells.

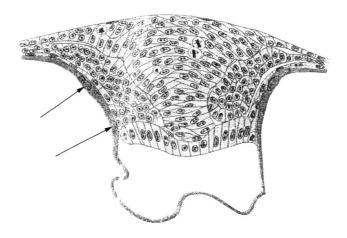

Fig. 14/3   Transverse section through *Belone acus* (Sobotta, 1898). Only roof of KV is cellular. Arrows point to *periblast nuclei*.

'Reis (1910)' found that in *Amiurus nebulosus* KV is formed as a closed or open fold and the lumen is created by a process of unfolding. However, in some embryos of the same species she observed at the location of KV vacuolised cells surrounded by a thick wall. She suggested that the lumen of KV is formed as a result of the coalescence of the vacuoles. The process of vacuolization may replace the folding process or simply serve to increase the lumen of KV. Each of these two processes has been reported to occur separately: In the bass, *Serranus atrarius*, KV arises by a process of folding, in Salmonidae its development may be construed as the hollowing of a solid thickening '(Wilson, 1889/1891)'.

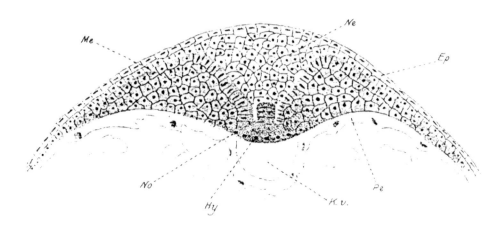

Fig. 14/4 Transverse section through herring embryo (Cunningham, 1885). Only roof of KV is cellular. Ep - *epidermis*; Hy - roof of KV; K.v. - Kupffer's vesicle; Me - *mesoderm*; Ne - neural *ectoderm*; No - notochord; Pe - *periblast* with *nuclei*.

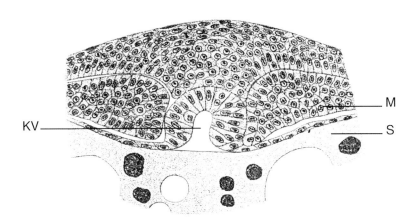

Fig. 14/5 Cross-section through *Coregonus* at 4 somite stage (Sobotta, 1898). Roof and sides of KV cellular. KV - Kupffer's vesicle; M - *mesoderm*; S - yolk *syncytium*.

In large-yolked eggs KV appears before the end of epiboly while in small-yolked eggs it appears after epiboly has been completed '(Sobotta, 1898; Brummett and Dumont, 1978)'. While there are obviously species-specific differences with regard to time of appearance of KV, its location is usually the same, namely anterior to the tailbud blastema. 'Kopsch (1900)' provided a review of KV morphology of a large number of early publications.

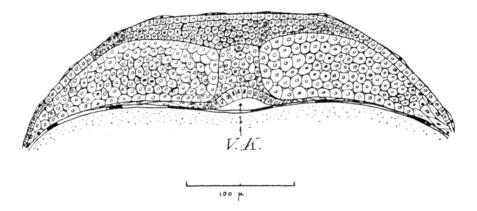

Fig. 14/6 Cross-section through perch (Chevey, 1926). Roof and sides of KV cellular. V.K. - Kupffer's vesicle.

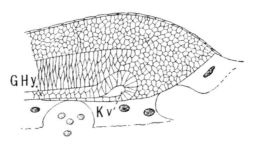

Fig. 14/7 Longitudinal section through embryo of the trout *Salvelinus* (Sumner, 1900a). The whole wall of KV is cellular. Kv - Kupffer's vesicle; Ghy - gut-hypoblast.

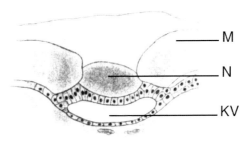

Fig. 14/8 Cross-section through *Salvelinus* (Berent, 1896a). The whole wall of KV is cellular. KV - Kupffer's vesicle; M - *mesoderm*; N - notochord.

As regards suggestions for the functional significance of KV, there have been many but none of them convincing. As one of the early researchers stated "Although we have been able to trace the entire history of KV in several species of ova, its significance remains as complete a puzzle as ever" '(Agassiz and Whitman, 1884)'. In contrast, 'Sumner (1900a)' from its position and its content concluded that KV furnishes material for the embryo, and later '(Sumner, 1904)' specifies that it 'probably represents a precociously formed region of the embryonic gut (probably post-anal), having a definite function to perform in the embryonic life (nutritive?). The cells destined to form its walls must be regarded as belonging to the hypoblast'. Similarly, 'Ziegler (1902)' suggested that KV facilitates the nutrient export to and the metabolism of the tail bud, which is characterized by active mitoses.

All of the early observations were carried out on either living embryos or sectioned material observed with light-microscopy, which has its limitations. Subsequently, SEM, supported with analyses by TEM, of KV of *Fundulus heteroclitus* revealed some unexpected details. Each columnar cell of the roof of KV is supplied with one long cilium emanating from its centre and extending into the cavity of the vesicle. The periblast floor of KV is characterized by the presence of numerous surface folds, ruffles, and microvillous projections. These results, too, would suggest that KV functions as an absorptive organ supplying nourishment from the yolk mass to the early embryo. Active movements of the cilia might act in stirring the partly digested yolk '(Brummet and Dumont, 1978)'.

'Melby *et al.* (1993)' reported that recent fate-mapping studies show that the epithelial cells lining KV will later form tail mesodermal derivatives, including notochord and muscle. 'Cooper and d'Amico (1996)' showed that at the tailbud stage of the zebrafish the cluster of so-called forerunner cells (see Chapter 11) form the roof of KV.

While the early investigators discussed the functional significance of KV without bringing forward any persuasive answer, the topic seems to have lost interest, i.e. it was no longer intensively discussed. This is reflected in the remark by ' Kimmel *et al.*, (1995)' 'We think God gave it [the KV] to fish embryos purely as a staging aid'!

Summary

Kupffer's vesicle is a vesicular structure of hitherto unknown function. It is situated at the end of the solid post-anal duct and is present only in teleosts. In all species analyzed its roof is made up of cells continuous with the hypoblast whereas the floor and sides are cellular in some species but in others bounded by the periblast covering the yolk.

Endnotes

[1] G. Kupffer'sche Endblase
   F. Vésicule de Kupffer
   I. Vesicula di Kupffer

Chapter fifteen

Ectodermal derivatives

'Discovery consists of seeing what everybody has seen and thinking what nobody has thought.'

Albert Szent-Gyorgy

At the end of gastrulation the outer layer of the embryo and yolksac consists of ectoderm[1]. In teleosts it is often referred to as 'epiblast' (to contrast it with the inner layer, the 'hypoblast'). The ectoderm covering the embryonic shield is greatly thickened compared with the extraembryonic area and it undergoes neurulation as described in Chapter 12. Eventually the whole embryo and the yolksac become covered by the one-layered epidermis topped by a flattened layer, the EVL or periderm (Chapter 11).

During neurulation the dorsal epiblast differentiates into neural ectoderm subjacent to the one-layered EVL. Cell labelling in the zebrafish embryo revealed that the lateral thickenings of the neural ectoderm at the stage of 90-100% epiboly contain a mixture of neural, neural crest and epidermis progenitor cells. At the 2-somite stage a clear separation between the neurogenic and epidermal progenitor cells occurs which involves for the latter a journey of about 400 microns within 1-2 hours. The neural keel remains in contact with the overlying epidermal cells until about the 14 somite stage when a distinct epidermal fusion in the midline is observed '(Schmitz *et al.*, 1993)' (see Chapter 12).

15.1 NEUROECTODERM

As mentioned above, early in development the neural ectoderm will separate from the putative epidermal ectoderm. During neurulation (Chapter 11) the neural cord and the brain are laid down. Their primordia consist of a solid cell mass and the lumina appear later.

15.1.1 Brain (encephalon)

The anterior part of the neurochord of the neurula is transitorily divided into 11 so-called neuromeres. The most anterior neuromeres represent the anlage of the brain and consist of two parts: 1) the prosencephalon [with the prospective (presumptive) telencephalon[2] and diencephalon[3]] and 2) the rhombencephalon (with the prospective mes-[4], met-[5] and myelencephalon[6]). However, other authors put forward as the early brain a 'three brain stage' consisting of forebrain (telencephalon and diencephalon), midbrain (mesencephalon) and hindbrain (rhombencephalon) (Figs 15/1). The early stage, be it considered bi- or tripartite, is followed by the 'five brain stage', consisting of 1) telencephalon 2) diencephalon 3) mesencephalon (tectum opticum), 4) metencephalon (cerebellum) 5) myelencephalon (medulla oblongata). The brain of the young embryo exhibits a cephalic flexure, between the mid- and forebrain (Fig. 15/2). The subsequent straightening will manifest a distinctive embryonic stage. The most conspicuous change observed in the teleost brain is the continued lateral extension (called valvula cerebelli) of the metencephalon and its progress backward over the mesencephalon until it almost covers the latter with its two broad lobes, which continue to increase in width (Fig. 15/3). The patterning of the zebrafish brain has been recently reviewed by 'Kimmel (1993)'.

The large sense organs of the head, i.e. olfactory organ, eye and auditory organ with its associated static organ, are primarily connected with integration centres in the brain. They start as solid thickenings, which subsequently develop lumina (Fig. 15/4).

The paired olfactory organ starts as a rostral placode situated on each side of the head. The placodes subsequently invaginate to olfactory pits and in the adult teleost they have achieved a dorsal position and possess one or two openings each. The nasal cavities are independent of the oral cavity.

The eye represents an externalized part of the brain (diencephalon) and will be dealt with separately in Chapter 17.

The 'ear' of teleosts is an organ of equilibrium; it consists only of the inner part, the labyrinth. Adult teleosts are characterized by having large statoliths (usually called otoliths); these are the utricolith (lapillus), sacculolith (sagitta) and lagenolith (asteriscus) (Fig. 15/5). The solid auditory placode first forms a thick-walled auditory vesicle. While the internal cavity gradually increases in size the walls themselves become concomitantly thinner. One and then two refringent bodies, the asterisk and sagitta, make their appearance (Fig. 18/30a,b). This is followed by the development of the labyrinth already visible at an early stage (Fig. 18/30c). The two vertical semicircular canals are formed first, followed by the horizontal one. It should be mentioned in this context that ostariophysi among the teleost fishes have a series of bones called Weberian ossicles, which acoustically couple the swimbladder to the inner ear. Ontogeny of the inner ear has been dealt with by 'Noakes and Godin (1988)' while the hearing organs of mature fishes have been recently reviewed by 'Bretschneider et al. (2001)'.

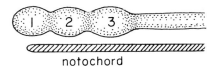

Fig. 15/1 Primary brain. 1 - Forebrain (*Prosencephalon*); 2 - Midbrain (*Mesencephalon*); 3 - Hindbrain (*Rhombencephalon*). Note that the notochord is shorter than the neural chord.

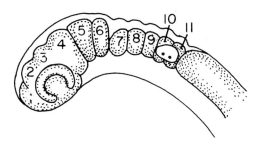

Fig. 15/2 Early brain of salmon (redrawn after Hill, 1900). The 11 neuromeres represent. 1 - *bulbus olfactorius* (part of *telencephalon*); 2, 3 - *prosencephalon* (*tel- and diencephalon*); 4, 5 - *mesencephalon*; 6 - anterior *rhombencephalon* (*metencephalon*); 7, 11 - posterior *rhombencephalon* (*myelencephalon*).

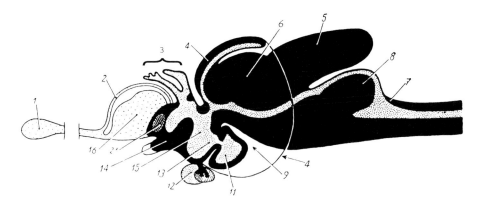

Fig. 15/3 Brain of adult teleost (Portmann, 1976, with kind permission of Benno Schwabe, Basel). 1 - *bulbus olfactorius*; 2 - hemisphere; 3 - *epiphysis*; 4 - roof of *mesencephalon*; 5 - *metencephalon*; 6 - *valvula cerebelli*; 7 - roof of *myelencephalon*; 8 - *lobus impar*; 9 - *plica ventralis*; 11 - *saccus vasculosus*; 12 - *hypophysis* (pituitary); 13 - *infundibulum*; 14 - optic chiasma; 15 - *diencephalon*; 16 - basal ganglion and *pallium*.

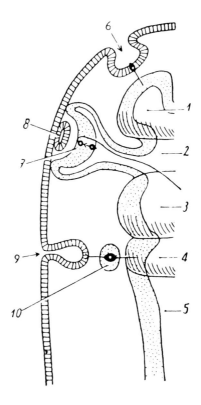

Fig. 15/4 Relationship between main sense organs and brain in the vertebrate embryo (in the teleosts the *anlagen* are solid) (Portmann, 1976, with kind permission of Benno Schwabe, Basel). 1 - *telencephalon*; 2 - *diencephalon*; 3 - *mesencephalon*; 4 - *metencephalon*; 5 - *myelencephalon*; 6 - olfactory anlage; 7 - eye cup; 8 - lens anlage; 9 - labyrinth anlage; 10 - ganglion of *nervus octavus*.

Summary

The early divisions in the anterior neural chord are referred to as neuromeres of which there are up to 11. The most anterior ones are substantially larger and correspond to primordia of the brain. The anlage of the brain is bi- or tripartite before it is made up of five parts. Sensory nerves carry impulses from the three main sensory organs, olfactory organ, eye and ear to the brain. In teleosts, the placodes of these sense organs, as well as the anlage of CNS, are first solid and only later cavitate.

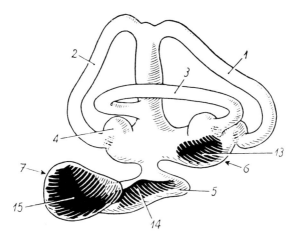

Fig. 15/5 Right labyrinth of an adult teleost (with large statoliths), viewed from the side (rostral part on the right hand side) (Portmann, 1976, with kind permission of Benno Schwabe, Basel). 1-3 - semicircular canals (1 = rostral); 4 - *ampulla* of canal; 5 - *sacculus*; 6 - *recessus utriculi*; 7 - *lagena*; 13 - *utriculolith* ('*lapillus*'); 14 - *sacculolith* = ('*sagitta*'); 15 - *lagenolith* = ('*asteriscus*').

(a) Epiphysis (Glandula pinealis) in the adult

A special posterior evagination of the diencephalic roof, the epiphysis cerebri[7], in teleosts contains the light-sensitive pineal organ[8] composed of a stalk and pineal vesicle (Fig. 15/6). 'Descartes (1644)' regarded the pineal organ as the seat of the soul. 'Von Frisch (1911)' was the first to report the pineal as an extraretinal photorecepor. He observed that pinealectomy of *Phoxinus phoxinus* resulted in a darkening of its skin. 'Scharrer (1928)' confirmed and extended his results. 'Breder and Rasquin (1947)' concluded that in the cave-dwelling blind characin *Anoptichthys* the pineal determines the directional sense of the light reaction. 'Hoar (1955)' reported that blinded smolts of *Oncorhynchus* show a light reaction, which is abolished after the destruction of the pineal. This so-called 'third eye' was subsequently shown by EM to contain photoreceptors similar to the retinal cones as well as supporting cells '(Oksche and Kirschstein, 1967)' (Fig. 15/7). The skull of teleosts overlying the pineal organ is largely cartilagenous. In some species the skin above the pineal organ is translucent (lacking in melanophores) and is known, therefore, as 'pineal window'. In some, there are chromatophores in the skin over the parietal foramen and, by their concentration and dispersion, they determine the amount of light reaching the pineal '(Nicol, 1963)'. 'Omura and Oguri (1969)' analyzed with histological methods the pineal organ of 15 species of teleosts and classified six types of pineals. Electrophysiological studies indicate that the pineal organ is directly photosensitive. It is above all a 'dusk receptor' and when exposed to sunlight, its main function is to detect a 'shadow' passing over the head '(Tamura and Hanyu, 1980)'. 'Meissl and Ekström (1986)' analyzed the pineal photoreceptors of the trout and reported that though pineal cells are morphologically similar to retinal cones, functionally they resemble rods. In contrast to the mammalian retina in which the photoreceptors are connected through bipolar cells to ganglion cells, the pineal photoreceptors of lower vertebrates are synaptically connected to second-order neurons projecting into the brain '(Omura and Oguri, 1971; Collin et al., 1986a,b)'.

In mammals the pineal organ produces melatonin which inhibits gonadal function. It is secreted according to a diurnal rhythm. 'Oguri and Omura (1973)' suggested that it is conceivable that the pineal organ of the fish also produces melatonin, which means that the pineal in fish might have a dual function. Light-entrained rhythmic melatonin biosynthesis in pineal organ culture of the pike, *Esox lucius*, '(Falcon, Guerlotte et al., 1987)' and the goldfish, *Carassius auratus*, '(Kezuka, Aida et al., 1989)' was reported while a rhythm of melatonin synthesis could not be detected in the pineal gland of *Salmo gairdneri* '(Max and Menaker, 1992)'. It has been suggested that there is a gradual shift from a direct photosensory function of the pineal in the more primitive vertebrates to a neuroendocrine role in the advanced forms. In other words, the pineal can be considered, even in teleosts, as an extraretinal photoreceptor, a biological clock and an endocrine gland.

'Mano et al. (1999)' and 'Kojima and Fukada (1999)' reported that a novel rhodopsin, called 'exo-rhodopsin' (extraocular rhodopsin), is expressed in the pineal and is responsible for the entrainment of the clock-phase in the zebrafish, the medaka and the European eel; it is absent in the retina. Its amino acid sequence is similar to, but only 74% identical with, that of the retinal rhodopsin.

(b) Epiphysis in the embryo

In embryos the pineal complex comprises a pineal body and, rostral to it, a 'dorsal sac' of unknown function and the parapineal organ[8], which arises from the anterior anlage of the diencephalic roof. The parapineal, which (represents an anterior evagination of the epiphysis), is said to be present only in embryonic or juvenile fishes where it is connected with adjacent brain centers by means of a tractus parapinealis. The 'dorsal sac' is another embryonic organ of hitherto unknown function. The velum, situated between the 'dorsal sac' and the paraphysis, is pronounced during ontogeny but is sometimes difficult to detect in the adult '(Holmgren, 1965)' (Figs. 15/3, 15/6). The earliest description of the embryonic development of the pineal- and parapineal organs in various teleosts was given by 'Rabl-Rückhardt (1883)'. This was followed by observations made by 'Hoffmann (1885)' and 'Hill (1891, 1894)'.

Ultrastructural evidence on developing cells in the teleostean pineal is very scant. It has been established that photoreceptors with outer segments are present in the embryonic pineal organ of teleosts before they appear in the retina. In *Gasterosteus* photoreceptor outer segments are present in the pineal organ at 72 hours postfertilization (hpf) and in the retina not before 144 hpf (hatching occurs between 120-144 hpf). Photoreceptors in the pineal organ of these embryos are thought not to be involved in visual functions but rather serve as photoneuroendocrine transducers '(Ekström et al., 1983; van Veen et al., 1984)'. Immunoreactive opsin in the outer segment appears first at 84 hpf in a few pineal gland cells of the stickleback embryo, and the number of immunoreactive outer segments increases during development '(Collin et al., 1986b)'. Similarly, fully developed photoreceptor outer segments are present in the pineal organ of the ayu, *Plecoglossus altivelis*, 11-15 days after artificial fertilization (hatching at 17 days) according to 'Omura and Oguri (1971). They suggest that, like the photoreceptors in the retina, those of the pineal undergo degeneration and formation of new outer segment lamellae. The initial appearance in the pineal of immunoreactivity to S-antigen, transducin, opsin and 5-HT was established in *Salmo salar* 30 days before any reaction was visible in the retina '(Oestholm et al., 1987)'.

'Kazimi and Cahill (1999)' studied the development of a circadian melatonin rhythm in embryonic zebrafish. Their data showed that a circadian oscillator regulating melatonin synthesis becomes responsive to light between 20 and 26 hpf. Since at this stage retinal photoreceptors have not yet begun to differentiate in contrast to pineal ones, it has been suggested that the first circadian rhythms are of pineal origin. Thus, 'ontogenetically speaking', maybe the 'third' eye should be called the 'first' eye. This is reminiscent of the elaborations discussions by 'Oksche (1989)' who mentioned this hypothesis, though mainly from a phylogenetical point of view.

In *Coregonus albus* 1 to 2 days after the development of the pineal organ, the anlage of the parapineal (anterior epiphyseal vesicle) appears in a rostral position and to the left of the median plane. In its early stage it is a small ovoid body, in a symmetrical position with the pineal. It degenerates later and will be found posteriorly to the mature pineal organ '(Hill, 1891, 1894)'. In *Salvelinus fontinalis* the parapineal loses the connection with the diencephalic roof at the 10millimetres stage when the wall of the pineal consists of 1-3 layers of undifferentiated cells. A well-developed saccus dorsalis (dorsal sac)[9], rostrally bordered by a deep velum transversum, is visible in front of the pineal organ. In the 23 mm embryo of *Salvelinus fontinalis* the pineal organ appears now folded, as observed in the adult trout. The cells have become radially arranged. The pineal still retains a funnel-like opening into the 3rd ventricle. The parapineal is now round and its wall made up of columnar ependymal cells surrounding a small lumen. The parapineal here, too, is lacking in the adult fish '(Holmgren, 1965)' (Fig. 15/6).

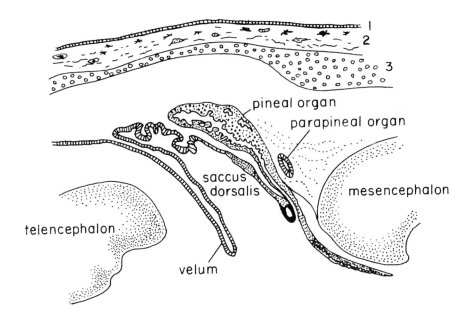

Fig. 15/6 Pineal body of a teleost (redrawn after Holmgreen, 1965). **1** - *epidermis*; **2** - *dermis* with *chromatophores*; **3** - cartilage.

Summary

The posterior evagination of the diencephalic roof develops into the pineal gland. In fishes it is a light-sensor with retinal cone-like photoreceptors. They appear before the photoreceptors of the retina. The pineal is most probably a component of the circadian system and contains a more or less self-sustained circadian oscillator. Two further dorsal diencephalic derivatives are the parapineal organ and the 'dorsal sac' of unknown function. The last two structures are lacking in the adult.

15.1.2 Neuromast organs

The solid anlage of the neuromasts which form the lateral line system, or occur as free neuromasts, is observed in the neighbourhood of, and maybe related to, the stato-acoustic system. The nerve fibres from both systems converge into a common acoustico-lateral area within the brain '(Herrick, 1962; T. Yamamoto, 1975; Noakes and Godin, 1988)'.

Neuromasts consist of bundles of cells, often with an appearance similar to that of taste buds. The neuromast cells are elongated, and each bears a projecting hair-like structure. Invariably there is present above the 'hairs', and enclosing their tips, a mass of gelatinous material. This is known as the cupula and is secreted by

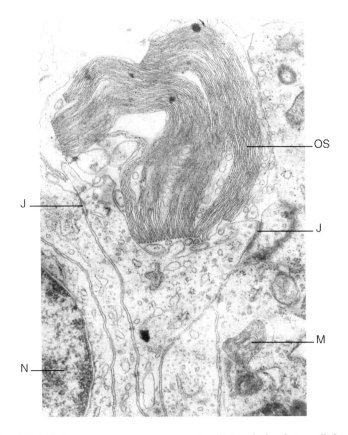

Fig. 15/7 TEM of photoreceptor outer segments of pineal gland. J - cellular junction; M - mitochondrion; N - nucleus; OS - outer segment saccules. × 15 400

the neuromast cells. It moves freely in the surrounding water. 'Schulze (1861)' was the first to observe that the cupula bends in response to slight motions of the water. It appears that the neuromasts respond to waves or disturbances in the water through movement of the cupula and consequent bending of the hair-like processes. This sense organ supplements vision by making the fish aware of nearby moving objects or obstructions to navigation' '(Romer and Parsons, 1977; Noakes and Godin, 1988; Helfman et al., 1997)'.

Many species of fish larvae have free neuromast organs. 'Iwai (1967)' reviewed earlier reports as well as his own work on *Tribolodon hakonensis*, *Tridentiger trigonocephalus*, and *Oryzias latipes*. These organs are usually situated in lateral rows along the body with the small hump of sensory cells being the main element visible especially in sea-caught or fixed specimens. Phase-contrast microscopy makes it possible to retain and see the cupulae, which are relatively enormous,

0.05 millimetre long in *Oryzias* but 0.1-0.4 millimetre long in some other species. In the adult it is likely that they are more often sunk into pits. The ability of postembryos to avoid capture before the eyes fully develop is almost certainly the result of the early presence of the neuromast or pineal organs.

Summary

Neuromast cells are part of the lateral line system. However, in many larvae they are present as free neuromast organs. In both cases they are covered by a gelatinous cupula which encloses the sensory 'hairs'. Disturbances in the water displace the cupula and stimulate sensory neurons. In postembryos neuromasts may be used as 'distant touch' receptors before the eyes are fully developed.

15.1.3 Neural crest

The neural crest emanates from the neural chord, initially as a wedge-shaped mass, which sends off the pluripotent neural crest cells. The neural crest, because of its origin from the ectodermal neural chord and its formation of mesenchymal cells, is referred to as 'ectomesodermal'. This is to distinguish it from the 'real mesoderm', which then is referred to as 'mesentoderm'. Though in teleosts — due to the unique formation of its neural tube — there is, strictly speaking, no neural 'crest', nevertheless, already early in development a population of mesenchymal cells is seen to segregate from the neuroepithelium and can be distinguished in the dorsal portion of the neural keel '(Eisen and Weston, 1993)'. These mesenchymal cells will generate various cell types including chromatophores, which at an early stage of the embryo migrate unto the yolksac and the body, where they are especially aggregated over the brain region. (The cells of the black pigment cell layer adjoining the retina are the only ones not of neural crest but of neurectodermal origin.) The mesenchymal cells of the head and the chondrocytes forming the embryonic skeleton are also of neural crest origin as are the sensory and sympathetic neurons and Schwann cells (rev. 'Selleck *et al.*, 1993'). The development of the neural crest in the zebrafish has been recently reviewed by 'Raible *et al.* (1992)' and 'Eisen and Weston (1993)'. Trunk neural crest cells have been investigated in living embryos of *Brachydanio rerio*. 'Raible and Eisen (1994)' labelled individual premigratory cells intracellularly with fluorescent vital dyes and recorded their complete lineages. It was concluded that some neural crest cells of the trunk are specified before reaching their final locations.

Summary

The neural crest, which runs along the neurochord, is mainly of ectodermal origin. Apart from forming neural tissue it is engaged in contributing mesodermal cells, hence the term of ectomesoderm given to the neural crest (as opposed to the term entomesoderm assigned to the mesoderm proper). The cells generated include sensory and sympathetic neurons, Schwann cells and pigment cells (retinal pigment epithelium cells excepted).

15.2 INTEGUMENT [10]

The functional significance of the skin of teleosts is the adaptation to various stresses, e.g. osmotic stress (represented by chloride cells and maybe pseudobranch cells) and protection from environmental hazards represented e.g. by the presence of mucus cells and microridges of the EVL.

15.2.1 EVL or Periderm

The EVL is the outermost layer (see Chapter 11). It displays a surface pattern of microridges. Individual cells are delineated by ridges, often unbroken, which circumscribe the perimeter of the cell. Further microridges coil towards the centre at an even distance from the adjacent ones (Fig.15/8). A few shorter segments are found in the centre of the cell. The arrangement of the microridges is often referred to as 'fingerprint pattern' but occasionally also as maze-like, filigree and, two-dimensional reticular pattern '(Yamada, 1968)'.

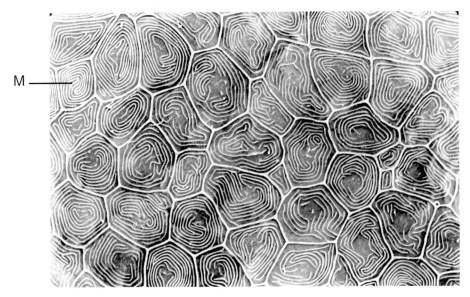

Fig. 15/8 SEM of skin surface (*periderm*, EVL) of two week old free-swimming *Oreochromis mossambicus*. × 1 780. Note microridges of EVL ('fingerprint pattern') and absence of hatching gland openings. M - microridges.

The cells of the EVL are sealed together by tight junctions (called 'terminal bars' at the light microscopical level and 'zonulae occludentes' at the EM level) which temporarily 'open and close'. Since the functional significance of tight junctions is to form a diffusion barrier between cells and also to create and maintain cell polarity, the EVL cells form a challenging object for molecular approaches to tight junctions. Recently, 'occludin' was suggested as a candidate for an adhesion molecule active at tight junctions '(Tsukita et al., 1996)'.

According to 'Rojo et al., (1997)' the external surface of the pavement cells (EVL) of *Salmo trutta* is folded into poorly developed microvilli [should be called microridges] which increase in height and number over time. EVL cells join each other by tight junctions apically and by desmosomes and interdigitations basally. It was observed that in *Salmo trutta* a few pavement cells, especially at the hatching time, undergo apoptosis. The data about the lifetime of the EVL vary. Some authors maintain that the EVL disappears before hatching and others that it lasts even into the adult life of the fish.

15.2.2 Epidermis

The histology of the skin of adult teleosts is depicted in Fig. 15/9. In the embryo the skin consists of a one-layered epidermis. During development it becomes three- or four-layered and remains covered by the EVL.

The whole surface of the early embryos is engaged in respiration. In some fishes this type of respiration continues throughout life (e.g. in the eel). However, in many pelagic species respiration of the body surface dominates only until the gills take over. In contrast, in most fishes developing under less favourable oxygen conditions, special respiratory sites, connected with the blood vascular system, are developed before the gills appear (Figs. 18/28, 18/30a,b).

The epidermis may contain transitory hatching gland cells (see Chapter 16), chloride cells, chromatophores, neuromasts (lateral line system), mucus glands and, in some nidicolous teleosts, adhesive glands (Figs. 1/3, 15/15, 16/1b). Transient dense hyaline plates in the outer epidermal layer were reported for the Pacific sardine, herring and anchovy by 'O'Connell (1981)'.

(a) Headfold

An epidermal headfold has been observed in embryos of ovoviviparous species. For instance, in the guppy, *Poecilia reticulata*, the one-layered epidermis over the head region invaginates (leaving the EVL behind) and makes possible the development of a pericardial hood (see Chapter 20 and Fig. 20/33).

(b) Finfold [11]

The epidermis, subtended by gel-filled dermal spaces, forms in the embryo the finfold in which later on the unpaired fins (dorsal, caudal, anal fin) will develop (Fig.

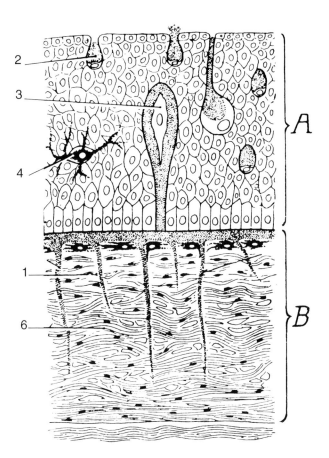

Fig. 15/9 Skin of adult teleost (Portmann, 1976, with kind permission of Benno Schwabe, Basel). **A** - *epidermis*; **B** - *cutis (dermis)*. 1 - *stratum germinativum*; 2 - unicellular *mucus* gland; 3 - unicellular goblet cells; 4 - chromatophore; 5 - *cutis*, soft (loose) part; 6 - *cutis*, compact (dense) part.

18/30). According to 'Ryder (1885b,c)' this fold at the time of hatching of the salmon contains 'simple embryonic rays', a term coined by 'A. Agassiz'. Because of their hairlike, unsegmented appearance 'Ryder' called them 'actinotrichia' (horny fibres of 'Balfour'). According to him they are formed from the mesoblast. The number of actinotrichia exceeds in number the permanent rays of the adult salmon at least 10-fold. Direct observation of mesenchyme cell movement in vivo in the teleostean fin bud has been reported by 'Wood and Thorogood (1984)'. They used the killifish *Aphyosemion scheeli*, because of the transparency of its fin structure. The actinotrichia are laid down in a highly ordered arrangement and mesenchymal cells migrate from the base of the fin bud between these fibrils using them as a substratum. These unsegmented collagenous fin rays of the embryo will be replaced later by the segmented bony fin rays (lepidotrichia).

(c) Chloride cells

Chloride cells, the principal sites of ion excretion, are another type of cell present in the epidermis '(Alderdice, 1988)'. They are characterized by the presence of numerous mitochondria, a ramifying system of branching smooth-walled tubules and an apical area exposed to the environment. Chloride cells in fishes are present in the gills and in the skin, especially in the integument of the yolksac as established first by 'Shelbourne (1857)'. It has been suggested that the chloride cells in the yolksac envelope may play a major role in ion transport before the gills become functional. While 'Jones et al. (1966)' were unable to find any chloride cells in the epidermis covering the head, yolksac and tail of herring reared at different salinities, 'Lasker and Threadgold (1968)' stressed that the organization and ultrastructure of the skin of the larval Pacific sardine, *Sardinops caerulea*, closely resembles that of the larval herring. However, a major difference is the presence of chloride cells in the sardine: they are found in all the areas overlying the yolk but are absent in the tail and head. On the other hand, 'Hølleland (1990)' observed chloride cells all over the surface of the yolksac of herrings. 'Katsura and Yamada (1986)' described the appearance and disappearance of chloride cells during the embryonic and postembryonic development of the goby, *Chaenogobius urotaenia*. 'Hwang (1988)' reported the distribution and density of the chloride cells in several postembryos. He showed that salinity stress alters the ultrastructure of the chloride cells but does not affect the surrounding skin cells.

The integumental chloride cells resemble ultrastructurally the chloride cells of the gills. Chloride cells either extrude Cl^- (in marine fish) or absorb Cl^- (in freshwater species). Chloride cell density varies between integumental regions of a fish embryo, and also between species. The cells may develop or degenerate at various ages of the embryo/postembryo, or following transfer between saltwater and freshwater stage, or shortly before hatching. They may disappear from the epidermis also after the yolk has been resorbed, presumably with concurrent development of branchial and other adult-type regulatory tissues. 'Shen and Leatherland (1978)' observed chloride cells on the yolksac and branchial epithelium of *Salmo gairdneri* embryos just before hatching and an increasing number in postembryos. However, when the embryos and postembryos were reared in different salinities, no difference in number of chloride cells was observed. According to 'O'Connell (1981)' in the anchovy chloride cells were not yet present at hatching but were found in the 3.3 millimetre long postembryo. As yolk recedes, chloride cells differentiate over much of the surface of the pericardium and the yolksac. Chloride-cells have also been reported for the surface layer of the yolk and pericardial membrane of the livebearing embryo of *Poecilia reticulata* by 'Dépêche (1973)' (see Chapter 20).

SEM analyses of the embryos of the mouthbrooding *Oreochromis* (*Sarotherodon*) *mossambicus* revealed openings, proposed to be apical orifices of the chloride glands, situated at the junction of three EVL cells '(Ayson *et al.*, 1994)'. The distribution of the openings is similar to the ones thought to be associated with hatching glands in the same species, *Sarotherodon* (*Oreochromis*) *mossambicus*, as well as in *Herotilapia multispinosa* (Figs. 16/1a,b, 16/5, 16/6). 'Foskett and Scheffey (1982)' analyzed the chloride cell in the opercular membranes and the branchial epithelium of the adult *Sarotherodon mossambicus* using the vibrating probe technique which proved definitely these cells as the extrarenal salt-secretory cells. 'Ayson *et al.* (1994)' stressed that in the adult fish the gill epithelium is the major site for chloride cells (mitochondria-rich cells) while in embryos and newly hatched postembryos, which possess only rudimentary gills, the yolksac envelope with its multitude of chloride cells seems to play a major role in ion transport. This is further documented by the findings of changes of activity in these cells in relation to changes in environmental salinity. Cortisol was shown to stimulate the activity of the chloride cells in the yolksac-membrane '(Ayson *et al.*, 1995)'. 'Shiraishi *et al.* (1997)' described multicellular complexes of chloride cells in the yolksac envelope of *Oreochromis mossambicus* (adapted to fresh water and sea water). Their ultrastructure is characterized by the presence of numerous mitochondria and extensive tubular systems continuous with the basolateral membrane. The multicellular complex consists of chloride cells and adjacent accessory cells sharing a common apical opening (see Fig. 15/12).

'Rojo *et al.* (1997)' found three different types of chloride cells in the gills of brown trout (*Salmo trutta*) embryos: immature, mature and apoptotic. The authors suggested that the immature chloride cell is a developing accessory cell. They had shown in a previous publication that rainbow trout hatch with only half of their chloride cells in a fully differentiated state '(González *et al.*, 1996)'. According to 'Morgan and Tovell (1973)'a wave of cellular death of chloride cells occurs during post-hatching stages of rainbow trout when there is a respiratory and iono-osmotic transition from the integument to the gills. At the same time, and in the same region, EVL degenerate and are renewed.

(d) Pseudobranch[12] (false gill, imitation gill, opercular gill)

Gills are formed from the ectodermal part of the gill slits; while the lining of the gill passages in any fish includes derivatives of both ecto- and entoderm '(Romer and Parsons (1977)'. The pseudobranch is a teleost-specific organ and was discovered by 'Broussonet (1782, 1785)' who called it a 'small gill' and ascribed to it a respiratory function. 'Müller (1839)' reported a detailed analysis of the pseudobranchial vascularization and the connection between it and the choroid body of the eye. He also noted a shunt between the two efferent pseudobranchial (or afferent ophthalmic) vessels. He suggested that the pseudobranch is responsible for controlling ocular blood pressure. 'Granel (1927)' insisted that the pseudobranch was not a rudimentary gill but a fully-developed organ. However, it is morphologically a hemibranch, i.e. the remnant of the first gill arch. [The normal gill bar carries a gill on both surfaces (holobranch) and hemibranch is the term used if a gill is present only on one surface of the gill bar.] It is interesting to note that in most elasmobranchs and in the chondrostei and holostei (*Acipenser* and *Polypterus*) the first gill pouch has been transformed into a spiracle.

A study of the teleostean blood circulation reveals that the pseudobranch receives oxygenated blood (from the hyoidean artery). The efferent vessel leaving the highly vascularized pseudobranch is the ophthalmic artery, which carries 'doubly oxygenated blood' to the eye (Fig. 15/10).

'Müller (1839)' supplied a list of 280 species of fish. Of these 39 had no pseudobranch. The remaining 241 with a pseudobranch included 198 with free, and 43 with a covered and glandular pseudobranch. This was followed by five further publications, in German and French, on the pseudobranch and vascularization of the eye. 'Maurer (1884)' repeated that the pseudobranchs can be covered/glandular (as in *Esox* and *Gadus*) or non-covered/free (as in *Alosa* and *Barbus*). A free pseudobranch is not covered by an epithelial fold and structurally resembles a normal gill except that it possesses only one row of filaments. The secondary lamellae are free of one another, as are the filaments though the pseudobranch is joined to the operculum for most of its length (Fig. 15/11). A covered pseudobranch shows different stages of coverage. It may be covered by the opercular epithelium, may show fusing of the filaments and lamellae and, because of a preponderance of acidophilic cells (cells with acidophilic granulations), is called a 'glandular pseudobranch'. It may even be buried deeply inside the body thereby completely loosing contact with the external medium '(Granel, 1927)'. 'Leiner (1939)' experimentally analyzed in detail the physiological role of the pseudobranch. He called it 'Augenkiemendrüse', a term which in itself indicates — in his words- that the pseudobranch constitutes an 'auxiliary gland for the respiration of the retina'. According to 'Bridges *et al.* (1998)' marine species tend to have larger and better developed pseudobranchs than freshwater species.

In the adult teleost the epithelium of the pseudobranch consists of chloride and acidophilic cells; the latter are commonly known as 'pseudobranch cells'. Other cells mentioned are supportive cells, rodlet cells and mucus cells. As an example, the pseudobranch of the bass, *Dicentrachus labrax*, and the smelt, *Osmerus esperlangus*, contain both pseudobranch and chloride cells, whereas the rainbow trout posesses only the former type '(Mattey *et al.*, 1978, 1979)'.

EM revealed that the pseudobranch cells of the lamella are closely applied to the basal lamina and endothelium of the blood spaces. Adjacent to the blood supply the mitochondria are closely packed but separated by long, parallel arrays of tubules that are arranged in an orderly fashion. As the tubules (with a relatively constant diameter) approach the basal (vascular) cell border they lose their parallel nature and form a reticular network that opens freely to the cell surface (Fig. 15/13 with inset). 'Rich and Philpott (1969)' described the presence of many fine, rounded projections from the luminal surface of the tubular membranes, which are not present on other membranes of the cell. 'Mendoza *et al.* (1976)'

Ectodermal derivatives   283

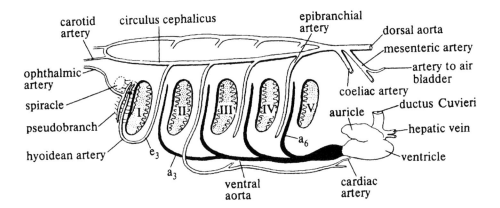

Fig. 15/10 Diagram of the branchial circulation of a teleost (Goodrich, 1909). There are five branchial slits. The spiracle is closed. The hyoidean artery supplies oxygenated blood to the vascularized pseudobranch. Its efferent vessel, the ophthalmic artery, carries 'doubly oxygenated' blood to the eye. I-V - branchial arches; a_3-a_6 - efferent vessels emanating from the ventral *aorta*; e_3 - efferent of the first branchial arch.

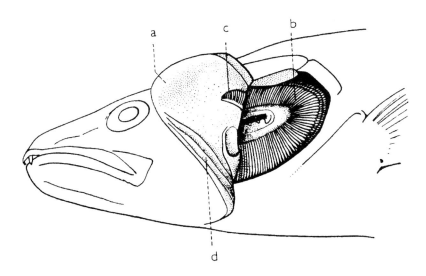

Fig. 15/11 Pseudobranch of *Lucioperca* (Rauther, 1910). This is an example of a teleost with an 'attached type' of pseudobranch, which becomes exposed when *operculum* is lifted. a - *operculum*; b - gill; c - pseudobranch; d - first holobranch.

using lanthanum nitrate showed that the tubules are completely open to the externum of the cell.

The ultrastructure of the chloride cell in the pseudobranch is similar to the one found in the epidermis and on the gill surface. It is rich in mitochondria, which are surrounded by a highly branching network of intercellular tubules, which connect with the narrow saccules near the vascular border. Mitochondria and tubules are absent in the apical region, where smooth surfaced vesicles are abundant. The apex is characterized by a pit. The chloride cell shows connection with the plasma membrane on all surfaces of the cells except at the extreme apical crypt end, while the pseudobranch cell, described above, shows connections only to the basal vascular border (Fig.15/12).

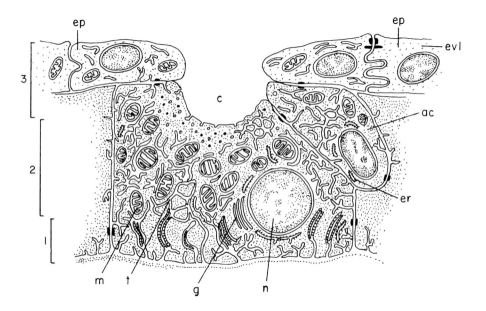

Fig. 15/12 Chloride cell of pseudobranch (modified from Dunel and Laurent, 1973, with kind permission of The Biology of the Cell, Paris). Schematic representation. The chloride cell communicates with the external *medium* with a crypt (c). The chloride cell is flanked by an accessory cell (ac). The *periderm* (evl) [a flattened *epithelium* (ep)] has retracted to give way to the crypt of the chloride cell. **1** - vascular region adjacent to fenestrated capillary *endothelium* (not shown). **2** - central region with *mitochondria* (m), anastomosing tubules (t), Golgi body (g), granular endoplasmic *reticulum* (er) and *nuclei* (n). **3** - Apical region with different types of vesicles.

The pseudobranch supplies the blood to the choriocapillaris and the falciform body of the eye (see Chapter 17). In other words, the ophthalmic artery is an efferent artery of the pseudobranch. 'Hyrtl (1838)' was the first to prove that the pseudobranch receives arterial blood. Therefore, since the pseudobranch receives already oxygenated blood from the adjacent gill and if it functions itself also as a gill, the blood it sends off to the eye is 'doubly-oxygenated' or 'super-oxygenated'. This result, seen in the salmon, has subsequently been confirmed by various authors. A correlation with vision was reported by 'Fairbanks et al. (1969)', 'Wittenberg and Haedrich (1974)' and 'Wittenberg and Wittenberg (1974)'. They showed with oxygen electrode measurements that the choroid rete behind the retina of the fish eye shows a high oxygen pressure (see Chapter 17 on eye). Finally, 'Bridges et al. (1998)' conclude that 'most evidence now indicates that the pseudobranch is integrally linked with the choroid rete and the control and supply of oxygen to the retina of the fish eye'. This should encourage work comparing the development of the pseudobranch with that of the eye.

Besides this particular function, various other roles have been attributed to the pseudobranch, including ion regulation, supplementary gas exchange, regulation of blood pressure in the eye, an endocrine function and sensory reception for regulation and ventilation [rev. 'Beatty (1977)', Laurent and Dunel-Erb (1984)', 'Ryan (1992)']. 'Bridges et al., (1998)', too, stress that the role of the pseudobranch in vision 'does not exclude other functions, and although some questions have been answered, many remain to be investigated'.

Unfortunately, the development of the pseudobranch has seldom been investigated. The eel possesses as larva a well-developed, uncovered pseudobranch, which is completely absent in the adult ('Laurent and Dunel-Erb, 1984'). In some teleosts the pseudobranch appears before the gills. It is known that the pseudobranch in the early fish embryo receives blood directly from the heart via the ventral aorta. It would be interesting to establish when the deoxygenated blood supply changes to the adult 'doubly oxygenated' type and compare and contrast probable changes in the structure/ultrastructure of the pseudobranch. An example of the ultrastructure of the pseudobranch cells of an embryo of the guppy (*Poecilia reticulata*) is given in Fig. 15/14.

'Parry and Holliday (1960)' suggested that the pseudobranch may produce or activate a hormone causing pigment aggregation in chromatophores and that the entry of such a hormone into the general circulation is regulated by the choroid gland of the eye.

(e) Adhesive organs[13]

Nidicolous fish, as the name suggests, do not move around after hatching (see Chapter 1). They still possess a large yolksac. In order to ensure efficient respiration, these types of embryos possess a number of structural adaptations, the adhesive organs being one of them. In cyprinids the adhesive organs are represented by unicellular glands, arranged in the skin of the head, in front of and below the eyes '(Soin, 1971)'. The hatchling of the pike, *Esox lucius*, attaches itself with paired adhesive organs to water plants or even to the water surface.

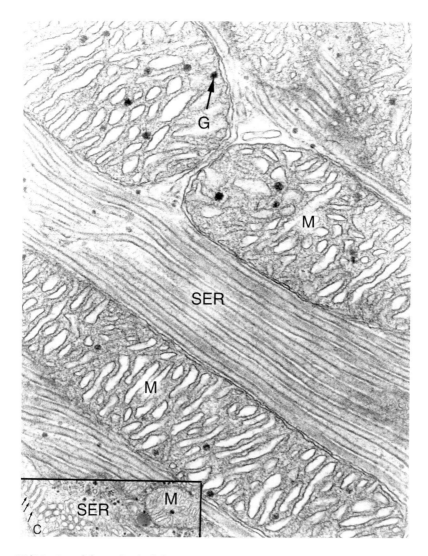

Fig. 15/13 Pseudobranch of adult guppy (TEM) (courtesy A. Yacob) × 51 800. Section of pseudobranch cell. Note that *intermitochondrial* space is almost entirely taken over by the tubules. Inset: shows the numerous basally located tubules. Some of them are seen to open unto the blood capillaries (arrows) × 17 000. C - capillary; G - *mitochondrial* granule; M - *mitochondrion*; SER - smooth endoplasmic *reticulum* (tubules).

Their adhesive glands consist of a group of unicellular mucus-producing cells, which lie cranio-ventrally in the region of the eye and nose. When the adhesive glands are at the height of their activity the fish tend to remain motionless in this position for a considerable time. Now and then they disengage, swim a short

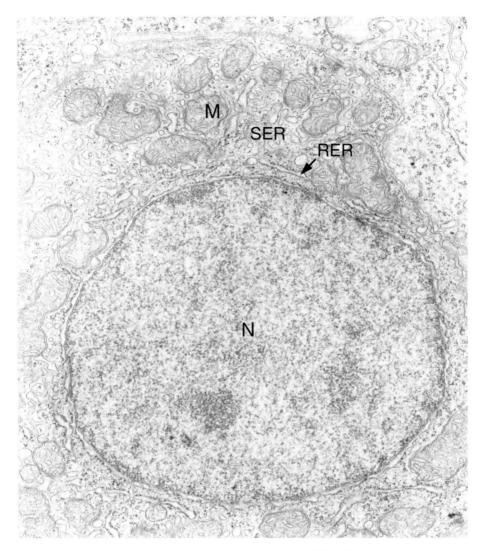

Fig. 15/14 TEM of an embryonic pseudobranch cell of *Poecilia reticulata* of stage 6 (courtesy A. Yacob) × 22 000. Numerous tubules are present but not yet regularly arranged. M - *mitochondrion*; N - *nucleus*; RER - rough endoplasmic *reticulum*; SER - smooth endoplasmic *reticulum*.

distance and attach themselves again '(Gihr, 1957)' (Fig. 15/15). In *Betta splendens* the adhesive organs are situated beneath the eyes, but in addition numerous glandular tubercles are distributed over the head. This fish, too, suspends itself from the water surface or adheres to plants '(Soin, 1971)'.

Three pairs of adhesive glands on the head are encountered for example in the cichlids *Pterophyllum scalare* (Fig. 18/30) and *Herotilapia multispinosa* (Fig. 16/1b). There is a single anterior pair, which has a forward facing aspect, whereas the other two pairs occur on the top of the head where they are very closely associated with each other. These glands are covered by the EVL, which covers the rest of the body (Fig. 15/16). The EVL is absent inside the gland, where numerous pits are revealed, through which the adhesive cement has been secreted (Fig. 15/17).

The sticky secretion of the adhesive glands attaches the postembryos to the substrate. They are similar to the adhesive glands described by 'Peters and Berns (1983)'. The adhesive phase ends with the degeneration of the adhesive glands and is followed by the so-called free phase, which starts with the 'swim-up' stage and the external uptake of food (see Fig.1/1).

Summary

The outermost cell layer of the integument is the EVL (periderm), which persists in the embryo or even longer, in some fishes up to the adult stage. On its surface a concentric pattern of microridges ('fingerprint pattern') is evident when viewed with electron microscopy. (15.2.1)

Subjacent to the EVL is the epidermis which in the embryo is one layer thick. The epidermis may invaginate at the anterior end to form a headfold (example guppy), which will later house the pericardial hood. The finfold, when capillarized at different sites, will assist in respiration and disposal of waste (see Chapter 18). The median dorsal and ventral finfolds are the sites where the unpaired dorsal, caudal and anal fins will develop. (15.2.2a,b)

The epidermis may contain chloride cells (probable sites of ion excretion) (15.2.2c) and hatching gland cells (see Chapter 16). It may further include cells forming the pseudobranch (false gill or opercular gill), an organ unique to teleosts (15.2.2d). Various functions have been attributed to it; the most likely one is a role in vision, since it sends doubly-oxygenated blood to the eye. The epidermis may also form adhesive glands which are made up of mucus-producing cells. With their help, newly hatched altricial fishes become attached to plants or suspended from the water surface. (15.2.2e)

Fig. 15/15 Adhesive gland of pike (Gihr, 1957, with kind permission of the director of the Muséum d'Histoire naturelle, Geneva). Drawing of vertical section through the adhesive gland area of the pike, *Esox lucius*. 1 - upper layer of *epidermis* engaged in secretion; 2 - basal layer of *epidermis*; 3 - 'normal' *epidermis*; 4a - columnar secretory cells;

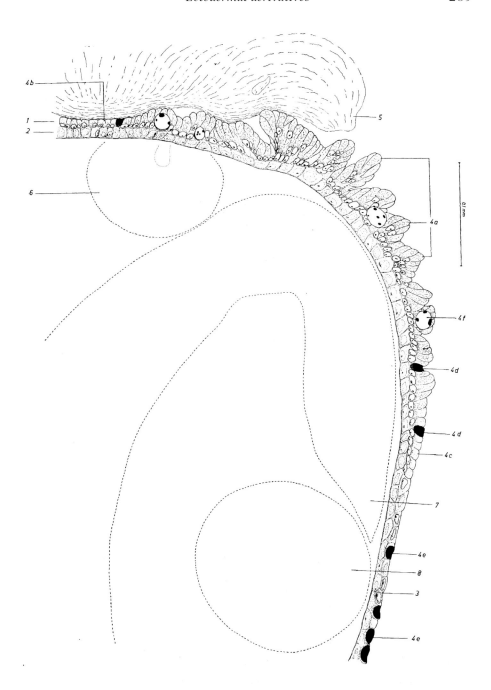

4b - cuboidal secretory cells (median location); 4c - cuboidal secretory cells (lateral location); 4d - calycal secretory cells; 4e - cuboidal normal (non-secretory) cells; 4f - vacuoles (?); 5 - secreted adhesive mass; 6 - olfactory placode; 7 - eye cup; 8 - lens placode.

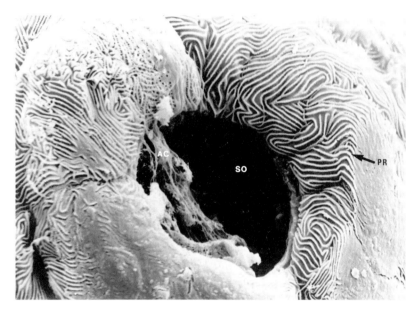

Fig. 15/16 SEM showing *periderm* (EVL) covering the adhesive gland of *Herotilapia multispinosa* 12 hours after hatching. × 3 500. AC - adhesive cement; PR - microridges of *periderm* (EVL); SO - orifice of adhesive gland.

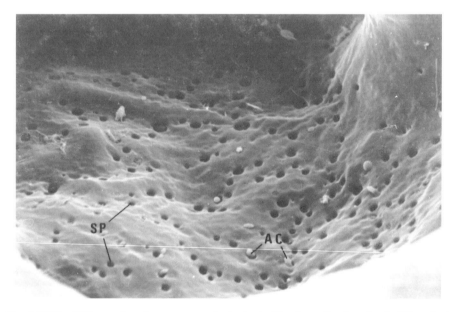

Fig. 15/17 SEM revealing the secretory pits in the inside of an adhesive gland of *Herotilapia multispinosa*, 12 hours after hatching. × 5 100. SP - secretory pores; AC - remnants of adhesive 'cement'.

Endnotes

1. G. Oberhaut, Hornblatt of 'Oellacher', Umhüllungshaut of 'Reichert'
2. Anterior forebrain
   G. Vorderhirn of 'Gegenbaur' and 'von Baer'
3. Posterior forebrain,
   G. Zwischenhirn of 'von Baer'
4. Midbrain.
   G. Mittelhirn of 'von Baer', Zwischenhirn of 'Gegenbauer'
5. Hindbrain or cerebellum
   G. Hinterhirn of 'von Baer', Mittelhirn of 'Gegenbaur'
6. Medulla oblongata
   G. Nachhirn of 'Gegenbaur'
7. from Gk. epi = upon and physis = growth, synonym of pineal gland
8. Glandula pinealis pineal from L. pinea = pine cone
   G. Zirbel, Zirbeldrüse
9. G. Zirbelpolster after 'Burckhardt', vorderes Parietalorgan after 'Studnicka, 1905', parencephalon after 'Kupffer', parapineal after 'Hill, 1891'. '(Holmgren, 1965)'
10. from L. integumentum = covering
11. G. Flossensaum
12. G. Augenkieme, Augenkiemendrüse
13. G. Haftdrüse

Chapter sixteen

Hatching

'There is no final experiment. Nature like a mirage, constantly challenges our perception.
Science, at best, is really an art of partial truths.'

Rajendra Raghow

Hatching is a process in which the developing fish embryo emerges from its envelope. In the words of 'Yamagami (1988)' it changes from an 'intracapsular' to a 'free-living' life. In different species hatching takes place at different developmental times of the embryo (Fig. 1/1). 'Yamagami' distinguished between mechanical and enzymatic hatching. Enzymatic shedding of the envelope is brought about mainly by the secretion of a proteolytic enzyme by unicellular hatching glands (UHG), which are of ecto- and/or endodermal origin. The hatching enzyme dissolves the inner layers of the envelope; this is followed by the lashing movements of the embryo, which result in breaking of the envelope.

'Moriwaki (1910)' was the first to report an enzymatic digestion of a teleostean egg envelope (*Oncorhynchus keta*). He reported that the perivitelline fluid derived from one embryo could digest more than 15 egg envelopes at a temperature as low as 8°C. He also discovered a large number of unicellular glands on the surface of the embryonic body 10 days before hatching. Since perivitelline fluid obtained before the time of hatching did not dissolve the egg envelope, he concluded that the secretions of the unicellular glands were instrumental in the digestion of the egg envelope. 'Wintrebert (1912a,b)' investigated the hatching of the rainbow trout and the goldfish. In order to establish if the movements of the embryo are the reason for the breaking of the egg envelope, he treated the embryos with chloretone, which renders them inert. He found that hatching was not prevented but simply retarded. This treatment had rendered the envelope extremely thin and fragile. When it was cut open a liquid oozed out which was present only at the time of hatching, not before. Most frequently the embryos hatched head-first. When the liquid was poured on unfertilized eggs, notable thinning of the envelope was observed after 3 hours, and after 6 hours it had become utterly fragile. By letting the tail and/or head of the embryo protrude through the envelope, it was concluded that the secretion stems from the body. 'White (1915)' described the movements of the embryo of *Salvelinus fontinalis* as the sole means of hatching. Of 23 cases observed, eight appeared head first, one tail first, and in 14 cases the yolksac broke through before the body. Hatching in this species lasted from 45 minutes to 5 or 6 hours.

Various studies dealing with the hatching mechanism followed, e.g. 'Chevey (1925)' for *Perca fluviatilis* and 'Armstrong (1936)' for *Fundulus heteroclitus*. 'Ishida (1944a,b)' suggested for *Oryzias latipes* that the current of the perivitelline fluid from the mouth to the gill openings is responsible for the breakdown of the hatching glands. This current starts just before hatching and is brought about by the beating of the pectoral fins, the rhythmic opening of the mouth and the opercular movements. 'Harder (1953)' supposed that the process of hatching is started, if not being maintained, by the pressure of the extending notochord. In the brook trout, 'hatching is initiated by movements starting at the head and later extending through the whole length of the body'. Normally, the head of the embryo emerges first. The embryos which hatch with the tail emerging first, normally die soon afterwards ('Lindroth, 1946)'. A thorough review of the hatching process in teleosts is given by 'Yamagami (1981, 1988)', 'Schoots (1982)', 'Oppen-Bernsten (1990)' and Luczinski *et al.* (1993)'.

16.1 DISTRIBUTION OF UNICELLULAR HATCHING GLANDS (UHG)

'Yanai (1966)' put forward a classification of UHG based on their distribution: 1) on body and yolksac, 2) body or yolksac, 3) in mouth cavity, pharyngeal cavity and gills, 4) distribution as for 1) and 3) combined. The distribution of UHG may change during ontogeny as observed in the pike: mature glands appear first in the head region and only later in the anterior half of the yolksac.

Examples are 1) *Salmo irideus* '(Wintrebret, 1912a)' and *Oreochromis mossambicus* '(McNeilly† and Kunz, 1992)' (Fig. 16/1a). 2) *Pterophyllum scalare* (Fig. 1/1) and *Herotilapia multispinosa* '(McNeilly† and Kunz, 1992)'(Fig. 16/1b)]. 3) *Oryzias latipes* '(T. Yamamoto, 1975)'. 4) A great number of embryos described belong to this category: *Oncorhychus keta* '(Moriwaki, 1910)', trout '(Bourdin, 1926a,b)', *Salmo gairdneri, S. trutta, Salvelinus fontinalis, S. pluvius* and *Oncorhynchus mykiss* '(Hagenmaier, 1974; Yokoya and Ebina, 1976; Ishida, 1985 and Blanquez *et al.*, 1996)', Coregoninae '(Hagenmaier, 1974a,b; Yokoya and Ebina, 1976)'.

'Blanquez *et al.* (1996)' stressed the positional information necessary for the UHG to develop at the appropriate site. Most of the UHG are irregularly scattered singly, but sometimes occur in pairs as in the pike and *Tilapia*. They also appear in clusters as in *Brachydanio rerio* on the yolksac covering the pericardium '(Kimmel *et al.*, 1995)' and in the trout '(Blanquez *et al.*, 1996)'. The pharyngeal and epidermic epithelia of the trout are multilayered at the time of hatching. Some of the UHG are in direct contact laterally while other neighbouring glands are separated by thin feet of other epithelial cells.

Hatching

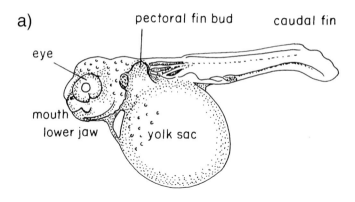

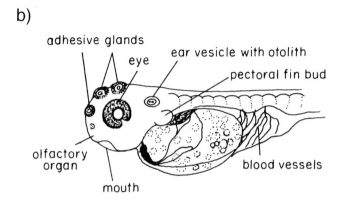

Fig. 16/1   Distribution of hatching glands one day posthatching. **a** - *Oreochromis mossambicus*: Openings of hatching glands on head and anterior yolksac indicated by half-circles cccc. **b** - *Herotilapia multispinosa*: Openings of hatching glands confined to yolksac. Indicated by circles with dotted periphery. (The dark circles represent yolk globules, which are found throughout the yolksac.)

16.2   HISTOLOGICAL STRUCTURE OF UHG

The UHG are easily distinguishable from mucus cells by the presence of secretory granules (Fig. 16/2). 'Yokoya and Ebina (1976)' described giant UHG in the skin of the head and yolksac and in the epithelium of the mouth and gills of salmonids. They contained granules with a strong affinity for toluidene blue and were seen to discharge these secretory granules at hatching. The granules of the endodermal UHG of *Oryzias latipes* are eosinophilic and also contain a small amount of haematoxylin-stained substance ('Ishida, 1944a').

16.3 ULTRASTRUCTURE OF UHG

The ultrastructure of UHG for *Oryzias latipes* was described by 'M. Yamamoto (1963, 1975)', for *Barbus schuberti* by 'Willemse and Denucé (1973)', for *Leuciscus hakuensis* by 'Ouji and Matsuno (1973)' and for *Salmo gairdneri, S. trutta, Salvelinus fontinalis* and *S. pluvius* by 'Yokoya and Ebina (1976)'. The cells contain a highly electrondense nucleus with a conspicuous nucleolus and many RER vesicles. In all the above species the secretory granules are thought to contain a hatching enzyme (or enzymes) associated with the Golgi body (Fig. 16/3).

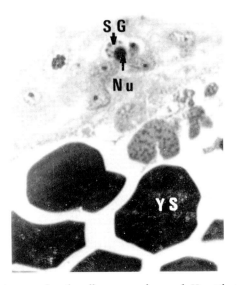

Fig. 16/2 Light micrograph of yolksac envelope of *Herotilapia multispinosa* fixed at hatching time. 1 μm section. × 500. Nu - *nucleus* of UHG; SG - secretory granules; YS - platelet in yolksac.

16.4 ULTRASTRUCTURAL DEVELOPMENT OF UHG

Differentiation of UHG from precursory endodermic cells already begins in the medaka, *Oryzias latipes*, some days after egg fertilization, at a time when the eye is pigmented '(Ishida, 1944a,b)'. The development of the UHG reveals that in the 3 day old embryos they are distinguishable from the other endodermal cells by an electrondense nucleus with a prominent nucleolus, highly developed flattened cisternae of the ER and a conspicuous Golgi body. In the 4-5 days old embryo the cisternae are dilated and contain a finely granular material. The putative secretion granules gradually replace the apical cisternae. When the glandular

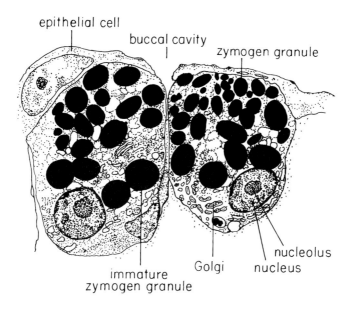

Fig. 16/3 Ultrastructure of unicellular hatching gland (UHG) in the buccal cavity of a prehatching stage embryo of *Oryzias latipes* (drawing modified after EM by M.Yamamoto et al., 1979, with kind permission of Academic Press, Orlando, Florida). × 3 000. Hatching cells are covered by a layer of EVL (epithelial cells).

cells burst and emit the granules into the perivitelline space, they leave behind crescent-shaped electrondense shells '(T. Yamamoto, 1975)'. The 'primitive' UHG display a large central nucleus which is round in *Salmo trutta* and *S. pluvius*, round or oval in *S. gairdneri* and crescent-shaped in *Salvelinus fontinalis*. Extensive RER, filling most of the cytoplasm and its dilated cisternae, contains flocculent material of medium density. Some ribosomes are free and polysomal in the cytoplasm. Mitochondria and a few small putative glycogen granules are scattered throughout the cytoplasm. The immature UHG contain some secretory granules of various size and density. During the course of maturation these granules come to occupy more than half of a cell. Immediately before hatching, granules fuse and the UHG form apical junctions with adjacent cells. At hatching, the granules with some cytoplasm are discharged (apocrine[1] secretion). However, the UHG of both *Oryzias latipes* and *Leuciscus hakuensis* have been reported to show holocrine[2] secretion '(Ishida, 1944a,b; Iga, 1959)'.

Maturation of the hatching gland apparatus involves two parallel processes: an increase in the number of hatching gland cells and in the amount of the enzyme chorionase. Embryos at the eyed-egg stage already possess immature UHG, which contain few secretory granules. The number of secretory granules increases while the UHG grow in size. The granules eventually dislocate the nucleus to the periphery. '(Luczynski and Kirklewska, 1984; Luczynski et al., 1993)'. According to 'Iuchi and Yamagami (1979)' and 'Schoots et al. (1982a)' synthesis of enzymatic vacuoles is observed before eye-pigmentation. However, hatching glands were not observed before eye-pigmentation in either *Oreochromis mossambicus* or *Herotilapia multispinosa* '(McNeilly† and Kunz, unpublished). Also in *Brachydanio rerio* the time of the first appearance of hatching enzyme granules coincides with that of eye-pigmentation '(Willemse and Denucé, 1973)'.

16.5 SECRETIONS BY UHG AND THEIR DEATH BY APOPTOSIS[4]

'Bourdin (1915)' regarded the hatching gland as morphologically merocrine[3] but functionally holocrine. The mature UHG are usually interspersed as a single layer between the EVL and the prospective epidermis. While the hatching enzyme is secreted, the apical part of the cell protrudes towards the perivitelline space and secretes its contents into it (so-called eccrine or merocrine secretion) as observed e.g. for the Coregoninae by 'Luczynski et al. (1993)'. This contrasts with reports by 'Ishida. (1944a,b)' and 'Iga (1959)' on *Oryzias latipes* and *Leuciscus hakuensis* respectively. They defined the secretion of the hatching glands in these embryos as holocrine (16.5). Similarly, in *Salmo gairdneri, S. trutta, Salvelinus fontinalis* and *S. pluvius* the glands detach themselves from the epithelium after releasing their granules '(Yokoya and Ebina, 1976)'. 'Schoots et al. (1982b)' suggested that this probably coincides with the shedding of the EVL around hatching time, reported for *Salmo trutta fario* L. by 'Bouvet (1976)'. In contrast, 'Schoots et al. (1982b)' observed that in the pike, *Esox lucius*, the posthatching glandular cells remain in the epidermis of the larvae and finally undergo apoptosis and are phagocytized by neighbouring cells. In *Salmo trutta*, 3 days after hatching, 'Rojo et al. (1997)' reported apoptosis of UHG scattered in the cephalic epidermis and the yolksac epithelium, which had not released the hatching enzyme. So-called 'classical apoptosis' was observed by the same authors in trout embryos, from at least eight days before hatching until three days after hatching, with a wave of cellular deaths occurring around hatching time. Subsequent infiltration of leucocytes, and in some instances, secondary necrosis of apoptotic UHG, was reported by the same authors.

Massive secretion of the enzyme shortly before the actual emergence of the embryo may be under endocrine control of prolactin '(Schoots et al., 1982a)'. Histological analyses of in vitro experiments on the breakdown of the envelope had suggested that the entire ZRI is digested during the hatching process '(Bell et al., 1969; M. Yamamoto and Yamagami (1975)'. The breakdown of two thin envelopes (pike and zebrafish) and a thick one (*Nothobranchius korthausae*) was analyzed by EM '(Schoots et al., 1982a)'. The thick ZRI of the last species is entirely digested by the hatching enzyme whereas the thinner ZRI of the pike and zebrafish decreases only by 30 and 15% respectively. In all three species it is possible to break down the layer completely while the ZRE remains intact. It has been suggested that in vivo this layer is digested by microorganisms present on the outer surface.

Immature UHG in *Oryzias latipes* are characterized by the presence of zymogen granules in the Golgi body. Electrically stimulated UHG swell and separate the joints of the overlying EVL cells so that the apical surface of the gland cell becomes exposed to the buccal cavity. A pattern of the openings of UHG, i.e. each orifice is located at the junctions of three EVL cells, was observed by 'Yamagami (1981)' and 'Schoots et al. (1982b)'. TEM of the UHG of *Oryzias latipes* revealed a coalescence of the zymogen granules, and a decrease in electron density of the substances '(Yamagami, 1981)'. Similarly, the SEM of *Oreochromis mossambicus* and *Herotilapia multispinosa* revealed orifices opening on the head and on the most anterior parts of the body of the former unto the surface of yolksac and on the base of the pectoral fin bud in the latter, separating the junction of three peridermal cells (Figs. 16/4). This is easily observable by following the course of the outermost microridges of the periderm (EVL) cells (Fig. 16/5). Higher magnification reveals a well-developed grid structure within the orifice. This structure may be the result of dried remnants after secretion of the hatching enzyme (Fig.16/6). However, the EM investigations of the EVL of the same two species were given different interpretations by 'Ayson et al., 1994; Shiraishi et al., 1997', Chapter 15). At the end of secretion UHG undergo apoptosis (programmed cell death) '(Blanquez et al., 1996; Royo et al., 1997)'.

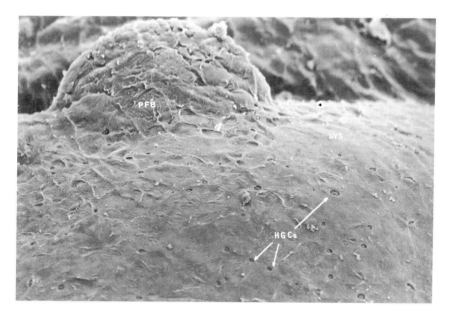

Fig. 16/4 Distribution of openings of unicellular hatching glands (UHG) on one day posthatching in *Herotilapia multispinosa*. SEM × 860. PFB - pectoral fin bud; HGCs - openings of unicellular hatching glands; YS - yolksac.

'Luczynski et al. (1993)' concluded that the following processes are involved in hatching: 1) enzymatic dissolution or softening of the proteinaceous layers of the envelope, 2) rupturing of the remnants by body movements, 3) mastication by the embryo, 4) anoxia.

16.6 TRIGGERING OF UHG SECRETION

As described in 16.4 the hatching enzyme (hatching proteinase, chorionase) is produced by UHG, which store the enzyme until the time of hatching. Different stimuli received by the embryo from its external and/or internal environments trigger the secretion of the enzyme thus initiating the process of hatching. 'Ishida (1944a,b)' observed that in *Oryzias latipes* opercular movements begin shortly before the breakdown of the UHG. Flushing of water from a capillary inserted into the mouth of the embryo broke down the cells. He concluded that the opercular movements of the embryo are necessary for the secretion of the UHG. Oxygen concentration and temperature exert a combined influence on hatching since oxygen uptake reaches its maximum immediately before hatching [reviewed by 'Luczynski et al. (1993)']. Moreover, electrical stimuli at various developmental stages can bring about hatching '(Iuchi and Yamagami, 1976; M. Yamamoto et al., 1979)'. Hatching time can also be controlled by increasing the environmental salinity or pH '(Luczynski et al., 1993)'. Since oxygen uptake by embryos reaches its maximum level immediately before hatching, temporary lowering of oxygen supply can cause precocious hatching '(DiMichele and Taylor, 1981; DiMichele and Powers, 1984)'. Shortage of O_2 also results in stimulation of the opercular movements, which in turn accelerate hatching '(Yamagami, 1981)'. Similarly, bubbling nitrogen through the water can stimulate rainbow trout eggs to hatch '(Hagenmaier, 1974a,b; DiMichele and Taylor, 1981; Iuchi et al., 1985)'.

16.7 HATCHING ENZYME

As mentioned in the introduction, 'Moriwaki (1910)' discovered the hatching enzyme (chorionase) in the chum salmon, *Oncorhynchus keta*. This was later followed by analyses of the enzyme for other fishes; *Oryzias latipes* by 'Ishida (1944a,b)'; *Fundulus heteroclitus* by 'Kaighn (1964)'; *Salmo gairdneri* by 'Hagenmaier (1974a)' and *Esox lucius* by 'Schoots and Denucé (1981)'.

The hatching enzyme of coregoninae embryos has a molecular weight (MW) of about 12,000. It also dissolves the envelopes of *Esox lucius* and *Salmo trutta fario* ('Luczinski et al., 1993'). The molecular weight of the hatching enzyme of *Salmo gairdneri* was reported as 24,000. In salmonids and the pike the hatching enzyme is a zinc metalloprotease '(Hagenmaier, 1974a,b; Schoots and Denucé, 1981; Hagenmeier and Lindemann, 1984)'. The biochemical properties of the hatching enzyme were analysed by various investigators '(K. Yamagami, 1988)'.

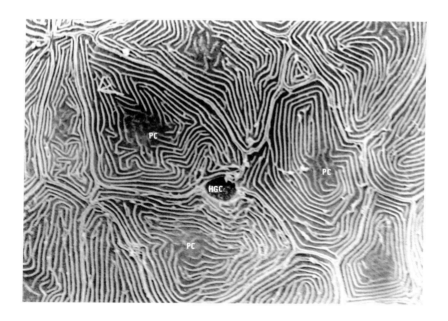

Fig. 16/5 *Oreochromis mossambica* at hatching time. Opening of hatching gland cells at the junction of three peridermal cells (note fingerprint-pattern). SEM x 4 000. HGC - opening of unicellular hatching gland; PC - *periderm* cell.

A central question has long been whether the hatching enzyme is a single enzyme or a system of multiple enzymes. The results of immunocytochemical analyses proved that there is a single hatching enzyme in the UHG of the pike, *Esox lucius*. The distribution of the glandular cells over the embryo changes during ontogeny. Mature cells first appear in the head region and only later also in the anterior half of the yolksac '(Schoots et al., 1982a,b)'. These authors suggested that the hexagonal pattern of the EVL seems to become interrupted where glandular cells are located. These authors also reported the persisistence of UGH remnants in a few days old postembryos.

The hatching enzyme of *Oryzias latipes* was purified as a single enzyme first but it later was found as a result of immunocytochemistry and immunoblotting to be an enzyme system composed of two distinct but similar proteases. They were named HCE and LCE (high and low choriolytic enzyme). They were found to be synthesized concurrently and packaged in the same secretory granule though in a discrete arrangement. The core of the secretory granule is occupied by HCE, while LCE is localized in the periphery. HCE is several times more abundant than LCE and the appearance of HCE seemingly precedes that of LCE. HCE had a molecular weight of 24,000 and LCE of 25,500 '(Yasumasu et al., 1992; L.Yamagami et al., 1993)'.

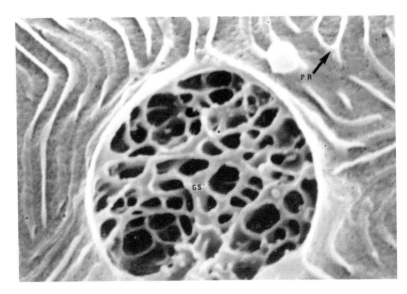

Fig. 16/6 *Herotilapia multispinosa* one hour posthatching. Opening of unicellular hatching gland displaying grid structure probably derived from remnants of the glandular secretion. SEM 20 500 ×. GS - grid structure; PR - and arrow ridge (part of fingerprint-pattern of peridermal cell).

Summary

The principal cause of hatching is due to secretions of unicellular hatching glands (UHG), which digest the envelope. Movements of the embryo play an accessory role in the process of rupturing the envelope. UHG are of ectodermal, and some of endodermal, origin. Hatching glands are located in different areas of the embryo according to species; their distribution may change during ontogeny. They have a function only during a very restricted period of time. The synthesis of the hatching enzyme begins at an early phase of development and continues until the time of its secretion.

Endnotes

[1] apocrine (during secretion portion of the cell is lost). From Gk apo = away, off and Gr krinein = to separate.
[2] holocrine (the whole cell is lost during secretion). From Gk holos = whole, entire and Gk krinein = to separate.
[3] merocrine (also eccrine) (none of the cytoplasm is lost during secretion). From Gk meros = part and Gk krinein = to separate.
[4] programmed cell death in contrast to necrosis (decay due to disease)

Chapter seventeen

Development of the eye

'The eye sees only what the mind is prepared to comprehend.'
Robertson Davis

The retina is of neuroectodermal origin, i.e. in reality a part of the brain. The role of the easily accessible retina for studies of the CNS [central nervous system] has been repeatedly emphasised and cogently expressed by 'Nowak (1988)': 'The isolated retina, this 'forgotten' part of the CNS, is a convenient model tissue for the CNS. It can be used not only for screening of various compounds, but also for biochemical, pharmacological and electrophysiological investigations of basic neural mechanisms, e.g. the functions and interactions of neurotransmitter candidates in the CNS. The processing of neuronal information, the role and mode of action of second messengers, and various synaptic and membrane phenomena may also be studied in this model'.

The teleostean eye displays several typical variations compared to the mammalian eye (Fig. 17/1). Therefore, a detailed description of the mature teleost eye is essential as an introduction. It should be emphasized that, since fishes and their eyes grow all through life, the ontogeny of the retina can be followed also with histological sections through the margin and ventral fissure of any eye of an 'adult' (sexually mature fish) (see 17.1.3c) (Figs. 17/23, 17/25) '(Kunz, 1987; Kunz et al., 1994; Olson et al., 1999)'.

17.1 EYE OF SEXUALLY MATURE TELEOST

While the majority of teleostean eyeballs are nearly spherical, some deep-sea fishes which are exposed to dim light conditions have so-called 'tube eyes' with the retina reduced to a patch (Fig. 17/2). Cave fishes, too, show optic degeneration. A most interesting example is provided by the cave fish, *Astyanax mexicanus*, from 29 Mexican caves. It exhibits various degrees of eye regression '(Jeffery and Martasian, 1998)'. The 'four-eyed fishes' Anablepidae float at the surface and with the help of two different corneas and retinal areas are able to see simultaneously both above and below the water-level '(Avery and Bowmaker, 1982)'.

The refractive index of the thin teleostean cornea is identical to that of water. The spherical lens is the only structure responsible for focussing. The eye at rest is set for near vision and accommodates for distant vision. Accommodation is achieved not by modification of the shape of the spherical lens but by backward movement of the lens in response to contraction of the ventrally attached musculus retractor lentis. A dorsal ligament functions as an antagonist to the retractor muscle.

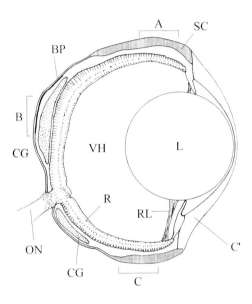

Fig. 17/1 Diagram of dorso-ventral section through teleost eye (Ennis and Kunz, 1984b, with kind permission of Academic Press, London). BP - black patch; C' - *cornea*; CG - *choroid* gland; L - lens; ON - optic nerve; R - *retina*; RL - *retractor lentis* muscle; SC - *sclera*; VH - vitreous humour; Darkened part of sclera = cartilage; Sections A, B, C indicate location of sections of Figs. 17/3 and 17/4).

The sclera which protects the eye is often entirely made up of collagenous fibres but may also contain a ring of cartilage (which in some fish is even ossified), usually situated above the ora serrata (Fig. 17/1, SC).

Subjacent to the sclera is the highly vascularized choroid. It contains a monolayer of hexagonal cells filled with black melanin granules of neural crest origin. The argentea, a silvery layer of the choroid, is found mainly in pelagic teleosts. In the viviparous guppy, *Poecilia reticulata*, the dorsal region of the argentea is made up of a layer of guanine crystals interspersed with chromatophores while in the ventral region the argentea is characterized by several layers of guanine crystals and the absence of black chromatophores. In this fish a silvery layer is absent in a region called 'black patch' situated above the exit of the optic nerve.

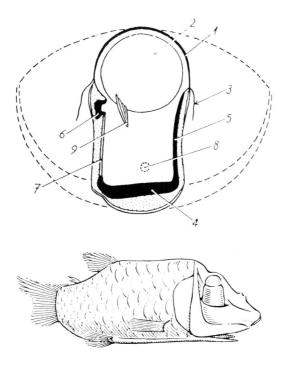

Fig. 17/2 Telescopic eye of *Opisthoproctus* (after Brauer and Chun, 1908). The deep sea fish is shown twice the natural size. The dotted line indicates the size of a normal teleostean eye, with identical size, and *retina*-lens distance. The eye is facing upwards whereas in some other deep sea fish the telescopic eyes face forward. 1 - *cornea*; 2 - lens; 3 - *integument* of body; 4 - main *retina*; 5 - accessory *retina*; 6 - specialized isolated part of the retina; 7 - non-*retinal epithelium*; 8 - exit of optic nerve; 9 - retractor lentis muscle.

It has been suggested that the absence of guanine crystals in this region allows for the retinomotor movements (see 17/34) of the 'tiered cones', present only in the retinal part underlying the 'black patch' '(Kunz and Wise, 1977; Kunz and Ennis, 1983)'(Fig.17/3 a,b, Fig. 17/4 a,b).

The iris in fish is not supplied with muscles. Therefore, in contrast to vertebrates generally, the teleostean pupil does not expand or contract (miosis) during dark- and light adaptation (though the iris of the catfish, *Hypostoma*, is mobile). Instead the retinomotor movements (also called photomechanical changes) of the pigment in the retinal pigment epithelium (RPE) and of the visual cells (rods and cones) regulate the amount of light reaching the photoreceptors. Rods extend sclerally during the day and contract vitreally in the dark. In contrast, the cones contract during the light phase and extend into the RPE during the dark. This rhythm has proved to be circadian in various teleosts including the trout and the guppy. (Figs. 17/20, 17/21, 17/31 f, 17/32 f).

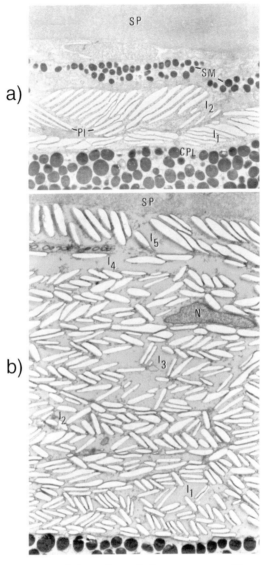

Fig. 17/3 Different parts of the *sclera* of the guppy eye (Kunz and Wise, 1977, with kind permission of Springer-Verlag, Heidelberg). **a** - EM of the dorsal black-iridescent region (area A) of Fig. 17/1. Part of a scleral cartilage plate (SP) covers the argentea, which is composed of a layer of stellate melanophores (SM) (called suprachoroidea in mammals), overlying two layers of iridophores (I_1 and I_2), and the choroid pigment epithelium (CPL). Pl, spaces (clefts) that contained the crystals. × 5 000. **b** - EM of the ventral silvery-iridescent region (area C of Fig. 17/1). The argentea contains five layers of iridophores (I_1-I_5). The suprachoroid layer of stellate melanophores observed in area A, is absent. SP, scleral cartilage plate; N, nucleus. × 5 000.

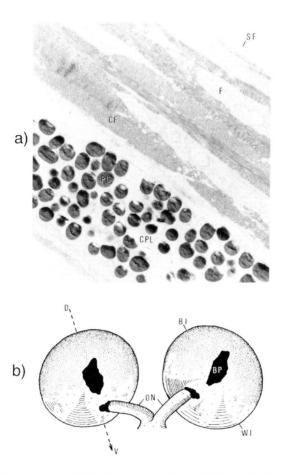

Fig. 17/4 Continuation of 17/3 (Kunz and Wise, 1977, with kind permission of Springer-Verlag, Heidelberg). **a** - EM showing part of the opaque black patch (area B of Fig. 17/1). Both the suprachoroidea and the iridophore layers are absent from this region. The dense fibrous sclera (SF) lies directly on the choroid pigment epithelium (CPL). × 7 500. CF - collagen fibres; CPL - choroidal pigment *epithelium*; F - fibroblasts; PG - pigment granule; SP - scleral plate. **b** - Drawing of whole excised eyes displaying the different regions of the *argentea*. × 17. BI - dorsal black-iridescent region of area A; BP - opaque black patch (area B); WI - ventral silvery-iridescent region (area C); D-V - dorso-ventral meridian along which the ultrathin sections of 17/3a,b and 17/4a were taken. ON - optic nerve.

The teleostean retina itself is not permeated by blood vessels. However, it shows a markedly elevated oxygen pressure. Blood entering the eye cup reaches the choriocapillaris, which, with its fenestrations directed towards the RPE, furnishes most of the blood-borne substances to the avascular RPE and retina.

External to the choriocapillaris there is often found a crescent-shaped choroid gland, which functions as a vascular countercurrent system (rete mirabile[1]) '(Copeland, 1980)'. It has been suggested that teleosts without a choroid gland possess a pseudobranch (Fig. 17/5a) '(Walls, 1967)'. Some teleosts possess additionally a vascular falciform (sickle shaped) process which receives its blood from the lentiform body (another countercurrent system, which in turn receives its blood from the choriocapillaris.) (Fig. 17/5c). The hyaloid artery sends blood to the ventral retractor lentis muscle used for moving the lens along the optic axis during accommodation. When the falciform process is absent or reduced, the hyaloid artery supplies a capillary network (hyaloid vessels) expanding over the internal surface of the retina [example: guppy in 'Kunz and Callaghan (1989)'] (Fig. 17/5b). A flow chart summarizing the entire possible vascular system of the eye has been displayed by 'Copeland (1980)' (Fig. 17/5c).

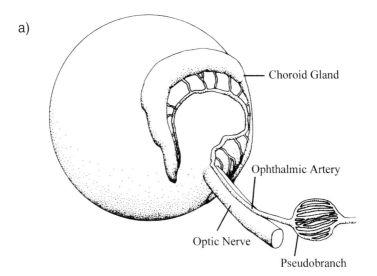

Fig. 17/5 Blood supply to teleost eye [drawn after Copeland (1980) with kind permission of Academic Press, New York] a) Doubly oxygenated blood emanating as ophthalmic artery from the pseudobranch (hemibranch) (Figs. 15/10-15/13) supplies blood to the choroid gland or body (made up of *retia mirabilia*). It is horseshoe-shaped, i.e. it appears as a circle, with a ventral opening around the exit of the optic nerve and rests on the *choriocapilllaris*. **b** - In the absence of a falciform body the hyaloid artery (which enters the vitreous cavity at the optic disc) supplies a capillary network to the inner *retina*. Example: bisected guppy eye displaying hyaloid vessels radiating from the exit of the optic nerve (ON) in centre. **c** - Diagrammatic flow chart displaying the vascular system of the pseudobranch to the eye in *Salmo*. The *lentiform* body receives blood from the *retinal* artery (LA) and passes it on (FA) to the falciform process (which supplies oxygen to the *retractor lentis* muscle). The lentiform body also receives oxygenated blood from the *choriocapillaris*, which it supplies to the

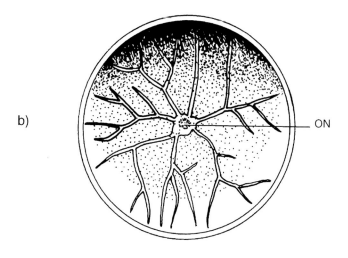

b)

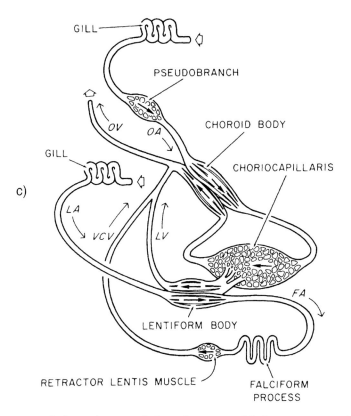

c)

falciform process and also to the ventral choroid vein. FA - falciform artery; LA - lentiform (*retina*l) artery; LV - lentiform vein; OA - ophthalmic artery; OV - ophthalmic vein.

The differentiated retina consists sclero-vitreally (from the outside to the inside) of the following layers: Bruch's membrane (BM); pigment layer (RPE); visual cell layer (VCL); outer nuclear layer (ONL); outer plexiform layer (OPL); inner nuclear layer (INL); inner plexiform layer (IPL); ganglion cell layer (GCL), and nerve fibre layer. The glial Müller cells (MC) span the whole width of the neural retina. Their scleral ends form a network called the external limiting membrane (ELM) and the vitreal end-feet form the internal limiting membrane (ILM) '(Walls, 1967)'. (Fig. 17/6).

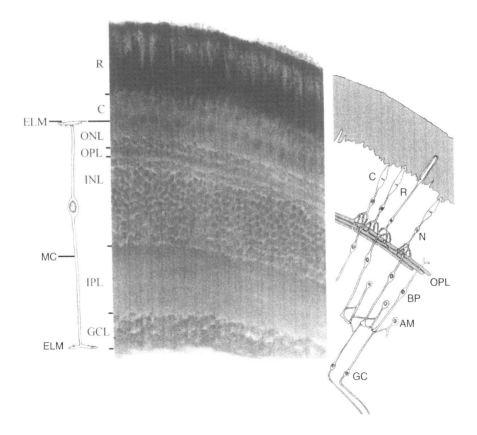

Fig. 17/6 Radial section through *retina* of *Poecilia reticulata* (Kunz, 1971b, with kind permission of the director of the Muséum d'Histoire naturelle, Geneva). On the left the Müller cell traversing the *retina* is drawn. The scleral end forms the ELM, the *nucleus* is situated in the INL and the vitreal end forms the ILM. On the right a diagram of the radial section is shown. A - amacrinecell; BC - bipolar cell; C - cones; ELM - external limiting membrane; GC - ganglion cell; GCL - ganglion cell layer; ILM - inner limiting membrane; INL - inner nuclear layer; IPL - inner plexiform layer; MC - Müller cell; N - *nucleus*; ONL - outer nuclear layer; OPL - outer plexiform layer; R - rods.

17.1.1 Bruch's membrane (BM)

BM was first described in 1884 as a 'structureless membrane' separating the RPE from the choriocapillaris. In adult vertebrates generally it consists of up to five sublayers: basal lamina, inner collagenous layer, elastic layer, outer collagenous layer and basal lamina of the choriocapillaris. EM studies revealed a strong bond between BM and the adjacent RPE with collagen fibres of the former merging with filaments from the basal lamina, which in turn merge with the plasma membrane of the RPE. However, in all teleosts so far tested, BM is three-layered, i.e. the elastic layer is missing. The functional significance of BM is the provision of structural support and attachment for the RPE. However, it also acts as a selective filter between retina and choriocapillaris. The transport of nutrients to the retina is facilitated by fenestrations in the basal lamina of BM and the adjacent endothelium '(Ennis and Kunz, 1984a; Dearry and Burnside, 1986)'.

17.1.2 Retinal pigment epithelium (RPE)

The RPE is made up of a single layer of cuboidal cells which in surface view or sectioned tangentially are hexagonally shaped (Fig. 17/7). As its name suggests, it contains pigment bodies. They occur as spherical and oblong (rod-shaped) melanin granules. The latter move vitreally during the day to shield the rods from light during daytime and again move sclerally at night; this process is known as retinomotor movements. The spherical granules do not change position (Figs. 17/20, 17/21). The RPE contains also parallel stacks of flattened discs, the myeloid bodies, which are in places continuous with the ER. They are easily differentiated from phagosomes because they lack an enclosing membrane and many have ribosomes associated with the outermost cisternae '(Ennis and Kunz, 1984b)'. Other inclusions are Golgi body, SER and RER, lipid droplets and the phagosomes, which are the shed tips of photoreceptor outer segments (see 17.2.3d). The RPE mediates above all the flow of ions, nutrients and waste products between the choriocapillaris and the retina and it is, therefore, often referred to as a 'wet nurse'. The lateral walls of the RPE cells are joined by zonulae adherentes and zonulae occludentes; the scleral cell membrane shows extensive infoldings and the vitreal membrane displays long processes. Between the regular cells there are interspersed differently shaped smaller cells with a denser cytoplasm. They contain aggregates of melanosomes bounded by a membrane. They are found both in the light- and dark-adapted eyes. (Fig. 17/8 and inset).

17.1.3 Photoreceptors (visual cells)

The retinal photoreceptors include rods, which function in dim light and different types of cones, which are engaged in colour vision. The retina of most teleosts contains rods and cones (duplex retina) (Fig. 17/9).

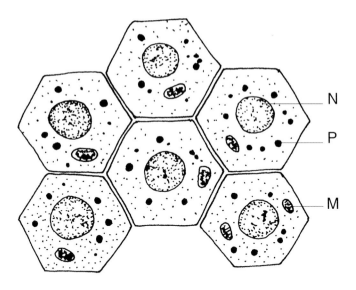

Fig. 17/7 Diagram to show the hexagonal surface pattern of the RPE. M - *mitochondrion*; N - *nucleus*; P - pigment granule.

Crepuscular fish have more rods than cones. A simplex retina (rods only) is encountered in eyes of deep-sea, cave- and nocturnal fishes (Fig. 17/2). Grouped rods have been observed in a number of deep-sea fishes '(Locket, 1970)'. In contrast, four species of deep-sea teleosts with visual cells showing the characteristic morphological features of cones, at both the LM and EM level, have been described '(cf. Munk, 1981)'. Extraretinal photoreceptors have been identified in the brain and in the pineal organ of teleosts (Chapter 16). Opsin found in the zebrafish horizontal cells is not considered to be photoreceptive '(Kojima and Fukada, 1999)'. In mammals the suprachiasmatic nucleus (SCN) of the hypothalamus is thought to be the circadian pacemaker. It has been suggested that the recently discovered photosensitive sets of ganglion cells (which innervate the SCN) in the rat may entrain the circadian clock; in other words, they may represent the primary photoreceptors for synchronizing the circadian clock to environmental time '[Berson *et al.* (2002) and Hattar *et al.* (2002)]'.

(a) Ultrastructure

The photoreceptors derive their name from the shape of their outer segments, i.e. rods have cylindrical and cones conical outer segments. There are single cones of different lengths (long, short, miniature) and double (equal and unequal) cones. Single and paired cones are often arranged in a mosaic (Fig. 17/9). Triple cones

Development of the eye 313

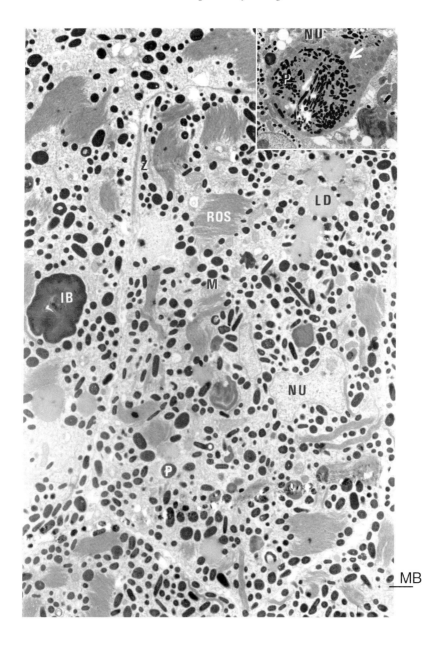

Fig. 17/8 TEM of a tangential section through the RPE of *Trachinus vipera* (Kunz *et al.*, 1985, with kind permission of Springer-Verlag, Heidelberg). × 5 900. Inset upper right: small pigment cell with membrane bound cluster of melanosomes (arrow) × 3 350. IB - inclusion body (phagosome?); LD - lipid droplet; M - *mitochondrion*; MB - myeloid body; NU - *nucleus*; P - pigment granule (*melanosome*); ROS - rod outer segment; Z - junction.

also have been observed in a number of species (Fig. 17/13 inset). In the trout, *Salmo trutta*, they are observed to be at the base of a branching radial cone line. Triple as well as quadruple cones have been observed in flatfish. According to 'Scholes (1975)' in the cyprinid, *Scardinius erythrophthalmus*, 50% of all cones are double cones. Rods, too, differ. The number of cone columns is larger in peripheral than in the fundic area because the retina is spherical. Some irregularities are inevitably introduced by the insertion of a new column of cone cells. The rods of the paddlefish *Polyodon spatula* and the catfish, *Ictalurus*, are long and very thick whereas in other species, e.g. the sole, they are long and thin. In deep-sea fish, but also in some shallow water species, the rods are arranged in tiers. In the adult fishes, rods are not generally arranged in patterns (except for the rim and embryonic fissure (Figs. 17/23-17/25) while a regular distribution is observable early in development.

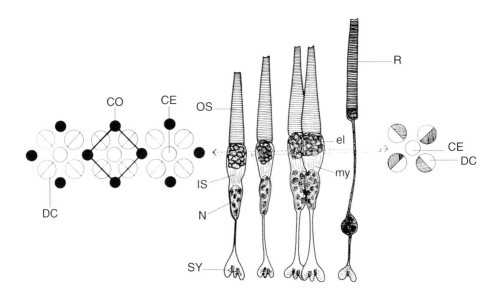

Fig. 17/9 Diagram of photoreceptors and cone mosaics of *Salmo trutta* (Bowmaker and Kunz, 1987, with kind permission of Elsevier Science, Oxford). Centre: diagram of radially sectioned photoreceptors showing from right to left a rod, a double cone, a long single and a short single cone. Left: diagram of a section through the ellipsoid region of the cone inner segments in postembryos and yearling trout to show the mosaic. Single cones are at the corners (CO) and the centre (CE) of the square units with the double cone (DC) forming the sides. Right: similar diagram, but from a two year old fish. The corner cones (putative UV-cones) are now absent. CE - central cone; CO - corner cone; DC - double cone; el - ellipsoid; IS - inner segment; my - myoid; N - *nucleus*; OS - outer segment; R - rod; SY - synapse.

The outer segments of rods are generally longer than those of cones. The outer segment of rods contains a column of free-floating discs separated from the enclosing membrane with the exception of the most basal discs. The 'discs' of the cone outer segment remain in continuity with the enclosing membrane and, therefore, should be called 'saccules'. The outer segment of both rods and cones is connected with the inner segment by a cilium. The distal centriole is called basal body and gives rise to microtubules, which form in a 9+0 pattern the 'backbone' of the so-called accessory outer segment (AOS) emanating from the cilium. The proximal centriole lies at right angles to the distal one. The AOS runs alongside the outer segment sensu stricto and EM analyses reveal that in both photoreceptors the AOS set of microtubules is continuous with that of the cilium. In cones the AOS is connected with the outer segment by a thin plasmabridge, whereas in the rods it lies embedded within the outer segment and in cross-sections appears as an incision [rev. 'Yacob et al., (1977)']. (Figs. 17/10 17/11, 17/12).

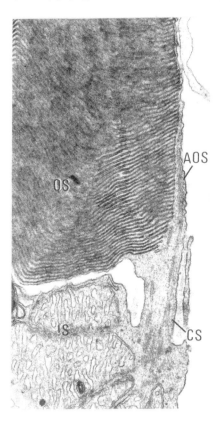

Fig. 17/10 TEM of longitudinally sectioned rod of guppy *retina* (Yacob et al., 1977, with kind permission of Springer-Verlag, Heidelberg). × 29 000. Ciliary stalk (CS) extends into outer segment (OS) as accesssory outer segment (AOS).

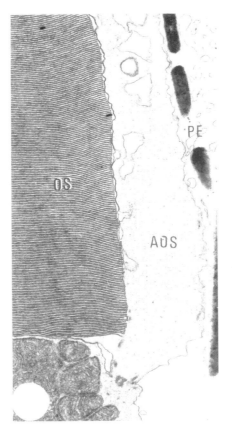

Fig. 17/11 TEM of longitudinally sectioned guppy cone showing accessory outer segment (AOS), closely appositioned to pigment *epithelium* (PE). OS-outer segment (Yacob et al., 1977), with kind permission of Springer-Verlag, Heidelberg). × 16 000.

The photoreceptor inner segment is divided into a scleral part called ellipsoid and a vitreal part, the myoid (Fig. 17/9). The outer membrane of the cone ellipsoid is often drawn out into so-called fins, which interlock with fins of adjacent cones (Fig. 17/13). The ellipsoid is filled with mitochondria while the myoid contains Golgi apparatus, ER and microfilaments. As its name suggests it is responsible for the contraction and extension of the photoreceptors (see (c) retinomotor movements). In the rods of guppy and trout the mitochondria are often orientated to point towards the cilium whereas in cones mitochondria display a vitreo-scleral size gradient (Fig. 17/12).

In structurally and ultrastructurally unequal double cones, the longer of the two closely apposed partners is called 'principal', and the shorter, 'accessory'. The double cones display along their apposing inner segments subsurface membranes (SS) arranged into cisternae, first described by 'Berger (1967)' (Fig. 17/13).

Development of the eye 317

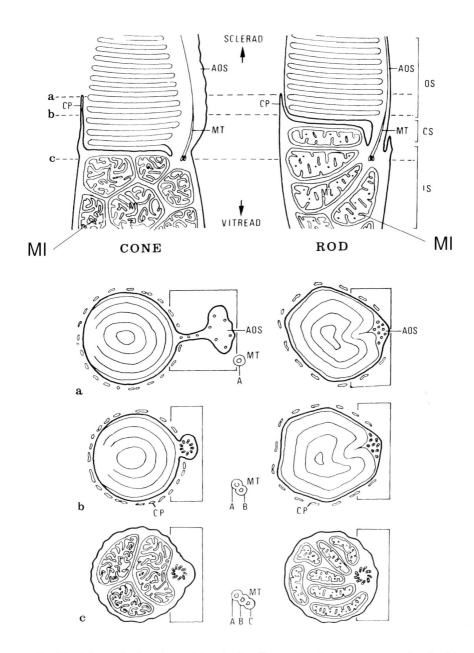

Fig. 17/12 Schematic drawings of longitudinally sectioned guppy cone and rod with cross-sections at three levels (a-c) (Yacob *et al.*, 1977, with kind permission of Springer-Verlag, Heidelberg). AOS - accessory outer segment. CP - calycal process; CS - ciliary stalk; IS - inner segment; MI - *mitochondrion*; MT - microtubule(s), A = single MT, AB = doublet MT, ABC = triplet MT; OS - outer segment.

This accounts for six membranes separating the two cone partners. 'Marchiafava *et al.,* (1985)' and 'Marchiafava (1986)' have shown that the two members of the double cones in the retina of the tench, *Tinca tinca*, and other fish are functionally coupled. Coupling persists even if the cones are deprived of their pedicles. However, no gap junctions between the inner segments of double cones of the tench and the goldfish, *Carassius auratus* have been found '(Cantino *et al.*, 1986)'. The apical part of the inner segment gives rise to the calycal processes (CP). They run along the outer segment and as the name suggests they enclose it like a cup (Fig. 17/12).

In the guppy, *Poecilia reticulata*, the arrangement of the outer segment discs of the unequal double cones is also different between members: In radial section the membranes are straight and close together in the longer partner and looser and wavy in the partner containing the ellipsosome (Fig. 17/17, 17/18). In equal double cones the discs are equal in both members, i.e. when cross-sectioned they are straight and close together.

(b) Visual pigments

The photopigment of the teleostean rods is either rhodopsin or porphyropsin, or both. The type of pigment of individual cones is established by microspectrophotometry (MSP), the direct measurement of visual pigment in individual photoreceptors. *Poecilia reticulata*, as most fish, is trichromatic '(MacNichol *et al.*, 1978; Levine and MacNichol, 1982)' (Fig. 17/14). Others are dichromatic such as *Trachinus vipera* '(Bowmaker and Kunz, 1985)' and tetrachromatic, including UV-receptors. '(Bowmaker, 1990)'. MSP of teleost retinae is greatly facilitated by the fact that the various types of teleost colour cones are also morphologically different. Many species possess two types of single cones (short and long) and structurally equal and/or unequal double cones. The short single cones usually absorb maximally at short wavelengths (blue), the very short ones at UV, the longer ones at longer wavelengths (green or red). The double cones are either equal or unequal with regard to photopigments (red/red or red/green) as well as ultrastructure. Equal cones are often referred to as twin cones.

As mentioned above, mitochondria in the cone ellipsoids increase in size in a vitreosclerad direction. The scleral-end dense mitochondrion in the shorter (accessory) member of the double cone of the guppy, *Poecilia reticulata*, was for a long time considered to represent an oil-droplet '(Müller, 1952)' (as depicted in Fig. 17/16a,b). Subsequently, 'oil-droplets', hitherto considered to be absent in teleosts, were observed in cones of all members of the cyprinodontoidei and the closely related exocoetoidei [cf. 'Kunz and Wise (1978)']. However, as shown by 'Kunz and Regan (1973)', the droplets in *Poecilia reticulata* do not stain with oil-soluble dyes. Subsequently, EM revealed that the 'droplets' have an organized structure (absent in oil-droplets) reminiscent of cristae of an unusually large mitochondrion '(Kunz and Wise, 1973, 1978)' (Fig. 17/17, 17/18). MSP of this dense spherical body in the guppy retina revealed that it contains a large concentration

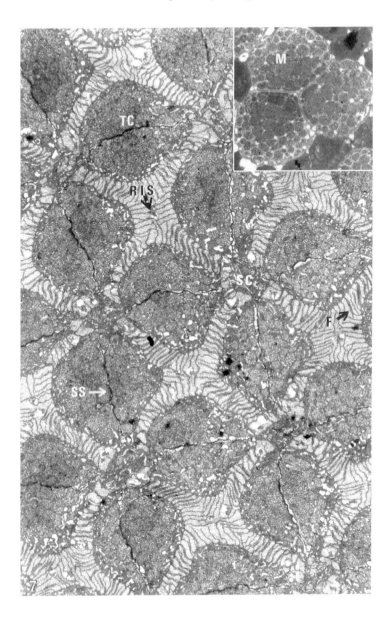

Fig. 17/13 TEM of transverse section at the level of ellipsoid of double cones of *Trachinus vipera*. Note square mosaic pattern and interlocking fins between double cone ellipsoids. In the interstices between the fins are a great number of cross-sectioned *microvilli* of the Müller cells (Kunz et al., 1985, with kind permission of Springer-Verlag, Heidelberg). × 4150. Inset: cross section of triple cone × 4 835. F - fins; M - ellipsoid of cone filled with *mitochondria*; Mi - *mitochondrion*; RIS - rod inner segment; SC - single cone; SS - subsurface cisterns; TC - double cone (twin cone).

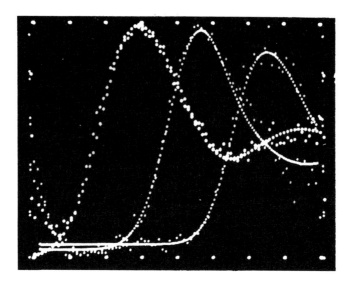

Fig. 17/14 Absorption spectra of three cone types in *Poecilia reticulata* [MacNichol et al. (1978) with kind permission of the American Association for the Advancement of Science]. Absorption maxima: From right to left: 408.5 nm (blue-absorbing small cone), 468.5 nm (green-absorbing long single cone and green partner of red-green double cone) and 546.5 nm (red double cones and red partner of red- and green-absorbing double cone). nm = nanometres

of a heme pigment resembling in its absorbance reduced cytochrome C with a maximum density sometimes greater than 1.0 at 415 nanometres (Fig. 17/15). It was given the name 'ellipsosome' and was shown to be invariably associated with the intermediate wavelength-absorbing (green) member of the double-cone of the guppy. Since this organelle lies immediately in front of the outer segment and is somewhat larger in diameter it should decrease considerably the violet light from reaching the outer segment (for which both red and green cones have about 25% of maximum absorptance). Ellipsosomes would thus increase colour contrast in the blue-violet portion of the spectrum '(MacNichol et al., 1978)'. Subsequently ellipsosomes were found in the cones of other fish (see e.g. 'Avery and Bowmaker, 1982' for *Anableps anableps*).

In the dorsal, fundic and upper 1/3 of the ventral retina of the guppy a square mosaic with short central cones and unequal double cones prevails. However, in the lower two-thirds the double cones are equal, with both members showing a vitreo-sclerad mitochondrial size gradient. This corresponds to the 'barrel-shaped' double cones described by 'Müller (1952)' (compare with Fig.17/16b). Short single cones are absent altogether and the centre of the square mosaic in this region is occupied by a long single cone. Moreover, in the ventral rim cones with a very dense ellipsosome-like body have been shown by EM. Their position

Development of the eye

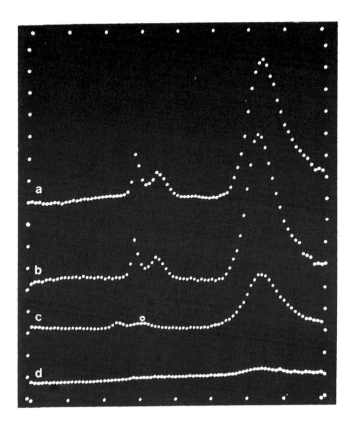

Fig. 17/15 Microspectrophotometric comparison of *ellipsosomes, erythrocytes* and ellipsoid containing *mitochondria* only [MacNichol *et al.* (1978) with kind permission of the American Association for the Advancement of Science]. a - *Fundulus ellipsosome*; b - *Poecilia ellipsosome*; c - *Poecilia erythrocytes*; d - *Poecilia* ellipsoid filled with *mitochondria* only. The alpha, beta and gamma bands of the heme pigments are in order from left to right. The spacing between dots on the left and right ordinate indicate an absorbance difference of 0.1.Ced Dots on the *abscissa* correspond to wavelengths of 741, 659, 594, 541, 496, 458, 425, 397 and 373 nanometres.

in the light-adapted retina is similar to that of rods. 'Endler (1980)' studied and discussed the basis of behavioural analyses and the complexity of the colour pattern polymorphism in the guppy. Detailed ultrastructural differences between the various region-specific cone types have been described by 'Ennis (1985)'. 'Archer and Lythgoe (1990)' observed in the same species three types of long-wavelength sensitive cones. Though different region-specific cones are encountered in the guppy retina at a LM and EM level, unfortunately MSP to measure visual pigments does not allow region-specific analyses in such a small eye.

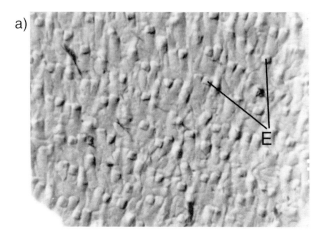

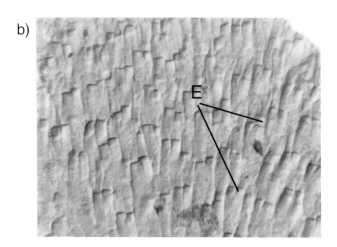

Fig. 17/16a,b Surface of excised guppy *retina* viewed under Nomarski. **a** - Dorsal region; **b** - Lower ventral region. See inset of 17/17b for subdivision of *retina*. Note different shapes of ellipsosomes in dorsal and ventral area. E - ellipsosmes.

A different way to establish spectral sensitivity is the so-called dorsal light response. This method was also used to analyze the colour vision of the guppy; its results correlated with the above-mentioned data on visual pigments '(Lang, 1967; Muntz et al., 1996)'. Additive colour experiments using a behavioural training technique have been used by 'Neumeyer (1992)' and established tetrachromatic colour vision in the goldfish (body length 6-7 centimetres).

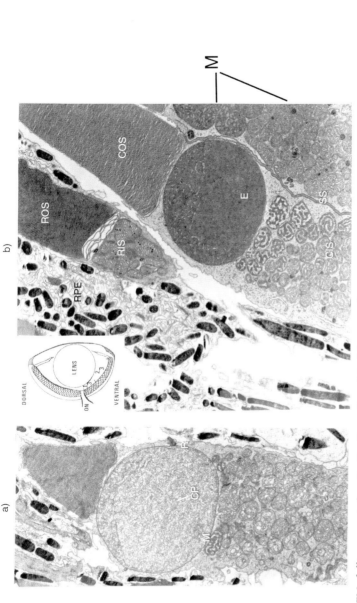

Fig. 17/17 TEM of ellipsosome in longitudinally sectioned double cones of guppy retina from different regions [Inset in b) shows diagrammatic representation of a dorso-ventral section indicating the subdivision of the ventral retina into 1/3 and 2/3]. [a, b from Y. Kunz and Wise (1978) with kind permission of Birkhäuser-Verlag, Basel]. **a** - Double cone and rod in the upper one- third of the ventral retina. × 7 000. Note cristate type of ellipsosome in the accesscry member. **b** - Double-cone of dorsal retina. × 7 000. Note highly electron-dense, so-called matrix type, of ellipsosome with peripheral *cristae* beside light-adapted rod. × 7 000. Inset: Diagram of dorso-ventral section through optic nerve of guppy eye. The longer (principal) cone is red-absorbing. It displays a vitreo-sclerad *mitochondrial* size gradient. The cross-sectioned saccules of the OS are straight. The outer segments of the shorter (accessory) green cone containing the 'ellipsosome' with peripheral cristae are 'wavy'. CIS - cone inner segment; COS - cone outer segment; CP - cristate type of ellipsosome; E - matrix-type of ellipssosome; M - mitochondria; ON - optic nerve; RIS - rod inner segment; ROS - rod outer segment; RPE - retinal pigment *epithelium*; SS - subsurface membranes.

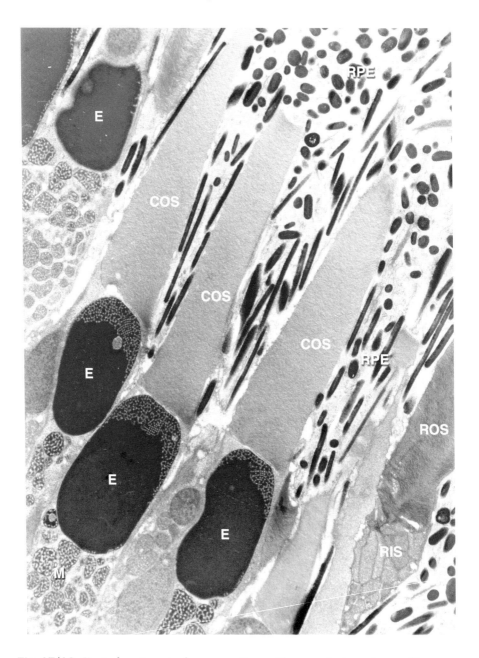

Fig. 17/18 Ventralmost part of guppy retina with very electron-dense ellipsosomes (Courtesy of Sara Ennis). × 13 500. COS - cone outer segment; E - ellipsosome; M - *mitochondrion*; R - rod; RIS - rod inner segment (ellipsoid); ROS - rod outer segment; RPE *retina*l pigment *epithelium*. × 9 500.

Development of the eye

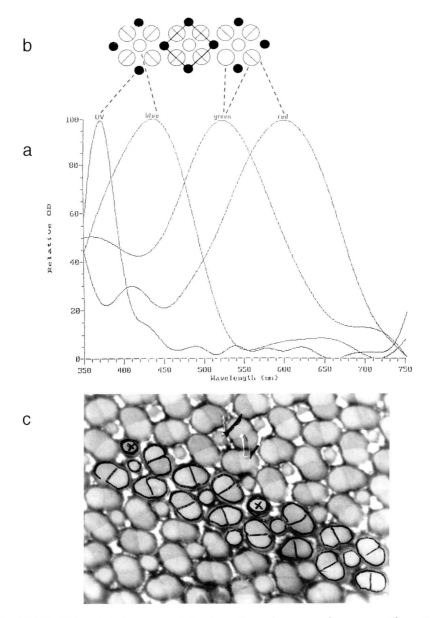

Fig. 17/19 Different 'colour cones' in the retina of young salmon. **a** - Absorption spectra obtained by msp of four cone types in young parr of *Salmo salar* (Loew and Kunz, unpublished). **b** - Diagram of square mosaics indicating that the corner cones are taken up by UV-cones. The single UV and blue cones intercalate in a row; double cones are arranged in zig-zag rows. Rare large green cones take up the position of a paired cone. **c** - Surface view of excised retina showing the position of large green cones (x) within the cone mosaic. × 3 000.

Some teleosts possess a pure-cone fovea (shallow depression to pronounced deepening), which usually lies in the temporal region of the retina, e.g.in blenniids, serranids and syngnathids '(Wagner, 1990)'. However, most fishes have instead a site of high cone density, called 'area', with an absence of rods. Salmonids possess an area in the ventro-temporal part of the retina, as do the perch, *Perca fluviatilis*, '(Ahlbert, 1976)', pike '(Bathelt, 1970)' and clupeids '(Walls, 1942)'. The guppy has two areae according to 'Vilter (1948)' and three areae (temporal, ventral and fundic), according to 'Müller (1952)'. EM analyses have shown an area with tiered cones in the fundic region of the eye of the guppy (Fig. 17/34) and the ventro-temporal eye of the zebrafish (Fig. 17/45a,b). In some deep-sea fishes pure-rod foveas are encountered '(Vilter, 1953, 1954; Munk, 1966)' while *Scopelosaurus lepidus* possesses a deep pure twin-cone fovea in the temporal part of the retina '(Munk, 1977)'.

(c) Retinomotor movements (photomechanical changes)

As mentioned above, retinomotor movements (RMM) are a substitute for pupillary movements to adjust for light- and dark-adaptation. In light the cones contract, the rods extend (both as a result of myoid contraction and extension) and the pigment granules in the RPE migrate vitreally. The reverse movements occur in the dark. (Figs. 17/20, 17/21). RMM follow diel changes '(Ali, 1975; Levinson and Burnside, 1981; McCormack et al., 1989)' (Fig. 17/32f).

(d) Shedding and renewal of photoreceptor outer segments

Rods shed the tip of their outer segments early in the morning and cones do so early at night '(LaVail, 1976; Besharse, 1982; McCormack et al., 1989) (Fig. 17/31f). The shed packets of discs, now called phagosomes, are taken up and degraded by the RPE (Fig. 17/20, 17/21) and new discs are added at the base of the outer segments (Fig. 17/22). The continual growth in length of the photoreceptors in teleosts is reflected by a predominance of disc (saccule) synthesis over degradation.

17.1.4 Outer nuclear layer and outer plexiform layer (Fig. 17/6)

The perikarya of rods and cones make up the outer nuclear layer. Usually, the cone nuclei are oval and stain only lightly while the rod nuclei are round and stain densely. Rod nuclei lie vitreally to the cone nuclei. In the outer plexiform layer rod and cone synaptic terminals make functional contact with processes of the bipolar and horizontal cells of the inner layer. Rod terminals are known as spherules (usually with one synaptic ribbon) and cone terminals as pedicles (with several synaptic ribbons) (Fig. 17/9). Coupling between photoreceptors also takes place by telodendria (fine processes extending laterally from receptor terminals).

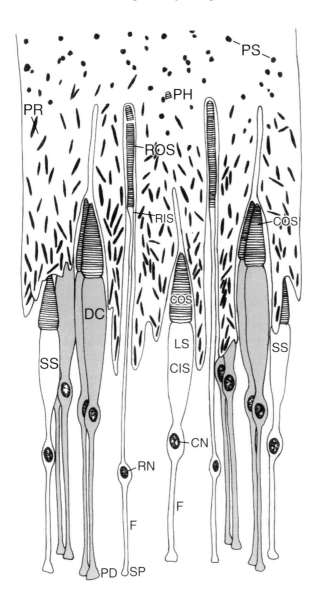

Fig. 17/20 Diagrammatic representation of light-adapted *retina*. Rods have extended, cones contracted and the rodshaped pigment granules of the RPE have moved vitreally. CIS - cone inner segment; CN - cone *nucleus*; COS - cone outer segment; DC - double cone; F - photoreceptor fibre; LS - long single cone; PD - pedicle (cone terminal); PH - *phagosome*; PR - rodshaped pigment granules; PS - spherical pigment granules; RIS - rod inner segment; RN - rod *nucleus*; ROS - rod outer segment; SP - spherule (rod terminal); SS - short single cone.

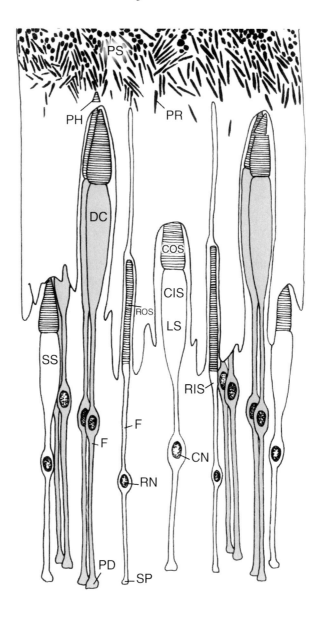

Fig. 17/21 Diagrammatic representation of dark-adapted *retina*. Rods have contracted, cones have extended and the rodshaped pigment granules have moved sclerally. CIS - cone inner segment; CN - cone *nucleus*; COS - cone outer segment; DC - double cone; F - photoreceptor fibre; LS - long single cone; PD - pedicle (cone terminal); PH - *phagosome*; PR - rodshaped pigment granules; PS - spherical pigment granules; RIS - rod inner segment; RN - rod *nucleus*; ROS - rod outer segment; SP - spherule (rod terminal); SS - short single cone.

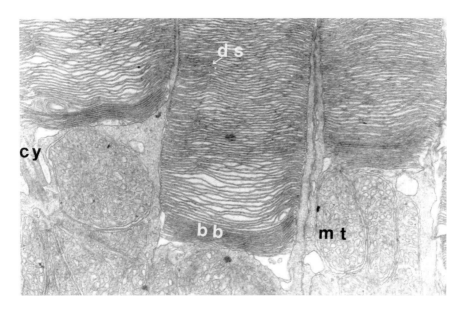

Fig. 17/22 Addition of basal discs in rod outer segments during daytime in *retina* of *Trachinus vipera*. Note that the basal 'discs' are not yet free-floating; they are still a continuation of the outer segment membrane (similar to saccules in the whole cone outer segment) × 19 000. bb - newly formed basal 'discs'; cy - connecting *cilium*; ds - rod outer segment discs; mt - *mitochondrion*.

17.1.5 Inner nuclear layer (Fig. 17/6)

Horizontal cells, arranged in tiers, take up the scleralmost position in this layer. Their dendrites are directed towards the photoreceptor terminals in the outer plexiform layer. This prompted 'Ramon y Cajal (1892)' to call the horizontal cells 'brush-like' (F. 'cellules en brosse'). Most horizontal cells possess a long axon, which extends for up to about 500 microns between bipolar and amacrine[2] cells and terminates in a tubular swelling. The bipolar cells connect photoreceptors to ganglion cells. However, they also receive synaptic input from horizontal cells, which is responsible for the receptive field organization (with chromatically antagonistic centre-surround response patterns). The bipolar cells are ultrastructurally easily recognized due to their scalloped outline, large mitochondria and homogeneous, dense distribution of clear vesicles. They contain synaptic ribbons, which, in contrast to those in the photoreceptor terminals, are shorter and pentalaminar. The Müller cells (MC) (radial fibres of Müller) span the full depth of the retina. In the guppy, their nuclei are holly-leaf shaped in radial sections and are more electrondense than the nuclei of the bipolar and amacrine cells. Fine radial processes of the MC cells invade the OPL and IPL and ganglion cell layer. The ELM (external limiting membrane) represents the linear arrangement of junctional complexes between the MC and photoreceptors, while the internal limiting membrane is formed by the basal processes of the MC, arranged in a mosaic fashion. The MC processes are easily distinguishable from those of the neighbouring cells by their high concentration of glycogen granules. Their mitochondria lack intramitochondrial granules and have their cristae restricted mainly to the periphery. Sclerally, long microvilli project into the interphotoreceptor space and interdigitate with the fins of the photoreceptors (Fig. 17/13). At the apex of the MC, below the ELM, a basal body has been observed, while a second centriole, normally associated with a solitary cilium, has not been found. The cilium always arises vitreally to the ELM '(Ennis and Kunz, 1986)'. Interplexiform cells, with their somata in the inner nuclear layer convey signals from the inner to the outer plexiform layer. Amacrine cells take up a vitreal position; some come to lie in the ganglion cell layer (displaced amacrine cells). They differ from conventional neurons by the apparent lack of an axon. They synapse with bipolar terminals, ganglion cell dendrites, and other amacrine cells. Thus, with the exception of photoreceptors, amacrine cells are in contact with every class of retinal neuron '(Wagner, 1990)'.

17.1.6 Inner plexiform layer (Fig. 17/6)

This layer represents the synaptic region which contains axons and telodendria of bipolar cells, processes of amacrine cells, interplexiform and Müller cells, and dendrites of ganglion cells.

17.1.7 Ganglion cell layer (Fig. 17/6)

As the name suggests, this layer is made up ganglion cells (so-called third order neurons). Their axons form the nerve fibre layer, which as optic nerve carries the visual information from the retina to the rest of the brain. In some fish, such as *Trachinus vipera*, the optic nerve is flattened into a tape which is folded into pleats '(Kunz *et al.*, 1985)'. This allows movemens of the eye for bipolar or unipolar vision.

Summary

The eye of the sexually mature teleost is protected on its outside by a coat. The part of the coat embedded in the head is the sclera. It is made up of collagenous tissue, which may contain additionally (in some fish ossified) cartilage. The anterior part of the coat is the transparent cornea. Since its refractive index is identical to that of water, it does not play any role in focusing. Adherent to the sclera is the vascular choroid, which contains melanin granules and, mainly in pelagic species, guanine crystals (hence the name 'argentea'). The choriocapillaris is often overlain by a crescent-shaped vascular countercurrent system, known as choroid gland. The choroid extends as iris and surrounds the pupil, which does not contract or expand during light and dark adaptation. The teleostean retina is avascular. Blood supply is effected by the choriocapillaris and in some also by the choroid gland. The latter expands ventrally into the falciform process, which supplies blood to the retractor lentis muscle. This muscle is engaged in accommodation of the eye. By contraction and relaxation it moves the spherical lens back- and forewards along the optical axis. At rest, the teleost eye is focused on nearby objects; contraction of the muscle results in bringing distant objects into focus. In species with a reduced or absent falciform body the hyaloid artery supplies a capillary network to the inner side of the retina.

Bruch's membrane is the outermost layer of the retina. Subjacent is the retinal pigment epithelium (RPE); the movements of its melanin granules (RMM) take part in light- and dark adaptation. The optical portion of the retina consists of various layers, as shown in Fig. 17/6. Retinal processing of visual stimuli (detection of brightness, contrast, colour and movement) is accomplished by the interaction of retinal neurons: photoreceptors (visual cells) composed of rods and cones; bipolar cells, horizontal cells, amacrine cells, interplexiform cells, and ganglion cells, the axons of which form the optic nerve. In addition, the glial cells of Müller traverse the whole retina.

Visual cells are represented by rods (for dim light vision) and cones (for colour vision). The various types of cones are long and short single, and double or twin-cones. They are organized into cone mosaics with the rods distributed at random.

Rods and cones undergo extension and contraction in a daily circadian rhythm. In the light melanin granules of the RPE move vitreally and rods contract into the RPE. In the dark, the melanin granules move sclerally and cones expand. These opposite movements, called retinomotor movements (RMM), follow a diurnal rhythm. The visual cells shed the tips of their light sensitive outer segment also according to a daily rhythm: Rods shed during the day and cones during the night. The packages of discs shed by the visual cells into the RPE are known as phagosomes.

The outer nuclear layer (ONL) is formed by the rod and cone nuclei. Their terminals synapse with bipolar neurons in the outer plexiform layer (OPL). The somata of the horizontal, bipolar, Müller, interplexiform and amacrine cells make up the inner cell layer (INL). In the inner plexiform layer (IPL) synapses between the amacrine and ganglion cells are found. The axons of the ganglion cell layer leave the eye as the optic nerve, which enters the brain.

Visual cells are organized into cone mosaics with the rods distributed at random. In the guppy, the cone mosaics are made up of 4 double cones, a short single central cone and long single cones at the corners. In the salmonids, the corner cones are missing (though present in younger stages) (see 17.2).

17.2 CONE MOSAICS IN SALMONIDS

'Bowmaker and Kunz (1987)' established that the 1 year-old brown trout, *Salmo trutta*, is tetrachromatic with cone visual pigments absorbing maximally at 600, 535, 441 and 355 nanometres. The double cones were red and green sensitive and the blue and UV pigment resided in single cones. Histological analyses revealed a cone mosaic with four double cones, a central and four corner single cones. The MSP analysis did not allow to differentiate morphologically between the two types of single cones. Based on histological investigations by 'Fürst (1904)' who had observed the loss of corner cones in the retina of older salmon versus younger salmon (no ages given), it was decided to analyze microspectrophometrically and histologically the retina of 20 months old fish. It was revealed that at this age the trout had no UV-pigment and had cone mosaics without corner cones (Fig. 17/19). Similarly, MSP analysis of Atantic salmon retinae revealed the presence of UV-cones in young parr and their absence in older fish ('Loew and Kunz', unpublished results, Fig. 17/19). Literature search revealed that other authors had described the 'disappearance' of corner cones in salmonids (Lyall, 1957a,b; Bathelt, 1970 and Ahlbert, 1976)'.

The above reported results on the Atlantic salmon might suggest a reduction of UV-cones during smoltification. Thyroid hormones have been shown to participate in the vast metabolic and developmental changes preparing the fish for life in the sea. Therefore, a relation was assumed to exist between increased iodine levels at smoltification and the disappearance of the UV-cones. The role of thyroid hormones in the migratory Atlantic salmon reviewed by 'Specker (1988)' revealed two peaks of thyroxine immediately preceding changes in the digestive tract and habitat of the fish: the first peak preceded the transition from the yolk-dependent stage to the predatory stage and the second peak was correlated with smoltification. Subsequently, 'Browman and Hawryshyn (1992)' reported that the addition of thyroxine to the non-migratory rainbow trout, *Oncorhynchus mykiss*', induced a precocious disappearance of the UV- cones. In contrast, retinal EM analyses of the developing salmon, *Salmo salar*, by 'Kunz et al. (1994)' (Fig. 17/9) revealed that from the time the postembryos begin exogenous feeding (average length 2.5 centimetres) to the parr of an average length of 4 centimetres, the corner cones (putative UV-cones) are regularly distributed over the whole retina. At a very early age of the parr, i.e. at 120 days, the corner cones begin to disappear first in the ventral half and are absent over the rest of the retina at day 220. In other words, in the Atlantic salmon, *Salmo salar*, the disappearance of UV cones occurs long before smoltification.

The disappearance of one-third of cones in the main retina of the trout, *Salmo trutta*, between the ages of 12-20 months and in the eye of 4-7.5 months old Atlantic salmon poses the further question as to how the loss of synaptic contact with, presumably, horizontal and bipolar cells affects the retinal circuitry, and, finally the retinal mapping in the optic tectum. While it is still a matter of debate as to what advantages UV-vision conveys to the young teleost, the functional sig-

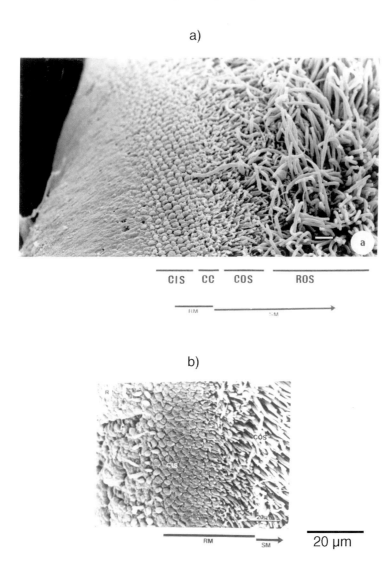

Fig. 17/23 Growth zone along the periphery and the ventral embryonic fissure of the retina of the two year old brown trout (Salmo trutta) and the Atlantic salmon (*Salmo salar*). (Kunz, 1987, with kind permission of Birkhäuser-Verlag, Basel). **a** - SEM of embryonic (choroid) fissure with growth zones. Bar = 10 microns. **b** - SEM of rim (margin) with growth zone. Bar = 20 microns. ROS are in a scleral position. They are not visible in the younger zones in surface view since they develop in a more vitreal position to that of the cones. — In the COS region, the peripheral row mosaic is followed by square mosaic, as seen below. CC - connecting cilium; CIS - cone inner segments; COS - cone outer segments; RM - row mosaic; ROS - rod outer segments; SM - square mosaic.

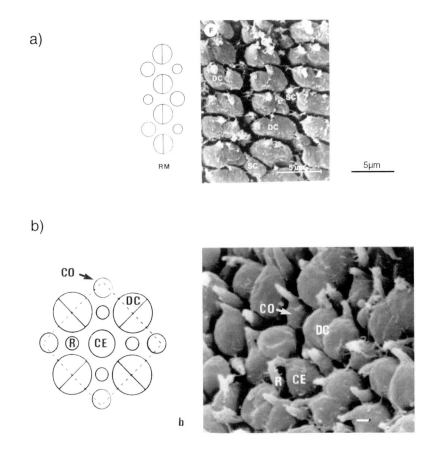

Fig. 17/24 Higher magnification than 17/21a,b to show arrangements of row and cone mosaic. in growth zones (Kunz, 1987) with kind permission of Birkhäuser-Verlag, Basel). **a** - Diagram and SEM of row mosaic; **b** - Diagram and SEM of square mosaic. Bar 1 micron CE - central cone; CO - corner cone; DC - double cone; R - rod; RM - row mosaic; SC - single cone; SM - square mosaic.

nificance of the gradual loss of the UV-cones with age over most of the retina is equally enigmatic. The loss of UV-stimulated cones may cause the fish to move to deeper waters where the penetration of UV-light is diminished. It has been shown that solar UV-radiation can cause epidermal necrosis in salmonids kept in surface waters. Such a skin trauma deprives the fish of its protective outer layer and subsequently exposes it to a variety of infection agents '(Bullock, 1982)'. Young *Perca fluviatilis* are equipped with UV-sensitive cones when they feed on zooplankton. According to 'Sandstroem (1999)' the UV-cones 'regress' after the fish change from a pelagic to a benthic littoral habitat.

The cone mosaic of surface prejuvenile rockfish, *Sebastes diplopora*, displays a square mosaic with central and corner cones. After migration to depths averaging several 100 metres, the rest of their life is spent near the bottom as juveniles and adults. The benthic juveniles lose their corner cones, beginning near the periphery and proceeding inward, and eventually also the central single cones disappear. This leaves the benthic fish with a retina with double cones only, arranged in rows '(Boehlert, 1978)'.

On the basis of MSP, age-related loss of UV-cones has since been reported for other teleosts '(Douglas *et al.*, 1987; rev. Bowmaker, 1990)' and ultraviolet vision in vertebrates, including teleosts, has been reviewed by 'Jacobs (1992)'.

It should be stressed here again that, unlike mammalian eyes, those of teleosts grow throughout life and new photoreceptors are continuously added in the growth zones of the margin: moreover, in fishes which have an open ventral fissure, photoreceptors are also continuously added along the fissure. Therefore, the question arose whether UV cones, when they are no longer present over the main retina, would still be formed in the growth zones, and if so, what their fate would be. 'Kunz (1987)' and 'Kunz *et al.* (1994)' analyzed with TEM and SEM the growth zones of the 2 year old brown trout and of Atlantic salmon older than 7.5 months. It emerged that in both salmonids corner cones (putative UV cones) are continuously added in both growth zones (margin/rim and ventral fissure) (Figs. 17/23-17/24a,b, 17/25). However, these cones were very short-lived and were never incorporated into the mosaic pattern of the main retina. In the Atlantic salmon the UV-cones were shown to die as a result of apoptosis (normal cell death), characterized by clumping and condensation of the nucleus and loss of cytoplasmic organelles. They subsequently became engulfed by macrophages and Müller cells. The spaces vacated by the dying corner cones were taken up by newly-formed rods. This is the only time rods form part of a regular mosaic. It was also shown that both in brown trout and Atlantic salmon in the most peripheral region of the circumferential growth zone and the embryonic fissure cones are arranged in a row mosaic pattern (row mosaics have the subsurface membranes of their double cones arranged in a straight line, while in the square mosaics they form a zig-zagging line). The rows of double cones are separated by rows of single cones corresponding to the central and corner cones of the square mosaic. Still within the growth zone, the row mosaic changes into a square mosaic with the shorter single cones taking up a corner position. The cone density in the growth zones is greater than in the main retina. (Figs. 17/23, 17/24a,b, 17/25).

Continuing retinal neurogenesis in the sexually mature fish results in a continuous making and breaking of retinotectal synapses and a constant shift of retinal terminals in the optic tectum. Such a changing pattern was established in the zebrafish '(Marcus *et al.*, 1999)' although it was previously reported that neurogenesis in the optic tectum of this fish ceases early in life '(Rahmann, 1968; Schmatolla and Erdmann, 1973)'.

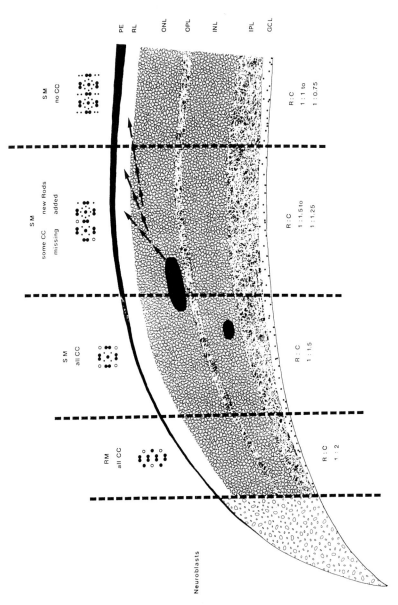

Fig. 17/25 The developmental areas in the rim of the eye of the salmon, *Salmo salar*. Diagrammatic representation. (Kunz *et al*. 1994 with kind permission of Elsevier Science, Oxford). Darkened patches and black arrows represent areas of new developing rods and their movements to replace corner cones and fill the region around the central cone. Note changes from rim to main *retina*: row mosaic changes to square mosaic. Rods take up special sites within the mosaic: first around the central cone, then taking up positions vacated by (putative UV) corner cones. C - cone; CC - corner cone; GCL - ganglion cell layer; INL - inner nuclear layer; IPL - inner plexiform layer; ONL - outer nuclear layer; OPL - outer plexiform layer; PE - pigment *epithelium*; R - rod; R:C - rod to cone ratio; RL - receptor layer; RM - row mosaic; SM - square mosaic.

17.3 DEVELOPING TELEOST EYE

It is generally stated that fishes grow throughout life and so do their eyes, by cell addition and stretching. However, the field of view does not change '(Fernald, 1993)'. An example to the contrary would be the guppy since the male stops growing at a length of about 1.8 centimetres while the female continues to grow.

It is the optic-cup of the embryo that induces the overlying ectoderm to form the lens. Additionally, the mesoderm contributes further structures of the eyeball: The retina becomes enclosed by the vascular choroid, which together with retinal layers forms much of the ciliary body and the iris. Subsequently, an external coat of connective tissue, the sclera, encloses the developing eye.

17.3.1 Eye primordium to optic cup

The vertebrate eye generally arises as an evagination of the diencephalon (Fig. 17/26a). It is called optic vesicle[2] and is lined by ciliated ependymal cells. It later folds in to form the optic cup '(Coulombre, 1961, 1965)'. In contrast, the primordium of the teleostean eye, as well as the rest of the CNS, are laid down as a solid mass of cells. Lumina appear only later (see Chapter 12). The teleostean equivalent to the optic vesicle, in vertebrates generally, is called optic lobe. The first lumen in the CNS is observed in the primordium of the eye and is called subretinal space; it later becomes confluent with the ventricles of the brain '(Kunz, 1971a)'. (Figs. 17/26a, 17/27a). The attachment of the eye to the brain is referred to as optic stalk, which later will become entirely replaced by the optic nerve.

While the optic cup is formed, both sides grow laterally and downward to meet below the optic stalk '(Walls, 1967)'. In many teleostean species these sides do not fuse but remain as the so-called ventral embryonic fissure. It is through this channel that early in development the blood vessels enter the developing eye (Fig 17/26b,c). The optic cup induces the overlying surface ectoderm to form the lens, which becomes detached from the epidermis (Fig. 17/26c). During development the lens diameter and the focal distance increase, which results in an increase in the size of image projected onto the retina.

While it is generally stated that teleostean eyes are characterized by an embryonic fissure, various stages between completely open and closed fissure have been described: In the trout, *Salmo trutta*, the fissure, through which a falciform process enters, is lined by the RPE and subjacent retinal layers and remains open. However, in the Mozambique tilapia, *Tilapia mossambica*, the fissure persists into adult life, too, but is lined with RPE only and is associated with a partially formed falciform process. In the guppy, *Poecilia reticulata*, the embryonic fissure closes completely during embryonic development. No falciform process develops; instead the hyaloid artery entering through the fissure supplies blood to the inner surface of the retina (hyaloid blood vessels) '(Kunz and Callaghan, 1989)' (17/5b). In the zebrafish, *Brachydanio (Danio) rerio*, the original anterior-posterior orientation of the eye primordium ultimately becomes the ventral-dorsal axis of the completed

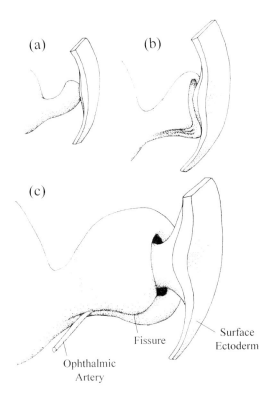

Fig. 17/26 Diagrams displaying the formation of a teleostean eye. **a** - evagination of the *diencephalon*, which is solid in fish. **b** - invagination to form the optic cup. This is known as the ventral embryonic fissure, choroid fissure or optic fissure. In many fish the cup fuses ventrally; **c** - The ophthalmic artery enters through the fissure, and the cup induces the overlying surface *ectoderm* to form a solid lens.

eyecup as a result of two waves of rotation. Therefore, during the invagination of the cup the choroid fissure comes to lie first in the anterior region of the cup and only concomitant with a further rotation of the eyes the fissure takes up its final ventral position '(Schmitt and Dowling, 1994)'.

The optic cup consists of two neuro-epithelial layers: The outer layer remains single and develops into the retinal pigment epithelium (RPE), while the inner layer, which is analogous to the ventricular zone of other parts of the developing brain, proliferates into the multilayered neural retina. The layers surrounding the eye are of mesodermal origin.

It should be stressed again that the RPE is made up of the only pigment cells in the animal body not derived from the neural crest. However, the pigmented layers, which cover the eye later (choroid), contain chromatophores stemming from the neural crest. The black granules of the RPE protect the developing eye long before the photoreceptors appear. Due to the presence of black pigment in the RPE, this early stage is universally referred to as 'eyed ovum' or 'eyed egg' in fishery parlance. It is at this stage that the eggs can be shipped and generally handled without disrupting their development. The sclera surrounding the eye is of mesodermal origin. Fishes that live at low light levels possess in their RPE fewer chromatophores but more reflecting particles which may develop into a tapetum lucidum.

17.3.2 Development of Bruch's membrane (BM)

In guppy and rainbow trout the BM develops vitreo-sclerad into a trilaminate structure. The three laminae are the basal lamina of the RPE, a collagenous layer and the basal lamina of the choriocapillaris '(Kunz and Ennis, 1984a)'. While in *Salmo gairdneri* BM measures 100 nanometres in width at hatching, it increases to 300 nanometres when exogenous feeding has started. This increase is due to an expansion of the collagenous layer.

17.3.3 Development of the retina

The anlage of the retina is first composed of pseudostratified columnar cells, called neuroblasts, with spindle-shaped nuclei at all levels. Initially all cells in the neuro-epithelium undergo division; the ganglion cells are the first to become postmitotic. Those in the fundic region are the first to cease mitosis and cell division becomes restricted to the annular margin (ora serrata) '(Hollyfield, 1972)'. At the same time, a spreading out (stretching) of existing cell layers takes place '(Lyall, 1957a; Johns, 1981)'. An additional site of neurogenesis has been observed in the salmonids. It is the ventral embryonic fissure (also called choroid fissure) through which blood vessels enter '(Ahlbert, 1976, Kunz, 1987)' (Fig. 17/26b,c). As mentioned above, e.g. in the guppy, *Poecilia reticulata*, the fissure closes during development; therefore, in this fish further growth of the retina is restricted to the margin.

Two developmental gradients are observed: one vitreal to scleral, i.e. from the ganglion cell layer to the visual cell layer, the other from the fundus to the periphery. '(Blaxter and Staines, 1970; Hollyfield, 1972; Grün, 1975b; Sharma and Ungar, 1980; Fernald, 1993)'. In fish with an open fissure, such as found in salmonids, a third growth region has to be taken into consideration.

It has been assumed that in all teleosts morphological development starts in the fundus as in vertebrates generally. However, the zebrafish seems exceptional in that differentiation begins in the part ventral to the exit of the optic nerve '(Kljavin, 1987)'. 'Schmitt and Dowling (1996)', using immunohistochemical markers on retinal wholemounts, describe that differentiation commences within the area ventro-nasal to the optic nerve and choroid fissure, in ganglion cells 38 hours post-fertilization (hpf), in double cones and rods at 50 hpf. The staining pattern for ganglion cells and double cones then progresses into the central and temporal retina to reach finally the nasal area. The development of the zebrafish retina has been recently reviewed by 'Malicki (1999)'.

For some time the opinion prevailed that teleosts with a duplex retina have larvae with pure cone retinae. In other words, the rods appear much later '(Ali, 1959; Blaxter and Staines, 1970; Wagner, 1974; Burckhardt et al., 1980; Sandy and Blaxter, 1980; Johns, 1982; Raymond, 1985)'. 'Fuiman and Delbos (1998)', who measured the changes in visual sensitivity of red drum, *Sciaenops ocellatus*, suggested that scotopic sensitivity is mediated by cones before the rods proliferate. However, EM analysis of the developing retina of both guppy and brown trout shows that, at least in these two species, rods can be identified already at the earliest stage. EM revealed that they develop at a more vitreal level than the cones, which must be the reason that they were not observed earlier when analyses were restricted to light microscopy '(Kunz, 1990)' (Fig. 17/27c, 17/28b). EM investigations of the retina of the zebrafish embryo revealed that in this fish, too, the rods and cones differentiate at the same time '(Kljavin, 1987)'.) As regards their number, rods lag behind that of cones in early development of both guppy and trout. However, later in development the number of rods begins to increase over the whole expanse of the retina. This seems to suggest that rods, in addition to their formation in the circumferential germinal area, emanate also from another site. It had been proposed that they do so from the base of the ONL '(Baburina, 1961; Ahlbert, 1973, 1976; Johns and Fernald, 1981)', or from the INL '(Lyall, 1957a; Blaxter and Jones, 1967; Johns, 1982)'. Earlier 'Müller (1954)' had observed that in the guppy the INL loses cells during development. It was on this basis that 'Kunz et al. (1983)' suggested that in this fish new rods emanate directly from the underlying INL. Later, autoradiographic analyses showed that in the retina of the salmon the progenitors of later generations of rods occur in the INL and subsequently in the ONL '(Goodrich and Kunz, 1992, unpublished results)'. 'Julian et al. (1998)' showed for the mature rainbow trout, *Oncorhynchus mykiss*, that cell birth and proliferation occur also in the inner nuclear layer, at least up to 2 years posthatching. They suggest that these proliferative cells migrate from the inner nuclear layer into the outer nuclear layer and once there differentiate into rods as illustrated in the diagram for *Salmo salar* (Fig. 17/25).

It should be emphasized that newly hatched larvae of many marine species such as the sole, *Solea solea*, the mackerel, *Scomber scombrus*, the whiting, *Merlangius merlangus*, the pilchard, *Sardina pilchardus* and the sardine, *Sardinops caerulea*, have unpigmented and, therefore, presumably non-functional eyes. However, other marine fish such as the plaice, *Pleuronectes platessa*, the cod, *Gadus morrhua*, and the herring, *Clupea harengus*, hatch with pigmented eyes. During their larval life flatfish engage in a pelagic life style and display a pure cone retina up to metamorphosis. The larvae are bilaterally symmetrical with their eyes located on each side of the head. They descend to the sea bed to metamorphose. During this process the bilateral body symmetry is lost so that the adults come to lie on one side leading a benthic lifestyle. Metamorphosis begins with the migration of one eye across the top of the skull to lie adjacent to the eye on the other side. Both eyes are now on the exposed side of the body. Retinomotor movements associated with dark and light adaptation develop once the rods have appeared. With the development of rods, shoal formation begins as does the development of the lateral line '(Blaxter, 1968)'. In some flatfish the eyes are on the left side (e.g. *Platichthys flesus*) while in others (e.g. *Scophthalmus maximus*) they are on the right side, and relatively few species can have the eyes on each side '(Medina *et al.*, 1993)'. As a result of the ocular displacement, the spatial relationships between labyrinths and eyes are altered. As mentioned above, the sole lacks eye pigmention at hatching (length of fish 2.5 millimetres). Eyes are pigmented at a length of about 3.5 millimetres, at the same time when active pectoral fin motion, a functional gut and the arrival of the larvae at the bottom of the sea are observed '(Flüchter, 1970)'. Both the splitnose rockfish, *Sebastes diploproa*, and the red sea bream, *Pagrus major*, move to deeper habitats once the retina has improved its sensitivity to low light levels '(Boehlert, 1979; Kawamura *et al.*, 1984)'. Early flatfish probably fixate their prey binocularly, and after eye migration the field of view changes completely. During this change a drop in feeding has been observed in the plaice '(Blaxter, 1988)'. The diencephalic projections to the retinae in two species of flatfishes, *Scophthalmus maximus* and *Pleuronectes platessa*, are discussed by 'Meyer *et al.*, (1993)'. Preliminary observations by 'Osse and Van der Boogart (1997)' of *Scophthalmus maximus* suggest a presence of retino-tectal projections already at the premetamorphic stage, i.e. prior to taking up benthic life.

17.3.4 Developmental sequence in the retina of salmonids and the guppy

If not stated otherwise, the results are taken from the following publications; on guppy: 'Kunz and Wise (1974), Kunz and Ennis (1983)', Kunz *et al.* (1983), Ennis and Kunz (1984a, 1986)'; and on rainbow trout: 'Schmitt (1987); 'Schmitt and Kunz (1989)'; on the brown trout: 'McCormack *et al.* (1989)'; on all three: 'Kunz (1990)'. The retina is divided into a dorsal, central (fundic) and ventral region. The optic nerve exits at the boundary between fundic and ventral region (Fig. 17/1, 17/17b inset).

(a) Retinal pigment epithelium (RPE)

At the stage A (hatching) of the rainbow trout and at stage 5 of the guppy the RPE is largely differentiated. In the trout the basal membrane of the RPE displays invaginations continuous with the ER. Apart from extensive RER, pinocytotic (coated) vesicles and numerous mitochondria have been observed. These infoldings and the presence of vesicles may be indicative of passage of material. The nuclei are oval, pale staining and with their long axis parallel to the plane of the RPE. The membrane-bound pigment granules (also called melanosomes) are spherical and this stage is often referred to as 'eyed ova'. Cell junctions (alternating zonulae adhaerentes and zonulae occludentes) are evident. (17/27b).

At stage B of the rainbow trout and stage 6 of the guppy embryo some nuclei of the RPE become radially oriented and show marginal heterochromatin (Fig. 17/27c). Myeloid bodies (parallel stacks of smooth flattened cisternae), which are continuous with the RER (and in the guppy with lipid bodies), are prominent in the basal region of the RPE (Fig. 17/27d). Rod-shaped, intermingled with spherical pigment granules are now evident. RPE microvilli interdigitate with the newly formed calycal processes of the cone inner segments (Fig. 17/27c).

Stage C of the rainbow trout is characterized by the appearance of membrane-bound clusters of small melanin granules and the presence of apical projections indicative of developing REP processes (Fig. 17/27e). Mitochondria are observed mainly in the basal zone and profuse RER and SER in the perinuclear region (17/28c).

At Stage D of the rainbow trout/stage 7 of the guppy, rod-shaped pigment granules are concentrated in the RPE processes which extend vitreally between the photoreceptor inner segments. Phagosomes (shed tips of photoreceptor outer segments) associated with myeloid bodies and RER are evident (Fig. 17/29). Myeloid bodies are distinguishable from phagosomes by their lack of an enclosing membrane (Fig. 17/27d). While myeloid bodies in the adult appear to be modifications of the SER, in the postembryo of trout (and the guppy) they are continuous with the RER. In the guppy MB were seen to arise from the ribosome-studded outer membrane of the nuclear envelope. Myeloid bodies in amphibians are thought to be linked with the appearance of phagosomes '(Nguyen-Legros, 1978; Yorke and Dickson, 1985)', while in guppy and rainbow trout they are observed before phagosomes appear (stage B). In the brown trout the number of large phagosomes changes cyclically, with a sharp rise and subsequent fall after sunset and a second peak after sunrise. The small phagosomes, too, show rhythmic fluctuations (Fig. 17/31a). The early morning peak is due to shedding from rods, while the peak early at night represents shedding from cones '(LaVail, 1976; Besharse, 1982)'. In the guppy, shedding of rod tips has just started during day time (Fig.17/30a).

At stage E in rainbow trout membrane-bound lipid droplets, in contact with the nuclear membrane and/or myeloid bodies, are evident for the first time, while in the guppy lipid droplets have already been observed at stage B.

Both in guppy (stage 8) and trout (stage E), retinomotor movements (RMM) have started in which pigment granules also take part; the rod-shaped pigment granules are concentrated in the elongated apical processes of the RPE, whereas the spherical granules have remained at the base of the RPE cells (Fig. 17/2, 17/21). Guppies have started to shed cone tips (Fig. 17/30a,b). In both trout and guppy, phagosomes are encountered in great numbers, some of them in the process of degeneration. In the brown trout the number of large as well as small phagosomes exhibit a two-peaked pattern over a 14 hour period (Fig. 17/31b).

At stage F in the trout, BM is now 300nm in width as a result of an expansion of the collagenous layer. Granules are observed in BM and within invaginations of the RPE basal membrane, in the region adjacent to fenestrations in the endothelium of the choriocapillaris. Pinocytotic (coated) vesicles are observed in the

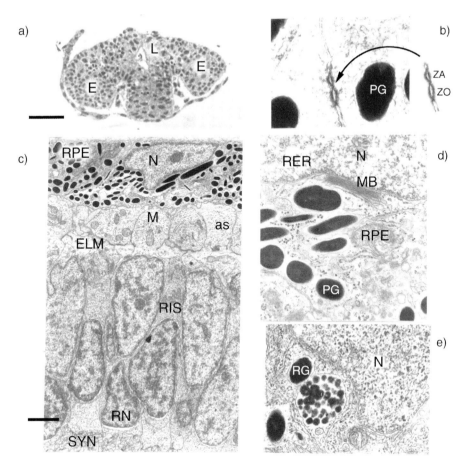

Fig. 17/27 a - Transverse section through head region of guppy on 5th embryonic day, 5-8 somites (stage 3). Solid eye primordium. **b** - Junctional complex with alternating *zonulae adhaerentes* (ZA) and *zonulae occludentes* (ZO) between cells of the RPE of the newly hatched trout (stage A) (Schmitt and Kunz, 1989, with kind permission of S. Karger, Basel). × 45 000. **c** - EM showing radially oriented photoreceptor *nuclei* and budding inner segments of both cones and rods of the guppy (stage 6) (Kunz *et al.*, 1983, with kind permission of Springer-Verlag, Heidelberg). Bar = 2 microns. **d** - Myeloid body (MB) in association with nuclear membrane and continuous with RER in the RPE of the guppy (Ennis and Kunz, 1984, with kind permission of Academic Press, London). Bar = 1 micron. **e** - Membrane-bound composite of small pigment granules in RPE of trout. (Schmitt and Kunz, 1989, with kind permission of S. Karger, Basel) × 20 000. accessory outer segment CIS - cone inner segment; CG - composite pigment granules; CN - cone *nucleus*; E - eye *primordium*; ELM - external limiting membrane; L - lumen; M - *mitochondrion*; MB - myeloid body; N - *nucleus*; PG - pigment granule; RER - rough endoplasmic *reticulum*; round pigment granules; RIS - rod inner segment; RN - rod *nuclei*; RPE - *retinal* pigment *epithelium*; SYN - synapse; ZO - Zonula adheseus; ZO - zonula occludens.

vicinity. MB are still continuous with the RER and closely associated with the nucleus, lipid droplet or phagosomes. The RPE processes now contact the apical microvilli of the Müller cells. In trout and guppy, RMM are pronounced with the rod-shaped pigment granules moving sclerad, i.e. leaving the apical processes in the dark. In the brown trout a peak in the number of phagosomes is observed at dawn and a smaller peak early in the night (Fig. 17/31d).

(b) Visual cell layer (VCL) including outer nuclear layer (ONL)

Guppy, salmon, brown and rainbow trout have a duplex retina, short and long single cones and double cones. Differentiation starts in the fundus (1/3 above the exit of the optic nerve) and progresses from there to the rest of the retina (Fig. 17/18 inset).

Stage A (hatching of rainbow trout and stage 5 of the guppy) reveals that the basal part of the ONL is taken up by rod nuclei which are smaller and more electron-dense than the cone nuclei situated above them. The inner segments of rods and cones are different. In rods they stain more deeply and their mitochondria are more elongated than in cones. Both rod- and cone inner segments contain abundant RER, free ribosomes, an elaborate Golgi body, microtubules and lamellated bodies. Both have connecting cilia developed, but at a different level. The ELM is observable at a light microscopical level. At EM level it is seen to be composed of zonulae adherentes connecting Müller fibres and photoreceptor inner segments. The developing cone inner segments project through the ELM while the rod inner segments are situated vitreally to it. At this stage prospective long single cones and double cones can be distinguished. The inner segments of the latter are apposed in pairs leaving a space filled with electrondense material between them (Fig. 17/27). This is in contrast to reports by other investigators who report that double cones appear much later than single cones '(Baburina, 1961; Ahlbert, 1969; Blaxter and Jones, 1967; Ahlbert, 1973; Grün, 1975b; Branchek and Bremiller, 1984)'.

At the trout stage B and guppy stage 6 apical microvilli, the future calycal processes, extend from the inner segment into the ventricular space. The apical connecting cilium of each cone has begun to enlarge into an accessory outer segment (see 'Yacob *et al.*, 1977)'. In some developing photoreceptors the formation of discs heralds the prospective outer segment. These infolding discs lie at a right angle to the microtubules of the cilium. In prospective cones the term 'disc' is really a misnomer since the infoldings remain continuous with the plasma membrane while in prospective rods the folds become free-floating discs; only in the most basal area do the infoldings persist (Fig. 17/28a). However, at this stage the rod outer segment reveals free floating discs in packages of two (Fig. 17/28b). Along the apposing inner segments of double cones subsurface cisternae form in a vitreo-scleral direction as a result of fusion of smooth endoplasmic reticulum vesicles. Subsurface membranes or cisterns seem to be exclusive to teleost cones; they are not encountered in amphibian or reptilian double cones. Rod nuclei form now a distinct layer and have become more heterochromatic than the cone nuclei (Fig. 17/27c).

At stage C (guppy late stage 6) the differentiation of the rods is ahead of that of the cones (in the guppy: average number of rod discs 110, of cone 'discs' 72) (Fig. 17/29). The free-floating outer segment discs are separated by the plasma membrane into packages of 9 discs (Fig. 17/33a). Division of inner segment into myoids with Golgi body and ER and ellipsoid with accumulated mitochondria (elongate in rods and round in cones) is distinct. Subsurface membranes along the apposing inner segments of double cones begin to fuse into cisterns. At the base of this region large basal mitochondria have been observed (guppy) as well as microtubules emanating from the whole appositional region (rainbow trout). The close proximity of mitochondria with developing cisterns has also been observed in *Tilapia leucosticta* '(Grün, 1975b)' and in neurons of the spinal cord '(Rosenbluth, 1962)'. Subsurface membranes seem to occur exclusively in neurons.

Stage D (guppy stage 7). The rods and the different type of cones are differentiated. The rods have attained a light-adapted position (Fig. 17/29). The number of rod discs in the trout now numbers 800 and that of cones 340. The cone to rod ratio in both species is 2:1. Scattered over the salmon and brown trout retina are encountered sparse large single cones (green sensitive). '(Loew and Kunz, unpublished results) (Figs. 17/19) and also some triple cones (similar to Fig. 17/13 inset).

In the guppy, fixed in ovario, a putative ellipsosome (modified scleral mitochondrion in the accessory member of double cones) is observed for the first time (Fig. 17/29). It is characterized by peripheral cristae and an electron-lucent lumen; there is no structural change of the ellipsosome taken from embryos of light- or dark-adapted females. The ellipsosome-bearing member of the double cone possesses already 'wavy' outer segment discs as observed in the sexually mature fish (Figs. 17/17a,b). Slight diel retinomotor movements of rods and cones and the pigment granules of the RPE have started over the whole retina.

In the brown trout, cyclical shedding of phagosomes is already rhythmic (Fig. 17/31).

Stage E (guppy stage 8). In the guppy the dorsal region and the upper one-third of the ventral region (inset Fig. 17/17b) are differentiated and exhibit the same type and arrangement of photoreceptors as found in the fundic region. Along the contact zone of the inner segments of double cones (DC) and long single cones (LS) are fenestrated subsurface cisternae. In contrast, they are absent in the contact zones of double cones and short single-cone ellipsoids. Additionally, the rod myoids, which are surrounded by double, long single and short single cones, display fenestrated subsurface cisternae in all contact zones (Fig. 17/35). However, in the lower two-thirds of the ventral region differentiation is still progressing. Here, the types of cones are very different from the rest of the retina. The short single cones are lacking and are replaced by long single cones. The double cones are structurally equal; they lack an ellipsosome but the mitochondria are still arranged along a vitreo-scleral size gradient and the spacing of the outer segment

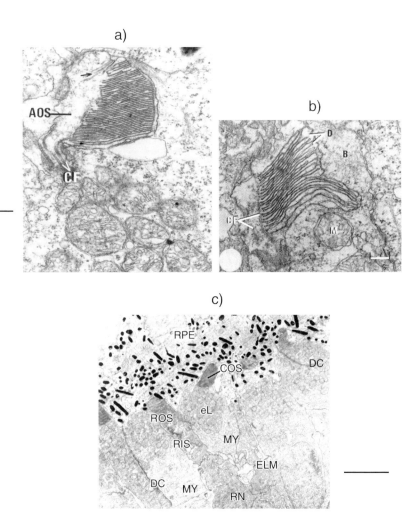

Fig. 17/28 a - Outer segment formation in guppy cone (stage 6) (Kunz et al., 1983, with kind permission of Springer-Verlag, Heidelberg). Bar = 0.4 microns. Arrow indicates tubular formation of future discs. **b** - Outer segment formation in a guppy rod. (stage 6) (Kunz et al., 1983, with kind permission of Springer-Verlag, Heidelberg). Bar = 0.1 micron. Some discs (D) are pinched off from the outer segment plasma membrane. **c** - RPE and photoreceptor layer in a trout of stage C (Schmitt and Kunz, 1988, with kind permission of S. Karger, Basel). RPE ensheathes photoreceptor outer segments. Rods continue to extend sclerally. RIS and CIS are divided into ellipsoid and myoid. Bar = 2 micron. AOS - accessory outer segment; B - bleb; CF - connecting *cilium*; COS - cone outer segment; D - discs; DC - double cone; eL - ellipsoid; ELM - external limiting membrane; M - *mitochondrion*; MY - myoid; RPE - *retinal* pigment *epithelium*; RIS - rod inner segment; RN - rod nucleus; ROS - rod outer segment; RPE - *retinal* pigment *epithelium*.

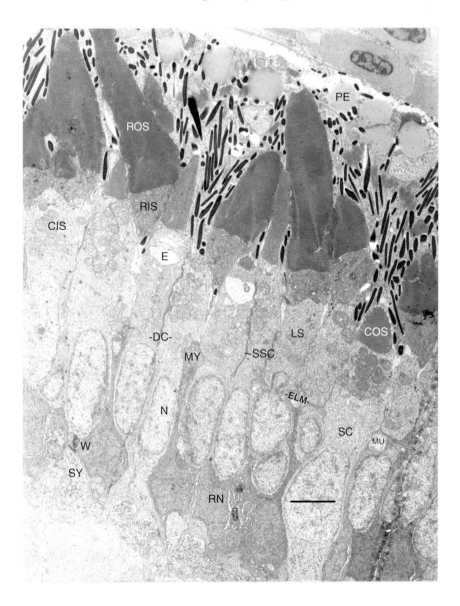

Fig. 17/29 Radial TEM section through pigment *epithelium* and photoreceptors of the fundic region of the guppy eye at stage 7. Rods are in a light-adapted position (Kunz *et al.*, 1983, with kind permission of Springer-Verlag, Heidelberg). Bar = 3 microns. CIS - cone inner segment; COS - cone outer segment; E - ellipsosome; ELM - external limiting membrane; EL - ellipsoid; L - lipid body; LS - long single cone; MB - myeloid body; MU - Müller cell; MY - myoid; N - cone *nucleus*; PE - pigment *epithelium*; RIS - rod inner segment; RN - rod *nucleus*; ROS - rod outer segment; SC - short single cone; SR - synaptic ribbon; SSC - subsurface *cisternae*; SY - synapse; W - lamellated body.

(OST) discs is the same in both partners. A view of the mosaic arrangement displays best the differences between fundic/upper third ventral and lower two-third ventral region. In the latter, four rods are grouped around a central long single cone (LS) and one rod adjacent to 'alternating' long single cones (LSS). In the contact zones between equal double cones and central long single cones fenestrated cisternae are detected. As opposed to the other regions, the subsurface cisternae always extend to the myoid level. At night the first indication of a change from square to row-mosaic is observed in the fundic region.

When the fixative is applied through the maternal cloaca, the ellipsosome of the embryo consists of a large vesicular structure with peripheral cristae both in the light- and dark-adapted state (Fig. 17/29). However, in embryos removed from the mother and subsequently kept in the light or dark, the ellipsosome is traversed by cristae, which show different regional arrangements and an electron-dense patch. (Fig.17/34). Subsurface membranes, in the region where double cones touch long single cones and rods, disappear in the dark. It has been suggested that they serve to stabilize the cone square mosaic in daylight (Fig. 17/35).

In the guppy, shedding of rod outer segments during the light-phase and of cone outer segments during the dark phase is shown in Figs. 17/30a,b. Some rod shedding has been already observed at late stage 7.

Retinomotor movements. Distinct diel retinomotor movements of rods, double and long single cones and the pigment granules in the RPE have started over the whole retina. In the light-adapted stage the rod outer segments of the guppy reach over the midline of the RPE. In the trout cone indices between night and day vary by about 20% and rod indices by about 30% (17/32b, 17/36a,b).

In the fundus, which is overlain by an opaque 'black patch' (Figs. 17/1, 17/4), the in situ dark-adapted double cones are in a tiered arrangement, which allows space for the dark-adapting rods (Fig. 17/34). The patch appears black due to the absence of guanine crystals in the argentea (Fig. 17/3). Therefore, this patch, devoid of rigidity, may bulge outwards in the dark. With the exception of the area of the 'black patch' a change from the square mosaic of the light-adapted guppy retina to a row mosaic in the dark is operational over the entire retina.

Stage F (stage 9 guppy). Invaginations of the rod outer plasma membrane separate free-floating discs into packages of 22 (Fig. 17/33b). Rod outer segments are now in a fully light-adapted position separated from the base of the RPE only by a junctional complex. In the trout, shedding of large phagosomes occurs as a burst at dawn and a smaller peak appears early in the night. The peaks of small phagosomes display a regular pattern with a peak each before dawn and in the middle of the day (Fig. 17/31d).

In the trout, cone inner segments display lateral fins interlocking with microvilli of Müller cells (similar to Fig. 17/13 for *Trachinus vipera*, whereas in the guppy fins are rudimentary and restricted to the myoid region. Nuclei of cones, except those of short single cones, protrude through the ELM.

Retinomotor movements in the guppy do not affect the short single cones and in the brown trout both types of single cones remain stationary. During dark-

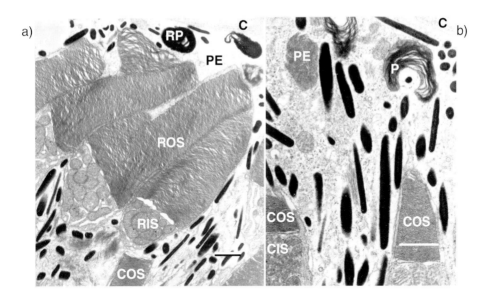

Fig. 17/30 Onset of shedding of photoreceptor outer segment tips in the guppy (TEM) (Kunz and Ennis, 1983, with kind permission of the director of the Muséum d'Histoire naturelle, Geneva). **a** - Rod outer segments shed during light-phase at stage 8. **b** - Cone outer segments shed during subsequent dark-phase. C - choroid gland; COS - cone outer segment; CIS - cone inner segment; P - cone phagosome; PE - pigment *epithelium*; RIS - rod inner segment; ROS - rod outer segment; RP - rod phagosome.

adaptation the cones of the guppy change from a square into a row mosaic. In the trout the cone outer segments do not enter the RPE but bend (Fig. 17/36b) and the accessory outer segment (AOS) now forms an incisure in the rod outer segment. In the guppy the ultrastructure of the ellipsosome of the infused embryos is still similar to the previous stages and in specimens, removed from the ovary and subsequently kept in a cyclic light regime, the ellipsosome is again traversed by cristae which enclose a dense patch. No differences were observed between light- and dark-adapted stages. In continuous darkness the eyes remain dark-adapted, which would suggest that circadian controls are not yet established although RMM have been developed for over 2 weeks.

Stage G and following stages:
Prepartum guppy embryos
Microscopical analyses of this stage had shown that the short single central cones (blue) are present in the dorsal, fundic and the upper one-third of the ventral region. In the lower two-thirds of the ventral part the short single cones are lacking and are replaced by long singles. As regards the mosaic arrangement, in the

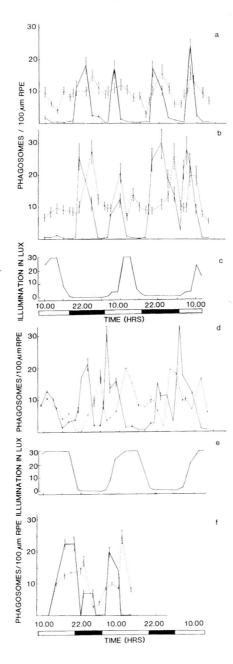

Fig. 17/31 Developmental pattern of phagosome count (i.e. shedding rhythm of rod outer segment tips) in brown trout (McCormack et al., 1989, with kind permission of S. Karger, Basel). **a** - stage D; **b** - stage E; **c** - ambient light conditions for a and b; **d** - stage F; **e** - ambient light conditions for d; **f** - one year old trout. Broken line refers to small and solid line to large phagosomes.

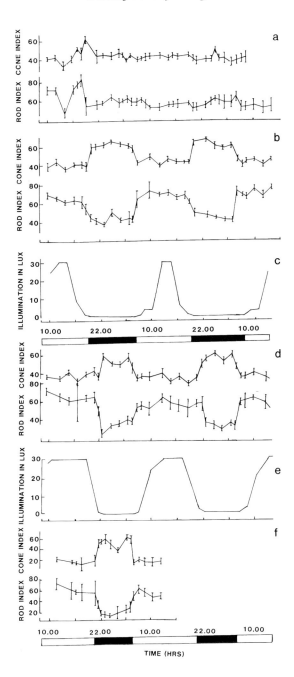

Fig. 17/32 Ontogeny of rod and double cone RMM in brown trout (McCormack *et al.*, 1989, with kind permission of S. Karger, Basel). **a** - stage D; **b** - stage E; **c** - ambient light conditions for a and b; **d** - stage F; **e** - ambient light conditions for d); **f** - one year old trout.

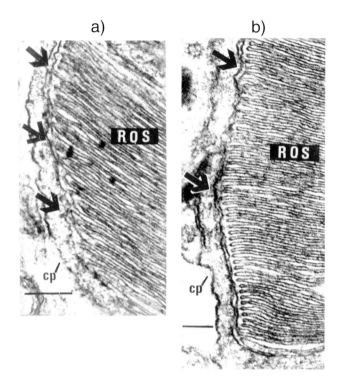

Fig. 17/33a At stage C of the trout the discs of the rod outer segments (ROS) are separated into packages of ~9 as a result of invagination of the plasma membrane (arrows) (TEM) (Schmitt and Kunz, 1989, with kind permission of S. Karger, Basel). Bar 0.25 microns.

Fig. 17/33b At stage F the packages of free-floating discs (arrows) in rod outer segments (ROS) of the trout have increased to ~22 discs (Schmitt and Kunz, 1989, with kind permission of S. Karger, Basel). Bar 0.1 microns. cp - calycal process.

upper one-third it is the same as described for the fundic and dorsal region: rows of zig-zagging unequal double cones are separated by rows of alternating short and long single cones, with the rods regularly arranged. In the lower two-thirds zig-zagging equal double cones (probably red/red) alternate with rows composed exclusively of long single cones (probably green). A region-specific cone distribution was referred to in the review by 'Levine and MacNichol (1982)'.

Neonate and 5 and 14 days old guppy
EM of the ellipsosome of light- and dark-adapted guppies at the above stages reveals cristae with a dark patch, both much more strongly stained than in the previous stages.

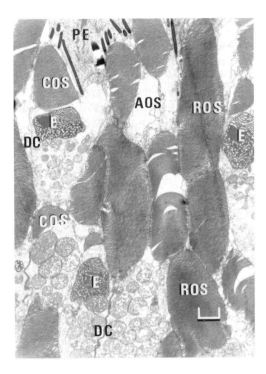

Fig. 17/34 Tiered double-cones in the 'black patch'-region of the guppy during dark-adaptation at stage 8 (TEM) (Kunz and Ennis, 1983, with kind permission of the director of the Muséum d'Histoire naturelle, Geneva). Bar 1 micron. Structure of 'black patch' (see Fig. 17/4a) allows for bulging of visual cell layer in the dark. AOS - accessory outer segment; COS - cone outer segment; DC - double cone; E - ellipsosome; ROS - rod outer segment.

(c) Outer plexiform layer (OPL) of the trout

At stage B electron-lucent pedicles expanding from cone axons and more densely staining spherules arising directly from the perinuclear region of the rods are observed. Both spherules and pedicles contain synaptic vesicles, neurofilaments, microtubules and free ribosomes.

At stage C cone pedicles and rod spherules contain small synaptic ribbons. Dyads as well as some triads have been observed. Adjacent to synaptic ribbons electrondense material fills the synaptic gaps. Growth cones of bipolar cells, running along Müller cells terminate in the OPL (Fig. 17/28c).

Stage D reveals that synaptic ribbons have increased in length (150nm in rods and 200nm in cones), and the number of triads has increased. Contiguous to the basal membrane of spherules and pedicles electron-dense zones on dendritic membranes representing presumptive basal contacts are observed.

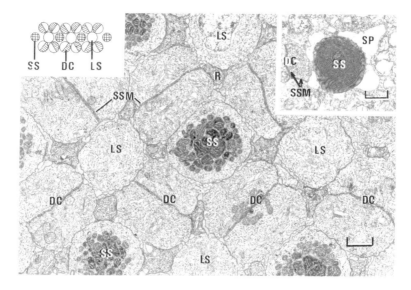

Fig. 17/35 TEM of dorsal square mosaic pattern during light-phase (late stage 8) of the guppy (Kunz and Ennis, 1983, with kind permission of the director of the Muséum d'Histoire naturelle, Geneva). Bar 1 micron. Note zig-zagging of the subsurface-membranes (SSM) separating the partners of the double-cones (DC) and additional SSM in the contact zones between DC and LS, DC and rods (R) and LS and rods. Inset upper right: Short single cone (SS), sectioned at level of their outer segment, show intercellular space ( ) surrrounding it. Subsurface membranes (SSM) in contact-zones of double cones (DC) and long single cones (LS) are also evident. Inset upper left: diagram of the square mosaic showing rows of alternating SS and LS and zig-zag arrangement of DC.

Stage E is characterized by the appearance of telodendrial contact between rod and cone terminals. These occur between a rod and a cone or between two cones. A halo of electronlucent cytoplasm now surrounds the synaptic ribbon. One ribbon and up to five invaginating dendrites are observed in spherules (rod terminals), whereas three ribbons and up to 15 invaginationg dendrites occur in pedicles (cone terminals). Additionally, presumptive synaptic contacts are observed between photoreceptors and horizontal and bipolar cell dendrites. These contacts resemble conventional synapses and are observed within synaptic terminals, along telodendria or the basal membrane. Spinules (finger like extensions with electron-dense contact zones) are for the first time observed in the contact zones between horizontal cell dendrites and apposing synaptic ribbons of pedicles. (Spinules occur in the light and disappear in the dark.) Gap-junctions between the somata of two external horizontal cells are shown in Fig. 17/37d).

At stage F the network of interreceptor contacts has increased dramatically. Telodendria are joined by electron-dense contact zones (Fig. 17/37e) or gap junc-

Development of the eye 355

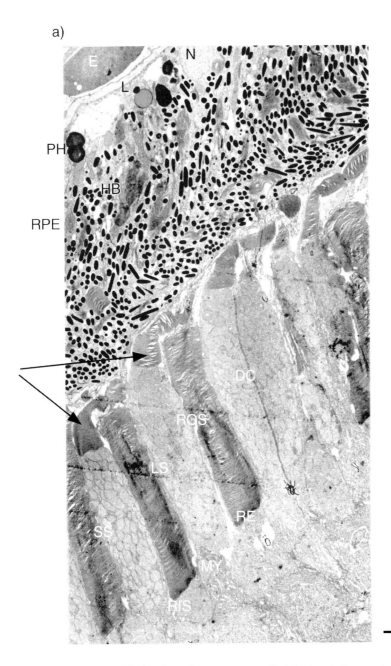

Fig. 17/36a TEM of dark-adapted *retina* at stage F of the trout. Cone outer segments are bent (arrows) at the RPE-*retina*l interface while in other teleosts they penetrate the RPE. Bar 5 microns.

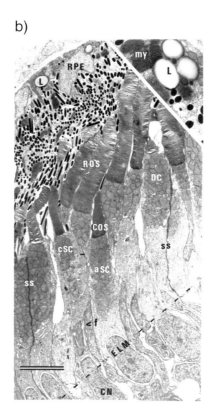

Fig. 17/36b Light-adapted *retina* at stage E of the trout. (Schmitt and Kunz, 1989, with kind permission of S. Karger, Basel) Bar 5 microns. Rod outer segments (ROS) reach to the midline of the RPE. Note lipid droplets. aSC - accessory (corner) cone; CN - cone *nucleus*; COS - cone outer segment; cSC - central single cone; DC - double cone; E - erythrocyte; ELM - external limiting membrane; L - lipid droplet; LS - long single cone; my - myeloid body; MY - myoid; N - nucleus; ph - phagosome; RIS - rodinner segment; ROS - rod outer segment; RPE - *retina*l pigment *epithelium*; ss - subsurface membranes. Bar 5 microns.

tions (Fig. 17/37f). They are also observed invaginating neighbouring pedicles which possess now synaptic ribbons 600-900nm long. Presumptive as well as mature conventional synapses between dendrites are observed. Gap junctions between the somata of external horizontal cells are pronounced (length of 1.0-2.0 microns).

The following changes are observed after 15 minutes dark adaptation in the OPL: synaptic vesicles are noted along the synaptic ribbon and congregated at the arciform density. Electron-dense spots, supposedly resulting from degenerating spinules, occur within the dendrites invaginating the pedicles.

(d) Inner nuclear layer (INL) of the trout

Stage A. The nuclei of the horizontal cells lie with their long axis parallel to the OPL. They are further differentiated than the vitreally adjoining bipolar cells, which lie sclerad to spindle-shaped neuroblasts in the midregion where fusiform nuclei of the Müller cells are also encountered. Their radial fibres are recognized by the presence of microtubules, glycogen granules, free ribosomes, RER and SER. Pyknotic cells are usually surrounded by these fibres. In the vitreal part presumptive amacrine[3] cells are identified on the basis of their large electron-lucent nuclei. Their perikarya contain mitochondria, RER, microtubules and a Golgi body. (Fig. 17/37a,b)

Stage B. The spindle-shaped neuroblasts of the midregion are no longer observed. Neural processes of the bipolar cells traverse the layer sclerovitread, parallel with light-staining Müller cells.

Stage E. The somata of external horizontal cells are joined by small gap junctions (Fig. 17/37d) and are more electron-dense than internal horizontal cells. Axons extend laterally and are joined by electron-dense contact zones. The vitreal part of the INL is traversed by large bundles of Müller cell fibres and bipolar cell axons containing numerous microtubules. (Fig. 17/37c)

(e) Müller cell of the guppy

In the stage 5/6 embryo, a diplosome is evident in the Müller cell scleral to its nucleus. An axoneme is not observed. The diplosomes are situated at the OPL. At stage 7 when the ELM is well developed, no centrioles or axonemes are observed in the Müller cell at the ELM level. They are situated further vitreally at the level of the ONL. A solitary cilium is observed in stage 8 embryos '(Ennis and Kunz, 1986)'.

(f) Inner plexiform (IPL) layer of the trout

At stage A, a network of fine neural processes of amacrine and ganglion cells is present containing microtubules, neurofilaments and sparse mitochondria. Fibers of the Müller cells traverse the IPL to reach the inner border of the retina where they form their 'endfeet' (inner limiting membrane, ILM).

At stage C, dendrites of amacrine and ganglion cells as well as those of bipolar cells become distinguishable. Interplexiform dendrites cannot be observed with normal EM analysis. Conventional synapses are plentiful in the midregion. Synapses (120-200 nm in length) occur between amacrine and ganglion cells or between two amacrine cells.

At stage D, the density of conventional synapses in the midregion has increased. Bipolar cells now form conventional synapses along their axons and at their terminals which harbour synaptic ribbons. They synapse with both amacrine and ganglion cells. Bipolar cells also display branched synapses.

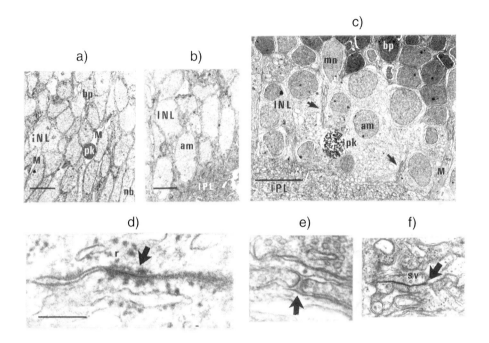

Fig. 17/37 TEM of inner retina of the trout (Schmitt and Kunz, 1989, with kind permission of S. Karger, Basel). Stage A (hatching): **a** - Differentiating bipolar cells lie sclerad to spindle-shaped neuroblasts. Pyknotic *nuclei* are surrounded by Müller cell fibres. Bar 5 microns. **b** - Differentiating amacrine cells, characterized by large electron-lucent *nuclei*. Perikarya contain *mitochondria*, RER, microtubules and a Golgi body. Bar 5 microns. Stage E ('swim-up stage'): **c** - *Nuclei* of bipolar, Müller and amacrine cells are easily distinguished. Dendrites of bipolar and amacrine cells (arrows) and a broad Müller fibre extend towards the IPL. Bar 12 microns. **d** - Gap junction (arrow) between two *somata* of external horizontal cells. Bar 0.1 micron. Stage F (external food uptake): **e** - Electron-dense zone (arrow) between two *telodendria*. × 100 000. **f** - Gap-junctions (arrow) between two *telodendria*. × 33 000.
am - amacrine cell; bp - bipolar cell; INL - inner nuclear layer; IPL - inner plexiform layer; M - fibre of Müller cell; mn - *nucleus* of Müller cell; nb - neuroblast; pk - pyknotic *nuclei*; r - ribosome; sv - synaptic vesicle

Stage E. At this stage the density of conventional synapses and electron-dense zones in the midregion has increased. The synaptic ribbon of the bipolar cells is now not only apposed to both amacrine and ganglion cells but also to two amacrine cells.

At stage F, a stratification within the IPL is evident as bipolar axons terminate in four strata. Multistratified axons are evident. In the midregion the density of conventional synapses has increased. Two or three synaptic ribbons occur within bipolar axon terminals. Serial synapses involving three consecutive conventional synapses between amacrine cell dendrites are observed. Amacrine cell dendrites may also be joined by small electron-dense zones, presumed to be gap junctions. Reciprocal synapses are observed at synaptic ribbons where amacrine cells are both pre- and post-synaptic to the bipolar cells.

Summary

Retinal pigment epithelium (RPE) of trout and guppy

The retinal pigment epithelium (RPE) first contains spherical and later also rod-shaped granules. The latter are contained within the RPE processes and undergo positional changes during dark-adaptation (retinomotor movements). The RPE of both guppy and rainbow trout is highly specialized before the onset of photoreceptor outer segment formation. Junctional complexes hold the lateral membranes of the cells together from stage A onward. As a result of the retinomotor movements, which start at stage D, these junctions become progressively restricted to the basal region. In the early stages, myeloid bodies are continuous with the rough endoplasmic reticulum (RER) (in contrast to observations on the amphibian eye) and establish continuity with the smooth endoplasmic reticulum (SER) only from stage E onwards. Myeloid bodies appear earlier than phagosomes, which display a rhythmic shedding pattern before retinomotor movements have become cyclical. Apart from their association with lipid droplets, some myeloid bodies are found to be continuous with the nuclear envelope in both trout and guppy. In guppies, rods shed phagosomes before cones do. (17.2.4a)

Visual cell layer (VCL), outer nuclear layer (ONL) of trout and guppy

In the fundus of the eye of guppy and trout rods can be distinguished already at stage A, which is in contrast to the contention of other investigators who maintain that rods appear only much later in development. Rod precursors are situated more vitreally than those of cones and their development proceeds faster. At stage B spherules and pedicles can be distinguished. Double cones appear at the same time as single cones in both trout and guppy, contrary to observations on other fish. After rods have achieved a light-adapted position, retinomotor movements (RMM) begin. In the guppy the short single cones and in the trout the short and long single cones do not participate in these movements. In the brown trout RMM start at stage D and a cyclical (dusk/dawn) bimodal rod and cone pattern is established at stage E.

At stage E the cones are arranged in a so-called square mosaic. In the guppy this pattern changes into a row mosaic during dark adaptation.

Phagosomes were observed for the first time at the previous stage D. In the brown trout they exhibit a circadian shedding pattern at stage F. In the guppy the ellipsosome of stages 7, 8 and 9 of the infused embryos (fixed in ovario), be they dark- or light adapted, reveal a structure with peripheral cristae and an electron-lucent lumen. In embryos removed from the mother and kept under cyclical light and dark conditions, the ellipsosome is traversed by cristae with different regional arrangement and an electrondense patch. This state persists in neonates and in 5 and 14 days old postembryos though the cristae and the patch are much more strongly stained than in earlier stages. Again, no differences between the dark- and light-adapted state are observed.

In the guppy the dorsal, fundic and upper one-third of the retina display the same types of cones forming the square mosaic, while in the lower two-thirds the short single central cones are absent and replaced by long single cones. Cones with intensely stained ellipsosomes are also observed in the ventralmost area. This is reminiscent of the 'colour pattern polymorphism' mentioned on the basis of behavioural and MSP studies.

In the guppy the subsurface membranes separating the apposing inner segments of the double cones are preceded by the presence of electron-dense material. Additional subsurface membranes in the appositional areas between double cones, long single cones and rods, 'disappear' in the dark.

Outer plexiform layer (OPL) of trout

In the OPL, at stage C, cone terminals (pedicles), rod terminals (spherules) and growth cones of bipolar cells are observed. At stage E, three synaptic ribbons and up to 15 invaginating dendrites occur in cone terminals and one ribbon and up to five dendrite invaginations in rod terminals. Telodendrial contacts between rod and cone terminals are evident. Spinules (fingerlike processes present in the light and absent in the dark), between horizontal cell dendrites and synaptic ribbons of pedicles, are for the first time observed. By stage F the network of above contacts has increased dramatically.

Inner nuclear layer (INL) of trout

At stage A the horizontal cells are arranged into two layers (external and internal) followed by the innermost layer of amacrine cells. By stage B the midregion has been differentiated revealing traversing Müller fibres and bipolar cells. At stage E the somata of the horizontal cells are seen to be connected by gap junctions and in the vitreal part nuclei of bipolar, Müller and amacrine cells can be differentiated.

Inner plexiform layer of trout

The number and type of synapses increases from stage C onwards.

17.3.5 Developmental sequence of the zebrafish eye and retina

Morphogenesis of the eye commences 11.5 hours post-fertilization (hpf). At ~13 hpf the eye anlage has taken up a wing-like shape. From ~14-15 hpf the optic primordia bend ventrally and rotate slightly in an anterior direction. Invagination commences at ~15-16 hpf and the choroid fissure becomes apparent. It increases in depth as the eye cup invaginates further and is completely formed by 24 hpf.

A further rotation of the eye in relation to the embryonic axis takes place between 24 and 36 hpf so that the choroid fissure has taken up a ventral position, as is typical for all other teleost embryos so far analyzed. The two eye rotations in the zebrafish embryo occur in parallel with the early and late stages of cephalic flexure '(Schmitt and Dowling, 1994)'.

During the period of 24 and 36 hpf the retina almost doubles in width. Axons of ganglion cells leave the eye between 32 and 34 hpf. While the pseudostratified retina is more than six cell layers across at hpf 36, at 50 hpf three nuclear layers (ganglion cell, inner nuclear and outer nuclear layer) are observed. At 72 hpf the majority of embryos have hatched. The basic synaptic pathways between photoreceptors and ganglion cells in all regions of the retina are completed between 70 and 74 hpf. The presence of ribbon synapses, though rare, within the OPL and IPL confirms that the basic 'vertical' pathway between photoreceptors, bipolar and ganglion cells has been established at that stage '(Kljavin, 1987; Burrill and Easter, 1995; Raymond et al., 1995; Schmitt and Dowling, 1994, 1996, 1999)'.

'Branchek and Bremiller (1984)' and 'Branchek (1984)' were the first to describe in detail the development of the photoreceptors, both structurally and functionally.

Retinal differentiation proceeds vitreosclerad with the exception of bipolar cells, which — as in other vertebrates — follow the differentiation of horizontal and visual cells. Synaptogenesis also takes place in a scleral direction with exception of bipolar cell synapses '(Grün, 1982)'.

While in most teleosts differentiation of the retina appears first in the fundic region, in the zebrafish it is first observed in the ventronasal region adjacent to the choroid fissure '(Kljavin, 1987; Raymond et al., 1995; Burrill and Easter, 1995; Schmitt and Dowling 1994, 1996, 1999)'. 'Kljavin (1987)' was the first to describe the early rod formation in this region.

The newly hatched zebrafish (from 60 hpf onwards) lacks vision. The retina becomes functional shortly before the swim-up stage and the onset of exogenous food uptake. A comparison of the differentiation of the zebrafish eye with retinotectal projections and the development of visual behaviour reported by other authors is supplied by 'Schmitt and Dowling (1999)'. These authors stress that at the onset of visual behaviour gap junctions between horizontal cells, interreceptor contacts, spinules within the cone pedicles, serial and reciprocal synapses within the IPL — all features present at the swimming up stage and feeding of the rainbow trout (see above) — are still lacking in the zebrafish.

The following detailed results on the retinal development are based on publications by 'Schmitt and Dowling (1994, 1996, 1999)', if not stated otherwise.

(a) Retinal pigment epithelium (RPE)

At 32 hpf, pigment granules are observed in the elongate flattened cells (Fig.17/38a); their area has increased by 40 hpf (Fig. 17/38b). At 50 hpf large irregular

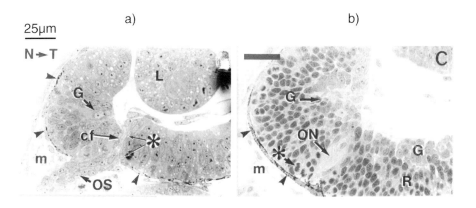

Fig. 17/38a,b Zebrafish eye. Horizontal sections along the nasal-temporal axis (N-T) (Schmitt and Dowling 1999, with kind permission of Wiley Liss Inc., a subsidiary of John Wiley & Sons, Inc.). **a** - At 32 hpf. Bar = 25 microns. At this stage the first ganglion cells (G) are observed ventro-nasally at the level of the optic stalk (OS), which is surrounded by mesenchymal cells (m). Pigment granules (arrowheads) do not extend into the choroid fissure (cf), which is lined by flattened cells (asterisk). **b** - At 40 hpf. Bar 25 microns. *Retina* has increased in thickness and cell density. Ganglion cells have formed a large patch within the nasal *retina* and two layers extend temporally to the optic nerve. Inreased area of pigmentation (arowheads). * points with arrow to patch of mitotic cells. cf - choroid fissure; G - ganglion cells; L - lens; m - mesenchymal cells; ON - optic nerve; OS - optic stalk; R - *retina*.

horizontally oriented nuclei as well as round and spindle shaped pigment granules are apparent (Fig. 17/39a,b,c,d).

At 60 hpf RPE processes containing RER, SER and free ribosomes, have begun to extend between the photoreceptors. Often pigment granules composed of many miniature granules are encountered (Fig. 17/42), as previously described for the salmonids by 'Schmitt and Kunz (1989)'. At 70-74 hpf the nuclei have elongated and small lipid droplets and membrane-bound clusters of pigment granules are observed (Fig. 17/43c,d).

Fig. 17/39a-d Zebrafish eye at 50 hpf (Schmitt and Dowling 1999, with kind permission of Wiley Liss Inc., a subsidiary of John Wiley & Sons, Inc.). **a** - Dorsoventral (D-V) section through optic nerve (ON). Bar 40 microns. Lamination observed within the ventral *retina* and from the optic nerve extending into the fundic and dorsal *retina*. * ventro-nasal patch. **b** - Magnified fundic part [to the right of optic nerve since photomicrograph is tilted by 90° with regard to a)]. Bar ... 12 microns. Small arrows point to presumably displaced amacrine cells within the inner plexiform layer. nb are presumptive bipolar cells within

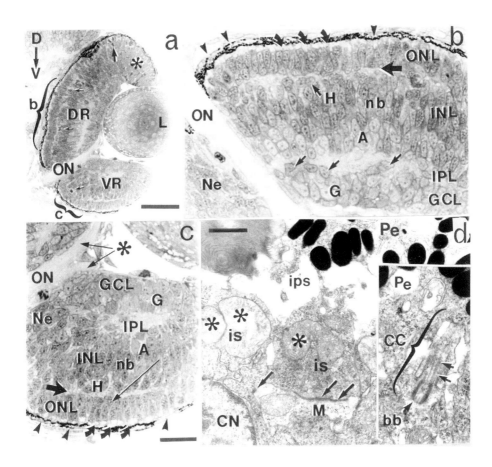

the inner nuclear layer. Curved arrows point to inner segments of visual cells and arrowheads to pigment granules of the RPE. **c** - Magnified ventral part, to the right of optic nerve. Arrowheads point to pigment granules. Thin arrow indicates presumptive rod *nuclei*. * fork of optic nerve. Bar 14 microns. **d** - EM shows two types of pigment granules in the REP and differentiation of photoreceptor inner segments with large *mitochondria* (asterisks) extending towards the pigment *epithelium* into the interphotoreceptor space. Müller cells project between cone *nuclei* and form *zonulae adherentes* (arrows) with the developing inner segments. Bar 1 micron. Inset: Microtubules (small arrows) extend from the basal body into the connecting cilium of the photoreceptor. Bar 0.5 microns. A - amacrine cells; b - dorsal region; bb - basal body; CC - connecting *cilium*; CN - cone *nucleus*; DR - fundic (central) *retina*; G - *ganglion* cell; GCL - *ganglion* cell layer; H - horizontal cells; INL - inner nuclear layer; IPL - inner plexiform layer; ips - interphotoreceptor space; is - photoreceptor inner segment; L - lens; M - Müller cell; nb - *neuroblast*; Ne - undifferentiated neuroepithelial cells; ON - optic nerve; ONL - outer nucleear layer; Pe - pigment *epithelium*; v - ventral region; VR - ventral area

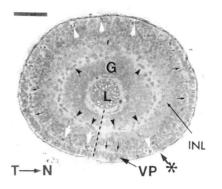

Fig. 17/40 Sagittal EM section along the temporal-nasal axis (T-N) through the mid-peripheral region at 54 hpf. (Schmitt and Dowling 1999, with kind permission of Wiley Liss Inc., a subsidiary of John Wiley & Sons, Inc.) *Nuclei* presumed to be of displaced amacrine cells within the interplexiform layer are indicated by black arrowheads. Bar 45 microns. * ventro-nasal patch. G - ganglion cell; INL - inner nuclear lay; L - lens; VP - ventronasal patch.

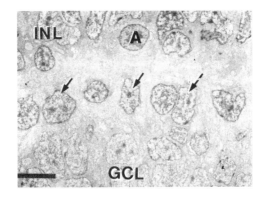

Fig. 17/41 EM of 50 hpf shows the *nuclei* (arrows) within the inner plexiform layer, presumed to be displaced amacrine cells which move from the inner plexiform layer to the ganglion cell layer between 50 and 60 hpf (Schmitt and Dowling 1999, with kind permission of Wiley Liss Inc., a subsidiary of John Wiley & Sons, Inc.) Bar 6 microns. A - amacrine cells; GCL - ganglion cell layer; INL - inner nuclear layer.

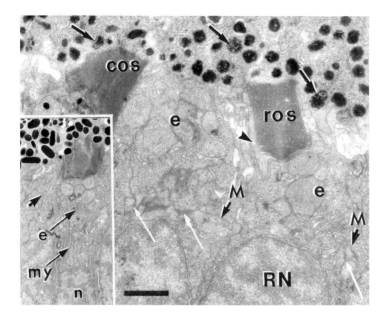

Fig. 17/42 TEM of zebrafish eye showing developing inner and outer segments in the ventronasal patch at 60 hpf (Schmitt and Dowling 1999. with kind permission of Wiley Liss Inc., a subsidiary of John Wiley & Sons, Inc.). Bar 1.5 microns. White arrows indicate *zonulae adherentes* between Müller cells and ellipsoids which form the outer limiting membrane. Black arrows point to pigment granules made up of small granules. Inset: Inner segment is now divided into ellipsoid and myoid. Black arrow points to developing connecting *cilium*. cos - cone outer segment; e - ellipsoid; M - Müller cell; my - myoid; n - *nucleus*; RN - rod *nucleus*; ros - rod outer segment.

(b) Visual cell layer (VCL) and outer nuclear layer (ONL)

At 50 hpf, retinal lamination is observed within the ventral and dorsal retina (Fig. 17/39a). At the same developmental stage a ventral view of whole mounts of eyes treated with FRet 11[4] revealed on a small percentage of embryos photoreceptors along the nasal side of the choroid fissure (Fig. 17/44A). Analysis of transverse sections along the dorsal-ventral axis showed in the ventral region small round nuclei (most probably those of rods) within the outer nuclear layer and adjacent to the outer plexiform layer, as well as photoreceptor inner segments scleral to the ELM (Fig. 17/39b,c). EM revealed in the 'ventronasal patch' and in the fundic region photoreceptor inner segments extending sclerally to the outer limiting

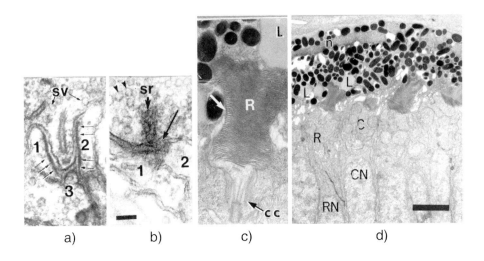

Fig. 17/43 Transverse section along the dorso-ventral axis at 74hpf (Schmitt and Dowling 1999, with kind permission of Wiley Liss Inc., a subsidiary of John Wiley & Sons, Inc.). **a** - Triad of photoreceptor synapse: two lateral horizontal processes (1,2) and a central bipolar process (3) opposing the ribbon. Synaptic vesicles and electron-dense postsynaptic zones (arrows) are observed. Bar 120 nm. **b** - Presumptive synaptic ribbon within a bipolar cell axon terminal in the IPL is opposed by two postsynaptic processes (1,2). Arrow heads point to synaptic vesicles and arrow to electron dense material associated with the postsynpatic membranes. Bar 90 nm. **c** - Rod outer segment discs (white arrow) have become discontinuous except for the most basal part. Bar 2 microns. **d** - TEM of dorsal periphery reveal rod *nuclei* vitreally to cone *nuclei*. Note outer segments of both rods and cones and the presence of lipid droplets in the pigment *epithelium*, the *nuclei* of which have become elongated. Bar 7.5 microns. C - cone; cc - connecting *cilium*; CN - cone *nucleus*; L - lipid droplet; n - *nuclei* in pigment *epithelium*; R - (white), rod outer segment; R - rod; RN - rod *nucleus*; sr - synaptic ribbon; sv - synaptic vesicle.

membrane and containing mitochondria, RER and abundant free ribosomes as well as microtubules extending from the basal body into the axoneme of the connecting cilium. Zonulae adherentes between Müller cell apical fibres (often with large mitochondria) projecting sclerally between heterochromatic cone nuclei and inner segments, the future ELM, were evident. (Fig. 17/39d and inset). At 55 hpf a small number of photoreceptor outer segments were observed. By 57 hpf the number of rods on the nasal side of the fissure had increased and rod staining extended to the temporal side (Fig. 17/44B).

'Larison and Bremiller (1990)' followed the cone mosaic during development by staining with a monoclonal antibody specific for the double cone phenotype.

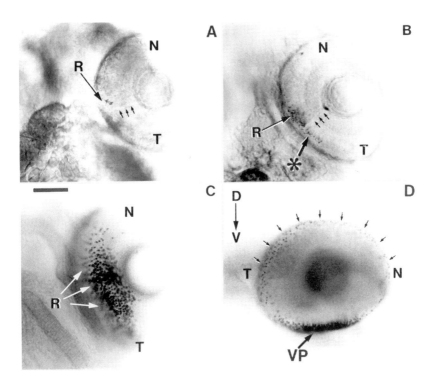

Fig. 17/44 A-D Topographical pattern of rod differentiation within the nasal and temporal regions of the *retina* in embryonic wholemounts (FRet 11[4] staining) (Schmitt and Dowling 1996, with kind permission of Wiley Liss Inc., a subsidiary of John Wiley & Sons, Inc.). **A** - 50hpf. Initial rods along the nasal side of the choroid fisssure (small arrows). **B** - 57hpf. Increased number of rods on nasal side of fissure (small arrows). Stained rods now extend temporally (asterisk). **C** - 72 hpf. Rods extend equally into both nasal and temporal regions. (white arrows). **D** - 72 hpf. Rods densely aggregated within the ventral patch (large arrow) and randomly scattered throughout nasal and temporal regions within the dorsal periphery (small arrows). D - dorsal; N - nasal; R - rods; T - temporal; V - ventral; VP - ventral patch. Bar = 75 µm in A-C, 100 µm in D.

They reported that an adult type row mosaic pattern is already established at 54 hpf'.

In the ventronasal patch large oval cone nuclei are interspersed by smaller round nuclei representing probably putative rod nuclei. In contrast to other teleosts (for trout see 'Kunz *et al.*, 1983', 'Schmitt and Kunz, 1989') there is no clear separation of the two types of photoreceptor nuclei, which was already referred to by 'Kljavin (1987)'.

EM revealed at 60 hpf small outer segments of both rods and cones in the ventral part. Their differentiation is most advanced in the ventronasal patch where the rod outer segment measures 3 microns in length (Fig. 17/42). Outer segments are less than one micron in length within the nasal region and still absent in the dorsal and temporal retina. The inner segment shows a distinct division into ellipsoid with large mitochondria and an elaborate Golgi body and SER and RER in the myoid region. The ELM composed of zonulae adherentes between Müller cells and photoreceptor ellipsoids is pronounced (Fig. 17/42 and inset). Instead of spreading nasally, as cones and ganglion cells do, rods have spread into the temporal-ventral region by 60 hpf. Their number increases rapidly within the ventral region while they are observed only sporadically in the dorsal part.

Analysis of whole mount eyes at 72 hpf showed a considerable equal increase of stained rods on the temporal and nasal regions of the ventral retina, with a large dense patch of staining across the centre of the ventral retina. There were scattered cells in the dorsal retina (Fig. 17/44 C,D). This matches the pattern demonstrated by in situ hybridization in zebrafish embryonic wholemounts using riboprobes specific for rhodopsin '(Raymond et al., 1995)'.

At 74 hpf, 1 micron and ultrathin sections display numerous rods within the ventronasal patch. On each side of the patch photoreceptors are less developed and dark-staining neuroblasts still lie within the INL. Cone outer segments measure maximally 3 microns whereas rod outer segments are 5 microns long and show the formation of free floating discs. Rod nuclei lie now vitreally to the cone nuclei. (Fig. 17/43 d). Rods are also scattered randomly throughout both the nasal and temporal regions in the dorsal periphery. They are not encountered in the fundic region (above the exit of the optic nerve) until 9-10 days pf. It has been suggested that retinoic acid regulates the differentiation of the rods: the application of exogenous retinoic acid to zebrafish during the initial stages of photoreceptor differentiation resulted in a precocious development of rods and an inhibition of cones '(Hyatt et al., 1996)'.

By 72 hpf the retina has become functional i.e. shortly before the postembryos become free-swimming and start feeding. By 74 hpf a so-called 'area' (area of high acuity) has arisen in the ventro-temporal region. Cones are arranged into three tiers causing a bulge of the eye. The outer segments of the most vitreal tier measure 3.5 microns, in the next tier 2.6 microns and in the most scleral one 2.0 microns. The pigment epithelium in this region is broader and the ONL is tightly packed with cone nuclei. Rods are rarely found. However, rods are encountered in considerable numbers ventrally to the area. These features become more pronounced at day 5 pf (Fig. 17/45 a,b) and persist at least through 21 days pf. (The problem of how the area of visual acuity maintains its relative topographical position during development was investigated by 'Zygar et al. (1999)' on the developing *Haplochromis burtoni*.)

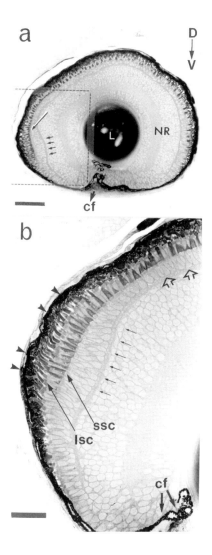

Fig. 17/45 a,b 'Area' (site of acute vision) in the ventro-temporal *retina*. (Schmitt and Dowling 1999, with kind permission of Wiley Liss Inc., a subsidiary of John Wiley & Sons, Inc.) **a** - Sagittal section at the level of the lens at 5 days postfertilization. The boxed area is characterized by an increased outer nuclear layer (small arrows), a tiering of cones (long arrow) and an absence of rods in the temporal area. Bar 60 microns. **b** - Higher magnification of boxed area in **a**. A tier of short single cones (ssc) lies vitread to a tier of longer single cones (lsc), which extend into the pigment *epithelium* (arrowheads). The outer nuclear layer in this region is made up of two to three layers of tightly packed *nuclei* (small arrows) in contrast to the more dorsally placed *nuclei* (open arrows). Bar 20 microns. cf - choroid fissure; DV - dorso-ventral; NR - nasal *retina*; ssc - short single cones; lsc - long single cones.

(c) Outer plexiform layer (OPL)

At 50 hpf this layer is less than 2 microns wide in the central and ventral retina, but not yet evident at the dorsal region. Despite the absence of photoreceptor terminals, many dendritic processes are observed. Due to the presence of glycogen granules, these processes are presumed to be those of horizontal cells. At 55 hpf the OPL has become more distinct temporally to the ventronasal patch but remains poorly formed in the dorsal periphery. At 60 hpf electron-lucent neural processes from the OPL invaginate the photoreceptor terminals both in the ventronasal patch and the region dorsal to the optic nerve. The terminals are characterized by dark-staining cytoplasm due to the presence of ribosomes and polyribosomes. At 62 hpf the first photoreceptor synaptic ribbons are observed in the ventronasal patch. Pedicles and spherules are not easily distinguished as is the case e.g. in the salmonids and the guppy '(Kunz *et al.*, 1983; Schmitt and Kunz, 1989)'.

At 70-74 hpf photoreceptor ribbon synapses form triads consisting of two lateral horizontal cell processes and a third central bipolar cell process opposing the ribbon (Fig. 17/43a).

(d) Inner nuclear layer (INL)

At 50 hpf, the INL is 25 microns wide. Horizontal cells are small and irregularly shaped. In the mid-region spindle-shaped neuroblasts considered to be presumptive bipolar cells are evident. Two to three light staining nuclei of amacrine cells are observed at the vitreal border. The differentiation of horizontal, bipolar and amacrine cells is most advanced in the ventronasal patch. (Fig. 17/39c) At 60 hpf the nuclei of the horizontal cells are elongated and lie parallel to the retinal surface. Dark-staining bipolar cells are distinguished in the mid-region. Along the midline fusiform nuclei of Müller cells and their light staining fibres are shown. The INL width has increased to 28-30 microns.

(e) Inner plexiform layer (IPL)

This layer is 6 micron thick at 50 hpf and includes a population of dark staining nuclei presumed to be displaced amacrine cells which will be incorporated later into the ganglion cell layer. No mature synapses are observed, only a sparse number of electron-dense zones, 100-150nm in length, which may represent the initial stages of conventional synapse formation.

At 60 hpf, the IPL is 8 microns wide. Presumptive bipolar cell axons and terminals are distinguished from smaller, electron-lucent processes of amacrine and ganglion cells. Amacrine processes are characterized by an irregular distribution of synaptic vesicles which are lacking in ganglion cell processes possessing ribosomes. A conventional synapse (small cluster of synaptic vesicles at the presynaptic membrane of an amacrine cell and electron-dense material coating the post-synaptic membrane of a ganglion cell) has been observed. In addition to the electrodense zones described for 50 hpf conventional synapses are now present both in the ventral and dorsal region. At 70-74 hpf presumptive synaptic ribbons within a bipolar cell axon terminal, opposed to two postsynaptic processes, are still rare (Fig. 17/43b). However, early stages of ribbons occur frequently. The number of conventional synapses has greatly increased.

(f) Ganglion cell layer (GCL)

At 32 hpf, a small number of ganglion cells with oval light-staining nuclei and prominent nucleoli are observed in the ventro-nasal patch. Their axons leave the eye via the embryonic fissure as seen in Fig. 17/38a. The optic lumen in the optic stalk is narrow but is continuous with the ventricle of the forebrain. By 40 hpf in addition to the large patch of ganglion cells within the nasal retina, two layers of ganglion cells extend temporally to the optic nerve (Fig. 17/38b). At 50 hpf the eye, sectioned along a median dorso-lateral axis is divisable into a central and dorsal third above the exit of the optic nerve and a ventral third below it. In all areas 2-3 ganglion cell layers are observed according to 'Hu and Easter (1999)' (Fig.17/ 39a).

At 50 hpf regularly spaced cells, presumed to be displaced amacrine cells from the inner plexiform layer, move to the ganglion cell layer (Fig. 17/39b). An EM of an embryo at 50 hpf displays their nuclei (Fig. 17/41). These cells are thought to move subsequently (i.e. from 50-60 hpf) into the ganglion cell layer as shown in a sagittal section (Fig. 17/40).

Summary

The developing zebrafish eye undergoes two rotations, parallel with cephalic flexions, until the embryonic (choroid) fissure takes up a ventral position as is typical for teleosts. While in most teleosts differentiation of the retina appears first in the fundic region above the exit of the optic nerve, in the zebrafish it occurs in the ventronasal region adjacent to the choroid fissure. Differentiation begins ~32 hours postfertilization (hpf) in the so-called 'ventronasal patch'. Differentiations of all cell types spreads from this area into the nasal and dorsal region before extending temporally. By about 70 hpf differentiation within all regions has attained a similar degree. An area (high density and tiering of cones) arises in the temporal retina. The zebrafish hatches at 60hpf and attains vision only shortly before the 'swimming up' stage to start exogenous food uptake.

Differentiation in the inner retina proceeds vitreosclerad, with the exception of bipolar cells, which follow differentiation of horizontal and visual cells. Synaptogenesis also takes place in a scleral direction with exception of bipolar cell synapses. A specialized area characterized by a high density and tiering of cones is observed in the temporal region.

17.4 DEVELOPMENT OF THE LACTATE DEHYDROGENASE (LDH) PATTERN IN THE TELEOST EYE

Studies of the energy metabolism in fish development stressed the role of the glycolytic enzyme lactate dehydrogenase (LDH; 1.1.1.27). Two loci of LDH have been observed in most vertebrates: The A-locus coding for the A_4 enzyme, predominantly expressed in anaerobic tissue such as white skeletal muscle and the LDH-B locus coding for the B_4 isozyme which occurs predominantly in aerobic tissue such as the heart muscle. There is a third gene, LDH-C, which is expressed in teleosts in liver and retina as two different isozymes, which are thought to be encoded by two different genes '(Leibel and Peairs, 1990)'. The retina-specific C_4 LDH was first observed in the eye of various species of teleost fish by 'Markert and Faulhaber (1965)' and 'Nakano and Whiteley (1965)'. It was referred to as E_4 by 'Massaro and Markert, (1968)' and later as C_4-LDH by 'Whitt et al (1975)'. It is also expressed in the optical part of the brain (Fig.17/46). The C_4 isozyme showed even more aerobic behaviour than the B_4 isozyme. 'Quattro *et al.* (1993)', who studied the evolutionary implications for the vertebrate LDH family, conclude that their phylogenetic analyses support previous hypotheses that teleost LDH is derived from a duplication of the LDH-B locus.

The LDH isozyme pattern of different organs of the adult *Poecilia reticulata*, obtained with starch gel electrophoresis and stained with nitroblue tetrazolium is shown in Fig. 17/46. At the bottom of the figure the C-isozymes of the eye were stained selectively as a result of prior application of 3 M urea as an inhibitor of A- and B-isozymes (Kunz, 1971b). Cryostat sections treated in the same way showed the staining most pronounced in the inner segments of all three types of cones. The cone outer segments stained as well though the rod outer segments seem to be devoid of LDH activity '(Kunz, 1971b)'.

Investigations into the development of the LDH pattern revealed that the C-bands are first evident in the stage 7 embryo (rods and cones differentiated) and already display the adult pattern (Fig. 17/47a). The retina of a stage 6 embryo (photoreceptors not yet differentiated) did not display any C-bands (Fig. 17/47b). Cryostat sections with and without the inhibitor yielded the same results. The LDH isozyme pattern at stage C/D shows already the adult distribution of intensity ('Kunz, 1971b)'.

Ontogenesis of the LDH pattern in two salmonids (*Salmo salar, S. trutta*), analyzed with starch-gel electrophoresis, revealed that in both species the eye isozymes are resolved shortly after hatching. In the early postembryo the eye-bands move faster in the salmon and faster in the postembryo than in the adult. By the end of the postembryonic phase the eyes of both salmon and trout possess the adult number and distribution of LDH activity '(Kunz, 1975)'.

'Whitt (1970)' had studied extensively the kinetic, physical and immunochemical properties of the C_4 isozyme and suggested that it may be specially suited for cells such as the photoreceptors, with a high constant aerobic metabolism. He also discussed a working hypothesis that this isozyme plays an important role in the regeneration of rhodopsin in the photoreceptors of the teleost retina.

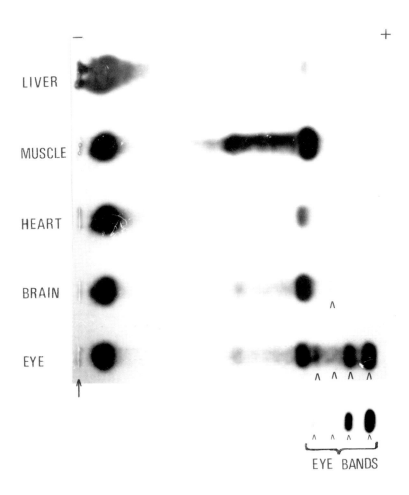

Fig. 17/46 Lactate dehydrogenase isozyme pattern of different organs of adult *Poecilia reticulata* (starch gel electrophoresis) (Kunz, 1971b, with kind permission of the director of the Muséum d'Histoire naturelle, Geneva). Arrow denotes point of application of sample. Below: Effect of inhibitor (3M urea) on lactate dehydrogenase pattern of the eye which selectively does not affect the eye bands. Arrow indicates site of application of sample. ^ indicates bands unique to eye and brain. — - cathode; + - anode.

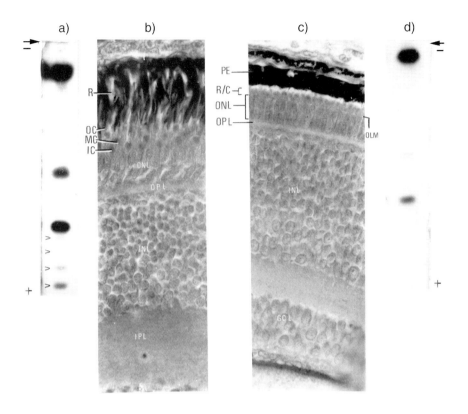

Fig. 17/47 a-d Lactate dehydrogenase isozyme pattern of guppy embryos (Kunz, 1971b, with kind permission of the director of the Muséum d'Histoire naturelle, Geneva). **a** - Starch gel electrophoresis of eye of embryo stage 7 showing adult pattern. **b** - Histological section of *retina* of stage 7 embryo showing that at this stage all three types of cones and rods are present and have assumed a light-adapted position. **c** - Histological section of *retina* of stage 6 embryo showing that rods, cones and pigment *epithelium* are present but not yet differentiated. **d** - Electrophoresis reveals that 'eye-bands' are not yet present. Arrow indicates site of application of sample. ^ indicates bands unique to eye and brain. — - cathode; + - anode; IC - inner cones; IPL - inner plexiform layer; MC - middle cones; OC - outer cones; OLM - outer limiting membrane; ONL - outer nuclear layer; OPL - outer plexiform layer; PE - pigment; R - rods; R/C - rods and cones.

Summary

Lactate dehydrogenase is a multilocus enzyme. The C_4-isozyme is primarily expressed in the retinal photoreceptors (and optical parts of the brain) of many teleosts. Developmental studies of the retina of *Poecilia reticulata, Salmo trutta* and *S. salar* revealed that the adult LDH-pattern with the specific C-isozymes is already present in the newly differentiated photoreceptors. It has been suggested that this isozyme may be specially suited for cells, such as the photoreceptors, which show a high aerobic metabolism and that it may play an important role in the regeneration of the visual pigment.

Endnotes

[1] rete mirabile (retia mirabilia),
G. Wundernetz
[2] F. vésicule oculaire
[3] [amacrine= Greek 'without a long fibre', name given by 'Ramon y Cajal (1892)']
[4] Fret 11 is a monoclonal antibody which labels rod outer segments in zebrafish postembryos.
[5] Enzyme that catalyzes the interconversion of pyruvate and lactate.

Chapter eighteen

Mesodermal derivatives

'Si, dans le domaine de l'embryologie ils sont les prix d'efforts plus pénibles, les fruits n'en sont aussi que plus doux de cueillir.'
Henneguy 1888

The mesoderm forms muscles, connective tissue, lateral plate[1] (coelom bounded exteriorly by the somatopleura[2] and interiorly by the splanchnopleura[2]), kidney, blood, blood vessels and mesenchyme. The material destined to form the anlage of the notochord is usually referred to as chorda-mesoderm. The notochord has been dealt with in Chapter 11.

Somites (metameric series of blocklike masses of cells) are constituted by the myotome forming the muscles, the dermatome yielding the cells for the dermis and the sclerotome forming the vertebrae and the sheath of the notochord and eventually vertebral cartilage (Fig. 18/1). Various aspects of the embryology of teleost skeletal muscle formation have been dealt with recently by 'Currie and Ingham (2001)'.

In teleosts the primary lateral plate gives rise to the intermediary cell mass[3] (ICM), the primary nephric ducts and the somato- and splanchnopleura of the secondary lateral plate. The term mesenchyme denotes loose cells which exist outside epithelial layers or are emitted from them (Fig. 18/1). Mesenchymal cells can migrate as already mentioned in the textbook by 'Ziegler (1902)'.

'Mizuno et al. (1996)' suggested that the yolk of the zebrafish has a mesoderm-inducing capacity. This was as a result of placing an animal pole of an embryo on a blastomere-free yolk of another one. 'Ober and Schulte-Merker (1999)' discussed this result and reasoned that the yolk used by these researchers was not completely devoid of blastomeres. To circumvent this problem, 'Ober and Schulte-Merker' used zebrafish recombinants between completely blastomere-free yolk and 'animal cap tissue' of a blastula representing presumptive ectoderm, which showed that the yolk is the source of the mesoderm-inducing signal. While the authors stress the difficulty of removing all blastomeres from the yolk, they do not seem to be concerned about the possible presence, in the cortex of the yolk, of the central periblast[5] nuclei, which are 'daughters' of nuclei from former marginal cells.

18.1 KIDNEY

As shown in a cross-section through the trunk, there is interposed between the somite (epimere) and the lateral plate (hypomere) an intermediate, segmentally-arranged region called nephrotome (mesomere). As the name suggests it gives rise to the kidney (nephros) (Fig. 18/1).

The assumed primordium of all kidney types is called holonephros[7]; its cranial part develops into the pronephros and its caudal part into the opisthonephros (Figs. 18/2, 18/3). The pronephros is situated behind the labyrinth. The pronephric segments are often modified into a pronephric chamber, in the wall of which blood vessels form a glomus (Fig. 18/4). In the amphibia 2-3 pronephric segments are functional in the larval stage only. In the selachians the pronephros is functional in the embryos only. However, in the early stages of teleosts 2-7 pronephric segments remain functional (e.g. in *Salmo salar* 3) while in some the pronephros keeps functioning as part of the kidney, even in mature life. In all so-called higher vertebrates the pronephros remains an embryonic structure.

In teleosts (and *Polypterus*) the connection between opisthonephros and the gonads (as observed in *Acipenser, Neoceratodus, Lepidosteus*) no longer takes place. The opisthonephros is referred to as mesonephros or metanephros in the anamniotes (cyclostomes, amphibia and fish) while in amniotes (reptiles, birds, mammals) the term metanephros is used exclusively. The kidney-duct of anamniota is referred to as archinephric duct.

The basic unit of the kidney is the nephron (tubule). Its connection with artery, afferent and efferent vein and duct is represented in Fig. 18/5. Arterial blood is supplied to the glomerulus in freshwater fishes. The glomerulus is made up of a net of arterial capillaries supplied by a segmental artery coming from the dorsal aorta. In contrast to the kidney of other vertebrates, the teleostean nephrons do not possess a nephrostomal connection with the coelom (Fig. 18/5B-D). The glomerulus is absent in the kidney of marine fishes (Fig. 18/5C,D). In a marine environment this aglomerular condition saves a great amount of water; salt excretion and disposal of nitrogenous waste are now largely taken over by the gills.

Besides waste-disposal and osmoregulation, pro- and mesonephros in teleostei are engaged also in blood formation (see 18.2.2).

Summary

The kidney is formed as a segmentally arranged structure in the nephrotome (mesomere), which is situated between the epi- and hypomere. The kidney of teleosts consists of pronephros and mesonephros, while the pronephros in most other vertebrates is only a transitory embryonic structure. The kidney of marine teleosts is aglomerous and, therefore, does not receive arterial blood, while the kidney of freshwater types contains a glomerulus. In marine fish, which have to conserve water, the excretion of salt and waste is largely taken over by the gills.

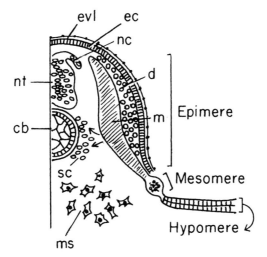

Fig. 18/1 Cross section through trunk region of a teleost. Three regions: Epimere (somites), Mesomere (nephrotome), Hypomere (lateral plate), cb - notochord; d - dermatome; ec - *ectoderm*; evl - envelope (*periderm*); m - myotome; ms - mesenchymal cell; nc - neural crest; nt - neural chord; sc - sclerotome.

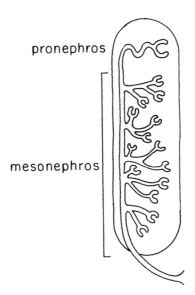

Fig. 18/2 *Pronephros* and *mesophros* of teleostean kidney with archinephric duct. (In teleosts the *testis* does not connect with the kidney and sperm is released by a separate sperm duct) (modified from Portmann, 1976, with kind permission of Schwabe-Verlag, Basel).

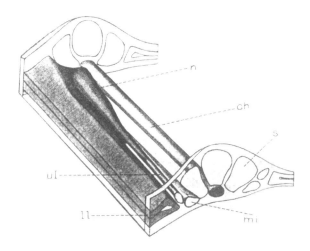

Fig. 18/3 Plastic model of the pronephric anlage in the trout (Felix, 1897). ch - notochord; ll - lateral plate; mi - intermediary cell mass; n - nephrotome; s - somite; uI - archinephric duct.

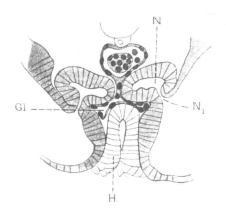

Fig. 18/4 Transverse section through the *pronephros* of the trout embryo (Swaen and Brachet, 1889). H - hypoblast (future *entoderm*); Gl - *glomerulus*; N - pronephritic chamber; N_1 - nephrotome.

The teleostean kidney does not possess a nephrostome, i.e. it does not connect with the coelom. In addition to excretory and osmoregulatory functions the teleostean kidney plays a part in blood-formation (haematopoiesis).

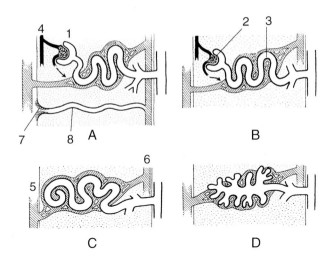

Fig. 18/5 Connection of a *nephron* with *coelom*, artery, vein and kidney duct (Portmann, 1976. with kind permission of Benno Schwabe, Basel). **A** *amphibia* (*anura*). Several nephrostomes (connection between tubule and *coelom*) connect secondarily with veins. **B, C, D** various types found in teleosts. None of them has a nephrostome. **B** normal type **C, D** nephrons without a *glomerulus*. Left = coelom, artery and vein carrying 'caudal blood'. Right = kidney duct and vein transporting blood craniad. 1 - Bowman's capsule; 2 - *glomerulus*; 3 - kidney tubule; 4 - artery; 5 - afferent vein; 6 - efferent vein; 7 - nephrostome; 8 - original kidney tubule.

18.2 HAEMATOPOIETIC SITES
(FOR RED BLOOD CELLS AND VASCULAR ENDOTHELIA)

In all vertebrates the red blood cells (erythrocytes) are the first to appear in circulation, followed much later by the other types of blood cells. In the mature fish blood-formation takes place in spleen, liver, pro- and mesonephros, intestine, pancreas and gonads. Of these, spleen and kidney are the blood forming organs most frequently observed '(Salvatorelli, 1971)'.

In the teleost embryo different haematopoietic sites have been observed. The so-called classical theory of blood formation in teleosts maintains that the first haematopoietic sites are situated intraembryonically in the intermediary cell mass (ICM)[3]. This mass is situated between the segmentally arranged somites and the lateral plates from which it has been separated (Figs. 18/6, 18/7). The finding of haematopoiesis at this location was unexpected since in the embryo of other meroblastic taxa, such as birds and mammals, the first blood cells in circulation are produced extra-embryonically, in the so-called blood islands[4]. However, it was discovered later that in some teleosts, too, the yolksac produces erythroid cells and, moreover, that in others blood is formed both by the ICM and the yolksac blood islands. The first haematopoietic sites, be they the ICM or extra-embryonic blood islands, or both, give rise only to erythroblasts (precursors of the red blood cells).

Before the mature sites come into operation, so-called secondary sites take over. One of these is the endocardium (the cell sheet lining the heart), where the detachment of haemocytoblasts has been observed (Fig.18/8). This is followed by haematopoiesis in the pro- and mesonephros (Fig. 18/2).

18.2.1 First haematopoietic sites

The early investigators realized that the first circulating 'blood' is only a serum[11], which does not carry any blood cells. 'Kupffer (1868)' reviewed critically the results of the early investigators. The haematopoietic sites they observed are the following:

(a) Surface of the yolk

The assumption that in teleosts the surface of the yolk participates in blood formation is very old '(Baumgärtner, 1830; Schultze, 1836; Fillipi, 1841 and Henneguy, 1888)'. 'Von Baer (1835)' thought that the blood cells in fish are formed from the plasma. Subsequently, 'Vogt (1842)' described that in *Coregonus palea* haematopoiesis takes place everywhere where vessels are formed and also at the interior surface of the heart. After consolidation of the walls of the heart and the vessels haematopoiesis becomes restricted to the 'couche hématogène' (haematogenic layer) lying immediately upon the yolk '(Kupffer, 1868; Oellacher, 1873; Henneguy, 1888; Rückert and Mollier, 1906)'. It was later reported by various authors that this 'couche hématogène' is nothing else but the periblast. 'Truman (1869)' mentioned that in the pike, *Esox lucius*, 'blood corpuscles are being detached from the 'yelk surface' where they had their origin and pass to the opening of the heart, even before it has commenced to beat'. 'Goette (1869)' reported for the trout that blood originates from a formation of free cells (periblast cells?) from the yolk. 'Klein (1876)' in his treatise on the trout mentioned that 'it is only natural to draw the inference that within the area of the latter, i.e. the yolksac, notorious for the development of blood and blood-vessels, the 'parablast[5]' is concerned in the formation of blood and blood-vessels'. 'Gensch (1882)' described that in the pike and *Zoarces viviparus* blood cells are formed by the periblast[5]. Similarly, Ryder (1882b, 1885)' reported that in the silver gar, *Belone longinostris*, red blood cells are formed from the periblast (also called hypoblast or secondary entoderm) on the yolk surface. 'Ryder (1884)', in his study on the development of the cod, *Gadus morrhua*, questioned again the function of the periblast with its free nuclei and suggests that it 'assimilates the constituents of the food-yelk in order to convert them into a form suitable for the growth of the cells, in other words, the multinucleated parablast assumes the role of provisional blood'. He continues: 'Inasmuch as we know that there are free nuclei imbedded in this plas-

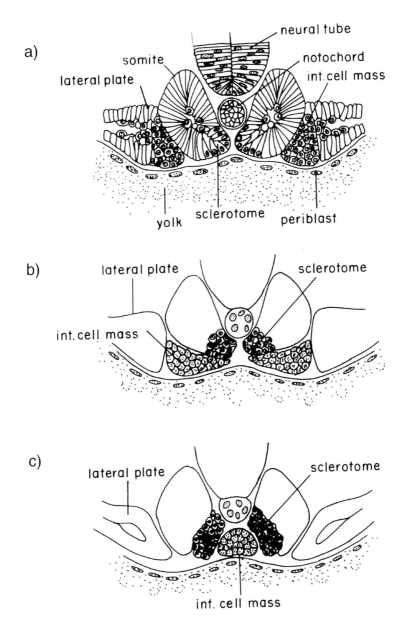

Fig. 18/6 Formation of intermediary cell mass (ICM) in the trout (redrawn from Swaen and Brachet, 1889). **a** - Cross-section through 6th somite of an 11-somite embryo. **b** - Cross-section through 6th somite of a 19-somite embryo. **c** - Cross-section through 6th somite of a 28 somite embryo. For clarity *entoderm* not included in the drawings.

modium or yelk-hypoblast, is it not possible that they may become the means of developing cells for the walls of the vitelline capillaries as well as blood corpuscles? On the inner side of the vessels, the blood cells are seen to lie in immediate contact with the plasmodium or yelk-hypoblast, and it is to be inferred that blood cells are budded off directly into them, the division of the free nuclei in the subjacent plasmodium probably multiplying and giving rise to these blood corpuscles'. In a following treatise 'Ryder (1887)', in the chapter of the common shad, *Clupea sapidissima*, stated 'the vascular network developed over the periblast and in intimate connection with it, in the ova of many species of fishes, is also splanchnopleural and homologous with the area vasculosa or omphalomesenteric meshwork developed over the yolk of higher forms. Of the correctness of this homology I think there can scarcely be any doubt whatever'. 'Mollier (1906)' decided that the view of the above authors who quote the periblast as the original site for haematopoiesis was no longer worth considering. However, it should be stressed here that at that time terms for layers were often confused: e.g. 'Ryder (1884)' maintained that the periblast is homologous with the splanchnopleura of large-yolked forms such as the salmonids, and 'Cunningham (1885b)' wrote that the periblast, after the formation of the intestine, gives rise to part of the splanchopleural mesoblast.

(b) Endocardium

'Aubert (1854, 1856)' observed in the pike, *Esox lucius*, at the time of heart formation, blood cells between the 'ventral plates' (lateral plates) and the yolk. He noted the detachment of the first blood corpuscles and their entrance into the heart (which at the time contracted 64-70 times per minute) and their subsequent exit into the aorta. The cells, which were still colourless, were first spherical, then took up the form of a circular and finally of an elliptical disc. Similar to 'Vogt (1842)' he suggested that blood cells detach themselves from the endocardium and, in fact, from blood vessels wherever these are forming. 'Lereboullet (1861)' maintained that although the first blood cells in the trout appear in the heart, they do not originate from it.

'Kupffer (1874-1876)' reported that in the embryo of the herring, *Clupea harengus*, no blood cells are observed, even after resorption of the yolk. The same results were noted by 'Hoffmann (1881)' for *Scorpaena, Julis* and *Fierasfer*. 'Ryder (1882/1884)' was in agreement with this view. Later, 'John (1932)' confirmed that even the larval stages of the herring, *Clupea harengus*, are devoid of blood. The eggs are deposited on the sea bottom and the nutritive material in the yolksac is absorbed through the colourless vascular system. When the yolk is absorbed, the larva measures 10 millimetres. At a length of 25 millimetres metamorphosis begins and the fish migrates through mid-water to the shore. Only at this stage (some months after hatching) haematopoiesis commences. At the first opaque stage there are only a few erythroblasts in the vascular fluid; at the second transitional stage, when the silvery sheen begins to appear, a large number of erythroblasts are observed and at the third stage, the silvery or 'whitebait' stage, the vascular system contains typical erythrocytes. It has been suggested that the blood cells are liberated from the endocardial lining, then multiply rapidly and become erythroblasts. Similarly, in the leptocephalus stage of the eel, *Anguilla vulgaris*, erythrocytes are completely absent and are only detected after metamorphosis when the by-now opaque larva has reached the shore. 'Hoffmann (1882a,b)' stressed that in the trout, *Salmo fario*, blood corpuscles are formed from the endothelium of the heart (Fig. 18/8). 'Henneguy (1888)', too, stated that the first blood cells of the trout are formed in the heart. 'Engen (1968)' who analyzed the organogenesis of the viviparous *Hyperprosopon argenteum* reasoned that 'since no vitelline circulation develops and the endothelial cells of the heart appear very similar to the early circulating blood cells the first blood cells must be derived from this source'.

(c) Intermediary cell mass (ICM) [2]

Early investigators

Haematopoiesis was reported to take place in the intermediary cell mass (ICM) discovered by 'Oellacher (1873)'. He described it as unsegmented and of mesodermal origin. It separates from the ventral part of the somites (called by him 'Urwirbel') of each side, then fuses in the midline and is absent in the most anterior and most posterior region of the embryo. (Others claimed that the ICM divides off the lateral plate instead.) It later grows towards the median line where it unites with similar cells from the other side resulting in a large median group of ICM. However, 'Oellacher' did not recognize the true functional significance of the ICM. He suggested, without delivering any evidence, that this mass 'sensu verbi plenitore' (in the fuller sense of the word) will form the stroma for the holonephros (kidney primordium[7]) and the intestine. As regards the formation of the kidney he was 'prophetic'. He also described that a small group of cells situated below the notochord may represent the anlage of the aorta. In the cross-sectioned middle trunk this anlage is made up of only 2 cells lying beside each other; more posteriorly, at the level of the otoliths, 3-4 cells are observed. It should be stressed here again that no other vertebrate taxa show a structure homologous to the ICM. 'Hoffmann (1882a,b)' mentioned that he did not understand the above mentioned observations by 'Oellacher'.

'Wenckebach (1885)', observed in the living *Perca fluviatilis* that at the stage of the eye-pigmentation the blood cells 'arise as a solid mass of tissue, situated in the region where afterwards the vena vertebralis (median vein, axial vein)[6] will be found. The central portion of this cylindrical mass of cells gradually loosens, the cells are carried away by the blood-fluid which has appeared in the meantime, acquire a yellow tint, and become the blood-cells. According to him the blood-vessels on the yolksac appear to arise first as lacunae. When he became aware of the PhD thesis of 'Ziegler (1882)', who had reported on 'Oellacher's' ICM in *Salmo* and realized its functional significance, 'Wenckebach (1886)' confirmed that the ICM is the site of origin for red blood cells of the embryos he had analyzed. He studied the origin of the blood cells in histological sections of *Belone acus* and confirmed 'Ziegler's' results in the salmon. In *Perca* he observed the axial vein (vena vertebralis posterior sine cardinalis) from which blood cells become detached. 'Von Kowalewski (1886b,c)' searched in vain for the blood-forming tissue but concluded that it must most probably develop 'at the expense' (G. auf Kosten) of the mesoderm, as 'Ziegler (1885)' had previously suggested.

'Ziegler (1887)' studied blood formation also in the pike. He found that in this species the ICM behind the pronephros forms the aorta while posterior to this region the ICM forms part of the aorta wall as well as blood cells. He did not observe in this fish any passage of blood cells onto the yolk. 'Reinhard (1888)' casually mentioned that the mesoderm moving forward between the epidermic stratum and the periblast furnishes the material for the formation of blood. 'Wilson (1889/91)' mentioned that the ICM is absent in *Serranus atrarius*. Below the subnotochord the aorta is formed by scattered mesoblastic cells the origin of which he could not determine.

As mentioned earlier, it was finally 'Ziegler (1882, 1887, 1892)' who concluded that the ICM in the salmon forms the axial vein/median vein[6] or the cardinal vein with a great number of blood cells. In other words, the peripheral cells of the ICM form the endothelial wall while the inner cells give rise to the blood cells. According to him the aorta is a product of mesenchymal cells of the sclerotome (Fig. 18/1). 'Ziegler' found in the salmon (on day 16) the ICM to extend from the pronephros to the anus. It is formed as a result of the fusion of two lateral cell strips. The fusion runs parallel with the separation of the entoderm from the notochord and subnotochordal cord (see Fig. 12/4). On day 17 the ICM starts to descend, i.e. it invades the surface of the yolk via the dorsal mesentery, migrates round the intestine and below it via the ventral mesentery unto the yolk. The blood cells can also, below the intestine, move from one side to the other (day 18 and 19). During all this time mitoses are frequent. On day 20, when the blood returns via the median vein into the body of the embryo, this vein is shown to send off small vessels unto the yolk.

'Hoffmann (1888)', seemingly unaware of the results of 'Ziegler', mentioned the presence of the ICM in the salmon, *Salmo salar*, and supposed that it is partly engaged in haematopoiesis. 'Henneguy (1888)' suggested that in the trout the ICM arises from the lateral plates and furnishes the cells for the blood vessels (aorta and veins); according to him the first blood cells are detached from the endothelium of the heart. 'Sobotta (1894)' reported that in the salmonids the anlagen of blood, heart, blood vessels and excretory system are evident only when the blastopore is closed. Blood and vessels originate from a 'peculiar cell mass situated under the notochord'. This mass is split off from the median corner of the somites (sclerotome); in the trunk of the embryo it lies between the notochord and the flat cells of the entoderm. Corresponding to its symmetrical origin the mass is seen to be made up of two portions. The outer cells will form the endothelia of the vessels, above all the two venae cardinales and the aorta; the remainder of the cells will be freed later and will form the main mass of blood cells. He stresses that this subchordal mesoderm mass (he does not use the term ICM) is unique within the vertebrates and represents a caenogenetic[8] formation. The entoderm of the head and the most posterior region has formed the gills and Kupffer's vesicle respectively, while in the haematopoietic middle region the entoderm is still represented as a flat layer. In the ventral mesentery the Vena subintestinalis has formed and in the head as well as on the yolksac vessels have appeared. It would seem to him that later any part of the mesoderm could yield blood and vessels. Much later the vascular 'tailbouquet' of 'Lereboullet (1861)' will connect up with the body vessels.

'Kaestner (1895)' in a treatise on the development of the muscles of the body and tail mentioned 'bloodcords' (cf. 'Sobotta, 1902'). 'Felix (1897)' agreed with 'Oellacher' and 'Henneguy' that the ICM originates from the lateral plate. He gave a detailed account of the ICM (called by him 'venous cords') in the trout and how it forms blood cells in a vessel with an endothelial lining (the cardinal vein). He suggested that the aorta is formed from cells of the somites (sclerotome) and that it contains no blood cells. He stressed that the (primary) lateral plate is made up of three parts: the medially situated venous cord, the pronephric duct and the lateral plate in the narrow sense (secondary lateral plate).

'Sobotta (1902)' published an extensively illustrated treatise on the formation of blood in the salmonids. He described the ICM, which he calls 'blood cords' (G. Blutstränge). These start at the level of the 8th somite and extend caudally. The ICM is visible at the 12-13 somite stage and is formed by the medial part of the primary lateral plates (in his earlier publication 1894 it was formed by the somites), remains unsegmented and takes up a subchordal position. The sclerotome will form first the aorta and then the venae cardinales which will enclose the blood cells of the ICM ('bloodcords'). In the ventral mesoderm the vena subintestinalis and in the head region some anterior vessels have been formed. 'Sobotta' suggested that these vessels do not arise in loco but are formed from the major vessels. Somewhat later, also on the yolksac near the embryo, the first extraembryonic vessels filled with blood were observed (their origin has not been established). The blood is not yet coloured. 'Sobotta' was not aware of the publications by 'Swaen and Brachet (1899, 1901)'.

In spite of the results of the above authors, the discovery of the fate of the ICM is usually attributed to 'Swaen and Brachet (1899, 1901)'. The following description will refer to their publication on the trout. The development of the intermediary lamella (1st - 5th somite) and of the ICM (5th somite to caudal end) has been illustrated by Fig. 18/6. The mesoderm in the middle and posterior regions of the trout embryo is made up on each side of two parts, a somite and ventrally to it a so-called primary lateral plate. The latter then divides off a portion immediately adjacent to the somites to form the ICM. Both anlagen of the ICM (Fig. 18/6a) move medially before the formation of the intestine and fuse into a median cell cord. (Fig. 18/6b,c). The fused ICM lies between notochord and intestine, and is laterally apposed by the sclero-myotomes. The lateral plate is now called 'secondary lateral plate'. It is smaller in the anterior region than in the middle region and is referred to as 'intermediary lamella'. In the region of the first 3 somites the ICM forms only the aorta, while the caudal ICM furnishes the aorta, the vena cardinalis (median vein, G. Stammvene) and the red blood cells. In the 26-somite embryo the remaining solid cell cord, situated ventrally to the aorta, moves in the region between the 14th and the 16th somite along both sides of the intestine unto the yolk where it expands under the splanchnopleura. The by-now loose spherical cells come to lie in unlined grooves of the yolk syncytial layer. The cells divide mitotically and become red erythrocytes, which stay in communication with the median vein within the embryo.

There is a strong possibility, as admitted by 'Swaen and Brachet (1901)', that in the trout wandering mesenchymal cells form blood cells on the yolksac. It should be stressed here that 'Swaen and Brachet' have also called attention to a mass of cells derived from the heart anlage lying beneath and outside the heart endothelium. This mass of cells has been claimed to wander from below the pericardium and give rise to vessels and blood on the yolksac. However, 'Stockard (1915a)' stressed that he has seen nothing in his studies, which would indicate that any cells left over from the heart formation had wandered upon the yolk or given rise to blood cells or vascular endothelia.

While 'Swaen and Brachet (1899)' claimed that in the trout the axial mass is continuous with the vascular layer in the head region and that it is from this layer of the ICM in close association with the splanchnic mesoderm that the heart, blood cells and vessels originate, 'Mahon and Hoar (1956)' were not able to establish the continuity between the ICM in the head region (yielding the heart) and the trunk region (yielding blood cells, aorta and other vessels) in the salmon, *Oncorhynchus keta*. 'Swaen and Brachet (1899)' stated that the head mesoblast can be separated into three portions 1) the ICM close to the top of the pharynx 2) the lateral plate split into two lamellae and 3) the general head mesoblast close around the brain. 'Gregory (1902/1903)' was unable to find the ICM in the trout, *Salmo salar, S. alsaticus*, carp and pike. He seems to interpret the mesoderm as entoderm.

'Swaen and Brachet (1901)' subsequently investigated the development of the ICM in *Leuciscus* and *Exocoetus* and found that the massively developed ICM furnishes not only the aorta and the axial veins but also the red blood cells. However, in the pelagic species *Clupea sprattus, Rhombus (spp.), Solea vulgaris, Pleuronectes microcephalus, Trachinus vipera, Caranx trachurus* and *Callionymus lyra* the ICM is very thin in the anterior trunk region. Sometimes only single cells are visible. In the middle and caudal part the cells are more numerous (Fig. 18/7a,b) but an ICM mass similar to that found in trout, *Leuciscus cephalus* and *Exocoetus volitans* is never encountered. Once the cells have moved medially, they form the aorta and the cardinal veins (Fig. 18/7c). No blood cells are encountered at this stage. The blood cells arise only after the embryo has hatched and is free-swimming. The late formation of blood in pelagic eggs had been already referred to earlier by 'Kupffer (1878)' and 'Cunningham (1891)' and was stressed again by 'Stockard (1915a)'. 'Brook (1884)' had observed in the pelagic eggs of *Trachinus vipera* that neither blood corpuscles nor a circulation are visible up to

3 or 4 days after hatching. Pelagic eggs are very small and develop rapidly. Their anterior ICM is relatively very small and gives rise to the aorta; the ICM of the middle and caudal part is larger. At the time of hatching the vessels are still hollow and the blood cells have not yet appeared. 'Ziegler (1887)' and 'Felix (1897)' differ from 'Swaen and Brachet (1901, 1904)' in that they derive the aorta from the sclerotome, which under the notochord forms a mesenchymal aortic string. 'Derjugin (1902)' reported that in the pelagic eggs of *Lophius* the cells forming the aorta and venae cardinales are given off by the sclerotomes [cf. 'Mollier (1906)']. The ICM in the pelagic *Belone acus* is very small according to 'Sobotta (1902)'; there is no cardinal (axial) vein and the blood returns via the subintestinal vein onto the yolk. In contrast, 'Borcea (1909)' described that the blood of *Belone* is formed in an ICM, which is highly developed.

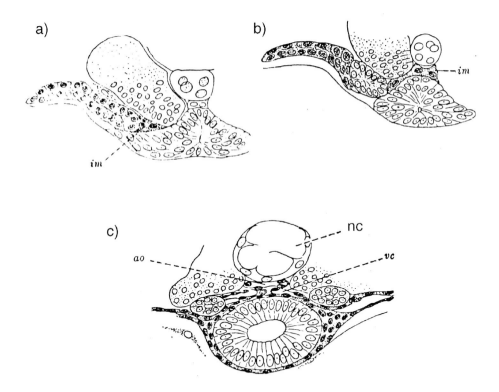

Fig. 18/7 Formation of the sparse intermediary cell mass (ICM) characteristic for a pelagic embryo (*Solea vulgaris*) Swaen and Brachet, 1901. **a** - Cross-section through 5th somite. **b** - Cross-section through 5th somite of an embryo older than a). **c** - Cross-section of the anterior trunk region of a newly hatched sole. ao - *aorta*; en - *entoderm* forming intestine; im - intermediary cell mass; nc - notochord; vc - *vena cardinalis*.

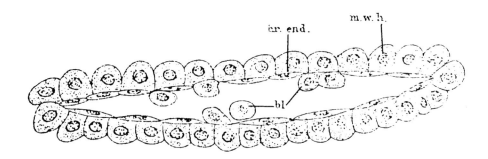

Fig. 18/8 Blood forming *endocardium* in primordial heart (Hoffmann, 1881). bl - blood cells; hr. end - heart *endothelium*; m.w.h. - myocardium.

'Marcus (1905)' wrote an essay on the blood formation in teleosts. His own observations were carried out on *Gobius capito*. He mentioned that in this species the highest development of the 'bloodmesoderm' is observed postanally. He also observed it in the same position in the tail of a 20 day old trout. [The postanal gut subsequently atrophies together with Kupffer's vesicle '(Wilson, 1889/1901)']. 'Stolz-Picchio (1933)' reviewed the results of early authors on the different haematopoietic sites with the exception of the blood islands.

Later investigators
'Gihr (1957)' in her thesis on the development of the pike, *Esox lucius*, mentioned that in the 35-45 somite stage the first blood corpuscles disperse over the whole yolk surface. She concluded on the basis of the results by 'Ziegler (1887)' that they originate from mesoderm cells left over after the formation of the endothelium of the heart. According to 'Ziegler' there is principally no difference between these cells and those originating from the ICM which wander onto the yolksac. Most of them will form pigment cells. 'Bielek (1974)' confirmed that in the pike the ICM separates from the primary lateral plates at the 30-35 somite stage. It forms the primary blood cells and the anlage of the venae cavae posteriores. At the stage of 30-40 somites blood cells migrate unto the yolksac. The onset of the blood circulation occurs at the 55-60 somite stage. She differentiated the various stages from the primitive blood cells to the erythrocytes (Table 2).

'Mahon and Hoar (1958)' traced the development of the blood cells, the aorta and other vessels from the ICM in the chum salmon, *Oncorhynchus keta*.

'Bielek (1974a)' studied the ontogenesis of haematopoiesis in the grayling, *Thymallus thymallus*. The primordial primitive blood cells are found in the ICM of the same somitic stage as in other salmonids. Their size decreases significantly from the 30 to 45 somite stage when they measure 9 x 7.5 microns. At the 50-60 somite stage typical haemoblasts with a strongly basophilic cytoplasm are

Table 2 Sequential appearance of red blood cells in the developing pike *Esox lucius* (modified after Bielek, 1975, with kind permission of Urban & Fischer Verlag, Jena)

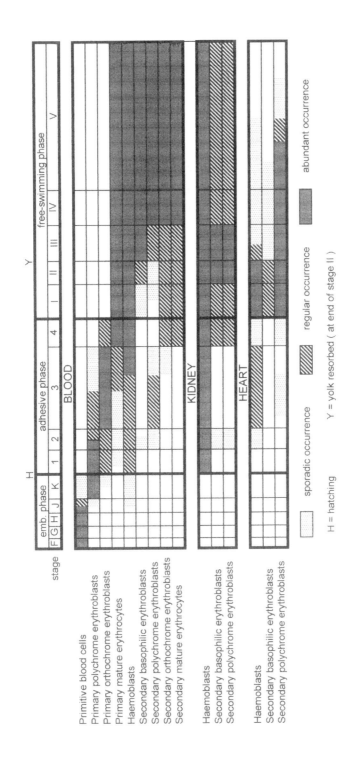

observed. Further development into primary erythrocytes, and in the caudal part also into haemoblasts, takes place in the now circulating blood. Shortly before hatching the majority of the erythrocytes display an orthochrome stainability whereas the erythrocytes are mature only shortly before the postembryos take up exogenous food. At this stage precursors of secondary erythrocytes begin to appear and immature erythroid cells become very numerous later (Table2).

In *Brachydanio rerio* (as well as in *Gasterosteus aculeatus* and *Pungitius pungitius*) the blood was made visible by the benzidine and hydrogen peroxide technique in the central part of the embryo in the middle somitic region by 'Colle-Vandevelde (1963b)'. Similarly, 'Rieb (1973)' observed that in *Brachydanio rerio* the first visible blood cells are erythroblasts along the axis of the embryo. They are not yet mobile and are situated below the notochord and are accumulated in the anal region. At the age of 23 hours erythroblasts suddenly move from the myelencephalon region onto the upper part of the yolksac. Since the heart has not yet started to beat, the author concluded that the erythroblasts actively migrate to colonize the yolksac. 'Al-Adhami and Kunz (1977)' agreed that the ICM is the first haematopoietic and vasculogenetic site in the zebrafish. Blood cell formation starts very early (12 somites), i.e. before the cells have reached the midline to form the ICM. This is in contrast to reports on any other ICM embryo. Proliferation within this site is very rapid; clusters of small lymphoid haemoblasts appear and no stem cells[9] (basophilic haematocytoblasts), fixed or free, are observed (though stem cells are readily observed in the subsequent haematopoietic sites). At the 12 somite stage also some amoebocytic cells are visible on the dorsal side of the anterior yolksac. These cells have migrated from the mesenchyme of the presomitic region as observed by 'Colle-Vandevelde (1963b).' At the 18-somite stage the posterior yolk-mass has become partially constricted. In the anterior trunk region (somites 1-5) the sparse ICM from each side have migrated towards the midline and come to lie between the hypochord (subnotochordal rod) and the undifferentiated entoderm (compare with Fig. 12/4). In the mid-trunk region (somites 6-16) the number of cells in the ICM has increased greatly. They show already signs of differentiating into primordial blood cells, prior to their mediad migration. They disaggregate and become round. Mitosis is commonly observed. These cells represent 'small lymphoid haemoblasts'(Fig. 18/9a). At the level of somites 16-18 the ICM have not yet separated from the lateral plate. Apart from the amoeboid cells mentioned at the 12-somite stage, no mesenchymal cells are encountered on the yolksac. At the 24-somite stage the posterior sac has elongated into a cylinder. Anteriorly (somites 1-7), the ICM have formed endothelia (future aortic roots). In the mid-trunk region (somites 8-19) endothelia of the future dorsal aorta and axial vein have appeared. Between the vessel primordia, clusters of haematopoietic cells are observed. The majority of these are in the small lymphoid haemoblast stage and are partially surrounded by endothelial cells (Fig, 18/9b). The pronephric duct has acquired a lumen. Scattered ICM are found at the level of somites 20-21. At the 28-somite stage the heart beats intermittently. In the mid-trunk region the dorsal aorta overlies the axial vein which is filled with primitive blood cells. Most of these are in the proerythroblast stage. Compared with the small lymphoid haemoblasts (Fig. 18/9a), their nucleo-cytoplasmic ratio has increased; their nucleus is more compact and its membrane more prominent (Fig. 18/21a). On the yolksac amoeboid cells are no longer apparent. There is no indication of a blood-island; the yolksac ectoderm directly overlies the periblast. In contrast, 'Kimmel *et al.* (1995)' in their extensive publication on the stageing of embryonic development of the zebrafish mentioned for the 20-somite stage the existence of a blood-island below the last three somites and above them the rudiment of the pronephric duct (still devoid of a lumen).

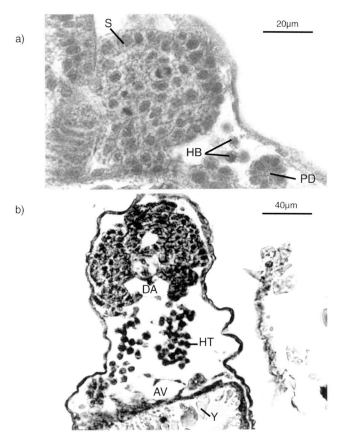

Fig. 18/9 Haematopoiesis in the zebrafish, *Brachydanio rerio* (Al-Adhami and Kunz, 1977, with kind permission of the director of the Muséum d'Histoire naturelle, Geneva). **a** - 18 somite embryo. Cross-section through 15th somite. Some of the IMC have differentiated into small lymphoid *haemoblasts* prior to migrating medially. **b** - 24 somite embryo. Transverse section through posterior trunk region. Dorsal *aorta* and axial vein are formed, with haematopoietic tissue between them. AV - axial vein; DA - dorsal *aorta*; HB - lymphoid *haemoblast*; HT - haematopoietic tissue; PD - pronephric duct; S somite.

Diagrammatic drawings depicting the above-mentioned developmental stages according to 'Al-Adhami and Kunz (1977)' are given in Fig. 18/10a-c.

EM investigations of early haematopoiesis and developing lymphoid organs in the zebrafish were reported by 'Willett *et al.* (1999)' from 5 somites through three weeks of development.

Brachydanio rerio has become a model organism for the genetic study of vertebrate development. In situ hybridization studies have followed the expression of early vascular and haematopoietic markers (the transcription factors GATA-1 and GATA-2) '(Zon, 1995; Detrich *et al.*, 1995)'. Fig. 18/11A-D displays the expression of GATA-1 during zebrafish ontogenesis. Epiboly in this species is very rapid: it is accomplished before the formation of the first pair of somites. Moreover, 'Kimmel *et al.* (1995)' considered that involution as a cell layer may not occur at all. Instead, cells near but usually not just at the margin move to deeper positions as individuals in a process called 'ingression'. Already at the 2-somite stage putative progenitors are expressed as two paraxial stripes flanking the posterior embryo (Fig.18/11A). The stripes have converged medially by the 5-somite stage (18/11B). They meet anteriorly, but posteriorly remain separate still at the 18-somite stage (compare Fig.18/11C with 18/9A). Complete fusion of the two stripes is present by the 24-somite stage (compare Fig. 18/11D with Fig. 18/9B).

The zebrafish as a vertebrate model of haematopoiesis has been reviewed by 'Bahary and Zon (1998)' and 'Amemiya (1998)'.

(d) Blood islands [4]

Other investigators concluded that the blood islands are the primary site. In the yolksac of teleosts the mesodermic layer (lateral plate) is largely, if not entirely, absent. The ectoderm lies directly on the periblast. However, between these two layers mesenchymal cells have been noticed, though these form groups and never seem to be organized into a continuous layer. According to 'Stockard (1915a)' the existence of the erythrocytes is very pronounced in the blood islands, where aeration is no doubt considerably better than in the ICM.

The observations of 'Kupffer (1868)' on the embryos of *Gasterosteus aculeatus* and *Spinachia* agreed with those of 'Aubert (1856)', that the future pigment and blood cells occur together, under the mesoderm, on both halves of the yolk. The increase in the number of blood cells is more pronounced on the right half of the yolk.

'Colle-Vandevelde (1962)' found that the first blood islands and vascular elements in *Pterophyllum scalare* appear on the yolksac around the closing blastopore (or terminal bud[10]). The ICM is poorly developed and furnishes only endothelia for blood vessels. Her observations terminated before the onset of heartbeat and the maturation of blood cells. Subsequently, 'Al-Adhami and Kunz (1976)' confirmed that in this species the first blood islands are found on the yolksac at the 6 somite stage (Fig.18/12) and form blood cells as well as endothelial cells while the ICM furnishes only axial blood vessels. They further observed that cells

Mesodermal derivatives 395

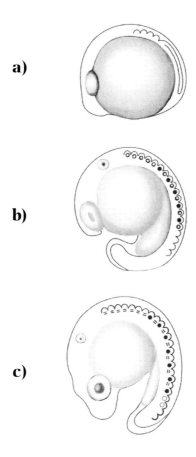

Fig. 18/10 Diagrammatic drawings to show fate of intermediary cell mass (ICM) in the zebrafish *embryo* (modified after Al-Adhami and Kunz, 1977, with kind permission of the director of the Muséum d'Histoire naturelle, Geneva). **a** - 5 somite stage, ICM not yet separated from lateral plate; **b** - 18 somite stage; **c** - 24 somite stage. oooo - undifferentiated; = = = - vasculogenetic •••• haematopoietic.

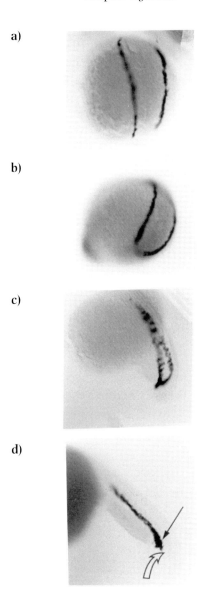

Fig. 18/11 Expression of GATA-1 mRNA (early vascular and haematopoietic markers) as a result of *in situ* hybridization studies in whole zebrafish embryos (Detrich *et al.*, 1995, with kind permission of the National Academy of Sciences, U.S.A.). **a-b** - two stripes of paraxial *mesoderm* of the posterior embryo expressing GATA-1. The two cell masses expressing GATA-1 converge medially between the 2 and 5 somite stage. **c** - At the 18 somite stage the stripes meet anteriorly. **d** - At the 24 somite stage they fuse completely. Closed arrow indicates posterior boundary of GATA-1 expression in the IMC. Open arrow points to undifferentiated, GATA-2 expressing progenitors in the posterior IMC.

of the ICM, in addition to producing axial blood vessels, appear to form, or at least to be incorporated into, the connective tissue coat of the alimentary canal. Blood islands on the yolksac of Pterophyllum scalare form only stem cells (basophilic haemocytoblasts) with finely-granular cytoplasm and a large vesicular nucleus (diameter 4.5 microns) containing 1-3 nucleoli. The cell shape is elliptic (5.5 microns by 8.5 microns). When the blood islands extend anteriorly, the posterior parts do not meet but are separated by a mesh of capillaries. In the anterior parts the cells destined to become endothelial cells (a) and those destined to differentiate into blood cells (b) may be distinguished: (a) The peripheral cells next to the ectoderm, as well as cells scattered within the island, have increased considerably in size (10-12 microns). They have a vesicular nucleus and faintly staining basophilic cytoplasm. Some of these cells contain several vacuoles, which may coalesce. The cells finally organize themselves into an epithelial sheet, with the nuclei positioned basally and the vacuoles apically, with respect to the forming capillaries. (b) This cell type is smaller (8x10 microns) with strongly basophilic granular cytoplasm and a spherical nucleus with a large central or slightly eccentric nucleolus. Mitotic figures are frequent. This is a free haematopoietic stem cell, which will eventually become surrounded by the capillary endothelium (a). In the posterior, older part of the island, capillary endothelia are now fully formed. They enclose free stem cells as well as proerythroblasts (which correspond to eosinophilic erythroblasts of mammals). The proerythroblasts are spherical cells with a round nucleus containing one or two nucleoli. The cytoplasm is less basophilic than that of free stem cells. Once released into the circulation, they differentiate and mature into spherical, disc-like erythrocytes [referred to as embryonic erythrocytes (erythrocytes-E)].

'Colle-Vandevelde (1963a,b)' studied the blood anlage with the aid of benzidine and hydrogen peroxide in various teleosts. In *Lepadogaster candollii* a ring-shaped area under and behind the terminal bud stained at the 11-somite stage. In the 11-somite stage of *Betta splendens* the blood anlage was exclusively on the anterior part of the yolksac surrounding the head and continuous with the heart primordium.

'Colle-Vandevelde (1963a,b)' studied also the vasculogenesis and haematogenesis in *Blennius gattorugine* on the basis of histological examination. With the application of benzidine to embryos in toto she visualized the presence of haemoglobin. The main vessels appear late in the embryo, i.e. at the 11-somite stage (closure of the blastopore at 2-somite stage). They are thought to take their origin from the sclerotomes since 'not a single embryo at anyone stage of development ever showed [a] trace of an ICM or a vascular layer between the dorsal and lateral mesoderm'. Blood-islands are exclusively extra-embryonic, in a crescent-shaped area on the yolksac, in the vicinity of the terminal bud. They are shown to contain haemoglobin at the 11-somite stage. At the same stage blood-cell differentiation extends laterally to the embryo in an anteriorly-directed movement upon the surface of the yolk up to the level of the second somite.

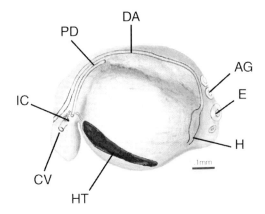

Fig. 18/12 Embryo of the angelfish *Pterophyllum scalare*. Reconstructed polystyrene model of a 16 somite embryo displaying location and extent of blood islands. (Heart transferred to right in drawing.) (Al-Adhami and Kunz, 1976, with kind permisision of Springer-Verlag, Heidelberg). AG - adhesive gland; CV - caudal vein; DA - dorsal *aorta*; E - eye; H - heart; HT - haematopoietic tissue; IC - intermediate cells; PD - pronephric duct.

(e) Intermediary cell mass (ICM) and blood islands

Several early workers mentioned that wandering mesenchymal cells do migrate from the embryo to the yolksac where they form isolated blood cells or small groups as e.g. 'Swaen and Brachet (1901)' mentioned above under (c). However, they considered this blood formation insignificant in amount as compared with haematopoiesis in the ICM.

'Stockard (1915a,b)' was the first to state definitely that the blood in a teleost may arise in two different localities, i.e. the ICM and the blood islands on the yolksac. He studied normal *Fundulus heteroclitus* embryos and also those in which the blood circulation had been arrested with chemicals. This treatment was based on the assumption that it is only safe to conclude that the sites, in which bloodcells are found, are the loci of their origin, in embryos in which the blood fails to circulate. The chief place of the formation of erythroblasts and vessels is the ICM, which extends from the level of the anterior portion of the kidney to the tail of the embryo, and the yolksac blood islands, which are formed by the migration of wandering cells at a very early period. He reasoned that since there is no true mesodermal layer in the yolksac wall of teleosts, the formation of blood islands must be brought about by wandering cells. When epiboly is almost complete, many cells migrate away from the caudal end of the embryo. The wandering cells include future black and brown chromatophores (Fig. 18/13 left) and elongate

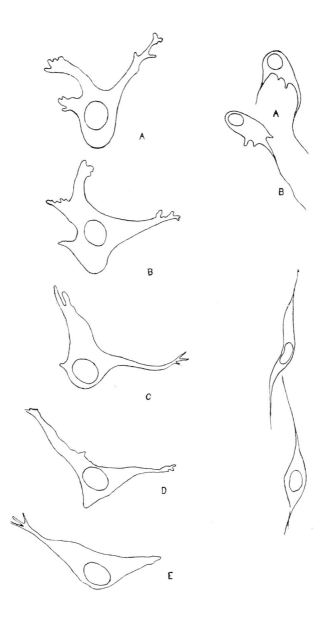

Fig. 18/13 Early migration of mesodermal cells from the edges of embryonic shield and from the germ ring of *Fundulus heteroclitus* embryos (Stockard, 1915c). Left: Camera outlines of one cell drawn at 5 minute intervals (A to E). The cell is a migrating future *chromatophore* in an embryo 50 hours old. Right: Camera outlines of wandering mesenchymal cells (48 hours old), all of the future endothelial type. A and B are two outlines of the same cell at a 6 minute interval.

spindle cells with thin cytoplasmic processes which are assumed to be putative vascular endothelial cells (Fig. 18/13 right). Prevention of blood circulation reveals that the blood cells which have developed on the yolksac blood islands never occur so far anteriorly on the ventral surface of the yolksac as to reach the venous part of the heart. According to 'Stockard' this phenomenon is not to be confused with the publication by 'Swaen and Brachet (1901)' who reported that cells are pushed out laterally from the embryo as branches from or portions of the ICM. In contrast, the 'true blood islands' as observed in *Fundulus*, usually occur at a great distance from the embryo.

'Stockard (1915b)' also observed that blood cells do not normally divide when completely enclosed by vascular endothelium. This fact, he concluded, is the key to the shifting series of haematopoietic organs during development. 'Stockard (1915c)' subsequently studied in normal embryos of *Fundulus heteroclitus* the wandering mesenchymal cells on the living yolksac and especially their development into chromatophores, vascular endothelia and blood cells (Fig. 18/14a,b). The wandering cells stemming from the ICM finally come to lie on the posterior and ventral surfaces of the yolksac where they form blood islands. Few, if any, reach the anterior regions before the circulation begins. The future endothelial cells wander out from the caudal end and side of the embryonic body. The mesodermal layer of the teleostean yolksac is, therefore, represented by numerous separately wandering mesenchymal cells. The putative blood cells are easily differentiated from other migratory cells on the yolksac. For instance, chromatophores are large amoeboid cells. In contrast, the putative blood cells are small and more or less circular in outline and move more slowly than the other types; they also display short thick protruding processes. Whereas the spindle shaped future endothelial cells and chromatophores wander away from the embryo along its entire lateral border as well as from the caudal end, the putative blood cells leave the embryo near the tail bud[10] (terminal bud). As the tail is formed, the cells emanate from the place of union between the ventral wall of the tail and the yolksac.

The presence of blood islands is easily observable in living *Fundulus* embryos, in normal specimens as well as in those with no circulation. The vascular net can be readily made out since pigment cells arrange themselves along the vessels (Fig. 18/14a,b). A change from endothelial cells to blood cell types, or vice versa, has never been observed.

The early yolksac is non-vascular i.e. none of the blood masses is covered by endothelium. Only later endothelial walls are formed around these masses and a vascular network is established. Regarding the formation of the vessels, 'Stockard (1915b)' discovered that the endothelium of the cardinal veins and aorta arises from mesenchymal cells surrounding the ICM (These cells are different in nature from the cells actually constituting the mass). Whether the aorta is derived from the ICM or from the sclerotome has not been settled.

A study of sections showed that the cord of mesoblast which has been designated as ICM leads caudad to the 'end-bud', which is 'well out in the tail'. The ventral cells from this mass wander away into the yolksac from the extreme

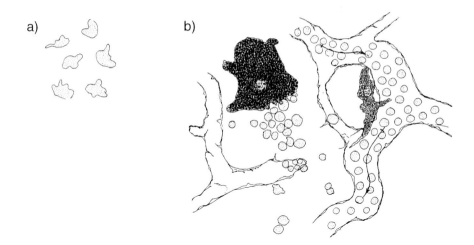

Fig. 18/14 Wandering cells on the yolksac of *Fundulus* embryo. (Stockard, 1915c). **a)** - A group of six early erythroblasts, not yet surrounded by *endothelium*, on the yolksac of a 90 hour *Fundulus* embryo. Short amoeboid processes project and the cells move very slowly. **b)** - Sketch of the yolksac of the same 90 hour embryo as in a). All four derivatives of the wandering mesenchymal cells are shown. The circulation is established in the vessel to the right and the current follows the direction of the arrows. The small vessel to the left is not yet connected with the current. Its wandering endothelial tip is approaching a group of erythroblasts not yet enclosed by *endothelium*. They lie near a large black *chromatophore*. A brown *chromatophore* is shown on the large vessel.

caudal position and other cells also wander away laterally from the ICM (Fig. 18/16). These cells group themselves into small clumps and are to give rise to erythroblasts or future red blood cells on the yolksac. Differentiation is very slow and just before the onset of the circulation the cells are seen to be circular erythroblasts. The complete change from the wandering state into typical haemoglobin-bearing blood cells may be followed on the living yolksac. The wall of the early vessels is very irregular with spaces between the component cells. Blood cells are often caught in these spaces, a condition which in sections would appear as though the blood cell actually formed a part of the endothelial wall and might incorrectly be interpreted as endothelial cells changing into blood cells. A conversion from endothelial to blood cell, or vice versa, has never been observed. The low number of cells emanating from the head-region is thought to give rise to either the heart or the pericardial wall (Fig. 18/15).

'Reagan (1915/1916)' was very critical of 'Stockard's' results. He reported that the head mesenchyme and the heart are haematopoietic. He did not mention the name of the species he analyzed and refers simply to 'an embryo'.

In contrast to the results presented for the rainbow trout by 'Sobotta (1902)', 80 years later 'Iuchi and Yamamoto (1983)' established in the same species the

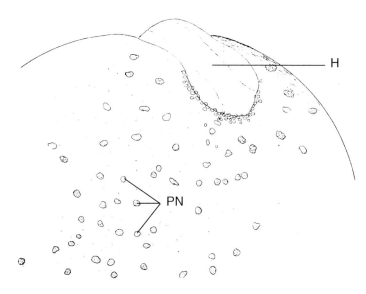

Fig. 18/15 Outline of the head end of a 56 hour *Fundulus* embryo. Scarcely any wandering mesenchymal cell is observed in this region. Large *periblast nuclei* are seen scattered over the yolk surface (Stockard, 1915c). H - head end; PN - *periblast nuclei*.

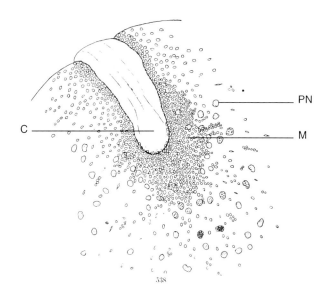

Fig. 18/16 The caudal end of the same embryo as above. Note the great contrast in the abundance of out-wandering mesenchymal cells (Stockard, 1915c). C - caudal end; M - mesenchymal cells; PN - *periblast nucleus*.

presence of erythropoietic organs by histochemical and immunochemical methods in both sites: The first erythropoietic site appears in blood islands scattered over the yolksac (7th embryonic day). Shortly afterwards (7.5-8th day) the region of the ICM becomes reactive but less so than the blood islands. The reactivity in both sites increases up to the 12th day, and is lost after the 15th day (5th day before hatching). So-called larval erythrocytes (Le) originate in both sites, mainly from the 7th to 12th day. They are disc-shaped and contain larval haemoglobin. Their increase in number during the pre-hatching phase is thought to be due to mitotic divisions in the peripheral blood from the 12th to the 15th day. Erythropoiesis for ImA (immature-adult)-erythrocytes starts in the kidney and the spleen one day before hatching and is correlated with haemoglobin switching.

'Colle-Vandevelde (1961, 1963b)' reported that at the 2-somite stage of the live-bearing *Lebistes reticulatus* (*Poecilia reticulata*) erythropoiesis takes place at the border of the blastodisc. It rapidly extends towards the posterior part of the embryo where it makes contact with the ICM. Haematopoiesis in the ICM is observed only at the 5-somite stage which yields an inverted Y anlage the long axis of which is constituted by the ICM and the branches by the blood islands on the yolksac. Blood circulation starts at the 15-17 somite stage. The data are based on histological analyses and treatment of living eggs with benzidine and hydrogen peroxide. 'Al-Adhami (1977)' in his thesis noted that in *Poecilia reticulata* before formation of somites has started future haematopoietic stem cells[9] are found extended over the yolksac to the level of the germ ring. At the 4-somite stage blood islands are formed, occasionally surrounded by endothelia. At the 9-somite stage the ICM has differentiated into the endothelium of the future dorsal aorta, the lumen of which is very small. By the 12-somite stage capillaries extend to the ventralmost part of the yolksac in the head region. Posteriorly, they are limited to the dorsal hemisphere. The capillaries enclose blood cells, which may be termed small lymphoid haemoblast-proerythroblasts. The yolksac capillary network is in intimate contact with the maternal ovarian circulation. By the 15 somite stage the haematopoietic activity of the yolksac has declined sharply and a well defined haematopoietic tissue has developed intraembryonally between the dorsal aorta and the newly formed axial vein posterior to the third somite (ICM). Circulating blood cells are at the erythroblast stage, of a spherical shape with a diameter of 6-9 microns. The spherical nucleus has a prominent nuclear membrane. Nucleoli vary from three to several and coarse chromatin particles are present. When the eyes become pigmented and pectoral fin buds are formed, haematopoiesis in the yolksac has ceased and continues in the embryo between the dorsal aorta and the axial vein. The circulating blood cells are at the early polychromatocyte stage.

(f) Extended blood mesoderm

'Seelbach and Kunz (1976)' from their observations of the development of *Poecilia reticulata* drew conclusions, which were different from those of 'Colle-Vandevelde (1961, 1963b)' and 'Al-Adhami (1977)'. According to 'Seelbach and Kunz' haematopoiesis takes place in three consecutive sites: 1) At the 0-2 somite stage praecranial blood islands are observed in the anterior germ ring (Fig. 18/17a). Their cells originate from the rostral mesenchymal cells within the head region of the embryo (Fig. 18/18a) and migrate in groups rostrally and laterally between yolk syncytium and extraembryonic ectoderm (Fig. 18/17c). These cells are proerythroblasts with a small nucleus and lightly basophilic cytoplasm. The migration of the cranial mesenchymal cells coincides with the onset of the formation of the headfold (see Chapter 20, Figs. 20/33a,b). 2) Once the headfold is formed, a migration of these cells from the head-region is no longer possible '(Kunz, 1971a)' (Figs. 18/18b, 20/33c). At the 3-somite stage posterior blood islands, containing cells differentiating into proerythroblasts, are found at the posterior germ ring on each side of the 'tail-bud' of the embryo (Figs. 18/17b, 18/19a,b,c). These islands are carried by the germ ring caudally and laterally (Fig. 18/17c), so that at the 6-7 somite stage they establish contact with the praecranial blood islands. They form now the 'ring of blood islands along the whole germ ring' as formerly described by 'Colle-Vandevelde (1961a,b)'. They also establish contact with the embryonic mesoblast. 3) At the 13-somite stage (18/17d) the posterior anlage, now surrounded by endothelia, merges with the posterior arms of the aorta. This corresponds to the Y-shaped anlage described by 'Colle-Vandevelde (1961a,b)' (Fig. 18/20a-c). 4) In the 13-somite embryo the ICM in the region of somite 1-6 has formed the aorta and from somite 6-13 the aorta appears filled with many blood cells. At this stage the aorta has not yet any connection with the heart. It is obvious that the blood formation in the embryo lags behind that of the blood islands. The results of 'Colle-Vandevelde' confirm this; she has shown with the benzidine-peroxidase reaction that at the 12-somite stage the embryo contains immature erythroblasts whereas mature blood cells are encountered on the yolksac. As soon as the blood circulation starts (14/15-somite stage according to 'Kunz (1964, 1971a)', the blood islands disappear from the yolk. At this stage the ICM forms, in addition to blood cells, the axial vein, which lies ventral to the aorta.

'Seelbach and Kunz (1976)' suggested that all three blood primordia originate from the peripheral mesoblast which, therefore, they call 'bloodmesoderm' (a term borrowed from 'Marcus (1905)'. The cells which involuted first will form the cranial mesenchyme, several cells of which leave the embryo again to form the praecranial blood islands (Fig. 18/17a) (and perhaps also the endothelium of the heart). This is followed by the involution of the cells which yield the ICM (Fig. 18/17b). Then follows the involution of the peripheral mesoderm cells that form the posterior blood islands which remain in contact with the involuted ICM. The last involuting cells will aggregate in the tailbud. The extension of one primordium into three differently located blood anlage is a striking example of physiological adaptation. The intra-ovarian development at high temperatures makes adjustments necessary which guarantee an adequate oxygen supply to the developing embryo. The extensive blood anlage is the first link in a chain of transitory organs (see 'Kunz, 1971a'). The situation in the guppy makes also possible a morphologi-

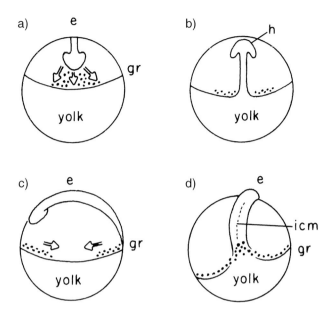

Fig. 18/17 Diagrammatic presentation of haematopoiesis in the extended blood-mesoderm of the guppy *Poecilia reticulata* (Seelbach and Kunz, 1976). **a** - praecranial blood islands in the anterior germ ring (0-2 somites); epiboly 30%. **b** - posterior blood anlage (3 somites); epiboly has almost reached equator. **c** - lateral view of *embryo*. Cells of a and b approach each other (6-7 somites); epiboly 50%. **d** - cells of the advancing germ ring establish contact with the ICM of the embryo now extending into the posterior arms of the *aorta*. (13 somites); epiboly 2/3. Y-shaped primordium of ICM fuses with putative blood cells of germ ring. This corresponds to Y-*primordium* reported by 'Colle-Vandevelde (1961a,b)'. **e** - embryo; gr - germ ring; h - head; icm - intermediary cell mass.

cal and tentatively physiological interpretation of teleost embryos, which have either the ICM or blood islands as sole haematopoietic sites. It also exemplifies how the variation of early blood circulation (see below) is closely linked with the location of haematopoietic sites.

On that basis, in the 'ICM embryos' (example: trout) all of the peripheral mesoderm enters the embryo to form the ICM, which produces the vessels of the body (30 somites at closure of blastopore). Subsequently, migratory cells are sent off onto the yolksac where they form blood and vessels. In 'blood-island embryos' (example: angelfish, *Pterophyllum scalare*) early on (6 somites) one part of the involuting mesoderm remains on the yolksac to contribute the blood islands while subsequently another part enters the embryo to form the ICM, which, however, supplies only endothelia for blood vessels. In 'mixed (ICM + blood island) embryos'

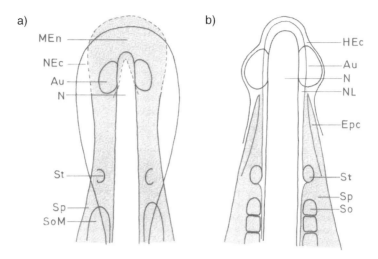

Fig. 18/18 Comparison of head region of presomitic and 3-somite guppy embryo (corresponding to diagrams 18/17a,b) (Seelbach and Kunz, 1976). **a** - Praesomitic embryo showing the extensive rostral *mesentoblast* (grey) from which putative blood cells are migrating to the germ ring. **b** - By the 3-somite stage (end of ectodermal headfold formation) the rostral *mesentoblast* in grey) has been drastically reduced. A - *aorta*; Au - eye *primordium*; Epc - *pericardium*; HEc - *ectoderm* (skin)/ beginning of heafold; MEn - *mesentoblast/mesentoderm*; N - neural chord; NEc - neural *ectoderm*; NL - neural crest; So - somite; SoM - unsegmented somite mesoblast; Sp - *splanchnopleura*; St - *primordium* of statocyst.

(example: *Fundulus heteroclitus*) the chief place of blood cell formation is the ICM, which extends from about the level of the anterior portion of the kidney to behind the anus and well into the tail of the embryo. In addition, ventral and dorsal blood islands have grown on the yolksac, most probably stemming from lateral plate mesoderm (ICM?).

Summary

The early investigators considered that the blood originates from the yolk surface, which has been given different names: e.g. periblast[5] or (primary) entoderm.

Some authors suggested that the endocardium is the primary haematopoietic site. However, the endocardium has been later put forward as a secondary site of haematopoiesis in several fish, see 18.2.2.

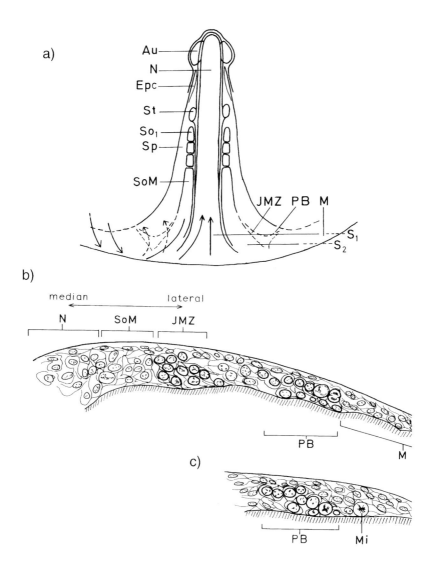

Fig. 18/19 Posterior blood anlage situated in the germ ring (3 somite guppy) (Seelbach and Kunz, 1976). **a** - External view of tailbud region. Arrows indicate the involuting peripheral mesoblast (M) (corresponding to diagram 18/17b,c). **b** - Cross-section of a) at the level of S_1. **c** - Cross-section of a) at the level of S_2. A - *aorta*; Au - eye *primordium*; Epc - *pericardium*; JMZ - intermediary cell mass; M - peripheral mesoblast; Mi - *mitosis*; N - neural chord; PB - posterior blood anlage; So_1 - first somite; SoM - unsegmented somite mesoblast; Sp - *splanchnopleura*; St - *primordium* of statocyst.

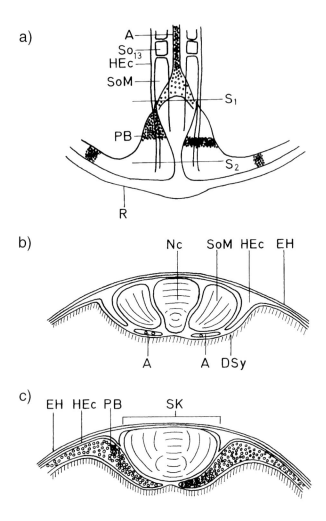

Fig. 18/20 Posterior blood anlage (PB) and intermediary cell mass (JMZ) of 13 somite guppy (corresponding to 18/17d) (Seelbach and Kunz, 1976). **a** - External view of the tailbud region. The intermediate cell mass has given rise to the aorta (A) and blood corpuscles. The caudal arms of the aorta are confluent with the posterior blood anlage (PB), which, in turn, merges with the germ ring (see Fig. 18/17d). **b** - Cross-section of a) at the level of S_1. **c** - Cross-section of a) at the level of S_2. A - *aorta*; DSy yolk *syncytium/periblast*; EH - envelope; HEc - *ectoderm* (skin)/ beginning of heafold; Nc - undifferentiated neural chord; PB - posterior blood anlage; R - germ ring; SK - tailbud; So_{13} - somite; SoM - unsegmented somite mesoblast.

As regards the role of the ICM, the early investigators differ 1) 'Ziegler', 'Wenckebach' and 'Hoffmann' observed in *Perca* and in the salmon, *Salmo salar*, blood-cells arising in the region of the ICM. 2) According to 'Swaen and Brachet' the blood-cells and the endothelia of the aorta and the cardinal veins are derived from the ICM of the lateral plate (salmonids, *Leuciscus* and *Exocoetus*). 3) 'Felix' maintained that the ICM of the lateral plate delivers only blood-cells and the endothelia for the cardinal vein. The endothelia of the aorta derive from the sclerotome. 4) 'Sobotta', in his later publication, stated that the ICM produces exclusively blood cells, while all endothelia (heart included) derive from the sclerotome. 5) In the pelagic fishes the ICM is very weakly developed and furnishes endothelia for the large vessels. Blood cells appear only very late. 6) Some authors described that the haematopoietic tissue arises from the sclerotome. 7) Various authors, such as 'Wenckebach', 'Ziegler' and 'Sobotta' mentioned the migration of blood-cells from the embryo onto the yolksac. 8) Later investigators confirmed the results reported for the salmonids. The ICM was also confirmed as the primary haematopoietic organ in the pike and grayling, *Thymallus thymallus*. 9) Many authors dealt with the zebrafish, *Brachydanio rerio*. It displays haematopoietic and vasculogenetic activity in the embryo even before the bilateral ICM has started to move ventrally ('Colle-Vandevelde', 'Rieb', 'Al-Adhami and Kunz'). The use of haematopoietic transcription factors GATA-1 and GATA-2 made visible the involution of the putative haematopoietic mesoderm and its incorporation into the zebrafish embryo ('Detrich et al., 1995; Zon, 1995').

'Blood islands on the yolksac are the primary sites of blood formation in *Gasterosteus aculeatus*, *Spinachia*, *Pterophyllum scalare*, *Lepadogaster condollii*, *Betta splendens* and *Blennius gattorugine*.

Haematopoiesis in both ICM and blood islands was reported for other species. The study of *Fundulus* embryos in which the blood circulation had been experimentally prevented, revealed the occurrence of blood islands on the yolksac as well as haematopoiesis in the ICM. Mesenchymal cells — future black and brown chromatophores, endothelial and blood cells — migrate away mainly from the caudal region. It has been concluded that the wandering cells, though they have the appearance of ordinary mesenchymal cells, are already made up of these four classes when they emerge from the embryo. Additionally, analyses of a salmonid (rainbow trout), which had already been studied by the early investigators, revealed that it, too, shows haematopoiesis in both the ICM and on the yolksac. *Lebistes reticulatus* (*Poecilia reticulata*) was diagnosed as another blood island/ICM embryo.

According to 'Seelbach and Kunz (1976)' in the embryo of *Poecilia reticulata* three consecutive parts of the originally peripheral mesoderm are haemo- and vasculogenetic: the praecranial mesoblast, the ICM and the posterior blood islands. The deliberations by the authors of the haematopoiesis observed in the guppy seem to be in agreement with the opinion of 'Mollier' who, as early as 1906, suggested that it is conceivable that in different teleosts the extent of the peripheral mesoderm is not the same.

18.2.2 Secondary haematopoietic sites in the embryo (endocardium and kidney)

The endocardium is the inner lining of the heart (Fig. 18/24). The teleostean kidney is made up of pro- and mesonephros (Fig. 18/2). The pronephros is present in all groups of vertebrates at some time during their development. 'Balfour (1882)' shows that the pronephros is a purely larval organ and never constitutes an active part in the excretory system of the sexually mature. In teleosts the kidney is thought not only to be erythropoietic but may also be a major site of the production of an erythropoiesis-stimulating factor 'Wickramasinghe et al. (1994)'.

'Mahon and Hoar (1956)' observed in *Oncorhynchus keta* (an ICM-embryo) that once the heart is formed, migratory embryonic blood cells from the pericardial cavity become distributed over the yolk. They suggested that there is a continuous extra-embryonic mesoderm present on the yolksac, at least in the part proximal to the embryo.

'Bielek (1974b)' investigated the sequence of haematopoietic sites in *Thymallus thymallus*. After haematopoiesis in the ICM, which produces primary blood cells, the appearance of 'swollen endothelial cells' in the endocardium, from the 50 to the 60 somite stage, may represent the genesis of haemoblasts. The 'swollen state' of the heart endothelium continues and haemoblasts, which divide into polychrome erythroblasts, as well as some large haemoblasts are observed. Shortly before hatching a shortlived production of secondary erythrocytes in the kidney is evident. Haematopoiesis on a large scale starts in the kidneys of postembryos shortly before the onset of exogenous feeding.

In *Pterophyllum scalare*, involution, which has left behind mesodermal cells (future blood-islands) on the yolksac, is rapid. Haematopoiesis in the blood-islands ceases at hatching. At closure of the blastopore, ICM have already formed in the trunk of the postembryo. Later they will fuse dorsally and form endothelial cells. The subsequent haematopoietic site is the endocardium, which shows many cells with basophilic cytoplasm resembling the fixed stem cells of the blood islands. This lasts until the third postembryonic day. Subsequently, haematopoiesis switches to the pronephros which remains haematopoietic to the mature stage. Erythropoiesis in this site produces elliptical, though still immature, erythrocytes (erythrocytes-ImA). Thus for the latter part of the postembryonic phase, until complete absorption of the yolk, there is a mixed erythrocyte population in circulation. During metamorphosis into the laterally compressed adult, the mature type of erythrocyte (erythrocyte-A) makes its first appearance. In the mature fish the pronephric tubules have degenerated and the site is filled with haematopoietic tissue and also contains strands of adrenal tissue. Additionally, the mesonephros is also haematopoietic, though to a much lesser degree than the pronephros '(Al-Adhami and Kunz, 1976)'.

In *Lebistes reticulatus* (*Poecilia reticulata*), at the stage of eye pigmention, intra-embryonic haematopoiesis has shifted to the glomeruli of the pro- and mesonephros and the circulating cells are embryonic erythrocytes (e-erythrocytes).

On the first postembryonic day, when the fish swims about actively and takes up food, the majority of the circulating erythrocytes are still of the embryonic type. However, a few immature adult-type erythrocytes (ImA-erythrocytes) which are flattened (diameter 6.0-8.7 microns and 2.7-5.7 microns) with a slightly spherical to oval nucleus have appeared. On the 5th postembryonic day, when traces of yolk are still present, the pronephric glomeruli have disappeared. The majority of the circulating erythrocytes are of the ImA type. A few erythrocytes of the adult (mature) type (A-erythrocytes) have made their appearance (diameter 6.7-8.7 microns and 4.0-5.0 microns) with an elongated nucleus. Six days after yolk absorption, the majority of the circulating erythrocytes are of the mature type (A-erythrocytes). In the mature guppy the mesonephros is the only blood forming organ '(Al-Adhami, 1977)'.

In *Brachydanio rerio* (primary site ICM) stem cells are readily seen in the subsequent haematopoietic sites: The first such site is the endocardium where on the second postembryonic day some endocardial cells resemble typical haemocytoblasts. They contain a homogeneous basophilic cytoplasm and a vesicular nucleus with a large central nucleolus. Transitional cells, between these and the typical endocardial cells, are also observed. While on the 5th postembryonic day (free swimming; onset of external food uptake) the endocardium continues to supply basophilic stem cells (Fig. 18/21b), free stem cells (basophilic haemocytoblasts) and stem cell-like endothelial cells appear in proximity to the glomera and the pronephric tubules (Fig. 18/21c). On the 8th postembryonic day stem cells are no longer visible in the endocardial wall. By the 12th day erythrocytes and leucocytes are formed in the pronephros and fragments of erythrocytes are discernible within the plasma of splenic cells. On the 30th postembryonic day the glomera are degenerated and mesonephric tubules have developed. There is no demarcation between pro- and mesonephros; the whole kidney is haematopoietic. The presence of intermediate stages of the erythropoietic series indicates that they are all formed in the kidney (Fig. 18/22a). Interrrenal tissue is present around the posterior cardinal veins. At the age of 6 months the pronephric tubules have degenerated into solid cords of cells which have the appearance of mammalian lymph nodes (Fig. 18/22b). The haematopoietic cell mass is mainly composed of coarse granulocytes and plasma cells '(Al-Adhami and Kunz, 1976)'. The mesonephroi are fused, except for the most posterior part. The haematopoietic tissue is most abundant in the anterior part '(Al-Adhami and Kunz, 1977)'. The presence of well-developed lymphoid tissue in the pronephros of other mature fish has been previously reported '(Smith *et al.*, 1967)'. Equally, on the basis of EM studies of *Cyprinus carpio* 'Smith *et al.* (1970)' had suggested that the pronephros in teleosts generally has a function closely resembling that of the mammalian lymphnode. In other words, the teleostean pronephros would be the primary source of cells for the immune stystem. On the basis of investigations into lymphoid organs and immunoglobulin producing cells in the Alantic cod, *Gadus morrhua*, 'Schroder *et al.*, (1998)' suggested that pronephros and spleen appearing at hatching are the first lymphoid organs, whereas the thymus is observed much later, i.e. in 9 millimetres long larvae.

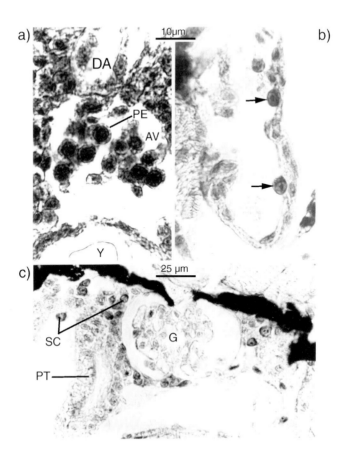

Fig. 18/21 Haematopoiesis in *Brachydanio rerio* (Al-Adhami and Kunz, 1976, with kind permission of the director of the Muséum d'Histoire naturelle, Geneva). **a** - 28 somite embryo. Transverse section of posterior trunk region. Blood cells (most of them at the *proerythroblast* stage) are enclosed by axial vein. **b** - 5th postembryonic day. Transverse section of heart. Some endocardial cells show resemblance to haematopoietic stem cells (arrows). **c** - 5th postembryonic day. Transverse section of *pronephros*. Haematopoietic stem cells between *glomus* and pronephric tubule. AV - axial vein; DA - dorsal *aorta*; G - glomus; PE - proerythroblast; PT - pronephric tubule; SC - Haematopoietic stem cell; Y - yolk.

'Willet *et al.* (1999)' maintained that viewed with the EM the 'stem cells' observed by LM in the endocardium of the zebrafish represent 'irregular endocardial cells joined together by desmosomes'. They also mentioned a discrepancy between their observations by EM and the LM reports by 'Al-Adhami and Kunz (1977)' with regard to the timing of appearance of haematopoietic lineages in the pronephros.

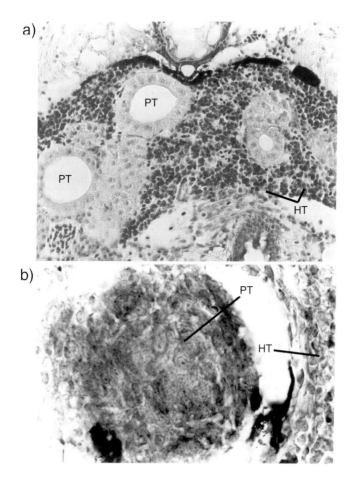

Fig. 18/22 Later haematopoiesis in *Brachydanio rerio* (Al-Adhami and Kunz, 1977 with kind permission of the director of the Muséum d'Histoire naturelle, Geneva). **a** - 30th postembryonic day. Transverse section of *pronephros*. Masses of haematopoietic tissue fill the subcapsular and intertubular spaces. HT - haematopoietic tissue; PT - pronephric tubules. **b** - Adult. Transverse section of *pronephros*. Degenerating solid pronephric tubule (PT) surrounded by haematopoietic tissue (HT), which consists mainly of plasma cells and coarse granulocytes.

In the rainbow trout, 1 day before hatching, erythropoiesis in the kidney and spleen begins and immature and mature erythrocytes (e.ImA and e.A) stain up. The shift in erythropoietic sites is correlated with a switch from larval to mature haemoglobins '(Iuchi and Yamamoto, 1983)'.

414   *Chapter eighteen*

Summary

Secondary sites of haematopoiesis in ICM- and bloodisland-embryos are the cardiac endothelium, followed by pronephros and mesonephros and, in the rainbow trout, also by the spleen. The kidney of some species later develops lymphoid tissue.

18.3   ERYTHROPOIETIC STAGES IN THE DEVELOPING PIKE, ESOX LUCIUS

An overview of the appearance of the different blood cells during the embryonic and postembryonic (adhesive and free-swimming) phase of the pike is given in Table 2, based on 'Bielek (1975)'.

On the basis of the results by 'Gihr (1957)' blood circulation starts at stage J (55-60 somites) and carries the still uncoloured blood cells unto the yolksac. According to 'Bielek (1975)' the size of the primitive blood cells decreases since their anlage at stage F of 'Gihr' (30-35 somites). At the onset of the blood circulation the majority of them has transformed into primary polychrome erythroblasts. At the early stage K (66 somites) large haemoblasts make their appearance. The first orthochrome erythroblasts appear around the time of hatching (stage 1) and some primary mature erythrocytes are observed at the beginning of the adhesive phase. At stage 2 the first secondary basophilic and early polychrome erythroblasts are observed for a short while only. During the second half of the adhesive phase (from stage 3 onwards) the peripheral blood consists mainly of orthochrome cells while the number of polychrome cells (probably of the second generation) is greatly diminished. The number of haemoblasts (medium and small sized) is increasing. The orthochrome erythroblasts gradually reach adult dimensions. Otherwise no essential qualitative and quantitative changes are observed to the end of the adhesive phase. The beginning of the free phase (stage I) is characterized by the increase of secondary erythroblasts (polychrome and basophilic forms) and mature secondary erythrocytes. Exogenous food uptake starts at 18-25 millimetres length of the postembryos, i.e. before all the yolk has been resorbed.

The development of bloodforming tissue in kidney and heart of the pike has also been described by 'Bielek (1975)'. The developing reticular kidney tissue encloses haemoblasts and increasingly venous sinuses from the beginning of the adhesive stage onwards. A short period of erythropoiesis is observed in the early adhesive phase (secondary basophilic and secondary polychrome erythroblasts). A substantial increase in these cell types takes place from the last stage of the adhesive phase to the end of the freeswimming period, with a substantial increase restricted to stages II and III. Haemoblasts and erythrob blasts occur in a ratio of 1:1. A weak haematopoiesis has been observed in the heart from stage 2 onwards. It is characterized by the presence of haemoblasts, secondary polychrome erythroblasts, whereas the ocurrence of secondary basophilic erythroblasts starts only at the beginning of the free-swimming phase. (Table 2).

18.4 HAEMOGLOBIN

Many fish possess multiple haemoglobins which differ in amino acid composition, electrophoretic mobility and solubility. Associated with the shift from early/larval to late type erythroblast is a shift in the chain structure of haemoglobin. Differences in the electrophoretic profiles between embryonic and mature haemoglobin of *Zoarces viviparus* were reported by 'Hjorth (1974)'. Developmental changes in the haemoglobin of *Salmo salar* have been described by 'Koch *et al.* (1964a,b)' and of *Clupea harengus* by 'Wilkins and Iles (1966)'. For *Brachydanio rerio*, globin expression was detected by the 15-somite stage '(Detrich *et al.*, 1995; Zon, 1995; Orkin and Zon, 1997)'. 'Chan *et al.* (1997)'characterized the globins expressed in the mature zebrafish which opens the possibility for the study of developmental switching of haemoglobins. Some antarctic fish have no haemoglobin.

18.5 HEART

18.5.1 Heart of sexually mature teleosts

The teleostean heart is made up of four parts: a thin-walled sinus venosus, atrium (auricle), a muscular ventricle, bulbus cordis (reduced to a muscular ring with two opposite valves), bulbus arteriosus, also called bulbus aortae (which does not contract rhythmically and is continuous with the ventral aorta). Only venous blood is passing through the teleostean heart. The muscular part (the myocardium) and the overlying epicardium form the walls of the heart. The heart cavity is lined by the endothelium. The whole heart is protected by the pericardium; the pericardial cavity (part of the coelom) is filled with a liquid (Fig. 18/23).

18.5.2 Development of the teleostean heart

The development of the general vertebrate heart is illustrated in Fig. 18/24. The 'heart cells' are given off by the splanchnopleura of each side; they then fuse and eventually form the endothelium lining the muscular heart (endocardium). The splanchnopleura also forms the epi- and myocardium and the somatopleura gives rise to the pericardium. The ventral mesocardium and later the dorsal mesocardium disappear.

The teleostean heart tube becomes divided into atrium and ventricle, with the atrium anterior to the ventricle. This arrangement is unique to fish and is opposite to the one observed in mammals and birds. The polarity switches during development, i.e. the sinus venosus comes to lie posteriorly and the arterial outflow anteriorly (Fig. 18/25).

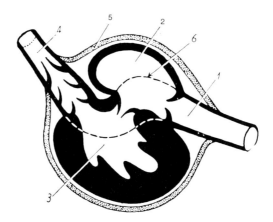

Fig. 18/23 Diagram of adult fish heart (Portmann, 1976, with kind permission of Benno Schwabe, Basel). 1 - *sinus venosus* with sinuatrial valves; 2 - thinwalled auricle (*atrium*), with atrioventricular valves; 3 - thickwalled ventricle; 4 - *bulbus arteriosus* with a single set of valves in most teleosts (though *Albula vulpes* and *Tarpon atlanticus* have two sets). This diagram displays several valves which are representative of the *conus ateriosus* of lower fish such as the sturgeon, dipnoids and *selachii*. 5 - *pericardium*; 6 - dashed line indicates the first *primordium* of the heart (heart tube).

Early investigators

'Rathke (1833)' working with *Blennius viviparus* was unable to establish the anlage of the heart. He assumed that it is formed by the 'Gefässblatt' (vesicular layer). However, he observed the subsequent formation of the ventricle and auricle. According to 'Colle-Vandevelde (1963a)' the endocardium and ventral aorta in this species begin to differentiate at the same time as the first pairs of somites appear, and the epimyocardium is formed by the splanchnic layer of the pericardium (onset of haematopoiesis on the yolksac only at 11-somite stage). The earliest mention of heart development is found in 'von Baer (1835)'. 'Vogt (1842)' observed the anlage of the heart as a 'heap of cells', which later acquires a lumen and is situated between the primordium of the eye and the labyrinth and above the 'couche hématogène'. The heart pulsates long before a cavity appears. 'Lereboullet (1854)' mentioned that in *Esox* the formation of the pericardium precedes that of the heart, while the opposite is true for *Perca*. 'Aubert (1856)' dealt with the formation of heart and pericardium in detail in his work on the pike. He observed that the heart is first a solid cylinder and subsequently acquires a lumen. He further described that the heart assumes an S-shape, that the first pulsations do not create any current, and that the first current does not contain blood but only a serous liquid. 'Reichert (1858)' could not follow the development of the pericardium but concluded that it must be established from a cleft in the meso-

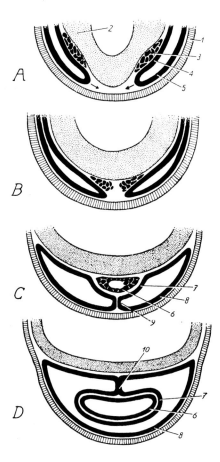

Fig. 18/24 A-D Development of paired heart *primordium* in holoblastic animals (frog). Only the ventral part of the cross-sectioned embryo is shown (Portmann, 1976, with kind permission of Benno Schwabe, Basel). 1 - *ectoderm*; 2 - wall of *pharynx*; 3 - heart cells; 4 - *splanchnopleura*; 5 - *somatopleura*; 6 - *endothelium* (heart tube); 7 - *myoepicardium*; 8 - *pericardium*; 9 - ventral *mesocardium*; 10 - dorsal *mesocardium*.

derm. He described the arterial and venous end of the heart, the acquisition of an S-form, and the division of the heart into two parts. 'Kupffer (1868)' put forward the suggestion that the pericardial cavity arises as a result of a split in the mesoderm. In his investigations on *Gasterosteus aculeatus* he noted that the pulsations start when a cavity appears. He agreed with 'Vogt', 'Lereboullet' and 'Aubert' that regular pulsations start before blood cells circulate. He suggested that the heart has been formed as a result of the splitting of the mesoderm. He maintained that the anlage of the heart is paired, with which 'Balfour (1883-1885)' agrees.

'Kupffer (1878)' mentioned that in the herring, *Clupea harengus*, the heart begins to contract long before it has a cavity.

'Truman (1869)' described in detail the development of the blood circulation in the pike. Colourless blood-corpuscles flow to the heart before it has commenced to beat! At this period the flow is independent of the suction of the heart. 'Oellacher (1873)' described the primordial heart in Salmonidae. He was the first to analyze from histological sections the development of the heart from a paired primordium to a single mass and the appearance of a cavity. 'Ryder (1884)' described in detail the development of the heart in *Gadus morrhua*, which is 'typical of that process in young osseous fish generally'. He noted that 'the contractility of the heart at an early stage is a remarkable phenomenon, in that it as yet contains no clearly marked spindle-shaped muscle cells or fibers. It pulsates very slowly and irregularly — once or twice a minute'. 'Ziegler (1885, 1887)' dealt with the development of the heart in the salmon and 'Wenckebach (1886)' with that of *Belone acus* and *Crenilabrus pavo*. 'Hoffman (1882a,b) described for *Salmo* that the primordium of the heart, i.e. the endothelium, consists of a mass of cells which are a product of the 'entoderm of the periblast' which later cavitates. Simultaneously, two lateral folds of splanchnic mesoblast pass down beneath the pharynx, and as a result of a dorsal and ventral union they surround the endothelium. As a result the tubular heart is now made up of an interior endothelium surrounded by the future myocardium. 'List (1887)' only briefly mentioned the development of the pericardium and the heart of *Ctenolabrus*. 'Brook (1884)' reported that the heart in *Trachinus vipera* is first solid but that a lumen appears when it has reached the floor of the pericardial space and that the heart is only a simple hollow cellular tube at the time it commences to pulsate. 'Henneguy (1888)' reported that the heart has a 'double primordium'; however, he thought it was made up of two endothelial tubes stemming from the splanchopleura. Before him 'Kupffer (1868)' and 'Balfour (1883-1885)' had reported that the heart primordium was double. 'M'Intosh and Prince (1890)' described the origin of the heart as a solid cylindrical structure and as 'a ventral outgrowth of the splanchnic mesoblast'. The heart is contractile while still a simple tube. The early pulsations, which in *Serranus atrarius* commence usually 1 or 2 days after the formation of the heart, do not bring about a current of haemal fluid. The heart beats for some time before blood cells appear in its lumen according to 'Felix (1897)'. 'Swaen and Brachet (1899)' described the median fusion of the left and right 'masses of heart cells' and the development of the pericardium. They suggested that not all the heart cells form the future endocardium; a great number of these cells move out over the yolksac and will form the endothelia of the yolksac vessels. 'Sobotta (1902)' described how in salmonids the heart anlage is formed as a result of the fusion in the midline of cells originating from the lateral plates of each side. The first pulsations are observed after the 60 somite stage. They are slow and irregular, they number 40-50 per minute and rise to 100-120 per minute. The heart pulsates for a whole day before there is a circulation; this is followed by the circulation of serum and finally blood cells. The author constructed four models displaying the development of the teleostean heart. 'Gregory (1902/03)' studied extensively the heart formation in carp, pike, trout, *Salmo salar* and *Salmo alsaticus*. He largely agreed with the observations of 'Swaen and Brachet (1899)' but interpreted the results differently. According to him the endocardium develops from an undifferentiated cellular mass — neither entoderm nor mesoderm — which he called (lateral) mesentoderm. 'Stockard (1915b)' mentioned that in *Fundulus*, as in other teleosts, the heart endothelium partially forms in loco. But he added that 'a few mesenchyme cells are found along the border of the head; these cells later take part in either the formation of the heart or pericardial wall'. It is the lateral lamella anterior to the first somite which gives rise to these cells.

Later investigators

'Mahon and Hoar (1956)' traced the development of the heart from a mesodermal lamella of the ICM in the head region of *Oncorhynchus keta*. When the epimyocardium enclosing the endocardium is formed haematopoiesis starts at this site as mentioned above. 'Colle-Vandevelde (1963a)' showed for *Blennius gattorugine* that the anlage of the heart is paired. This is already visible at the 2-somite stage while blood is formed only at the 11-somite stage.

The primordium is a tube which later bends into an S- shape (often called looping) in the sagittal plane between the atrium and the ventricle in such a way that the sinus venosus and the atrium become dorsally located with respect to the ventricle and the bulbus cordis. The latter is reduced to a muscular ring with two opposite valves. The ventral aorta starts off with an arterial bulb (not to be confused with the cardiac bulbus) which is supplied with a smooth muscular wall. The positional changes of the parts of the heart and the Ductus Cuvieri during development are shown for *Pterophyllum scalare* in Fig. 18/25 '(Kunz, 1964)'.

The heart of the zebrafish, *Brachydanio rerio*, at the beginning a contractile tube, is contained in an enormous pericardial cavity. At first the contractions are only 1 every 2 seconds. At 25 hours the extraembryonic erythrocytes are the first to enter the heart '(Rieb, 1973)'. The formation of the zebrafish heart tube was examined by injection of single blastomeres with a lineage tracer dye followed by serial sectioning of the immunostained embryos. At the early blastula stage most cardiac progenitors were found in a marginal zone between 90° longitude (midway between the future dorsal and ventral axis), which gives rise to myocardial cells only, and 180° longitude (the future ventral axis), which yields myocardial, endothelial, including endocardial and blood cells. Cardiac progenitors were among the first mesodermal cells to involute '(Warga and Kimmel, 1990; Stainier et al., 1993)'. According to the last authors 'these cardiogenic precursors appear to migrate medially as part of the lateral plate mesoderm, and to generate a pair of tubular primordia, one on each side of the midline. These tubular primordia then fuse to generate the definitive heart tube, which consists of an inner, endocardial tube and an outer, myocardial tube'. This fusion occurs in the zebrafish at the 21-27 somite stage. According to 'Al-Adhami and Kunz (1977)' there is a development in the pulsation pattern: the heart starts to beat intermittently at the 28-somite stage and rhythmically at the 31-somite stage.

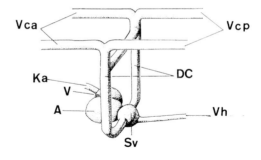
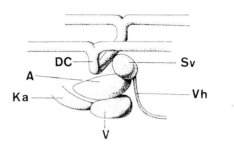

Fig. 18/25 *Ductus Cuvieri* and its positional changes during development of the postembryo of *Pterophyllum scalare* (Kunz, 1964, with kind permission of the director of the Muséum d'Histoire naturelle, Geneva). Diagram above: 1st to 3rd postembryonic day. Diagram below: from 4th postembryonic day onwards. A - auricle (*atrium*); DC - *Ductus Cuvieri*; Ka - gill artery; Sv - *sinus venosus*; V - ventricle; Vca - anterior cardinal vein; Vcp - posterior cardinal veins; Vh - hepatic vein.

Summary

The heart arises from a solid paired primordium from the lateral plate mesoderm, which subsequently fuses and forms the endocardium. The cells for the epi- and myocardium are supplied by the splanchnopleura. A cavity appears and the tube assumes an S-form. The differentiation of the heart tube into regions is a gradual process; the first part to be marked off is the dilated anterior end or sinus venosus, then auricle, ventricle and conus arteriosus. The first heart contractions do not involve any liquid. This is followed by the circulation of plasma and subsequently blood. The different parts of the heart change position during development. The teleostean heart receives only venous blood.

18.6 BLOOD CIRCULATION

18.6.1 Sexually mature fish

A general plan of the blood circulation of a mature fish is given in Fig. 18/26. The dorsal aorta is the main route of transport of oxygenated blood from the gills to the rest of the body. Teleosts possess only 4 complete pairs of aortic arches. The first arch, mandibular arch, disappears completely and the second arch, hyoid arch, subsists in certain species and supplies oxygenated blood to the pseudobranch (see Figs. 15/10, 15/11).

18.6.2 Yolksac circulation

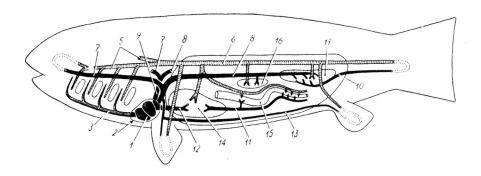

Fig. 18/26 Primary blood circulation of a fish (Portmann, 1976, with kind permission of Benno Schwabe, Basel). 1 - heart; 2 - *pericardium*; 3 - gill artery; 5 - aortic root; 6 - *aorta*; 7 - anterior cardinal vein; 8 - posterior cardinal vein; 9 - *Ductus Cuvieri*; 10 - *vena caudalis* (*vena renalis advehens*); 11 - *vena subintestinalis* (*vena portae*); 12 - *vena hepatica*; 13 - *vena lateralis*; 14 - liver; 15 - intestine; 16 - gonad; 17 - kidney.

The teleostean yolksac circulation is at the height of its development exclusively venous; it forms part of the liver portal system: The vena subintestinalis functions as a vena portarum and the reticulum of blood vessels on the yolk are part of the venae hepaticae. The functions of the yolksac circulation are gaseous exchange, excretion and uptake and distribution to the embryo of nutrients contained in the yolk.

As already the early investigators ['von Baer (1835)', 'Vogt (1842)', 'Aubert (1856)' and 'Lereboullet (1861)'] remarked, the first circulation on the yolk involves only the liquid part of the blood, which they called serum[11], and the 'vessels' are represented by grooves which only later acquire endothelial walls. Even when the circulating blood contains cells, the path of the circulation changes during development. Apart from species-specific differences there may also be different patterns between individuals of the same species.

'Forchhammer (1819)' described with figures the yolksac circulation of the embryo of *Blennius vipiparus*. He mentioned 300 as the number of embryos carried to term by one mother. This topic was followed by 'Rathke (1832)' who illustrated in colour the development of the embryonic and yolksac circulation of the same species. He called the axial vein[6] 'vena vertebralis posterior sine cardinalis'. The onset of the circulation was first observed by 'Vogt (1842)' in *Coregonus palea* and by 'Aubert (1856)', 'Lereboullet (1861)' and 'Truman (1869)' in the pike. 'Vogt' noted that the venae cardinales appear late and that in the trunk they are fused into one vessel (now known as median or axial vein) as far as the site of the liver. 'Rathke (1832)' and 'Goette (1873)' described two veins instead of an axial vein in *Blennius viviparus*. 'Hoffmann' mentioned in *Salmo salar* two venae cardinales posteriores, with one of them aborting later. Before blood cells appear, the circulation is different and the heart is at that time only a slightly bent tube.

'Wenckebach (1886)' studied the circulation in *Blennius, Gobius niger, Syngnathus* and *Belone acus*; he maintained that the blood vessels on the yolk are formed by cells migrating mainly from the head region but also from other sites of the embryo. However, most of the migrating cells form chromatophores. Three main yolk vessels lead into the sinus venosus. He observed the detachment of blood cells from the dorsal median vein. 'Lereboullet' also noticed the absence of any cellular walls in the first yolk vessels. 'Ziegler (1887)' found that early on the caudal vein leads into the vena subintestinalis, which furnishes the yolksac circulation; later on the caudal vein severs its connection with the vena subintestinalis and continues into the unpaired axial vein ('Stammvene') of the trunk. The subintestinal vein now passes into the liver, which sends venae hepaticae over the yolksac. The blood of the yolk 'vessels' is taken up by the sinus venosus of the heart. This pattern was observed in *Perca fluviatilis, Salmo,* and *Esox*. 'His'' results on *Syngnathus* and *Belone* were not complete. He observed that in both fish the vena subintestinalis leads over the yolksac. 'Borcea (1909)' studied *Belone acus* and *Syngnathus acus*; he referred to 'Ziegehagen (1896)' and 'Wenckebach (1886)' whose results on *Belone*, according to him, were rather incomplete. 'Borcea' further described, with illustrations, the circulation in *Gobius capito, Exocoetus volitans* and *Uranoscopus scaber*. 'Anthony (1918)' reported that in *Gasterosteus gymnurus* the yolksac circulation is provided by an artery. Because of the many, often wrong and incomplete results and interpretations of these early investigators, 'Portmann (1927)' decided to analyze additional species; this allowed him to draw general conclusions as to the singularity of the teleostean yolksac circulation. He analyzed first the development of *Cottus bubalis* and found that contrary to earlier observations on teleosts [vide 'Ziegler (1887)'] the first circulation in this fish is characterized by the inclusion, in the body, of the axial vein (which receives its blood from the caudal vein) without any circulation on the yolksac. The same first circulation was observed in *Gobius capito, Gasterosteus gymnurus* and *Cottus bubalis* (Fig. 18/27a). 'Portmann' showed that in *Cottus* the first vessel of the yolksac is a large hepatic vein, which flows into the sinus venosus (Fig. 18/27b). This is followed by the occurrence of several hepatic veins. (Fig. 18/27c).

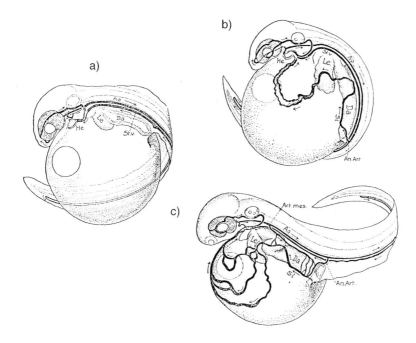

Fig. 18/27 a-c Development of the vitelline circulation in *Cottus bubalis* (Portmann, 1927). An.art - anal artery; Art.mes - mesenteric artery; Ao - *aorta*; Da - intestine; He - heart; Le - liver; Si - subintestinal vein; St.v - median (axial) vein.

Summarizing his and earlier results 'Portmann (1927)' divided the yolksac circulation into two groups: In type 1 only venous blood emanating from the liver reaches the yolk; the vena subintestinalis runs along the intestine — avoiding the yolksac — directly into the liver which gives off one or several hepatic veins. Species of this type are *Cottus bubalis, Gobius*, maybe *Gasterosteus, Perca* and *Salmo, Uranoscopus* and *Pholis*. Type 2 includes fish with a yolksac circulation contributed by veins of the liver and the vena subintestinalis. The arteria mesenterica sends one branch into the liver and the other along the dorsal part of the intestine where it meets within the capillary net of the anal artery. This blood is collected by the vena subintestinalis. Examples include *Belone, Exocoetus, Cristiceps*. In addition, the Ductus Cuvieri capillarize on the yolksac.

Some time later, 'Kryzanowski (1934)', divided the yolksac circulation into 6 categories according to the vessels leading unto the yolksac: 1) Vena subintestinalis (*Esox, Haplochilus*), 2) Vena hepatica (*Salmo, Coregonus, Perca, Acerina*), 3) Vena subintestinalis and Vena hepatica (*Gymnarchus*), 4) Ductus Cuvieri (Cyprinidae), 5) a combination of 1-4 (*Xiphophorus*), 6) Vena hepatica and Arteria vitellina (*Cottus, Cottomorphus*). Subsequently, 'Brachet (1935)' abstained from proposing any classification because of the great variety encountered.

'Kunz (1964)' investigated other teleostean species and found that *Coregonus alpinus* and *Perca fluviatilis* belong to type 1 of 'Portmann' and group 2 of 'Kryzanowski'. (Fig. 18/29d). However, *Pterophyllum scalare* (Fig. 18/29a-c) and *Poecilia reticulata* (*Lebistes reticulatus*) (Fig. 18/28a-e) could not be included in any of these groupings. In both species the Vena caudalis first leads over the yolk and only later runs through the trunk. Therefore, they display the first circulation of the early investigators. In *Poecilia* the Vena subintestinalis carries venous blood from the intestine into the liver. This viviparous embryo has a very distinct blood-supply: the V. hepatica (temporarily double), Dd. Cuvieri, V. caudalis, the transitory V. hepatica posterior (posterior intestinal vein) which leads into the V. caudalis before the latter spreads over the yolksac (Fig. 18/28b). The embryo is characterized by a greatly enlarged pericardium which forms a hood. At the so-called 'pericardial hood stage' (see Chapter 20) the yolksac circulation is mainly supplied by the Dd. Cuvieri. They branch on the yolksac and send a main vessel dorso-craniad and another one dorso-caudad. The former forms also the vascular net on the hood. The V. caudalis sends blood to the most caudal part of the yolksac (Fig. 18/28c). In many embryos the blood does not return into the trunk below the aorta as V. caudalis dorsalis but as V. caudalis ventralis; in some specimens both veins may be encountered together. The V. hepatica leads into the dorso-caudal main branch of the left D. Cuvieri. The yolksac circulation changes decisively when the 20th embryonic day has been reached. The Dd. Cuvieri supply blood to the most anterior part of the yolksac and to the 'neckstrap' (reduced pericardial hood). The caudal part of the yolksac receives blood from the V. caudalis which before the descent unto the yolksac receives venous blood from the area of the anal fin. The left half of the yolksac is vascularized by the V. hepatica which may appear double. The yolksac vessels usually reunite into three ventral main branches, which open into the sinus venosus. Immediately before hatching the hood splits dorsally and subsequently disappears; only the ventral part of the yolksac remains vascularized by the Dd. Cuvieri and the Vena hepatica (Fig. 18/28d,e).

In *Pterophyllum scalare* the Vena caudalis profunda capillarizes in the ventral finfold ('postanal fin'). It divides again into many branches before it enters the yolksac ('preanal fin'), which also receives blood from the Arteria mesenterica at around hatching time. On the 2nd postembryonic day the preanal fin is addition-

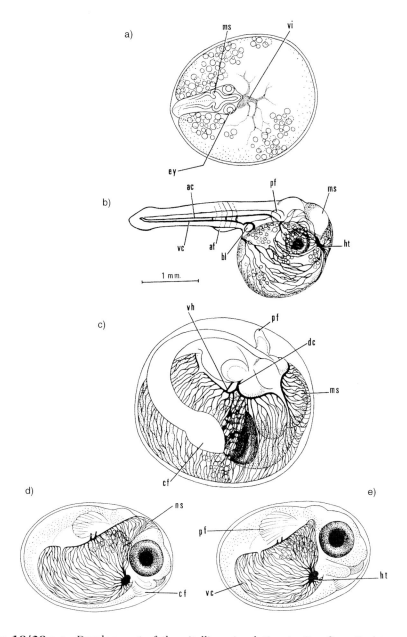

Fig. 18/28 a-e Development of the vitelline circulation in *Poecilia reticulata* removed from ovary (Kunz, 1964, 1971a, with kind permission of the director of the Muséum d'Histoire naturelle, Geneva). **a** - stage 4 (6th embryonic day); **b** - stage 5 (10th embryonic day); egg envelope removed; **c** - stage 6 (15th embryonic day); **d** - stage 7 (20th embryonic day); **e** - stage 8 (25th embryonic day). ac - *arteria caudalis*; af - anal fin; bl - urinary bladder; cf - caudal fin; dc - *Ductus Cuvieri*; ey - eye; h - heart; ms - *mesencephalon*; ns - neckstrap; pf - pectoral fin; vc - *vena caudalis*; vh - *vena hepatica*; vi - vitelline circulation.

ally capillarized by the vena subintestinalis, and the blood supply from the liver unto the yolksac is increased (Fig. 18/29a-c). The Ductus Cuvieri have moved frontally and ventrally to the oesophagus; the two descending limbs are connected by a crossing vessel before they empty into the sinus venosus (Fig. 18/25a). On the 4th postembryonic day the capillarization of the 'postanal fin' has disappeared, and the blood now returns via the Vena caudalis profunda into the trunk. The preanal fin is also greatly reduced and receives blood only from the vena subintestinalis. Most of the blood on the yolksac is now supplied by the many hepatic veins (Fig.18/29c). Due to the change of positions of the ventricle and auricle of the heart, the cross-branch of the Dd. Cuvieri now enters directly the dorsally situated sinus venosus (Fig. 18/25b). The transient vascularization of post- and preanal fins is interpreted as an additional area for gas exchange before the gills take over (Fig. 18/29a-c).

At hatching *Coregonus alpinus* displays a yolk circulation taken over entirely by hepatic veins (Fig. 18/29d). A similar picture is displayed by *Perca fluviatilis* approaching hatching time '(Kunz, 1964)'.

In *Brachydanio rerio* the caudal vein follows the course reported for ICM-embryos generally. In the early embryo this vein bypasses the yolksac to lead directly into the trunk, where haematopoiesis is taking part in and around the axial vein. According to 'Rieb (1973)' the first circulation is established at 25 hrs. Somewhat behind the junction of the dorsal aortic roots the axial vein divides into two posterior cardinal veins. At the beginning only one of the posterior Vv. cardinales, either the right or the left, is functional and sends blood over the yolksac, which is collected by the Dd. Cuvieri. Finally the erythrocytes disperse themselves as in a fan over the yolk. They move in lacunae between the periblast[5] and the ectoderm. Towards the stage of 27 hrs (eyes pigmented) the embryonic circulation extends into the anterior region of the embryo. The blood on the yolksac reaches the sinus venosus of the heart via the Dd. Cuvieri (still devoid of an endothelium along most of their course). The functional significance of this direct contact of the blood cells with the yolksac syncytium may be to enhance respiration, excretion and yolk absorption. The erythrocytes are spherical (10 microns diameter) and when they become oval they measure 9.5 x 4.8 microns. The Vena subintestinalis appears only on the 5th day when the yolk is practically absorbed. It follows the intestine and the stomach and enters the liver. The hepatic vein empties directly into the sinus venosus or the Dd. Cuvieri. Nothing new is added in recent literature on the zebrafish. '(Kimmel et al., 1995)' showed with Nomarski optics at the 21-somite stage the pronephric duct (still closed) with accumulated blood cells dorsally to it. By age 24 hrs the duct has acquired a lumen and opens to the outside.

'Soin (1971)' dealt extensively with the development and structural types of the teleostean vitelline circulation and discussed the phylogenetical links between the different types.The following functional significance has been put forward on morphological grounds: The vascular system of the yolksac is believed to be mainly respiratory. The intake of perivitelline fluid through the mouth and the exit through the operculum in the late embryo, assisted by the movements of

Mesodermal derivatives 427

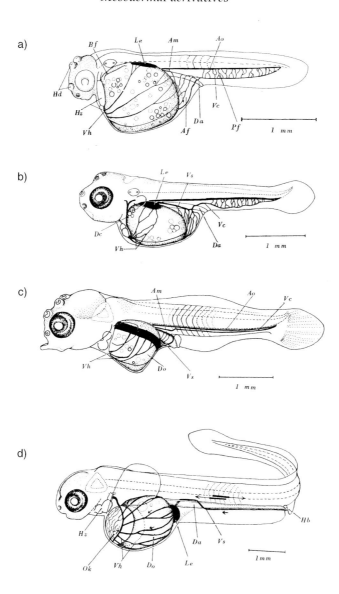

Fig. 18/29 a-d Vitelline circulation in *Pterophyllum scalare* and *Coregonus alpinus* (Kunz, 1964, with kind permission of the director of the Muséum d'Histoire naturelle, Geneva). **a** - *Pterophyllum* hatching (2nd embryonic day); **b** - *Pterophyllum* 2nd postembryonic day; **c** - *Pterophyllum* 4th postembryonic day; **d** - *Coregonus alpinus* hatching (60th embryonic day). In drawings a)-c) only some oil droplets are shown. Af - anal fin; Am - mesenteric artery; Ao - *aorta*; Bf - pectoral fin; Da - intestine; Dc - *Ductus Cuvieri*; Do - yolk; Hb - urinary bladder; Hd - adhesive glands; Hz - heart; Le - liver; Ok - oil globule; Pf - postanal fin; Vc - *vena caudalis*; Vh - *vena hepatica*; Vs - *vena subintestinalis*.

the pectoral fins, is called a respiratory pump. These ventilatory movements are first infrequent and gradually become rhythmical. The vitelline circulation is the first embryonic vascular system to develop. It may be followed by other vascular surfaces such as the finfold and pericardium, until the gills are taking over. The vitelline circulation is also engaged in the absorption of yolk though many eggs, especially pelagic ones, are not provided with a yolksac circulation and still absorb their yolk. An excretory function is also subserved by the early capillarized surfaces. The ontogeny of respiration has been reviewed by 'Rombough (1988)'.

A review on what guides early embryonic blood vessels in vertebrates, including teleosts, was given to a peer-reviewed forum by 'Weinstein (1999)'. Vasculogenesis as the primary means of generating the first major vascular tracts during early embryogenesis, including yolksac and trunk axial vessels, was discussed. It was stressed also that in the zebrafish axial mesoderm (notochord) is necessary for formation of the dorsal aorta, but is not required to form the posterior cardinal vein, and that major intraembryonic vessels need to be positioned precisely with respect to adjacent organs and reproducibly and exactly 'plumbed together' with other major vessels.

Summary

Since in all cases the liver is in contact with the yolksac, the Vv. hepaticae, either exclusively, or, at least to some degree, take part in the yolksac portal system. In the zebrafish the liver is formed only at the end of the yolk absorption, and the Dd. Cuvieri are the only yolksac vessels (though mainly devoid of walls). The Dd. Cuvieri, though highly capillarized, form yolksac vessels also in the guppy, *Poecilia reticulata* (*Lebistes reticulatus*). In many groups also the Vena subintestinalis (V. portae), and in some species the V. caudalis, flow over the yolksac. In other words, it can be stated safely that in the teleosts all the great veins emanating from the body can supply blood to the yolksac.

The vascularized pericardial hood in *Poecilia* and the transient capillarization of the pre- and postanal fins in *Pterophyllum* represent additional surfaces for gaseous exchange. The buccal and opercular pump, assisted by the beating of the pectoral fins, cause the flow of perivitelline fluid through the anterior part of the embryo. After hatching, the beating of the pectoral fins will assist in moving water over the gills.

Endnotes

[1] G. Seitenplatte
   F. parois latéral
[2] In the U.S. literature the term somatopleura includes the adjacent ectoderm and the term splanchnopleura the adjacent entoderm.

³ G. intermediäre Zellmasse, Blutstränge of 'Sobotta', Venenstränge of 'Felix'.
  F. Masse intermédiaire
  I. Massa intermedia
⁴ G. Blutinseln
  F. îlots sanguins, ébauche vasculaire
  I. isole sanguine
⁵ see endnote[14] of Chapter 10
⁶ axial vein, median vein
  G. Stammvene
⁷ G. Urniere
⁸ The development in embryos or postembryos of functional adaptations not present in the mature animal.
⁹ G. Stammzellen
  F. précurseur commun, cellule primitive
  I. cellule capostipiti
¹⁰ G. Endknospe
   F. bouton terminal
¹¹ The term 'plasma' would be used nowadays since the term 'serum' refers to the liquid remaining after the material of the protein clot (fibrin) has been removed from clotted blood plasma.

Chapter nineteen

Entoderm and its derivatives

'Pour bien savoir les choses il faut en savoir le détail.'
La Rochefoucauld

The entoderm (endoderm, enteroderm) is the inner germ layer of the gastrula; it furnishes the lining epithelia of the digestive system (pharynx, intestine), as well as the parenchymal cells of liver[1], acinar cells of the pancreas, also part of the gills, the pseudobranch (Chapter 15, Figs. 15/10-15/14), the swim bladder, Kupffer's vesicle (Chapter 14) and part of the pituitary, the adenohypophysis. The hypochord is also thought to be of entodermal origin and is dealt with in Chapter 12. (Fig. 12/4). However, in a recent publication on zebrafish entoderm it was decided that 'this peculiar tissue...has no obvious association with the gut... but our data show that hypochord is not derived from cells near the margin and implies perhaps that hypochord should not be considered endodermal' '(Warga and Nüsslein-Volhard, 1999)'.

Most writers on fish embryology have thought that the entoderm arises as a differentiation from the inflected layer of the germ ring. The involuted cells (called primary hypoblast[2]) were held to separate sooner or later into two layers, the lower of which was the entoderm (secondary hypoblast), the upper one the mesoderm. In some fish the primordium of the entoderm lies superficially and is a separate layer from its inception and subsequent inflection. (Figs. 11/13, 11/14, 12/2-3, 13/1-13/6).

The term 'primary entoderm' refers to the 'lower layer' (the primitive hypoblast of 'Goette') which later differentiates into notochord, mesoderm and enteroderm (also called 'secondary entoderm' or 'intestinal epithelium'). In teleosts notochord, mesoderm and entoderm are difficult to distinguish at an early stage due to the absence of lumina. The median cells are prospective notochord and, at each side of the notochord, the thin layers represent the future intestine with its glands, which is now called secondary entoderm or enteroderm, or simply 'entoderm'. The differentiation into chorda, mesoderm and entoderm is observable first in the trunk of the embryo, later in the head, while at the posterior end the prospective layers are still fused '(Hoffmann, 1881; Ziegler, 1902)'.

The entoderm in the head region will form, in association with the ectoderm, the gills and pseudobranch. Posterior entodermic organs are the oesophagus, the swimbladder, the intestine and the liver. The entoderm generally forms only the lining of the intestine and its glands. However, it provides entodermal material in bulk to the putative liver (and to a lesser degree to the pancreas). The teleostean liver is made up of anastomosing liver cords and tubules, the meshes of which become filled with 'inwandering' mesenchyme³ cells. Lobes are observed in some species whereas lobules — as encountered in the mammalian liver — are absent. The gall bladder is formed as a dilation of the liver duct lined with a single layer of columnar cells, which gradually flatten. The ductus choledochus (common bile duct) empties into the intestine. Although the liver cells perform a multitude of reactions they look alike when viewed with LM.

While in all other meroblastic vertebrates during their yolksac phase the intestine is open to the yolk mass, which suggests a direct absorption of the yolk, in teleosts the intestinal tube remains distinct from the yolk mass and no umbilical cord is present (Figs. 19/7a,b, 19/16C). [However, 'von Baer (1838)' and 'Lereboullet (1861)' expressed an opposite opinion.].

As shown in Chapter 11 on gastrulation, the teleostean yolk-syncytium with the overlying ectoderm epibolize, i.e. spread to enclose the entire yolk mass. The entoderm does not follow this epibolic movement and consequently the yolk is not enclosed by an entodermal layer, as already observed by 'Rusconi (1836)'. Absorption of yolk, therefore, in contrast to the situation in other vertebrates, seems to occur without any involvement of the gut '(Bachop and Schwartz (1974)'.

'Hoffmann (1881)' in his review insisted never did he see an involution in the eggs of the many Mediterranean fish eggs he investigated. 'Sumner (1990b)' gave an up-to-date historical account, mentioning that 'nearly all writers on fish embryology have thought this germ-layer [entoderm] to arise as a differentiation from the inflected layer of the germ ring. The cells of the latter were held to separate sooner or later into two layers, the lower of which was the entoderm, the upper the mesoderm. The only writers, as far as I know, who have maintained the existence of a distinct entoderm rudiment are 'von Kowalewski', 'Berent' and 'Reinhard'. 'Von Kowalewski (1886a,b,c)' described in *Gobius* sp. and *Carassius auratus* that the entoderm derives from a few entoblastic cells situated at the posterior margin of the blastodisc, between the mesoderm and the epithelial envelope (EVL). At its beginning the entoderm is solid and triangular in cross section with its basis facing the EVL and its apex along the longitudinal axis of the embryo and pointing towards its head. 'Reinhard (1888)' suggested that the entire hypoblast is a derivative of the walls of Kupffer's vesicle (Chapter 14) and 'Berent (1896a)' contended that in trout, *Salmo fario*, and salmon, *Salmo salar*, the entoderm arises as a separate rudiment (Fig. 13/6). 'Sumner (1900a,b)' observed that during a certain period of the development of *Muraena*, at the posterior margin, a direct continuity is observable between the so-called prostomal thickening and a thin layer of cells representing what he called the gut-hypoblast. More than a year earlier, in a paper before the American Morphological Society, he had described for the catfish blastoderm a pronounced thickening of the pavement layer (EVL) on the embryonic (posterior) margin of the blastoderm as a 'new factor'. In a subsequent paper, 'Sumner (1904)' undertook a morphological as well as an experimental analysis of early fish development. In other words, he subjected to experimental tests, as well as re-examined, former preparations. He emphasized that 'after a thorough reexamination of my former preparations, as well as of many new ones, I still feel absoutely certain of the existence of the 'New Factor' previously discussed by me. But the theoretical conclusions then offered were, I now think, obscurely stated and in part false'. As regards the prostomal thickening, he analyzed this process in further fish, such as *Salvelinus fontinalis, Abramis crysoleucas,* two species of Siluridae, an unknown species of *Muraena*. He found strong indications of this collection of cells in *Fundulus heteroclitus* and *Ctenolabrus (Tautogolabrus) adspersus* but he 'cannot speak with the same degree of certainty in the case of these latter fishes'. He still had no doubt about the continuity of the superficial layer of the epiblast with the prostomal thickening; however, its continuity with the gut-hypoblast needed clarification. He stressed that a middle layer is formed by the inflection of the margin of the blastoderm (exclusive of the EVL) around its entire circumference; from this layer (primary hypoblast of most authors) chorda and lateral plates will be formed. A lower layer, derived from an independent collection of cells (called prostomal thickening in the earlier paper of 1900) first appears as a thickening of the EVL of the blastoderm, on its posterior margin. This lower layer (referred to by him as enteroblast) gives rise to the gut epithelium. 'Lanzi (1909)' and 'Reis (1910)' agree with 'Sumner'. Similarly, 'Boeke (1903)' maintained that in the muraenoids the entoderm develops independently of the layer which is destined to form the mesoderm and chorda. However, with some other earlier investigators, he suggested that the entoderm takes its origin from the periblast of the yolk.

Cross-sections of the salmonid gastrula stage display a continuous entoderm layer below the notochord and the mesoderm as mentioned earlier by 'Goette (1878)' and 'Henneguy (1888)' for the trout (Figs. 12/1-12/3, 19/1). In contrast, the entodermal cells of *Ctenolabrus* grow medially and meet under the notochord as observed by 'Agassiz and Whitman (1885)'. Similarly, in the sea bass, *Serranus atrarius*, the entoderm, chorda and mesoderm cells differentiate immediately following involution. The entoderm flanks both sides of the notochord and eventually grows together and fuses '(Wilson, 1889/1891)' (Fig.19/2).

The entoderm consists of one layer according to 'Romiti (1873)', 'Goette (1873)' 'Hoffmann (1881)', 'Agassiz and Whitman (1884)', while 'Oellacher (1873)' insisted that the entoderm in the trout is two or three layered and 'Henneguy (1880)' mentions at least two strata. 'Ziegler (1882)' described that the entoderm consists of one or more layers. These different results may be dependent on the time of observation since the entoderm gradually accumulates in the midline to form the intestinal primordium. 'Kupffer (1884)' declared that he agrees with calling the periblast and the yolk together 'entoblast'.

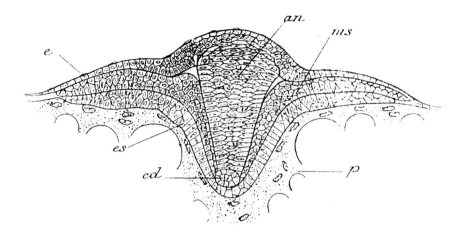

Fig. 19/1 Cross-section of trout embryo, posterior to the auditory placode, to show a continuous *entoderm* below notochord and neural chord (Henneguy, 1888). an - neural axis; cd - notochord; e - *ectoderm*; es - secondary *entoderm*; ms - *mesoderm*; p - *periblast*. Periderm is indicated, but not labelled.

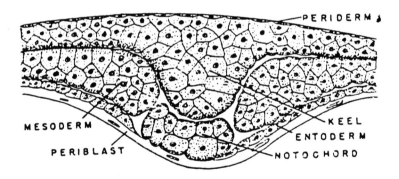

Fig. 19/2 Cross-section of embryo of sea bass to show discontinuous *entoderm*, below notochord (Wilson, 1891).

Later investigators reported again that before gastrulation in some teleosts the entodermal cells are not exposed to the surface at the caudal portion of the blastodisc, e.g. trout '(Pasteels, 1936)'. He specified that the entoderm is not formed by the EVL but involutes at its site of contact. 'Oppenheimer (1936a)' described for *Fundulus* that the putative entodermal cells lie at the surface and involute separately from the mesodermal cells. All agree that the lumen in the intestine appears later than in Kupffer's vesicle (Chapter 14). 'Brummet (1968)' was engaged in deletion-transplantation experiments of the posterior embryonic shield of *Fundulus heteroclitus*. As regards entoderm formation she concluded that 'endoderm formation in the teleost embryos remains an intriguing and completely unsolved problem'.

The theory of involution was again challenged from 1966 to 1982 (see Chapter 11). But subsequently 'Wood and Timmermans (1988)', using Nomarski differential interference contrast microscopy, established that involution does take place in the rosy barb and 'Warga and Kimmel (1990)' employing double-label methods for clonal analysis in zebrafish followed directly the inflection of the hypoblast. It emerged that the entoderm involutes before the mesoderm and that the order in which cells of a clone involute corresponds to their subsequent order along the anterior-posterior axis of the embryo. However, the above authors did not extend their investigations to the formation of the intestine. Subsequently, 'Warga and Nüssslein-Volhard (1999)' showed, by labeling with lineage tracer dye single cells at the margin of the early blastula, that endodermal and mesodermal precursors are distributed throughout the margin. They further established that the progeny of marginal clones differentiates into either only endoderm, or only mesoderm, or a combination of both. The structures of entodermal origin are derived predominantly from the more dorsal and lateral cells of the margin (see also Chapter 13 on fatemaps).

Ontogeny of endocrine cells of the intestine and rectum of the sea bass, *Dicentratus labrax* L.), was investigated at an ultrastructural level by 'Hernanez et al. (1994)'. Different endocrine cells develop parallel with the differentiation of the intestine and rectum.

The alimentary canal of adult fish consists mostly of an intestine. There is a very short oesophagus. There is practically no distinction between it and the stomach, which is straight and usually without function. Many teleosts show an increase of the intestinal surface, the so-called pyloric caeca or appendices. There may be just five or more than 900 of these (Fig. 19/5a). Other fish, such as the cyprinids, have no real stomach but an elongated intestine instead (Fig. 19/5b).

19.1 FORMATION OF THE PHARYNX

In the head region the gill cavity is laid down when the ectoderm forms a fold, which at several sites comes in contact with the entoderm; this is called a pharyngeal pouch (Fig. 19/3). At these sites a fusion of ecto- and entoderm follows which results in a gill cleft '(Hoffmann, 1878; Henneguy, 1888; Ziegler, 1902)'. A lateral view of an embryo of *Salmo salar* with several gill clefts is shown in Fig. 19/4.

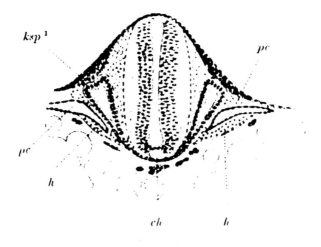

Fig. 19/3 Cross-section through head region of salmon. The *primordium* of the first gill cleft is indicated (Ziegler, 1902). Magnification × 78. ch - notochord; h - heart cells; ksp[1] - *primordium* of first gill cleft; pc - *pericardium*.

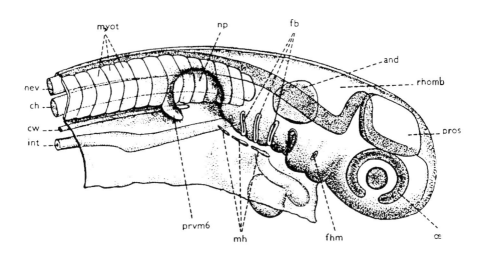

Fig. 19/4 Lateral view of older *salmo* embryo (length 8 mm) with several clefts (Ziegler, 1902, after Harrison, 1895). and - auditory *placode*; ch - notochord; cw - pronephric duct; fb - gill clefts; fhm - 1st gill cleft; int - intestine; mh - myotome processes; myot - myotomes; nev - neural chord; np - pectoral fin; oe - eye; pros - *prosencephalon*; prvm6 - ventral process of 6th somite; rhomb - *rhombencephalon*.

19.2 FORMATION OF THE THYROID [4]

The thyroid gland is a derivative of the pharynx and was first described and given its name in 1656 by Thomas Wharton, a London physician. In mammals, birds and amphibians it is a highly vascularized shield-like paired gland and in reptiles a single midline gland. The appearance of the thyroid in teleosts is unusual in that it is not an encapsulated organ but consists of scattered follicles extending along the ventral aorta and often into the branchial arches. Heterotopic (ectopic) thyroid tissue may be found also in the kidneys (pronephros) and other non-pharyngeal sites such as eye, brain, spleen '(Leatherland, 1994)'. In all fishes the thyroid appears early in embryonic development. The presence of the thyroid in fishes was reported first by 'Simon (1844)' and 'Maurer (1886)' '(cf. Leatherland, 1994)'.

The histological structure of the thyroid is very characteristic: the walls of its follicles are composed of a secretory epithelium, with tight junctions at their apices, surrounding a lumen containing colloid. The follicles vary their appearance according to the activity of the gland. The follicular cells are concerned a) with the formation of thyroglobulin in the lumen by taking up iodide from the extracellular space, transporting it into the follicular lumen where thyroglobulin (T_4 and T_3) is formed. and b) with reabsorption of the thyroid hormones and their transmission into the blood circulation. T_3 is generally considered to be the biologically active form of the hormone, with T_4 acting as a precursor or prohormone. In unstimulated thyroid follicles their epithelial cells are squamous or cuboidal with a high nucleo-cytoplasmic ratio and enclosing a lumen filled with a usually homogenous colloid. Following stimulation by the pituitary, the follicles tend to be smaller, the epithelial cells columnar, the nucleo-cytoplasmic ratio smaller and the colloid in the lumen partly or wholly depleted.

The literature on thyroid hormones and early ontogeny in teleost fishes was reviewed by 'Leatherland (1994)': While immersion of embryos of *Cyprinus carpio*, *Sarotherodon mossambicus* and *S. niloticus* in T_4 increased their rate of yolk absorption, paradoxically, immersion of both denuded embryos and embryos with their envelopes intact of *Osseochromis mossambicus* in a solution of either T_4 or T_3 delayed the secretion of hatching enzymes '(Lam, 1980; Nacario, 1983; Lam and Sharma, 1985; Reddy and Lam, 1991, 1992)'. For Salmonidae and for a wide spectrum of freshwater and marine teleosts high levels of T_4 were reported for eggs, embryos and larvae. It was shown that the content of thyroid hormone in the egg was related to maternal plasma thyroid hormone levels. The study of the T_4 and T_3 contents of whole salmonid embryos showed a clear decrease in the thyroid hormone content between the egg and early embryo and the time of hatching '(Leatherland *et al.*, 1989)'.

19.3 FORMATION OF THE INTESTINE
(SEE ALSO CHAPTER 11 ON GASTRULATION)

'Goette (1869, 1873)' observed in the trout the early differentiation of the entoderm, (which he called 'Darmblatt' = intestinal leaf) at the stage when the notochord begins to form. The entoderm arises not by cleavage from the upper layer but grows forward and under from the germ ring (by a 'wheeling movement' = involution) and extends with its free ends to the limits of the embryonic shield. However, 'Ryder (1882 b)' declared that in teleosts the intestine is formed 'from behind forwards by splitting of the hypoblast'. In the cod, *Gadus morrhua*, the hypoblast is 'confounded or blended with the mesoblast up to the time when the muscular and the peritoneal layers are differentiated'. The intestine is primitively solid and the appearance of a lumen is by separation or retreat of its cells from its axis '(Ryder, 1884)'. 'Agassiz and Whitman (1885)' and 'Henneguy (1888)' described how in the trout the first two germ layers arise as a result of inflection at the germ ring. The intestine is formed as a thickening caused by the fusion of lateral folds and the lumen arises as a result of the separation of the interior cells from each other. 'Von Kowalewski (1886a,b)' described and followed the development of the entoderm cells in *Gobius, Carassius* and the trout. He noted that cells involute at the posterior end of the future embryo to form the mesoderm and suggested that during involution the entoderm destined to form the intestine is left behind. This entodermal rudiment will subsequently form a solid cord lying on the periblast and getting thinner towards the head. He later modified the interpretation of his results and concluded that the primordium of the entoderm he had described earlier was only part of the entoderm and was destined to form Kupffer's vesicle (see Chapter 14). He now observed, anterior to Kupffer's vesicle, and continuous with it, a solid cellular rod forming the axis of the embryo. This rod on each side is delimited by mesoderm and will form from its dorsal cells the notochord and from 1-2 ventral cell layers the intestine '(von Kowalewski, 1886c)'. The work of 'von Kowalewski' met at the time with little acceptance.

The origin of the intestine as a closed entodermal fold has been suggested by 'Goette (1878)'(Fig. 12/1-12/3), 'Agassiz and Whitman (1884)', 'Hoffmann (1883)', 'Wilson (1889/1891)', 'Henneguy (1888)' and 'Lwoff (1894)', 'Swaen and Brachet (1889) (Fig. 18/4)'. In contrast, the origin of the intestine as a solid rod (thickening of the entodermal lamella), however of short duration, has been proposed by 'Ziegler (1882)', 'Cunningham (1885a,b)' and 'Berent (1896a,b)' for *Salmo salar* and the trout, *Salmo fario*. According to 'Kingsley and Conn (1883)' and 'Ryder (1884)' in pelagic eggs generally, the alimentary canal is reported to originate as a solid thickening of the entodermal lamella. None of the early investigators suggested how exactly the solid band is converted into a tube. In passing, they simply mention a 'separation of cells'. Later investigators confirmed that the intestine of teleost embryos and newly hatched postembryos is solid and nonfunctional '(Rombout *et al.*, 1978; Timmermans, 1987)'. However, 'Mangor-Jensen and Adoff (1987)' and 'Tyler and Blaxter (1988a)' observed that drinking occurs in the cod, *Gadus morrhua*, and the halibut, *Hippoglossus hippoglossus*, respectively. Moreover, 'Tyler and Blaxter (1988b)' have shown that the herring, plaice and cod larvae change their drinking rate in response to changes in ambient salinity: in other words, the function of the intestine at this stage is osmoregulatory.

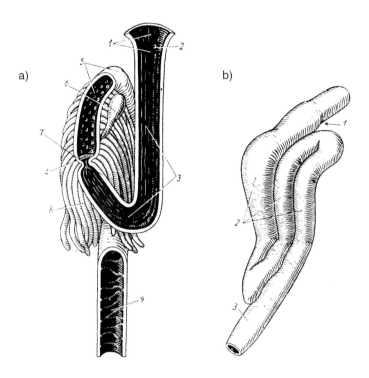

Fig. 19/5 Intestine of adult teleosts (Portmann, 1976, with kind permission of Benno Schwabe, Basel). **a** - intestine of salmon. 1 - *oesophagus*; 2 - duct of swimbladder 3 - stomach; 4 - *pylorus*; 5 - middle intestine; 6 - gall bladder; 7 - duct of gall bladder; 8 - appendices of pylorus; 9 - rectum. **b** - intestine of a cyprinid without a real stomach. 1 - *pylorus*; 2 folds of middle intestine; 3 - *rectum*.

19.4 POSTANAL GUT

The postanal gut is a transitory organ of unknown function. In the sea bass, *Serranus atrarius*, at a stage when the tail has barely began to develop and when KV has disappeared, the postanal gut has become a prominent feature (Fig. 19/6a,b). It increases in size until it almost vanishes in the caudal mass of cells.

Immediately after this stage, atrophy of the postanal gut starts to set in, beginning at the anterior limit of the tail. 'Schwarz (1889)' described in detail the histological changes in the tailend of the postembryo of salmon and pike. The neural chord, notochord and intestine which were separated anteriorly merge into the indifferent cell mass of the tailbud. Within the postanal gut and at the end of the notochord KV is visible. 'Mahon and Hoar (1956)' reported similar observations in *Oncorhynchus keta*.

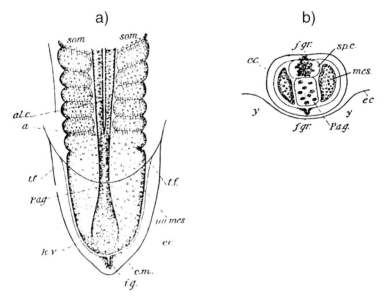

Fig. 19/6a Surface view of tail end of *Serranus atrarius* embryo, viewed from below (Wilson, 1889). a - *anus*; al.c - alimentary canal; cm - caudal mass; ec - *ectoderm*; f.g. - fin groove; k.v - Kupffer's vesicle; P.a.g - postanal gut; som - somite; un.mes - undivided *mesoderm* in tail; t.f. - (tailfold), marks line along which the *ectoderm* covering the tail bends round over the yolk.

Fig. 19/6b Transverse section through embryo of the same stage as above (Wilson, 1889). ec - *ectoderm*; f.gr - fin groove; mes - *mesoderm*; P.a.g. - postanal gut; s.pc - spinal cord; y - yolk.

The salmon, aged 2 days after closure of the blastopore, shows a protruding tail. The pharynx in the gill region represents a flat compressed tube. The oesophagus and the stomach show a small lumen. In the region of the liver, where the anlage of the intestine had shown the most massive development, the intestine is still compact while the posterior part displays a slit-like lumen as far as the anus. A solid postanal gut is evident posterior to the anus.

Summary

It seems that the location of the prospective entoderm is species-specific. This is already evident when contemplating the differences in fate-maps (Chapter 13). The investigators suggest that the entodermal cells, once involuted to the inside, grow centripetally and aggregate to form the intestine. Its anlage was suggested to be a solid or a folded cord. All authors seem to agree that the lumen in the intestine appears late regardless of the original morphology (solid rod or folded cord). No neurenteric canal is formed since both the primordia of the intestine and the neural cord lack a lumen. The postanal gut is a mere streak of entoderm of hitherto unknown function.

19.5 FORMATION OF THE LIVER

In *Serranus atrarius* the primordial liver is visible one day after hatching. It arises as a solid outgrowth from the dorsal wall of the intestine, which has already a lumen. The liver cells soon become arranged into a reticulum. The liver grows ventrally between the ectoderm and the yolksac '(Wilson, 1889/1891)'. Its further development will be dealt with below under liver/yolksac contact.

'Ryder (1884)' remarked that the secretion of bile into the gall bladder, which precedes the ingestion of food in *Cottus, Salmo* and others, is not observed in *Gadus morrhua*. He suggested that the secretion of bile in the former is probably analogous to the meconium discharged by recently-born infants.

Publications dealing with the development of the liver in vertebrates in general are for instance those by 'Brachet (1896)' and 'Siwe (1929)'. Special accounts of the teleostean liver, essentially with regard to the phylogenesis of the vertebrate liver in general, were given by 'Segerstrale (1910)', 'Riggert (1922)', 'Siwe (1929)', 'Smallwood and Derrickson. (1933)' and 'Williams (1939)'. 'Segerstrale (1910)' supplied a general account of the teleostean liver. 'Siwe (1929)' described the liver in the trout as two cranially-directed sacs arising from a posterior common part. 'Smallwood and Derrickson (1933)' investigated the liver development in the carp, *Cyprinus carpio*. 'Williams (1939)' gave a detailed account of the liver and the liver-yolksac contact. The liver anlage of the winter-flounder, *Pseudopleuronectes americanus*, can be distinguished 10 days after fertilization as a median thickening of the ventral intestinal wall. After hatching the cells rearrange themselves into anastomosing cords and tubules. In older larvae the spaces of this reticulum are taken up by mesenchymal cells, which will give rise to the connective tissue and the endothelia of the adult liver. The outside diameter of the liver duct measures 15 microns at hatching and increases to 75 microns in older larvae. The wall of the duct is made up of a single layer of columnar cells, which gradually become flattened. The primordial liver of the cod, *Gadus callarias*, is similar to that of the flounder in that it arises about ten days after fertilization as a proliferation of the ventral part of the intestine. However, the liver of the cod is bilobed and its anlage is not solid. Cavities and clefts are evident but they are no longer distinguishable in the 12 to 15 day old embryo. At hatching the liver cells are arranged into a reticulum made up of cords and tubules. Again, mesenchyme is seen to invade the spaces to form the connective tissue framework and walls of the blood vessels. The gall bladder arises as a diverticulum of the liver duct. The initially columnar cells lining the gall bladder and the duct become flattened during development.

It would seem that the liver of all teleostean embryos is always in contact with the yolksac.

Summary

The teleostean liver arises as a solid outgrowth from the intestine. It may contain lobes but is not made up of lobules as is the case e.g. for the mammalian liver. It is traversed by connective tissue cells and contains a large gall bladder.

The teleostean liver is in contact with the yolksac: for instance dorsally on the right or left side in salmonids, caudally in Coregonus. This contact may have implications for the uptake of yolk (see 19.6).

19.6 LIVER-YOLKSAC CONTACT

As mentioned above, all vertebrates, with exception of the teleosts, show the digestive tube open to the yolksac and the entoderm surrounding it (Fig. 19/7a). In teleosts the originally solid intestine later acquires a lumen and becomes flanked by the ventrally progressing lateral plate (somatopleura and splanchnopleura) (Fig. 19/7b).

In most teleosts investigated so far (with the exception of the zebrafish) the liver maintains a very close association with the yolk and eventually envelops it completely.

'Ryder (1883a)' already remarked that in spite of the close association 'there is no connection of any kind between the intestine and the yelk sac at any time, such as has been described as connecting the yolk and the intestine in some embryo sharks.' He went on to note: 'The manner in which the yelk of fish ova is absorbed or incorporated into the body of the young fish, especially in those forms in which no vessels traverse the yelk bag, was for a long time a puzzle to me'. In contrast to the above, 'von Baer (1835)' had mentioned earlier a 'direct communication' between yolksac and intestine immediately in front of the liver. Subsequently, 'Wilson (1889/1891)' observed that in *Serranus atrarius* the prominent liver is connected with the final disappearance of the yolk and periblast and concluded that 'it is very evident that the liver is absorbing the yolk and periblastic protoplasm' (Fig. 19/8a). 'Forchhammer (1819)' in his thesis on the viviparous blenny had suggested that the liver takes up yolk in the contact area. Similarly, 'Chevey

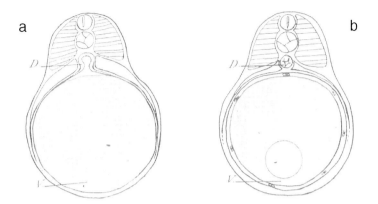

Fig. 19/7 Diagrammatic cross-sections through anterior trunk region (Chevey 1924, 1925). **a** - through a vertebrate except a teleost. Intestine is open to yolk. **b** - through *Perca fluviatilis* and *Serranus atrarius* to demonstrate the separation of intestine and yolk. D - intestine; V - yolk.

(1924, 1925)' as a result of his observations on *Perca fluviatilis* described the intimate contact between liver cells and yolksac in this fish (Fig. 19/8b,c). The largest area of contact (called by him 'primary contact') is seen along the oil globule which is finally completely surrounded by the liver. 'Chevey' concluded that 'le foie joue donc un rôle embryonnaire transitoire bien spécial, rôle de résorption du vitellus' (thus the liver plays a transient but special embryonic role, namely that of absorbing the yolk). He showed that the periblast cytoplasm penetrates the yolk and eventually diffuses entirely through it, absorbing the yolk on its inner side while it is itself being converted into hepatic tissue on its outer side. This happens from one to two days before hatching to 5 or 6 days after.

The above-mentioned results on the perch led 'Portmann and Metzner (1929)' to analyze other teleost embryos. They found that in *Salmo salar* the liver is originally separated from the yolk by two layers of the splanchnopleura (one layer surrounding the intestine, the other the liver) (Fig. 19/9a). The splanchnopleura gradually thinned to one layer (Fig. 19/9b) and then disappearrd yielding what they called a 'secondary direct contact' between liver and yolk (Fig. 19/9c). In many places the boundaries of the liver cells became indistinct. This would indicate a complete contact of cytoplasm of the liver with the cytoplasm of the periblast. The area of contact increased during development. In the salmon the liver-yolk-sac contact decreased again as a result of the expanding lateral plate. However, the close contact between liver and yolksac persisted for a long time; Fig. 19/9d shows schematically a cross-section of a fish 40 days after the onset of hatching. The same authors observed in the pike a close contact between liver and yolk while the contact zone between pancreas and yolk syncytium is characterized

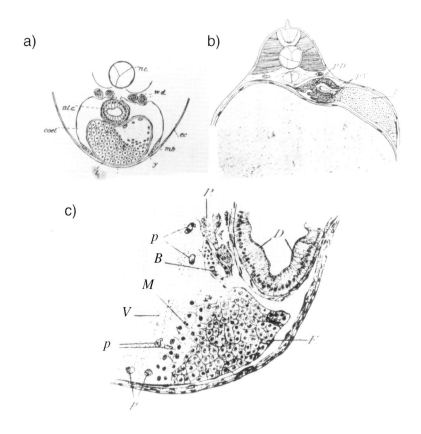

Fig. 19/8a Intimate contact between liver cells and periblast of yolk in the postembryo of *Serranus atrarius* (Wilson, 1889). al.c. - alimentary canal; coel - *coelom*; ec - *ectoderm*; l - liver; mb - muscle band; nc - neural chord; wd - Wolff'ian duct; y - yolk.

Fig. 19/8b Liver-yolk contact of 9 day old *Perca fluviatilis* embryo. Yolk resorbed by vitelline circulation (Chevey, 1924). F - liver; PD - pancreatic duct; PV - ventral pancreas.

Fig. 19/8c Intimate contact between liver cells and periblast of yolk of newly hatched *Perca fluviatilis*. Yolk resorbed by liver (Chevey 1924). B - gall bladder; D - intestine; F - liver; M - *mitosis*; P - pancreas; p - *periblast nuclei*; V - *vitellus*.

by the presence of the splanchnopleura and chromatophores (Fig. 19/10a,b). However, no direct contact was evident between yolk and liver in *Cottus* and *Gobio*. 'Pohlmann (1939)' observed the decomposition of the splanchnopleural layers between the liver and yolksac periblast in *Trichopodus trichoperus* and *Cichlasoma bimaculata*. However, her observation of pigment cells between liver cells and the periblast would indicate the presence of an endothelial layer.

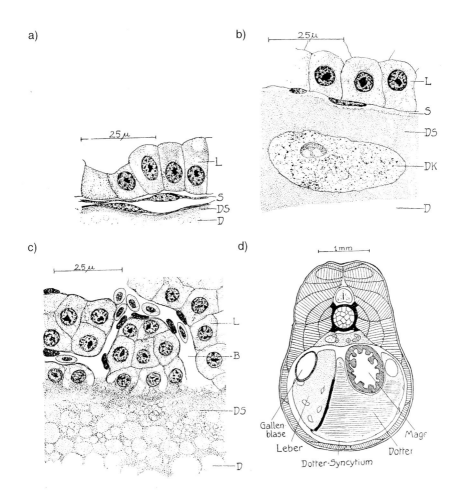

Fig. 19/9 Liver and yolk contact in the salmon (Portmann and Metzner, 1929). **a** - In the young salmon postembryo, up to two weeks after hatching, the liver is separated from the yolksac *periblast* by two layers of the sp*lanchnopleura*. **b** - 30 days after hatching the two *splanchnopleurae* are pressed tightly together so that they appear as one layer. **c** - This displays an area of the same specimen showing a large area of intimate contact not only between liver cells and *periblast* but also between the lumen of two capillaries and the *periblast*. **d** - Schematic drawing showing the area of intimate liver/yolksac contact in *Salmo* 40 days after hatching. Abbreviations for a-c: B - blood vessel; D - yolk; DK - *periblast nuclei*; DS - yolk *syncytium (periblast)*; L - liver cell; S - *splanchnopleura*. Labelling for d: Dotter - yolk; Dotter-Syncytium - yolk *syncytium (periblast)*; Gallenblase - gall bladder; Leber - liver; Magen - stomach.

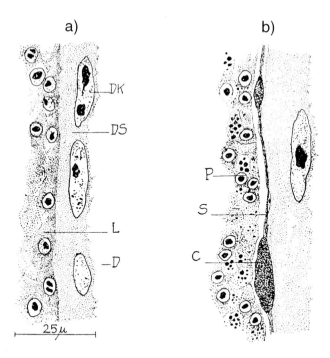

Fig. 19/10 Two neighbouring longitudinal sections of the pike *Esox* lucius (Portmann and Metzner, 1929). **a** - taken through the area of liver: contact between liver cells and *periblast* is evident. **b** - through the region of the *pancreas*. A separating splanchopleural layer is clearly visible, accentuated by the presence of chromatophores. C - chromatophore; D - yolk; DK - *periblast nuclei*; DS - yolk syncytial layer; L - liver; P - *pancreas*; S - *splanchnopleura*.

'Williams (1939)' devoted a whole thesis to the development of the liver in teleosts with special reference to the liver and yolk-periblast contact. He observed an intimate connection between the developing liver and periblast in the flounder *Pseudopleuronectes americanus*. The connection was so close as to provide an optical continuity between the liver cell and periblast cytoplasm. The area of contact was at its maximum 'during a few days of larval life'. The results obtained with the yellow perch, *Perca flavescens*, were inconclusive. The few observations on the cod, *Gadus callarias*, seemed to agree with those on the flounder. The results on the shad, *Alosa sapidissima*, were again similar to those of the flounder. The author concluded that the developing liver in some teleosts is active in yolk-periblast absorption.

The contact between liver and yolk was another initial object of a thesis '(Kunz, 1964)'. She confirmed the results of 'Chevey (1924)' showing that in *Perca* the liver lies directly upon the yolksac without separation of splanchnopleural layers (called 'primary contact') (Fig. 19/11a). In contrast, in the eggs of *Lebistes reticulatus* (*Poecilia reticulata*), *Coregonus alpinus* and *Pterophyllum scalare* the liver is initially separated from the yolksac syncytium by a double layer of splanchnopleura. This is followed by a change into a single-cell layer. In none of these species was a complete disappearance of the splanchnopleura (secondary contact) observed (Figs. 19/11bc; 19/12a,b; 19/13a,b; 19/14a,b).

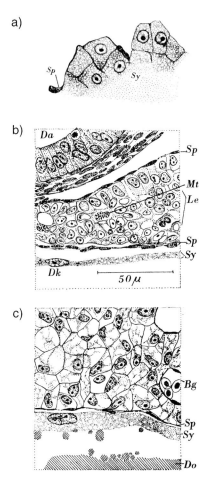

Fig. 19/11 Liver-yolk contact in *Perca fluviatilis* and *Coregonus alpinus* (Kunz, 1964, with kind permission of the director of the Muséum d'Histoire naturelle, Geneva). **a** - intimate liver/yolksac contact on 4th postembryonic day of *Perca fluviatilis*. **b** - *Coregonus alpinus* on 30th embryonic day. Two layers of *splanchnopleura* separate liver from yolk. **c** - *Coregonus alpinus* on 60th embryonic day (hatching). Only one layer of *splanchnopleura* is left. Liver cells are highly vacuolized. Da - intestine; Do - yolk; Bg - capillary; Dk - *periblast nucleus*; Le - liver; Mt - *mitosis*; Sp - *splanchnopleura*; Sy - *periblast*.

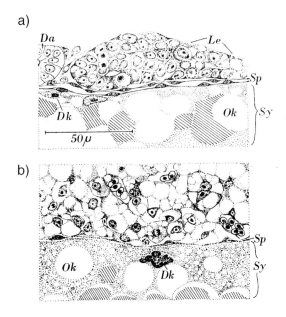

Fig. 19/12 Liver-yolk contact in *Pterophyllum scalare* (Kunz, 1964, with kind permission of the director of the Muséum d'Histoire naturelle, Geneva). **a** - on the day of hatching two layers of *splanchnopleura* separate liver from yolk. **b** - on the 4th postembryonic day the *splanchnopleura* in the contact zone is reduced to one layer. Da - intestine; Dk - *periblast nuclei*; Le - liver; Ok - oil globules; Sp - *splanchnopleura*; Sy - syncytial layer of yolk.

Periblast nuclei are much larger than those in the liver or in teleoestean organs generally (Fig. 19/14a). Later in development they divide amitotically and take up a radial position (Figs. 19/14b,c,d).

Based on all the above observations, the results regarding liver-yolksac contact so far obtained can be assigned to two categories '(Kunz, 1964)':

1) The liver is laid down before the mesoderm of the lateral plate starts to grow into the extraembryonic region. As a result the liver parenchyma lies in direct contact with the periblast. This would correspond to a 'primary direct contact' (examples: *Serranus atrarius, Perca fluviatilis, Pseudopleuronectes americanus, Gadus callarias, Alosa sapidissima*) (Fig. 19/15a).

2) The lateral plate surrounds the yolksac before the liver is laid down (Fig. 19/15b). As a result the intestine with its derivatives as well as the yolksac are completely surrounded by the splanchnopleura. The periblast surrounding the yolk is separated from the liver by two layers of splanchnopleura, i.e. the yolksac- and liver-splanchnopleura (Fig. 19/15c).

a) No reduction of the splanchnopleura layers in the contact zone is later observed (examples: *Cottus* and *Gobio*) (Fig. 19/15c).

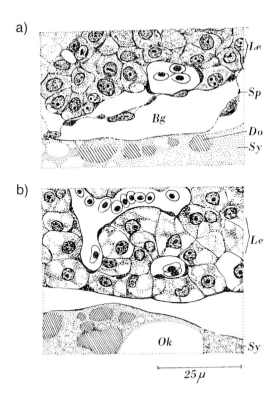

Fig. 19/13 Liver-yolk contact in *Poecilia reticulata* (*Lebistes reticulatus*) (Kunz, 1964 with kind permission of the director of the Muséum d'Histoire naturelle, Geneva). **a** - two layers of *splanchnopleura* separate liver from yolk on the 10th embryonic day. **b** - the *splanchnopleura* is reduced to one layer on the 30th embryonic day. Bg - blood capillary; Do - yolk; Le - liver; Ok - oil globules; Sp - *splanchnopleura*; Sy - syncytial layer of yolk.

b) The originally two-layered splanchnopleura is reduced to one layer. At least over most of the contact area this remains so to the end of the yolksac resorption (examples: *Lebistes reticulatus* (*Poecilia reticulata*), *Coregonus alpinus*, *Pterophyllum scalare*, *Trichopodus trichopterus* and *Cichlasoma bimaculatum* (Fig. 19/15d).
c) The two splanchnopleural layers are reduced to one layer, which later disappears so that — as in case 1) — the liver cells lie directly on the yolksac periblast. This phenomenon is called 'secondary direct contact' (examples: *Salmo*, *Esox*, *Pseudopleuronectes americanus*, *Alosa sapidissima*, *Gadus callarius*) (Fig. 19/15a).

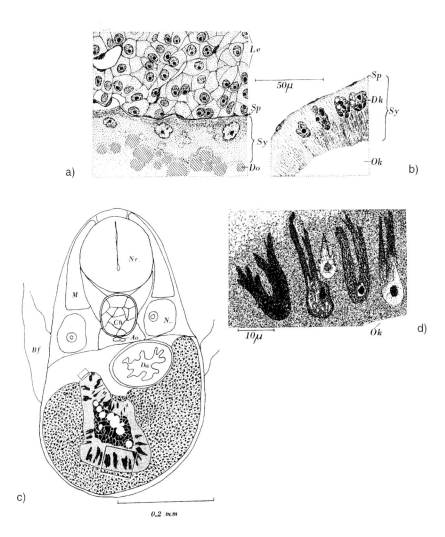

Fig. 19/14a-d Size and fate of *periblast nuclei* (Kunz, 1964, with kind permission of the director of the Muséum d'Histoire naturelle, Geneva). **a** - Note the difference between *nuclei* of *periblast* surrounding the yolk and those of the adjacent liver (*Coregonus*, 15th postembryonic day). **b** - The *periblast nuclei* adjacent to the very large oil globule of the same specimen have assumed a radial direction and divide amitotically. **c** - Centripetal orientation of *periblast nuclei* surrounding the yolk remnants of *Pterophyllum scalare* on its 8th postembryonic day. **d** - Detail of square box of c) showing radial 'comet-like' *periblast nuclei*. Ao - *aorta*; Bf - pectoral fin; Ch - notochord; Da - intestine; Dk - *periblast nuclei*; Do - yolk; Le - liver; M - somite; N - kidney; Nr - neural chord; Ok - oil globules; Sp - *splanchnopleura*; Sy - syncytial layer of yolk or oil-globule Ok.

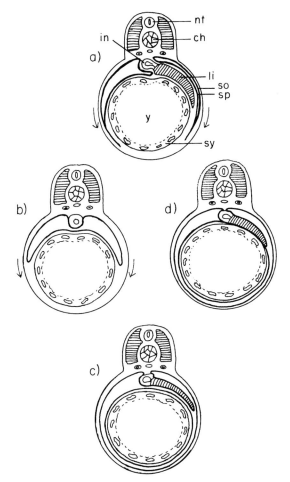

Fig. 19/15 Summary of liver-yolksac contact. Schematic representation of the presence of *splachnopleura* layer(s) in three different groups of teleosts (redrawn from Portmann and Metzner, 1929 and Kunz, 1964, with kind permission of the director of the Muséum d'Histoire naturelle, Geneva). **a** - Coelom grows to surround the yolk sphere **after** the liver has been laid down. Result: there is no *splachnopleura* present in the contact area (as observed in *Serranus atrarius* and in the perch). Compare with Fig. 19/8b,c. **c** - Coelom grows around yolk sphere **before** the liver is laid down which results in two layers of *splachnopleura* separating liver and yolksac as observed in salmon (19/9a), *Coregonus* (19/11b). *Pterophyllum* (19/12a) and *Poecilia* (19/13a). **d** - The contact zone is further reduced to one layer as observed in all species mentioned in c). However, in the salmon and the pike the single layer disappears, which yields a 'direct contact' similar to a). This has been called a 'secondary contact'. ch - notochord; in - intestine; li - liver; nt - neural tube; so - *somatopleura*; sp - s;planchnopleura; sy - *periblast (syncytium)*; y - yolk.

It has been suggested that the functional significance of the 'primary' and 'secondary direct contact zone' is the uptake of yolk into the embryo at a time or in regions where the blood circulation on the yolk is absent. This gave rise to the hypothesis that, since the liver is an outgrowth of the intestine, this represents entodermal absorption of yolk via a detour and brings the teleosts back into the fold of meroblastic vertebrates. It should be mentioned in this context that the yolk-periblast of the winter flounder, *Pseudopleuronectes americanus*, becomes constricted at its middle and eventually separates the anterior from the posterior part. Both parts become absorbed but only the anterior one has remained in contact with the liver. Additionally, in late postembryonic stages of many fish, a considerable remnant of yolk is found at a great distance from the liver and it, too, becomes eventually absorbed. Moreover, yolk absorption by the liver need not be prevented by one or two splanchnopleural layers. So rather than trying to squeeze the teleost embryo into its 'proper phylogenetic place', on morphological grounds, one should leave it to modern molecular scientists to shed light on the problem of yolk absorption at a time and in sites where the yolksac circulation is absent.

In order to appreciate the differences between the development of teleosts and other vertebrates a schematic comparison is supplied by Figs. 19/16A-C and Fig. 19/17. (It should be emphasized that this comparison does not imply a relationship, in an ancestrtal sense, to amphibia.)

Fig. 19/16A shows a cross-section of a frog embryo, which is an example of a macrolecithal type of embryo. The incision in the intestine, the ventral blood forming mesoderm and the ectoderm is meant to explain the situation of the telolecithal elasmobranchs (B) and teleosts (C).

Fig. 19/16 B demonstrates the entoderm/yolk contact observed in the elasmobranchs. In these fish the intestine remains in contact with the yolk by an umbilical cord (Fig. 19/17).

Fig. 19/16C displays the situation in the teleosts. The main difference is clearly the missing intestine-yolk contact. However, other differences encountered are, for example, the fact that in teleosts the extra-embryonic blood-vessels are not present in the splanchnopleura but rather lie between both layers of the lateral plate. However, the blood cell formation in teleosts is unique in many other ways (see Chapter 18).

Summary

In different teleosts the liver is found in various positions on the yolksac (anterior dorsal left and right, lateral left and right, caudal).

The morphological type of the hepato[1]-vitelline connection is thought to reveal the possibility of yolk uptake by the liver; 1) In the zones of 'primary direct contact' the cytoplasm of the periblast surrounding the yolk is continuous with the liver cell cytoplasm in the contact zones. 2) In other types very early on two splanchnopleural layers intervene between liver and yolk. Subsequently, they either a)

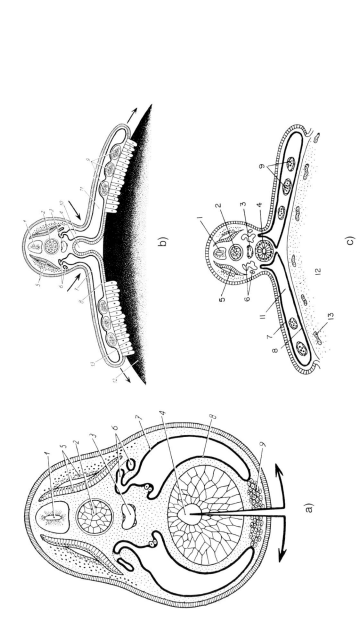

Fig. 19/16 Comparison of schematic transverse sections through the trunk region of **a** - holoblastic (frog), **b** - meroblastic (elasmobranch) and **c** - meroblastic (teleost) embryos. Figs. a) and b) are with kind permission of Benno Schwabe, Basel). The body plan is characterized by structures found at equivalent locations. 1 - neural tube; 2 - notochord; 3 - blood vessels; 4 - intestine; 5 - somite; 6 - kidney primordium; 7 - *somatopleura*; 8 - *splanchnopleura*; 9 - ventral blood cells (blood islands in b) and c)); 10 and 11 - intra- and extraembryonic coelom; 12 - yolk; 13 - yolk *entoderm* in b), *periblast* with *nuclei* in c). The holoblastic embryo a) has been cut ventrally and the arrows indicate the extension of the cut part over the yolk mass of b). The diagram of c clearly shows that in teleosts there is no continuity between the intestine and the yolk sphere.

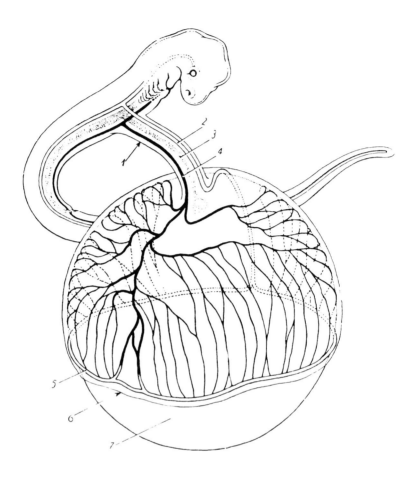

Fig. 19/17 Embryo of a shark displaying connection between intestine and yolk via 'umbilical chord' (Portmann, 1976, with kind permission of Benno Schwabe, Basel). 1 - umbilical cord; 2 - vitelline artery; 3 - yolk duct; 4 - vitelline vein; 5 - marginal blood vessel of yolk sac; 6 - margin of epiboly; 7 - free yolk.

remain intact, or b) become reduced to one layer or c) they degenerate altogether. As a result of c) optical continuity between the periblast and liver cell cytoplasm is displayed, a condition which has been called 'secondary direct contact'.

In the case of 'primary' or 'secondary direct contact' it has been suggested that the periblast cytoplasm resorbs the yolk on its inner side while it is being converted into hepatic tissue on its outer side. However, it should be stressed that one or two layers (2a, 2b) of splanchonpleura cells do not, from a physiological point of view, prevent yolk uptake. Teleosts are unique in that they do not show a connection between yolk and intestine as observed in all other macrolecithal vertebrates. Since the liver is formed by the intestine, it has been suggested that teleost embryos conform basically to the mode of yolk absorption common to lower fish, amphibians, birds and reptiles. However, it should be stressed that from a morphological point of view the liver is not the only agent that could be active in yolk absorption (see for instance, the situation in the flounder, *Pseudopleuroneces americanus*, in which the posterior part of the yolksac becomes separated from the liver).

Endnotes

[1] Gk. hepar
[2] rudiment of both meso- and entoderm (as opposed to epiblast), see Chapter 11
[3] loose mesodermal cells, see Chapter 18
[4] from Gk. thureos = shield, and eidos = form

Chapter twenty

Viviparity

> 'Most of the available information is contained in anatomical and morphological studies in which function is deduced from structure. It is fortunate that these deductions are probably valid.'
>
> Wourms (1981)

Viviparity in teleosts has been defined as 'a process in which eggs are fertilized internally and are retained and undergo development in the maternal reproductive system' '(Wourms *et al.*, 1988)'.' Balon (1981, 1990)' classified reproductive styles into different ecological groups and he called viviparous fish 'internal live bearers'.

There has been controversy about which author gave the first account of live-bearing fishes: the names and dates are given as follows: 'LeConte (1851)', 'Lord (1852)', 'Jackson (1852)', 'Gibbons (1852)',' Webb (1852)', 'Agassiz (1853, 1854)' (cf. 'Eigenmann, 1892'). However, the fact that *Zoarces viviparous* produces live young ones was noted already in 1624 by 'Schonevelde'. He mentioned 300 embryos in a brood.

'Grove and Wourms (1994)' stressed that while, in teleosts generally, the follicle wall plays an important role in regulating oogenesis and transport of metabolites and yolk components to the developing oocyte, it has some important additional functions in some groups of viviparous teleosts: 'First, spermatozoa must cross the follicle wall for fertilization to occur. Second, the site of gestation must be compartmentalized from the rest of the maternal organism to maintain the appropriate environment for embryonic development and to protect the embryo from the maternal immune system. Third, there must be metabolic exchange between the embryo and the mother. Finally in those species where embryonic nutrition is matrotrophic, nutrients must cross the follicular epithelium.'

Within vertebrates, internal fertilization followed by viviparity first arose among the fishes. It is prevalent in the chondrichthyes but less so in teleosts where it occurs in 14 families belonging to the orders gadiformes, ophiiformes, atheriniformes, cyprinodontiformes, scorpaeniformes, perciformes and beloniformes. Only 510 of an estimated total of more than 24 000 teleost species are livebearing.

Subsequent development takes place either within the follicle (follicular/intrafollicular) or within the lumen of the ovary (intralumenal/intraluminal) or body cavity of the female. The latter is the most prevalent mode in viviparous teleosts.
1) Intrafollicular gestation has been observed in Cinidae, Poeciliidae, Anablepidae and Labriosomidae. These embryos leave the follicle only immediately before birth.
2) Intralumenal gestation has been described for some embryos of the Perciformes, e.g. Zoarcidae, Brotulidae, Clinidae, Embiotocidae, Scorpaenidae, Comephoridae and the Cyprinodontiformes, e.g., Goodeidae and Jenynsiidae; in the Gadiformes, e.g., Brotulidae and Ophidioidae and in the Beloniformes in one family only, the Hemirhamphidae.
3) There is an in-between group, in which intrafollicular gestation is prolonged to an embryonic stage, while hatching and completion of development occurs in the ovarian lumen (e.g. Jenynsia, the ophidioid Dinematichthys and the hemirhamphid *Novorhamphus hageni*).

According to the maternal-embryonic relationship, we call embryos which depend exclusively on the yolk stored in their yolksac lecithotrophic[1] (ovoviviparous) while embryos which, apart from their own yolk reserves, rely essentially on the supply of maternal nutrients are known as matrotrophic[2] (truly viviparous). The first type of embryos undergoes a decrease in embryonic weight during gestation whereas the second type exhibits a considerable weight increase. In the latter group the number of young born is greatly reduced. All embryos of livebearing teleosts depend on the female for gas exchange, respiration, and osmoregulation-excretion.

Teleosts are characterized by a cystovarian ovary, i.e. a single hollow ovary lined by the germinal epithelium. In livebearing teleosts this epithelium is continuous with the gonoduct, which opens to the exterior as a genital pore. The investment of the gonoduct is a continuation of the muscular wall of the ovary and, therefore, is not homologous with the gonoduct of other vertebrates which is formed by the Müllerian duct '(Wourms, 1994)'. In some forms, such as *Sebastodes rubrovinctus*, the ovary is double '(Eigenmann, 1892)' and in some others, such as *Dermogenys pusillus*, *Lucifuga subterraneus* and *Stygocola dentata*, the ovary is partially fused. But in all these cases only a single gonoduct has been observed. Lobulated ovigerous folds and, in some goodeids, a median septum protrudes into the ovarian cavity. Both structures contain highly vascularized connective tissue and are limited externally by a layer of circular muscle. The whole ovary is surrounded by the peritoneal epithelium (rev. 'Amoroso, 1960; Wourms *et al.*, 1988). (Figs. 20/1a,b,c; 20/22))

In most livebearing fishes internal fertilization is guaranteed by a remodeling of the male anal fin into a copulatory organ. In some teleosts the sperms are stored in ovarian folds which allows for superfoetation (also superembryonation)[4]. In the most extreme cases (*Heterandria formosa* and several species of *Poeciliopsis* and

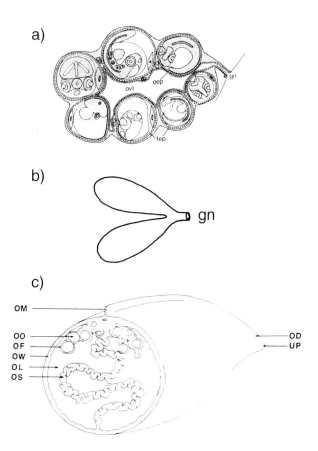

Fig. 20/1 Types of ovaries of livebearing teleosts. **a** - Single ovary. Diagram of longitudinal section through a poeciliid ovary. It is lined with a single layered *epithelium*, which continues into the gonoduct. The developing embryos are surrounded by a follicular *epithelium* and embedded in connective tissue (Wourms *et al.*, 1988, with kind permission of Academic Press, London). fep - follicular *epithelium*; gn - gonopore; ovl - ovarian *lumen*; oep - ovarian *epithelium*. **b** - Paired ovary. For detail refer to Fig. 20/22 showing ovary of *Anableps anableps*. **c** - Partially fused ovary of *Ameca splendens*. Both the anterior third of the ovary and the embryos have been deleted (Lombardi and Wourms, 1985, with kind permission of Wiley Liss Inc., a subsidiary of John Wiley & Sons, Inc.). OF - follicular wall; OL - ovarian *lumen*; OM - *mesorchium*; OO - *oogonia*; OS - ovarian *septum*; OW - ovarian wall; IE - inner ovarian *epithelium*; OL - ovarian lumen; OM - ovarian mesentery; OO - oocyte; OD - oviduct; OS - ovarian *septum*; OW - ovarian wall; UP - urogenital pore.

Aulophallus) as many as nine broods, with their members in different stages of development, are found in the ovary at the same time.

The oocytes of viviparous teleosts are surrounded by a reduced envelope (ZR). In the viviparous zoarcids and scorpaenids ovulation precedes fertilization. In most other groups fertilization is intrafollicular. In some the two events occur so closely together that it is difficult to determine the sequence '(Stuhlmann, 1887; Mendoza, 1940; Bretschneider and DeWit, 1947; Moser, 1967a,b; Kristofferson et al., 1973, Wourms et al., 1988)'.

In his review on 'viviparity in teleost fishes' 'Turner (1947)' concluded that there seems to be no relation between the development of viviparity and ecological conditions. The livebearing habit is found in both marine and fresh-water fishes. It occurs in fishes living in the cold, lightless depths of the sea, in dark, fresh-water caves, in shore habitats, in the tepid waters of tropical lagoons, in swift-running streams, and in the lakes of elevated sun-drenched plateaus. Some live bearers are bottom-feeding, others are top-swimmers and surface-feeders, and still others are swift, efficient carnivores.

Recent extensive reviews of viviparity in fish have been published by 'Wourms (1981)', 'Balon (1985)', 'Wourms et al (1988)', 'Schindler and Hamlett (1993)', 'Wourms (1994)' and 'Greven (1995)'.

In the following the topic is divided according to the maternal-embryonic relationship:

20.1 MATROTROPHIC SPECIES

A maternal contribution of nutritive material is evident if during gestation an increase in net dry weight of the embryos is observed. Since maternal nutrients reach the embryo via different epithelial tissues, a division into dermotrophic uptake (via general body surface, finfolds, gill filaments, buccal epithelium, yolk-sac and pericardial sac) and enterotrophic uptake (via gut and its derivatives such as trophotaeniae) has been suggested by 'Wourms and Lombardi (1985)'. 'Balon (1985)' uses the term 'obligate internal bearers' for matrotrophic fish.

The term oophagy refers to ingestion of eggs periodically ovulated during gestation and adelphophagy to eating of other siblings. Both terms are listed under 'intrauterine embryonic cannibalism'. Embryonic nutrients are known as embryotrophs[3]. When they are of maternal origin, they are also called histotrophs (histogenic when secreted and histocytic when derived from degenerated cells).

Zoarcidae (eelpouts)

The perciform *Zoarces viviparus* (G. Aalmutter) is a littoral marine fish, which inhabits the intertidal zone. However, Zoarcidae are also found in brackish waters and river mouths. As mentioned above, *Zoarces viviparus* was the first livebearing fish reported '(Schoenevelde, 1624; cf. Stuhlmann, 1887)'. 'Schoenevelde' was a medical general practitioner; he named the fish he observed *Mustela vivipera* and published his findings in Ichthyologia (Hamburg). These findings were subsequently quoted but nothing of major importance on viviparous fish was published over almost 200 years. 'Forchhammer (1819)' described in his thesis the development of this fish which he called *Blennius viviparus* (viviparous blenny) (Fig. 20/7).

This was followed by a publication by 'Rathke (1824)' and an embryological monography by the same author in 1833. He stressed that amazingly the embryo develops at the same site where the ovum was formed, i.e. in the ovary. *Blennius viviparus* hatches after 3 weeks and remains in the ovarian cavity for about 4 months. The continuous growth of the postembryo and the gradual decrease of the ovarian fluid indicate matrotrophy. A cursive description of the ovary of this fish was given by 'M'Intosh (1885)'. 'Stuhlmann (1887)' mentioned that from 110 up to 405 young ones are born to this fish (referred to as *Zoarces viviparus*) at the same time. The number depends, as in oviparous fish, on the age of the mother. 'Van Bambeke (1888)' reviewed the number of embryos per fish communicated by the early investigators. He stressed that the size of the mother, as well as the geographical location of the fish, influence the numbers. Almost a century later, 'Goetting (1976)' again stressed that the number of embryos depends on the length of the mother; he quoted a number of '100 or little more'. The adult male possesses a small cutaneous papilla located behind the vent. During the breeding season this papilla is transformed into a copulatory organ. The female possesses a small aperture behind the anus. At the time of copulation, the mature oocytes leave the follicles to enter the ovarian cavity where fertilization occurs '(Soin, 1968)'.

The ovarian wall of *Zoarces viviparus* is highly vascularized. Dorsal arteries originate directly from the aorta descendens and ventrally the arteria mesenterica supplies the blood. Venous drainage is accomplished by the dorsal vena renalis and ventral veins, which with the Vena mesenterica enter the liver. During oogenesis highly vascularized ovigerous processes containing developing follicles extend into the lumen of the ovary (Figs. 20/2-4). The LM analyses of 'Goetting (1976)' confirmed what 'Stuhlmann (1887)' had already described. The process called by the latter an 'eggsac' (G. Eiersack) the former named a 'funiculus'. A 'liquor' originating from the theca folliculi surrounds the oocytes, which are large (2.8-3.2 millimetres) and contain numerous small oil droplets. The follicular vessels bring them nutrients and carry out gas exchange and waste disposal. After ovulation fertilization takes place in the ovarian cavity. The follicle does not undergo atresia but while retaining its capillary network becomes transformed into a long (1.5-2 millimetres) 'epithelial outgrowth' (G. Zotte), called 'calyx nutritius simplex' by 'Bretschneider and DeWit (1947)' (Fig. 20/5,6). These protruberances, together with the vascularized epithelial cells of the ovary wall, secrete nutrients into the ovarian lumen. Thus the ovulated eggs fall into a viscous liquid, which is further enriched by degenerated unfertilized eggs. 'Soin (1968)' noted that the ovarian wall is characterized by the presence of thin protuberances (G. 'Zotten'), which can attain a length of 1.5 centimetres and which carry the developing eggs. The layers of the *Zoarces* oocytes are the same as in oviparous fish; however, they are thinner and the vascularization of the follicle is much more pronounced. 'Goetting (1974)' compared by freeze etching the ultrastructure of the oocytes of various teleosts, which revealed cytomorphological differences as regards the envelope. *Zoarces* showed a decrease in the number as well as in the diameter of layers compared with oviparous eggs.

'Kristofferson et al (1973)' agreed with 'Stuhlmann (1878)' and 'Bretschneider and DeWit (1947)' that after ovulation the numerous large and 'empty' calyces (which they call villi) of the ovary do not degenerate but remain functional and produce the embryotrophe[3] (histiotrophe), also called 'uterine milk', for the postembryos and are also engaged in gas exchange and excretory function. The vascular system of such a protuberance is shown in Fig. 20/5. After ovulation, too, small protuberances carrying the eggs of the next generation are observed (Fig.20/6). They are also engaged in gas exchange and excretion. Chemical analyses revealed that the concentration of inorganic ions is roughly equal in plasma and histiotrophe. However, the latter contains very little protein, or in some cases even none. It differs widely from the surrounding brackish water which is hypotonic to the plasma '(Veith, 1980)'. According to 'Lombardi and Wourms (1985a)' 'in terms of function, the ovarian calyces of *Zoarces* are equivalent to the maternal portion of a mammalian epithelio-chorional placenta, whereas in terms of morphology, they resemble the chorionic villi of the mammalian chorioallantoic placenta'.

The embryos hatch after 3 weeks and remain in the ovary for 2-4 months ('Rathke, 1833)'. According to 'Soin (1968)', hatching occurs at the approximate age of 2 months and at a length of 17.1 millimetres. 'Bretschneider and DeWitt (1947)' reported that the zygote shows a weight of 20 milligrams and that after about 4 months the fish is born at a weight of 240 milligrams and a length of 44 millimetres.

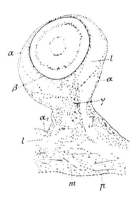

Fig. 20/2 Oocyte in follicular protuberance of *Zoarces viviparus* (Stuhlmann, 1887) × 650. a - outer vascular layer; l - lymph spaces; a_1 - vascular stalk developing; m - *muscularis*; ß - inner vascular layer; p - peritoneal membrane; y - vascular stalk.

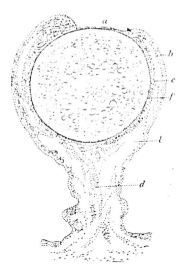

Fig. 20/3 Egg in follicle of pregnant female of *Zoarces viviparus*. Vessels are developed (Stuhlmann, 1887). × 55. a - depression; d - vascular stalk; b - outer vascular layer; f - follicular *epithelium*; c - inner vascular layer; l - lymph space.

Fig. 20/4 Part of a pregnant ovary of *Zoarces viviparus* with protuberances and eggs (Stuhlmann, 1887). Natural size.

'Stuhlmann (1887)', and before him 'Forchhammer (1819)' and 'Rathke (1833)', observed that, immediately after hatching, the free embryos in the ovarian cavity swallow ovarian liquid (nowadays referred to as embryotrophe or histiotrophe) consisting of ovarian secretions, desquamated ovarian cells and non-viable embryos. 'Kolster (1905)' provided histological evidence that the nutritive material in the early embryo is at first derived from the yolk and that later the postembryo relies mainly on the intake and absorption of embryotrophe.

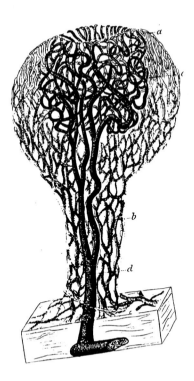

Fig. 20/5 A protuberance 3-4 months after the birth of the embryos of *Zoarces viviparus*.. Slightly schematic (Stuhlmann, 1887). × 4. Doubly injected (*arteria mesenterica, vena mesenterica*). a - polar contraction (depression) c - central vesicular Knäuel. b - outer vascular layer; d - central arterial stalk.

Fig. 20/6 A protuberance 3-4 months after parturition. An egg for the next generation of embryos of *Zoarces viviparous* is shown at the base of the stalk (Stuhlmann, 1887) × 4.

'Stuhlmann (1887)' analyzed the contents of the embryonic intestinal system. The middle part of the intestine was filled with erythrocytes. They were in the process of decaying whereas the ones in the middle of the lumen were intact. Folded membranes were also observed which might represent the membranes shed by the embryos when hatching. This might explain why 'Forchhammer (1819)', not finding any evidence of them, concluded that the egg membranes are absorbed by the embryos. 'Forchhammer (1819)', 'Rathke (1824, 1833)' and 'Stuhlmann (1887)' described and showed figures of the extremely large rectum in the early embryo of Zoarces viviparus (Figs. 20/7-20/12). 'Stuhlmann' could not detect any blood cells in this area and concluded that they are absorbed at an accelerated rate by the folded rectal wall described by all three authors (Fig. 20/8). The hatched embryos lie almost motionless in the viscous ovarian liquid. Since respiration by the gills and body surface might not be sufficient for them, 'Stuhlmann' suggested that the absorbed blood cells might supply the necessary oxygen to the embryos.

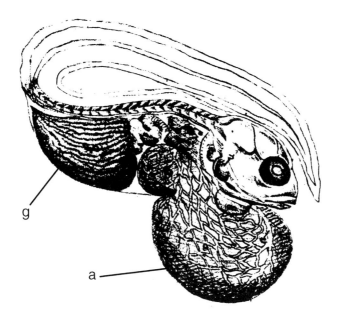

Fig. 20/7 Embryo of *Blennius viviparous* 15 or 20 days old. Part of the yolk already absorbed (Forchhammer, 1819). a - yolk; g - *rectum*.

'Amoroso (1960)' stressed that the female *Zoarces* secretes a fatty mucous substance into its ovarian cavity but he mistakenly suggested that the embryotrophe is taken up by the free embryos through their skin. 'Soin (1968)' — as suggested already before him by 'Forchhammer', 'Rathke' and 'Stuhlmann' mentioned above — noted in the free embryos a significantly modified hindgut with longitudinal folds, which would indicate uptake of food by the mouth.

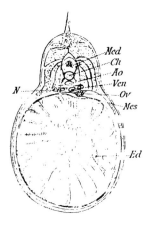

Fig. 20/8 Cross-section of the abdomen of an embryo of *Zoarces viviparus* at the level of the *rectum* displaying deep folds (Stuhlmann, 1887). × 10. Ao - *aorta*; Med - neural chord; Ch - notochord; Mes - mesentery; Ed - rectal folds; Ov - ovary; N - kidney.

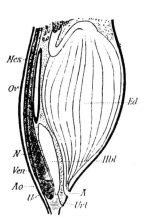

Fig. 20/9 Scheme of abdominal organs of a mature embryo of *Zoarces viviparus*. Longitudinal section (Stuhlmann, 1887). A - *anus*; N - kidney; Ao - *aorta*; Ov - ovary; Ed - *rectum*; U - *ureter*; Hbl - urinary bladder; Urt - mouth of *ureter*; Mes - mesentery; Ven - vein.

Mixed feeding (yolk and embryotrophe) starts in the *Zoarces* embryo when the yolksac is still quite large. The yolksac remains large for a lengthy period due to an increase in its water content. Thus, while yolk is being continuously absorbed, the large yolksac with its increasing capillary net ensures an adequate respiratory surface '(Soin, 1968)'. It had been noted that while the hindgut had decreased in diameter the urinary bladder was greatly enlarged '(Stuhlmann, 1887)'. 'Rathke (1833)' showed the decrease of the yolksac parallel with the development of the hindgut, an indication of uptake of embryotrophe by the gut (Figs. 20/10-12). In contrast, 'Korsgaard (1986)', who studied the maternal-embryonic and –postembryonic relationship in *Zoarces viviparus*, found that during the period from fertilization to hatching no change in dry weight was observed in the embryo. He, therefore, concluded that during this early period only the yolk provides nutrition for the embryo. However, an extensive nutritive relationship between mother and postembryos commences immediately after hatching. EM analyses revealed that the intestinal epithelium of the postembryo is organized into villi, the apical surfaces of which display microvilli. The presence of microvilli at an EM level corresponds to the brush border at LM level and is indicative of an absorptive surface '(Kristoffersson *et al.*, 1973; Wourms, 1981)'.

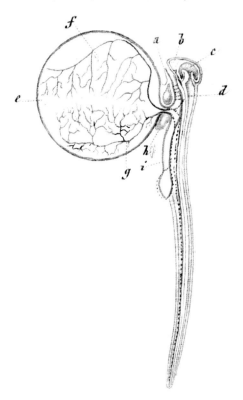

Fig. 20/10 Median section of embryo of *Blennius viviparus*, seen from the left (Rathke, 1833). × 16. a - lower jaw; b and c - brain; d - neurochord; e - yolksac; f - arteries of yolksac; g - veins of yolk sac; h - liver; i - intestine.

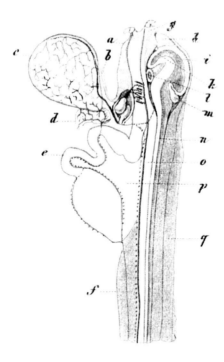

Fig. 20/11 Median section of embryo of *Blennius viviparus*, seen from the left, with decreased yolk sac and increased hindgut (Rathke, 1833). × 16. a - lower jaw; b - wall of yolksac; c - yolksac (pituitary?); d - liver; e - intestine; f - anal fin; g - upper jaw; h,i,k - brain; l - vesicular protuberance of brain; m - *metencephalon*; n - neural chord; o - onset of formation of vertebrae; p - *pancreas*; q - dorsal fin.

All the authors seem to agree that the respiratory surface over the yolksac is maintained by the highly developed yolksac circulation (while 'Stuhlmann', in addition, speculates about a possible uptake of ingested blood cells by the embryo). It has to be borne in mind that the developing fish are dependent on the oxygen concentration in the maternal ovarian fluid which itself is dependent on the oxygen concentration of the environment '(Soin, 1968)'. *Zoarces* is an inhabitant of northern arctic waters. These have a high oxygen content, which provides favourable conditions for the embryos developing in the ovary. Electrophoretic patterns showed that haemoglobins of embryos and adult *Zoarces* differ '(Hjorth. 1974)'.

The newborn *Zoarces* greatly resembles the adult fish. However, as mentioned earlier, it still carries remnants of yolk while it has started to take up external food. The paired and median fins are fully formed and provided with rays.

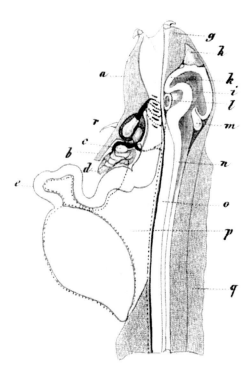

Fig. 20/12 Median section of embryo of *Blennius viviparus* on its last day, seen from the left. Hepatic vein flows over remnants of yolk sac into the heart. Note greatly enlarged hindgut (Rathke, 1833). × 16. a - lower jaw; b - wall of yolksac; c - yolksac; d - liver; e - intestine; f - anal fin; g - upper jaw; h,i,k - brain; l - vesicular protuberance of brain (pituitary?); m - *metencephalon*; n - neural chord; o - onset of formation of vertebrae; p - *pancreas*; q - dorsal fin.

Clinidae (clinid blennies)

The first to suggest that *Clinus superciliosus* is viviparous was 'Johnson (1883) cf. 'Breder and Rosen (1966)'. This was followed by 'Smith (1950)' who described 28 species of Clinidae and remarked: 'Males have an intromittent organ for copulation... Large number of young being born in August-October'. Superfoetation[4] (also called superembryonation) is very pronounced with up to 12 simultaneous broods '(Veith and Cornish, 1986)'. As in most teleosts, brood size varies with the size of the mother: in clinids between 50 upwards to 1 000 young, at different stages of development, have been reported '(Gunn and Tresher, 1991)'.

Viviparous clinids possess two ovaries enclosed by a single bilobed sac, which opens by means of a single genital pore into the cloaca. The left ovary is smaller than the right one '(Veith and Cornish, 1986)' (Fig. 20/1b).

Intrafollicular gestation of *Clinus superciliosus* and *C. dorsalis* has been described by 'Veith (1979a,b, 1980)'. In both species embryotrophe rich in amino acids and lipids is taken up. There is a shift in absorptive sites from the skin to the gut. On the basis of autoradiographic studies using ^{3}H-thymidine it was demonstrated that the tracer injected into gravid *C. superciliosus* is concentrated first in the follicular epithelium and subsequently absorbed by the embryos. Absorption by their epidermis (most active in ventral yolksac and pericardium) is observed first. It reaches a peak in embryos of 6 to 8 millimetres. Radioactivity in the epidermis of the fins was most pronounced in the dorsal fin. The surface of the pericardial sac, the yolksac and the tail are extensively folded which provides an increase in surface area for the uptake of nutrients, excretion of waste products and respiration. The epidermis covering the trunk and the dorsal, anal and caudal fins displays an intricate network of microridges covered by microvilli and seems to lack an EVL, so characteristic for oviparous species. The embryonic 'trophodermic skin' is in close contact with the secretory portion of the follicle and this relationship is referred to as pseudoplacenta. [The cells do not interdigitate as described for *Anableps* where the contact zone is referred to as a placenta '(Knight et al., 1985)']. While reduction in this epidermal complexity takes place, the mouth and anus open. The embryonic gut increases in size and structure, which is indicative of nutrient uptake. By the time the embryo has reached the length of 12.5 millimetres nutrient uptake has been completely taken over by the gut.

Only part of the follicular epithelium hypertrophies and is secretory. The remainder is engaged in oxygen supply to the embryos via a rich capillary network. With the above-mentioned epidermal changes, respiration has switched almost entirely from the surface area of the epidermis (now with reduced ridges and microvilli) to the gills. Gill activity is judged by the rate of opercular movements, which show a mean rate of 44 per minute in the 12 millimetre embryo and increase to 83 per minute in the 20 millimetre embryo.

In *Clinus dorsalis*, lipid concentrations in the embryotrophe are higher than in *C. superciliosus* and, therefore, the gut is thought to be the principal site of nutrient uptake. No pseudo-placental arrangement was observed in this species '(Veith and Cornish, 1986)'. *Heteroclinus perspicillatus* and an undescribed but common *Heteroclinus* species known as Scott's Weedfish, show a similar follicular development to *Clinus superciliosus*. In both, portions of the ovarian follicular wall are hypertrophied and release the embryotrophe by apocrine secretion. The remainder consists of a flattened epithelium, which covers the maternal vascular system and is in close contact with the embryo. While in *C. superciliosus* epidermal microridges are in close contact with follicular hypertrophied cells, such an arrangement was not observed in *H. perspicillatus* by 'Veith and Cornish (1986)'.

The increase in mass from fertilization to full-term embryo or newborn was found to be 34,370% for *C. superciliosus*, 20,125% for *H. perspicillatus*, 5,744% for Scott's Weedfish and 3,140% for *H. heptaeolus*. Brood size varied from 50 to 1,000. Superfoetation has been observed in 9 out of 12 species of the subfamily Clininae.

The mode of nutrition in *Heteroclinus perspicillatus* and Scott's Weedfish shifts from lecithotrophy during early development to trophodermal matrotrophy and, eventually, intestinal nutrient transfer. Microvilli and microridges covering the body (including the fins) and the yolksac have been described by 'Gunn and Thresher (1991)' in contrast to the observations reported by 'Veith and Cornish (1986).

Cyprinodontidae (Killifishes)

This family is considered to be oviparous in habit. However, 'Harrington (1961)' demonstrated that a population of *Rivulus marmoratus* contained hermaphroditic individuals capable of fertilizing their own eggs. He wrote, 'Although the species is basically oviparous, *Rivulus marmoretus* shows a trend toward ovoviviparity — a disposition foreshadowed by an instant egg retention by *Oryzias*' '(Breder and Rosen (1966)'. According to 'Balon (1985)' *Rivulus marmoratus* and *Oryzias latipes* belong to the ecological group of 'facultative internal bearers', which retain the fertilized eggs rarely beyond the cleavage phase. It is worth stressing that retention of fertilized (activated) eggs is for a relatively short interval, comparable to the total duration of embryonic development

Poeciliidae (Livebearers)

All poeciliids are viviparous with the exception of *Tomeurus gracilis* '(Dépêche, 1976)'. Most are of the lecithotrophic type (see 20.2.2). The matrotrophic mosquito fish *Heterandria formosa* is the smallest livebearer. The eggs are minute and the yolk is exhausted very early in gestation. The number of young per birth varies between one and six. The embryos develop to term in the follicle and undergo an increase in dry mass during their development, which indicates that exogenous nutrients are taken up by the embryo. Monosaccharides, amino acids, and even some macromolecules, pass from the mother across the follicular epithelium and the intact ZR '(Wourms, 1981)'. While the embryonic dry weight increases by 3,900% in *Heterandria formosa*, the change is less pronounced in *Poeciliopsis turneri* (1,800%) '(Scrimshaw, 1945; Thibault and Schultz, 1978; Grove and Wourms, 1991, 1994)'. 'Trexler (1985)' suggested that *Poecilia latipinna*, which is normally lecithotroph, may be capable of 'facultative matrotrophy'.

In the area of the yolksac and the pericardial hood[6] (see 'Kunz', Chapter 18) there is subjacent to the embryonic epithelium an extensive vascular network, the vitelline portal system. The contact zone between the embryonic and the follicular tissue is known as follicular placenta or pseudoplacenta. It was analyzed under LM by 'Turner (1940a), 'Fraser and Reader (1940)' and 'Scrimshaw (1944, 1945)' and under EM by 'Dépêche (1973)'.

While *Heterandria formosa* has a highly developed extraembryonic pericardium (Fig. 20/14) and a reduced yolksac, an undescribed species of *Poeciliopsis* (*Poeciliopsis* B), a more advanced specialized undescribed species (*Poecilistes* D) and *Aulophallus elongatus* display a follicular placenta and, in addition, a highly expanded coelom (belly sac) (Figs. 20/15, 20/16). Superfoetation is exhibited to a high degree in the latter two species '(Turner, 1940a)'. The pericardial sac of *Heterandria* develops very early and rapidly and finally encloses the whole embryo leaving only the tailbud outside. The sparse yolk is not involved in this hood formation. Later the anterior portion of the head pushes through, which leaves the pericardial sac as a 'neckstrap'[7]. This situation is foreshadowed in the oviparous poeciliids, such as *Fundulus heteroclitus*, in which the lateral and the anterior pericardial folds partially enclose the head and in which the outer part of the pericardial wall (somatopleura) becomes vascular. In contrast, *Fundulus majalis*, displayed in Fig. 20/13, was described as matrotrophic by 'Ryder (1885c)'. Since the development in *Heterandria* parallels the amniotic folds in amniotes, the outer vascular layer in this fish is referred to as pseudo-chorion and the inner non-vascular layer as pseudo-amnion. The extraembryonic pericardial cavity corresponds to the extra-embryonic coelom in mammals. However, since the inner ectodermal and mesodermal layers of the extraembryonic pericardium adhere tightly to the embryo (Fig. 20/14), no equivalent of the mammalian amnion is formed. Judging from the early exhaustion of the yolk supply in *Heterandria* the vascular system of the pericardium and the coelomic cavity must carry maternal nutrients to the embryo, which is expressed in the increase of dry-mass of the embryo mentioned above '(Turner, 1940a; Grove and Wourms, 1991, 1994)'. *Poeciliopsis B* is characterized by a large 'belly sac' covered by the portal system (Fig. 20/15) while in the drawing depicting *Poecilistes D* (Fig. 20/16) the portal circulation is not displayed. Instead, a part of the body wall has been removed to display the large coelom and pericardium '(Turner, 1940a)' (Fig. 20/16).

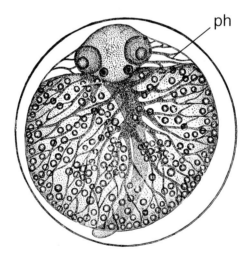

Fig. 20/13 *Fundulus majalis*, which displays a vascularized pericardial hood (ph), was described as matrotrophic (Ryder, 1885). × 32.

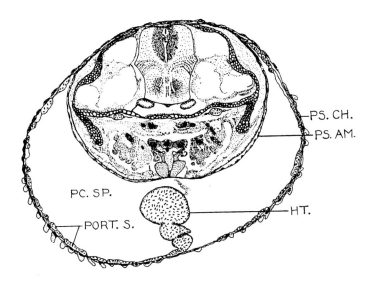

Fig. 20/14 Cross section of 4 mm embryo of *Heterandria formosa*, one of the smallest of the poeciliids, illustrating increased pericardium (PC.SP.) enclosing the heart (HT.), *pseudochorion* (PS.CH.) containing the portal blood system (PORT.S.) and *pseudoamnion* (PS.AM.) (Turner, 1940a).

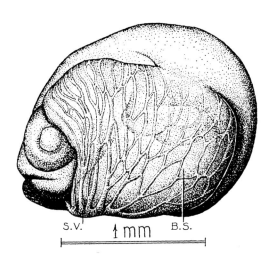

Fig. 20/15 *Poeciliopsis B* (an undescribed species of Poeciliopsis). Yolk sac almost completely absorbed. 'Belly sac', which contains anteriorly the *pericardium* and posteriorly the expanded *coelom*, is covered by portal network. Head has eruptd through anterior part of percardium (Turner, 1940a). B.S. - belly sac; S.V. - *sinus venosus*.

474 Chapter twenty

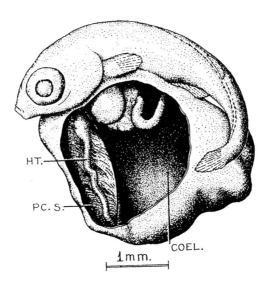

Fig. 20/16 Another undescribed species of Poeciliopsis (*Poecilistes D*). Length of embryo: 5 mm. Portal network covering the 'belly sac' is not shown. Coelom and pericardial sac opened showing the gut and liver in the former and the heart in the latter. The head does not become enveloped by the pericardium, which is simply drawn up along the sides of the head (Turner, 1940a). H.T. - heart; COEL. - *coelom*; PC.S. - pericardial sac.

The follicular wall surrounding the early embryo of *Heterandria formosa* is shown in Fig. 20/17. At this early stage the inner layer is a syncytium and the outer layer is fibrous and only slightly vascular. An undescribed species still more advanced than *Poeciliopsis B* and *Poecilistes D* is called *Poeciliopsis C*. Its belly sac is more developed, the follicle more vascular and the villi longer. Its villous follicular wall is shown in cross-section in Fig. 20/18 and in surface view in Fig. 20/19. While the villi are branched in *Poeciliopsis C*, they are finger-like and unbranched in *Aulophallus elongatus* (Fig. 20/20).

'Amoroso (1960)' posed the question as to whether the amnion arose as a cushion in viviparity, or as an aquarium in terrestrial life. 'Grove and Wourms (1991, 1994)' reported that during early development the entire embryonic surface, including a pericardial hood, functions in absorption of nutrients. This is based on the observations by SEM and TEM which revealed that the surface cells possess apical microvilli as well as coated pits, vesicles, and a well-developed RER in their cytoplasm. While the absorptive epithelium covering the pericardial and yolksac persists into older embryonic stages, the cells of the remaining embryonic areas become non-absorptive, i.e. they lack microvilli and show a paucity of other organelles. Cells with this morphology are characteristic of the embryonic surface of oviparous species that are not engaged in maternal-embryonic nutrient transfer.

Viviparity

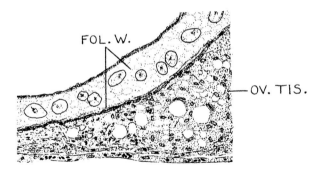

Fig 20/17 Section of follicular wall surrounding the early *embryo* of *Heterandria formosa*. The inner layer is a *syncytium* while the outer layer is only slightly vascular at this early stage (Turner, 1940a). FOL.W. - follicular wall; OV.TIS - ovarial tissue.

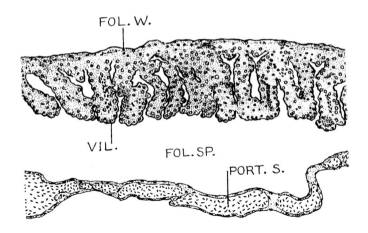

Fig. 20/18 Section of follicular wall with *villi* and vascular belly sac of the embryo of *Poeciliopsis* C (Turner, 1940a). FOL.SP. - follicular space; FOL.W - follicular wall; PORT.S. - portal system; VIL. - *villus*.

Eventually all the surface cells lose their microvilli and develop microplicae typical of epidermal cells. Studies with tracers have indicated that these cells are non-absorptive '(Grove, 1985)'.

Chloride cells were also observed in the surface epithelium of the embryo of *Heterandria*. They are characterized by the possession of numerous mitochondria and a network of tubular channels which communicate with the cell surface

Fig. 20/19 Internal surface of follicle of *Poeciliopsis* C with low, branched *villi* (Turner, 1940a).

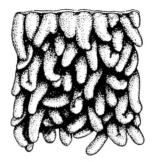

Fig. 20/20 Internal surface of follicle surrounding the early embryo of *Aulophallus elongatus*. *Villi* are fingerlike and unbranched (Turner, 1940a).

and thus resemble branchial chloride cells of adult teleosts and also chloride cells of other teleost embryos '(Grove and Wourms, 1991)'. The chloride cells in *Heterandria* closely resemble those of the lecithotrophic *Poecilia reticulata*. These cells are presumed to mediate osmoregulation.

At LM level 'Turner (1940a)' described the maternal contribution to the placenta as a follicular (inner) vascular epithelium which proliferates into a syncytium, and a fibrous capsule (outer epithelium) which is thin and additionally vascular. Towards the end of gestation the syncytium flattens and eventually disappears. Concomitantly, the fibrous capsule thins and its vascularity increases greatly. Subsequently the development of the maternal component of the follicular placenta of *Heterandria formosa* was studied by EM '(Grove and Wourms, 1991, 1994)'. The post-fertilization follicular cells display several ultrastructural features suggestive of a maternal-embryonic nutrient transfer. Junctional complexes between the cells are evident which were not observed in the prefertilization stage. This would indicate a barrier to the free movement of large molecules and ions and would immunologically protect the embryo. Since maternal-embryonic nutrient transfer does not seem to be important at the early embryonic stage, the authors suggested that at this stage the follicular cells may be engaged in the regulation of the follicular fluid content and/or may be involved in the remodeling of the egg envelope which decreases in thickness during embryonic development. When the follicular placenta becomes established, it is characterized by the presence of numerous apical microvilli and other apical projections, numerous mitochondria and extensive folding of the basal plasma membrane. An increase in this surface area would enhance the uptake and transport of low molecular weight molecules such as amino acids, monosaccharides and fatty acids. Additionally, numerous coated and smooth-surfaced pits were observed along the basal surface of the follicle cells indicating endocytosis and protein trafficking. The authors suggest also that the follicle cells may absorb low molecular substances from the follicular fluid, and coated vesicles at the apical surface of the cells may indicate receptor-mediated endocytosis of macromolecules.

The embryos of *Xiphophorus helleri* receive some maternal nutrients during their development '(Haas-Andela, 1976)' since in vitro development is successful only if glucose and casein are added to the culture medium.

Anablepidae (Foureyed Fishes)

All of these fishes are found in estuarine habitats and are viviparous. Impregnation takes place in the follicle where the embryos are retained until parturition. The average brood number is nine. In *Anableps dowi*, brood size is a function of body size as is generally true for oviparous fishes. 'Knight *et al.* (1985)' examined 102 gravid females and found the number of embryos ranging from 1 in a 15.1 centimetres long female to 35 in a 23.3 centimetres long female. Superfoetation was uncommon. The small eggs (0.7-1.0 millimetres diameter) contain only a small amount of yolk, which is absorbed by the 5 millimetres stage '(Turner, 1938, 1940a)'. According to 'Wourms (1981)' the weight of *Anableps anableps* increases by 298,000 % and that of *A. dowi* by 842,900 % by the end of gestation.

'Artedi (1705-1735)' was the first to mention *Anableps*, calling it '*A. artedi*'. This was followed by a drawing of the species by 'Seba (1758)' and another one by 'Gronovius', copied later by 'Linné', who called the fish *Cobitis anableps*. These authors were interested in the 'four-eyed condition' of the fish. The first to realize that the Anablepidae are viviparous was 'Bloch (1794)' who provided a detailed description. This was followed with reports by 'Gill (1861)' and 'Dow (1861)' who referred to the specimens as *Anableps dowi* Gill. 'Wyman, 1850-1857' found that early embryos were enclosed in the ovary in thin-walled fluid filled sacs which did not communicate with each other. The embryos were free within the sacs (Figs. 20/1b, 20/22). He concluded that since the embryo continued to increase in size after the yolk was consumed, 'it is nourished by a fluid secreted by the walls of the sac (covered with papillae) in which it is lodged in the earlier stages, or by the parietes of the general ovarian cavity, in which the foetuses are received towards the end of gestation. The high degree of vascularity of the sac is favourable to this supposition'. 'Garman (1895)' verified the observations of the previous author, described the vascularization of the papillae and suggested that in addition to their absorptive function the 'bulbs' facilitate oxygenation of the blood. 'Ryder (1885c)' reported that 'Wyman's memoir on *Anableps* contains such valuable observations and reflections on the viviparity of fishes that I will here reproduce the most important parts of his paper'. This was followed by a discussion on the results of the four authors mentioned above to which he added his own observations on the embryonic intestine. It was 'Garman' who dealt with the systematics of these fish and differentiated among three species: *Anableps anableps, A. microlepis* and *A. dowi*. 'Röthlisberger (1921)' wrote a thesis on *Anableps tetrophthalmus*. He mentioned that the embryonic development takes place within the follicle. The follicular epithelium is extremely thin and the embryos get their nourishment via the arteria mesenterica and the aorta descendens of the ovary or from the decaying follicular cells. 'Turner (1938, 1940c)' reported that the Anablepidae possess an enlarged pericardial sac and a highly developed follicular pseudoplacenta. The yolksac is transient and replaced very early by an elongated and coiled midgut in *Anableps anableps* or a distended posterior intestine in *A. dowi* (Fig. 20/21a,b), which are engaged in digesting the follicular nutrients taken up into the alimentary canal. The pseudoplacenta is made up of vascular bulbs of the 'belly sac' (pericardium) in close proximity to vascularized villi of the follicular epithelium. Both bulbs and villi develop and recede in parallel. His extensive observations at the histological level were not followed by other investigators until 'Knight *et al.* (1985)' embarked on an ultrastructural study of anablepid development.

'Knight *et al.* (1985)' mention that the largest specimen of *A. anableps* examined at term was 51 millimetres long with a dry mass of 149 milligrams, which represents a postfertilization mass increase of 298,000%. The largest term embryo of *A. dowi* measured 77 millimetres at term, with a weight of 910 milligrams and a postfertilization increase of 843,000%. The follicular placenta is formed by a close apposition of the follicular epithelium and the pericardial trophoderm. This term corresponds to the 'belly sac' referred to by 'Turner (1938, 1940a)' which is a ventral ramification of the extraembryonic pericardial sac replacing the yolksac during early gestation. The highly vascularized trophoderm is characterized by hemispherical projections which 'Knight *et al.*' term vascular bulbs. Within each bulb, the vascular plexus expands into a blood sinus. Following the maximal expansion of the pericardial trophoderm there is a gradual reduction as the embryos approach full term. Ultrastructural analysis showed that at the late stage (60 millimetres) of *A. dowi* the cells of the external surface of the bulbs possess microplicae while microvilli, characteristic of absorptive cells, are absent. It has been reported that the pericardial trophoderm of *A. dowi* is resorbed after, and in *A. anableps* prior to, parturition '(Turner, 1938; Knight *et al.*, 1985)'.

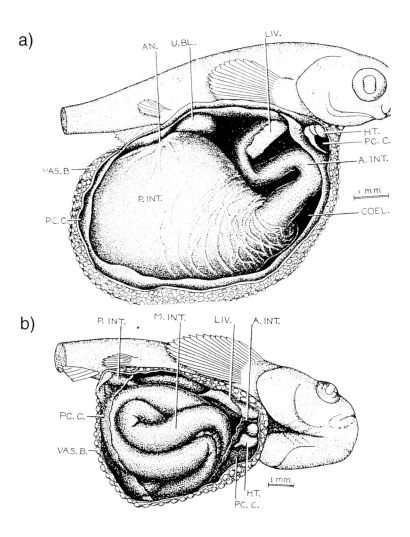

Fig. 20/21 Embryo of *Anableps* (Turner, 1940c, 1947). **a** - 10 mm stage of *A. dowi* with an expanded posterior portion of the intestine, which occupies most of the 'belly sac'. **b** - 21 mm stage of *A. anableps*. Gut fills most of the 'belly sac'. AN. - *anus*; A.INT - anterior intestine; COEL. - *coelom*; HT. - heart; LIV. - liver; M.INT - middle intestine; PC.C. - pericardial cavity; P.INT. - posterior intestine; U.BL. - urinary bladder; VAS.B. - vascular bulb.

During middle to late gestation (length ca. 60 millimetres) EM reveals differentiation in the adjacent follicular epithelium into two regions: one of shallow, pitlike depressions of the follicular epithelium within which vascular bulbs interdigitate in a 'ball and socket' manner; the second comprised of villous extensions of the hypertrophied follicular epithelium in contact with the dorsal and lateral surfaces of the embryo. It would seem that in both regions the follicular cells constitute a transporting rather than a secreting epithelium. From a morphological point of view *Anableps* is regarded as the most efficient adaptation for maternal embryonic nutrient transfer in teleosts (Fig. 20/22) '(Wourms, 1994)'.

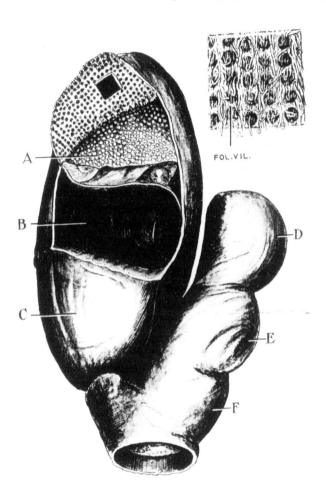

Fig. 20/22 Paired ovary of *Anableps anableps* containing 21 mm long embryos (Turner, 1938). (see diagram 20/1b). **A** - Follicle with ovarian wall taken off and flap of the follicle wall laid back to show relation of yolksac and yolksac bulbs to the follicle wall and the *villi*. **B** - Space from which an embryo and follicle have been removed. **C, D, E, F** - Embryos enclosed within intact follicles and ovarian wall. Inset. Internal surface of follicle wall greatly enlarged [window] to show follicular villi (FOL.VIL.) and depressions caused by contact with yolksac.

Little ultrastructural information is available regarding the role of the hypertrophied mid- and hindgut of *Anableps*. During the last period of gestation the gut is probably a supplemental site for the absorption of maternally derived nutrients. As to whether the pericardial sac and the gut are engaged simultaneously or sequentially or alternately in other physiological exchanges (e.g. respiration, eliminating of waste products) cannot be solved at a morphological level. The egg envelope is absent in *Anableps* '(Knight et al., 1985)'.

A tubular anal fin is used to transfer sperm between male and female. In *A. dowi* partial spermatozeugmata[9] and in *Anableps anableps* free spermatozoa are transferred '(Grier et al., 1981)'.

Goodeidae (Mexican topminnows)

Goodeidae are fresh-water fishes, which are isolated geographically i.e. confined to the Mexican plateau and adjacent streams. According to 'Hubbs and Turner (1939)' all goodeids are viviparous viz. approximately 40 species representing 18 genera.

The eggs do not exceed 1 millimetre in diameter and are supplied only with little yolk. However, the embryos of some species grow to a length of 26 millimetres. Neither spermatophores[8] nor storage of sperm has been observed '(Turner, 1933)'. The anal fin of the males does not possess an elongated gonopodium, as e.g. observed in the *Poeciliidae*. However, the anterior portion of the anal fin is partially separated and has short, stiff rays. This structure in itself is not adequate for the transmission of sperms. Transfer is effected through a urogenital organ consisting of a pearshaped muscular mass perforated by the vas deferens and urinary canal and lying between the anus and the anal fin [rev. 'Wourms (1981)]'. Transfer of spermatozeugmata[9] in Goodeidae has been reported by 'Grier (1981)'.

The goodeids are said to have the most reduced egg envelope. The thickness in *Xenophorus captivus* is 0.5 microns and in *Ameca splendens* 1.5 microns '(Riehl and Greven, 1993)'. Their envelope can be ultrastructurally resolved into layers of different compactness and electron density. Heterogeneity was also observed in its physicochemical composition: Lectin-binding experiments revealed asymmetry in the distribution of carbohydrate moieties, while the entire chorion contains glycoconjugates '(Schindler and Vries, 1989)'.

Fertilization in goodeids takes place within the follicle '(Turner, 1933, 1937, 1940d; Amoroso, 1960; Lombardi and Wourms, 1985a and Schindler and de Vries, 1989)' (Fig. 20/1c). However, the young embryo remains in the follicle only for a short time, during which the small amount of yolk is being resorbed. This is followed by a discharge of the embryo into the ovarian cavity where further development takes place. From one-half to two-thirds of the embryos are absorbed in the ovary, e.g. in an ovary containing 60 embryos at hatching only 30 or fewer are born '(Turner, 1933)'. The surviving postembryos in the ovarian cavity display nutritive rectal (or anal) processes called trophotaeniae (Fig. 20/23). They are considered to be derivatives of the embryonic hindgut and are thought to take up nutrients discharged into the ovarian cavity (Fig. 20/24). The trophotaeniae consist of a core of mostly highly vascularized loose connective tissue covered by an epithelium, which is continuous with the epithelium of the hindgut. The trophotaeniae arising in the anus are species-specific as regards form, number and histological structure. They are lost shortly after birth. The contact zone of the trophotaenial epithelium with the ovarian epithelium has been termed 'trophotaeneal placenta'. 'Turner (1933, 1937, 1940d)' suggests that the Goodeidae should be classified as 'ovo-viviparous with a superimposed viviparity' or, as expressed by 'Wourms (1981)', as progressing from strict lecithotrophy to extreme matrotrophy.

The massive increase in embryonic dry mass from fertilization to birth (in *Goodea atripinnis* 68-times and in *Ameca splendens* 150-times) is also attributed to the function of the trophotaeniae. The histology of the trophotaeniae of *Girardinichthys innominatus* and *Lermichthys* is shown in Fig. 20/25a,b.

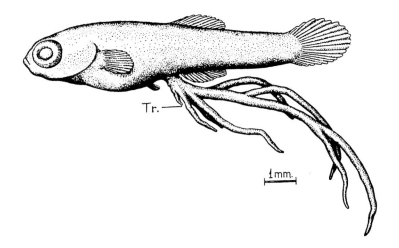

Fig. 20/23 Goodeid with *trophotaeniae* (Turner, 1933). Tr. *trophotaeniae*.

After all the yolk reserves have been taken up, the ovarian fluid is absorbed in the alimentary canal. Thus, the functional significance of the ovarian fluid has become manyfold, i.e. respiratory, excretory and nutritive.

Ataeniobius toweri has been described by 'Turner (1940d)' as a goodeid, which, as all other members of this family, shows a gonopodium formed by a modified anal fin. The species is labelled as a primitive goodeid since its postembryos lack the proctodeal trophotaeniae so characteristic of goodeids. A vascular network instrumental for respiration covers the yolksac and the pericardium. (Fig. 20/26).

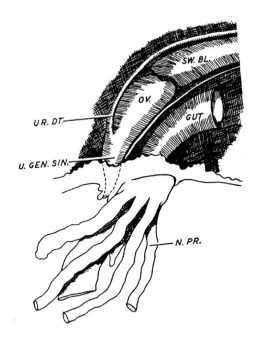

Fig. 20/24 Relationship of the nutritive *trophotaeniae* with the gut in the goodeid *Basaldichthys whitei* (Turner, 1933). N.PR. - nutritive processes; O.V. - ovary; SW.BL. - swim bladder; UR.DT. - urinary duct; U.GEN.SIN. - urogenital *sinus*.

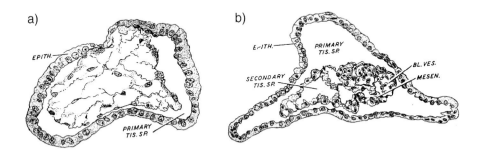

Fig. 20/25 Cross-sections of *trophotaeniae* (Turner, 1933). **a** - *trophotaenia* of 4.5 mm *Lermichthys multiradiatus* embryo. × 500. Note undifferentiated and loose mesenchymal type of core covered by a *plexus* of capillaries. **b** - *trophotaenia* of 13 mm *Girardinichthys innominatus* embryo. × 333. Core with large secondary tissue spaces and a considerable number of large internal blood vessels. BL.VES. - blood vessel; EPITH. - *epithelium*; MESEN. - *mesenchyme*; PRIMARY TIS.SP. - primary tissue space; SECONDARY TIS.SP. - secondary tissue space.

The vascular networks of the pericardium and the yolksac decrease simultaneously and respiration is shifted to the gills which are now exposed and bathed in the ovarian fluid, which is drawn into the alimentary canal by the opercular movements of the embryo (Fig. 20/27). The genus *Goodea* displays a considerably more expanded pericardial sac (Fig. 20/28) and is characterized by rosette shaped trophotaeniae (Fig. 20/29a).

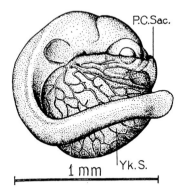

Fig. 20/26 Early stage of *Ataeniobius toweri* embryo displaying pericardial sac and yolksac covered by vitelline circulation (Turner, 1933). PC.Sac. - pericardial sac; Yk.S. - yolksac.

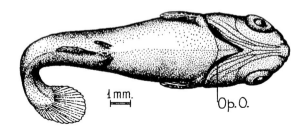

Fig. 20/27 Older embryo of *Ataeniobius toweri* illustrating wide opercular openings and absence of *trophotaeniae* (Turner, 1933). Op.O. - opercular opening.

The capillarized pericardial sac of *Ataenobius toweri* covers the head only to some degree, similar to the situation in *Fundulus heteroclitus*. In contrast, *Lermichthys multiradiatus* shows a very early development of the trophotaeniae, before the pericardial hood has fully expanded, which would suggest that in this species the voluminous trophotaeniae have taken over the functional role of both yolksac and pericardial hood '(Turner, 1940d; Amoroso, 1960; Wourms, 1981)'.

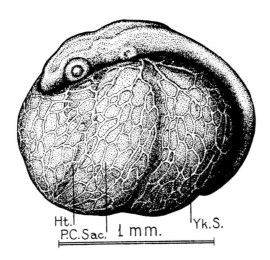

Fig. 20/28 Early embryo of *Goodea luitpoldii* showing relatively small yolk sac, expanded pericardial sac and portal network (Turner, 1933). Ht. - heart; P.C. Sac - pericardial sac; Yk.S. - yolk sac.

Trophotaeniae exist as different types according to species but are also to some extent different even within the same species. There are essentially three types: rosette, sheathed and unsheathed ribbon-type.

They have a core of vascular and loose connective tissue covered by a simple epithelium. The rosette-type, as the name suggests, consists of short lobulated processes gathered at their base; they are connected to the posterior end of the gut. They are found in *Goodea, Allotoca, Neoophorus, Ameca* and *Xenoophorus* (Fig. 20/29a,b). In the elongated sheathed type the long and ribbon shaped trophotaeniae are not pressed against the ovarian wall but are surrounded by ovarian fluid. An example of this type would be found in *Balsadichthys, Ilyodon, Neotoca, Skiffia, Ollentodon, Lermichthys* and *Girardinichthys* (Fig. 20/30a). The elongated unsheathed type contains a spongy stroma with many tissue spaces. The cells of the covering epithelium are of irregular shape and varying in height. These trophotaeniae, too, are suspended in the ovarian cavity. This type is encountered in *Alloophorus, Xenotoka, Chapalichthys, Zoogeneticus Hubbsina* (Fig. 20/30b) '(Turner, 1937?, 1940d; Amoroso, 1960; Wourms, 1988)'.

Structural analyses of the trophotaeniae revealed substantial species-specific diversity in the organization of the covering epithelium, which is continuous with the absorptive epithelium of the hindgut (Fig. 20/24, 20/25a,b). The trophotaeniae of *Allophorus, Ameca, Ilyodon, Skiffia, Xenoophorus, Xenotoca* and *Zoogeneticus* display ultrastructural features associated with endocytosis of macromolecules and have been shown to endocytose rapidly, in vitro and in vivo, the tracers horseradish peroxidase and ferritin '(Hollenberg and Wourms, 1994)'. All were classified by 'Hubbs and Turner (1939)' as having ribbon-type trophotaeniae except *Xenoophorus*, which possesses rosette-types. The two genera *Goodea* and *Gerardinichthys* lack an endocytotic apparatus. The trophotaeniae of the former are of the short rosette-type whereas those of the latter are of the ribbon-type. Electrophoretic analyses comparing ovarian fluid with maternal serum indicated that both contain essentially identical proteins. It has been suggested that the protein-absorbing trophotaeniae represent reduced forms of the protein absorbing epithelium of the gut. Goodeidae do not possess a stomach. Several studies have shown that the intestines of stomach-less teleosts are ultrastructurally divided into an anterior, mainly lipid absorbing and a posterior mainly protein absorbing segment. In some species a short terminal segment is specialized in osmoregulation '(Hollenberg and Wourms, 1994)'.

The developmental changes in the trophotaeniae of *Neoteca bilineata* were described by 'Mendoza (1939)'. He later concluded, in contradiction to 'Turner, 1940c', that superfoetation is not a characteristic feature of this species '(Mendoza, 1939)'. The structural details of the trophotaeniae of *Hubbsina turneri* were described by 'Mendoza (1956)'. He subsequently suggested that in *Goodea luitpoldi* the larval fin fold may be engaged in the absorption of nutrients '(Mendoza, 1958)'.

The ultrastructure of the absorptive trophotaenial epithelium of *Characodon* (*Xenotoca*) was described by 'Mendoza (1972)'. As the cells become more active they change from closely apposed columnar cells to cells separated by fluid-filled spaces. However, none of these spaces was seen to open into the ovarian lumen; in other words, there remains a completely apically closed epithelial surface. Eventually the spaces enlarge so much that they remain attached to each other only by a few persisting points identified as desmosomes. The cells contain apical microvilli (single, bifurcated or even arborescent) at their free surface and different kind of vesicles which would point to a pinocytotic activity. The cells eventually fragment and are lost immediately after birth '(Wourms and Cohen, 1975)'.

'Lombardi and Wourms (1985a)' described SEM and TEM analyses of the internal ovarian epithelium of the goodeid *Ameca splendens*. Matrotrophic embryonic development takes place within ovarian chambers formed by a highly folded longitudinal ovarian septum (Fig. 20/1c). During gestation the highly vascularized maternal component of the trophotaenial placenta is in direct contact with the body surfaces and the trophotaenial epithelia of the developing embryos. TEM and SEM of *Ameca* trophotaeniae revealed that the morphology of their epithelial cells is nearly identical with that of the intestinal absorptive cells of mammalian embryos. For instance the surface of the cuboidal trophotaenial cells possess apical microvilli (called brushborder at the LM level) '(Lombardi and Wourms, 1985b)'. According to 'Turner (1933)' the embryos are born in an advanced stage in which the male and female gonads are already differentiated.

According to 'Lombardi and Wourms (1988)' the development of trophotaeniae, externalized embryonic gut derivatives, in *Ameca splendens* and *Goodea atripinnis* can be differentiated into 5 stages which are characterized as follows:

1) Formation of anus and perianal lips. 2) Dilation of anus and, during lecithotrophic phase, enlargement of the perianal lips, which now protrude outward resulting in the formation of the absorptive epithelium of a trophotaenial peduncle (called trophotaenial stalk by 'Mendoza, 1956'). 3) Marked hypertrophy and lateral expansion of the perianal lips resulting in short trophotaenial processes. 4) Continuation of outward expansion of the inner mucosal surface of the trophotaenial peduncle resulting in its eversion and lobulation. Externalized absorptive surface responsible for dramatic increase of embryonic dry mass = placental function established. 5) Axial elongation and dichotomous branching of the processes. Rosette (e.g. of *Goodea atripinnis*) and ribbon trophotaeniae (e.g. in *Ameca splendens*) differ in the degree of axial elongation (Figs. 20/29, 29/30).

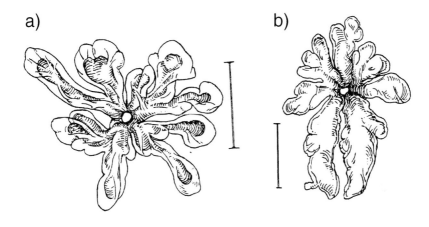

Fig. 20/29 Rosette-type *trophotaeniae* (Turner, 1937). a - of 16 mm embryo of *Goodea luitpoldi*. Bar = 1 mm; b - of 5 mm embryo of *Neoophorus diazi*. Bar = 1 mm.

Jenynsiidae (Jenynsiids)

This family is confined to South America. The anal fin of the male is modified into a true tube (gonoduct). According to 'Grier *et al.*, (1981)' free sperms are transferred in *Jenynsia lineata*.

The egg of Jenynsiidae, similar to that of the Goodeidae, has only a very small yolk mass. Degeneration of embryos which serve as nutrients to the survivors is, therefore, common '(Scott, 1928; Turner, 1940d; Amoroso, 1960)'. The embryos develop in the follicle until the completion of cleavage. After rupture of the follicle further development takes place in the ovarian cavity. The portal capillary

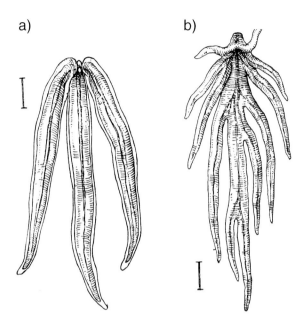

Fig. 20/30 Elongated *trophotaeniae* (Turner, 1937). **a** - Sheathed *trophotaeniae* of 11 mm embryo of *Neotoca bilineata*, just before birth. Bar = 1 mm. The number of *trophotaeniae* is reduced to three, with a single median posterior process attached to the rectal membrane and a pair attached laterally. **b** - Unsheathed *trophotaeniae* of 8 mm embryo of *Zoogeneticus quitzeoensis*. In a typical specimen (as shown here) there are 11 *termini*, arranged as illustrated. Bar = 1 mm.

network on the small yolksac is, after the disappearance of the latter, shifted to the extraembryonic pericardial sac which in the meantime has hypertrophied. The pericardial sac measures 3 millimetres in diameter in embryos of 5 millimetres length. At the 6 millimetre stage both the pericardial sac with its network and the remains of the yolksac have disappeared. At 7 millimetres the opercula have opened and the ovarian fluid is drawn into the pharyngeal cavity (Fig. 20/31a). Thus the embryo has started to feed on ovarian secretions and thriving also on 'intraovarian cannibalism'. Later, the ovarian epithelium regresses and is replaced by highly vascular folds protruding into the ovarian cavity. These come into contact with the gills of the embryos; sometimes they even enter the opercular cleft and the mouth (Fig. 20/31b). The folding of the ovarian flaps within the pharyngeal and oral cavities signifies a great increase in surface. This intimate association between the bucco-pharyngeal tissue of the embryo and the maternal ovarian tissue is called a branchial placenta '(Turner, 1947; Amoroso, 1960;

Richter *et al.*, 1983; Wourms *et al.*, 1988)'. 'Turner (1940b)' had suggested that while respiration and elimination of waste products are taking place between the maternal blood and the gills of the embryos, nutritive substances from the ovarian fluid are swallowed and absorbed by the gut. 'Richter *et al.* (1983)' reported a 24,000% increase in dry mass during gestation of *Jenynsia lineata* embryos.

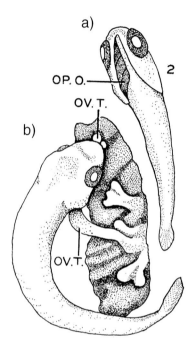

Fig. 20/31 Yenynsiid embryos (Turner, 1940b). **a** - embryo in late gestation with open opercular clefts. **b** - embryo in late gestation showing ovarian tissue extending through the opercular opening into pharyngeal cavity. A small portion is protruding from the mouth. OP.O. - opercular openings; OV.T. - ovarian tissues.

Parabrotulidae (parabrotulids)

This family of marine fish were first included in the Brotulidae, then Zoarcidae until they were assigned a family of their own.

The embryos of *Parabrotula dentiens* pass their later stages in the ovarian cavity. They exhibit trophotaeniae arranged around and, for a short distance, in front of the anus. These external processes are covered by a syncytial layer. It is assumed that they have a nutritive as well as respiratory function '(Turner, 1936, 1947)'. 'Wourms (1981)' reported that Turner's specimens were probably not well pre-

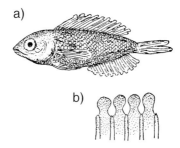

Fig. 20/32 Embiotocid embryo (Eigenmann, 1892). **a** - The male is already sexually mature at this stage. **b** - Flat extensions of soft tissue between the ends of the rays of the vertical fins. These vascular protrusions are assumed to take up nutrients and to facilitate respiration.

served and an unpublished study by 'Lombardi and Wourms (1979)' on well-preserved museum specimens revealed that the surface of the trophotaeniae is taken up by a cellular epithelium with apical microvilli which would suggest that the processes are gut derivatives and would be engaged in absorption. Trophotaeniae in *Parabrotula plagiophthalmus* have been described by 'Turner (1936)' and 'Wourms and Lombardi (1979)'.

Both *Oligopus longhursti* and *Microbrotula randalli* are characterized by the possession of trophotaeniae emanating from the anal area '(Turner; 1933, 1937; Cohen and Wourms, 1976; Wourms, 1981)'. Each trophotaenia may be longer than the embryo itself and is histologically and ultrastructurally identical with others and those of many Goodeidae. The cells display microvilli at their proximal surface and are, therefore, thought to increase the surface area for absorption of maternal nutrients. During their development in the ovarian lumen postembryos are said to grasp with their mouth processes of the ovarian wall ('ovigerous bulbs'). This close apposition is known as 'oral or buccal placenta' because the ovarian tissue is presumed to supply nourishment to the embryo '(Wourms and Cohen, 1975; Wourms *et al.*, 1988)'. 'Suarez (1975)' had observed late-term embryos of *Ogilbia cayorium* with ovarian nipples (ovigerous bulbs) in their mouth. Similar observations had been reported for *Lucifuga* by 'Lane (1909)' [cf. 'Wourms (1981)'].

Bythididae and Aphyonidae (Order Ophidiiformes)

Both families of ophidioid fishes were previously referred to as Brotulidae. They comprise 33 genera containing about 100 species.

The sperm packets are encapsulated as spermatophores[8] '(Gardiner, 1978)'. Explicit information about the copulatory apparatus in *Dinematichthys* and *Ogilbia cayorum* is given by 'Turner (1946)' and 'Suarez (1975)'. The latter also describes the female copulatory apparatus. In *Ogilbia cayorum* ovulation and hatching coincide and follow shortly after fertilization, while in *Dinematichthys* follicular gestation is prolonged to the mid-tailbud stage 'Wourms (1981)'.

In *Cataetyx memoriabilis* 'Meyer-Rochow (1970)' and *Diplacanthopoma rivers-Andersoni* 'Alcock (1895)' the ovulated egg retains its envelope during much of ovarian gestation. Finally, the event of hatching may approximate parturition, e.g. in *Cataetyx*, or precede it by a significant interval, e.g. at mid-late term in *Diplacanthopoma*.

A buccal placenta has been observed in two species of ophidioids. It shows a close association between the maternal ovarian lumenal epithelium and either the embryonic branchial and pharyngeal epithelium or the buccal epithelium ['Wourms and Cohen (1975); Lombardi and Wourms (1978); Cohen and Wourms (1976); Lombardi and Wourms (1985a,b) and Wourms et al., (1988)'.

'Alcock (1895)' suggested that the epithelium of the hypertrophied vertical finfold of *Diplacanthopoma* acts as an absorptive surface (similar to the Embiotocidae; see below for *Cymatogaster*). Since in *Diplacanthopoma* there is no vascular connection on the foetal side, he suggests 'I am inclined to think that the nutrient material is absorbed not so much from the thin tough ovarian capsule as from the ovary itself, namely from those ova in which no trace of a germinal vesicle can be found'. Adelphophagy is encountered in the embryos of the deep-water *Cataetyx memoriabilis* (*C. messieri*) '(Gilchrist, 1905, and Meyer-Rochow, 1970)'.

Embiotocidae (surfperches)

This family, which is mostly marine, seems to be composed entirely of viviparous fishes. 'Breevort (1856)' records a specimen of viviparous fish discovered by Dr. John L. LeConte in 1851. The fact that this family is viviparous was subsequently almost simultaenously described in 1852 by 'Jackson', 'Lord', 'Gibbons' and 'Webb' (cf. 'Eigenmann, 1892'). 'Agassiz (1853)' published the first account of these fishes and, as early as 1858, 200 species were known. 'Lord (1852)' [cf. 'Eigenmann (1892)'] wrote about the pregnant ovary: 'The fish are arranged to economize space; when the head of a young fish points to the head of its mother, the next to it is reversed, and looks towards the tail. However, a regular arrangement of the embryos within the ovary has been discussed and denied by 'Eigenmann (1892)'. The male reproductive organ transmits sperms in packets containing each about 600 spermatozoa. These have a long rod-shaped head in place of a globular one as is usual in fishes '(Eigenmann, 1892)'. EM and tritium autoradiography suggest that the spermatophore[8] binding material is a proteineous substance synthesized in the rough endoplasmic reticulum (RER) of the efferent sperm duct. In contrast to the results obtained for ophidioid fishes studied at LM level, the sperm packets of *Cymatogaster* do not seem to be surrounded by an extracellular capsule and, therefore, are called spermozeugma or spermatozeugmata[9], as opposed to spermatophora[8]. However, EM studies revealed an extracellular capsule '(Gardiner, 1978)'. After penetration the sperms are stored for about 6 months within pockets of the ovarian epithelium. '(Eigenmann, 1892; Gardiner, 1978); Lee and Lee, 1989)'. 'Eigenmann (1892)' described the egg of *Cymatogaster* aggregatus, which is very small (diameter of 200-230 microns)]. Fertilization by sperms, previously packaged into spermatophores '(Gardiner, 1978)' is intrafollicular and ovulation takes place immediately afterwards, when the embryo is still at an early cleavage stage. Hatching occurs at the time of ovulation and further development takes place within the ovarian cavity. Gestation time in the ovarian cavity lasts probably 5 to 6 months. The number of young varies from 3-50 depending on the size of the female '(Lee and Lee, 1989)'. Hypertrophied ovigerous folds separate the embryos into different compartments. The cells of the ovary, apart from being engaged in gas exchange, provide the food (embryotrophe /histotrophe) for the enclosed embryos. This is made evident by the dramatic increase, during gestation, in size (up to 40 millimetres) and mass (240 fold) of the surfperch postembryos. Various ways of nutrient uptake have been suggested: 1) highly vascularized vertical fins with their spatulate extensions e.g. in *Cymatogaster aggregatus* '(Ryder, 1893; Amoroso, 1960)' (Fig. 20/32a,b). 2) embryonic gill tissue closely apposed to the ovigerous folds (branchial placenta) as observed in *Cymatogaster* and *Rhacochilus* '(Turner, 1952; Wourms et al., 1988)'. 3) Later in development the hypertrophied hindgut is involved in absorption of nutrients '(Eigenmann, 1892; Amoroso, 1960)'. It was also observed in *Neoditrema ransonneti* by 'Igarashi (1962)' and 'Dobbs (1974)'. 'Ryder (1885c)' had already described the hypertrophied hindgut 'fitted with villi of the most extraordinary length'. He concluded that these structures have some important function, namely the digestion of the fluids secreted by the walls of the ovary. SEM of the hindgut of *Micrometrus minimus* reveals microvilli at the apical end of the cells which may function as 'internal trophotaeniae' '(Wourms, 1981)'. Similar results have been obtained by 'Lee and Lee (1989)' analyzing the Japanese surfperch *Ditrema temmincki*. According to 'Amoroso (1960)' and 'Turner (1947)' the males are sexually mature already at birth (Fig. 20/32a).

Hemirhamphidae (Halfbeaks)

According to 'Greven (1995)' the 'incomplete results' of 'Brembach (1991)' suggested that both *Dermogenys montanus* and *Nomorhamphus hageni* are matrotrophic.

Since other species analyzed are definitely lecithotrophic, general results regarding this family are treated below under 2) lecithotrophic fishes.

Scorpaenidae (Scorpionfishes, Rockfishes)

Most of these demersal marine fish are oviparous. Four genera (*Sebastes/Sebastodes, Sebasticus, Helicolenus* and *Hozukius*) are viviparous. They are characterized by a large brood size. According to 'Turner (1947)' they number thousands, or exceeding two millions in *Sebastes paucispinis* '(Moser, 1967a,b; Phillips, 1964; Wourms, 1981)'.

Fertilization and development take place in the ovarian lumen '(Ryder, 1886; Eigenmann, 1892)'. In late embryos of various species of *Sebastes* the hindgut is the site of nutrient uptake. The cells lining the hindgut of *S. schlegeli* exhibit microvilli with associated tubular invaginations and small round vesicles typical of cells engaged in endocytosis. Apart from ovarian secretions moribund eggs may be a nutrient source too '(Boehlert and Yoklavich, 1984; Boehlert et al., 1986; Wourms et al., 1988)'. Hatching occurs prior or just after parturition '(Moser, 1967a)'. *Helicolenus papillosus* extrudes hundreds of thousands of developing embryos in a mass of jelly-like material which may provide nutrients or oxygen for them though they still possess a yolksac '(Wourms, 1981; Erickson and Pikitch (1993)'.

20.2 LECITHOTROPHIC SPECIES

These embryos are characterized by undergoing a dry weight decrease of about 35% during development, the decrease ranging from 25-55%. This value is similar to the weight loss observed in developing embryos of oviparous fish.

Hemirhamphidae (halfbeaks)

All forms of hemirhamphids — marine pelagic, fresh and brackish water — have developed a viviparous habit '(Breder and Rosen, 1966)'.

The males of *Dermogenys*, *Hemirhamphodon*, *Zenarchopterus* and *Nomorhamphus* display peculiar modifications of their anal fin [rev. 'Wourms (1981)']. The ovaries are paired and converge posteriorly. The eggs are fertilized in the follicle, where they remain encapsulated until just before birth '(Riehl and Graven, 1993)'. Judging from the ultrastructure of the egg envelope, 'Wourms (1981)' concluded that the first three mentioned species are lecithotrophic. 'Greven (1995)' reported that the average weight of mature eggs of *Dermogenys pusillus* is around 1.09 milligrams while the dry weight of the newborn lies between 0.47 and 0.8 milligrams. This decrease in weight of 20-50% confirms that this species is lecithotrophic.

Poeciliidae (live-bearing toothcarps)

As mentioned before (20.1.4), all Poeciliidae are viviparous with the exception of *Tomeurus gracilis* which displays a so-called 'facultative viviparity' '(Gordon, 1955)'. The degree of viviparity encountered among the Poeciliidae varies with the amount of yolk present. Most are of the lecithotrophic type.

The Poeciliidae form one of the dominant teleost groups in the fresh and brackish waters of Middle America and the West Indies. The phylogeny of the Poeciliidae has been of major interest, due to their many different features, some anatomical and behavioural, others physiological and biochemical or genetic and developmental.

Follicular gestation of *Poecilia* was first reported by 'Duvernoy (1844)' and of *Gambusia* by 'Ryder (1885c)'. They noted that the embryos emerge into the lumen of the ovary only just prior to the actual birth. 'Tavolga and Rugh (1947)' reported that in the viviparous Poeciliidae both fertilization and development of the embryos take place within the ovarian follicle. Gravid females from eleven species of the subfamily poeciliopsinae revealed superfoetation '(Turner, 1940a)'.

The eggs of the guppy and the mosquitofish *Gambusia affinis* are yellow, as mentioned by 'Soin (1971)' when he put forward the theory of the respiratory importance of carotenoids contained in teleostean eggs. The guppy *Poecilia reticulata* Peters (*Lebistes reticulatus* Regan) is an euryhaline species, popular as a freshwater fish with many aquarists who often refer to it as millionfish. These fish are adaptable also to laboratory conditions. Over the years this species has been referred to by 13 different names '(Dépêche, 1976)'[5]. The guppy is also eurythermic; it has been encountered in thermal water of 29-33°C and tolerates low temperatures down to 16°C.

Poecilia reticulata and *Gambusia affinis* are considered lecithotrophic since their dry weight decreases from the zygote up to parturition by about 35% '(Scrimshaw, 1945; Dépêche, 1976; Thibault and Schultz, 1978)'.

Sperm transfer occurs as spermatozeugmata according to 'Grier (1981)'. After insemination the sperms may be retained in the ovarian pockets of the female guppy for several months and will fertilize sequential broods (time between the parturition of sequential broods is at 18° C ca. 35 days). The eggs contain enough yolk to bring the development of the embryos to term. The follicular epithelium of this species, which is considered to act only as 'a pathway for respiratory exchanges', is called by 'Turner (1940a)' a 'pseudoplacenta' (in contrast to the 'follicular placenta' encountered in the matrotrophic poeciliid *Heterandria formosa*).

The following development of *Poecilia reticulata* is based on investigations by 'Kunz, 1963, 1964, 1971a, 1980)'. At the 3-somite stage (stage 1,2) the head of the guppy embryo is buried in the yolk. Before the eye cups are formed, ectodermal headfolds develop and continue to grow in a ventral direction until they meet (Fig. 20/33 a-c) and in a caudal direction to the posterior region of the mesencephalon, near the first gill pouch (stage 3) (Fig. 20/33h). The ectodermal folds are the prequisite for the formation of the pericardial hood (Fig. 20/33 d-f). First the bilateral pericardial primordia meet ventro-medially to form the endocardial tube (stage 4) (Fig 20/33d). They subsequently start to be vascularized and greatly extend in the anterior diriection (stage 5) (Figs. 20/33e) and become drawn over the entire head to form the hood (stage 6) (Figs.20/33f, 20/34a). The outer wall of the whole headfold (somatopleura) becomes vascularized and the blood vessels are seen to be continuous with the yolksac circulation. The apices of

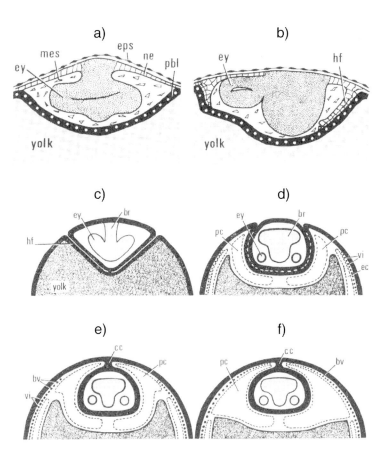

Fig. 20/33 Diagram showing the formation of the ectodermal headfold in *Poecilia reticulata*, followed by the formation and regression of the pericardial hood (Kunz, 1971a, with kind permission of the director of the Muséum d'Histoire naturelle, Geneva). **a-f** - cross sections; **a** - stage 1 (4th embryonic day); **b** - stage 2 (early 5th embryonic day) This oblique section shows the cranio-caudal gradient (right half with small lumen in eye *primordium* is cranial to the left half).; **c** - stage 3 (late 5th embryonic day), headfold completed.; **d** - stage 4 (6th embryonic day), development of pericardial hood.; **e** - stage 5 (10th embryonic day), hood meets dorsally; begin of vascularization.; **f** - stage 6 (15th embryonic day), pericardial hood fully vascularized.; br - brain; cc - chorioamniotic connection/somatopleural bridge; bv - blood vessel of pericardial headfold; ec - extraembryonic *coelom*; eps - epidermic *stratum*; ey - eye *primordium*; hf - ectodermal headfold; mes - *mesenchyme*; ne - neural *ectoderm*; pb l - *periblast*; pc - *pericardium*; vi - vitelline blood circulation.

the headfolds do not fuse, i.e. a somatopleural bridge persists along the mid-dorsal line (Fig. 20/33f).

Vascularization of the pericardial hood at stage 6 is brought about by the anterior branch of the Ductus Cuvieri (Fig. 20/34a). At stage 7 the head has elongated and its tip has pushed forward through the hood by separating the folds. These have become reduced to a so-called neckstrap which remains supplied with blood from the anterior branches of the Ductus Cuvieri (Figs. 20/33g, 20/34b). Pectoral fins beat now rhythmically thereby producing a vigorous circulation of the perivitelline fluid. The mouth movements begin and perivitelline fluid is drawn into the pharyngeal cavity. Two to three days later opercular movements start which result in a dorsal splitting of the neckstrap and the sliding down of its remnants on each side of the head. Perivitelline fluid taken in by the mouth can now leave the gill cavity freely (Fig. 20/34c). Stage 8 is characterized by the pronounced absorption of the yolk. However, the size of the yolksac does not decrease, since its lumen is encroached upon by the continually enlarging bilobed urinary bladder (Fig. 20/35bc, compared with 20/35a).

Immediately before hatching, stage 9, the remnants of the pericardial neckstrap have disappeared. The urinary bladder has been emptied and has collapsed which resulted in the uptake of the yolk remnants into the body. Hatching of the 'slimmed embryo' now occurs which is followed by parturition. (Fig.20/34d). The withdrawal of the yolk into the body must be the reason why several authors mistakenly claim that the newborn *Poecilia* has used up all its yolk (note incorporated yolk remnants in Fig. 20/36a,b). This must be due to the fact that as a result of dissection or fixation of the gravid female, the embryos instantaneously contract their body muscles which causes an immediate release of the contents of the bladder. The collapse of the bladder of the embryos can only be avoided by a deep narcosis of the female. Birth of a guppy is shown in Fig. 20/37a-c. The newborn feeds immediately and swims to the surface to fill the swim bladder (characteristic of a nidifugous fish) (Figs. 20/37d; 1/2). Therefore, this stage is also referred to as 'swim up stage'. The urinary bladder has been restored. Its normal size is seen in Figs. 20/37d; 1/2) and its normalized wall in Fig. 20/36b.

Several authors described a 'large bladder' in other poeciliids but did not refer to it as urinary bladder ['Duvernoy (1844)' for *Poecilia surinamensis*, 'Ryder (1885c)' for *Fundulus majalis*, 'Bailey (1933)' for *Xiphophorus helleri* and 'Turner (1940a)' who talks about a 'belly sac']. A (normally sized) bilobed urinary bladder was mentioned by a few authors working on Poeciliidae ['Ryder (1885)' for *Gambusia patruelis*, 'Oppenheimer (1937)' for the oviparous *Fundulus heteroclitus*, 'Tavolga and Rugh (1947)' for *Platypoecilus*]. However, 'Turner (1947)' mentions for *Heterandria formosa* that at an early stage the urinary bladder becomes expanded and pushes into the pericardial and peritoneal cavities. He mentions the fact that this temporary large bladder may store urinary wastes like the allantoic sacs of some of the higher vertebrates.

Both the embryonic and maternal adaptations of *Poecilia reticulata* have been described on the basis of LM level investigations '(Turner, 1940a; Kunz, 1963,

Viviparity

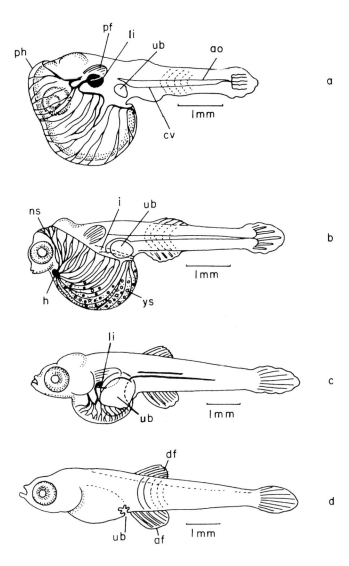

Fig. 20/34 Embryos of *Poecilia reticulata* removed from envelope to show development of urinary bladder (combined drawings from Kunz, 1963, 1964, with kind permission of the director of the Muséum d'Histoire naturelle, Geneva). **a** - stage 6, 15th embryonic day. Pericardial hood and normal urinary bladder. **b** - stage 7, 20th embryonic day. Hood reduced to pericardial neckstrap. Urinary bladder increased in size. **c** - stage 8, 25th-30th embryonic day. Neckstrap split and receded due to onset of opercular respiratory movements. Bilobed urinary bladder greatly enlarged. **d** - Newlyborn. Urinary bladder collapsed. Remnant of yolk withdrawn into body. ao - *aorta*; af - anal fin; cv - caudal vein; df - dorsal fin; h - heart; i - intestine; li - liver; ns - neckstrap; pf - pectoral fin; ph - pericardial hood; ub - urinary bladder; ys - yolk sac.

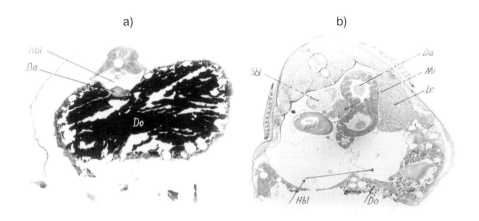

Fig. 20/35 Cross-section through trunk to show changes in size of bilobed urinary bladder ('Kunz, 1963, with kind permission of the director of the Muséum d'Histoire naturelle, Geneva). **a** - stage 6, 15th embryonic day. Normally sized urinary bladder above intestine. **b** - stage 8, 25th-30th embryonic day. Large bilobed urinary bladder. Da - intestine; Do - yolk; Hbl - urinary bladder; Mi - spleen; Sbl - swim-bladder; Le - liver.

1964, 1971a; Dépêche, 1964, 1970, 1973, 1976)'. EM analyses showed that the ectoderm enclosing the yolksac and the pericardial hood is bistratified. The lower part includes the chloride cells which appear early in development; most of them are no longer present in the postembryo '(Dépêche, 1970)'.

The close contact between the vitelline/pericardial blood circulation and the maternal ovarian circulation in *Poecilia reticulata* is, to use a mammalian term, 'epithelio-chorial'. This type of placenta is encountered in mammals with an extensive placenta (as opposed to those with a massive placenta). Mammals with an extensive placenta (e.g. ruminants and pig) are characterized by the possession of a voluminous allantois. As this allantois contains very little urine, it is not considered to be a storage place for excretory matter; its main function is said to press fetal against maternal circulation '(Portmann, 1976)'. Similarly, the function of the greatly enlarged urinary bladder in *Poecilia reticulata* may be to provide an intimate contact between yolksac circulation and ovarian circulation during the recession of the yolk material. The similarity can be drawn further when, after 'Felix (1898)', one considers the urinary bladder of teleosts to be of entodermal origin, as is the allantois of higher vertebrates. According to 'Tavolga (1949)' the bladder of *Xiphophorus helleri* is formed from Kupffer's vesicle, which itself is an outpushing of the postanal gut. In other fish, however, the urinary bladder appears a long time after the disappearance of Kupffer's vesicle; it arises as a dorsal evagination of the cloaca to which later mesodermal elements are added (e.g. in *Serranus atrarius* after 'Wilson (1891)' and in *Esox lucius* after 'Gihr (1957)'.

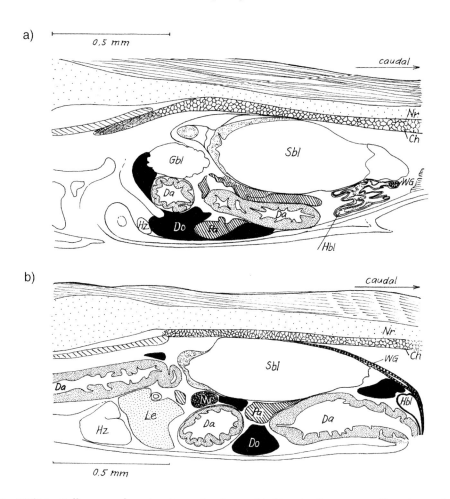

Fig. 20/36 Collapse and restoration of urinary bladder in the guppy. Drawings of histological sections (Kunz, 1964, with kind permission of the director of the Muséum d'Histoire naturelle, Geneva). **a** - Diagram of longitudinal median section of a newborn guppy. The collapse of the urinary bladder (Hbl) has caused the withdrawal of the yolk remnants (Do) into the body. **b** - Diagram of median section of a guppy a few days old. The urinary bladder is restored to its normal size. Remnants of yolk still present. Ch - notochord; Da - intestine; Do - yolk; Gbl - gall bladder; Hbl - urinary bladder; Hz - heart; Le - liver; Mi - spleen; Nr - neural chord; Pa - *pancreas*; Sbl - swim bladder; WG - Wolffian duct.

Thus, there seems to exist not only an analogy but also a homology between the allantois and the teleostean urinary bladder; both are of entodermal origin: and moreover, in some mammals, the allantois is finally incorporated into the urinary bladder.

SEM on the ectodermal surface of yolksac and pericardial hood of the guppy was performed by 'Dépêche (1973, 1976)'. The yolksac reveals an abundant network of microridges, 0.3 to 0.5 microns high, reminiscent of the ultrastructure of the EVL (see Chapter 15). The hexagonal cells are delimited by long microridges and by apical zonulae occludentes (terminal bars at LM level). The long axis of the hexagon measures 40 microns at the beginning of epiboly and due to the stretching of the cells subsequently increases to over 80 microns. Granules are observed on top of the microridges, which are thought to represent secreted mucopolysaccharides. There are zones with less microridges but displaying circular 'microcraters'. The surface of the pericardial sac looks different. The microridges are shorter and straighter. Smooth intercalating zones may be an expression of the regression of the pericardial hood. 'Microcraters' are absent. On both the vitelline and pericardial surface free apices of chloride cells are seen at the contact zone of 2-3 superficial cells. This is an indication of a bistratified ectoderm. The number of chloride cells on the vitelline surface amounted to 80-100 per millimetre^2 and on the pericardial hood to 60-80 per millimetre^2. The presence of chloride cells is an expression of the salinity tolerance of these fish. It was shown experimentally that osmotic stress induces an overlapping of adjacent superficial epithelial cells.

'Dépêche (1962, 1976)' studied in detail the in vitro development of the guppy. He concluded that only water, oxygen and a certain number of mineral ions are indispensable for its embryonic development. Phosphate ions were necessary for a complete in vitro development (a lack of phosphates leads to oedematous pericardia). In vitro embryos were retarded in length and dry weight compared with in vivo embryos. Embryos taken at advanced stages developed to term in solutions lacking in an ionic equilibrium. In other words, the older embryos are able to osmoregulate. They no longer differ in weight and length from normal embryos. Tritiated L-leucine is taken up by the in vitro embryo especially during the first days of development. Radio-iodide (I^{131}) absorption in ovario and in vitro rapidly diminishes from fertilization onwards. It seems, therefore, that maternal nutrients are furnished during the first week of the gestational period, but are not essential for the development of the embryo.

At the LM level the maternal contribution of the 'pseudoplacenta' in *Poecilia reticulata* is made up of 1) a follicular (inner) avascular epithelium in contact with the vitelline envelope 2) a fibrous capsule (outer epithelium) which is thinner than the previous layer and becomes vascular following fertilization '(Turner, 1940a; Takano, 1964)'. At the EM level the mature follicle, from within outward, consists of 1) a (bilaminar) ZR interna composed of amorphous material and perforated by pore canals enclosing microprojections of the ovum which penetrate the 2) fibrous ZR externa and become tightly packed into recesses of the 3) simple, low columnar (inner) follicular epithelium, which is separated by 4) a supporting basal membrane from 5) an outer capsule (internally made up of stratified squamous epithelium but externally of fibers). Rarely are capillaries observed; they lie between the inner and outer envelopes, i.e. within a split of the intervening basement membrane. The outermost layer, external to the collagen fibre bundles, is continuous with the ovarian capsule '(Jollie and Jollie, 1964a)'.

Fig. 20/37 Birth of *Poecilia* reticulata (Stoeckli, 1957). **a** - Female guppy, ready to give birth, has moved to the bottom of the *aquarium*. **b** - Birth of a guppy ('breach presentation'). **c** - The newborn rests on the gravel. **d** - The newborn swims up to fill its swim bladder and to take up food; compare with Fig. 1/2. Restored urinary bladder visible. Example of extreme 'nestfleer'.

Following fertilization the changes observed at the LM level are 1) disappearance of the (inner) follicular epithelium, 2) increase in vascularity of the (outer) fibrous capsule, and 3) an increase in fibre content of the outer covering '(Turner, 1940a)'.

The following EM observations are restricted to an area called the pseudoplacental barrier. This shows 6-7 layers. The layers from the maternal to the embryonic side consist of 1) maternal capillary endothelium, 2) maternal capillary basement membrane, 3) inner follicular epithelium, 4) a fertilization membrane which is absent three and four weeks after fertilization, 5) embryonic yolksac epithelium, 6) embryonic capillary basement membrane and 7) embryonic capillary endothelium, separated from the surface of the embryo by the perivitelline space. While follicular capillaries are increased both in number and size, the only ultrastructural evidence of a possible trans-pseudoplacental exchange are micropinocytotic vesicles abundant in all cellular layers of the barrier '(Jollie and Jollie. 1964b)'.

Contrary to LM reports ultrastructural studies show that the inner follicular epithelium remains cellular and persists. Additionally, fluctuations in glycogen content of this layer are observed. The outer envelope does not become more fibrous, as described by LM; instead the cells become more heavily keratinized. As recently as 1994 'Wourms' stressed that in the lecithotrophic *Poecilia* the follicle cells are not specialized for nutrient transport, but may be engaged in gas exchange.

Belonesox belizanus

This is another lecithotrophic species. Dry mass falls by 30% from egg to the newborn '(Greven, 1995)'.

Summary

Viviparity is a highly successful reproductive strategy. Internal fertilization is secured by penetration of a species-specific male copulatory organ (modified ventral fin rays). Superfoetation, i.e. simultaneous development of multiple batches of eggs or embryos at different stages of development, is often observed. This is made possible because of storage of sperm in the ovarial folds.

Embryonic development of viviparous species occurs either in the ovarian lumen (intralumenal) or in the follicle (intrafollicular). Zoarcidae and Scorpaenidae are the only two families in which ovulation precedes fertilization. Lecithotrophic embryos depend exclusively on yolk reserves laid down during oogenesis. Therefore, they exhibit a considerable decrease (30-40%) in dry mass during gestation. However, data on the lecithotrophic embryo of *Poecilia reticulata* indicate only a progressive settlement of trophic autonomy during the first days of the developmental period. Early on radioactive lecithin and iodide are taken up. However, other in vitro experiments were successful in bringing embryos to term in solutions without the addition of nutrients. Therefore, viviparity cannot be explained, at least in this fish, by special nutritional requirements of the developing embryo.

Matrotrophy denotes dependence on maternal nutrients and, therefore, is differentiated on the basis of increasing gestational weight. Matrotrophic embryos display a wide variety of structural specializations. Dermotrophic transfer sites include general body surface, gill filaments, buccal epithelium, finfolds, epaulettes, yolksac, pericardial sac, pericardial amniochorion and pericardial trophoderm. Enterotrophic transfer sites include gut, branchial portion of the branchial placenta and trophotaeniae. The greatly hypertrophied hindgut of developing embryos appears to be the principal site of nutrient uptake. The free embryos of *Zoarces vivipara*, first described in 1762, are typical representatives. They ingest nutrients supplied by the ovary (histiotrophe) in a greatly expanded hindgut (enterotrophy). In the Anablepidae, too, a distended posterior gut absorbs the ingested food. The embryonic gut is a major site of nutrient absorption also in Embiotocidae, *Sebastes* and some ophioids. The brood of *Clinus superciliosus* changes from trophodermy or dermotrophy (nutrient uptake through the epidermis) to enterotrophy. The embryos of the matrotrophe poeciliid *Heterandria formosa* are, except for the tail bud, enclosed by the greatly enlarged pericardium. Its cells are provided with apical microvilli and together with an adapted follicular layer represent a follicular pseudoplacenta.

Postembryos of Goodeidae possess extensive rectal processes called trophotaeniae which absorb nutrients dissolved in the ovarian fluid. The modified contact between the trophotaeniae and the ovarian epithelium is known as trophotaenial placenta. Parabrotulidae, Bythitidae and Aphyonidae, too, have postembryos with trophotaeniae. In the last two families trophotaeniae can be longer than the postembryo.

The Jenynsiidae feed on ovarial secretions taken up by the gut. The yolksac and pericardial vascularization vanish early and are replaced by highly vascularized ovarian folds. These extend into the gills, opercular clefts and oral cavity and are referred to as branchial placenta and buccal placenta. Both yolksac, pericardial vascularization and branchial placenta are thought to be engaged in respiration, elimination of waste products and osmoregulation.

Embiotocidae take up nutrients through their vertical fins. This is followed by a branchial placenta and in prepartum-free embryos most of the absorption is taking place through the hindgut. The postembryonic hindgut is also a major site of nutrient absorption in the Scorpaenidae.

Other physiological requirements of embryos and postembryos are, apart from nutrition, gaseous exchange, osmoregulation and excretion. These are the sole maternal contributions in lecithotrophic fish. A popular representative of this category is *Poecilia reticulata*. Its embryos display a greatly increased pericardium the outer wall of which is vascularized. This upper part of the pericardium is drawn over the entire head within an ectodermal headfold and called a hood. This is reminiscent of an amniotic fold (however, without an amniotic cavity). During further development the hood regresses to a neckstrap and when gill respiration starts it is dorsally split and slides ventrally. *Poecilia* is also characterized by a bilobed urinary bladder; its size increases concomitantly with the reduction of the yolk and the yolksac circulation It may be the function of this greatly enlarged bladder to store waste products and additionally to maintain a close contact between the decreasing yolksac circulation and the ovarian circulation.

Endnotes

[1] from Gr. lekithos = yolk; troph Gr.= nourishment
[2] from L. mater = mother; troph Gr.= nourishment

3. Embryotrophe = embryonic nutrients,
   G. foetale Nährstoffe.
4. Superfoetation (superfetation). Mother carrying multiple broods of young in different stages of development.
5. *Poecilia reticulata*, Peters, 1859; *Lebistes poecilioides*, de Philippi, 1861; *Girardinus guppyi*, Gunther, 1866; *Girardinus reticulatus*, Gunther, 1866; *Poecilioides reticulatus*, Jordan et Gilbert, 1833; *Heterandria guppyi*, Jordan, 1887; *Acanthophacelus reticulatus*, Eigenmann, 1907; *Girardinus vandepolli*, Meek, 1909; *Acanthophacelus guppyi*, Eigenmann, 1910; *Girardinus poecilioides*, Boulanger, 1912; *Poecilia poeciliodes*, Langer, 1913; *Lebistes reticulatus*, Regan, 1913; *Glaridichthys (Girardinus) reticulatus*, Milewski, 1920.
6. G. Kapuze des Pericardiums
   F. capuchon péricardique
7. G. Kopfband
   F. Lanière nucale
8. see endnote[1] in Chapter 8
9. see endnote[2] in Chapter 8

Chapter twenty-one

Synthesis

'Everything flows and nothing abides everything gives way and nothing stays fixed.'

Heraklit

Embryology (now called developmental biology) of teleosts was intensively analyzed from the beginning of the 19th century onwards. It is, therefore, most intriguing if not to say incomprehensible, that there are textbooks and even atlases of the development of amphibia, birds and various mammals widely available, but so far none on teleost fish. This was the compelling reason for the writing of this book.

'Wourms and Whitt (1981)' in a publication relating to a Symposium on Developmental Biology of Fishes (December 1979 in Tampa, Florida) discussed various reasons for the lack of any 'historical evolution of developmental biology [of fishes]' and mentioned, inter alia, difficulties in breeding and keeping the fish alive in captivity. As a thriving industry, based on the breeding of tropical fish, has existed for some time, they conceded that 'obviously, for many species, this supposed difficulty was more perceived than real'. They continued to argue that fish developmental biologists have tended to work with a number of diverse groups of fishes rather than settling on a few species and that there was no emergence of a piscine analogue of *Drosophila melanogaster*. In the meantime, however, the emphasis of molecular research has included the zebrafish, *Danio (Brachydanio) rerio*. The choice of this fish may be due to the transparency of its eggs and the fast course of development. Notwithstanding the immense importance of this research, it has to be hoped that it will not be at the expense of abandoning investigations on other species. Moreover, the study of developing teleostei presents a unique topic for comparative embryology, ecomorphology, behaviour and evolutionary considerations. It is self-explanatory that changes achieved by molecular geneticists can be perceived only if the normal state of the respective organ or tissue is known. In other words, this discipline has to rely on the availability of a thorough descriptive analysis of the normal morphology/histology/cytology of the developing tissue or organ investigated.

Teleosts are the most speciose vertebrate taxon. They contain more than 24,000 species compared with just 25,000 species of all tetrapods. The classification of teleosts in this book is almost wholly one of convenience. For an exact classification the reader is referred to the publication of 'Greenwood, Rosen, Weitzman and Myers (1966)'.

It should be emphasized that the present treatise covers a long period of time (beginning of the 19th to the beginning of the 21st century) during which names of genus and species may have changed. Moreover, in many of the early papers of considerable length the scientific name of the species described was not mentioned, as for instance for pike, salmon and trout, to name just a few. Common and scientific names used in this book are those given in the original citations.

In the compiling of this book, the vast literature dealing with the descriptive morphology, histology and cytology of teleost development has been combed and integrated and many results obtained by the author and her team have been included. Meaningful results of the very early investigators have been recorded in all chapters. The book is divided into 21 chapters, starting with the egg and embryonic development up to hatching. This is followed by a description of ectodermal, mesodermal and entodermal derivatives and the development of various organs. Some chapters are subdivided according to taxonomic groups.

Phylogenetic considerations or speculations within the teleost taxon are outside the scope of this work. However, the main features of gastrulation are compared diagrammatically with those of amphibia, in an effort to stress the apparent differences between the two groups, on the basis that teleostan development is meroblastic and that of amphibian holoblastic. The problem of yolk absorption in teleosts is highlighted by showing the unique lack of contact between the intestine and the yolk, as compared with the situation in birds and elasmobranchs. In short, most of the special features of fish ontogeny may be understood better when compared with those of other vertebrates.

Since the eggs of salmonids and the pike can be obtained in very large numbers and can be incubated under strictly controlled conditions, they provide a most suitable material for investigating the precise nature of teleost ontogeny. Their slow development may be a drawback although, within a small range, it can be accelerated or slowed down by changing the water temperature. Short development can be observed in warm-water or tropical fish, reared with relative ease in aquaria. An added advantage is that their eggs and the early embryos are transparent.

Teleostean eggs are generally spherical, not egg-shaped. Most of our knowledge derives from demersal eggs and ova obtained from fisheries. In addition, eggs of mouthbrooders and viviparous fish are easily obtained and reared in university laboratories. Fortunately, teleostean eggs are not included in a staple human diet though the eggs of the sturgeon are eaten, but only as a delicacy (as already observed by Shakespeare's Hamlet, who addressed some players with the words "'twas caviar to the general – a good thing unappreciated by the ignorant".)

Teleostean eggs are protected by envelopes, with a structure and thickness according to their habitat. The variations of the outermost layer, produced by the follicle cells, are extraordinarily manifold. It may consist of simple viscous material which glues eggs to the substratum (lake/sea bottom, floating plants). In contrast, the viscous mass of demersal eggs may be expressed in filaments of different lengths, some adhesive, others non-adhesive, covering the same egg. They can occur in bundles, sometimes of different lengths, display ornaments as shapes of anchors or mushrooms or make up different patterns of nets or rafts. They adhere to filaments of other eggs of the same species or are arranged as a netlike tube with an opening at each end, as observed in the perch. And within this tube, the hundreds of eggs are organized in such a way that the micropyles (see below) face the lumen of the tube; this organization provides them with the greatest chance to attract sperms floating in the water. Pelagic eggs usually reveal a pattern of hexagons when viewed with light- and electron-microscopy.

Below the viscous membrane, referred to in this book under 'accessory structures', a layer with radial canals is observed (zona radiata). These canals harbour microvilli from the egg and macrovilli from the ovarian follicle cells, which is indicative of exchange of material. The villi retreat to allow for fertilization. Following fertilization the radial canals become plugged or disappear altogether and this accounts for their late discovery or claimed absence in some fishes.

Fertilization in teleosts is also very unusual. The sperms do not contain an acrosome (present in all other vertebrates for enzymatic digestion of the egg envelope); they enter the egg through the micropyle, an opening into a canal of species-specific shape. An egg contains only one micropyle and fertilization has to be fast since the micropyle of the newly-ovulated egg becomes plugged after a short period in water. Fertilization is internal in the viviparous species. The male fish of *Poecilia reticulata* is provided with a gonopodium formed from modified anal fin rays while intromittent organs are either not known in some species, or are not specifically modified anal fins (e.g., rockfish, coelacanth).

Their embryos develop either for the whole time in the follicle, or reach the ovarian lumen after a certain period and stay there, or in the body cavity, until parturition. The embryos of viviparous species are provided with a minimum of yolk and some with different structures for the uptake of food supplied by the mother.

The yolk of the eggs is divided into lipid and proteid yolk and the various ways of their genesis are described. Oil-drops of various size and distribution are observed. One would expect them to be responsible for floating, but they are also present in demersal eggs. The distribution of yolk in teleost ova is usually quoted as being telolecithal (not permeated by cytoplasm as, for instance, in amphibia). However, it has been shown since early times that in many fishes the eggs become telolecithal only after activation caused by contact with water. The yolk is covered with a cytoplasmic cortex, which at the onset of gastrulation becomes invested with nuclei resulting in the so-called syncytial periblast. Its nuclei do not divide mitotically but amitotically, grow to enormous sizes and have been shown to be polyploid. Research with modern techniques into these nuclei is still lacking; it would prove very promising since all the yolk (food for all early embryos) has to pass the periblast to reach the embryo proper.

The cytoplasm surrounding the yolk is more pronounced at the apical pole of the egg and is known as blastodisc. Within this structure cleavage is taking place; in contrast to avian development, early cleavage is very regular, looking like a developing chess-board. What an advantage, not to say invitation, for any cellular biologist! The blastodisc is made up of so-called deep cells (DC) covered by a thin layer of envelope cells (EVL).

The unique fatemap of teleosts allows the displacement of cells to be followed during gastrulation. The process of invagination is totally lacking. Only a thin covering layer (periderm, EVL) and the underlying periblast undergo epiboly. It has been proposed that the arrangement and rearrangement of the periderm cells during epiboly represent one of the most precise of all morphogenetic movements known. The DC are contributing to epiboly by thinning and spreading the blastodisc. This is achieved by radial movement of the DC, intercalating among more superficial DC but not among EVL cells. The blastodisc displays a thickened rim, the germ ring. Involution of mesodermal cells occurs along the whole ring, but is most pronounced at the future posterior end of the embryo. This process is known as convergent-extension and is the result of mediolateral intercalations of the DC of the blastodisc. The ectodermal cells are left behind; they will give rise to the nervous system and to the skin covering the developing embryo and its yolksac. The entoderm, its position on the fatemap and its movement into the embryo, are still a matter of debate. A large body of evidence favours the theory that the marginal periblast contributes material to the blastoderm before and during cleavage.

Neurulation (brain, eye and neurochord formation) and the development of the intestine take place without the formation of an interior lumen; cavitation occurs later. However, a most disconcerting fact to embryologists is the presence of a very large vesicle located at the posterior end of the early embryo. It is called Kupffer's vesicle and its function has eluded the researchers up to now. The vesicle disappears before the tail starts to grow.

The skin of embryos and the covering of the yolksac are devoid of scales, so prominent in adults, but contain chromatophores eminated by the neural crest and special cells which secrete hatching enzymes. Hatching takes place at a species-specific stage of development. Altricial embryos hatch at a stage when they are helpless and would not survive without parental care; precocial embryos are independent soon after hatching or parturition.

The eye is part of the brain and the retina is, therefore, a model tissue for investigations of the CNS (central nervous system). Moreover, the study of vision of teleosts has attracted many investigators outside the club of ichthyologists, since these fish display colour vision and some also have UV-sensitive cones in their retina. An additional advantage for eye research is the fact that the different colour receptors of the teleost retina are also morphologically different and, additionally, are mostly arranged in regular mosaic patterns; this greatly facilitates various types of investigations on colour vision. The fact that the fish and their eyes continue growing throughout their life allows 'embryonic' data to be collected from the growth zones of adult fish, thereby avoiding the trouble of rearing embryos for specific investigations.

Formation of red blood cells, too, is unique in that the first haematopoetic sites, expected to be blood islands on the yolk as in other meroblastic taxa, have been observed in this location only in a few species. In others it takes place within the embryo in the so-called intermediate cell mass, and in others again at both embryonic and extraembryonic sites.

As mentioned earlier, the site of the appearance of the first entodermal cells differs according to genus or family. In all specimens examined, the intestine is initially solid. One part of the entoderm, the liver, has received much attention. Its original position is in contact with the yolk and it has been suggested on morphological grounds, not only recently, but for the first time in 1819, that it is engaged in the uptake of yolk before and/or during the vascularization of the yolksac. Since the liver is entodermal, this would bring the teleosts back into the fold of the other developing vertebrates.

Apart from being most interesting specimens for developmental biologists, from Aristotle to present-day molecular researchers, fish have also fascinated other disciplines:

Amongst some of the Salish Indians of the West Coast of today's Canada it was believed that fish evolved from humans who lived according to the type of fish they changed into: sockeye, coho, humpback, spring and dog salmon. It is because of this belief that their sought-after art work (prints and wood carvings) on salmon shows human faces or forms in the body of the fishes.

A conference on 'Alternative Life-History Styles of Animals'[1] held in 1987 in South Africa displayed a mystical orientation towards Taoism, which sees all changes in nature as manifestations of the dynamic interplay between the polar opposites yin and yang. The logo of the conference was a coelacanth embryo with a yin yang yolksac. This picture is also displayed on some of the issues of Guelph Ichthyology Reviews.

One of the myths of the Celtic race recounts 'The Coming of Finn' (the Fair One), who became an Irish hero. 'He himself went to learn the accomplishments of poetry and science from an ancient sage and Druid named Finegas, who dwelt on the river Boyne. Here, in a pool of this river, under boughs of hazel from which dropped the Nuts of Knowledge on the stream, lived Fintan, the Salmon of Knowledge, which whoso ate of him would enjoy all the wisdom of the ages. Finegas had sought many a time to catch this salmon, but failed until Finn had come to be his pupil. Then one day he caught it, and gave it to Finn to cook, bidding him eat none of it himself, but to tell him when it was ready. When the lad brought the salmon, Finegas saw that his countenance was changed. "Hast thou eaten of the salmon?" he asked. "Nay," said Finn, "but when I turned it on the spit my thumb was burnt, and I put it to my mouth.". "Take the Salmon of Knowledge and eat it", then said Finegas, "for in thee the prophecy is come true. And now go hence, for I can teach thee no more." '(Rollestone, 1949)'.

In pagan Graeco-Roman Europe fishes were used purely as decorative signs. However, by the first decades of the second century they were depicted as a Christian symbol already in engravings on Roman monuments and in frescoes of the catacombs. It has been suggested that the popularity of fish among Christians was principally due to the famous acronym consisting of the five Greek letters forming the word for fish, ιχθυs (ichthys), as Iesous Christos Theou Yios Soter (Jesus Christ, Son of God, Saviour) '(Kant, 1993)'.

Endnote

[1] Alternative Life-History Styles of Animals (1989), Conference held in Grahamstown, South Africa, June 1987 (ed M.N. Bruton), Kluwer Academic Publishers, Dordrecht, The Netherlands.

REFERENCES

Able, K.W., Markle, D.F. and Fahay, M.P. (1984) *Cyclopteridae*: Development, in *Ontogeny and Systematics of Fishes* (ed American Society of Ichthyologists and Herpetologists), Allen Press Inc., Lawrence, KS, U.S.A. pp. 428-437.

Able, K.W. (1984) Cyprinodontiformes: Development, in *Ontogeny and Systematics of Fishes* (ed American Society of Ichthyologists and Herpetologists), Allen Press Inc., Lawrence, KS, U.S.A. pp. 362-368.

Abraham, V.C., Gupta, S. and Fluck, R.A. (1993a) Ooplasmic segregation in the medaka (*Oryzias latipes*) egg. *Biological Bulletin* **184,** 115-124.

Abraham, V.C., Hilge, V., Riehl, R. and Iger, Y. (1993b) The muco-follicle cells of the jelly coat in the oocyte envelope of the sheathfish (*Silurus glanis* L.). *Journal of Morphology* **217**, 37-43.

Afzelius, B.A. (1970) Thoughts on comparative spermatology, in *Comparative Spermatology* (ed B. Baccetti), Academic Press Rome, pp. 565-572.

Afzelius, B.A. (1978) Fine structure of the garfish spermatozoon. *Journal of Ultrastructure Research* **64**, 309-314.

Agassiz, A. (1879) I. On the young stages of bony fishes. II. Development of the flounder. *Proceedings of the American Academy of Arts and Sciences* **14**, 1-25.

Agassiz, A. (1882) On the young stages of some osseous fishes. Part III. *Proceedings of the American Academy of Arts and Sciences* **17**, 271-303.

Agassiz, A. and Whitman, C.O. (1884) Contributions from the Newport Marine Laboratory communicated by Alexander Agassiz. XIV. On the development of some pelagic fish eggs. Preliminary Notice. *Proceedings of the American Academy of Arts and Sciences* **20**, 23-75. *Journal of the Royal Microscopical Society* **2**, 863-865.

Agassiz, A. and Whitmann, C.O. (1889) The development of osseous fishes II. The pre-embryonic stages of development. *Memoirs of the Museum of Comparative Zoology, Harvard College* **14,** No. 1, part II, 1-40.

Agassiz, J.L.R. (1853) Extraordinary fishes from California, constituting a new family (Holoconoti or Embiotocidae). *American Journal of Science* **16**, 389-390.
Also *Archiv der Naturgesellschaft Berlin*, (1854) **20**, **1**, 149-163.

Agassiz, J.L R. (1842) Histoire naturelle des Poissons d'eau douce de l'Europe centrale. *Mémoires de la Société des Sciences Naturelles, Neuchâtel,* Pepitpierre, Institut Lithographique de H. Nicolet, Neuchâtel.

Agassiz L. and Vogt, C. (1842) Anatomie des Salmones. *Mémoires de la Société des Sciences Naturelles, Neuchâtel* **3**, 1-196.

Ahlbert, I.B. (1969) The organization of the cone cells in the retinae of four teleosts with different feeding habits (*Perca fluviatilis* L., *Lucioperca lucioperca* L., *Acerina cernua* L. and *Coregonus albula* L. *Ark. Zool. Ser. 2*, **22**, 445-481.

Ahlbert, I.B. (1973) Ontogeny of double cones in the retina of perch fry (*Perca fluviatilis*). *Acta Zoologica* **54**, 241-251.

Ahlbert, I.B. (1976) Organisation of the cone cells in the retina of salmon (*Salmo salar*) and trout (*Salmo trutta*) in relation to their feeding habits. *Acta Zoologica* **57**, 13-35.

Ahlstrom, E.H. and Moser, H.G. (1980) Characters useful in identification of. pelagic marine fish eggs. *Rep. Calif. Coop. Oceanogr. Fish Invest.* **21**, 21-31.

Aketa, K (1954) The chemical nature and the origin of the cortical alveoli in the egg of the medaka, *Oryzias latipes. Embriologia* **2**, 63-66.

Al-Adhami (1977) A comparative study of the haematopoiesis in developing teleosts. PhD thesis, National University of Ireland, Dublin. 95 pp., 101 figs.

Al-Adhami, M.A. and Kunz, Y.W. (1976) Haematopoietic centres in the developing angelfish, *Pterophyllum scalare* (Cuvier and Valenciennes). *Wilhelm Roux's Archives* **179**, 393-401.

Al-Adhami, M.A. and Kunz, Y.W. (1977) Ontogenesis of haematopoietic sites in *Brachydanio rerio* (Hamilton-Buchanan) (teleostei). *Development, Growth and Differentiation* **19**, 171-179.

Alcock, A. (1895 On a new species of viviparous fish, diplacanthopoma, Günther, Family Ophidiidae. *Annals and Magazine of Natural History. Ser. 6*, **16**, 144-146.

Alderdice, D.F. (1988) Osmotic and ionic regulation in teleost eggs and larvae, in *Fish Physiology* **11A**, (eds W.S. Hoar and D.J. Randall), Academic Press, pp.163-251.

Alderdice, D.F., Jensen, J.O.T. and Velsen, F.P.J. (1984) Measurement of hydrostatic pressure in salmonid eggs. *Canadian Journal of Zoology* **62**, 1977- 1987.

Ali, M.A. (1959) The ocular structure, retinomotor and photobehavioural responses of juvenile Pacific salmon. *Canadian Journal of Zoology* **37**, 965-996.

Ali, M.A. (1964) Stretching of the retina during growth of salmon (*Salmo salar*). *Growth* **28**, 83-89.

Ali, M.A. (1975) Retinomotor responses, in *Vision in Fishes* (ed M.A.Ali.), Plenum Press, New York, pp. 313-355.

Amanze, D. and Iyengar, A. (1990) The micropyle: a sperm guidance system in teleost fertilization. *Development* **109**, 495-500.

Amemiya, E. (1928) (in Japanese) cf. 'T. Yamamoto (1931)'.

Amemiya, C.T. (1998) The zebrafish and haematopoietic justice. *Journal of Nature Genetics* **20**, 222-223.

American Society of Ichthyologists and Herpetologists (ed) (1984) *Ontogeny and Systematics of Fishes. An International Symposium.* Special Publication No.1. Allen Press Lawrence, KS 66044, U.S.A.

Amoroso, E.C. (1960) Vivipariy in fishes, in *Hormones in Fish* (ed I. Chester Jones), Zoological Society of London, pp. 153-181.

Anderson, E. (1967) The formation of the primary envelope during oocyte differentiation in teleosts. *Journal of Cell Biology* **35**, 193-212.

Anderson, E. (1968) Cortical alveoli formation and vitellogenesis during oocyte differentiation in the pipefish, *Syngnathus fuscus*, and the killifish, *Fundulus heteroclitus*. *Journal of Morphology* **125**, 23-31.

Anderson, E. (1974) Comparative aspects of the ultrastructure of the female gamete. *International Review of Cytology* **4** Suppl., 1-61.

André, M.J. (1875) Sur la préparation du micropyle dans la coque des oeufs de truite. *Journal de l'Anatomie et de la Physiologie*, pp. 197-202.

Anthony, R. (1918) Recherches sur le développement de la circulation chez l'Epinoche (*Gasterosteus gymnurus* Cuv.). *Archive de Zoologie expérimentale et générale* **57**.

Antila, A. (1984) Steroid conversion by oocytes and early embryos of *Salmo gairdneri*. *Annales Zoologici Fennici* **21**, 465-471.

Archer, S.N. and Lythgoe, J.N. (1990) The visual pigment basis for cone polymorphism in the guppy, *Poecilia reticulata*. *Vision Research* **30**, 225-233.

Aristotle, '*History of Animals*', Book VI, chapter 14, 568[a].

Armstrong, P.B. (1936) Mechanism of hatching in *Fundulus heteroclitus*. *Biological Bulletin (Woods Hole, Mass.)* **71**, 407.

Armstrong, P.B. (1980) Time-lapse cinemicrographic studies of cell motility during morphogenesis of the embryonic yolk sac of *Fundulus heteroclitus* (Pisces: Teleostei), *Journal of Morphology* **165**, 13-29.

Armstrong, P.B. (1962) Stages in the development of *Ictalurus nebulosus*. Syracuse University Press, Syracyse, New York.

Armstrong, P.B. and Child, J.W. (1965) Stages in the normal development of *Fundulus heteroclitus*. *Biological Bulletin* **128**, 143-168.

Arndt, E.A. (1956) Histologische und histochemische Untersuchungen über die Oogenese und bipolare Differenzierung von Süsswasser-Teleosteern. *Protoplasma* **47**, 1-36.

Arndt, E.A. (1960a) Untersuchungen über die Eihüllen von Cypriniden. *Zeitschrift für Zellforschung und Mikroskopische Anatomie* **52**, 315-327.

Arndt, E.A. (1960b) Ueber die Rindenvakuolen der Teleostieroocyten. *Zeitschrift für Zellforschung* **51**, 209-224.

Artedi, P. (1705-1735) *Bibliotheca Ichthyologia sive opera omnia piscibus* (Wisshoff, Leiden)

Assheton, R. (1907) Certain features characteristic of teleostean development. *Guy's Hospital Report*, pp. 345-388.

Aubert, H. (1856) Beiträge zur Entwicklungsgeschichte der Fische. II. Die Entwicklung des Herzens und des Blutes im Hechteie. *Zeitschrift für wissenschaftliche Zoologie* **7**, 345-364.

Auerbach, M. (1904) Die Dotterumwachsung und Embryonalanlage vom Gangfisch und der Aesche im Vergleich zu denselben Vorgängen bei der Forelle. *Verhandlungen der Naturwissenschaftlichen Vereinigung Karlsruhe* **17**, 57-82.

Avery, J.A. and Bowmaker, J.K. (1982) Visual pigments in the four-eyed fish, *Anableps anableps*. *Nature* **298**, 62-63.

Ayson, F.G., Kaneko, T., Hasegawa S. and Hirano, T. (1994) Development of mitochondrion-rich cells in the yolksac membrane of embryos and larvae of tilapia, *Oreochromis mossambicus*, in fresh water and seawater. *Journal of Experimental Zoology* **270**, 129-35.

Ayson, R.G., Kaneko, T., Hasegawa S. and Hirano, T. (1995) Cortisol stimulates the size and number of mitochondrion-rich cells in the yolksac membrane of embryos and larvae of tilapia (*Oreochromis mossambicus*) *in vitro* and *in vivo*. *Journal of Experimental Zoology* **272**, 419-425.

Babin, P.J. and Vernier, J.M. (1989) Review. Plasma lipoproteins in fish. *Journal of Lipid Research* **30**, 467-490.

Baburina, E.A. (1961) Development of eyes and their function in the embryos and larvae of *Lucioperca lucioperca*. *Trudy Institute Norf Zhivot* **33**, 151-171.

Baccetti, B. (ed) (1970) *Comparative Spermatology*. Accademia Nazionale dei Lincei, Rome; Academic Press, New York, London.

Baccetti, B. (1985) Evolution of the sperm cell, in *Biology of Fertilization. Vol. 2 Biology of the sperm*, (eds D. Mazia and A. Tyler), McGraw-Hill, New York, pp. 91-115.

Bachop, W. (1965) Size and shape, number and distribution of periblast nuclei in the muskellunge embryo (clupeiformes: *Esocidae*). *Transactions of the American Microscopical Society* **84**, 80-86.

Bachop, W. and Price, J.W. (1971) Giant nuclei formation in the yolk sac syncytium of the muskellunge, a bony fish (Salmoniformes: Esocidae: *Esox* masquinogy). *Journal of Morphology* **135**, 239-246.

Bachop, W. and Recinos, A. (1973) Scanning microdensitometry of giant *nuclei* in the yolk sac syncytium of the zebrafish embryo (cyprinoformes: cyprinidae: *Brachydanio rerio*). *Bulletin of the South Carolina Academy of Science* **35**, 99.

Bachop, W.E. and Schwartz, F.J. (1974) Quantitative nucleic acid histochemistry of the yolk sac syncytium of oviparous teleosts: Implications for hypotheses of yolk utilization, in *The Early Life History of Fish*, (ed J.H.S. Blaxter), Springer, Berlin, Heidelberg, New York, pp. 345-353.

Baer von, K.E. (1835) *Untersuchungen über die Entwickelungsgeschichte der Fische, nebst einem Anhange über die Schwimmblase*. F.C.W. Vogel, Leipzig, 53 pp.

Bahary, N. and Zon, L.I. (1998) Use of the zebrafish (*Danio rerio*) to define hematopoiesis. *Stem Cells* **16**, 89-98.

Bailey, R.E. (1933) The ovarian cycle in the viviparous teleost, *Xiphophorus helleri*. *Biological Bulletin* **64**, 206-225.

Bailey, R.E. (1957) The effect of estradiol on serum calcium, phosphorus and protein of goldfish. *Journal of Experimental Zoology* **136**, 455-469.

Baker, J. A., Foster, S.A. and Bell, M.A. (1995) Armor morphology and reproductive output in threespine stickleback, *Gasterosteus aculeatus*. *Environmental Biology of Fishes* **44**, 225-330.

Balfour, F.M. (1875) A comparison of the early stages in the development of vertebrates. *Quarterly Journal of Microscopical Science* **14**, 323-364.

Balfour, F.M. (1880-81) *A Treatise on Comparative Embryology (1st and 2nd edition)*, in two volumes. Macmillan and Co., London.

Balfour, F.M. (1882) On the nature of the organ in adult teleosteans and ganoids, which is usually regarded as the head-kidney or pronephros. *Quarterly Journal of Microscopical Science n.s.* **30**, 12-16.

Balfour, F.M. (1883-1885) *Traité d'embryologie et d'organogenèse comparée*. Paris. A treatise on Comparative Embryology, in two volumes. Macmillan and Co., London

Ballard W.W. (1966) Origin of the hypoblast in *Salmo*, II. Outward movement of deep central cells. *Journal of Experimental Zoology* **161**, 211-220.

Ballard, W.W, (1966a) The role of the cellular envelope in the morphogenetic movements of teleost embryos. *Journal of Experimental Zoology* **161**, 193-200.

Ballard, W.W. (1966b) Origin of the hypoblast in *Salmo*. I. Does the blastodisc edge turn inward? *Journal of Experimental Zoology* **161**, 201-210.

Ballard, W.W. (1968) History of the hypoblast in *Salmo*. *Journal of Experimental Zoology* **168**, 257-272.

Ballard W.W. (1973a) A re-examination of gastrulation in teleosts. *Revue Roumaine de Biologie, Série Zoologie,* **18**, 115-135.

Ballard, W.W. (1973b) A new fate map for *Salmo gairdneri. Journal of Experimental Zoology* **184**, 49-74.

Ballard, W.W. (1973c) Morphogenetic movements in *Salmo gairdneri* Richardson. *Journal of Experimental Zoology* **184**, 27-48.

Ballard, W.W. (1973d) Normal embryonic stages for salmonid fishes, based on *Salmo gairdneri* Richardson and *Salvelinus fontinalis* (Mitchill). *Journal of Experimental Zoology* **184**, 7-26.

Ballard, W.W.(1976) Problems of gastrulation: Real and verbal. *Bioscience* **26**, 36-39.

Ballard, W.W. (1981) Morphogenetic movements and fate maps of vertebrates. *The American Zoologist* **21**, 391-399.

Ballard, W.W. (1982) Morphogenetic movements and fate map of the cypriniform teleost, *Catostomus commersoni (Lacepede). Journal of Experimental Zoology* **219**, 301-321.

Ballard, W.W. and L.M. Dodes (1968) The morphogenetic movements at the lower surface of the blastodisc in salmonid embryos. *Journal of Experimental Zoology* **168**, 67-84.

Ballowitz, E. (1890) Untersuchungen über die Struktur der Spermatozoen. Teil III: Fische, Amphibien und Reptilien. *Archiv für Mikroskopische Anatomie* **36**, 225-90.

Ballowitz, E. (1915) Ueber die Samenkörner der Forellen. *Archiv für Zellforschung* **14**, 185-192.

Ballowitz, E. (1916) Ueber die körnige Zusammensetzung des Verbindungsstückes der Samenkörper der Knochenfische. *Archiv für Zellforschung* **14**, 355-358.

Balon, E.K. (1975a) Reproductive guilds of fishes: a proposal and definition. *Journal of the Fisheries Research Board of Canada* **32**, 821-864.

Balon, E.K. (1975b) Terminology of intervals in fish development. *Journal of the Fisheries Research Board of Canada* **32**, 1663-1670.

Balon, E.K. (1977) Early ontogeny of *Labeotropheus* Ahl, 1927 (Mbuna, Cichlidae, Lake Malawi), with a discussion on advanced protective styles in fish reproduction and development. *Environmental Biology of Fishes* **2**, 147-176.

Balon, E.K. (1981) Additions and amendments to the classification of reproductive styles in fishes. *Environmental Biology of Fishes* **6**, 377-389.

Balon, E.K. (1984) Patterns in the evolution of reproductive styles in fishes, in *Fish Reproduction* (eds G.W. Potts and R.J. Wootton), Academic Press, London, pp. 35-51.

Balon, E.K. (1985) Early life history of fishes. New developmental, ecological and evolutionary perspectives, in *Developments in Environmental Biology of Fishes* **5**, Dr. W. Junk Publishers, Dordrecht, 1-280.

Balon, E.K. (1989) The Tao of life: from the dynamic unity of polar opposites to self-organization, in *Alternate Life-History Styles of Animals,* (ed M.N. Bruton), Kluwer Academic Publishers, Dordrecht. pp. 7-40.

Balon, E.K. (1989) Classification of reproductive styles, in *Alternate Life-History Styles of Animals,* (ed M.N. Bruton), Kluwer Academic Publishers, Dordrecht, pp 20-28.

Balon, E.K. (1989) The 'alprehost': creation of new forms or systems, in *Alternate Life-History of Animals,* (ed M.N. Bruton), Kluwer Academic Publishers, Dordrecht, pp. 29-32.

Balon, E.K. (1989) Towards the evolutionary synthesis, in *Alternate Life-History of Animals,* (ed M.N. Bruton), Kluwer Academic Publishers, Dordrecht, pp. 33-42.

Balon, E.K. (1990a) Epigenesis of an epigeneticist: the development of some alternative concepts of the early ontogeny and evolution of fishes. *Guelph Ichthyology Reviews* **1**, 1-48.

Balon, E.K. (1990b) 4. Classification of reproductive styles. 5. The 'alprehost': creation of new forms or systems. *Environmental Biology of Fishes* **44**, 20-428.

Balon, E.K. (2001) The sensory canal systems of the living coelacanth, *Latimeria chalumnae*: a new instalment. *Environmental Biology of Fishes* **61**, 117-124.

Balon, E.K. and Liem, K.F. (1995) Prelude to ecomorphology of fishes. *Environmental Biology of Fishes* **44**, 7-8.

Bathelt, D. (1970) Experimentelle und vergleichend morphologische Untersuchungen am visuellen System von Teleostiern. *Zoologisches Jahrbuch für Anatomie* **87**, 402-470.

Battle, H.I. (1944) The embryology of the Atlantic salmon (*Salmo salar Linnaeus*). *Canadian Journal of Research* **22**, 105-125.

Baumann, M. and Sander, K. (1984) Bipartite axiation follows incomplete epiboly in zebrafish embryos treated with chemical teratogens. *Journal of Experimental Zoology* **230**, 363-376.

Baumgärtner (1830) Beobachtungen über die Nerven und das Blut. Freiburg.

Baylis, J.F. (1981) The evolution of parental care in fishes, with reference to Darwin's role of male sexual selection. *Environmental Biology of Fishes* **6**, 223-251.

Bazzoli, N. and Godinho, H.P. (1994) Cortical alveoli in oocytes of freshwater neotropical teleost fish. *Bolletino di Zoologia* **61**, 301-308.

Beams, H.W. and Kessel, R.G. (1973) Oocyte structure and early vitellogenesis in the trout, *Salmo gairdneri*. *American Journal of Anatomy* **136,** 105-122.

Beams, H.W., Kessel, R.G., Shih C.Y. and Tung, H.N. (1985) Scanning electron microscope studies on blastodisc formation in the zebrafish *Brachydanio rerio*. *Journal of Morphology* **184**, 41-50.

Beatty, D.D. (1977) The role of the pseudobranch and choroid *rete mirabile* in fish vision, in *Environmental Physiology of Fishes,* (ed M. Ali.), Plenum Press, New York and London. pp. 673-678.

Becher, H. (1928) Beitrag zur feineren Struktur der *Zona radiata* des Knochenfischeies und über ein durch die Struktur der Eihülle bedingtes optisches Phänomen. *Zeitschrift für mikro-anatomische Forschung* **13**, 591-624.

Becker, K.A. and Hart, N.H. (1996) The cortical actin cytoskeleton of unactivated zebrafish eggs: spatial organization and distribution of filamentous actin, nonfilamentous actin, and myosin. *Molecular Reproduction and Development* **43**, 536-547.

Becker, K.A. and Hart, N.H. (1999) Reorganization of filamentous actin and myosin-II in zebrafish eggs correlates temporally and spatially with cortical granules exocytosis. *Journal of Cell Science* **112**, 97-110.

Begovac, P.C. and Wallace, R.A. (1988) Stages of oocyte development in the pipefish, *Syngnathus scovelli. Journal of Morphology* **197**, 353-69.

Begovac, P.C. and Wallace, R.A. (1989) Major vitelline envelope proteins in pipefish oocytes originate within the follicle and are associated with the Z3 layer. *Journal of Experimental Zoology* **251**, 56-73.

Behrens, G. (1898) Die Reifung und Befruchtung des Forelleneies. *Anatomische Hefte* **10**, 227-285.

Bell, G.R., Hoskins, G.E. and Bagshaw, J.W. (1969) On the structure and enzymatic degradation of the external membrane of the salmon egg. *Canadian Journal of Zoology* **47**, 146-148.

Bell, M.A. and Foster, S.A. (1994) The evolutionary biology of the threespine stickleback. Oxford University Press, Oxford.

Bell, M.A. and Foster, S.A. (1995) Intraspecific systematics of *Gasterosteus aculeatus* populations: Implications for behavioral ecology. *Behaviour* **132**, 1131-1152.

Bennet, M.V.L. and Trinkaus, J.P. (1970) Electrical coupling between embryonic cells by way of extracellular space and specialized junctions. *Journal of Cell Biology* **44**, 592-610.

Bennett, M.V.L., Spira, M.E. and Spray, D.C. (1978) Permeability of gap junctions between embryonic cells of *Fundulus*: A reevaluation. *Developmental Biology* **65**, 419-435.

Bennett, M.V.L., Spray, D.C. and Harris, A.L. (1981) Electrical coupling in development. *The American Zoologist* **21**, 413-427.

Berent, W. (1896a) Zur Kenntnis des Parablastes und der Keimblätterdifferenzierung im Ei der Knochenfische. *Jena Zeitschrift für Naturwissenschaften* **30**, 291-349.

Berent, W (1896b) Ueber die ersten Entwickelungsprozesse der Knochenfische. *Zeitschrift für wissenschaftliche Zoologie* **43**, 434-480.

Berger, E.R. (1966) On the mitochondrial origin of oil drops in the retinal double cone inner segments. *Journal of Ultrastructural Research* **14**, 143-157.

Berger, E.R. (1967) Subsurface membranes in paired cone photoreceptor inner segments of adult and neonatal *Lebistes* retinae. *Journal of Ultrastructural Research* **17**, 226-232.

Berson, D.M., Dunn, F.A. and Takao, M. (2002) Phototransduction by retinal ganglion cells that set the circadian clock. *Science* **295**, 1070-1073.

Bertin, L. (1958) Ovuliparité, oviparité, viviparité. *Traité de Zoologie* **13**, II, 1791-1812.

Besharse, J.C. (1982) The daily light-dark cycle and rhythmic metabolism in the photoreceptor and pigment epithelial complex. *Progress in Retinal Research* **1**, 81-118.

Betchaku, T. and Trinkaus, J.P. (1978) Contact relations, surface activity and cortical microfilaments of marginal cells of the enveloping layer and of the yolk syncytial and yolk cytoplasmic layers of *Fundulus* before and during epiboly. *Journal of Experimental Zoology* **206**, 381-426.

Betchaku,T. and Trinkaus, J.P. (1982) Membrane internalization plays an important rôle in teleostean epiboly. *Journal of Cell Biology* **95**, Part 2,96A.

Bianco, L. S. (1931-1933) 38. Monografia: Uova, larve e stadi giovanili di teleostei, in *Fauna e Flora del Golfo di Napoli*, (eds G. Bardi, F. Friedländer & Sohn), Stazione Zoologica di Napoli, pp. 1-176.

Bianco, LS. (1969) The above (1931-1933) translated into English by Israel Program for Scientific Translations, Jerusalem.

Bielek, E. (1974a) Observations on the ontogenesis of haemopoiesis in teleosts. I. Development of the blood picture of the grayling (*Thymallus thymallus* L.) In German with English summary. *Zoologisches Jahrbuch der Anatomie* **93**, 243-258.

Bielek, E. (1974b) Observations on the ontogenesis of haemopoiesis in teleosts. II. Development of haemopoietic organs in the grayling (*Thymallus thymallus* L.). German with English summary. *Zoologisches Jahrbuch der Anatomie* **93**, 259-271.

Bielek, E. (1975) Observations on the ontogenesis of haemopoiesis in teleosts. III. Ontogenesis of blood picture and haemopoietic organs in the pike (*Esox lucius* L.). German with English summary. *Zoologisches Jahrbuch der Anatomie* **95**, 193-205.

Billard, R. (1970) Ultrastructure comparée de spermatozoïdes de quelques poissons téléostéens, in *Comparative Spermatology*, (ed B. Baccetti), Academic Press, New York, London, pp. 71-79.

Billard, R. and Roubaud, P. (1985) The effect of metals and cyanide on fertilization in rainbow trout (Salmo gairdneri). *Water Research* **19**, 209-214.

Blanc, H. (1894) Note préliminaire sur la maturation et la fécondation de l'oeuf de la truite. *Bulletin de la Société Vaudoise des Sciences Naturelles* **27**, 272-275.

Blanchard, R. (1878) La fécondation dans la série animale (Impregnation of fishes). *Journal de l'Anatomie et de Physiologie* **14**, 737-739.

Blanquez, M.J., Rojo, C., González, E. and Illanes, J. (1996) Hatching gland cells of trout embryos as the target of positional information. *Anatomia, Histologia, Embryologia* **25**, 301-309.

Blaxter, J.H.S. (1968) Light intensity, vision and feeding in young plaice. *Journal of Experimental marine Biology and Ecology* **2**,293-307.

Blaxter, J.H.S. (1969) Development, eggs and larvae, in *Fish Physiology* **3**. (eds W.S. Hoar and D.J. Randall), Academic Press, New York. pp.177-252.

Blaxter, J.H.S. (1975) The eyes of larval fish, in *Vision in Fishes*, (ed M.A. Ali), Plenum Press, New York and London, pp. 427-443.

Blaxter, J.H.S. (1988) Pattern and variety in development, in *Fish Physiology* **11A**, (eds W.S Hoar and D.J. Randall), Academic Press, London, 58 pp.

Blaxter, J.H.S. and Jones, M.P. (1967) The development of the retina and retinomotor responses in the herring. *Journal of the Marine Biological Association U.K.* **47**, 677-697.

Blaxter, J.H.S. and Staines, M. (1970) Pure cone retinae and retinomotor responses in larval teleosts. *Journal of the Marine Biological Association U.K.* **50**, 449-460.

Bloch, M.E.(1794) *Allgemeine Naturgeschichte der Fische* 12 Vols. and atlas, vol. 4, Berlin, Hr. Hesse und Buchhandlung der Realschule, Berlin.

Blumer, L.S. (1979) Male parental care in the bony fishes. *Quarterly Review of Biology* **54**, 149-161.

Boeck, A. (1871) Om Silden og Sidlefiskerierne navnlig om det norske Vaarsildfiske.

Christiana 1871.

Boehlert, G.W. (1978) Intraspecific evidence for the function of single and double cones in the teleost retina. *Science* **202**, 309-311.

Boehlert, G.W. (1979) Retinal development in postlarval through juvenile *Sebastes diploproa*: adaptations to a changing photic environment. *Revue Canadienne de Biologie* **38**, 265-280.

Boehlert, G.W. (1984) Scanning electron microscopy, in *Ontogeny and Systematics of Fishes*, an Ahlstrom Symposium, (ed American Society of Ichthyologists and Herpetologists), Allen Press Inc., Lawrence, KS, U.S.A., pp. 43-48.

Boehlert, G.W. and Yoklavich, M.M. (1984) Reproduction, embryonic energetics and the maternal-fetal relationship in the viviparous genus *Sebastes*. *Biological Bulletin (Woods Hole, Mass.)* **167**, 354-370.

Boehlert, G.W., Kusakari, M., Shimizu, M. and Yamada M. (1986) Energetics during embryonic development in kurosoi, *Sebastes schlegeli*). *Journal of Experimental Marine Biology and Ecology* **101**, 239-556.

Boeke, J. (1902) On the development of the entoderm of Kupffer's vesicle, of the mesoderm of the head and of the infundibulum in Muraenoids. *Versl. Vergad. Wis-. Natuurk Afdeel. Konink Acad. Wetensch., Amsterdam* **4**, 442-448.

Boeke, J. (1903) On the early development of the weever fishes (*Trachinus vipera* and *Trachinus draco*) Tjidschrift der nederlandsche dierkundige vereeniging, 2. Serie, **8**, 148-157.

Borcéa, I.(1909) Observations sur la circulation embryonnaire chez les téléostéens. *Annales scientifiques de l'Université de Jassy* **6**, 84-100.

Boulekbache H. and Devillers, Ch. (1977) Biologie du développement. — Etude par microscopie électronique à balayage des modifications de la membrane des blastomères au cours des premiers stades du développement de l'oeuf de truite (*Salmo irideus* Gibb.). *Comptes Rendus de l'Académie des Sciences de Paris* **285**, 917-920.

Boulekbache, H., Darrière, Joly, C., Boucaut, J.C. and Thiery, J.P. (1984) Immunolocalization of fibronectin in fish embryo; involvement in gastrulation and epiboly. *Journal of Embryology and Experimental Morphology* **82**, 25a.

Bourdin, J. (1926a) Le mécanisme de l'éclosion chez les téléostéens. II. Evolution histologique des cellules séreuses cutanées provoquant l'éclosion, chez la truite. *Comptes Rendus de la Société Biologique (Paris)* **46**, 1183-1186.

Bourdin, J. (1926b) Le mécanisme de l'éclosion chez les téléostéens. III Morphologie et répartition des glandes séreuses du tégument. *Comptes Rendus de la Socété Biologique (Paris)* **46**, 1939-1941.

Bouvet, J. (1976) Enveloping layer and periderm of the trout embryo (*Salmo trutta fario* L.) *Cell and Tissue Research* **170**, 367-382.

Bowmaker, J.K. (1990) Visual pigments of fishes, in *The Visual System of Fish*, (eds R.H. Douglas and M.B.A. Djamgoz,), Chapman and Hall, London, pp. 81-107.

Bowmaker, J.K. and Kunz, Y.W. (1985) The visual pigments of the weever fish, *Trachinus vipera*: a microspectrophotometric study. *Experimental Biology* **44**, 139-145.

Bowmaker, J.K. and Kunz, Y.W. (1987) Ultraviolet receptors, tetrachromatic colour vision and retinal mosaics in the brown trout (*Salmo trutta*): age dependent changes. *Vision Research* **217**, 2102-2108.

Bozhkova, V. and Voronov, D. (1997) Spatial-temporal characteristics of intercellular junctions in early zebrafish and loach embryos before and during gastrulation. *Development, Genes and Evolution* **207**, 115-126.

Bozhkova, V., Potapova, T.V. and Chailakhyan, L.M. (1980) A study of intercellular exchange in loach embryos by means of fluorescent dye injection. *Soviet Journal of Developmental Biology* **11**, 363-369.

Bozhkova, V.P., Gedevanishvili, M.Sh., Kvavilashvili, I.Sh. and Voronov, D.A. (1992) Cyclic changes of membrane conductivity in fertilized and activated eggs of teleost (*Misgurnus fossilis*) and their relation to the cell shape. *International Journal of Developmental Biology* **36**, 579-582.

Bozhkova, V., Palmbakh, L.R., Khariton, V.Y. and Chaylakhyan, L.M. (1983) Organization of the surface and adhesive properties of cleavage furrows in the loach (Mis*gurnus fossilis*) eggs. *Experimental Cell Research* **149**, 129-239.

Brachet, A. (1896) Die Entwickelung und Histogenese der Leber und des Pankreas. *Ergebnisse der Anatomie und Entwicklungsgeschichte* **6**, 739-799.

Brachet, A. (1931/1935) Traité d'embryologie des vertébrés. Paris.

Brancheev, L.M. and Razumovskii, A.M. (1937) Relationship between the biological properties of the eggs of *Coregonus lavaretus baeri* and their pigmentation (in Russian). *Trudy Leningradskogo Obshchestva Estestvoispytatelei* **66**, nos. 1-2, issue 3.

Branchek, T. (1984) The development of photoreceptors in the zebrafish, *Brachydanio rerio*. II. Function. *Journal of Comparative Neurology* **224**, 116-122.

Branchek, T. and Bremiller, R. (1984) The development of photoreceptors in the zebrafish, *Brachydanio rerio*. I. Structure. *Journal of Comparative Neurology* **224**, 107-115.

Branchek, T. and Streisinger (1980) Photoreceptor development in zebrafish. *Society of Neuroscience Abstracts* **6**, 194.

Braum, E. (1978) Ecological aspects of the survival of fish eggs, embryos and larvae, in *Ecology of Freshwater Fish Production,* (ed S.D. Gerking), Blackwell Scientific Publications Oxford, London, Edinburgh, Melbourne, pp. 102-136.

Breder, C.M. Jr. and Rasquin, P. (1947) Comparative studies in the light sensitivity of blind characins from a series of Mexican caves. *Bulletin of the American Museum of National History* **89**, 325-351.

Breder, Ch.M. and Rosen, D.E. (1966) *Modes of Reproduction in Fishes,* Natural History Press, Garden City. New York.

Breevort, J.C. (1856) Notes on some figures of Japanese fish, taken from recent specimens by the artists of the U.S. Japan Expedition. 33rd congress, 2d session, House of Representatives, Ex. Doc. No. 97.

Brembach, M. (1991) *Lebendgebärende Halbschnäbler.* Natur und Wissenschaft, Solingen.

Bretschneider, L.H. and DeWitt, J.J.D. (1947) *Sexual endocrinology of non-mammalian vertebrates.* Elsevier, Amsterdam. 146 pp.

Bridges, C.R., Berenbrink, M., Müller, R. and Waser, W. (1998) Physiology and biochemistry of the pseudobranch: an unanswered question? *Comparative Biochemistry and Physiology* **119A**, 67-77.

Brivio, M.F. (1989) Structure, isolation and composition analysis of teleostean fish egg chorion. *PhD thesis, University of Milan, Italy.*

Brivio, M.F., Bassi, R. and Cotelli, F. (1991) Identification and characterization of the major components of the *Oncorhynchus mykiss* chorion. *Molecular Reproduction and Development* **28**, 85-93.

Brock, J. (1878) Beiträge zur Anatomie und Histologie der Geschlechtsorgane der Knochenfische. *Morphologisches Jahrbuch* **4**, 505-572.

Brook, G. (1884) Preliminary account of the development of the lesser weever-fish, *Trachinus vipera*. *Journal of the Linnean Society (Zoology)* **18**, 274-291.

Brook, G. (1885a) On the development of the herring, Part I *Report of the Fisheries Board of Scotland*, Appendix F, No. 1, 32-50.

Brook, G. (1885b) On the origin of the hypoblast in pelagic teleostean ova. *Quarterly Journal of microscopical Science* **25**, 29-36. In French: Sur l'origine de l'hypoblaste dans les oeufs des téléosteeéns pélagiques. *Archives de Zoologie expérimentale et générale* **3**, 21-23.

Brook, G. (1885c) On some points in the development of *Motella* mustela, L. *Journal of the Linnean Society, Zoology (London)* **18**, 298-307.

Brook, G. (1887a) On the relation of yolk to blastoderm in teleostean fish ova. *Proceedings of the Royal Physiological Society Edinburgh* **9**, 187-193.

Brook, G. (1887b) The formation of the germinal layers in teleostei. *Transactions of the Royal Society Edinburgh* **30**, part I, 199-239.

Broussonet, A. (1782) *Ichthyologia, sistens piscium descriptiones et icones. Decas i. London.* **IV**, 41 pp.

Broussonet, A. (1875) Mémoire pour servir à l'histoire de la respiration des poissons. *Journal de Physiologie* **31**, 289-304.

Browman, H.I. and Hawryshyn, C.W. (1982) Thyroxine induces a precocial loss of ultraviolet photosensitivity in rainbow trout (*Oncorhynchus mykiss*, teleostei). *Vision Research* **31**, 2101-2108.

Bruch, C. (1855) Ueber die Befruchtung des Thiereies, Mainz 1855.

Bruch, C. (1856) (in a letter dated December 28 1854 to C.Th.von Siebold) Ueber die Mikropyle der Fische. *Zeitschrift für wissenschaftliche Zoologie* **7**: 172-176

Brummet, A.R. (1955) The relationships of the germ ring to the formation of the tailbud in *Fundulus* as demonstrated by the carbon marking technique. *Journal of Experimental Zoology* **125**, 447-485.

Brummet, A.R. (1968) Deletion-transplantation experiments on embryos of *Fundulus heteroclitus*. I. The posterior embryonic shield. *Journal of Experimental Zoology* **169**, 316-334.

Brummet, A.R. (1969) Deletion-transplantation experiments on embryos of *Fundulus heteroclitus*. II. The anterior embryonic shield. *Journal of Experimental Zoology* **172**, 443-464.

Brummett, A.R. and Dumont, J.N. (1978) Kupffer's vesicle in *Fundulus heteroclitus*: a scanning and transmission electron microscope study. *Tissue and Cell* **10**, 11-22.

Brummett, A.F. and Dumont, J.N. (1979) Initial stages of sperm penetration into the egg of *Fundulus heteroclitus*. *Journal of Experimental Zoology* **77**, 417-434.

Brummett, A.R. and Dumont, J.N. (1981) Cortical vesicle breakdown in fertilized eggs of *Fundulus heteroclitus*. *Journal of Experimental Zoology* **216**, 63-79.

Brummett, A.F., Dumont, J.N. and Richter. C.S. (1985) Later stages of sperm penetration and second polar body and blastodisc formation in the egg of *Fundulus heteroclitus*. *Journal of Experimental Zoology* **224**, 423-39.

Brussonet, P.M.A. (1782) Mémoire pour servir à la l'histoire de la respiration des poissons. *Mémoires de l'Académie Royale des Sciences* 174-96 and *Ichthyologia*, London.

Bruton, M.N. (1989) The ecological significance of alternative life-history styles, in *Alternate Life-History of Animals*, (ed M.N. Bruton), Kluwer Academic Publishers, Dordrecht. pp. 503-550.

Buchholz, R. (1863) Ueber die Mikropyle von *Osmerus eperlanus*. *Archiv für Anatomie, Physiologie und wissenschaftliche Medicin*. pp. 71-8 and 367-372.

Budd, P.L. (1940) Development of the eggs and early larvae of six California fishes. *Dept. Nat. Sci., Div. Fish and Game. Bur. of Mar. Fish. State of Calif. Fish Bull. No.* **56** 1-53.

Buerano, C.C., Inaba, K., Natividad, F.F. and Morisawa, M. (1995) Vitellogenins of *Oreochromis niloticus*: Identification, isolation, and biochemical and immunochemical characterization. *Journal of Experimental Zoology* **273**, 59-69.

Bullock, A. (1982) in *The Role of Solar Ultraviolet Radiation in Marine Ecosystems*, (ed J. Calkins), NATO Conf. Ser. IV, p. 409.

Burckhardt, D.A., Hassin, G., Levine J.S. and MacNichol, E.F. (1980) Electrical responses and photopigments of twin cones in the retina of the walleye. *Journal of Physiology (London)* **209**, 215-228.

Burnett W., and Agassiz (1857) On the signification of cell segmentation. *Proceedings of the American Academy of Arts and Sciences* **3**, 43. (Meeting of June 21st 1853).

Burrill, J.D. and Easter, Jr. S.S. (1995) The first retinal axons and their microenvironment in zebrafish: Cryptic pioneers and the pretract. *Journal of Neuroscience* **15**, 2935-2947.

Burzawa-Gerard, E., Baloche, S., Leloup-Hatey, J., LeMenn, F., Messaouri, H., Nunez-Rodriguez, J., Peyon, P. and Roger, E. (1994) Ovogenèse chez l'anguille (*Anguilla anguilla* L.); ultrastructure de l'ovaire à différents stades de développement et implication des lipoprotéines au cours de la vitellogenèse. *Bulletin Français de Pêche et Pisciculture* **335**, 213-233.

Busson-Mabillot, S. (1973) Evolution des enveloppes de l'ovocyte et de l'oeuf chez un poisson téléostéen, *Journal de Microscopie* **18**, 23-44.

Busson-Mabillot, S. (1977) Un type particulier de sécrétion exocrine: celui de l'appareil adhésif de l'oeuf d'un poisson téléostéen. *Biologie Cellulaire* **30**, 233-244.

Calberla, D. (1877) Zur Entwicklung des Medullarrohres und der Chorda dorsalis der Teleostier und der Petromyzonten. *Morphologisches Jahrbuch* **3**, 226-270.

Callen J.C., Dennebouy, N. and Mounolou, J.C. (1980) Kinetic analysis of entire oogenesis in *Xenopus laevis*. *Development, Growth and Differentiation* **22**, 599-610.

Cameron, I.L. and Hunter, K.E. (1984) Regulation of the permeability of the medaka fish embryo chorion by exogeneous sodium and calcium ions. *Journal of Experimental Zoology* **231**, 231-454.

Cantino, D., Marchiafava, P.L., Strettoi, E. and Strobbia, E. (1986) Subsurface cisternae in retinal double cones. *Journal of Submicroscopical Cytology* **18**, 559-566.

Carbonnier, P. (1874) Sur le mode de respiration de diverses espèces de poissons à pharyngiens labyrinthiformes. *Comptes Rendus de l'Académie des Sciences Paris* **78**, 501-502.

Carnevali, O., Mosconi, G., Roncarati, A., Belvedere, P., Romano, M. and Limatola, E.

(1992) Changes in the electrophoretic pattern of yolk proteins during vitellogenesis in the gilthead sea bream, *Sparus aurata* L. *Comparative Biochemistry and Physiology* **103B**, 955-962.

Carranza, J. and Winn, H.E. (1954) Reproductive behavior of the blackstripe topminnow, *Fundulus notatus. Copeia* **1954, No. 4**, 273-278.

Carter, C.A. and Wourms, J.P. (1991) Cell Behavior during early development in the South American annual fishes of the genus *Cynolebias. Journal of Morphology* **210**, 247-266.

Carter, C.A. and Wourms, J.P. (1993) Naturally occurring diblastodermic eggs in the annual fish *Cynolebias*: Implications for developmental regulation and determination. *Journal of Morphology* **250**, 301-312.

Cavolini, F. (1787) Memoria sulla generazione dei pesci e dei granchi. Napoli. Translation into German by E.A.W. Zimmermann, Berlin, 1792).

Chambers, C. and Leggett, W.C. (1996) Maternal influences on variation in egg sizes in temperate marine fishes. *The American Zoologist* **36**, 180-196.

Chambolle, P. (1962) Recherches sur les facteurs physiologiques de la réproduction des poissons 'ovovivipares'. Analyse expérimentale sur *Gambusia spec. Biological Bulletin of the Marine Biological Laboratory, Woods Hole* **107**, 27-101.

Chan, F.Y., Robinson, J., Brownlie, A., Shivdasani, R.A., Donovan, A., Brugnara, J.K., Lau, B.C., Witkovska, H.E. and Zon, I. (1997) Characterization of adult alpha- and ß-globin genes in the zebrafish. *Blood* **89**, 688-700.

Chaudry, H.S. (1956) The origin and structure of the *zona pellucida* in the ovarian eggs of teleosts. *Zeitschrift für Zellforschung* **43**, 478-485.

Cherr, G.N. and Clark, W.H. (1982) Fine structure of the envelope and micropyles in the eggs of the white sturgeon, *Acipenser transmontanus* Richardson. *Development, Growth and Differentiation* **24**, 341-352.

Chevey, P. (1924) La connexion hépato-vitelline chez l'alevin de la perche (*Perca fluviatilis* L.). *Bulletin de la Société de Zoologie de la France* **49**, 136-145.

Chevey, P. (1925) Recherches sur la perche et le bar. *Bulletin Biologique de la France et de la Belgie* **59**, 145-292.

Clapp, C.M. (1891) Some points in the development of the toad-fish *(Batrachus Tau). Journal of Morphology* **5**, 494-501.

Clark, F.N. (1929) The life history of the Californian jack smelt, *Atherinopsis californiensis. Californian Department of Fish Game Bulletin* **16**, 1-22.

Clarke, S.C. (1883) December 15. *American Angler*.

Cohen, D.M. and Wourms, J.P. (1976) *Microbrotula randalli*, a new hybrid, whose embryos bear trophotaeniae. *Proceedings of the Biological Society of Washington* **89**, 81-96.

Collazo, A., Bolker, J.A. and Keller, R. (1994) A phylogenetic perspective on teleost gastrulation. *The American Naturalist* **144**, 133-152.

Collette, B.B., McGowen, G.E., Parin, N.V. and Mito, S. (1984) Beloniformes: Development and relationships, in *Ontogeny and Systematics of Fishes*, Special Publication Number 1 (eds The American Society of Ichthyologists and Herpetologists), Allen Press Inc. Lawrence, KS, U.S.A., pp. 335-354.

Colle-Vandevelde, A. (1961) Sur l'origine du sang et des vaisseaux chez *Lebistes reticulatus* (téléosteen) *Journal of Embryology and experimental Morphology* **9**, 68-76.

Colle-Vandevelde, A. (1962) Sur l'origine du sang et des vaisseaux chez *Pterophyllum scalare* (Téléosteen). *Annales de la Société Royale Zoologique de Belgique* **92**, 133-139.

Colle-Vandevelde, A. (1963a) Sur la vasculogénèse chez *Blennius gattorugine* (Téléostéen). *Publicazione della stazione zoologica di Napoli* **33**, 197-205.

Colle-Vandevelde, A. (1963b) Blood anlage in teleostei. *Nature* **198**, 1223.

Collin, J.P., Brisson, P., Falcon, J. and Voisin P. (1986a) Multiple cell types in the pineal: functional aspects, in *Pineal and retinal relationships*, (eds P.J. O'Brien and D.C. Klein), Academic Press, New York. pp.15-32.

Collin, J.P., Falcón, J., Voisin, P. and Brisson, P. (1986b) The pineal organ: ontogenetic differentiation of photoceptor cells and pinealocytes, in *The pineal gland during development*, (eds D. Gupta and R.J. Reiter), Croom-Helm Ltd., London.

Cooper, J.E. (1978) Identification of eggs, larvae and juveniles of the rainbow smelt, *Osmerus mordax*, with comparisons to larval alewife, *Alosa pseudoharengus* and gizzard shad, *Dorosoma cepedianum*. *Transactions of the American Fisheries Society* **107**, 56-62.

Cooper, M.S. and D'Amico, L.A. (1996) A cluster of noninvoluting endocytic cells at the margin of the zebrafish blastoderm marks the site of embryonic shield formation. *Developmental Biology* **180**, 184-198.

Copeland, D.E. (1974a) The anatomy and fine structure of the eye of teleost. I. The choroid body in *Fundulus grandis*. *Experimental Eye Research* **18**, 547-561.

Copeland, D.E. (1974b) The anatomy and fine structure of the eye in teleost. II. The vascular connections of the lentiform body in *Fundulus grandis*. *Experimental Eye Research* **19**, 583-589.

Copeland, D.E. (1975) The anatomy and fine structure of the eye in teleost. III. The structure of the lentiform body in *Fundulus grandis*. *Experimental Eye Research* **21**, 515-521.

Copeland, D.E. (1976) The anatomy and fine structure of the eye in teleosts. V. Vascular relations of choriocapillaris, lentiform body and falciform process in rainbow trout (*Salmo gairdneri*). *Experimental Eye Research* **23**, 15-27.

Copeland, D.E. (1980) Functional vascularization of the teleost eye, in *Current Topics in Eye Research* **3**, (eds J.A. Zadunaisky and H. Davson), Academic Press, New York, London, pp. 219-331.

Copeland, P.A., Sumpter, J.P., Walker, T.K. and Croft, M. (1986) Vitellogin levels in male and female trout, *Salmo gairdneri* Richardson, at various stages of the reproductive cycle. *Comparative Biochemistry and Physiology* **83B**, 487-493.

Cordier, R. (1941) Le transit des colorants acides et les phénomènes concomitants chez l'alevin de la truite. *Archives de Biologie, Liege* **52**, 361.

Coste, P. (1848) Nidification des épinoches et des épinochettes. *Mémoires des Savants étrangers à l'Académie des Sciences de l'Institut National de France* **10**, 574-588.

Coste, P. (1847-1859). Histoire générale et particulière du développement des corpes organisés. Paris (two plates on the development of *Gasterosteus*).

Coste, P. (1850) Origine de la cicatricule ou du germe chez les poissons osseux. *Comptes rendus de l'Académie des Sciences de Paris* **30**, 692-693.

Cotelli, F., Andronico, F., Brivio, M. and Lamia C.L. (1988) Structure and composition of the fish egg chorion (*Carassius auratus*). *Journal of Ultrastructure and Molecular Structure Research* **99**, 70-78.

Cotelli, F., Andronico, F., Bassi, M., Brivio, C., Ceccagno, C., Denis-Domini, M.L. La Rosa, S.C. and Dorin, L.L. (1986) Studies on the composition, structure and differentiation of fish egg chorion. *Cell Biology International Reports* **10**, 471.

Coulombre, A.J. (1961) Cytology of the developing eye. *International Review of Cytology* **11**, 161-194.

Coulombre, A.J. (1965) The eye, in *Organogenesis,* (eds R.L. DeHaan, and H. Ursprung), Holt, Rinehart, Winston, New York, pp. 219-251.

Coulombre, A.J. (1979) Rôles of the retinal pigment epithelium in the development of ocular tissue, in *The retinal pigment epithelium*, (eds M.F. Marmor and K.M. Zinn), Harvard University Press, pp. 53-57.

Cousins, K.L. and Jensen, J.O.T. (1994) The effects of temperature on external egg membranes in coho salmon (*Oncorhynchus kisutch*). *Canadian Journal of Zoology* **72**, 1855-1857.

Covens, M., Covens, L., Ollevier, F. and De Loof, A. (1987) A comparative study of some properties of vitellogenin (Vg) and yolk proteins in a number of freshwater and marine teleost fishes. *Comparative Biochemistry and Physiology* **88B**, 75-80.

Craik, J.C.A. and Harvey, S.M. (1984) Egg quality in rainbow trout: The relation between egg viability, selected aspects of egg composition, and time of stripping. *Aquaculture* **40**, 115-134.

Cunningham, J.T. (1885a) On the nature and significance of the structure known as Kupffer's'vesicle in teleostean embryos. *Proceedings of the Royal Society of Edinburgh* 13, 4-14.

Cunningham, J.T. (1885b) On the relations of the yolk to the gastrula in teleosteans, and in other vertebrate types. *Quarterly Journal of the Microscopical Society* **26**, 1-38.

Cunningham J.T. (1885c) Significance of Kupffer's vesicle and remarks on other questions of vertebrate morphology. *Quarterly Journal of Microscopical Science.* **25**, 1-15.

Cunningham, J.T. (1886a) The reproductive organs of *Bdellostoma*, and a teleostean ovum from the West Coast of Africa. *Transactions of the Royal Society of Edinburgh* **30**, 247-250.

Cunningham, J.T. (1886b) On the mode of attachment of the ovum of *Osmerus eperlanus*. *Proceedings of the Zoological Society of London* **4**, 292-295.

Cunningham, J.F. (1888) The eggs and larvae of teleosteans. *Transactions of the Royal Society of Edinburgh,* **33** Part I., 97-136

Cunningham, J.T. (1891) On some disputed points in teleostean embryology. *Annals and Magazine of Natural History* **7**, 203-121.

Cunningham, J.T. (1898) On the histology of the ovary and of the ovarian ova in certain marine fishes. *Quarterly Journal of the Microscopical Society* **40**, 101-163.

Curless, W.W. (1979) Early development and larval biology of the California topsmelt, *Atherinops affinis. The American Zoologist* **19**, 949 (Abstract)

Currie, P.D. and Ingham, P.W. (2001) Induction and patterning of embryonic skeletal muscle cells in the zebrafish, in *Muscle development and growth,* (ed J.A. Johnston), Academic Press, London, pp. 1-17.

Czeczuga, B. (1979) Carotenoids in fish. XIX.carotenoids in the eggs of *Oncorhynchus keta* (Walbaum) *Hydrobiologia* **63**, 45-47.

Dalcq, A.M. (1928) Les bases physiologiques de la fécondation et de la parthogénèse. *Les presses universitaires de France, Paris*. 174 pp.

Dalcq, A.M (1941) *L'oeuf et son dynamisme organisateur*. Albin Michel, Paris. 582 pp.

Dalcq, A.M. (1965) The cytochemical reaction for acid phosphatase as applied to sea-urchin eggs studied in whole mounts. *Archive de Biologie* (Liège) **76**, 439-546.

D'Ancona, U. (1931) *Exocidae*, in *Fauna e flora del golfo di Napoli*. 38.Monografia, (eds G. Bardi and R. Friedländer & Son), Stazione Zoologica di Napoli.

Dasgupta, J.D. and Singh, U.N. (1981) Early differentiation in zebrafish blastula: Role of yolk syncytial layer. *Wilhelm Roux's Archives* **190**, 358-60.

Davenport, J., L., Lønning and Kjørsvik, E. (1986) Some mechanical and morphological properties of the chorions of marine teleost eggs. *Journal of Fish Biology* **29**, 289-301.

Dean, B. (1895) On the gastrulation of teleosts. *Science N.S.* **3**, p. 60.

Dearry, A. and Burnside, B. (1986) Dopaminergic regulation of cone retinomotor movement in isolated teleost retina: I. Induction of cone contraction is mediated by D2 receptors. *Journal of Neurochemistry* **46**, 1006-1021.

Delsman, H.C. (1931) Fish eggs and larvae from the Java Sea.17. The genus *Stolephorus*. *Treubia* **13**, 217-243.

Dépèche, J. (1962) Le développement *in vitro* des oeufs de *Poeciliidae* (*Cyprinodontes vivipares*). *Comptes Rendus des Séances de l'Académie des Sciences de Paris* **255**, 2670-2672.

Dépèche, J. (1964) Les rapports anatomiques materno-embryonnaires au cours de la gestation des *poeciliidae* (*Cyprinodontes vivipares*). *Bulletin de l'Association des Anatomistes*, 49e *Réunion*, pp.536-544.

Dépèche, J. (1970) Osmorégulation embryonnaire et adaptation à l'eau de mer de *Lebistes reticulatus* (cyprinodonte vivipare). *Comptes Rendus de l'Académie des Sciences de Paris* **259**, série D, 908-910.

Dépèche, J. (1973) Infrastructure superficielle de la vésicule vitelline et du sac péricardique de l'embryon de *Poecilia reticulata* (Poisson Téléostéen). *Zeitschrift für Zellforschung* **141**, 235-253.

Dépèche, J. (1976) Acquisition et limites de l'autonomie trophique embryonnaire au cours du développement du poisson téléostéen vivipare *Poecilia reticulata*. *Bulletin Biologique de la France et de la Belgique* **110**, 45-97.

Derjugin, K.M. (1902) Ueber einige Stadien in der Entwickelung von *Lophius piscatorius* (Text in Russian). *Trav. Soc. St. Petersburg* **32**, 1-45.

Descartes, René (1644) Principia philosophiae (Mind Body Relationship). Amsterdam.

Detlaf, T.A. (1959) Die Bedeutung der Ca-ionen für die Aktivierung der Eier von Salmoniden (in Russian) *Z. obsc. Biol. (USSR)* **20**, 184.

Detrich III, H.W., Kieran, M.W., Chan, F.Y., Barone, L.M., Yee, K., Rundstadler, J.A., Pratt, S., Ransom, D. and Zon, L.I. (1995) Intraembryonic hematopoietic cell migration during vertebrate development. *Proceedings of the National Academy of Science U.S.A.* **92**, 10713-10717.

Devillers, Ch. (1951a) Symétrisation et régulation du germe chez la truite. *Comptes rendus de l'Association des Anatomistes* **38**, 418-424.

Devillers, Ch. (1951b) Les mouvements superficiels dans la gastrulation des poissons. *Archives d'Anatomie, Microscopie et Morphologie expérimentales* **40**, 298-309.

Devillers, Ch. (1956) Les aspects caractéristiques de la prémorphogenèse dans l'oeuf des téléostéens. L'origine de l'oeuf télolécithique. *Annales Biologiques* **32**, 437-456.

Devillers, Ch. (1961) Structural and dynamic aspects of the development of the teleostean egg, in *Advances in Morphogenesis,* (eds M. Abercrombie and J. Brachet), Academic Press, NewYork, **1**, pp. 379-428.

Devillers, Ch. (1965) Respiration et morphogenèse dans l'oeuf des téléostéens. *Annales Biol. France* **4**, 157-186.

Devillers, Ch. et Raichman, L. (1959) Quelques données sur l'utilisation du vitellus au cours de la gastrulation dans l'oeuf de *Salmo irideus*. *Comptes Rendus de l'Académie des Sciences, Paris* **247**, 2033-2229.

Devillers, Ch., Thomopoulos, A. and Colas, J. (1953) Différention bipolaire et formation de l'espace perivitellin dans l'oeuf de *Salmo irideus*. *Bulletin de la Société Zoologique de la France* **78**. 462-470.

Devillers, Ch., Domurat J. and Colas, J. (1959) *Comptes Rendus de l'Académie des Sciences à Paris* **249**, 2229.

DiMichele, L. and Powers, D.A. (1984) The relationship between oxygen consumption rate and hatching in *Fundulus heteroclitus*. *Physiol. Zool.* **57**, 46-51.

DiMichele, L. and Taylor, M.H. (1981) The mechanism of hatching in *Fundulus heteroclitus*: development and physiology. *Journal of Experimental Zoology* **217**, 73-79.

Dobbs, G.H. (1974) Scanning electron microscopy of intraovarian embryos of the viviparous teleost *Micrometrus minimus* (Gibbons) (*Perciformes, Embioticidae*). *Journal of Fish Biology* **7**, 209-214.

Donovan, M. and Hart, N. (1982) Uptake of ferritin by the mosaic egg surface of *Brachydanio*. *The Journal of Experimental Zoology* **223**, 299-304.

Donovan, J.J. and Hart, N.H. (1986) Cortical granule exocytosis is coupled with membrane retrieval in the egg of *Brachydanio*. *Journal of Experimental Zoology* **237**, 391-405.

Douglas, R.H., Bowmaker, J.K. and Y.W. Kunz (1987) Ultraviolet vision in fish, in *Seeing Contour and Colour,* (eds J.J. Kulikowski, C.M. Dickinson and I.J. Murray) Pergamon Press, Oxford, pp. 601-616.

Dow, J.M. (1861) Notice of a viviparous fish (*Anableps dowi*) from the bay of La Union, state of San Salvador. *Annals of the Natural History* **7**, 420.

Doyère, M. (1849) Sur un micropyle dans des oeufs du Loligo et *Syngnathus Ophidium*. *Bulletin de la Société Philomathique de Paris*. Dec. 15th, 1849.

Droller, M.J. and Roth, T.F. (1966) An electron microscope study of yolk formation during oogenesis in *Lebistes reticulatus* Guppy. *Journal of Cell Biology* **28**, 209-232.

Dumont, J.N. and Brummet, A.R. (1980) The vitelline envelope, chorion, and micropyle of *Fundulus heteroclitus* eggs. *Gamete Research* **3**, 25-44.

Dunel, S. and Laurent, P. (1973) Ultrastructure comparée de la pseudobranchie chez les téléostéens marins et d'eau douce. *Journal de Microscopie* **16**, 53-74.

Duvernoy, G.L. (1844) Observations pour servir à la connaissance du développement de la Poecilie de Surinam (*Poecilia surinamensis* Val.), précédées d'une esquisse historique des principaux travaux sur le développement des poissons aux deux premières époques de la vie. *Comptes Rendus de l'Académie des Sciences de Paris* **18**, 667-679.

Easter, S.S., Wilson, W.S., Ross, L.S. and Burrill, J.D. (1992). Tract formation in the brain of the zebrafish embryo, in *The Nerve Growth Cone*, (eds P.C. Letourneau, S.B. Kater and E.R. Macagno), Raven Press, New York.

Eddy, F.B. (1974) Osmotic properties of the perivitelline fluid and some properties of the chorion of Atlantic salmon eggs (*Salmo salar*). *Journal of Zoology* **174**, 237-243.

Egami, N. (1959) Record of the number of eggs obtained from a single pair of *Oryzias latipes* kept in laboratory aquarium. *Journal of the Faculty of Science Tokyo University, sect. 4, Zoology* **8**, 521-538.

Eggert, B. (1929) Entwicklung und Bau der Eier von *Salarias flavo-umbrinus* Rupp. *Zoologischer Anzeiger* **83**, 241.

Eggert, B. (1931) Die Geschlechtsorgane der *Gobiiformes* und *Blenniiformes*. *Zeitschrift für wissenschaftliche Zoologie* **139**, 249-558.

Eigenmann, C.H. (1890) On the egg membranes and micropyle of some osseous fishes. *Bulletin of the Museum of Comparative Zoology at Harvard College* **19**, 129-154.

Eigenmann, C.H. (1892) On the viviparous fishes of the Pacific coast of North America. *Bulletin of the U.S. Fish Commission* **12**, 381-478.

Eigenmann, C.H. and Shafer, G.D. (1890) The mosaic of single and twin cones in the retina of fishes. *American Naturalist* **34**, 109-118.

Einarsson, J.M. (1992) Cyclical changes, in the behaviour and localization of C_4-lactate dehydrogenase in the retina of the cichlid fish, *Oreochromis mossambicus*. PhD thesis submitted to the National University of Ireland.

Einarsson, J.M., Joyce, P. and Y.W. Kunz (1995) Kinetic and immunological differences between the retinal specific C_4- and B_4-lactate dehydrogenase of the cichlid fish *Oreochromis mossambicus*. *Comparative Biochemistry and Physiology* **112B**, 589-598.

Eisen, J.S. and Weston, J.A. (1993) Review. Development of the neural crest in the zebrafish. *Developmental Biology* **159**, 50-59.

Ekström, P., Borg, B. and van Veen, Th. (1983) Ontogenetic development of the pineal organ, parapineal organ, and retina of the three-spined stickleback, *Gasterosteus aculeatus* L. (Teleostei). *Cell and Tissue Research* **233**, 593-609.

Emeliyanova, N.G. and Makeyeva, A.P. (1985) Spermatozoon ultrastructure of some cyprinids (*Cyprinidae*) (in Russian). *Vopr. Ikthiol.* **25**, 459-468.

Endler, J.A. (1980) Natural selection on color patterns in *Poecilia reticulata*. *Evolution* **34**, 76-91.

Engen, P.C. (1968) Organogenesis in the walleye surfperch, *Hyperprosopon argenteum* (Gibbons). *Calif. Fish and Game* **54**, 156-169.

Ennis, S. (1985) Ultrastructural study of the outer retinal layers in the teleost *Poecilia reticulata* (Peters) — Adult, juvenile, neonate and embryonic stages. PhD thesis submitted to The National University of Ireland.

Ennis, S. and Kunz, Y.W. (1984a) Myeloid bodies in the pigment epithelium of a teleost embryo, the viviparous *Poecilia reticulata*. *Cell Biology International Reports* **8**, 1009-1011.

Ennis, S. and Kunz, Y.W. (1984b) Ageing of the collagenous layer in Bruch's membrane of the teleost *Poecilia reticulata*. *Cell Biology International Reports* **8**, 902.

Ennis, S. and Kunz, Y.W. (1986) Differentiated retinal Müller glia are ciliated — ultrastructural evidence in the teleost *Poecilia reticulata* P. *Cell Biology International Reports*

10, 611-622.

Erhardt, H. (1976) Licht- und elektronenmikroskopische Untersuchungen an den Eihüllen des marinen Teleosteers *Lutjanus synagris*. *Helgoländer wissenschaftliche Meeresuntersuchungen* **28**, 90-105.

Erhardt, H. and K.J. Götting (1970) Licht- und elektronenmikroskopische Untersuchungen an Eizellen und Eihüllen von *Platypoecilus maculatus*. Light and electron microscopic studies in egg cells and envelopes of *Platypoecilius* maculatus. *Cytobiologie* **2**, 429-440.

Erickson, D.L. and Pikitch, E.K. (1993) A histological description of shortspined thornyhead, *Sebastolobus alascanus*, ovaries: structures associated with the production of gelatinous masses. *Environmental Biology of Fishes* **36**, 273-282.

Erickson, D.L. and Trinkaus, J.P (1976) Microvilli and blebs as source of reserve surface membrane during cell spreading. *Experimental Cell Research* **99**, 375-384.

Evans, B.I. and Fernald, R.D. (1990) Metamorphosis and fish vision. *Journal of Neurobiology* **21**, 1037-1052.

Evans, B.I. and Fernald, R.D. (1993) Retinal transformation at metamorphosis in the winter flounder (*Pseudopleuronectes americanus*). *Visual Neuroscience* **10**, 1055-1064.

Eyclesheimer, A.C. (1895) The early development of *Amblystoma* with observations on some other Vertebrates. *Journal of Morphology* **10**, 343-418.

Fairbanks, M.B., Hoffert, J.R. and Fromm, P.O. (1969) The dependence of the oxygen-concentrating mechanism of the teleost eye (*Salmo gairdneri*) on the enzyme carbonic anhydrase. *Journal of General Physiology* **54**, 203-211,

Falcon, J., Guerlotte, J. F., Voison, P. and Collin, J.P. (1987) Rhythmic melatonin biosynthesis in a photoreceptive pineal organ. A study in the pike. *Neuroendocrinology* **45**, 479-486.

Farias, G., González and Maccioni, R.B. (1995) Tubulin and microtubule-associatied protein pools in unfertilized and fertilized eggs of the trout *Oncorhynchus mykiss*. *Journal of Experimental Zoology* **271**, 253-263.

Fausto, A.M., Carcupino, M., Scapigliati, G., Taddei, A.R. and Mazzini, M. (1994) Fine structure of the chorion and micropyle of the sea bass egg *Dicentrarchus labrax* (teleostea, perichthydae). *Bolletino di Zoologia* **61**, 129-133.

Favard, P. and André, J. (1970) The mitochondria of spermatozoa, in *Comparative Spermatology* (ed B. Baccetti), Academic Press, New York, London, pp. 415-429.

Felix, W. (1897) Beiträge zur Entwickelungsgeschichte der Salmoniden. *Anatomische Hefte, Wiesbaden* **8**, 249-466.

Fernald, R.D. (1985) Growth of the teleost eye: novel solutions to complex constraints. *Environmental Biology of Fishes* **13**, 113-123.

Fernald, R.D. (1989a) Retinal rod neurogenesis, in *Development of the Vertebrate Retina*, (eds B.L Finlay and D.R. Sengelaub), Plenum, New York. pp. 31-42.

Fernald, R.D. (1989b) Fish vision, in *Development of the vertebrate retina*, (eds B.L. Finlay and D.R. Sengelaub), Plenum, New York. pp. 247-265.

Fernald,R.D. (1991) Teleost vision; seeing while growing. *The Journal of Experimental Zoology, Supplement* **5**, 167-180.

Fernald, R.D. (1993) Vision, in *The Physiology of Fishes*, (ed. D.H. Evans), CRC Press, Boca Raton, pp.161-190.

Filippi, Filippo de (1841) Memoria sullo sviluppo del Ghiozzo d'aqua dolce (*Gobius fluviatilis*). *Annali Univ. di Medicina, Milano, compilati dal dottore Omodei.- Rev. Zool.* **1842**. p. 45.

Fioroni, P. (1987) *Allgemeine und vergleichende Embryologie der Tiere*, Springer, Berlin, Göttingen, Heidelberg. 429 pp.

Fishelson, L.) 1978) Oogenesis and spawn-formation in the pigmy lion fish *Dendrochirus brachypterus* (pteroidae) *Marine Biology* **46**, 341-348.

Fisher, K.C. (1963) The formation and properties of the external membrane of the trout egg. *Transactions of the Royal Society of Canada* **1**, Series IV, Section III, 323-332.

Flegler, C. (1977) Electron microscopic studies on the development of the chorion of the viviparous teleost *Dermogenys pusillus* (*Hemirhamphidae*). *Cell and Tissue Research* **179**, 255-270.

Flegler-Balon, C. (1989) Direct and indirect development in fishes — examples of alternative life-history styles, in *Alternative Life-History Styles*, (ed M.N. Bruton), Kluwer Academic Publishers, Dordrecht. pp, 71-100.

Fleig, R. (1993) Embryogenesis in mouth-breeding cichlids (*Osteichthyes, Teleostei. Roux's Archives for Developmental Biology* **203**, 124-130.

Flüchter, J. and Rosenthal, H. (1965) Beobachtungen über das Vorkommen und Laichen des Blauen Wittlings (*Micromesistius poutasson* Risso) in der Deutschen Bucht. *Helgoländer wissenschaftliche Meeresuntersuchungen* **12**, 149-155.

Fluck, R.A. (1978a) Acetylcholine and acetylcholinesterase activity in early embryos of the medaka *Oryzias latipes*, a teleost. *Development, Growth and Differentiation* **20**, 17-25.

Fluck, R.A. (1978b) Acetylcholine and aceteylcholinesterase activity in early embryos of a teleost, the Japanese medaka, *Oryzias latipes*. *Journal of General Physiology* 6a, 7a.

Fluck, R.A. (1982) Localization of acetylcholinesterase activity in young embryos of the medaka *Oryzias latipes*. *Comparative Biochemistry and Physiology* **72C**, 59-64.

Fluck, R.A. and Killian, C.E. (1982) Cellular basis of contraction of the enveloping layer of the medaka (*Oryzias latipes*), a teleost. *Journal of Cell Biology* **95**, 2, part2, 311A.

Fluck, R. Gunning, R., Pellegrino, J., Barron, T. and Panitch, D. (1981) Rhythmic contractions of the blastoderm of the medaka *Oryzias latipes*, a teleost. *Journal of General Physiology* **78**, 16a and *Journal of Experimental Zoology* **226**, 245-53.

Fluck, R.A., Killian, C.E., Miller K., Dalpe, J.N. and Shih, T.M. (1984). Contraction of an embryonic epithelium, the enveloping layer of the medaka (*Oryzias latipes*), a teleost. *Journal of Experimental Zoology* **229**, 127-142.

Fluck, R.A., Miller, A.L. and Laffe, L.F. (1991) Slow calcium waves accompany cytokinesis in medaka fish eggs. *Journal of Cell Biology* **115**, 1259-1266.

Flügel, H. (1964a) On the fine structure of the *zona radiata* of growing trout oocytes. *Die Naturwissenschaften* **51**, 542.

Flügel, H. (1964b) Electron microscopic investigations on the fine structure of the follicular cells and the *zona radiata* of trout oocytes during and after ovulation. *Die Naturwissenschaften* **51**, 564-565.

Flügel, H. (1964c) Desmosomes in the follicular epithelium of growing oocytes of the Eastern brook trout *Salvelinus fontinalis* (electron microscopic investigations). *Die Naturwissenschaften* **51**, 566.

Flügel, H. (1967a) Elektronenmikroskopische Untersuchungen an den Hüllen der Oozyten und Eier des Flussbarsches *Perca fluviatilis*. *Zeitschrift für Zellforschung* **77**, 244-256.

Flügel, H. (1967b) Licht- und elektronenmikroskopische Untersuchungen an Oozyten und Eiern einiger Knochenfische. *Zeitschrift für Zellforschung* **83**, 82-116.

Flügel, H. (1970) Zur Funktion des Golgi-Apparatus in den Follikelzellen von *Coreguonus albula* L.. *Cytobiologie* **4**, 450-459.

Forchhammer, T. (1819) De *Blenii vivipari formatione et evolutione observationes*. Kiliae, (ed C.F. Mohr, Kiel).

Foskett, J.K. and Scheffey, C. (1982) The chloride cell: definitive identification as the salt-secretory cell in teleosts. *Science* **215**, 164-166.

Foster, S.A. and Baker, J.A. (1995) Evolutionary interplay between ecology, morphology and reproductive behavior in threespine stickleback, *Gasterosteus aculeatus*. *Environmental Biology of Fishes* **44**, 213-223.

Franz, K. (1897) Ueber die Entwicklung von *Hypochorda* und *Ligamentum longitudinale ventrale* bei Teleostiern. *Morphologisches Jahrbuch* **25**, 143-169.

Fraser, E. and Reader, D. (1940) Observation on the breeding and development of the viviparous fish *Heterandria formosa*. *Quarterly Journal of microscopical Science* **8**, New Series, Part 4, No. 324.

Fribourgh, J.H., McClendon, D.E. and Soloff, B.L. (1970) Ultrastructure of the goldfish, *Carassius auratus (cyprinidae)*, spermatozoon. *Copeia* **1970, No. 2**, 274-279.

Frisch von, K. (1911) Ueber das Parietalorgan der Fische als funktionierendes Organ. *Sitzungsberichte der Gesellschaft für Morphologie und Physiologie (München)* **27**, 16-18.

Fuiman, L.A. (1984) Ostariophysi: Development and relationships, in *Ontogeny and Systematics of Fishes*, (ed American Society of Ichthyologists and Herpetologists), Allen Press Inc. Lawrence, KS 66074, U.S.A.

Fuiman, L.A. (1997) What can flatfish ontogenies tell us about pelagic and benthic lifestyles? *Journal of Sea Research* **37**, 257-267.

Fuiman, L.A. and Delbos, B.C. (1998) Development changes in visual sensitivity of red drum, *Sciaenops ocellatus*. *Copeia* **1989** (4), 936-943.

Fujinami N. and Kageyama, T. (1975) Circus movements in dissociated embryonic cells of a teleost, *Oryzias latipes*. *Journal of Cell Science* **19**,169-182.

Fulton, T.W. (1898) On the maturation of the pelagic eggs of teleostean fishes. *Zoologischer Anzeiger* **21**, 245-252.

Fürst, C.M. (1904) Zur Kenntnis der Histogenese und des Wachstums der Retina. *Acta universita Lund* **40**, 1-45.

Fusari, R. (1892) Sur les premières phases du développement des téléostéens. *Archivio Italiano di Biologia* **18**, 204-239.

Gardiner, D.M. (1978) The origin and fate of spermatophores in the viviparous teleost *Cymatogaster aggregata* (Perciformes: Embiotocidae). *Journal of Morphology* **155**, 157-172.

Garman, S. 1895) The cyprinodonts. *Memoirs of the Museum of Comparative Zoology* **19**, 1-179.

Gegenbaur, C. (1861) Ueber den Bau und die Entwickelung der Wirbelthiereier mit partieller Dottertheilung. *Müeller's Archiv für Anatomie und Physiologie,* **1861**, 491-529.

Geiger, W. (1956) Quantitative Untersuchungen über das Gehirn der Knochenfische, mit besonderer Berücksichtigung seines relativen Wachstums. *Acta anatomica* **26**, 121-163.

Gensch, H. (1882) Das sekundäre Entoderm und die Blutbildung beim Ei der Knochenfische. PhD Thesis, Königsberg.

Gerbe, Z. (1872) Recherches sur la segmentation de la cicatricule et la formation des produits adventifs de l'oeuf des Plagiostomes et particulièrement des Raies. *Journal de l'Anatomie et de la Physiologie* **8**, 609-616.

Gerbe, Z. (1875) Du lieu ou se forma la cicatricule chez les poissons osseux. *Journal de l'Anatomie et de la Physiologie* **11**, 329-333.

Gevers, P. and Denucé, J.M. (1993) Effects of retinoic acid on epiboly, convergence and gastrulation of embryos of the medaka, *Oryzias latipes* (Teleostei, Cyprinodontidae) *Roux's Archives of Developmental Biology* **203**, 169-174.

Gevers, P. and Timmermans, L.P.M. (1991) Dye-coupling and the formation and fate of the hypoblast in the teleost fish embryo, *Barbus conchonius*. *Development* **112**, 431-438.

Gevers, P., Coenen, A.J.M., Schipper, H., Stroband, H.W.J. and Timmermans, L.P.M. (1993) Involvement of fibronectin during epiboly and gastrulation in embryos of the common carp, *Cyprinus carpio*. *Roux's Archives of Developmental Biology* **202**, 152-158.

Gihr, M. (1957) Zur Entwicklung des Hechtes. *Revue Suisse de Zoologie* **64**, 355-474.

Gilch, G. (1957) Vergleichende Untersuchungen eines hydrostatischen Apparates larvaler Labyrinthfische. *Zoologisches Jahrbuch, Abteilung Anatomie und Ontogenie* **76**, 1-62.

Gilchrist, J.D.F. (1905) The development of South African fishes, Part 2. *Marine investigations in South Africa* **3**, 131-167.

Giles, N. (1984) Implications of parental care of offspring for the anti-predator behaviour of adult male and female three-spined sticklebacks, *Gasterosteus aculeatus* L., in *Fish Reproduction* (eds G.W. Potts and R.J. Wootton), Academic Press, London, pp. 275-89.

Gilkey, J.C. (1981) Mechanisms of fertilization in fishes. *The American Zoologist* **21**, 359-375.

Gilkey, J.C., Jaffe, L.F., Ridgway, E.E. and Reynolds, G.T. (1978) A free calcium wave traverses the activating egg of the medaka, *Oryzias latipes*. *The Journal of Cell Biology* **76**, 448-466.

Ginsburg, A.S. (1961) The block to polyspermy in sturgeon and trout with special reference to the role of cortical granules (alveoli). *Journal of Embryology and experimental Morphology* **9**, part l, 173-790.

Ginsburg, A.S. (1963) Sperm-egg association and its relationship to the activation of the egg in salmonid fishes. *Journal of Embryology and experimental Morphology* **11**, 13-33.

Ginsburg, A.S. (1968) *Fertilization in Fishes and the Problem of Polyspermy*. (ed. T.A. Detlaf, Moskau). Translated from Russian. Israel Program for Scientific Translations, Jerusalem, 1972. 366 pp.

Ginsburg, A.S. (1972) Translation of 1968 publication, mentioned above.

Ginzburg, A.S. see Ginsburg

Goette, A. (1869) Zur Entwicklungsgeschichte der Wirbelthiere. *Centralblatt für Medicinische Wissenschaft* **26**, 404-409.

Goette, A. (1873) Beiträge zur Entwicklungsgeschichte der Wirbelthiere. I. Der Keim des Forelleneies. *Archiv für Mikroskopische Anatomie* **9**, 679-708.

Goette, A. (1878) Beiträge zur Entwickelungsgeschichte der Wirbelthiere. III. Ueber die Entwickelung des Central-Nervensystems der Teleostier. *Archiv für Mikroskopische*

Anatomie **5**, 139-199.

Gomendio, M. and Roldan, E.R.S. (1993) Coevolution between male ejaculates and female reproductive biology in eutherian mammals. *Proceedings of the Royal Society of London B* **252**, 7-12.

González, M.E., Blánquez, M.JK. and Royo, C. (1996) Early gill development in *Oncorhynchus mykiss*: a light and scanning electron microscopy study. *Journal of Morphology* **229**, 201-217.

Gordon, M. (1955) Those puzzling 'little toms'. *Animal Kingdom*, p. 50-55.

Goronowitsch, N. (1885) Studien über die Entwicklung des Medullarstranges bei Knochenfischen, nebst Beobachtungen über die erste Anlage der Keimblätter und der Chorda bei Salmoniden. *Morphologisches Jahrbuch* **10**, 376-445.

Götting, K.-J. (1961) Beiträge zur Kenntnis der Grundlagen der Fortpflanzung und zur Fruchtbarkeitsbestimmung bei marinen Teleosteern. *Helgoländer wissenschaftliche Meeresuntersuchungen* **8**, 1-41.

Götting, K.-J. (1964) Entwicklung, Bau und Bedeutung der Eihüllen des Steinpickers (*Agonus cataphratus* L.). *Helgoländer wissenschaftliche Meeresuntersuchungen* **11**, 1-12.

Götting, K.-J. (1965) Die Feinstruktur der Hüllschichten reifender Oocyten von *Agonus cataphractus* L. (Teleostei, Agonidae). *Zeitschrift für Zellforschung* **66**, 405-414.

Götting, K.-J. (1966) Zur Feinstruktur der Oocyten mariner Teleosteer. *Helgoländer wissenschaftliche Meeresuntersuchungen* **13**, 118-170.

Götting, K.-J. (1967) Der Follikel und die peripheren Strukturen der Oocyten der Teleosteer und Amphibien. *Zeitschrift für Zellforschung* **79**, 481-491.

Götting, K.-J. (1969) Zur Feinstruktur der Dotterkerne in den Oocyten mariner Teleosteer. Verhandlungen der Deutschen Zoologischen Gesellschaft 1968. *Zoologischer Anzeiger*, **Supplement 32**, 161-168.

Götting, K.J. (1970) Zur Darstellung der Ultrastruktur des Teleosteer–Follikels mittels der Gefrierätztechnik. *Micron* **1**, 356-372.

Götting, K.J. (1974) Oocyte ultrastructure of oviparous and of ovoviviparous teleosts as revealed by freeze etching. *International Congress of Electron microscopy Canberra* **8**, 668-669.

Götting, K.J. (1976) Fortpflanzung und Oocyten-Entwicklung bei der Aalmutter (*Zoarces viviparus*) (Pisces, Osteichthyes). *Helgoländer wissenschaftliche Meeresuntersuchungen* **28**, 71-89.

Granel, f. (1927) La pseudobranchie des poissons. *Archive d'Anatomie et Microscopie* **23**, 175-317.

Grassi, G.B. (1914) Funzione respiratoria delle cosidette pseudobranche dei teleostei ed altri particolari intorno ad esse. *Bios. Genova* **2**,1-16.

Gray, J. (1932) The osmotic properties of the egg of the trout (*Salmo fario*). *Journal of experimental Biology* **9**, 277-299.

Greeley, M.S., Calder, D.R. Jr and Wallace, R.A. (1986) Changes in teleost yolk protein during oocyte maturation: correlation of yolk proteolysis with oocyte hydration. *Comparative Biochemistry and Physiology* **84B**, 1-9.

Gregory, E.H. (1902-03) Beiträge zur Entwicklungsgeschichte der Knochenfische. *Anatomische Hefte, 1. Abteilung,* **20**, 151-229.

Greven, H. (1995) Viviparie bei Aquarienfischen (*Poeciliidae, Goodeidae, Anablepidae, Hemiramphidae*). *Fortpflanzungsbiologie der Aquarienfische*, ed Schmettkamp, Bornheim. pp. 141-158.

Grier, H.J. (1975) Spermiogenesis in the teleost *Gambusia affinis* with particular reference to the role played by microtubules. *Cell and Tissue Research* **165**, 89-102.

Grier, H.J. (1981) Cellular organization of the testis and spermatogenesis in fishes. *The American Zoologist* **21**, 345-357.

Grier, H.J. and L.R. Parent (1994) Reproductive biology and systematics of phallostethid fishes as revealed by gonad structure. *Environmental Biology of Fishes* **41**, 287-299.

Grier, H.J., Burns, J.R. and Flores, J.A. (1981) Testis structure in three species of teleosts with tubular gonopodia. *Copeia* **1981**, 797-801.

Griffin, F.J., Vines, C.A., Pillai, M.C., Yanagimachi, R. and Cherr, G.N. (1996) Sperm motility initiation factor is a minor component of the pacific herring egg chorion. *Development, Growth and Differentiation* **38**, 193-202.

Grodzinski, Z. (1954) The yolk of bitterling *Rhodeus amarus* BL. *Folia Morphol. (Warsaw)* **13**, 13-26.

Gronovius, L.T. (1754-56) *Museum ichthyologicum* and *Zoophylacum*. (Theodor Haak, Leiden) (cf. Schneider-von Orelli. M., 1907).

Groot, E.P. and Alderdice, D.F. (1985) Fine structure of the external egg membrane of five species of Pacific salmon and steelhead trout. *Canadian Journal of Zoology* **63**, 552-566.

Gross, M.R. and Shine, R. (1981) Parental care and mode of fertilization in ectothermic vertebrates. *Evolution* **35**, 775-793.

Grove, B.D. (1985) The structure, function and development of the follicular placenta in the viviparous poeciliid fish, *Heterandria formosa*. PhD. Thesis, Clemson University.

Grove, B.D. and Wourms, J.P. (1991) The follicular placenta of the viviparous fish *Heterandria formosa*. I. Ultrastructure and development of the embryonic absorptive surface. *Journal of Morphology* **209**, 264-284.

Grove, B.D. and Wourms, J.P. (1994) Follicular placenta of the viviparous fish, *Heterandria formosa*: II. Ultrastructure and development of the follicular epithelium. *Journal of Morphology* **220**, 167-184.

Grün, G. (1975a) Electron microscopic study of receptor outer segment differentiation in the retina of *Tilapia leucosticta* (*cichlidae*). *Verhandlungsbericht der Deutschen Zoologischen Gesellschaft* **1974**, 167-170.

Grün, G. (1975b) Structural basis of the functional development of the retina in the cichlid, *Tilapia leucosticta (teleostei)*. *Journal of Embryology and experimental Morphology* **33**, 243-257.

Grün, G. (1980) Developmental dynamic in synaptic ribbons of retinal receptor cells (*Tilapia, Xenopus*). *Cell and Tissue Research* **207**, 331-339.

Grün, G. (1982) Development dynamics of synapses in the vertebrate retina. *Progress in Neurobiology* **80**, 257-274.

Gudger, E.W. (1905) The breeding habits and the segmentation of the egg of the pipefish (*Siphostoma floridae*). *Proceedings of the U.S. Natural Museum* **29**, 447-500.

Gudger, E.W. (1918) Oral gestation in the gaff-topsail catfish, *Felichthys felis*. *Carnegie Institute of Washington Pulications* **252**, 25-52.

Gudger. E.W. (1927) The nest and the nesting habits of the butterfish or gunnel *Pholis gunnellus*. *Natural History New York* **27**, 65-71.

Guitel, F. (1891) Sur les moeurs du *Gobius minutus*. *Comptes Rendus Hebdomadaires des Séances de l'Académie des Sciences* **113**, 292-296.

Guitel, F. (1892) Sur l'ovaire et l'oeuf du *Gobius* minutus. *Comptes Rendus Hebdomadaire des Séances de l'Académie des Sciences* **114,** 612-616.

Guitel, F. (1893) Observations sur les moeurs de trois blenniidés, *Clinus argentatus, Blennius montagui* et *Blennius* sphynx. *Archive de Zoologie Expérimentale et Générale 3*, série **1**, 325-384.

Gunn, J.S. and Thresher, R.E. (1991) Viviparity and the reproductive ecology of clinid fishes (*Clinidae*) from temperate Australian waters. *Environmental Biology of Fishes* **31**, 323-344.

Guraya, S.S. (1965) A comparative histological study of a fish (*Canna marulus*) and an amphibian (*Bufo stomaticus*). *Zeitschrift der Zellforschung* **65**, 662-700.

Guraya, S.S.(1982) Recent progress in the structure, origin, composition and function of cortical granules in animal eggs. *International Review of Cytology* **78**, 257-360.

Guraya, S.A. (1986) *The Cell and Molecular Biology of Fish Oogenesis*, (ed H.W. Sauer), Karger, Basel, 223 pp.

Haas-Andela, H. (1976) *In vitro* Kultur und Aufzucht von Embryonen lebendgebärender Zahnkarpfen der Gattung *Xiphophorus*. *Zoologischer Anzeiger* **197**, 1-5.

Haeckel, E. (1855) Ueber die Eier der *Scomberesoces*. *Archiv für Anatomie, Physiologie und wissenschaftliche Medicin* **1885 No. 4,** 23-31.

Haeckel (1875) Die Gastrula und die Eifurchung der Thiere. *Jenaische Zeitschrift für Naturwissenschaft* **9**, 402-508.

Hagenmaier, H.E. (1969) Der Nucleinsäure- bzw. Ribonucleoproteid-Status während der Frühentwicklung von Fischkeimen (*Salmo irideus* und *Salmo trutta fario*). *Wilhelm Roux's Archiv für die Entwicklungsmechanik der Organismen* **162**, 19-40.

Hagenmaier, H.E. (1972) Zum Schlüpfprozess bei Fischen. II. Gewinnung und Charakterisierung des Schlüpfsekretes bei der Regenbogenforelle (*Salmo gairdneri* Rich.) *Experientia* **28**, 1214-1215.

Hagenmaier, H.E. (1973) The hatching process in fish embryos. — III. The structure, polysaccharide and protein cytochemistry of the chorion of the trout egg, *Salmo gairdneri* (Rich.) *Acta Histochemica* **47**, 61-69.

Hagenmaier, H.E. (1974a) The hatching process in fish embryos. IV. The enzymological propertiers of a highly purified enzyme (chorionase) from the hatching fluid of the rainbow trout, *Salmo gairdneri* Rich.. *Comparative Biochemistry and Physiology* **49B**, 313-324.

Hagenmaier, H.E. (1974b) The hatching process in fish embryos. V. Characterization of the hatching protease (chorionase) from the perivitelline fluid of the rainbow trout, *Salmo gairdneri* (Rich.). as a metalloenzyme. *Wilhelm Roux's Archiv für die Entwicklungsmechanik der Organismen* **175**, 157-162.

Hagenmeier, H.E., Schmitz, I. and Foles, J. (1976) Zum Vorkomen von Isopeptidbindungen in der Eihülle der Regenbogenforelle *Salmo gairdneri* Rich., as a metalloenzyme. *Hoppe Seyler's Zeitschrift für Physiologische Chemie* **357**, 1435-1438.

Hagenmaier, H.E. and Lindemann, K. (1984) The hatching process in fish embryos. VII. Isolation of hatching enzyme of the rainbow trout (*Salmo gairdneri* Rich.) by a Percoll Density Gradient Centrifugation. *Zoologisches Jahrbuch der Physiologie* **88**, 447-452.

Hagström, B.E. and Lönning, S. (1968) Electron microscopic studies of unfertilized and fertilized eggs from marine teleosts. *Sarsia* **33**, 73-80.

Hall D.D., and Miller, F.J. (1968) A qualitative study of courtship and reproductive behavior in the male gourami, *Trichogaster leeri* (Bleeker). *Behaviour* **32**, 70-84.

Hamano S. (1964) A time-lapse cinematographic study on gastrulation in the zebrafish. *Acta Embryologiae et Morphologiae Experimentalis* **7**, 42-48.

Hamazaki, T., Iuchi, I. and Yamagami, K. (1984) Chorion glycoprotein-like immunoreactivity in some tissues of adult female medaka. *Zoological Science* **1**, 148-150.

Hamazaki, T., Iuchi, I. and Yamagami, K. (1985) A spawning female-specific substance reactive to anti-chorion (egg envelope) glycoprotein antibody in the teleost, *Oryzias latipes*. *Journal of Experimental Zoology* **235**, 269-279.

Hamazaki, T.S., Iuchi, I. and Yamagami, K. (1987a) Production of a "Spawning Female-Specific Substance" in hepatic cells and its accumulation in the ascites of the estrogen-treated fish, *Oryzias latipes*. *Journal of Experimental Zoology* **242**, 325-332.

Hamazaki, T.S., Iuchi, I. and Yamagami, K. (1987b) Purification and identification of vitellogenin and its immunohistochemical detection in growing oocytes of the teleost, *Oryzias latipes*. *Journal of Experimental Zoology* **242**, 333-342.

Hamazaki, T.S., Iuchi, I. and Yamagami, K. (1987c) Isolation and partial characterization of a "Spawning Female-Specific Substance" in the teleost, *Oryzias latipes*. *The Journal of Experimental Zoology* **242**, 343-49.

Hamazaki, T.S., Nagahama, Y., Iuchi, I. and Yamagami, K. (1989) A glycoprotein from the liver constitutes the inner layer of the egg envelope (*zona pellucida interna*) of the fish, *Oryzias latipes*. *Developmental Biology* **133**, 101-110.

Hancock, A. (1854) Observations on the nidification of *Gasterosteus aculeatus* and *G. spinachia*. *Transactions of the Tyneside Naturalists' Field Club* **2** (1851-54).

Harb, J.M. and Copeland, D.E. (1969) Fine structure of the pseudobranch of the flounder *Paralichthys lethostigma*. *Zeitschrift für Zellforschung* **101**, 167-174.

Harder, W. (1953) Zum Formwachstum des Herings. *Zeitschrift für Morphologie und Oekologie* **42**, 209-224.

Hárosi, F.I. and Hashimoto, Y. (1983) Ultraviolet visual pigment in a vertebrate: A tetrachromatic cone system in the Dace. *Sience*, **222**, 1021-1032.

Harrington, R.W. (1961)' Oviparous hermaphroditic fish with internal self-fertilization. *Science* **135**, 1749-50.

Harris, R.K. (1973) Cell suface movements related to cell locomotion, in *Locomotion of Tissue Cells*, [Ciba Foundation Symposium 14 (new series)], Elsevier, Excerpta Medica, North-Holland, Amsterdam, pp. 3-25.

Hart, N.Y. (1990) Fertilization in teleost fishes: Mechanisms of sperm-egg interactions. *International Review of Cytology* **121**, 1-66.

Hart, N.Y. and Collins, G.C. (1987) Morphological features of the plasma membrane during activation of the *Brachydanio* egg. *Journal of Cell Biology* **105**, part 2, 254a.

Hart, N.Y. and Donovan, M. (1983) Fine structure of the chorion and site of sperm entry in the egg of *Brachydanio*. *Journal of Experimental Zoology* **227**, 277-296.

Hart, N.Y. and Wolenski, J.S. (1988) Actin in the cortex of the teleost egg. *Journal of Cell Biology* **107,** 174a.

Hart, N.Y. and Yu, S.F. (1980) Cortical granule exocytosis and cell surface reorganization in eggs of *Brachydanio*. *Journal of Experimental Zoology* **213**, 137-159.

Hart, N.H., Becker, K.A. and Wolenski, J.S. (1992) The sperm entry site during fertilization of the zebrafish egg: Localization of actin. *Molecular Reproduction and Development* **32**, 217-228.

Hart, N.H., Pietri, R. and Donovan, M. (1984) The structure of the chorion and associated surface filaments in *Oryzias* — Evidence for the presence of extracellular tubules. *Journal of Experimental Zoology* **273**, 273-296.

Hart, N.Y., Wolenski, J.S. and Donovan. M.J. (1987) Ultrastructural localization of lysosomal enzymes in the egg cortex of *Brachydanio rerio*. *Journal of Experimental Zoology* **244**, 17-32.

Hart, N.H., Yu, S.F. and Greenhut, V.A. (1977) Observations on the cortical reaction in eggs of *Brachydanio rerio* as seen with the scanning electron microscope. *Journal of Experimental Zoology* **201**, 325-332.

Hartmann, M. (1944) Befruchtungsstoffe (Gamone) bei Fischen (Regenbogenforelle). *Die Naturwissenschaften* **32,** 231.

Hartmann, M., von Medem F.G., Kuhn K. and Bielig, H.-J. (1947a) Ueber die Gynogamone der Regenbogenforelle. *Die Naturwissenschaften* **34**, 25-26.

Hartmann, M., von Medem, F.G., Kuhn, R. and Bielig H.-J. (1947b) *Untersuchungen über die Befruchtungsstoffe der Regenbogenforelle*. Zeitschrift für Naturforschung **2B**, 330-349.

Hassett, MM. (1909) The Catholic Encyclopedia, vol. 6. Robert Appleton Company.

Hatta, K. and Kimmel, Ch.B. (1993) Midline structures and central nervous system coordinates in zebrafish. *Perspectives on Developmental Neurobiology* **1**, 257-268.

Hatta, K., Püschel, A.W. and Kimmel, Ch.B. (1994) Midline signaling in the primordium of the zebrafish anterior central nervous system. *Developmental Biology* **191**, 2061-2065.

Hatta, K., Kimmel, Ch.B., Ho, R.K. and C. Walker (1991) The cyclops mutation blocks specification of the floor plate of the zebrafish central nervous system. *Nature* **350**, 339-341.

Hattar, S., Liao, H.-W., Takao, M., Berson, D.M. and Yau, K.-W. (2002) Melanopsin-containing retinal ganglion cells: architecture, projections, and intrinsic photosensitivity. *Science* **295**, 1065-1070.

Hawkes, J.W. (1974) The structure of fish skin. I. General organization. *Cell and Tissue Research* **149**, 147-158.

Hawryshyn, C.W. (1998) Vision, Chapter 4 in *The Physiology of Fishes*, 2nd edn (ed D.H., Evans), CRC Press, Boca Raton, New York, pp. 345-374.

Hayes, F. (1942) The hatching mechanism of salmon eggs. *Journal of exerimental Zoology* **89**, 357-73.

Hayes, F. and Armstrong, F. (1942) Physical change in the constituent part of developing salmon eggs. *Canadian Journal of Research, sect. D* **20**, 99-114.

Hearne, M.E. (1984), in *Ontogeny and Systematics of Fishes*. Special Publication Number1. (eds American Society of Ichthyologists and Herpetologists), pp. 153-155.

Helfman, G.S., Collette, B.B. and Facey, D.E. (2nd edition 1999) *The Diversity of Fishes*. Blackwell Science, Malden, Mass. U.S.A.

Heming, T.A. and Buddington, R.K. (1988) Yolk absorption in embryonic and larval fishes, in *Fish Physiology* 11A, (ed P.W. Hoar), Academic Press, San Diego, p. 414-430.

Hempel, C. (1979) *Early life history of fish: The egg stage*. University of Washington Press, Seattle.

Henneguy, L.F. (1878) Procédé technique pour l'étude des embryons de poissons. *Bulletin de la Société Philomatique de Paris* **3**, 75-77.

Henneguy L.F. (1880) Note sur quelques faits relatifs aux premiers phénomènes du développement des poissons osseux. *Bulletin de la Société Philomatique de Paris* **4**, 132-135.

Henneguy, L.F. (1885) Sur la ligne primitive des Poissons osseux. *Zoologischer Anzeiger* **8**, 103-108.

Henneguy, L.F. (1887) Sur le mode d'accroissement de l'embryon des poissons osseux. *Comptes Rendus de l'Académie des Sciences* **104**, 85-87.

Henneguy, L.F. (1888) Recherches sur le développement des Poissons Osseux. Embryogénie de la Truite. *Journal de l'Anatomie et de la Physiologie* **24**, 413-617.

Hensel, R.F. (1870) The freshwater fishes of Southern Brazil. *Archiv für Naturgeschichte (Wiegmann)* **36**, 50-91.

Hensel, K. and Balon, E.K. (2001) The sensory canal systems of the living coelacanth, *Latimeria chalumnae*: a new instalment. *Environmental Biology of Fishes* **61**, 117-240.

Henshall, J.A. (1888) On some peculiarities of the ova of fishes. *Journal of the Cincinatti Society of Natural History* **11**, 81-85.

Hernández, G.M.P., Lozano, M.T. and Agulleiro, B. (1994) Ontogeny of the endocrine cells of the intestine and rectum of sea bass (*Dicentrarchus labrax* L.): an ultrastructural study. *Anatomy and Embryology* **190**, 529-539.

Herrick, C.J. (1962) *Neurological foundation of animal behavior*. Hafner Publ. Comp., New York and London.

Hertwig, O. (1975) Beiträge zur Kenntnis der Bildung, Befruchtung und Theilung des tierischen Eies. *Morphologisches Jahrbuch* **1**, 1-88.

Hertwig, O. (1906) Die Lehre von den Keimblättern. *Handbuch der vergleichenden und experimentellen Entwicklungslehre der Wirbeltiere*, **1, part 1.** Fischer, Jena.

Hildebrand, S.F. and Schroeder, S.F. (1928) Fishes of the Chesapeake Bay. *Bulletin of the U.S. Bureau of Fisheries* **43**, pt.1, doc.1024, 366 pp.

Hill, C. (1891) Development of the epiphysis in *Coregonus albus*. *Journal of Morphology* **5**, 503-510.

Hill, C. (1894) The epiphysis in teleosts and *Amia*. *Journal of Morphology* **9**, 237-268.

Hirai, A. (1987) Techniques for studying the early life history of fishes. 9-Identification of fish eggs using the micropyle. (In Japanese) *Aquabiology* **9**, 177-179.

Hirai, A. (1988) Fine structure of the micropyles of pelagic eggs of some marine fishes. *Japanese Journal of Ichthyology* **35**, 351-358.

Hirai, A. (1993) Fine structure of the egg membranes in four species of *Pleuronectinae*. *Japanese Journal of Ichthyology* **40**, 227-235.

Hirai, A. and Yamamoto, T.S. (1986) Micropyle in the developing eggs of the anchovy, *Engraulis japonica*. *Japanese Journal of Ichthyology* **33**, 62-66.

Hirose, K. (1972) The ultrastructure of the ovarian follicle of medaka, *Oryzias latipes*. *Zeitschrift für Zellforschung* **123**, 316-329.

His, W. (1873) *Untersuchungen über das Ei und die Eientwicklung bei Knochenfischen*, F.C.W. Vogel, Leipzig.

His, W. (1876) Untersuchungen über die Entwickelung von Knochenfischen besonders über diejenige des Salmens. *Zeitschrift für Anatomie und Entwicklungsgeschichte* **1**, 1-40.

His, W. (1878) Untersuchungen über die Bildung des Knochenfischembryo (Salmen). *Archiv für Anatomie und Entwicklungsgeschichte (Anatomische Abtheilung)*. 180-221.

His, W. (1898) Ueber Zellen- und Syncytienbildung: Studien am Salmonidenkeim. *Abhandlungen der Sächsischen Gesellschaft der Wissenschaften, mathematische-physikalische Classe.* **24**, 401-468.

Hisaoko, K.K. and Battle, H.I. (1958) The normal developmental stages of the zebrafish, *Brachydanio rerio* (Hamilton-Buchanan). *Journal of Morphology* **102**, 311-326.

Hjorth, J.P. (1974) Genetics of *Zoarces* populations. VII. Fetal and adult hemoglobins and polymorphism common in both. *Hereditas* **78**, 69-72.

Ho, R.K. (1992a) Axis formation in the embryo of the zebrafish *Brachydanio rerio*. *Seminars in Developmental Biology* **3**, 53-64.

Ho, R.K. (1992b) Cell movements and cell fate during zebrafish gastrulation. *Development 1992 Supplement*, 65-73.

Ho, R.K. and Kimmel, Ch.B. (1993) Commitment of cell fate in the early zebrafish embryo. *Science* **261**, 109-111.

Hoar, W.S. (1955) Phototactic and pigmentary responses of sockeye salmon smolts following injury to the pineal organ. *Journal of the Fisheries Research Board of Canada* **12**, 178-185.

Hofer, B. (1909) *Die Süsswasserfische von Mitteleuropa*. Leipzig.

Hoffmann, C.K. (1880) II. Wissenschaftliche Mittheilungen. 1. Vorläufige Mitteilung zur Ontogenie der Knochenfische. *Zoologischer Anzeiger* **3**, 607-610.

Hoffmann, C.K. (1881) Zur Ontogenie der Knochenfische. *Naturk. Verhandelingen der Koninklike Akademie voor Wetenschappen te Amsterdam* **21**, 1-168.

Hoffmann, C.K. (1883) Zur Ontogenie der Knochenfische. (Continuation of above 1881 publication).*Verhandelingen der Koninklike Akademie voor Wetenschappen te Amsterdam* **23**, 1-60.

Archiv für mikroskopische Anatomie **23**, 45-104.

Hoffmann, C.K. (1885) Epiphyse und Parietalauge, in *Bronn's Klassenordnungen des Tierreiches* **6**, 1981-1993.

Hoffmann, C.K. (1888) Ueber den Ursprung und die Bedeutung der sogenannten "freien" Kerne in dem Nahrungsdotter bei den Knochenfischen. *Zeitschrift für wissenschaftliche Zoologie (Leipzig)* **46**, 517-548.

Hogan, C.J. and Trinkaus, J.P. (1977a) I. Modifications de l'oeuf non fécondé après la ponte. II. Premières phases du développement. *Mémoires couronnés et mémoires des savants étrangers de l'Académie des Sciences Belgique* **40**, 1-66.

Hogan, J.C. and Trinkaus, J.P. (1977b) Intercellular junctions, intramembraneous particles and cytoskeletal elements of deep cells of the *Fundulus* gastrula. *Journal of Embryology and experimental Morphology* **40**, 125-141.

Holder, C.F. (1904) The boy anglers. p. 175. New York.

Holder, N. and Xu, Q. (1997) The Zebrafish — An overview of its early development, in *Methods in Molecular Biology* **97**, (eds P.T. Sharpe and I. Mason), Humana Press, Inc., Totowa, N.J. pp. 431-439.

Hølleland, T. (1990) The distribution of chloride cells in embryonic and larval stages of herring *Clupea harengus*, in *Program 'Developmental and Aquaculture of Marine Fish Larvae'*, Bergen, Norway. Abstract P45.

Hollenberg, F. and Wourms, J.P. (1994) Ultrastructure and protein uptake of the embryonic *trophotaeniae* of four species of goodeid fishes (*teleostei: Atheriniformes*). *Journal of Morphology* **219**, 105-129.

Hollyfield, J.G. (1972) Histogenesis of the retina of the killifish, *Fundulus heteroclitus*. *Journal of Comparative Neurology* **144**, 373-380.

Holmdahl, D.E. (1932) Die zweifache Bildungsweise des zentalen Nervensystems bei den Wirbeltieren. *Wilhelm Roux's Archiv der Entwicklungsmechanik* **129**, 206-254.

Holmgren, U. (1965) On the ontogeny of the pineal- and parapineal organs in teleost fishes. *Progress in Brain Research* **10**, 172-182.

Holt, E.W.L. (1890) On the ova of *Gobius*. *Annals and Magazine of Natural History (London)*. Serie 6, **6**, 34-40.

Holt, E.W.L. (1893) On the eggs and larval and postlarval stages of teleosteans (in Survey of fishing grounds, west coast of Ireland, 1890-1891). *Scientific Transactions of the Royal Dublin Society*, 2. ser. **5**, 5-121.

Hosokawa, K. (1979) Scanning electron microscopic observations of the micropyle in *Oryzias latipes*. *Japanese Journal of Ichthyology* **26**, 90-94.

Hosokawa, K. (1985) Electron microscopic observation of chorion formation in the teleost, *Navodon modestus*. *Zoological Science* **2**, 513-522.

Hosokawa, K.T., Fusimi, T. and Matsusato, T. (1981) Electron microscopic observations of the chorion and micropyle apparatus of the porgy, *Pagrus major*. *Japanese Journal of Ichthylogy* **27**, 339-343.

Howe, E. (1987) Breeding behaviour, egg surface morphology and embryonic development in four Australian species of the genus *Pseudomugil* (Pisces: Melanotaeniidae). *Australian Journal of Marine and Freshwater Research* **38**, 885-895.

Hu. M. and Easter, S.S. Jr. (1999) Retinal neurogenesis: The formation of the initial central patch of postmitotic cells. *Developmental Biology* **207**, 309-321.

Hubard, J.W. (1894) The yolk nucleus in *Cymatogaster aggregatus* Gibbons. *Proceedings of the American Philosophical Society* **33**, 74-83.

Hubbs, C.L. and Turner, C.L. (1939). Studies of the fishes of the order *cyprinodontes*. XIV A revision of the *goodeidae*. *Miscellaneous Publications of the Museum of Zoology* **42**, 3-80.

Hurley, D.A. and Fisher, K.C. (1966) The structure and development of the external membrane in young eggs of the brook trout, *Salvelinus fontinalis* (Mitchill). *Canadian Journal of Zoology* **44**, 173-190.

Huver, C.W. (1960) The stage at fertilization of the egg of *Fundulus heteroclitus*. *Biological Bulletin* **119**, 320.

Huver, C.W. (1962) A study of the site and origin of the teleost blastodisc. PhD Thesis Yale University, New Haven.

Huver, C.W. (1964) Comparative studies of blastodisc formation in teleosts. *The American Zoologist* **4**, 319-320.
Hwang, P.P. (1988) Ultrastructural study on multicellular complex of chloride cells in teleosts. *Bulletin of the Institute of Zoology Acad. Sinica* **27**, 225-234.
Hwang, P.P. (1989) Distribution of chloride cells in teleost larvae. *Journal of Morphology* **200**, 1-8.
Hyatt, G.A., Schmitt, E.A., Fadool, J.J. and Dowling, J.E. (1996) Retinoic acid alters photoreceptor development *in vivo*. *Proceedings of the National Academy of Sciences, U.S.A.* **93**, 13298-13303.
Hyllner, J.S. (1994) Isolation, partial characterization, induction, and the occurrence in plasma of the major vitelline envelope proteins in the Atlantic halibut (*Hippoglossus hippoglossus*) during sexual maturation. *Canadian Journal of Fisheries and Aquatic Science* **51**, 1700-1707.
Hyllner, S.J., Silversand, Ch. and Haux, C. (1994) Formation of the vitelline envelope precedes the active uptake of vitellogenin during oocyte development in the rainbow trout *Oncorhynchus mykiss*. *Molecular Reproduction and Development* **39**, 166-175.
Hyllner, S.J., Oppen-Berntsen, D.O., Helvik, U.V., Walther, B.T. and Haux, C. (1991) Oestradiol-17ß induces the major vitelline envelope proteins in both sexes in teleosts. *Journal of Endocrinology* **131**, 229-236.
Hyrtl, J. (1838) Beobachtungen aus dem Gebiet der vergleichenden Gefässlehre. II. Ueber den Bau der Kiemen der Fische. *Medicinisches Jahrbuch* **15**, 232-248.
Iga, T. (1959) Development of the hatching gland in the teleost, *Leuciscus hakuensis* Günther *(In Japanese) Science Report (Nat.Sci) Shimane University* **9**, 64-68, cited by 'Yokoya and Ebina, 1976'.
Igarashi, T. (1962) Morphological changes of the embryo of a viviparous teleost, *Neoditrema ransonneti* Steindachner during gestation. *Bulletin of the Faculty of Fisheries, Hokkaido University* **13**, 47-52.
Ignatieva, G.M. (1974) Relative duration of the same periods of early embryogenesis in teleosteans. *Ontogenesis* **5**, 427-436.
Ignatieva, G.M. (1976a) Regularities of early embryogenesis in teleosts as revealed by studies of the temporal pattern of development. I. The duration of the mitotic cycle and its phases during synchronous cleavage divisions. *Wilhelm Roux's Archives* **179**, 301-312.
Ignatieva, G.M. (1976b) Regularities of early embryogenesis in teleosts as revealed by studies of the temporal pattern of development. II. Relative Duration of corrsponding periods of development in different species. *Wilhelm Roux's Archives* **179**, 313-325.
Ignatieva G.M. (1991 The rainbow trout *Salmo gairdneri*, in *Animal species for Developmental Studies, Vol. 2, Vertebrates*, (eds T.A. Dettlaff and S.G. Vassetzky), Consultants Bureau, New York and London.
Ishida, J. (1944a) Hatching enzyme in the fresh-water fish, *Orizyas latipes*. *Annotationes Zoologicae Japonenses* **22**, 137-154.
Ishida, J. (1944b) Further studies on the hatching enzyme of the freshwater fish, *Oryzias latipes*. *Annotationes Zoologicae Japonenses* **22**, 155-164.
Ishida, J. (1948) 'Oguma Commemorative Volume on Cytology and Genetics'.

Ishida, J. (1985) Hatching enzyme: past, present and future. *Zoological Science* **2**, 1-10.

Iuchi, I. and Yamagami, K. (1976) Major glycoproteins solubilized from the teleostean egg membrane by the action of the hatching enzyme. *Biochimica and Biophysica Acta* **453**, 240-249.

Iuchi, I. and Yamagami, K. (1979) Induction of a precocious secretion of the hatching enzyme in the rainbow trout by electric stimulation. *Zool. Mag. (Tokyo)* **85**, 273-277, cited by 'Yamagami, K. (1988)'.

Iuchi, I. and Yamamoto, M. (1983) Erythropoiesis in the developing rainbow trout, *Salmo gairdneri irideus*: Histochemical and Immunochemical detection of erythropoietic organs. *Journal of Experimental Zoology* **226**, 409-417.

Iuchi, I., Hamazaki, T. and Yamagami, K. (1985) Mode of action of some stimulants of the hatching enzyme secretion in fish embryos. *Development, Growth and Differentiation* **27**, 573-581.

Ivanenkov, V.V, Minin, A.A., Meshcheryajov, V.N. and Martynova, L.E. (1987) The effect of local microfilament disorganization on ooplasmic segregation in the loach (*Misgurnus fossilis*) egg. *Cell Differentiation* **22**, 19-28.

Ivankov, V.N. and Kurdyayiva, V.P. (1973) Systematic differences and the ecological importance of the membranes in fish eggs. *Journal of Ichthyology* **13**, 864-873.

Iwai, T. (1967) Structure and development of lateral line in teleost larvae, in *Lateral Line Detectors. Proceedings of a symposium. New York.* Indiana University Press, Bloomington, Indiana and Condon. pp. 27-44.

Iwamatsu, T. (1965) Effect of acetone on the cortical changes at fertilization of the egg of the medaka, *Oryzias latipes. Embryologia* **9**, 1-12.

Iwamatsu, T. (1968) Structural change in the egg surface after fertilization in the fish, *Oryzias latipes. Annotationes Zoologicae Japonenses* **41**, 148-153.

Iwamatsu, T. (1969) Changes of the chorion upon fertilization in the medaka, *Oryzias latipes. Bull. Aichi Univ. Educat.* **18** (Nat. Sci.) 43-56.

Iwamatsu, T. (1973) On the mechanism of ooplasmic segregation upon fertilization in *Oryzias latipes. Japanese Journal of Ichthyology* **20**, 73-78.

Iwamatsu, T. (1983) A new technique for dechorionation and observations on the development of the naked egg in *Oryzias latipes. Journal of Experimental Zoology* **228**, 83-89.

Iwamatsu, T. (1994) Stages of normal development in the medaka, *Oryzias latipes. Zoological Science* **11**, 825-839.

Iwamatsu, T. (1998) Studies on fertilization in the teleost. I. Dynamic responses of fertilized medaka eggs. *Development, Growth and Differentiation* **40**, 475-483.

Iwamatsu, T. and Keino, H. (1978) Scanning electron microscopic study on the surface change of eggs of the teleost, *Oryzias latipes*, at the time of fertilization. *Development, Growth and Differentiation* **20**, 237-250.

Iwamatsu, T. and Ohta, T. (1976) Breakdown of the cortical alveoli of medaka eggs at the time of fertilization, with a particular reference to the possible role of spherical bodies in the alveoli. *Roux's Archives of Developmental Biology* **180**, 297-309.

Iwamatsu, T. and Ohta, T. (1978) Electron microscopical observation on sperm penetration and pronuclear formation in the fish egg. *Journal of Experimental Zoology* **205**, 157-180.

Iwamatsu, T. and Ohta, T. (1981) Scanning electron microscopic observations on sperm penetration in teleostean fish. *Journal of Experimental Zoology* **218**, 261-277.

Iwamatsu, T., Ishijima, S. and Nakashima, S. (1993a) Movement of spermatozoa and changes in micropyles during fertilization in medaka eggs. *Journal of Experimental Zoology* **266**, 57-64.

Iwamatsu, T., Nakashima, S. and Onitake, K. (1993b) Spiral patterns in the micropylar wall and filaments on the chorion in eggs of the medaka, Oryzias latipes. *Journal of Experimental Zoology* **267**, 225-232.

Iwamatsu, T., Shibata, Y. and Kanie T. (1995) Changes in chorion proteins induced by the exudate released from the egg cortex at the time of fertilization in the teleost, Oryzias latipes. *Development, Growth and Differentiation* **37**, 747-759.

Iwamatsu, T., Tada, Y. and Nishiyama, Y. (1982) On sperm nuclei and development in polyspermic eggs of the teleost, Oryzias latipes. *Annotationes Zoologicae Japonenses* **55**, 91-99.

Iwamatsu, T., Yoshimoto, Y. and Hiramoto, Y. (1988a) Mechanism of Ca^{2+} release in medaka eggs microinjected with inositol 1,4,5-Triphosphate and Ca^{2+}. *Developmental Boiology* **129**, 191-197.

Iwamatsu, T., Yoshimoto, Y. and Hiramoto, Y. (1988b) Cytoplasmic Ca^{2+} release and effects of microinjected divalent cations on Ca^{2+} sequestration and exocytosis of cortical alveoli in the medaka egg. *Developmental Biology* **125**, 451-457.

Iwamatsu, T., Yoshizaki, N. and Shibata, Y. (1997) Changes in the chorion and sperm entry into the micropyle during fertilization in the teleostean fish, Oryzias latipes. *Development, Growth and Differentiation* **39**, 33-41.

Iwamatsu T., Ohta, T., Oshima, E. and Sakai N. (1988) Oogenesis in the medaka Oryzias latipes — Stages of oocyte development. *Zoological Science* **5**, 353-373.

Jablonowski, J. (1898) Ueber einige Vorgänge in der Entwickelung des Salmonidenembryos nebst Bemerkungen über ihre Bedeutung für die Beurteilung der Bildung des Wirbeltierkörpers. *Anatomischer Anzeiger* **14**, 532-551.

Jablonowski, J. (1899) Ueber die Bildung des Medullarstranges beim Hecht. *Abhandlungen & Berichte des K. Zool. und Anthrop. Ethn. Museum Dresden Festschrift no.* **8**, 1-18.

Jacobs, G.H. (1992) Ultraviolet vision in vertebrates. *The American Zoologist* **32**, 544-554.

Jamieson, B.G.M. and Grier, H.J. (1993) Influences of phylogenetic position and fertilization biology on spermatozoal ultrastructure exemplified by exocoetoid and poeciliid fish. *Hydrobiologia* **271**, 11-25.

Jamieson, B.G.M. and Leung, K.-P. (1991) Introduction to fish spermatozoa and the micropyle. Chapters 5 and 15, in *Fish Evolution and Systematics; Evidence from Spermatozoa*, (ed B.G.M. Jamieson), Cambridge University Press, Cambridge, pp. 56-72, 167-179.

Janosik, J. (1885) Partielle Furchung bei den Knochenfischen. *Archiv für Mikroskopische Anatomie* **24**, 472-474.

Jeffery, W.R. and Martasian, D.P. (1998) Evolution of eye regression in the cavefish Astynax: Apoptosis and the Pax-6 gene. *The American Zoologist* **38**, 685-96.

Jesuthasan, S. and Straehle, U. (1997) Dynamic microtubules and specification of the zebrafish embryonic axis. *Current Biology* **7**, 31-42.

Jin, Z.X., Inaba, K., Manaka, K., Morisawa, M. and Hayashi, H. (1994) Monoclonal antibodies against the protein complex that contains the flagellar movement-initiating phosphoprotein of *Oncorhynchus keta*. *Journal of Biochemistry* **115**, 885-890.

John, C.C. (1932) The origin of erythrocytes in the herring (*Clupea harengus*). *Proceedings of the Royal Society, Ser. B*. **110**, 112-119.

Johns, P.R. (1977) Growth of the adult goldfish eye. III. Source of the new retinal cells. *Journal of Comparative Neurology* **176**, 343-357.

Johns, P.R. (1981) Growth of fish retinas. *The American Zoologist* **21**, 447-458.

Johns, P.R. (1982) Formation of photoreceptors in larval and adult goldfish. *Journal of Neuroscience* **2**, 176-198.

Johns, P.R. and Fernald, R.D. (1981) Genesis of rods in teleost fish retina. *Nature* **293**, 141-142.

Johnson, E.Z. and Werner, R.G. (1986) Scanning electron microscopy of the chorion of selected freshwater fishes. *Journal of Fish Biology* **29**, 257-265.

Jollie, W.P. and Jollie, L.G. (1964a) The fine structure of the ovarian follicle of the ovoviviparous poeciliid fish, *Lebistes reticulatus* I. Maturation of follicular epithelium. *Journal of Morphology* **114**, 479-501.

Jollie, W.P. and Jollie, L.G. (1964b) The fine structure of the ovarian follicle of the ovoviviparous poeciliid fish, *Lebistes reticulatus* II. Formation of follicular pseudoplacenta. *Journal of Morphology* **114**, 503-525.

Jones, M.P., Holliday, F.G.T. and Dunn, A.E.G. (1966) The ultrastructure of the epidermis of larvae of the herring (*Clupea harengus*) in relation to the rearing salinity. *Journal of the Marine Biological Association* **46**, 235-239.

Julian, D., Ennis, K. and Korenbrot, J.I. (1998) Birth and fate of proliferative cells in the inner nuclear layer of the mature fish retina. *Journal of Comparative Neurology* **394**, 271-283.

Kaestner, S. (1892) Ueber die allgemeine Entwickelung der Rumpf- und Schwanzmusculatur bei Wirbelthieren, mit besonderer Berücksichtigung der Selachier. *Archiv der Anatomie und Physiologie (Anat. Abth.)* **1892**, 153-222.

Kagan, B.M (1935) The fertilizable period of the egg of *Fundulus heteroclitus* and some associated phenomena. *Biological Bulletin, Wood's Hole* **69**, 185-201.

Kagawa, H. and Takano, K. (1979) Ultrastructure and histochemistry of granulosa cells of pre- and post-ovulatory follicles in the ovary of the medaka *Oryzias latipes*. *Bulletin of the Faculty of Fisheries Hokkaido University*. **30**, 191-204.

Kageyama, T. (1977) Motility and locomotion of embryonic cells of the medaka, *Oryzias latipes*, during early development. *Development, Growth and Differentiation* **19**, 103-110.

Kageyama, T. (1980) Cellular basis of epiboly of the enveloping layer in the embryo of medaka, *Oryzias latipes*. I. Cell architecture revealed by silver staining method. *Development, Growth and Differentiation* **22**, 659-668.

Kageyama, T. (1985) SEM observation of the external yolk syncytial layer in blastopore closure of the medaka, *Oryzias latipes*. *Development, Growth and Differentiation* **27**, 633-638.

Kageyama, T. (1986) Mitotic wave in the yolk syncytial layer of embryos of *Oryzias latipes* originates in the amplification of mitotic desynchrony in early blastomeres. *Developmental Biology* **3**, 1046.

Kageyama, T. (1996) Polyploidization of nuclei in the yolk syncytial layer of the embryo of the medaka, *Oryzias latipes*, after the halt of mitosis. *Development, Growth and Differentiation* **38**, 119-127.

Kaighn, M.E. (1964) A biochemical study of the hatching process in *Fundulus heteroclitus*. *Developmental Biology* **9**, 56-80.

Kane, D.A. (1999) Cell cycles and development in the embryonic zebrafish, in *Methods in Cell Biology*, (eds H.W. Detrich, M. Westerfield, L.I. Zon), Academic Press, New York, pp. 11-26.

Kane, D.A. and Kimmel, C.B. (1993) The zebrafish midblastula transition. *Development* **119**, 447-456.

Kanoh, Y. (1949) Ueber den japanischen Hering (*Clupea pallasii* Cuvier et Valenc.) I. Morphologie der reifen Eier. *Cytologia* **15**, 138-144.

Kanoh, Y (1952) Ueber das Ei und einige seiner Charakteristika bei einem japanischen Knochenfisch, Ayu (*Plecoglossus altivelis*). *Japanese Journal of Ichthyology* **3**, 147-155.

Kanoh, Y. (1953) Ueber den japanischen Hering (*Clupea pallasii*) II. Veränderung im Ei bei Befruchtung und Aktivierung. *Cytologia* **18**, 67-79

Kanoh, Y. and Yamamoto, T.S. (1957) Removal of the membrane of the dog salmon egg by means of proteolytic enzymes. *Bulletin of the Japanese Society of Science in Fisheries* **23**, 166-172.

Kant, L.H. (1993) *The interpretation of religious symbols in the graeco-roman world: a case study of early christian fish symbols*. PhD thesis, Yale University, 505 pp.

Karnaky (1986) Structure and function of the chloride cell. *The American Zoologist* **26**, 209-224.

Katssura, K. and Yamada, K. (1986) Appearance and disappearance of chloride cells throughout the embryonic and postembryonic development of the goby, *Chaenogobius urotaenia*. *Bulletin of the Faculty of Fisheries Hokkaido University* **37**, 95-100.

Kawamura, G., Tsuda, R., Kumai, H. and Obashi, S. (1984) The visual cell morphology of *Pagrus major* and its adaptive changes with shift from pelagic to benthic habitats. *Bulletin of the Japanese Society of Scientific Fisheries* **50**, 1975-1980.

Kawase, H. and Nakazono, A. (1995) Predominant maternal egg care and promiscuous mating system in the Japanese filefish, *Rudarius ercodes* (*Monoacanthidae*). *Environmental Biology of Fishes* **43**, 241-254.

Kazimi, N. and Cahill, G.M. (1999) Development of a circadian melatonin rhythm in embryonic zebrafish. *Developmental Brain Research* **117**, 47-52.

Keenleyside, M.H.A. (1979) Diversity and adaptation in fish behaviour. *Zoophysiology* **11**, 1-208.

Keller, R.E. and Trinkaus, J.P. (1987) Rearrangement of enveloping layer cells without disruption of the epithelial permeability barrier as a factor in *Fundulus* epiboly. *Developmental Biology* **120**, 12-24.

Kemp, N.A. and Allen, M.D. (1956) Electron microscopic observations on the development of the chorion of *Fundulus*. *Biological Bulletin* **11**, 293.

Kendall, A.W. Jr., Ahlstrom, E.H. and Moser, H.G. (1984) Early life history stages of fishes and their characters, in *Ontogeny and Systematics of Fishes*, (eds H.G. Moser *et al.*), Allen Press, Lawrence, Kansas, pp. 11-22.

Kessel, R.G. (1960) The role of cell division in gastrulation of *Fundulus heteroclitus*. *Experimental Cell Research* **20**, 277-282.

Kessel, R.G., Beams, Y.W., Tung, H.N. and Roberts, R. (1983) Unusual particle arrays in the plasma membrane of zebrafish spermatozoa. *Journal of Ultrastructure Research* **84**, 268-274.

Kessel, R.G., Tung, H.N., Roberts, R. and Beams, H.W. (1985) The presence and distribution of gap junctions in the oocyte-follicle cell complex of the zebrafish, *Brachydanio rerio*. *Journal of Submicroscopical Cytology* **17**, 239-253.

Kezuka, H. Aida, K. and Hanyu, I. (1989) Melatonin secretion from goldfish pineal gland in organ culture. *General Comparative Endocrinology* **75**, 217-221.

Kim, E.D (1981) Specificities of changes in aminoacids and lipids in oocytes of some fish during maturation, in *Ontogenetic Diversity in Fish*, (ed Naukova Dumka, Kiev), pp. 61-84 (in Russian)

Kimmel, C.B. (1989) Genetics and early development of zebrafish. *Trends in Genetics* **5**, 283-288.

Kimmel, C.B. (1993) Patterning the brain of the zebrafish embryo. *Annual Review of Neuroscience* **16**, 707-732.

Kimmel, C.B. and Law, R.D. (1985) Cell lineage of zebrafish blastomeres. II. Formation of the yolk syncytial layer. *Developmental Biology* **108**, 86-93.

Kimmel, C.B. and Warga, R.M. (1986) Tissue-specific cell; lineages originate in the gastrula of the zebrafish. *Science* **231**, 365-368.

Kimmel, Ch.B., Spray, D.C. and Bennett, M.V.L. (1984) Developmental uncoupling between blastoderm and yolk cell in the embryo of the teleost *Fundulus*. *Developmental Biology* **102**, 483-487.

Kimmel, Ch.B., Warga, R.M. and Kane D.A. (1994) Cell cycles and clonal strings during formation of the zebrafish central nervous system. *Development* **120**, 265-276.

Kimmel, C.B., Warga, R.M. and Schilling, T.F. (1990) Origin and organization of the zebrafish fate map. *Development* **108**, 581-594.

Kimmel, C.B, Ballard, W.W, Kimmel, B., Ullmann, T., Westerfield, M. (1995) Stages of embryonic development of the zebrafish, in *The Zebrafish Book*, (ed M. Westerfield). pp. 27-56

Kimmel, C.B., Ballard, W.W., Kimmel, S.R., Ullmann, B. and Schilling, T.F.D. Ullmann, B. and G. Schilling (1995) Stages of Embryonic Development of the Zebrafish. *Developmental Dynamics* **203**, 253-310.

Kingsley J.S. and Conn H.W. (1883) Some observations on the embryology of the teleosts. *Memoirs of the Boston Society of Natural History* **6**, 183-211.

Kiselev, I.V. (1980) The biological background of fertilization and incubation of fish eggs (in Russian). *Naukova Dimka. Kiev*. 296 pp.

Kjesbu, O.S. and Kryvi, H. (1993) A histological examination of oocyte final maturation in cod (*Gadus morhua* L.), in *Physiological and Biochemical Aspects of Fish Development*, (ed B.T. Walther and H.-J. Fyhn), University of Bergen, pp. 86-93.

Klein, E.E. (1876) Observations on the early development of the common trout (*Salmo fario*). *Quarterly Journal of microscopical Science, N.s.* **16**, 113-131.

Kljavin, I. (1987) Early development of photoreceptors in the ventral retina of the zebrafish embryo. *Journal of comparative Neurology* **260**, 461-471.

Knight, F.M., Lombardi, J., Wourms, J.P. and Burns, J.R. (1985) Follicular placenta and embryonic growth of the viviparous four-eyed fish (*Anableps*). *Journal of Morphology* **185**, 131-142.

Kobayakawa, M. (1985) External characteristics of Japanese catfishes (*Silurus*). *Japanese of Journal of Ichthyology* **32**, 104-106.

Kobayashi, W. (1985) Communications of oocyte-granulosa cells in the chum salmon ovary detected by transmission electron microscopy. *Development, Growth and Differentiation* **27**, 553-561.

Kobayashi, W. and Yamamoto, T.S. (1981) Fine structure of the micropylar apparatus of the chum salmon egg, with a discussion of the mechanism for blocking polyspermy. *Journal of Experimental Zoology* **217**, 265-275.

Kobayashi, W. and Yamamoto, T.S. (1985) Fine structure of the micropylar cell and its change during oocyte maturation in the chum salmon, *Oncorhynchus keta*. *Journal of Morphology* **184**, 263-276.

Kobayashi, W. and Yamamoto, T.S. (1987) Light and electron microscopic observations of sperm entry in the chum salmon egg. *Journal of Experimental Zoology* **243**, 243-322.

Koch, H.J., Bergström E. and Evans, J. (1946a) The multiple haemoglobins of *Salmo salar* L.. *Reports of the Swedish Salmon Research Institute* **6**, 1-7.

Koch, H.J., Bergström E. and Evans, J. (1946b) The microelectrophoretic separation on starch gel of the haemoglobins of *Salmo salar* L.. *Mededel Vlaamse Acad. Kl. Wet.* **26**, 1-32.

Koch, H.J., Bergström, E., Bodarwé-Schmitz and Evans, J.C. (1968) The separation of haemoglobins of *Salmo salar* L. by means of isoelectric focusing. *Mededel Vlammse Acad. Kl. Wet.* **30**, 1-16.

Kodjabachian, L., Dawid, I.B. and Toyama, R. (1999) Gastrulation in zebrafish: What mutants teach us. *Developmental Biology* **213**, 211-245.

Koehler, J.K. (1970) Freeze-etching studies on spermatozoa with particular reference to nuclear and post-nuclear cap structure, in *Comparative Spermatology*, (ed. B. Baccetti), Academia Nazionale dei Lincei, Rome, pp. 515-522.

Kojima, D. and Fukada Y. (1999) Non-visual photoreception by a variety of vertebrate opsins in *Rhodopsins* and phototransduction. *Novartis Foundation Symposium* 224, pp 265-282. Wiley, Chichester.

Kolessnikow, N. (1878) Ueber die Entwicklung bei Batrachiern und Knochenfischen. *Archiv für Mikroskopische Anatomie* **15**, 382-414.

Kölliker, A. (1858) Untersuchungen zur vergleichenden Gewebelehre, angestellt in Nizza im Herbste 1856. *Verhandlungen der Physikalisch-Medicinischen Gesellschaft in Würzburg* **8**, 80-109.

Kolster, R. (1905) Ueber die Embryotrophe, speziell bei *Zoarces viviparus* Cuv. *Festschrift für Palmén*, no, 4, pp. 1-46.

Kopsch, Fr. (1898) Die Entwicklung der äusseren Form des Forellen-Embryo. *Archiv für Mikroskopische Anatomie*, **51**, 181-213.

Kopsch, Fr. (1900) Homologie und phylogenetische Bedeutung der Kupffer'schen Blase. *Anatomischer Anzeiger* **17**, 497-509.

Kopsch, Fr. (1904) Untersuchungen über Gastrulation und Embryobildung bei den Chordaten. I. Die morphologische Bedeutung des Keimhautrandes und die Embryobildung bei der Forelle. Georg Thieme, Leipzig. pp.1-166.

Kopsch, Fr. (1911) Die Entstehung des Dottersackentoblast und die Furchung bei der Forelle (*Salmo fario*). *Archiv für Mikroskopische Anatomie* **78**, 618-659.

Korfsmeier, K.H. (1966) Zur Genese des Dottersystems in der Oocyte von *Brachydanio rerio*. Autoradiographische Untersuchungen. *Zeitschrift für Zellforschung* **71**, 283-296.

Korschelt, E. und Heider K. (1936) *Vergleichende Entwicklungsgeschichte der Tiere* **1**, 78-81.

Korschelt, E. und Heider, K. (1936) *Vergleichende Entwicklungsgeschichte der Tiere*. **2**, 1167-1205.

Korsgaard (1986) Trophic adaptations during early intraovarian development of embryos of *Zoarces viviparus* (L.). *Journal of Experimental Marine Biology and Ecology* **98**, 141-158.

Koshi, M. and Ogawa, Y. (1951) Studies of fertilization. Separation and chemical formula of activation and chemotaxis substance of sperm from the *Clupea pallasii* eggs. (in Japanese). *Medicine and Biology* **18**, 184-187.

Kostomarova, A.A. (1991) The loach *Misgurnus fossilis*. Chapter 5 in: *Animal Species for Developmental Studies* **2**, *Vertebrates*, (eds T.A. Dettlaff and S.G. Vassetzky), Consultants Bureau, London, New York. pp. 125-144.

Kostomarova, A.A. and Ignat'eva, G.M. (1968) Relationship of karyo- and cytotomy processes during the period of synchronous division in cleavage of loach (*Misgurnus fossilis* L.). Translated from *Doklady Akademii Nauk SSSR Biological Sciences Section* **183**, 636-638.

Kovàc, V. (1993) Early development of ruff, *Gymnocephalus cernuus*. *Folia Zoologica* **42**, 269-280.

Kovalevskaya, N.V. (1965) The eggs and larvae of synentognathous fishes (Beloniformes, Pisces) of the Gulf of Tonkin. *Tr. Okeanol. Akad. Nauk. SSSR* **80**, 124-146. (In Russian with English summary).

Kowalewski von, M. (1886a) Ueber die ersten Entwicklungsprozesse der Knochenfische. *Zeitschrift für wissenschaftliche Zoologie* **43**, 434-480.

Kowalewski von, M. (1886b) Ueber Furchung und Keimblätteranlage der Teleostier. *Sitzungsberichte der Physikalisch-medicinischen Societät zu Erlangen* **18**, 1-6.

Kowalewski von, M. (1886c) Die Gastrulation und die sogen. Allantois bei den Teleostiern. *Sitzungsberichte der Physikalisch-medicinischen Societät zu Erlangen* **18**, 31-36.

Kowalska-Dyrcz, A. (1979) Teleost gastrulation in the new light (in Polish). *Przeglad Zoologiczny* **23**, 115-125.

Koya, Y., Munehara, H. and Takano, K (1995) Formation of egg adhesive material in masked greenling, *Hexagrammos octogrammus*. *Japanese Journal of Ichthyology* **42**, 45-52.

Kraft von, A. and Peters, H.M. (1963) Vergleichende Studien über die Oogenese in der Gattung *Tilapia (Cichlidae, Teleostei)*. *Zeitschrift für Zellforschung* **61**, 434-485.

Kramer, D.L. (1873) Patental behaviour in the blue gourami, *Trichogaster trichopterus* (Pisces; Belontiidae) and its induction during exposure to various numbers to conspecific eggs. *Behaviour* **47**, 14-31.

Krigshaver, M.P. and Neifakh a.A. (1968) Protein synthesis in the blastoderm of the loach embryo. Doklady Akademii Nauk SSSR **180**, 1259-1261. (Translated from Russian)

Kristoffersson, R., Broberg S. and Pekkarinen, M. (1973) Histology and physiology of embryotrophe formation. Embryonic nutrition and growth in the eel-pout, *Zoarces viviparus* (L.). *Ann. Zool. Fennici* **10**, 467-477.

Kryzhanovskii, S.G. (1934) Die Atmungsorgane der Fischlarven (*Teleostomi*). *Zoologisches Jahrbuch* **58**, 21-60.

Kryzhanovskii, S.G. (1948) Ecological groups of fish and patterns of their development (in Russian). *Izvestiya Tikhookeanskogo Nauchno-Issledovatel'-skogo Instituta Rybnogo Khozyaistva i Okeanografii.* **47.**

Kryzhanovskii, S.G. (1949) Morphological patterns of development in the *cyprinidae, cobitidae* and *siluridae* (in Russian). *Trudy Inst. Morph. Zhvotn. Akad. Nauk S.S.S.R., Moscow* **1**, 1-332.

Kryzhanovskii, S.G. (1953) Features of the ripe eggs of teleosts (in Russian) *Voprosy Ikhtiologii,* No. l.

Kryzhanovskii, S.G. (1960) On the significance of the lipid inclusions in fish eggs. *Zool. Zh.* **39**, 111-123 (in Russian).

Kuchnow, K.P. and Scott, J.R. (1977) Ultrastructure of the chorion and its micropyle apparatus in the mature *Fundulus heteroclitus* (Walbaum) ovum. *Journal of Fish Biology* **10**, 197-201.

Kudo, S. (1976) Ultrastructural observations on the discharge of two kinds of granules in the fertilized eggs of *Cyprinus carpio* and *Carassius auratus*. *Development, Growth and Differentiation* **18**, 167-176.

Kudo, S. (1978) Enzymo-cytochemical observations on the cortical change in the eggs of *Cyprinus carpio* and *Carassius auratus*. *Development, Growth and Differentiation* **20**, 133-142.

Kudo, S. (1980) Sperm penetration and the formation of a fertilization cone in the common carp egg. *Development, Growth and Differentiation* **22**, 403-414.

Kudo, S. (1982) Ultrastructure and ultracytochemistry of fertilization envelope formation in the carp egg. *Development, Growth and Differentiation* **24**, 327-339.

Kudo S. (1983a) Ultracytochemical modifications of surface carbohydrates in fertilized eggs of the common carp. *Development, Growth and Differentiation* **25**, 85-97.

Kudo, S. (1983b) Response to sperm penetration of the cortex of eggs of the fish, *Plecoglossus altivelis*. *Development, Growth and Differentiation* **25**, 163-170.

Kudo, S. (1991) Fertilization, cortical reaction, polyspermy-preventing and anti-microbial mechanisms in fish eggs. *Bulletin of the Institute of Zoology Academy of Science Monograph* **16**, 313-340.

Kudo, S. (1992) Enzymatic basis for protection of fish embryos by the fertilization envelope. *Experientia* **48**, 277-281.

Kudo, S. and Inoue M. (1986) A bactericidal effect of fertilization extract from fish eggs. *Zoological Sci.* **3**, 323-329.

Kudo, S. and Inoue, M. (1989) Bactericidal action of fertilization envelope extract from eggs of the fish *Cyprinus carpio* and *Plecoglossus altivelis*. *Journal of Experimental Zoology* **250**, 219-228.

Kudo, S. and Sato, A. (1985) Fertilization cone of carp eggs as revealed by scanning electron microscopy. *Development, Growth and Differentiation* **27**, 121-128.

Kudo, S., Sato A. and Inoue, M. (1988) Chorionic peroxidase activity in the eggs of the fish *Tribolodon hakonensis. Journal of Experimental Zoology* **245**, 63-70.

Kügel, B., Hoffmann, R W. and Friess, A. (1990) Effects of low pH on the chorion of rainbow trout, *Oncorhynchus mykiss, and* brown trout, *Salmo trutta f. fario. Journal of Fish Biology* **37**, 301-310.

Kuntz, A. (1915) Notes on the embryology and larval development of five species of teleostean fishes. *Bulletin of the U.S. Bureau of Fisheries* **34**, 405-429.

Kuntz A. and Radcliffe, L. (1917) Notes on the embryology and larval development of twelve teleostean fishes. *Bulletin of the U.S. Bureau of Fisheries* Bull **1915/1961**, pp 87-134.

Kunz, Y. (1963) Die embryonale Harnblase von *Lebistes reticulatus. Revue Suisse de Zoologie* **70**, 291-297.

Kunz, Y. (1964) Morphologische Studien über die embryonale und postembryonale Entwicklung bei Teleostiern mit besonderer Berücksichtigung des Dottersystems und der Leber. *Revue Suisse de Zoologie* **71**, 445-552.

Kunz, Y.W. (1971a) Histological study of greatly enlarged pericardial sac in the embryo of the viviparous teleost *Lebistes reticulatus. Revue Suisse de Zoologie* **78**, 187-207.

Kunz, Y.W. (1971b) Distribution of lactate dehydrogenase (and its E-isozymes) in the developing and adult retina of the guppy (*Lebistes* reticulatus) *Revue Suisse de Zoologie* **78**, 761-776.

Kunz, Y.W. (1975) Ontogenesis of lactate dehydrogenase isozyme patterns in two salmonids (*Salmo salar* and *S. trutta*). *Experientia* **31**, 152-153.

Kunz, Y.W. (1980) Cone mosaics in a teleost retina: Changes during light and dark adaptation. *Experientia* **36**, 1371-1374.

Kunz, Y.W. (1987) Tracts of putative ultraviolet receptors in the retina of the two-year-old brown trout (*Salmo trutta*) and the Atlantic salmon (*Salmo salar*). *Experientia* **43**, 1202-1204.

Kunz, Y.W. (1990) Ontogeny of retinal pigment epithelium-photoreceptor complex and development of rhythmic metabolism under ambient light conditions. *Progress in Retinal Research* **9**, 135-196.

Kunz, Y.W. and Callaghan, E. (1989) The embryonic fissure in developing teleosts. *ASF Transactions* **189**, 195-202.

Kunz, W. and Ennis, S. (1983) Ultrastructural diurnal changes of the retinal photoreceptors in the embryo of a viviparous teleost (*Poecilia reticulata* P.). *Cell Differentiation* **13**, 115-123.

Kunz, Y.W. and Regan, C. (1973) Histochemical investigations into the lipid nature of the oil-droplet in the retinal twin-cones of *Lebistes reticulatus* (Peters). *Revue Suisse de Zoologie* **80**, 699-703.

Kunz, Y.W. and Wise, C. (1973) Ultrastructure of the "oil-droplet" in the retinal twin-cone of *Lebistes reticulatus* (Peters). Preliminary results. *Revue Suisse de Zoologie* **80**, 694-698.

Kunz, Y.W. and Wise, C. (1974) Development of the photoreceptors in the embryonic retina of *Lebistes reticulatus* (Peters). Electron microscopical investigations. *Revue Suisse de Zoologie* **81**, 697-701.

Kunz, Y.W. and Wise, C. (1977) Regional differences of the *argentea* and *sclera* in the eye of

Poecilia reticulata P. (*teleostei: cyprinodontidae*). A Light- and Electron microscopic study. *Zoomorphologie* **87**: 203-215.

Kunz, Y.W. and Wise, C. (1978) Structural differences of cone oil-droplets in the light and dark adapted retina of *Poecilia reticulata* P.. *Experientia* **34**, 246-248.

Kunz, Y.W., Ennis, S. and Wise, C. (1983) Ontogeny of the photoreceptors in the embryonic retina of the viviparous guppy, *Poecilia reticulata* P. (Teleostei). *Cell and Tissue Research* **230**, 469-486.

Kunz, Y.W., Ni Shuilleabhain, M. and Callaghan, E. (1985) The eye of the venous marine teleost *Trachinus vipera* with special reference to the structure and ultrastructure of visual cells and pigment epithelium. *Experimental Biology* **43**, 161-173.

Kunz, Y.W., Wildenburg, G., Goodrich, L. and Callaghan, E. (1994) The fate of ultraviolet receptors in the retina of the Atlantic salmon (*Salmo salar*). *Vision Research* **34**, 1375-1383.

Kupffer, C.W. (1866) Untersuchungen über die Entwicklung des Harn- und Geschlechtssystem. *Archiv für Mikroskopische Anatomie* **2**, 473-488.

Kupffer, C.W. (1868) Beobachtungen über die Entwicklung der Knochenfische. *M. Schultze's Archiv für Mikroskopische Anatomie* **4**, 209-272.

Kupffer, C.W. (1878) Die Entwicklung des Herings im Ei. *Jahresbericht der Commission zur wissenschaftlichen Untersuchung der deutschen Meere in Kiel für die Jahre 1874-76* **4-6**, 175-224.

Kupffer, C.W. (1879) Ueber die Entstehung der Allantois und die Gastrula der Wirbelthiere. *Zoologischer Anzeiger* **2**, 520-522, 593-597.

Kupffer, C.W. (1884) Die Gastrulation an den meroblastischen Eiern der Wirbelthiere und die Bedeutung des Primitivstreifs. *Archiv für Anatomie und Physiologie* pp.1-40.

Kusa, M (1954) The cortical alveoli of salmon egg. *Annotationes Zoologicae Japonenses* **27**, 1-6.

Laale, H.W. (1980) The perivitelline space and egg envelopes of bony fishes: a review. *Copeia* **1980(2)**, 210-226.

Laale, H.S. (1984) Naturally occurring diblastodermic eggs in the zebrafish, *Brachydanio rerio*. *Canadian Journal of Zoology* **62**, 386-390.

Lam, T.J. (1980) Thyroxine induces larval development and survival in *Sarotherodon* (*Tilapia*) *mossambicus* Ruppel. *Aquaculture* **21**, 287-291.

Lam, T.J. and Sharma, R. (1985) Effects of salinity and thyroxine on larval survival, growth and development of carp, *Cyprinus carpio*. *Aquaculture* **44**, 201-212.

Lane, H.H. (1909) On the ovary and ova in *Lucifuga* and *Stygicola*, in Eigenmann, Carl H., *Cave vertebrates of America*, pp 226-32, Washington.

Lang, H.-J. (1967) Ueber das Lichtrückenverhalten des Guppy (*Lebistes reticulatus*) in farbigen und farblosen Lichtern. *Zeitschrift für vergleichende Physiologie* **56**, 296-340.

Lange, R.H., Richter, H-P., Riehl, R., Zierold, D.K., Trandaburu, T. and Magdowski, G. (1983) Lipovitellin-Phosvitin crystals with orthorhombic features. Thin-section electron microscopy, gel-electrophoresis and microanalysis in teleost and amphibian yolk-platelets and a comparison with other vertebrates. *Journal of Ultrastructure* **83**, 122-140.

Lanzi, L. (1909) Recherches sur les premiers moments de développement de quelques téléostéens, spécialement par rapport à la valeur de ce qu'on apelle épaississement prostomal. *Archivio Italiano di Anatomia ed Embriologia* **8**, 292-306.

Larison, K. and Bremiller, R. (1990) Early onset of phenotype and cell patterning in the embryonic zebrafish retina. *Development* **109**, 567-576.

Lasker, R. and Threadgold, L.T. (1968) "Chloride cells" in the skin of the larval sardine. *Experimental Cell Research* **52**, 382-390.

Laurent, P. and Dunel-Erb S. (1984) The pseudobranch: Morphology and function, in *Fish Physiology* **10B**, (eds W.S. Hoar and D.J. Randall), pp. 285-323.

LaVail, M.M. (1976) Rod outer segment disc shedding in relation to cyclic lighting. *Experimental Eye Research* **23**, 277-280.

LaVail, M.M. and Mullen, R.J. (1976) Role of the pigment epithelium in inherited retinal degeneration analyzed with experimental chimaeras. *Experimental Eye Research* **23**, 27-45.

Leatherland, J.F. (1994) Reflections on the thyroidology of fishes: from molecules to humankind. *Guelph Ichthyology Reviews* **2**, 1-67.

Leatherland, J.F., Lin, L., Down, N.E. and Donaldson, E.M. (1989) Thyroid hormone content of eggs and early developmental stages of five *Oncorhynchus* species. *Canadian Journal of Fisheries and Aquatic Science* **46**, 2140-2145.

Lebal, W.S. and Peairs, K. (1990) Molecular homology of coexpressed eye and liver specific lactate dehydrogenase isozymes from the Basketmouth cichlid assessed by oxamate-sepharose and blue dextrose-sepharose-affinity chromatography. *Journal of Experimental Zoology* **253**, 107-114.

Lee, J.S. and Lee, T.Y. (1989) Reproductive cycle and embryonic development within the maternal body of the viviparous teleost, *Ditrema temmincki* (Bleeker). (Japanese) *Bulletin Nat. Fish University of Pusan* **29**, 37-51.

Lee, K.B.H., Lim, E.H. and Ding, J.L. (1992) Vitellogenin diversity in Perciformes. *Journal of Experimental Zoology* **264**, 100-106.

Leiner, M. (1939) Die Augenkiemendrüse (Pseudobranchie) der Knochenfische. Experimentelle Untersuchungen über ihre physiologische Bedeutung. *Zeitschrift für vergleichende Physiologie* **26**, 416-466.

Leis, J.M. and Moyer, (1985) Development of eggs, larvae and pelagic juveniles of three Indo-Pacific ostraaciid fishes (Tetradontiformes): *Ostracion meleagris, Lactoria fornasini* and *L. diaphana. Japanese* Journal *of Ichthyology* **32**. 189-202.

Lele, Z. and Krone, P.H. (1996) The zebrafish as a model system in developmental toxico-logical and transgenic research. *Biotechnology Advances* **14**, 57-72.

Lentz T. and Trinkaus J.P. (1967) A fine structural study of cytodifferentiation during cleavage, blastula and gastrula stages of *Fundulus heteroclitus. Journal of Cell Biology* **32**, 121-138.

Lereboullet, M. (1854) Résumé d'un travail d'embryologie comparée sur le développement du brochet, de la perche et de l'écrevisse. *Annales des Sciences Naturelles (Zoologie) 4. Série,* **1**, 237-289.

Lereboullet, M. (1861) Recherches d'embryologie comparée sur le développement de la truite, du lézard et du limnée. *Annales des Sciences Naturelles (Zoologie),* **16,** 4th série, 113-196.

Lereboullet, M. (1862) Recherches d'embryologie comparée sur le développement du brochet, de la perche et de l'écrevisse. *Mémoire couronné par l'Académie des sciences de Paris (Mémoires présentés par divers savants de l'Académie des sciences de l'Institut impérial de France; sc. mathém. et physiques,* **17**, 447 805.
Lesseps, R.J., Kessel, van A.H.M.G. and Denuce, J.M. (1975) Cell patterns and cell movements during early development of an annual fish, Nothobranchius neumanni. *Journal of Experimental Zoology* **193**, 137-146.
Lessman, Ch.A., and Huver, Ch.W. (1981) Quantification of fertilization-induced gamete changes and sperm entry without egg activation in a teleost egg. *Developmental Biology* **84**, 218-224.
Leukart, R. (1855) Ueber die Mikropyle bei Insekteneiern (with appendix on the micropyle of fishes). *Müller's Archiv*1855, p. 257.
Levine, J.S. and MacNichol, E.F. (1982) Visual pigments in teleost fishes: Effect of habitat, microhabitat and behaviour on visual system evolution. *Sensory Proceedings* **3**, 95-131.
Levinson, G. and Burnside, B. (1981) Circadian rhythm in teleost retinomotor movements: A comparison of the effects of circadian rhythm and light condition on cone length. *Investigative Ophthalmology* **20**, 294-303.
Lewis, W.H. and Roosen-Runge E.C. (1942) The formation of the blastodisc in the egg of the zebrafish, B*rachydanio rerio,* illustrated with motion pictures. *Anatomical Record* **84**, 463-464.
Lewis, W.H. and Roosen-Runge E.C. (1943) The formation of the blastodisc in the egg of the zebrafish, *Brachydanio rerio. Anatomical Record* **85**, 326.
Leydig, Fr (1857) *Lehrbuch der Histologie der Menschen und der Thiere*, Frankfurt
Leydig, Fr. (1883) *Untersuchungen zur Anatomie und Histologie der Thiere.* Emil Strauss.
Lichtenfeld, J., Viehweg, J., Schützenmeister, J. and Naumann, W.W. (1999) Reissner's substance expressed as a transient pattern in vertebrate floor plate. *Journal of Anatomy and Embryology (Berlin)* **200**, 161-174.
Lindroth, A. (1946) Zur Biologie der Befruchtung und Entwicklung beim Hecht. *Mitteilungen der Anstalt für Binnenfischerei bei Drottningholm, Stockholm,* pp. 1-174.
Linehart, O., Kudo, S., Billard, R., Slechta, V., Mikodina, E.V. (1995) Morphology, composition and fertilization of carp eggs: a review. *Aquaculture* **129**, 75-93.
Linné, C. von (1744) Systema naturae. Ed. IV of *Opera varia, in quibus continentur fundamenta botanica, sponsalia plantarum et systema naturae.*
List, J.H. (1887) Zur Entwicklungsgeschichte der Knochenfische (*Labridae*). *Zeitschrift für wissenschaftliche Zoologie* **45**, 595-645.
Lo Bianco, S. (1931-1933) Fauna e flora del golfo di Napoli, (eds G. Bardi, Rome and R. Friedländer & Son, Berlin), Stazione Zoologica di Napoli, 886 pp.
Locket, N.A. (1970) Deep-sea fish retinas. *British Medical Bulletin* **26**, 107-111.
Locket, N.A. (1980) Variation of architecture with size in the multiple-bank retina of a deep-sea-teleost, *Chauliodus sloani. Proceedings of the Royal Society of London (Biol.)* **208**, 223-242.
Lombardi, J. and Wourms, M.P. (1985a) The trophotaenial placenta of a viviparous goodeid fish. I. Ultrastructure of the internal ovarian epithelium, the maternal component. *Journal of Morphology* **184**, 277-292.

Lombardi, J. and Wourms, M.P. (1985b) The trophotaenial placenta of a viviparous goodeid fish. II. Ultrastructure of trophotaeniae, the embryonic component. *Journal of Morphology* **184**, 293-309.

Lombardi, J. and Wourms, M.P. (1988) Embryonic growth and trophotaenial development in goodeid fishes (*teleostei; atheriniformes*). *Journal of Morphology* **197**, 193-208.

Long, W.L. (1980a) Analysis of yolk syncytium behavior in *Salmo* and *Catostomus*. *Journal of Experimental Zoology* **214**, 323-331.

Long, W.L. (1980b) Proliferation, growth, and migration of nuclei in the yolk syncytium of *Salmo* and *Catostomus*. *Journal of Experimental Zoology* **214**, 333-343.

Long, W.L. (1984) Cell movements in teleost fish development. *BioScience* **34**, 84-88.

Lönning, S. (1972) Comparative electron microscopic studies of teleostean eggs with special reference to the chorion *Sarsia*. **49**, 41-48.

Lönning, S. (1981) Comparative electron microscope studies of the chorion of the fish egg. *Rapports et Procès verbaux des Réunions de Conseil International pour l'Exploration de la Mer* **178**, 560-564.

Lönning, S. and Hagstrom, B.E. (1975) Scanning electronmicroscope studies of the surface of the fish egg. *Asarte* **8**, 17-22.

Lönning, S., Kjörsvik, E. and Davenport, J. (1984) The hardening process of the egg chorion of the cod. *Journal of Fish Biology* **24**, 505-522.

Lönning, S., Kjörsvik, E. and Frank-Petersen, I.B. (1988) A comparative study of pelagic and demersal eggs from common marine fishes in northern Norway. *Sarsia* **73**, 49-60.

Luczynski, M. and Kirklewska, A. (1984) Dependence of *Coregonus albula* embryogenesis rate on the incubation temperature. *Aquaculture* **42**, 43-55.

Luczynski, M., Brzuzan, P. and Czerkies, P. (1993) The hatching process in *coregoninae* embryos. A review. *Fish Ecotoxicology and Ecophysiology* 199-208.

Ludwig, H. (1874) Ueber die Eibildung im Thierreiche. Arbeiten aus dem Zoologischen Institut (Würzburg) **1**, 287-510.

Lum, J.B., Lawrence, W.C. and Cameron, I.L. (1983) A blastula-specific requirement for Na^+ influx into the blastocoel for continued cell proliferation. *Cell and Tissue Kinetics* **16**, 523a.

Luther, W. (1935) Entwicklungsphysiologische Untersuchungen am Forellenkeim: Die Rolle des Organisationszentrums bei der Entstehung der Embryonalanlage. *Biologisches Zentralblatt* **55**, 114-137.

Luther, W. (1966) Entwicklungsphysiologie der Fische. *Fortschritte der Zoologie* **17**, 313-340.

Lwoff, B. (1894) Die Bildung der primären Keimblätter und die Entwicklung der Chorda und des Mesoderms bei den Wirbeltieren. *Bulletin de la Société Nationale de Moscou* **8**, 56-137,160-256.

Lyall, A.H. (1957a) The growth of the trout retina. *Quarterly Journal of Microscopical Science* **98**, 101-110.

Lyall, A.H. (1957b) Cone arrangements in teleost retina. *Journal of Microscopical Science* **98**, 189-201.

MacNichol, E.F., Kunz, Y.W., Levine, J.S., Harosi, F.I and Collins, B.A. (1978) Ellipsosomes: Organelles containing a cytochrome-like pigment in the retinal cones of certain fishes.

Science **200**, 549-552.

Mahon, E.F. and Hoar, W.S. (1956) The early development of the chum salmon, *Oncorhynchus keta* (Walbaum). *Journal of Morphology* **98**, 1-47.

Maienschein, J. (1997) Changing conceptions of organization and induction. *The American Zoologist* **37**, 220-228.

Malicki, J. (1999) Development of the retina. Chapter 15 in: *Methods in Cell Biology* **59**. *The Zebrafish: Biology*, (eds H.W. Detrich, III, M. Westerfield, L.I. Zon). Academic Press, New York.

Malone, T.E. and Hisaoka, K. (1963) A histochemical study of the formation of deutoplasmic components in developing oocytes of the zebrafish, *Brachydanio rerio*. *Journal of Morphology* **112**, 61-75.

Mangor-Jensen, A. (1987) Water balance of cod larvae. *Fish Physiology and Biochemistry* **3**, 17-24.

Manner, H.W., VanCura, M. and Muehleman, H. (1977) The ultrastructure of the chorion of the fathead minnow, *Pimephales proleas*. *Transactions of the American Fish Society* **106**, 110-114.

Mano, H., Kojima, D. and Fukada, Y. (1999) Exo-rhodopsin: a novel rhodopsin expressed in the zebrafish pineal gland. *Molecular Brain Research* **73**, 110-118.

Mansueti, A.J. (1964) Early development in the yellow perch, *Perca flavescens*. *Chesapeake Science* **5**, 46-66.

Marchiafava, P.L. (1986) Cell coupling in double cones of the fish retina (Plate 1). *Proceedings of the Royal Society of London, Ser. B* **226**, 311-315.

Marchiafava, P.L., Strettoi, E. and Alpigiani, V. (1985) Intracellular recording from single and double cone cells isolated from the fish retina (Tinca tinca). *Experimental Biology* **44**, 13-80.

Marcus, H. (1905) Ein Beitrag zur Kenntnis der Blutbildung bei Knochenfischen. *Archiv für Mikroskopische Anatomie* **66**, 333-354.

Marcus, R.C., Delaney, C.L. and Easter, S.S. (1999) Neurogenesis in the visual system of embryonic and adult zebrafish (*Danio rerio*). *Visual Neuroscience* **16**, 417-424.

Mark, E.L. (1890) Studies on *Lepidosteus*. *Bulletin of the Museum of Comparative Zoology, at Harvard College*, **19**, 67-127.

Markert, C.L. and Faulhaber, I. (1965) Lactate dehydrogenase isozyme patterns in fish. *Journal of Experimental Zoology* **159**, 319-332.

Massaro, E. and Markert, C.L. (1968) Isozyme patterns of salmonid fishes. Evidence for multiple cistrons for lactate dehydrogenase polypeptides. *Journal of Experimental Zoology* **168**, 223-238.

Matarese. A.C. and Sandknop, E.M. (1984) Identification of fish eggs, in *Ontogeny and Systematics of Fishes*, (ed American Society of Ichthyologists and Herpetologists), Alan Press Inc., Lawrence, U.S.A., pp. 27-31.

Matsubara, T. and Sawano, K. (1995) Proteolytic cleavage of vitellogenin and yolk proteins during vitellogenic uptake and oocyte maturation in barfin flounder (*Verasper moseri*). *Journal of Experimental Zoology* **272**, 34-45.

Mattei, X. (1970) Spermiogenèse comparée des poissons, in *Comparative Spermatology*, (ed B. Baccetti), Accademia Nazionale dei Lincei, Rome; Academic Press, New York,

London, pp. 5-69.

Mattei, X. (1991) Spermatozoon ultrastructure and its systematic implications in fishes. *Canadian Journal of Zoology* **69**, 3038-3055.

Mattei, C. and Mattei, X. (1978) La spermiogenèse d'un poisson téléostéen (*Lepadogaster lepadogaster*). La spermitide. *Biology of the Cell.* **32**, 257-266.

Mattei, X. and Boisson, C. (1966) Le complexe centriolaire du spermatozoïde de *Lebistes reticulatus*. *Comptes Rendus Hebdomadaires des Séances de l'Académie des Sciences* **D 262**, 2620-2622.

Mattei, X., Boisson, C., Mattei, C. and Reizer, C. (1967) Spermatozoïdes aflagelés chez un poisson: *Gymnarchus niloticus* (téléostéen, Gymnarchidae). *Comptes Rendus Hebdomadaires des Séances de l'Académie des Sciences* **D 265**, 2010-2012.

Mattey, D.L., Moate, R. and Morgan, M. (1978) Comparison of 'pseudobranch' type and 'chloride' cells in the pseudobranch of marine, freshwater and euryhaline teleosts. *Journal of Fish Biology* **13**, 535-542.

Mattey, D.L., Morgan, M. and Wright (D.E.) (1979) Distribution and development of rodlet cells in the gills and pseudobranch of the bass, *Dicentrachus* labrax (L). *Journal of Fish Biology* **15**, 363-370.

Maurer, F. (1884) Ein Beitrag zur Kenntnis der Pseudobranchien der Knochenfische. *Morphologisches Jahrbuch* **9**, 229-251.

Max, M. and Menaker, M. (1992) Regulation of melatonin production by light, darkness and temperature in the trout pineal. *Journal of Comparative Physiology* A, **170**, 479-489.

McCormack, C.A., Hayden T.H. and Kunz, Y.W. (1989). Ontogenesis of diurnal rhythms of cAMP concentration, outer segment disc shedding and retinomotor movements in the eye of the brown trout *Salmo trutta*. *Brain, Behavior and Evolution* **34**, 65-72.

McGowan and Berry, F.H. (1984) *Clupeiformes*: Development and relationships in *Ontogeny and Systematics of Fishes*, (ed American Society of Ichthyologists and Herpetologists), Allen Press Inc., Lawrence, KS 66040, U.S.A., pp. 108-125.

McKenzie, J.A. (1974) The parental behavior of the male brook stickleback, *Calaea inconstans* (Kirtland). *Canadian Journal of Zoology* **52**, 649-652.

McNeilly, S.† and Kunz, Y. (1992) The hatching glands and adhesive apparatus of a mouthbrooding and a substrate spawning cichlid — a light microscopical and scanning electron microscopical study. (Unpublished).

Meckel von Helmsbach, H. (1852) Die Bildung der für partielle Furchung bestimmten Eier der Vögel, im Vergleich mit dem Graafschen Follikel der Decidua des Menschen. *Zeitschrift für wissenschaftliche Zoologie* **3**, 420-433.

Medina, M., Repérant, J., Ward, R., Rio, J.-P. and Lemire, M. (1993) The primary visual system of flatfish; an evolutionary perspective. *Anatomy and Embryology* **187**, 167-191.

Meijer, H.A. and G. Te Kronnie (1996) Manipulation of gap junctional communication between the yolk syncytial layer and blastoderm cells and its effect on embryonic development of *Cyprinus carpio*. *Netherlands Journal of Zoology* **46**, 304-316.

Meissl, H. and Ekström, P. (1986) Antwortmuster pinealer Photorezeptoren der Forelle. *Verhandlungen der Deutschen zoologischen Gesellschaft*. Gustav Fischer Verlag, Stuttgart, p. 227.

Melby, A.E., Ho, R.K. and Kimmel, C.B. (1993) An identifiable domain of tail-forming cells in the zebrafish gastrula. *Society of Neuroscience Abstracts* **19**, 445.

Mendoza, G. (1939) The reproductive cycle of the viviparous teleost, *Neoteca bilineata*, a member of the family Goodeidae. I. The breeding cycle. *Biological Bulletin (Woods Hole, Mass.)* **76**, 359-370.

Mendoza, G. (1940) The reproductive cycle of the viviparous teleost, *Neotoca bilineata*, a member of the family Goodeidae. II The cyclic changes in the ovarian soma during gestation. *Biological Bulletin (Woods Hole, Mass.)* **78**, 349-365.

Mendoza, G. (1956) Adaptations during gestation in the viviparous teleost, *Hubbsina turneri*. *Journal of Morphology* **99**, 73-89.

Mendoza, G. (1958) The fin fold of *Goodeidea luitpoldii*, a viviparous cyprinodont teleost. *Journal of Morphology* **103**, 539-560.

Mendoza, G. (1972) The fine structure of an absorptive epithelium in a viviparous teleost. *Journal of Morphology* **136**, 109-130.

Mendoza, A., Zamora, A and Garcia, H. (1976) Structure fine des pseudobranchies chez les téléostéens. *Archive de Biologie (Bruxelles)* **87**, 215-224.

Merriman, D. (1940) Morphological and embryological studies on two species of marine catfish, *Bagre marinus* and *Galeichthys felis*. *Zoologica, New York* **25**, 221-248.

Methven, D.A. and Brown, J.A. (1991) Time of hatching affects development, size, yolk volume, and mortality of newly hatched *Macrozoarces americanus* (Pisces: Zoarcidae). *Canadian Journal of Zoology* **69**, 2162-2167.

Metscher, B.D. and Ahlberg, R.E. (1999) Review. Zebrafish in context: uses of laboratory model in comparative studies. *Developmental Biology* **210**,1-14.

Meyer-Rochow, V.B. (1970) *Cataetix memoriabilis* new sp. — ein neuer Tiefsee-*Ophidiidae* aus dem südostlichen Atlantik. *Abhandlungen Verhandlungen der Naturwissenschaften (NF)* **14**, 37-53.

Meyer, A., Biermann, C.H. and Orti, G. (1993) The phylogenetic position of the zebrafish (*Danio rerio*), a model system in developmental biology: an invitation to the comparative method. *Proceedings of the Royal Societey of London B* **252**, 231-236.

Meyer, D.L., Lara, J., Malz, C.R. and Graf, W. (1993) Diencephalic projections to the retinae in two species of flatfishes (*Scophthalmus maximus* and *Pleuronectes platessa*). *Brain Research* **601**, 308-312.

Miescher, F. 1874 Das Protamin, eine neue organische Basis aus den Samenfäden des Rheinlachses. *Berichte der deutschen chemischen Gesellschaft* **7**, 376-379.

Miescher, F. (1878) Die Spermatozoen einiger Wirbelthiere. Ein Beitrag zur Histochemie (after lectures given in April 1872 and November 1873). *Verhandlungen der Naturforschenden Gesellschaft in Basel* **1878, VI**, 138-208.

Miescher, F. (1896) Physiologisch-chemische:Untersuchungen über die Lachsmilch. *Archiv für experimentelle Pathologie und Pharmakologie* **37**. 100-155.

Mikodina, Ye.V. and Makeyeva, E.P. (1980) The structure and some properties of egg membranes in pelagophilous freshwater fishes. *Journal of Ichthyology* **20**. 86-94.

Miller, T.J., Herra, T. and Leggen, W.C. 1995) An individual-based analysis of the variability of eggs and their newly hatched larvae of Atlantic cod (*Gadus morrhua*) on the Scotian Shelf. *Canadian Journal of Fisheries and Aquatic Science* **52**, 1083-1093.

Minot, Sch.S. The concrescence theory of the vertebrate embryo. *The American Naturalist* **24**, 501-516.

M'Intosh, W.C. (1885) Notes from St. Andrews marine laboratory II. On the spawning of certain marine fishes. *Annals and Magazine of Natural History*. 5. ser., **15**, No. 9, 429-435.

M'Intosh, W.C. and Prince, E.E. (1890) On the development and life histories of the teleostean food- and other fishes. *Transactions of the Royal Society Edinburgh* **35**, 665-946.

Mito, S. (1963) Pelagic eggs from Japanese waters. IX. *Escheneida* and *Pleuronectida*. (in Japanese). *Japanese Journal of Ichthyology* **81**, 81-102.

Miyayama Y. and Fujimoto, T. (1977) Fine morphological study of neural tube formation in the teleost, *Oryzias latipes*. *Okajimas Fol. Anat. Jap.* **54**, 97-120.

Mizuno, T., Yamaha, E., Wakahara, M., Kuroiwa, A. and Takeda, H. (1996) Mesoderm induction in zebrafish. *Nature* **383**, 131-132.

Mollier, S. (1906) Die erste Entwickelung des Herzens, der Gefässe und des Blutes: Teleostier, in *Handbuch der Entwickelungslehre der Wirbeltiere*, Jena, Lief. **27** & **28**, 1125-1154.

Mommsen, T. and Walsh, P.J. (1988) Vitellogenesis and oocyte assembly, in Fish Physiology, **11A**, (eds W.S. Hoar and D.J. Randall), Academic Press, New York, London, pp. 347-406.

Mooi, R.D. (1990) Egg surface morphology of pseudochromoids (perciformes: percoidei), with comments on its phylogenetic implications. *Copeia* **1990(2)**, 455-475.

Mooi, R.D., Winterbottom, R. and Burridge, M. (1990) Egg surface morphology, development and evolution in the congrogadinae (pisces: perciformes: pseudochromidae). *Canadian Journal of Zoology* **68**, 923-934.

Morgan, M. (1974) Development of secondary lamellae of the gills of the trout, *Salmo gairdneri* (Richardson). *Cell and Tissue Research* **151**, 509-523.

Morgan, T.H. (1893) Experimental studies on the teleost eggs. *Anatomischer Anzeiger* **8**, 803-814.

Morgan, T.H. (1895) The formation of the fish embryo. *Journal of Morphology* **10**, 419-470.

Morgan, M. M. and Tovell, P.W.A. (1993) The structure of the gill of the trout, *Salmo gairdneri* (Richardson). *Zeitschrift der Zellforschung und Mikroskopischen Anatomie* **142**, 147-162.

Morisawa M. (1985) Initiation mechanism of sperm motility at spawning in teleosts. *Zoological Science* **2**, 605-615.

Moriwaki, I. (1910) The mechanism of escape of the fry out of the egg chorion in the dog salmon (in Japanese). *The 3rd Report of the Hokkaido Fisheries Research Station* (cited from 'Yamagami, 1988').

Moser, H.G. (1967a) Reproduction and development of *Sebastodes paucispinis* and comparison with other rockfishes off California. *Copeia* **1967**, 773-779.

Moser, H.G. (1967b) Seasonal histological changes in the gonads of *Sebastodes paucispinis* Ayers, an ovoviviparous teleost (family *scorpaenidae*). *Journal of Morphology* **123**, 333-336.

Mosher, C. (1954) Observations on the spawning behaviour and the early larval development of the sargassum fish, *Histria histria* (Linnaeus). *Zoologica* **39**, 141-152.

Müller, J. (1839) Weitere Mitteilungen über die Wundernetze zu dem comparativen Theil der vergleichenden Anatomie der Myxinoiden. *Berichte der Akademischen Wissenschaften*,

Berlin **1839**, 272-287.

Müller, H. (1952) Bau und Wachstum der Netzhaut des Guppy (*Lebistes reticulatus*). *Zoologisches Jahrbuch, Abteilung allgemeine Zoologie und Physiologie* **63**, 276-324.

Müller, H. (1954a) Die Dunkeladaption beim Guppy (*Lebistes reticularis* P.) *Zeitschrift für vergleichende Physiologie* **37**, 1-18.

Müller, J. (1854b) Ueber zahlreiche Porencanäle in der Eicapsel der Fische. *Müller's Archiv für Anatomie und Physiologie* **1854**, 186-190.

Müller, H. and Sterba, B. (1963) Elektronenmikroskopische Untersuchungen über Bildung und Struktur der Eihüllen bei Knochenfischen II Die Eihüllen jüngerer und älterer Ozyten von *Cynolebias belotti* Steindachner (Cyyprinodontidae). *Zoologisches Jahrbuch der Anatomie* **80**, 469-488.

Munehara, H. and Shimazi, K. (1989) Annual maturation changes in the ovaries of masked greenling *Hexagrammos octogrammus*. *Nippon Suisan Gakkaishi* **55**, 423-430.

Munk, O. (1966) Ocular anatomy of some deep-sea teleosts. *Dana Report* No. **70**, 63 pp.

Munk, O. (1977) the visual cells and retinal tapetum of the foveate deep-sea fish *Scopelosaurus lepidus* (teleostei) *Zoomorphologie* **87**, 21-49.

Munk, O. (1981) On the cones of the mesopelagic teleost *Trachypterus trachypterus* (Gmelin, 1789). *Vidensk. Meddr dansk naturh. Foren* **143**, 101-111.

Muntz, W.R.A., Partridge, J.C., Williams, S.R. and Jackson, C. (1996) Spectral sensitivity in the guppy (*Poecilia reticulata*) measured using the dorsal light response. *Marine and Freshwater Behav. Physiology* **28**, 163-176.

Murata. K., Iuchi, I. and Yamagami, K. (1994) Synchronous production of the low- and high-molecular-weight precursors of the egg envelope subunits, in response to estrogen administration in the teleost fish *Oryzias latipes*. *General and Comparative Endocrinology* **95**. 231-239.

Murata, K., Hamazaki, T.S., Iuchi, I. and Yamagami, K. (1991) Spawning female-specific egg envelope glycoprotein-like substances in *Oryzias latipes*. *Development, Growth and Differentiation* **33**, 553-562.

Murata. K., Sasaki, T., Yasumasu, S., Iuchi, I., Enami, J., Yasumasu, I. and Yamagami, K. (1995) Cloning of cDNAs for the precursor protein of a low-molecular-weight subunit of the inner layer of the egg envelope (chorion) of the fish *Oryzias latipes*. *Developmental Biology* **167**, 9-17.

Murie, J. (1868) On the supposed arrest of development of the salmon when retained in fresh water. *Proceedings of the Zoological Society* pp.247-54.

Nacario, J.F. (1983) The effect of thyroxine on the larvae and fry of *Sarotherodon niloticus* L. (*Tilapia nilotica*). *Aquaculture* **34**, 73-83.

Nagahama, Y., Clarke, W.C. and Hoar, W.S. (1978) Ultrastructure of putative steroid-producing cells in the gonads of coho (*Oncorhynchus kisutch*) and pink salmon (*Oncorhynchus gorbuscha*). *Canadian Journal of Zoology* **56**, 2508-2519.

Nagler, J.J. and Idler, D.R. (1989) Ovarian uptake of vitellogenin and another very high density lipoprotein in winter flounder (*Pseudopleuronectes americanus*) and their relationship with yolk proteins. *Biochemistry and Cell Biology* **68**, 330-335.

Nagler, J.J., Tyler,C.R. and Sumpter, J.P. (1994) Ovarian follicles of rainbow trout (*Oncorhynchus mykiss*) cultured within lamellae survive well and sequester and process vitellogenin. *Journal of Experimental Zoology* **269**, 45-52.

Nakano, E. (1956) Changes in the egg membrane of the fish egg during fertilization. *Embryologia* **3**, 89-103.

Nakano, E. (1969) Fishes, Chapter 7, in *Fertilization, Comparative Morphology, Biochemistry and Immunology, Volume II*, (eds Ch. B. Metz and A. Monroy), Academic Press, New York and London, pp. 295-324.

Nakano, E. and Whitely, A.H. (1965) Differentiation of multiple molecular forms of four dehydrogenases in the teleost. *Oryzias latipes. Journal of Experimental Zoology* **159**, 167-180.

Nakashima, S. and Iwamatsu, T. (1989) Ultrastructural changes in micropylar cell and formation of micropyle during oogenesis in the medaka *Oryzias latipes*. *Journal of Morphology* **202**, 339-349.

Nakashima, S. and Iwamatsu, T. (1994) Ultrastructural changes in micropylar and granulosa cells during in vitro oocyte maturation in the medaka, *Oryzias latipes. The Journal of Experimental Zoology* **270**, 547-556.

Nakatsuji, T., Kitano, T., Akiyama, N. and Nakatsuji, N. (1997) Ice Goby (Shiro-uo), *Leucopsarion petersii*, may be a useful material for studying teleostean embryogenesis. *Zoological Science* **14**, 443-448.

Nakazono, A. and Kawase, H. (1993) Spawning and biparental egg-care in a temperate filefish, *Paramonacanthus japonicus* (Monocanthidae). *Environmental Biology of Fishes* **37**, 245-256.

Nelsen, O.E. (1953) *Comparative Embryology of the Vertebrates*, McGraw Hill, London, New York.

Neumeyer, C. (1992) Tetrachromatic color vision in goldfish: evidence from color mixture experiments. *Journal of Comparative Physiology A* **171**, 639-649.

Ng, T.B. and Idler, D.R. (1983) Yolk formation and differentiation in teleost fishes, in *Fish Physiology* **9B**, (eds W.S. Hoar, A.J. Randall, E.M. Donaldson), Academic Press, New York.

Nguyen-Legros, J. (1978) Fine structure of the pigment epithelium in the vertebrate retina. *International Review of Cytology (suppl.)* **7**, 287-328.

Nicander, L. (1970) Comparative studies on the fine structure of vertebrate spermatozoa, in *Comparative Spermatology*, (ed B. Baccetti), Accademia Nazionale dei Lincei Rome; Academic Press, New York, London, pp. 47-55.

Nicol, J.A.C. (1963) Some aspects of photoreception and vision in fishes. *Advances in Marine Biology* **1**, 121-208.

Noakes, D.L.G. (1978) Ontogeny of behavior in fishes: A survey and suggestions, in *The Development of Behavior: Comparative and Evolutionary Aspects*, (eds G.M. Burghardt and M. Bekoff), Garland STPM Press, New York and London, pp103-125.

Noakes, D.L.G. (1981) Comparative aspects of behavioral ontogeny: a philosophy from fishes. Chapter 18, in *Behavioral Development*, The Bielefeld Interdisciplinary Project, (eds K. Immelmann, G.W. Barlow, L. Petrinovich and M. Main), Cambridge University Press, Cambridge, pp 491-508.

Noakes, D.L.G. and Godin, J.J. (1988) Ontogeny of behavior and concurrent developmental changes in sensory systems in teleost fishes, in *Fish Physiology*, **11B**, (ed W.S. Hoar and

D.J. Randall), Academic Press, San Diego, Ca, pp. 345-395.
Norberg, B. and Haux, C. (1985) Induction, isolation and a characterization of the lipid content of plasma vitellogenin from two *Salmo* species: Rainbow trout (*Salmo gairdneri*) and sea trout (*Salmo trutta*). *Comparative Biochemistry and Physiology* **81B**, 860-876.
Nosck, J. (1984) Biogenesis of cortical granules in fish oocytes. *Histochemical Journal* **16**, 435-437.
Nowak, J.Z. (1988) The isolated retina as a model of CNS in pharmacology. *Trends in Pharmacological Science* **9**, 80-82.
Nuccitelli, R. (1980) The electrical changes accompanying fertilization and cortical vesicle secretion in the medaka egg. *Developmental Biology* **76**, 483-498.
Nuccitelli, R. (1987) The wave of activation current in the egg of the medaka fish. *Developmental Biology* **122**, 522-534.
Nuccitelli, R. (1991) How do sperm activate eggs?, in *Current Topics in Developmental Biology*, Vol. 25, (ed H.R. Bode), Academic Press, Inc., pp. 1-16.
Ober, E.A. and Schulte-Merker (1999) Signals from the yolk cell induce *mesoderm, neurectoderm*, the trunk organizer and the notochord in zebrafish. *Developmental Biology* **215**, 167-181.
Ochi, H. (1993a) Maintenance of separate territories for mating and feeding by males of a maternal mouthbrooding cichlid, *Gnathochromis pfefferi* in Lake Tanganyika. *Japanese Journal of Ichthyology* **40**, 173-182.
Ochi, H. (1993b) Mate monopolization by a dominant male in a multi-male social group of a mouthbrooding cichlid, *Ctenochromis horei*. *Japanese Journal of Ichthyology* **40**, 290-218.
O'Connell, Ch.P. (1981) Development of organ systems in the Northern anchovy, *Engraulis mordax*, and other teleosts. *The American Zoologist* **21**, 429-446.
Oellacher, J. (1872) Beiträge zur Entwicklungsgeschichte der Knochenfische nach Beobachtungen am Bachforelleneie. I. Das unbefruchtete reife und das befruchtete Forellenei vor der Furchung. II. Die Furchung im Forellenkeim. *Zeitschrift für Wissenschaftlichen Zoologie* **22**, 373-385 and 386-421.
Oellacher J. (1873) Beiträge zur Entwicklungsgeschichte der Knochenfische nach Beobachtungen am Bachforelleneie. III Von der Bildung des Hornblatttes bis zum Auftreten der Rückenfurche. IV. Bildung des Medullarstranges, der Chorda, der Peritonealplatten, Urwirbelplatten und Urwirbel; erste Anlage des Auges, des Ohres und der Kiemenhöhle.V. Weitere Ausbildung der Urwirbel, des Darmes und der Kiemenhöhle, des Auges und Ohren, Entwickelung der Peritoneal- und Pericardialhöhle; der Kiemenspalten; des Herzens und der Urnierengänge *Zeitschrift für Wissenchaftliche Zoologie* **23**, pp 1-37, pp 37-66, pp 66-115.
Oestholm, T., Brännäs, E. and van Veen, T. (1987) The pineal organ is the first differentiated light receptor in the embryonic salmon, *Salmo salar* L. *Cell and Tissue Research* **249**, 641-646.
Oguri M. and Omura, Y. (1973) Ultrastructure and functional significance of the pineal organ of teleost, in *Response of Fish to Environmental Changes*, (eds W. Chavin, W. and S. Thomas) Springfield.
Ohta, H. (1984) Electron microscopic study on adhesive material of Pacific herring (*Clupea*

pallasi) eggs. *Japanese Journal of Ichthyology* **30**, 404-411.
Ohta, T. (1985) Electron microscopic observations on sperm entry and pronuclear formation in naked eggs of the rose bitterling in polyspermic fertilization. *Journal of Experimental Zoology* **234**, 273-281.
Ohta, T. (1986) Electron microscopic observations on the process of polar body formation in artificially activated eggs of the rose bitterling. *Journal of Experimental Zoology* **237**, 263-270.
Ohta, T. and Iwamatsu, T. (1983) Electron microscopic observations on sperm entry into eggs of the rose bitterling, *Rhodeus ocellatus*. *Journal of Experimental Zoology* **227**, 109-119.
Ohta, T. and Takano, K. (1982) Ultrastructure of micropylar cells in the pre-ovulatory follicles of Pacific herring, *Clupea pallasi* Valanciennes. *Bulletin of the Faculty of Fisheries Hokkaido University* **33**, 57-64.
Ohta, T. and Nashirozawa, C. (1996) Sperm penetration and transformation of sperm entry site in eggs of the freshwater teleost *Rhodeus ocellatus ocellatus*. *Journal of Morphology* **229**, 191-200.
Ohta, T. and Takano, (1982) Ultrastructure of micropylar cells in the pre-ovulatory follicles of Pacific herring, *Clupea pallasi* Valenciennes. *Bulletin of the Faculty of Fisheries Hokkaido University* **33**, 57-64.
Ohta, T. and Tenarishi, T. (1982) Ultrastructure and histochemistry of *granulosa* and micropylar cells in the ovary of the loach, *Misgurnus anguillacaudatus* (Cantor) *Bulletin of the Faculty of Fisheries Hokkaido University* **33**. 1-8.
Ohta, T., Takano, K., Izawa, T. and Yamauchi, K. (1983) Ultrastructure of the chorion and the micropyle of the Japanese eel, anguilla japonica. *Bulletin of the Japanese Society of Fisheries* **49**, 501.
Ohta, T., Kato, K.H., Abe, T. and Takeuchi, T. (1993) Sperm morphology and distribution of intramembranous particles in the sperm heads of selected freshwater teleosts. *Tissue and Cell* **25**, 725-735.
Ohtsuka, E. (1957) On the hardening of the chorion of the fish egg after fertilization, I. Role of the cortical substance in chorion hardening of the egg of *Oryzias latipes*. *Sieboldia* **2**, 19-29.
Ohtsuka, E. (1960) On the hardening of the chorion of the fish egg after fertilization. III. The mechanism of chorion hardening in *Oryzias latipes*. *Biological Bulletin (Woods Hole, Mass.)* **118**, 120-128.
Oksche, A. (1989) Pineal complex –the 'third' or 'first' eye of vertebrates: a conceptual analysis. *Biomedical Research* **10**, 187-94.
Oksche, A. and Kirschstein, H. (1967) Die Ultrastruktur der Sinneszellen im Pinealorgan von *Phoxinus laevis* L. *Zeitschrift für Zellforschung* **78**, 151-166.
Olson, A.J., Picones, A., Julian, D. and Korenbrot, J.I. (1999) A developmental time line in a retinal slice from rainbow trout. *Journal of Neuroscience Methods* **93**, 91-100.
Olt, A. (1893) Lebensweise und Entwicklung des Bitterlings. *Zeitschrift für wissenschaftliche Zoologie* **55**, 543-575.
Omura, Y. and Oguri, M. (1969) Histological studies on the pineal organ of 15 species of teleosts. *Bulletin of the Japanese Society of Scientific Fisheries* **35**, 991-1000.

Omura, Y. and Oguri, M. (1971) The development and degeneration of the photoreceptor outer segment of the fish pineal organ. *Bulletin of the Japanese Society of Scientific Fisheries* **37**, 851-860.

Oppen-Berntsen D.O. (1990) Oogenesis and hatching in teleostean fishes with special reference to egg shell proteins. PhD Thesis, University of Bergen, Norway. Cited by Luczynski, M. *et al.* (1993)'.

Oppen-Berntsen, D.O., Gram-Jensen, E. and Walther, B.T. (1992) Zona radiata proteins are synthesized by rainbow trout (*Oncorhynchus mykiss*) hepatocytes in response to oestradiol-17ß. *Journal of Endocrinology* **135**, 293-301.

Oppen-Berntsen, D.O., Helvik, J.V. and Walther, B.T. (1990) The major structural proteins of cod (*Gadus morrhua*) eggshells and protein crosslinking during teleost egg hardening. *Developmental Biology* **137**, 258-265.

Oppen-Berntsen, D.O., Hyllner, S.J., Haux, C., Helvik, J.V. and Walther, B.T. (1992a) Eggshell *zona radiata*-proteins from cod (*Gadus morhua*): extraovarian origin and induction by estradiol-17ß. *International Journal of Developmental Biology* **36**, 247-254.

Oppenheimer, J.M. (1935) The development of transplanted fragments of *Fundulus* gastrulae. *Proceedings of the National Academy of Science* **39**, 1149-1152.

Oppenheimer, J.M. (1936a) Processes of localization in developing Fundulus. *Journal of Experimental Zoology* **73**, 405-444.

Oppenheimer, J.M. (1936b) The Development of isolated blastoderms of *Fundulus heteroclitus*. *Journal of Experimental Zoology* **72**, 247-269.

Oppenheimer, J.M. (1937) The normal stages of *Fundulus heteroclitus*. *The Anatomical Record* **68**, 1-8.

Oppenheimer, J.M. (1938) Potencies for differentiation in the teleostean germ ring. *Journal of Experimental Zoology* **79**, 185-212.

Oppenheimer, J.M. (1947) Organization of the teleost blastoderm. *Quarterly Review of Biology* **22**,105-118

Oppenheimer, J.M. (1979) Fifty years of *Fundulus*. *Quarterly Review of Biology* **54**, 385-393.

Orkin, S.H. and Zon, L.I. (1997) Genetics of erythropoiesis: induced mutations in mice and zebrafish. *Annual Review of Genetics* **31**, 33-60.

Orsi, J.J. (1968) The embryology of the English sole, *Parophrys vetulus*. *Calif. Fish and Game* **54**, 133-155.

Osanai, K. (1956) On the ovarian eggs of the loach, *Lefus eschogonia*, with special reference to the formation of the cortical alveoli. *Scientific Reports Tohoku University Series 4*, **22**, 181-188.

Osanai, K. (1977) Scanning electron microscopy of the envelopes surrounding the chum-salmon oocytes. *Bulletin of the marine Biological Station Asamushi* **16**, 21-25.

Osse, J.W. and Van den Boogaart, M. (1997) Size of flatfish *larvae* at transformation, functional demands and historical constraints. *Journal of Sea Research* **37**, 229-239.

Ota, D., Marchesan, M. and Ferrero, E.A. (1996) Sperm release behaviour and fertilization in the grass goby. *Journal of Fish Biology* **49**, 246-256.

Ouji, M. and Matsuno, A. (1973). The structure of the hatching gland granules in the fresh-water teleost, *Leuciscus hakuensis*. *Memoirs of the Faculty of Literature and Science, Shimane University, Natural Science* **6**, 39-45.

Owsiannikow (Ovsyannikov), Ph. (1873) Ueber die ersten Vorgänge der Entwicklung in den Eiern des *Coregonus lavaretus*. *Bulletin de l'Académie Impériale des Sciences de St. Petersbourg* **19**, 225-235.

Owsiannikow (Ovsyannikov) Ph. (1885) Studien über das Ei, hauptsächlich bei Knochenfischen. *Mémoires de l'Académie des Sciences de St. Pétersbourg, Série 7*, **33**, 1-54.

Ozato, K. and Wakamatsu, Y. (1994) Review: Developmental genetics of medaka. *Development, Growth and Differentiation* **36**, 437-443.

Padmanabhan, K.G. (1955) Breeding habits and early embryology of *Macropodus cupanus* Cuvier and Valanciennes. *Bulletin of the Central Research Institute, University Travancore, Ser. C Naat. Sci. 4*.

Padoa, E. (1956) Famiglia: Callionymidae, in *Uova, larva e stadi giovanili dei Teleostei, Fauna Flora Golfo Napoli Monograph* **38**, (eds Salvatore lo Bianco, G. Bardi and R. Friedländer), pp. 697-708.

Pan, M.L., Bell, W. and Telfer, W.H. (1969) Vitellogenic blood protein synthesis by insect fat body. *Science* **165**, 393-394.

Papan, C. and Campos-Ortega, J.A. (1994) On the formation of the neural keel and neural tube in the zebrafish *Danio (Brachydanio) rerio*. *Roux's Archive of Developmental Biology* **203**, 178-186.

Parihar, R.P. (1979) Studies on the cleavage and gastrulation in *Cyrrhinus mrigala (Ham.)*. *The Annals of Zoology* **15**, 43-57.

Parry, G. and Holliday, F. (1960) An experimental analysis of the function of the pseudobranch in teleosts. *Journal of Experimental Biology* **37**, 344-353.

Pasteels, J. (1934) Répartition des Territoires et Mouvements Morphogénétiques de la Gastrulation de l'oeuf de Truite (*Salmo irideus*). *Comptes Rendus de l'Association anatomiaques (Bruxelles)* **29**, 451-458.

Pasteels, J. (1936) Etudes sur la gastrulation des vertébrés méroblastiques téléostéens. *Archive de Biologie (Liège)* **47**, 205-308.

Pasteels, J. (1940) Un aperçu comparatif de la gastrulation chez les chordés. *Biological Review* **15**, 59-106.

Patzner, R.A. (1984) The reproduction of *Blennius pavo (teleostei, blenniidae)*. II. Surface structures of the ripe egg. *Zoologischer Anzeiger (Jena)* **213**, 44-50.

Pecio, A. and Rafinski, J. (1999) Spermiogenesis of *Mimagoniates barberi (Characidae)*, an oviparous internally fertilizing fish. *Acta Zoologica (Stockholm)* **80**, 35-45.

Pelluet, D. (1944) Criteria for the recognition of developmental stages in the salmon (*Salmo salar*). *Journal of Morphology* **47**, 395-407.

Perry, D.M. (1984) Post-fertilization changes in the chorion of winter flounder, *Pseudopleuronectes americanus* Walbaum, eggs observed with scanning electron microscopy. *Journal of Fish Biology* **25**, 83-94.

Peter, K. (1947) *Grundlagen einer funktionellen Embryologie*, Johann Ambrosius Barth Verlag, Leipzig.

Peters von H.M. and Berns, S. (1983) On the larval adhesive apparatus of substrate brooding cichlids (teleostei). *Zoologisches Jahrbuch der Anatomie* **109**, 59-80).

Phillips, J.B. (1964) Life history studies on ten species of rockfish (Genus *Sebastodes*). *California Department of Fish and Game, Fish Bulletin* **126**, 1-70.

Pilliai, M.C., Yanagimachi, R. and Cherr, G.N. (1994) *In vivo* and *in* vitro initiation of sperm motility using fresh and cryopreserved gametes from the Pacific herring, *Clupea pallasi*. *Journal of Experimental Zoology* **269**, 62-68.

Pillai, M.C., Shields, T.S., Yanagimachi, R. and Cherr, G.N. (1993) Isolation and partial characterization of sperm motility initiation factor from the eggs of the Pacific herring, *Clupea pallasi*. *Journal of Experimental Zoology* **265**, 336-342.

Pohlmann J.H.L (1939)'Ontogenie und mikroskopischer Bau der Leber einiger Fische. *Arch. neerl. Zool.* **3**, 64-140.

Pommeranz, T. (1974) Resistance of plaice eggs to mechanical stress and light, in *The Early Life History of Fish*, (ed H.J.S. Blaxter), Springer, Berlin, pp.397-416.

Portmann, A. (1927) Die ersten Stadien des Blutkreislaufs bei Teleostierembryonen und die Ausbildung der Dottersackzirkulation. *Verhandlungen der Naturforschenden Gesellschaft Basel* **38**, 416-426.

Portmann, A. (1976) *Einführung in die vergleichende Morphologie der Wirbeltiere*. 4th edition, Schwabe, Basel.

Portmann, A. and Metzner, G. (1929) Die Verbindung von Leber und Dottersack bei Teleostierlarven. *Verhandlungen der Naturforschenden Gesellschaft Basel* **40**, 271-279.

Potts, G.W. (1984) Parental Behaviour in temperate marine teleosts with special reference to the development of nest structures, in *Fish Reproduction*, Chapter 5, (eds G.W. Potts and R.J. Wootton), Academic Press, London, pp. 223-44.

Price, J.W. (1934a) The embryology of the whitefish *Coregonus clupeaformis* (Mitchill). *Ohio Journal of Science* **34**, part I, 287-305.

Price, J.W. (1934b) The embryology of the white fish *Coregonus clupeaformis* (Mitchill). *Ohio Journal of Science* **34**, part II, 399-414.

Price, J.W. (1935) The embryology of the white fish *Coregonus clupeaformis* (Mitchill). *Ohio Journal of Science* **35**, part III, 40-53.

Prince, E. E. (1887) The significance of the yolk in the eggs of osseus fishes. *The Annals and Magazine of Natural History*, **5th series, No. 15**, 1-8.

Purser, G.L. (1940) Reproduction in *Lebistes reticulatus*. *Quarterly Journal of Microscopical Science* **81**, 151-157.

Quattro, J.M., Woods, H.A and Powers, D.A. (1993) Sequence analysis of teleost retina-specific lactate dehydrogenase C: Evolutionary implications for the vertebrate lactate dehydrogenase gene family. *Proceedings of the National Academy of Science U.S.A.* **90**, 242-246.

Rabl-Rückhardt, H. (1883) Das Grosshirn der Knochenfische und seine Anhangsgebilde. *Archiv für Anatomie und Physiologie (Anat. Abh.)* **1883**, 279-322.

Rabl, H. (1888) Ueber die Bildung des Mesoderms. *Anatomischer Anzeiger* **3**, 23-25.

Raffaele, F. (1888) Le uova galleggianti e le larve dei teleostei nel golfo di Napoli. *Mittheilungen aus der Zoologischen Station zu Neapel* **8**,1-83.

Raffaele, F. (1897) Osservazioni sul foglietto epidermico superficiale degli embrioni dei pesci ossei. *Mittheilungen aus der Zoologischen Station zu Neapel* **12**, 169-207.

Rahmann, H. (1968) Autoradiographische Untersuchungen zum DNS –Stoffwechsel (Mitosehäufigkeit) im ZNS von *Brachydanio rerio*. Ham. Buch. *Journal der Hirnforschung* **10**, 279-840.

Raible, D.W. and Eisen, J.S. (1994) Restriction of neural crest cell fate in the trunk of the embryonic zebrafish. *Development* **120**, 495-503.

Raible, D.W., Wood, A., Hodsdon, W., Henion, P.D., Weston, J.A. and Eisen, J.S. (1992) Segregation and early dispersal of neural crest cells in the embryonic zebrafish. *Developmental Dynamics* **195**, 29-42.

Ramon y Cajal (1892) G. Die Retina der Wirbeltiere. German translation by R. Greef. Wiesbaden 1894. French: la rétine des vertébrés. *Cellule* **9**, 121-225.

Ransom, W.H. (1852/1855) Letter to Dr. Allan Thomson Article on ovum. *Todd's Cyclopaedia of Anatomy and Physiology* **5** (suppl.), pp. 1-80 (1852), pp 81-142 (1855).

Ransom, W.H. 1854 On the impregnation of the ovum in the stickleback. *Proceedings of the Royal Society of London* **7**: 168-172.

Ransom, W.H. (1855) Further observations on the structure of the ova of fishes, with especial reference to the micropyle and the phenomena of their fecundation. *Report of the British Association for the Advancement of Science*, 25th meeting, Part 2, p.131.

Ransom, W.H. (1856) On the impregnation of the ovum in the stickleback. *Proceedings of the Royal Society of London* **7**, 168-172.

Ransom, W.H. (1867) On the conditions of the protoplasmic movements in the eggs of osseous fishes. *Journal of Anatomy and Physiology* **1**, 237-245.

Ransom, W.H. (1867/1868) Observations on the ovum of osseous fishes. *Philosophical Transactions of the Royal Society of London* **157**, 431-501.

Rathke, M.H. (1824) Beiträge zur Geschichte der Thierwelt. Neue Schrift. *Naturforschende Gesellschaft Danzig*, **Heft 2,3.**

Rathke, H. (1833) Bildungs- und Entwickelungsgeschichte des *Blennius viviparus* oder des Schleimfisches, in *Abhandlungen zur Bildungs- und Entwickelungs-Geschichte des Menschen und der Thiere*, F.C.W. Vogel, Leipzig, *Part 2*, pp 1-68.

Rauber, A. (1883) Neue Grundlagen zur Kenntnis der Zelle. *Morphologisches Jahrbuch* **5**, 661-705.

Rauther, M. (1910) Die akzessorischen Atmungsorgane der Knochenfische. *Ergebnisse und Fortschritte der Zoologie, Jena.* **2**, 517-585.

Raymond Johns, P. (1982) Formation of photoreceptors in larval and adult goldfish. *Journal of Neuroscience* **2**, 178-198.

Raymond, P.A. (1985) Cytodifferentiation of photoreceptors in larval goldfish delayed maturation of rods. *Journal of Comparative Neurology* **236**, 90-105.

Raymond, P.A. and Rivlin P.K. (1987) Germinal cells in the goldfish retina that produce rod photoreceptors. *Developmental Biology* **122**, 120-38.

Raymond, P.A., Barthel, L.K. and Curran G.A. (1995) Developmental pattern of rod and cone photoreceptors in embryonic zebrafish. *Journal of Comparative Neurology* **359**, 537-550.

Reagan, F.P. (1915-1916) A further study of the origin of blood vascular tissues in chemically treated teleost embryos, with especial reference to haematopoesis in the anterior mesenchyme and the heart. *Anatomical Records* **10**, 99-129.

Reddy, P.K. and Lam, T.J. (1991) Effect of thyroid hormones on hatching in the tilapia, Oreochromis mossambicus. *General and Comparative Endocrinology* **81**, 484-491.

Reddy, P.K. and Lam, T.J. (1992) Role of thyroid hormones in tilapia larvae (*Oreochromis mossambicus*). I. Effects of the hormones and an antithyroid drug on yolk absorption, growth and development. *Fish Physiology and Biochemistry* **9**, 473-486.

Reichenbach, A., Schaaf, P. and Schneider, H. (1990) Primary neurulation in teleosts — evidence for epithelial genesis of central nervous tissue as in other vertebrates. *Journal der Hirnforschung* **2**, 152-158.

Reichert, K.B. (1848) Beobachtungen über die ersten Blutgefässe und deren Bildung, sowie über die Bewegungen des Blutes in denselben bei Fischembryonen. *Studien des Physiologischen Instituts zu Breslau. Leipzig.*

Reichert, K.B. (1856) Ueber die Mikropyle der Fischeier und über einen bisher unbekannten, eigenthümlichen Bau des Nahrungsdotters reifer und befruchteter Fischeier (Hecht). *Müller's Archiv für Anatomie und Physiologie* **1856**, 83-124.

Reichert, K.B. (1856) Ueber die Müller-Wolffschen Körper bei Fischembryonen und über die sogenannten Rotationen des Dotters im befruchteten Hechteie. *Müller's Archiv für Anatomie und Physiologie* **1856**, 125-43.

Reichert, K.B. (1857) Der Nahrungsdotter des Hechteies — eine kontraktile Substanz. *Müller's Archiv für Anatomie und Physiologie* **1857**, 46-51.

Reichert, K.B. (1858) Beobachtungen über die ersten Blutgefässe und deren Bildung, sowie über Bewegung des Blutes in denselben bei Fischembryonen. *Studien des physikalischen Institutes zu Breslau*, Leipzig.

Reighard, J. (1890) The development of the wall-eyed pike, *Stizostedion vitreum* Raf. *Michigan State Board Fish Commission Bulletin* **1**, 95-158

Reinhard, W. (1888) Entwicklung der Keimblätter, der Chorda und des Mitteldarmes bei den Cyprinoiden. *Zoologischer Anzeiger* **11**, 648-655.

Reinhard, W. (1898) Die Bedeutung des Periblastes und der Kupffer'schen Blase in der Entwickelung der Knochenfische. *Archiv für Mikroskopische Anatomie* **52**, 793-820.

Reis, C. (1910) Untersuchungen über die embryonale Entwicklung der Knochenfische. *Bulletin International de l'Académie des Sciences, Cracovie.* **Série B**, 521-554.

Remak, R. (1854) Ueber Eihüllen und Spermatozoen. *Müller's Archiv* p. 252-56. (Hoffmann 1881 p. 22, Zeichnung Tafel II).

Renard, P., Flechon, B., Billard, R. and Christen, R. (1990) Biochemical and morphological changes in the chorion of the carp (*Cyprinus carpio*) oocyte, following the cortical reaction. *Journal of Applied Ichthyology* **6**, 81-90,

Repiachoff, W. (1883) Bemerkungen über die Keimblätter der Wirbelthiere. *Zoologischer Anzeiger* **7**, 148-152.

Retzius, M.G. (1905) Die Spermien der Leptokardier, Teleostier und Ganoiden. *Biologische Untersuchungen N.F., Stockholm* n.s. **12**, 103-15.

Retzius, M.G. (1910) Weitere Beiträge zur Kenntnis der Spermien mit besonderer Berücksichtigung der Kernsubstanz. *Biologische Untersuchungen, Stockholm* n.s. **15**, 63-82.

Retzius, M.G. (1912) Zur Kenntnis der Hüllen und besonders des Follikelepithels an den Eiern der Wirbeltiere. I. Bei den Fischen. *Biologische Untersuchungen, Stockholm* n.s. **17**, 1-29.

Reznick, D., Callahan, H. and Llauredo, R. (1996) Maternal effects on offspring quality in poeciliid fishes. *The American Zoologist* **36**, 147-156.

Rich, R. and Philpott, C. (1969) Repeating particles associated with an electrolyte-transporting membrane. *Experimental Cell Research* **55**, 17-24.

Richard, W.J. and Leis. J.M. (1984) *Labroidei*: development and relationships, in *Ontogeny and Systematics of Fishes* (ed American Society of Ichthyologists and Herpetologists), Allen Press Inc., Lawrence, KS 66044, U.S.A., pp 542-547.

Richardson, L.R. (1939) The spawning behaviour of *Fundulus diaphanus* (Le Sueur). *Copeia* **1939, No. 3**, 165-167.

Richter, J., Lombardi, J. and Wourms, J.P. (1983) Branchial placenta and embryonic growth in the viviparous fish *Jenynsia*. *The American Zoologist* **23**, Abstr. 808, p. 1017.

Ridgway, E.B., Gilkey, J.C. and Jaffe, L.F. (1977) Free calcium increases explosively in activating medaka eggs. *Proceedings of the National Academy of Sciences U.S.A.* **74**, 623-627.

Ridley, M. (1978) Paternal care. *Animal Behaviour* **26**, 904-932.

Ridley, M. and Rechten, C. (1981) Female sticklebacks prefer to spawn with males whose nests contain eggs. *Behaviour* **96**, 152-161.

Rieb, J.P. (1973) La circulation sanguine chez l'embryon de *Brachydanio rerio* (téléostéens, cyprinidae). *Annales d'Embryologie et de Morphogenèse* **6**, 43-54.

Riehl, R. (1976) A special attaching-mechanism between the *cortex radiatus externus* and the follicle epithelium of the oocytes of *Gobio gobio*. *Zeitschrift der Naturforschung* **31C**, 628-629.

Riehl, R. (1977a) Histochemical and ultrahistochemical investigations of the oocytes of *Noemacheilus barbatulus* (L.) and *Gobio gobio* (L.) (Pisces, Teleostei) *Zoologischer Anzeiger Jena* **198**, 328-354.

Riehl, R. (1977b) Parakristalline Körper in jungen Oocyten der Schmerle, *Noemacheilus barbatulus* (L.) (*Teleostei, Cobitidae*), Zeitschrift für Naturforschung **32c**, 305-306.

Riehl, R. (1978a) Die Oocyten der Grundel *Pomatoschistus minutus*. I. Licht- und elektronenmikroskopische Untersuchungen an Eihülle und Follikel. *Helgoländer wissenschaftliche Meeresuntersuchungen* **31**, 314-332.

Riehl, R. (1978b) Elektronenmikroskopische und autoradiographische Untersuchungen an den Dotterkernen in den Oocyten von *Noemacheilus barbatulus* (L.) und *Phoxinus phoxinus* (L.) (Pisces, Teleostei). *Cytobiologie* **17**, 137-145.

Riehl, R. (1978c) Licht- und elektronenmikroskopische Untersuchungen an den Oocyten von *Noemacheilus barbatulus* (L.) und *Gobio gobio* (L.) (Pisces, Teleostei). *Zoologischer Anzeiger* **201**, 199-219.

Riehl, R. (1978d) Bestimmungsschlüssel der wichtigsten deutschen Süsswasser-Teleosteer anhand ihrer Eier. *Archiv für Hydrobiologie* **83**, 200-212.

Riehl, R. (1980a) Ultracytochemical localization of Na+, K+-activated ATPase in the oocytes of *Heterandria formosa* Agassiz, 1853 (Pisces, Poeciliidae) Reprod. Nutr. Development **20A**, 191-196.

Riehl, R. (1980b) Micropyle of some salmonins and coregonins. *Environmental Biology of Fish* **5**, 59-66.

Riehl, R. (1991) Die Struktur der Oocyten und Eihüllen oviparer Knochenfische — eine Uebersicht. *Acta Biologica Benrodis* 3: 27-65.

Riehl, R. (1993) Surface morphology and micropyle as a tool for identifying fish eggs by scanning electron microscopy. *Microscopy and Analysis*, 29-31.

Riehl, R. and Götting, K.J. (1974) Zu Struktur und Vorkommen der Mikropyle an Eizellen und Eiern von Knochenfischen. *Archiv der Hydrobiologie* **74**, 393-402.

Riehl, R. and Götting, K.J. (1975) Bau und Entwicklung der Mikropylen in den Oocyten einiger Süsswasser-Teleosteer. *Zoologischer Anzeiger* **195**, 363-365.

Riehl, R. and Greven, H. (1993) Fine structure of egg envelopes in some viviparous goodeid fishes, with comments on the relation of envelope thinness to viviparity. *Canadian Journal of Zoology* **71**, 91-97.

Riehl, R. and Kock, K.H. (1989) The surface structure of Antarctic fish eggs and its use in identifying fish eggs from the southern ocean. *Polar Biology* **9**, 197-203

Riehl, R. and Kokoschka (1993) A unique surface pattern and micropylar apparatus in the eggs of *Luciocephalus* sp. (Perciformes, Luciocephalidae). *Journal of Fish Biology* **43**, 617-620.

Riehl, R and Patzner, R.A. (1991) Breeding, structure and larval morphology of the catfish *Sturisoma aureum* (Steindachner) (Teleostei, Loricariidae). *Journal of Aquariculture and Aquatic Sciences* **6**, 1-6.

Riehl, R. and Patzner, R.A. (1992) The eggs of native fishes. 3. Pike — *Esox lucius* L., 1758, *Acta biologica Benrodis* **4**, 135-139.

Riehl, R. and Patzner, R.A. (1998) Minireview: The modes of egg attachment in teleost fishes. *Italian Journal of Zoology* **65** Suppl., 415-420.

Riehl, R. and Schulte, E. (1977a) Licht- und elektronenmikroskopische Untersuchungen an den Eihüllen der Elritze (*Phoxinus phoxinus* [L.]; Telcostei, Cyprinidae). *Protoplasma* **92**, 147-162.

Riehl, R. and Schulte, E. (1977b) Vergleichende rasterelektronenmikroskopische Untersuchungen an den Mikropylen ausgewählter Süsswasser-Teleosteer. *Archiv für Fischerei Wissenschaften* **28**, 95-107.

Riehl, R. and Schulte, E. (1978) Bestimmungsschlüssel der wichtigsten deutschen Süsswasser-Teleostier anhand ihrer Eier. *Archiv. Biol.* **83**, 200-212.

Rieneck, D.R. (1869) Ueber die Schichtung des Forellenkeims.*M. Schultze's Archiv für Mikroskopische Anatomie* **5**, 356-366.

Riggert, F. (1922) Beiträge zur Anatomie der Fischleber. *Veterinär-medizinische Dissertation.* Hannover.

Rijnsdorp, A.D. and Vingerhoed, B. (1994) The ecological significance of geographical and seasonal differences in egg size in sole *Solea solea* (L.) *Netherlands Journal of Sea Research* **32**, 255-270.

Robertson, D.A. (1981) Possible functions of surface structure and size in some planktonic eggs of marine fishes. *New Zealand Journal of Marine and Freshwater Research* **15**, 147-153.

Rojo, M.C., Blánquez, M.J. and Gonzáles, M.S. (1997) Ultrastructural evidence for apoptosis of pavement cells, chloride cells, and hatching gland cells in the developing branchial area of the trout *Salmo trutta*. *Journal of Zoology London* **243**, 737-751.

Rolleston, T.W. (1949) *Myths and Legends of the Celtic Race, VI. Tales of the Ossianic Cycle.* George G. Harrap & Co. Ltd., London, pp. 252—308.

Rombough, P.J. (1988) Growth, aerobic metabolism and dissolved oxygen requirements of embryos and alevins of steelhead, *Salmo gairdneri*. *Canadian Journal of Zoology* **66**, 651-660.

Romer, A.S. and Parsons, T.S. (1977) *The Vertebrate Body (Saunders, Philadelphia, London, Toronto)*

Romiti, G. (1873) Studi di Embriogenia. I. Contribuzione allo studio dei foglietti embrionali. II. Sullo sviluppo del canale centrale della midolla spinale. *Rivista Clinica di Bologna* 363-369.

Roosen-Runge, E.C. (1938) On the early development — bipolar differentiation and cleavage — of the zebrafish, *Brachydanio rerio*. *Biological Bulletin* **75**, 119-133.

Rosen E.D. and Bailey, R.M. (1963) The poeciliid fishes (cyprinodontiformes), their structure, zoogeography, and systematics. *Bulletin of the American Museum of Natural History* **126**, 1-176.

Rosenbluth, J. (1962) Subsurface cisterns and their relationship to the neuronal plasma membrane. *Journal of Cell Biology* **13**, 405-421.

Röthlisberger, H. (1921) Entwicklungsgeschichte von *Anableps tetrophthalmus*. PhD Thesis. *Art. Füssli, Zürich.*

Rothschild, L. (1958) Fertilization in fish and lampreys. *Biological Reviews* **33**, 372-392.

Rott, N.N. and Sheveleva, G.A. (1968) Changes in the rate of cell divisions in the course of early development of diploid and haploid loach embryos. *Journal of Embryology and Experimental Morphology* **20**, 141-150.

Roubaud, P. and Pairault, C. (1980) Membrane differentiation in the pregastrula of the teleost, *Brachydanio rerio* Hamilton-Buchanan (Teleostei: Cyprinidae). A scanning electron microscope study. *Réproduction, Nutrition et Développement* **20**, 1515-1526.

Rouse, G.W. and Jamieson, B.G.M. (1987) An ultrastructural study of the spermatozoa of the polychaetes *Eurythoe complanta* (*Amphnionomidae*), *Clymenella* sp. and *Micromaldane* sp. (*Maldanidae*), with definition of sperm types in relation to reproductive biology. *Journal of Submicroscopic Cytology* **19**, 573-584.

Rückert, J. and Mollier, S. (1906) Die erste Entstehung der Gefässe und des Blutes bei Wirbeltieren, Chapter 5 in *Handbuch der vergleichenden und experimentellen Entwicklungslehre der Wirbeltiere,* (ed O. Hertwig), Gustav Fischer, Jena, pp. 1019-1278.

Rusconi, M. (1835) Lettera al Signor Weber sopra la fecondazione artificiale ne' pesci, e sopra le metamorfosi a cui soggiacciono l'uova de' pesci innanzi di prender forma di embrione. *Bibl. Ital.*, **79**, 250-257.

Rusconi, M. (1836) Ueber die Metamorphosen des Eies der Fische vor der Bildung des Embryo. *Archiv der Anatomie (Müller)* pp. 278-290.

Rusconi, M. (1840) Sopra la fecondazione artificiale nei pesci, e sopra alcune nuove sperienze intorno alla fecondazione artificiale nelle rane. (4th letter to Weber). *Müller's Archiv für Anatomie und Physiologie,* pp. 185-193.

Russel, F.S. (1976) The eggs and planktonic stages of British marine fishes. Academic Press, London, New York and San Francisco.

Ryan, B. (1992) The pseudobranch and gill of two teleost species — a structural and ultrastructural study. PhD thesis. The National University of Ireland. 162 pp.

Ryder, J.A. (1856) Embryology. Why do certain fish ova float? *The American Naturalist (Philadelphia)* **20**, 985-987.

Ryder, J.A. (1882a) The micropyle of the egg of the white perch. *Bulletin of the United States Fish Commission* (1881) **1**, 282.

Ryder, J.A. (1882b) Development of the silver gar (*Belone longirostris*), with observations on the genesis of the blood in embryo fishes, and a comparison of fish ova with those of other vertebrates. *Bulletin of the United States Fish Commission* (1881) **1**, 283-301.

Ryder, J.A. (1883a) Observations on the absorption of the yelk, the food, feeding, and development of embryo fishes comprising some investigations conducted at the central hatchery, armory building, Washington, D., in 1882. *Bulletin of the United States Fish Commission* **2**, 179-205.

Ryder, J.A. (1883b) Preliminary notice of the development and breeding habits of the Potomac catfish, *Ameiurus albidus*. *Bulletin of the U.S. Fish Commission* **3**, 225-230.

Ryder, J.A. (1884) A contribution to the embryography of osseus fishes with special reference to the development of the cod (*Gadus morrhua*). *Report U.S. Fish Commission. (Washington)* (1882) **10**, 455-605.

Ryder, J.A. (1885a) On certain features of the development of the salmon. *Proceedings of the U.S. National Museum (Washington)* **8**, 156-162.

Ryder, J.A. (1885b) On the formation of the embryonic axis. *The American Naturalist* **19**, 614-615.

Ryder, J.A. (1885c) On the development of viviparous osseus fishes. *Proceedings of the U.S. National Museum (Washington)* **8**, 128-156.

Ryder, J.A. (1886) The development of *Fundulus heteroclitus*. *The American Naturalist* **20**, 824.

Ryder, J.A. (1887) On the development of osseus fishes, including marine and freshwater forms. *Annual Report of the, Commissioner of Fish and Fisheries* (1885) **13**, 488-604.

Ryder, J.A. (1893) The vascular respiratory mechanism of the vertical fins of the viviparous Embiotocidae. *Proceedings of the Academy of Natural Science Philadelphia.* **1893**, 95-99.

Sakai, Y.T. (1961) Method for removal of chorion and fertilization of the naked egg in *Oryzias latipes*. *Embryologia* **5**, 357-568.

Sakai, Y.T. (1964a) Studies on the ooplasmic segregation in the egg of the fish *Oryzias latipes*. *Embryologia* **8**, 129-134.

Sakai, Y.T. (1964b) Studies on the ooplasmic segregation in the egg of the fish *Oryzias latipes*. II Ooplasmic segregation of the partially activated egg. *Embryologia* **8**, 135-145.

Sakai, Y.T. (1976) Spermiogenesis of the teleost, *Oryzias latipes*, with special reference to the formation of flagellar membrane. *Development, Growth and Differentiation* **18**, 1-13.

Salvatorelli, G. (1971) Osservazione sopra l'ematopoiesi in *Anguilla anguilla*. *Accademia delle Scienze di Ferrara* **48**, 59-68.

Samassa, P. (1896) Studien über den Einfluss des Dotters auf die Gastrulation und die Bildung der primären Keimblätter bei Wirbelthieren. *Verhandlungen der Deutschen Zoologischen Gesellschaft* **5**, 130-142.

Sander, K., Dollmetsch, K. and Vollmar, H. (1984) Zebrafish epiboly: wheeling movement of deep cells at the blastodisc rim. *Journal of Embryology and Experimental Morphology* **82**, 214.

Sandstroem, A. (1999) Visual ecology of fish — a review with special reference to percids. *Fiskeriverk. Rapp* **No. 2**, 45-80.

Sandy, J.F., Blaxter, J.H.S. (1980) A study of retinal development in larval herring and sole. *Journal of the Marine Biological Association U.K.* **60**, 59-71.

Sanzo, L. (1933) Macruridae. Fauna Flora Golfo Napoli; Monogr. 38. Uova, larva e stadi giovanili di teleostei. (ed S. Lo Bianco), Stazione Zoologica di Napoli. Pt **2**, 255-265.

Sargent, R.C. and Gross, M.R. (1986) Williams' principle: an explanation of parental care in teleost fishes, in *The Behaviour of Teleost Fishes* (ed T.J. Pitcher), Croom Helm, London & Sydney, pp. 275-293.

Sars, G.O. (1876) On the spawning and development of the codfish. *Report of the U.S. Fisheries Commission for 1873-75.* **3**, 213-222.

Satchell, BG.H. (1971) Circulation in Fishes. Chapter 5 in '*The Blood*', Cambridge University Press, London. 52-61

Scapigliati, G., Carcupino, M., Taddei, A.R. and Mazzini, M. (1994) Characterization of the main egg envelope proteins of the sea bass *Dicentrarchus labrax L (Teleostea, Serranidae)*. *Molecular Reproduction and Development* **38**, 48-53.

Schalkoff, E. and N.H. Hart (1986) Effects of A 23187 upon cortical granule exocytosis in eggs of *Brachydanio*. *Roux's Archives of Developmental Biology* **195**, 39-48.

Schantz, A.R. (1985) Cytosolic free calcium-ion concentration in cleaving embryonic cells of *Oryzias latipes* measured with calcium-selective microelectrodes. *Journal of Cell Biology* **100**, 947-954.

Schapringer, A. (1871) Ueber die Bildung des Medullarrohrs bei den Knochenfischen. *Sitzungsberichte der K.K. Akademie der Wissenschaften zu Wien, math. Naturw. Classe* **64**, Part III, 653-658.

Scharrer, E. (1928) Die Lichtempfindlichkeit blinder Elritzen (Untersuchungen über das Zwischenhirn der Fische). *Zeitschrift für vergleichende Physiologie* **7**, 1-38.

Scheuermann, H. (1979) The genus *Tropheus*, Part I. *Buntbarsche Bulletin* **70**, 5-14.

Scheuring, L. 1924) Biologische und physiologische Untersuchungen an Forellensperma. *Archiv für Hydrobiologie*, Suppl. B. 4, L. **2**, 181-318.

Schindler, J.F. and de Vries, U. (1989) Polarized distribution of binding sites for concavalin A and wheat-germ agglutinin in the zona pellucida of goodeid oocytes (teleostei). *Histochemistry* **91**, 413-417.

Schindler, J.F. and de Vries U. (1990) Effects of ammonia, chloroquine, and monensin on the vacuolar apparatus of an absorbtive epithelium. *Cell and Tissue Research* **259**, 283-292.

Schindler, J.F., and Hamlett, W.C. (1993) Maternal-embryonic relations in viviparous teleosts. *Journal of Experimental Zoology* **266**, 378-393.

Schlenk,W. and Kahmann, H. (1935) Ein Verfahren zur Messung der Spermatozoenbewegung. *Pflüger's Archiv der gesamten Physiologie* **236**, 398-404.

Schlenk, W. and Kahmann, H. (1938) Die chemische Zusammensetzung des Spermaliquors und ihre physiologische Bedeutung. *Biochemische Zeitschrift* **295**, 283-301.

Schmatolla, E. and Erdmann, G. (1973) Influence of retino-tectal innervation on cell proliferation and cell migration in the embryonic teleost tectum. *Journal of Embryology and Experimental Morphology* **26**, 697-712.

Schmitt, E.A. (1987) An investigation of the structural and ultrastructural development and of the effect of constant light and constant darkness in the post-embryonic retina of the rainbow trout, *Salmo gairdneri*. Ph.D. Thesis, National University of Ireland.

Schmitt, F. (1902) Ueber die Gastrulation der Doppelbildungen der Forelle, mit besonderer Berücksichtigung der Concrescenztheorie. *Verhandlungen der Deutschen Zoologischen Gesellschaft* **12**, 64-83.

Schmitt, E.A. and Dowling, J.E. (1994) Early eye morphogenesis in the zebrafish, *Brachydanio rerio*. *Journal of Comparative Neurology* **344**, 532-542.

Schmitt, E.A. and Dowling, J.E. (1996) Comparison of topographical patterns of ganglion and photoreceptor cell differentiation in the retina of the zebrafish, *Danio rerio*. *Journal of Comparative Neurology* **371**, 222-234.

Schmitt, E.A. and Dowling, J.E. (1999) Early retinal development in the zebrafish, *Danio rerio*: A light and electron microscopical analysis. *Journal of Comparative Neurology* **404**, 515-536.

Schmitt, E. and Kunz, Y.W. (1989) Retinal morphogenesis in the rainbow trout, *Salmo gairdneri*. *Brain, Behavior and Evolution* **34**, 48-64.

Schmitz, B., Papan, C. and Campos-Ortega, J.A. (1993) Neurulation in the anterior trunk region of the zebrafish *Brachydanio rerio*. *Roux's Archive of Developmental Biology* **202**, 250-259.

Schoenwolf, G.C. and Alvarez, I.S. (1992) 4. Role of cell rearrangement in axial morphogenesis, in *Current Topics in Developmental Biology*, Vol. **27**, (ed R.A. Pedersen), Academic Press Inc., pp. 129-173.

Scholes, J.H. (1975) Color receptors and their synaptic connections in the retina of a cyprinid fish. *Phil. Transactions of the Royal Society* **270**, 61-118.

Schonevelde, S.A. (1624) Uitvoerige en natuurkundige Beschrijving der Visschen, volgens het zamenstel van C. Linnaeus. **2**, p. 257 Or

Schonevelde, S.A. (1624) Ichthyologia et nomenclature animalium marinorum, fluviatilium, lacustrium quae in florentissimis ducatibus Slesvici et Holsatiae et Emporio Hamburgo occurrunt triviales; ac plerumque hactenus desideratorum imagines, breves descriptiones et explicationes. Hamburgi, 85 p.

Schoots, A.F.M. (1982) Enzymatic hatching of fish embryos. PhD Thesis, Katholieke Universiteit, Nijmegen. Cited by 'Luczynski M. *et al.* (1993)'.

Schoots, A.F.M. and Denucé, JM. (1981) Purification and characterization of hatching enzyme of the pike (*Esox lucius*). *International Journal of Biochemistry* **13**, 591-602.

Schoots, A.F.M., Stikkelbroeck, J.M., Bekhuis J.F. and Denucé J.M. (1982a) Hatching in teleostean fishes: Fine structural changes in the egg envelope during enzymatic breakdown *in vivo* and *in vitro*. *Journal of Ultrastructural Research* **80**, 185-196.

Schoots, A.F.M., Opstelten, R.J.G. and Denucé, J.M. (1982b) Hatching in the pike *Esox lucius* L.: Evidence for a single hatching enzyme and its immunocytochemical localization in specialized hatching gland cells. *Developmental Biology* **89**, 48-55.

Schoots, A.F.M., De Bont, R.G., Van Eys, G.J.J.M. and Denucé (1982c) Evidence for a stimulating effect of prolactin on teleostean hatching enzyme secretion. *Journal of Experimental Zoology* **219**, 129-132.

Schoots, A.F.M., Sackers, P.S.G. and Denucé J.M. (1983) Hatching in the teleost *Oryzias latipes*. Limited proteolysis causes egg envelope swelling. *Journal of Experimental Zoology* **226**, 93-100.

Schroder, M.B., Villena, A.J. and Jorgensen, T.O. (1998) Ontogeny of lymphoid organs and immunoglobulin producing cells in Atlantic cod (*Gadus morhua* L.). *Developmental and Comparative Immunology* **22**, 507-517.

Schultze, M. (1836) Das System der Cirkulation in seiner Entwickelung durch die Tierreihe und im Menschen. *Cotta, Stuttgart and Tübingen.*

Schulze, F.E. (1861) Ueber die Nervenendigung in den sogenannten Schleimcanälen der Fische und über entsprechende Organe der durch Kiemen athmenden Amphibien. *Archiv für Anatomie und Physiologie Med. (Reichert)* **1861**, 759-769.

Schwarz, D. (1889) Untersuchungen des Schwanzendes bei den Embryonen der Wirbeltiere. Nach Beobachtungen an Selachiern, Knochenfischen und Vögeln vergleichend dargestellt. *Zeitschrift für wissenschaftliche Zoologie* **49**, 191-223.

Scott, M.I.H. (1928) [cf. 'Schindler and Hamlett (1993)'].

Scott, W.B. and Crossman, E.J. (1973) Freshwater fishes of Canada. *Bulletin of the Fisheries Research Board of Canada* **184**, 966 pp.

Scrimshaw, N.S. (1944) Embryonic growth on the viviparous fish *Heterandria formosa*. *Biological Bulletin* **87**, 37-51.

Scrimshaw, N.S. (1945) Embryonic development in poeciliid fishes. *Biological Bulletin* **88**, 233-246.

Scuigna, C., Fluck, R. and Barber, B. (1988) Calcium dependence of rhythmic contractions of the *Oryzias latipes* blastoderm. *Comparative Biochemistry and Physiology* **89C**, 369-374.

Seba, A. (1758) *Thesaurus rerum naturalium*. 4 vols. In Latin and French.

Seelbach, V. and Kunz, Y.W. (1976) Haematopoiesis in the extended blood-mesoderm of the guppy *Poecilia reticulata*. Dedicated to Professor Adolf Portmann on the occasion of his 80th birthday of (unpublished).

Seelely, H.G. (1886) The freshwater fishes of Europe; a history of their genera, species, structure, habits and distribution. London. 444 pp.

Segerstrahle, E. (1910) Kenntnis der Teleostierleber. *Anatomische Hefte* **41**, 1910.

Selleck, M.A.J., Scherson, T.Y. and Bronner-Fraser, M. (1993) Review. Origins of neural crest diversity (1993). *Developmental Biology* **159**, 1-11.

Selman, K., Wallace, R.A. and Barr, V. (1986) Oogenesis in *Fundulus heroclitus*. IV. Yolk-vesicle formation. *The Journal of Experimental Zoology* **239**, 277-288.

Selman, K., Wallace, R.A. and Barr, V. (1988) Oogenesis in *Fundulus heroclitus*. V. The relationship of yolk vesicles and cortical alveoli. *Journal of Experimental Zoology* **246**, 42-56.

Shackley, S.E. and King, P.E. (1977) Oogenesis in a marine teleost, *Blennius pholis* L.. *Cell and Tissue Research* **181**, 105-128.

Shackley, S.E. and King, P.E. (1978) Protein synthesis in *Blennius pholis*. *Journal of Fish Biology* **13**, 179-193.

Shackley, S.E. and King, P.E. (1979a) Amino acid incorporation by isolated oocytes of the marine teleost *Blennius pholis* L. *Journal of Fish Biology* **14**, 375-380.

Shackley, S.E. and King, P.E. (1979b) Lipid yolk synthesis in the marine teleost, *Blennius pholis* L. *Cell and Tissue Research* **204**, 507-512.

Shackley, S.E., King, P.E. and Gordon, S.M. (1981) Vitellogenesis and trace metals in a marine teleost. *Journal of Fish Biology* **18**, 349-352.

Shanklin, D.R. and Armstrong, P.B. (1952) The osmotic behavior and anatomy of the *Fundulus* chorion. *Biological Bulletin* **103**, 295.

Sharma, S.C. and Ungar, F. (1980) Histogenesis of the goldfish retina. *Journal of Comparative Neurology* **191**, 373-382.

Shaw, E.S. (1955) The embryology of the seargent major, *Abdudefduf saxatilis*. *Copeia* **1955 no. 2**, 85-89.

Shelbourne, J.E. (1857) Site of chloride regulation in marine fish larvae. *Nature* **180**, 920-922.

Shelton, W.L. (1978) Fate of the follicular epithelium in *Dorosoma petenense* (Pisces: Clupeidae) *Copeia* **1978**. 237-244.

Shen, A.C.Y., and Leatherland, J.F. (1978) Structure of the yolksac epithelium and gills in the early developmental stages of rainbow trout (*Salmo gairdneri*) maintained in different ambient salinities. *Environmental Biology of Fishes* **3**, 345-354.

Shimizu M and Yamada, J (1980) Ultrastructural aspects of yolk absorption in the vitelline syncytium of the embryonic rockfish, *Sebastes schlegeli*. *Journal of Ichthyology*. **27**, 56-63.

Shiraishi, K., Kaneko, T., Hasegawa, S. and Hirano, T. (1997) Development of multicellular complexes of chloride cells in the yolksac membrane of tilapia (*Oreochromis mossambicus*) embryos and larvae in seawater. *Cell Tissue Research* **288**, 583-590.

Siewing, R. (1969) *Lehrbuch der vergleichenden Entwicklungsgeschichte der Tiere*, Parey, Berlin, Hamburg. 531 pp.

Sire, M-F., Babin, P-J. and Vernier, J-M. (1994) Involvement of the lysosomal system in yolk protein deposit and degradation during vitellogenesis and embryonic development in trout. *Journal of Experimental Zoology* **269**, 69-83.

Siwe, S.A. (1929) On the earliest liver anlagen in vertebrates and on the origin and development of the gallbladder. *Archive de Biologie* **39**, 479-510.

Small, J.V., Stradal, T., Vignal, E. and Rottner, K. 2002) The lamellipodium: where motility begins. Review. *Trends in Cell Biology* **12**, 112-120.

Smallwood, W.M. and Derrickson, M.B. (1933) The development of the carp, *Cyprinus carpio*. II. The development of the liver-pancreas, the islands of Langerhans and the spleen. *Journal of Morphology* **55**, 15-26.

Smith, J.L.B, (1950) The seafishes of Southern Africa. *Central News Agency Ltd. South Africa*. Second impression. **16**, 550 pp.

Smith, S. (1958) Yolk utilization in fishes, in *Embryonic Nutrition* (ed Dorothea Rudnick) the Developmental Biology Conference Series, 1956, Chicago: The University of Chicago Press. pp. 33-107.

Smith, M.S., Potter, M. and Bruce-Merchant, E. (1967) Antibody-forming cells in the pronephros of the teleost *Lepomis macrochirus*. *The Journal of Immunology* **99**, 876-882.

Smith, M.S., Wivel, N.S. and Potter, M. (1970) Plasmacytopoiesis in the pronephros of the carp (*Cyprinus carpio*). *The Anatomical Record* **167**, 351-370.

Sobotta, J. (1894) Ueber Mesoderm, Herz, Gefäss- und Blutbildung bei Salmoniden. *Anatomischer Anzeiger* **9**, 77-84.

Sobotta, J. (1896) Zur Entwicklung von *Belone acus*. *Anatomischer Anzeiger (Verhandlungen*

der Anatomischen Gesellschaft) **12**, 108-111.
Sobotta, J. (1898) Die morphologische Bedeutung der Kupffer'schen Blase. Ein Beitrag zur Gastrulation der Teleostier. *Verhandlungen der Phys. Medizinischen Gesellschaft zu Würzburg* **32**, 1-17.
Sobotta, J. (1902) Ueber die Entwickelung des Blutes, des Herzens und der grossen Gefässtämme der Salmoniden. *Anatomische Hefte* **19**, Part l, 579-688.
Soin, S.G. (1968) Some features in the development of the blenny [*Zoarces viviparus* (L.)] in relation to viviparity. *Probl. Ichthyol.* **8**, 222-229.
Soin, S.G. (1971) *Adaptional features in fish ontogeny*, 77 pp. Translated from Russian, (U.S. Department of Commerce and the National Science Foundation, Washington D.C).
Soin, S.G. (1981) A new classification of mature eggs of fishes according to the ratio of yolk to ooplasm. *Soviet Journal of Developmental Biology* **12**, 13-17.
Solnica-Krezel, L. and Driever, A. (1994) Microtubule arrays of the zebrafish yolk cell: organization and function during epiboly. *Development* **120**, 2443-2445.
Specker, J. (1988) Preadaptive role of thyroid hormones in larval and juvenile salmon. Growth, the gut and evolutionary considerations. *The American Zoologist* **28**, 61-118.
Spek, J. (1933) Die bipolare Differenzierung des Protoplasmas des Teleosteer-Eies und ihre Entstehung. *Protoplasma* **18**, 497-545.
Srivastava, S.J. and Srivastava, S.K. (1994) Seasonal changes in liver and serum proteins, serum calcium, inorganic phosphate and magnesium levels in relation to vitellogenesis in a freshwater catfish, Heteropneustes fossilis (Bloch). *Annales d'Endocrinologie (Paris)* **55**, 197-202.
Stahl, A. and Leray, C. (1961) L'ovogenèse chez les poissons téléostéens, I. Origine et signification de la *zona radiata* et de ses annexes. *Archives d'Anatomie Microscopique* **50**, 251-268.
Stainier. D.Y.R., Lee, R.K. and Fishman, M.C. (1993) Cardiovascular development in the zebrafish. I. Myocardial fate map and heart tube formation. *Development* **119**, 31-40.
Starck, D. (1975) Embryologie. *Ein Lehrbuch auf allgemein biologischer Grundlage*, 3rd edition, Thieme, Stuttgart.
Stehr, S.M. (1982) The development of the hexagonally structured egg envelope of the C-O sole (*Pleuronichthys coenosus*). M.S. thesis. University of Washington. 58 pp.
Stehr, C.M. and Hawkes, J.W. (1979) The comparative ultrastructure of the egg membrane and associated pore structures in the starry flounder, *Platichthys stellatus* (Pallas), and pink salmon, *Oncorhynchus gorbuscha* (Walbaum). *Cell and Tissue Research* **20**, 347-356.
Stehr, C.M. and Hawkes, J.W. (1983) The development of the hexagonally structured egg envelope of the C-O sole (*Pleuronichthys coenosus*). *Journal of Morphology* **178**, 267-284.
Stein, H. (1981) Licht- und elektronenoptische Untersuchungen an den Spermatozoen verschiedener Süsswasserknochenfische (*Teleostei*). *Zeitschrift für angewandte Zoologie* **68**, 183-198.
Sterba, G. (1958) Die Eihüllen des Schmerleneies *(Noemacheilus barbatula)*. *Zeitschrift für mikroskopisch-anatomische Forschung* **63**, 581-588.
Sterba, G. and Müller, H. (1962) Elektronenmikroskopische Untersuchungen über Bildung und Struktur der Eihüllen bei Knochenfischen, I. Die Hüllen junger Oozyten von *Cynolebias belotti* Steindachner (Cyprinidontidae). *Zoologisches Jahrbuch der Anatomie*

80, 65-80.

Stockard, C.R. (1915a) An experimental study of the origin of blood and vascular endothelium in the teleost embryo. *Anatomical Record* **9**, 124-127.

Stockard, C.R. (1915b) The origin of blood and vascular endothelium in embryos without a circulation of the blood and in the normal embryo. *American Journal of Anatomy* **18**, 227-327.

Stockard, C.R. (1915c) A study of wandering mesenchymal cells on the living yolksac and their developmental products: Chromatophores, vascular endothelium and blood cells. *The American Journal of Anatomy* **18**, 525-594.

Stockley, P., Gage, M.J.G., Parker, G.A. and Møller, A.P. (1996) Female reproductive biology and the coevolution of ejaculate characteristics in fish. *Proceedings of the Royal Societey of London, B.* **263**, 451-458.

Stoeckli, P. (1957) Die Geburt eines Guppy. *Leben und Umwelt* **14**, 49-51.

Stolz-Picchio (1933) Le nostre conoscenze sulla ematogenesi embrionale e larvale nei pesci. *Rendiconti 1st Lombardo* **66**, 575-589.

Strähle, U. and Blader, P. (1994) Early neurogenesis in the zebrafish embryo. *The Faseb Journal* **8**, 692-698.

Strähle, U. and Jesuthasan, S. (1993) Ultraviolet irradiation impairs epiboly in zebrafish embryos: evidence for a microtubule-dependent mechanism of epiboly. *Development* **119**, 909-919.

Strehlow D. and Gilbert, W. (1993) A fate map for the first cleavages of the zebrafish. *Nature* **361**, 451-453.

Stricker, S. (1865) Untersuchungen über die Entwickelung der Bachforelle. *Sitzungsberichte der k. Akademie der Wissenschaften mathematisch naturwissenschaftlicher Classe (Wien)* **51**, part 2, 546-554.

Stuhlmann, F. (1887) Zur Kenntnis des Ovariums der Aalmutter (*Zoarces viviparus* Cuv.). *Abhandlungen des Naturwissenschaftlichen Vereins in Hamburg* **10**, 1-48.

Suarez, S.S. (1975) The reproductive biology of *Olgilbia cayorum*, a viviparous brotulid fish. *Bulletin of Marine Science* **25**, 143-173.

Sumner, F.B. (1900a) Kupffer's vesicle and its relation to gastrulation and concrescence. *Memoirs of the New York Academy of Sciences.* **2**, **part 2**, 47-84.

Sumner, F.B. (1900b) The teleost gastrula and its modifications. *Science n.s.* **11**,169.

Sumner, F.B. (1904) A study of early fish development. *Archiv für Entwicklungsmechanik der Organismen* **17**, 92-149.

Suzuki, R. (1958) Sperm activation and aggregation during fertilization in some fishes. I. Behavior of spermatozoa around the micropyle. *Embryologia* **4**, 93-102.

Suzuki, R. (1961) Sperm activation and aggregation during fertilization in some fishes. VII. Separation of the sperm-stimulating factor and its chemical nature. *Japanese Journal of Zoology* **13**, 79-100.

Swaen, A. and Brachet, A. (1899) Etude sur les premières phases du développement des organes dérivés du mésoblaste chez les poissons téléosteens. *Archive de Biologie Liège et Paris* **16**, 173-311.

Swaen, A. and Brachet, A. (1901) Etude sur les premières phases du développement des organes dérivés du mésoblaste chez les poissons Téléostéens. *Archive de Biologie, Liège et*

Paris **18**, 73-198.
Swaen, A. and Brachet, A. (1904) Etude sur la formation des feuillets et des organes dans le bourgeon terminal et dans la queue des embryons des poissons téléostéens. *Archive de Biologie, Liège et Paris* **20**, 461-610.
Swarup, H. (1958) Stages in the development of the stickleback *Gasterosteus aculeatus* (L.). *Journal of Embryology and experimental Morphology* **6**, 373-383.
Szöllösi, D. and Billard, R. (1974) The micropyle of trout eggs and its reaction to different incubation media. *Journal de Microscopie* **21**, 55-62.
Takano, K. (1964) On the egg formation and the follicular changes in *Lebistes reticulatus*. *Bulletin of the Faculty of Fisheries Hakkaido University* **15**, 147-156.
Takano, M. and Onitake, K. (1989) On the mode of sperm entry into the micropylar canal in the medaka *Oryzias latipes* (abstract). *Zoological Science* **6**, 1169.
Takano, K. and Ohta, H. (1982) Ultrastructure of micropylar cells in the ovarian follicles of the pond smelt, *Hypomesus transpacificus nipponensis*. *Bulletin of the Faculty of Fisheries Hokkaido University* **33**, 65-78.
Takeda, H. and Miyagawa, T. (1994) Commitment of cell fate in embryo of the zebrafish, *Danio rerio*. *The Fish Biology Journal MEDAKA* **6**, 33-45.
Tamura, T. and Hanyu, I. (1980) Pineal photosensitivity in fishes, in *Environmental Physiology of Fishes* (ed Ali, A.) Plenum Press, New York and London, pp. 477-566.
Tavolga, W.N. (1949) Embryonic development of the platyfish (*Platypoecilus*) The swordtail (*Xiphophorus*) and their hybrids. *Bulletin of the American Museum of Natural History* **94**, 161-229.
Tavolga, W.N. and Rugh, R. (1947) Development of Platyfish, *Platypoecilus maculatus*. *Zoologica (New York)* **32**, 7-12.
Tchou, S. and C. Chen (1936) Fertilization in goldfish. *Contributions of the Institute of Zoology, National Academy Peiping* **3**, 35-55.
Tesoriero, J.V. (1977a) Formation of the chorion (*zona pellucida*) in the teleost, *Oryzias latipes*. I. Morphology of early oogenesis. *Journal of Ultrastructura Research* **59**, 282-291.
Tesoriero, J.V. (1977b) Formation of the chorion (*zona pellucida*) II. Polysaccharide cytochemistry of early oogenesis. II. Polysaccharide cytochemistry of early oogenesis. *Journal of Histochemistry and Cytochemistry* **25**, 1376-1380.
Tesoriero, J.V. (1978) Formation of the chorion (*zona* pellucida) in the teleost, *Oryzias latipes*. III. Autoradiography of [^{3}H]proline incorporation. *Journal of Ultrastructura Research* **64**, 315-326.
Tesoriero, J.V. (1980) The distribution and fate of ^{3}H-Glucose and ^{3}H-Galactose in oocytes of *Oryzias latipes*. *Cell and Tissue Research* **209**, 117-129.
Thiaw, O.T. and Mattei, X. (1996) Ultrastructure of the secondary egg envelope of cyprinodontidae of the genus *Epiplatys* Gill, 1862 (Pisces, Teleostei). *Acta Zoologica (Stockholm)* **77**, 161-166.
Thibault, R.E. and Schultz, R.J. (1978) Reproductive adaptations among viviparous fishes (Cyprinodontiformes: *Poeciliidae*). *Evolution* **32**, 320-333.
Thomas, R.J. (1968a) Yolk distribution and utilization during early development of a teleost embryo (*Brachydanio rerio*). *Journal of Embryology and Experimental Morphology* **19**, 203-215.

Thomas, R.J. (1968b) Cytokinesis during early development of a teleost embryo: Brachydanio rerio. *Journal of Ultrastructural Research* **24**, 232-238.

Thomas R.G. and Waterman, R.E. (1978) Gastrulation in the teleost, Brachydanio rerio. *Scanning Electron Microscopy* **17**, 531-540.

Thomopoulos, A. (1953a) Sur l'oeuf de *Perca fluviatilis* L.. *Bulletin de la Société de Zoologie de la France* **78**, 106-114.

Thomopoulos, A. (1953b) Sur l'oeuf de l'épinoche (*Gasterosteus aculeatus*). *Bulletin de la Société de Zoologie de la France* **78**, 142-158.

Thomopoulos, A. (1954) Sur l'oeuf de l'équille (*Ammodytes tobianus*). *Bulletin de la Société de Zoologie de la France* **79**, 112-118.

Thomson, A. (1855) Ovum. *The Cyclopaedia of Anatomy and Physiology* (eds Robert B. Todd) **5**, 81-142.

Timmermans, L.P.M. (1987) Early development and differentiation in fish. *Sarsia* **72**, 331-339.

Topczewski, J. and Solnica-Krezel (1999) Cytoskeletal dynamics of the zebrafish embryo, in *Methods in Cell Biology*, (eds J.W. Detrich, M. Westfield and L.I. Zon), Academic Press New York, pp 205-226.

Toshimori, K. and Ysuzumi, F. (!979) Gap junctions between microvilli of an oocyte and follicle cells in the teleost (*Plecoglossus altivelis*). *Zeitschrift für mikroskopisch-anatomische Forschung* **93**, 458-464.

Trexler, J.C. (1985) Variation in the degree of viviparity in the sailfin molly, *Poecilia latipinna*. *Copeia* **1985**, 999-1004.

Trinkaus, J.P. (1951) A study of the mechanism of epiboly in the egg of *Fundulus heteroclitus*. *Journal of Experimental Zoology* **118**, 269-319.

Trinkaus, J.P. (1963) The cellular basis of *Fundulus* epiboly. Adhesivity of blastula and gastrula cells in culture. *Developmental Biology* **7**, 513-532.

Trinkaus, J.P. (1965) Mechanism of morphogenetic movements, in *Organogenesis*, (eds R.L. DeHaan and H. Ursprung), pp. 55-101.

Trinkaus, J.P. (1966) Morphogenetic cell movements, in *Major Problems in Developmental Biology*, (ed M. Locke), Academic Press, New York and London, pp. 125-176.

Trinkaus, J.P. (1973a) Modes of cell locomotion *in vivo*, in *Locomotion of Tissue Cells* (Ciba Foundation Symposium 14 (new series), Elsevier, Excerpta Medica, North-Holland, Amsterdam. pp. 233-49.

Trinkaus, J.P (1973b) Surface activity and locomotion of *Fundulus* deep cells during blastula and gastrula stages. *Developmental Biology* **30**, 68-103.

Trinkaus, J.P. (1976) in *The Cell Surface in Animal Embryogenesis and Development*, (eds G. Poste and G.L. Nicholson), North-Holland, Amsterdam, New York, Oxford, pp. 233-303.

Trinkaus, J.P. (1980) Formation of protrusions of the cell surface during tissue cell movement, in *Tumor Cell Surfaces and Malignancy*, (eds R.O. Hynes and C.F. Fox), Alan R. Liss, New York, pp. 887-906.

Trinkaus, J.P. (1984) Mechanism of *Fundulus* epiboly — A current view. *The American Zoologist* **24**, 673-688.

Trinkaus, J.P. (1990) Some contributions of research on early teleost embryogenesis to general problems of development, in *Experimental Embryology in Aquatic Plants and*

Animals (ed H.-J. Marty), Plenum Press, New York, pp. 315-327.

Trinkaus, J.P. (1992) The midblastula transition, the YSL transition and the onset of gastrulation in *Fundulus. Development* **1992, suppl.** 75-80.

Trinkaus, J.P. (1993) The yolk syncytial layer of *Fundulus*: Its origin and history and its significance for early embryogenesis. *Journal of Experimental Zoology* **265**, 258-284.

Trinkaus, J.P. (1998) Gradient in convergent cell movement during *Fundulus* gastrulation. *Journal of Experimental Zoology* **281**, 328-335.

Trinkaus, J.P and Erickson, C.A. (1981) Locomotion of *Fundulus* deep cells during gastrulation. *The American Zoologist* **21**, 401-411.

Trinkaus, J.P. and Erickson, C.A. (1983) Protrusive activity. Mode and rate of locomotion, and pattern of adhesion in *Fundulus* deep cells during gastrulation. *Journal of Experimental Zoology* **228**, 41-70.

Trinkaus, J.P. and Lentz, T.L. (1967) Surface specializations of *Fundulus* cells and their relation to cell movements during gastrulation. *The Journal of Cell Biology* **32**, 139-155.

Trinkaus, J.P., Trinkaus, M. and Fink, R.D. (1991) An *in vivo* analysis of convergent cell movements in the germ ring of *Fundulus,* in *Gastrulation: Movements, Patterns and Molecules,* (eds R. Keller, W. Clark Jr and F. Griffin), Plenum Press, New York, pp 121-134.

Trinkaus, J.P., Trinkaus, M. and Fink, R.D. (1992) On the convergent cell movements of gastrulation in *Fundulus. Journal of Experimental Zoology* **261**, 40-61.

Truman E.B. (1869) Observations on the development of the ovum of the pike. *Monthly Microscopical Journal* **2**, 185-204.

Tsukahara, J. (1971) Ultrastructural study on the attaching filaments and villi of the oocyte of *Oryzias latipes* during oogenesis. *Development, Growth and Differentiation* **13**, 173-180.

Tsukamoto, Y. and Kimura, S. (1993) Development of laboratory-reared eggs, larvae and juveniles of the atherinid fish, *Hypoatherina tsurugae,* and comparison with related species. *Japanese Journal of Ichthyology* **40**, 261-267.

Tsukita, S., Furuse, M. and Itoh, M. (1996) Molecular dissection of tight junctions. *Cell Structure and Function* **21**, 381-385.

Turner, C.L. (1933) Viviparity superimposed upon ovoviviparity in the *goodeidae*, a family of cyprinodont teleost fishes of the Mexican plateau. *Journal of Morphology* **55**, 207-251.

Turner, C.L. (1936) The absorptive processes in the embryos of *Parabrotula dentiens*, a viviparous deep-sea brotulid fish. *Journal of Morphology* **59**, 313-324.

Turner, C.L. (1937) The *trophotaeniae* of the *goodeidae*, a family of viviparous cyprinodont fishes. *Journal of Morphology* **61**, 495-523.

Turner, C.L. (1938) Adaptations for viviparity in embryos and ovary of *Anableps anableps. Journal of Morphology* **12**, 323-349.

Turner, C.L. (1940a) Pseudoamnion, pseudochorion, and follicular pseudoplacenta in poeciliid fishes. *Journal of Morphology* **67**, 59-79.

Turner, C.L. (1940b) Adaptations for viviparity in jensynsiid fishes. *Journal of Morphology* **67**, 291-296.

Turner, C.L. (1940c) Follicular pseudoplacenta and gut modifications in anablepid fishes. *Journal of Morphology* **67**, 91-105.

Turner, C.L. (1940d) Pericardial sac, trophotaeniae, and alimentary tract in embryos of goodeid fishes. *Journal of Morphology* **67**, 271-283.

Turner, C.L. (1946) Male secondary sexual characteristics in *Dinematichthysm iluocoeeteoides*. *Copeia* **1946**, 44-96.

Turner, C.L. (1947) Viviparity in teleost fishes. *Science Monthly* **65**, 508-518.

Turner, C.L. (1952) An accessory respiratory device in embryos of the embiotocid fish, *Cymatogaster aggregata* during gestation. *Copeia* **1952**, 146-147.

Twelves, E.L. and Bachop, W.E. (1979) Giant nuclei in the yolk sac syncytium of the antarctic teleost *Notothenia coriiceps neglecta*. *Copeia* **4**, 909-911.

Tyler, Ch. (1993) Electrophoretic patterns of yolk proteins throughout ovarian development and their relationship to vitellogenin in the rainbow trout, *Oncorhynchus mykiss*. *Comparative Biochemistry and Physiology* **106B**, 321-329.

Tyler, Ch.P., Sumpter, J.P. and Bromage, N.R. (1988a) *In vivo* ovarian uptake and processing of vitellogenin in the rainbow trout, *Salmo gairdneri*. *Journal of Experimental Zoology* **246**, 171-179.

Tyler, Ch.P., Sumpter, J.P. and Bromage, N.R. (1988b) Selectivity of protein sequestration by vitellogenic oocytes of the rainbow trout, *Salmo gairdneri*. *The Journal of Experimental Zoology* **248**, 199-206.

Tyler, Ch.P., Sumpter, J.P. and Campbell, P.M. (1991) Uptake of vitellogenin into oocytes during early vitellogenic development in the rainbow trout, *Oncorhynchus mykiss* (Walbaum). *Journal of Fish Biology* **38**, 681-689.

Tytler, P. and Blaxter, J.H.S. (1988a) Drinking in yolksac larvae of the halibut, *Hippoglossus hippoglossus* (L.). *Journal of Fish Biology* **32**, 493-494.

Tytler, P. and Blaxter, J.H.S. (1988b) The effects of external salinity on the drinking rates of the larvae of herring, plaice and cod. *Journal of Experimental Biology* **138**, 1-15.

Ulrich, E. (1969) Etude des ultrastructures au cours de l'ovogenèse d'un poisson téléosteen, le Danio, *Brachydanio rerio* (Hamilton-Buchanan). *Journal de Microscopie* **8**, 447-478.

Utsugi, K. (1993) Motility and morphology of sperm of the ayu, *Plecoglossus altivelis*, at different salinities. *Japanese Journal of Ichthyology* **40**, 273-278.

Van Bambeke, Ch. (1872) Premiers effets de la fécondation sur les oeufs de poissons: sur l'origine et la signification du feuillet muqueux ou glandulaire chez les poissons osseux. *Comptes Rendus de l'Académie de Science Paris*, **74**, 1056-1060.

Van Bambeke, Ch. (1876) Recherches sur l'Embryologie des Poissons Osseux. I. Modifications de l'oeuf non fécondé après la ponte. II. Premières phases du développement. *Mémoires Couronnés et Mémoires des Savants étrangers de l'Académie des Sciences de Belgique* **40**, 1-66.

Van Bambeke, Ch. (1888) Remarques sur la reproduction de la Blennie vivipare (*Zoarces viviparus* Cuv.). *Bulletin de l'Académie Royale de Belgique* **15**, 92-117.

Van Beneden, M.E. (1877) Contribution à l'histoire du développement embryonnaire des téléostéens. *Bulletin de l'Académie Royale des Sciences, des Lettres et des Beaux Arts de Belgique* **44**, 742-770.

Van Beneden, M.E. (1878) A contribution to the history of the embryonic development of the teleosteans. *Quarterly Journal of Microscopical Science* **18**, 41-57.

Van der Ghinst, M. (1935) Mise en évidence de ferments dans le syncytium vitellin de la truite (*Salmo irideus*). *Bull. Histol. Appl. Physiol. et Pathol. et Tech. Microscop.* **12**, 257.

Van Haarlem, R. (1981) Cell behavior during early development of annual fishes. PhD thesis, Catholic University of Nijmegen, the Netherlands.

Van Haarlem, R. (1983) Early ontogeny of the annual fish genus *Nothobranchius*; Cleavage plane orientation and epiboly. *Journal of Morphology* **176**, 31-42.

Van Haarlem, R., Konings, J.G.B. and Van Wijk, R. (1983) Analysis of the relationship between the variation in intercleavage times and cell diversification during the cleavage stages of the teleost fish *Nothobranchius guentheri*. *Cell Tissue Kinetics* **16**, 167-176.

Van Veen, Th., Ekström, P., Nyberg, L., Borg, B., Vigh-Teichmann I., and Vigh, B. (1984) Serotonin and opsin immunoreactivities in the developing pineal organ of the three-spined stickleback, *Gasterosteus aculeatus* L. *Cell and Tissnue Research* **237**, 559-564.

Veit, O. (1923) Alte Probleme und neuere Arbeiten auf dem Gebiete der Primitiventwicklung der Fische. *Ergebnisse der Anatomie und Entwicklungsgeschichte* **24**, 414-490.

Veith, W.J. (1979a) Reproduction in the live-bearing teleost *Clinus superciliosus*. *South African Journal of Zoology* **14**, 208-211.

Veith, W.J. (1979b) The chemical composition of the follicular fluid of the viviparous teleost, *Clinus superciliosus*. *Biochem. Physiol.* **63A**, 37-40.

Veith, W.J. (1980) Viviparity and embryonic adaptations in the teleost *Clinus superciliosus*. *Canadian Journal of Zoology* **58**, 1-12.

Veith. W.J. and Cornish, D.A.(1986) Ovarian adaptations in the viviparous teleost *Clinus superciliosus* and *Clinus dorsalis* (*Perciformes: Clinidae*). *South African Journal of Zoology* **21**, 343-347.

Verma, P. (1970) Normal stages in the development of *Cyprinus carpio* var. *communis*. L. *Acta biologica Academy of Science, Hungary* **21**, 207-218.

Vigh-Teichmann, I., Szel, A., Röhlich P. and Vigh, B. (1990) A comparison of the ultrastructure and opsin immunocytochemistry of the pineal organ and retina of the deep-sea fish *Chimaera monstrosa*. *Experimental Biology* **48**, 361-371.

Vilter, V. (1948) Existence d'une double area dans la rétine du *Lebistes reticulatus* et significance fonctionnelle des structures aréales. *Comptes rendus de la Société de Biologie* **142**, 292-294.

Vilter, V. (1953) Existence d'une rétine à plusieurs mosaïques photoréceptrices chez un poisson abyssal bathypélagique, *Bathylagus benedicti*. *Comptes rendus des Séances de la Société de Biologie* **147**, 1937-1941.

Vilter, V. (1954) Différenciation fovéale dans l'appareil visuel d'un poisson abyssal, le *Bathylagus benedicti*. *Comptes rendus des séances de la Société de Biologie* **148**, 59-63.

Virchow, H. (1894) Ueber das Dottersyncytium und den Keimhautrand der Salmoniden. *Verhandlungen der Anatomischen Gesellschaft. Anatomischer Anzeiger* **10**, 66-77.

Vogt, C. (1842) Embryologie des Salmones, with atlas, **volume I** in *Histoire naturelle des poissons d'eau douce de l'Europe centrale, part 2* (ed L. Agassiz). Pepitpierre, Institut Lithographique de H. Nicolet, Neuchâtel. 328 pp.

Wagner, H.-J. (1974) Development of the retina of *Nannacara anomala* with special reference to regional variations of differentiation. *Zeitschrift der Morphologie der* Tiere **79**, 113-131.

Wagner, H.-J. (1990) Retinal structure of fishes, in *The visual system of fish*, (eds R. Douglas

and M. Djamgoz), Chapman and Hall, London, pp. 111-57.
Waldeyer, W. (1870) *Eierstock und Ei. Ein Beitrag zur Anatomie und Entwickelungsgeschichte der Sexualorgane*. Leipzig, W.Engelman, pp. 174.
Waldeyer, W. (1883) Archiblast und parablast. *Archiv für Mikroskopische Anatomie* **22**, 1-77.
Wallace, R.A. (1981) Cellular and dynamic aspects of oocyte growth in teleosts. *The American Zoologist* **21**, 325-343.
Wallace, R.A. and Begovac, P.C. (1985) Phosvitins in *Fundulus* oocytes and eggs: preliminary chromatographic and electrophoretic analyses together with biological considerations. *Journal of biological Chemistry* **260**, 11268-11274.
Wallace, R.A. and Selman (1979) Physiological aspects of ogenesis in two species of sticklebacks, *Gasterosteus aculeatus* L. and *Apeltes quadracus* Mitchill. *Journal of Fish Biology* **14**, 551-564.
Wallace, R.A. and Selman (1981) The reproductive activity of *Fundulus heteroclitus* females from Woods Hole, Massachusetts, as compared with more southern locations. *Copeia* **1981**, 212-213.
Wallace, R.A. and Selman, K. (1985) Major protein changes during vitellogenesis and maturation of *Fundulus* oocytes. *Developmental Biology* **68**, 172-182.
Walls, G.L. (1942, 1967) *The vertebrate eye and its adaptive radiation*. Hafner Publishing Co., New York.
Ward, C.R. and Kopf, G.S. (1993) Review. Molecular events mediating sperm activation. *Developmental Biology* **158**, 9-34.
Warga, R,M and Kimmel, C.B. (1990) Cell movements during epiboly and gastrulation in zebrafish. *Development* **108**, 569-580.
Warga, R.M. and Nüsslein-Volhard, C. (1999) Origin and development of the zebrafish endoderm. *Development* **126**, 827-838.
Warington, R. (1856) Observations on the habits of the stickleback. *The Zoologist, London* **14**, 4948-4950.
Wartenberg, H. (1964) Experimentelle Untersuchungen über die Stoffaufnahme durch Pinocytose während der Vitellogenese der Amphibienoocyten. *Zeitschrift der Zellforschung* **63**, 1004-1019.
Webb, T.A., Kowalski, W.J. and R.F. Fluck (1995) Microtubule-based movements during ooplasmic segregation in the medaka fish egg (*Oryzias latipes*). *Biological Bulletin* **188**, 146-156.
Wegmann, I. and Götting, K.J. (1971) Untersuchungen zur Dotterbildung in den Oocyten von *Xiphophorus helleri* (Heckel, 1948) (Teleostei; Poeciliidae). *Zeitschrift für Zellforschung und Mikroskopische Anatomie* **119**, 405-433.
Weil, C. (1872) Beiträge zur Kenntniss der Entwicklung der Knochenfische. *Sitzungsbericht der K. Akademie der Wissenschaften in Wien* **65**, Abth. 3, 171-179.
Weinstein, B.M. (1999) What guides early embryonic vessel formation? *Developmental Dynamics* **215**, 2-11.
Weliky, M. and Oster, G. (1990) The mechanical basis of cell rearrangement. 1. Epithelial morphogenesis during *Fundulus* epiboly. *Development* **109**, 373-386.
Weliky, M. and Oster, G. (1991) Dynamic models for cell. rearrangement during morpho-

genesis, in *Gastrulation, Movements, Patterns, and Molecules*. (eds R. Keller, Clark, W.H. and F. Griffin), Plenum Press, New York and London. pp. 135-146.

Welling, S.R. and Brown, G.A. (1969) Larval skin of the flathead sole, *Hippoglossoides elassodon*. *Zeitschrift für Zellforschung* **100**, 167-179.

Wenckebach, K.F. (1885) The development of the blood-corpuscles in the embryo of *Perca fluviatilis*. *Journal of Anatomy and Physiology* **19**, 231-236.

Wenkebach, K. (1886) Beiträge zur Entwicklungsgeschichte der Knochenfische. *Archiv für Mikroskopische Anatomie* **28**, 225-251.

Wessels and Schwartz (1953) Relation of the micropyle to cortical changes at fertilization in the egg of *Fundulus heteroclitus*. *Anatomical Record* **117**, 557-558.

White, G.M. (1915) The behavior of brook trout embryos from the time of hatching to the absorption of the yolk sac. *Journal of Animal Behaviour* **5**, 44-59.

White, B.N., Lavenberg, R.J. and McGowen, G.E. (1984) in *Ontogeny and Systematics of Fishes* (ed American Society of Ichthyologists and Herpetologists, Special Publication Number 1), Allen Press Inc., Lawrence KS, U.S.A. pp. 355-362.

Whitt, G.S. (1970) Developmental genetics of the lactate dehydrogenase isozymes of fish. *Journal of Experimental Zoology* **175**, 1-36.

Whitt. G.S. and G.M. Booth (1970) Localization of lactate dehydrogenase activity in the dells of the fish (*Xiphophorus helleri*) eye. *Journal of Experimental Zoology* **174**, 215-224.

Whitt, G.S., Shaklee, J.B. and Markert, C.L. (1975) Evolution of the lactate dehydrogenase isozymes of fishes, in *Isozymes IV: Genetics and Evolution,* (ed Markert, C.L.), Academic Press, New York. pp 381-400,

Wickler, W. (1956) Der Haftapparat einiger Cichliden-Eier. *Zeitschrift für Zellforschung* **45**, 304-327.

Wickler, W. (1957) Das Ei von *Blennus fluviatilis Asso* (=*Blennius vulgaris Poll.*) *Zeitschrift für Zellforschung* **45**, 641-648.

Wickler, W. (1959) Weitere Untersuchungen über Haftfäden an Teleosteer-Eiern, speziell an *Cyprinodon variegatus*. *Zoologischer Anzeiger* **163**, 90-107.

Wickramasinghe, S.N., Shiels, S. and Wickramasinghe, P.S. (1994) Immunoreactive erthrypoietin in teleosts, amphibians, reptiles and birds, in *Molecular, Cellular, and Developmental Biology of Erythropoiesis,* (eds I.N. Rich and T.R.J. Lappin), The New York Academy of Sciences, New York. pp. 366-376.

Wilkins, N.P. and Iles, T.D. (1966) Haemoglobin polymorphism and its ontogeny in herring (*Clupea harengus*) and sprat (*Sprattus sprattus*). *Comparative Biochemistry and Physiology* **17**, 1141-1158.

Willemse, H.Th.M. and Denucé J.M. (1973) Hatching glands in the teleost, *Brachydanio rerio, Danio malabaricus, Moenkhausia oligolepis* and *Barbus schuberti*. *Development, Growth and Differentiation* **13**, 169-178.

Willett, C.E., Cortes, A., Zuasti, A. and Zapata, A.G. (1999) Early haematopoiesis and developing lymphoid organs in the zebrafish. *Developmental Dynamics* **214**, 323-336.

Williams, L.L. (1939) A comparative study of the development of the liver in teleost fishes with special reference to the relation between liver and yolk-periblast. Ph.D. thesis. University of North Carolina, Chapel Hill. 92 pp.

Willmott, H.E. and Foster, S.A. (1995) The effects of rival male interaction on courtship and parental care in the fourspine stickleback, *Apeltes quadratus*. *Behaviour* **132**, 1107-1029.
Wilson, E.B. (1927) *The cell in development and heredity*. MacMillan, New York.
Wilson, H.V. (1889/1891) The embryology of the sea bass *(Serranus atrarius)*. *Bulletin U.S. Fish. Comm.* **9**, 209-277.
Wilson, E.T., Helde, K.A. and Grunwald D.J. (1993) Something's fishy here — rethinking cell movements and cell fate in the zebrafish embryo. *Perspectives* **9**, 348-352.
Wintrebert, P. (1912a) Le méchanisme de l'éclosion chez la truite arc-en-ciel. *Comptes Rendus des Séances de la Société de Biologie* **6**, 724-727.
Wintrebert, P. (1912b) Le déterminism de l'éclosion chez le cyprin doré (*Carassius auratus* L.). *Comptes Rendus des Séances de la Société de Biologie* **73**, 70-73.
Wirz-Hlavacek, G. and Riehl, R. (1990) Fortpflanzungsverhalten und Eistruktur des Piranhas *Serrasalmus nattereri* (Kner, 1860). *Acta Biologica Benrodis* **2**, 19-38.
Wittenberg, J.B. and Haedrich, R.L. (1974) The choroid *rete mirabile* of the fish eye. II. Distribution and relation to the pseudobranch and to the swimbladder *rete mirabile*. *Biological Bulletin Marine Biological Laboratory Woods Hole* **146**, 137-156.
Wittenberg, J.B. and Wittenberg, B.A. (1974) The choroid *rete mirabile* of the fish eye. I. Oxygen secretion and structure: Comparison with the swimbladder *rete mirabile*. *Biological Bulletin Marine Biological Laboratory Woods Hole* **146**, 116-136.
Wolenski, J. and Hart, N.H. (1987) Scanning electron microscope studies of sperm incorporation into the zebrafish (*Brachydanio*) egg. *Journal of Experimental Zoology* **243**, 259-273.
Wolenski, J. and Hart, N.H. (1988a) Effects of cytochalasins B and D on the fertilization of zebrafish (*Brachydanio*) eggs. *Journal of Experimental Zoology* **246**, 202-215.
Wolenski, J. and Hart, N.H. (1988b) The fertilization cone develops in the presence of actin filament cappers. *Journal of Cell Biology* **107**, 452a.
Wolpert, L. (1969) Positional information and the spatial patterning of cellular differentiation. *Journal of Theoretical Biology* **25**, 1-47.
Wolpert, L. (1978) Gap junctions: channels for communications in development, in *Intercellular Junctions and Synapses*, (eds J. Feldman, N.B. Gilula and J.D. Pitts), Chapman and Hall, London, pp. 81-96.
Wood, A. and Thorogood,:P, (1984) Contact guidance in vivo — an analysis of mesenchymal cell movements in the teleost fin bud. *Journal of Embryology and Experimental Morphology* (Special issue) **82**, 124. Abstract, European Developmental Biololgy Congress, Southampton UK 2-7 September, 1984.
Wood, A. and Timmermans, L.M. (1988) Teleost epiboly: a reassessment of deep cell movement in the germ ring. *Development* **102**, 575-585.
Wooton, R.J. (1976) *The biology of the sticklebacks*. Academic Press, London. 387 pp.
Wooton, R.J. (1990) *Ecology of teleost fishes*. Chapman and Hall, London.
Wourms, J.P. (1972a) Developmental biology of annual fishes. I. Stages in the normal development of *Austrofundulus myersi* Dahl. *Journal of Experimental Zoology* **182**, 143-168.
Wourms, J.P. (1972b) The developmental biology of annual fishes. II. Naturally occurring

dispersion and reaggregation of blastomeres during the development of annual fish eggs. *Journal of Experimental Zoology* **182**, 169-200.

Wourms, J.P. (1972c) Developmental biology of annual fishes. III. Pre-embryonic and embryonic diapause of variable duration in the eggs of annual fishes. *Journal of Experimental Zoology* **182**, 389-414.

Wourms, J.P. (1976) Annual fish oogenesis. I. Differentiation of the mature oocyte and formation of the primary envelope. *Developmental Biology* **50**, 338-354.

Wourms, J.P. (1981) Viviparity: The maternal-fetal relationship in fishes. *The American Zoologist* **21**, 473-515.

Wourms, J.P. (1994) The challenge of of piscine viviparity. *Israel Journal of Zoology* **40**, 551-568.

Wourms, J.P. (1997) The rise of fish embryology in the nineteenth century. *The American Zoologist* **37**, 269-310.

Wourms, J.P. and Cohen, D.M. (1975) Trophotaeniae, embryonic adaptations, in the viviparous ophidioid fish, *Oligopus longhursti*: a study of museum specimens. *Journal of Morphology* **147**, 385-401.

Wourms, J.P. and Lombardi, J. (1985) Prototypic trophotaeniae and other placental structures in embryos of the pile perch, *Rhacochilus vacca* (embriotocidae). *The American Zoologist* **25**, 95A.

Wourms, J.P. and Sheldon, H. (1976) Annual fish oogenesis, II. Formation of the secondary egg envelope. *Developmental Biology* **50**, 355-366.

Wourms, J.P. and Whitt, G.S. (1981) Future directions of research on the developmental biology of fishes. *The American Zoologist* **21**, 597-604.

Wourms, J.P., Grove, B.D. and Lombardi, J. (1988) The maternal-embryonic relationship in viviparous fishes, in *Fish Physiology* **11B,** (eds W.S. Hoar and D.J. Randall), Academic Press London. pp. 1-134.

Wülker, W. (1953) Bewegungsrhythmen im Teleostier-Ei (Zeitrafferfilm-Untersuchung). 1. *Esox lucius, Salmo trutta, S. fontinalis, S. irideus*. Zoologisches Jahrbuch, Abteilung Anatomie, Ontogenese **73**, 1-35.

Wülker, W. (1954) Bewegungsvorgänge im Aeschen-Ei (*Thymallus vulgaris*). *Archiv für Hydrobiologie* **20 Suppl.** 524-536.

Wylie, C. (ed) (1996) *Zebrafish Issue*, in *Development,* **123,** Company of Biologists Limited, Cambridge. 481 pp.

Wyman, J. (1850-57) Observations on the development of *Anableps Gronovii* (Cuv. & Val.). *Boston Journal of Natural History* **6**, 432-443.

Yacob, A., Wise, C. and Kunz, Y.W. (1977) The accessory outer segment of rods and cones in the retina of the guppy, *Poecilia reticulata* P. (teleostei). An electron microscopical study. *Cell and Tissue Research* **171**, 181-191.

Yamada, J. (1963) The normal developmental stages of the pond smelt, *Hypomesus olidus* (Pallas). *Bulletin of the Faculty of Fisheries Hokkaido University* **14**, 121-126.

Yamada, J. (1968) A study on the structure of surface cell layers in the epidermis of some teleosts. *Annotationes Zoologicae Japonenses* **41**, 1-8.

Yamagami, K. (1970) A method for rapid and quantitative determination of the hatching enzyme (chorionase) in the medaka, *Oryzias latipes. Annotationes Zoologicae Japonenses*

43, 1-9.

Yamagami, K. (1975) Relationship between two kinds of hatching enzymes in the hatching liquid of the medaka, *Oryzias latipes*. *Journal of Experimental Zoology* **192**, 127-132.

Yamagami, K. (1981) Mechanism of hatching in fish: Secretion of hatching enzyme and enzymatic choriolysis. *The American Zoologist* **21**, 459-471.

Yamagami, K. (1988) Mechanisms of hatching in fish, in *Fish Physiology*, **11A** (eds W.S. Hoar, D.J. Randall and J.R. Brett) Academic Press, pp. 447-499.

Yamagami, K., Hamazaki, T.S., Yasumasu, S., Masuda, K. and Iuchi, I. (1992) Molecular and cellular basis of formation, hardening, and breakdown of the egg envelope in fish. *International Review of Cytology* **136**, 51-92.

Yamagami, L., Yasumasu S. and Iuchi, I. (1993) Choriolysis by the medaka hatching enzyme, an enzyme system, in *Physiological and Biochemical Aspects of Fish Development*, (eds B.T. Walther and H.J. Fyhn), University of Bergen, Norway, pp. 104-111.

Yamamoto, K. (1952) Studies on the fertilization of the egg of the flounder. II. The morphological structure of the micropyle and its behavior in response to sperm-entry. *Cytologia* **16**, 302-306.

Yamamoto, K. (1956a) Studies on the formation of fish eggs. VII. The fate of the yolk vesicle in the oocyte of the herring, *Clupea pallasii*, during vitellogenesis. *Annotationes. Zoologicae Japonenses* **29**, 91-96.

Yamamoto, K. (1956b) Studies on the formation of fish eggs. VIII. The fate of the yolk vesicle in the oocyte of the smelt, *Hypomesus japonicus*, during vitellogenesis. *Embriologia* **3**, 131-138.

Yamamoto, K. (1956c) Studies on the formation of fish eggs. IX. The fate of the yolk vesicle in the oocyte of the flounder, *Liposetta obscura*, during vitellogenesis. *Bulletin of the Faculty of Fisheries, Hokkaido University* **12**, 208-212.

Yamamoto, K. (1963) Cyclical changes in the wall of the ovarian lumen in medaka, *Oryzias latipes*. *Annotationes Zoologicae Japonenses* **35**, 179-186.

Yamamoto, K. and Oota, I. (1967a) An electron microscope study of the formation of the yolk globule in the oocyte of zebrafish, *Brachydanio rerio*. *Bulletin of the Faculty of Fisheries, Hokkaido University* **17**, 165-174.

Yamamoto, K. and Oota, I. (1967b) Fine structure of yolk globules in the oocyte of the zebrafish, *Brachydanio rerio*. *Annotationes Zoologicae Japonenses* **40**, 20-27.

Yamamoto, M. (1963) Electron microscopy of fish development. I. Fine structure of the hatching gland of embryos of the teleost, *Oryzias latipes*. *Journal of the Faculty of Science, Tokyo IV*, **10**, 115-122.

Yamamoto, M. (1975) **5**, Hatching gland and hatching enzyme, in *Medaka (killifish), Biology and Strains*. (ed T. Yamamoto), Keigaku Publishing Company, Tokyo, pp. 73-79.

Yamamoto, M. and Yamagami, K. (1975) Electron microscopic studies on choriolysis by the hatching enzyme (chorionase) of the teleost, *Oryzias latipes*. *Developmental Biology* **43**, 313-321.

Yamamoto, M., Iuchi, I. and Yamagami, I. (1979) Ultrastructural changes of the teleostean hatching gland cell during natural and electrically induced precocious secretion. *Developmental Biology* **68**, 162-174.

Yamamoto, T. (1931) Studies on the rhythmical movements of the early embryos of *Oryzias* latipes. *Journal of the Faculty of Science of the Imperial University, Tokyo* **2**, 147-162.

Yamamoto, T. (1939) Changes of the cortical layer of the egg of *Oryzias latipes* at the time of fertilization. *Proceedings of the Imperial Academy of Tokyo* **15**, 269-271.

Yamamoto, T. (1944a) Physiological studies on fertilization and activation of fish eggs. I. Response of the cortical layer of the egg of *Oryzias latipes* to insemination and to artificial stimulation. *Annotationes Zoologicae Japonenses* **22**, 109-125.

Yamamoto, T. (1944b) Physiological studies of fertilization and activation of fish eggs. II. The conduction of the fertilization waves in the egg of *Oryzias latipes. Annotationes Zoologicae Japonenses* **22**, 109-136.

Yamamoto, T. (1954) Cortical changes in eggs of the goldfish (*Carassius auratus*) and the pond smelt (*Hypomesus elidus*) at the time of fertilization and activation. *Japanese Journal of Ichthyology* **3**, 162-170.

Yamamoto, T. (1956) The physiology of fertilization in the medaka (*Oryzias latipes*) *Experimental Cell Research* **10**, 387-393.

Yamamoto, T. (1958) The physiology of fertilization in fish eggs (in Japanese) in *Developmental Physiology* (eds K. Dan and T. Yamada), Bahukan, Tokyo. p.73.

Yamamoto, T. (1961) Physiology of fertilization in fish eggs. *International Review of Cytology* **12**, 361-405

Yamamoto, T. (1962) Mechanism of breakdown of cortical alveoli during fertilization in the medaka *Oryzias latipes. Embriologia* **7**, 228-251.

Yamamoto, T. (1975) *Medaka (Killifish) Biology and Strains.* **3**, Stages in Development, pp. 31-50. **4**, Rhythmical contractile movements, pp 59-72. **6**, Fertilization in *Oryzias* egg pp 80-96, **7**, Activation and conduction of fertilization wave pp 97-108. Kaigaku Publishing Company, Tokyo.

Yamamoto, T.S. (1955) Morphological and cytochemical studies on the oogenesis of the freshwater fish, medaka (*Oryzias latipes*) (in Japanese) *Japanese Journal of Ichthyology* **4**, 170-181.

Yamamoto, T.S. (1963) Eggs and ovaries of the stickleback, *Pungitius tymensis* with a note on the formation of a jelly-like substance surrounding the egg. *Journal of the Faculty of Science Hokkaido University* **15**, 190-199.

Yanagimachi, R. (1953) Effect of environmental salt concentration on fertilizability of herring gametes. *Journal of the Faculty of Science Hokkaido University (Zool.)* **11**, 139-144.

Yanagimachi, R. (1956) The effect of single salt solutions on the fertilizability of the herring egg. *Journal of the Faculty of Science Hokkaido University (Zool.)* **12**, 317-324.

Yanagimachi, R. (1957) Some properties of the sperm-activating factor in the micropyle area of the herring egg. *Annotationes Zoologicae Japonenses* **30**, 114-119.

Yanagimachi, R. (1988) Sperm-Egg Fusion, Chapter 1 in *Current Topics in Membranes and Transport* **32**, Academic Press, pp. 1-43.

Yanagimachi, R. and Kanoh, Y. (1953) Manner of sperm entry in herring egg, with special reference to the role of calcium ions in fertilization. *Journal of the Faculty of Science, Hokkaido University* **11**, 487-494.

Yanagimachi, R., Cherr, G.N., Pilai, M.C. and Baldwin, J.D. (1992) Factors controlling sperm entry into the micropyles of salmonid and herring eggs. *Development, Growth and Differentiation* **34**, 447-461.

Yanai, T. (1966) Hatching, in *Vertebrate Embryology* (in Japanese) (ed M. Kume). Bifukan Publications Co., Tokyo, pp. 49-58.

Yasumasu, S., Katow, S., Hamazaki, T.S., Iuchi, I. and Yamagami, K. (1992) Two constituent proteases of a teleostean hatching enzyme: Concurrent syntheses and packaging in the same secretory granules in discrete arrangement. *Developmental Biology* **149**, 349-356.

Yokoya, S. and Ebina, Y. (1976) Hatching glands in salmonid fishes, *Salmo gairdneri, Salmo trutta, Salvelinus fontinalis* and *Salvelinus pluvius. Cell and Tissue Research* **172**, 529-540.

Yorke, M. A. and Dickson, D.H. (1985) A cytochemical study of myeloid bodies in the retinal pigment epithelium of the newt, *Notophthalmus viridescens. Cell and Tissue Research* **240**, 541-548.

Yoshimoto, Y., Iwamatsu, T., Hirano, K. and Hiramoto, Y. (1986) The wave pattern of free calcium release upon fertilization in medaka and sand dollar eggs. *Development, Growth and Differentiation* **28**, 583-596.

Zahndt, J.P. and Porte, A. (1966) Signes morphologiques du transfer de matériel nucléaire dans le cytoplasme des ovocytes de certaines espèces de poisson. *Comptes Rendus de l'Académie des Sciences de Paris* **262D**, 1977-1987.

Ziegenhagen, P. (1896) Ueber Entwickelung der Circulation bei Teleostiern, insbesondere by Belone. *Verhandlungen der Anatomischen Gesellschaft. Anatomischer Anzeiger* **12**, 100-108.

Ziegler, H.E. (1882) *Die embryonale Entwicklung von Salmo salar.* Ph.D. thesis. Freiburg im Breisgau.

Ziegler, H.E. (1885) Ueber die Entstehung der Blutkörperchen bei Knochenfischembryonen. *Tageblatt der 58. Versammlung deutscher Naturforscher und Aerzte in Strassburg. No. 4.*

Ziegler, H.E. (1887) Die Entstehung des Blutes bei Knochenfischembryonen. *Archiv für Mikroskopische Anatomie* **30**, 596-687.

Ziegler, H.E. (1892) Ueber die embryonale Anlage des Blutes bei den Wirbelthieren. *Verhandlungen der Deutschen Zoologischen Gesellschaft*, 18-30.

Ziegler, H.E. (1894) Ueber das Verhalten der Kerne im Dotter der meroblastischen Wirbelthiere. *Berichte der Natuforschenden Gesellschaft in Freiburg i.B.* **8**, 192-209.

Ziegler, H.E. (1902) *Lehrbuch der vergleichenden Entwicklungsgeschichte der niederen Wirbeltiere*, Gustav Fischer, Jena.

Zon, L.I. (1995) Review article: Developmental biology of hematopoiesis. *Blood* **86**, 2876-2891.

Zotin, A.I. (1953) Initial stages of the hardening process of salmonid eggs. (in Russian) *Doklady Akad. Nauk. S.S.S.R.* **89**, 573-576.

Zotin, A.I. (1958) The mechanism of hardening of the salmonid egg membrane after fertilization or spontaneous activation. *Journal of Embryology and experimental Morphology* **6**, 546-568.

Zygar, C.A., Lee, M.J. and Fernald, R.D. (1999) Nasotemporal asymmetry during teleost retinal growth; Preserving an area of specialization. *Journal of Neurobiology* **41**, 435-442.

Species index

A

Abdudefduf saxatalis 99
Abramis brama 18, 89, 259
Abramis crysoleucas 433
Acanthorhodeus asmussi 6
Acerina 53, 105, 111, 424
Acerina cernua 105, 111, 114, 151, 259
Acerina vulgaris 20, 53, 72, 114, 149
Acheilognathus lanceolata 156, 157
Acheilognathus rhombeus 134, 135
Acheilognathus tabira 156, 157
Acipenser 109, 185, 282, 378
Acromonas hydrophila 176
Adioryx vexillarius 5
Adrianichthyidae 94
Aelurichthys 16, 106
Agonus cataphractus 50, 61, 62, 102
Albula vulpes 136
Alburnus alburnus 89, 142
Alburnus lucidus 42, 89
Allophorus 485
Allotoca 485
Alosa 13, 36, 87, 178, 226, 282
Alosa pseudoharengus 87
Alosa sapidissima 11, 446, 448, 449
Aluterus punctatus 136

Ameca 485, 486
Ameca splendens 459, 481, 482, 486, 487
Ameiurus albidus 15
Ameiurus nebulosus 222, 224
Amia 185
Amiurus catus 62
Amiurus nebulosus 38, 262
Amphiprion clarki 99
Anableps 470, 477, 478, 480, 481
Anableps anableps 320, 459, 477-481
Anableps artedi 477
Anableps dowi 18, 477, 478, 479, 481
Anableps microlepis 478
Anableps tetrophthalmus 478
Anarrhichas lupus 19
Anchoa guineensis 136, 139
Anchoviella 17
Anguilla 20
Anguilla anguilla 32, 34, 59, 63
Anguilla japonica 34, 109, 121
Anguilla vulgaris 385
Anoptichthys 271
Anoptichthys jordani 5
Antennarius 11, 14, 106
Antennarius senegalensis 139
Apeltes quadracus 15, 85

Aphyosemion scheeli, 279
Aphyosemion gardneri, 145
Aristichthys nobilis 62, 81
Arius 16
Arius commersonii 16, 19
Arrhamphus 137, 144
Arrhamphus sclerolepis 137
Aspius alburnus 111, 114
Aspredo 15
Astyanax mexicanus 303
Ataenobius toweri 482, 484
Atherinopsis affinis 96, 97
Atherinopsis californiensis 96
Atherion elymus 96
Aulophallus 459
Aulophallus elongatus 471, 474, 476
Austrofundulus 6
Austrofundulus myersi 17, 93, 222
Austrofundulus transilis 65, 71

B

Bagre marinus 19
Balistes forcipatus 136, 138, 142
Balsadichthys 485
Barbatula barbatula 29, 30
Barbina 5
Barbus 282
Barbus barbus 142
Barbus conchonius 117, 122, 157, 202, 220, 221
Barbus schuberti 296
Basaldichthys whitei 483
Bathygobius 17
Bathygobius soporator 17
Batrachus tau 187, 222
Belone 90, 91, 389, 422, 424
Belone acus 187, 200, 201, 262, 385, 389, 418, 422
Belone longirostris 12, 57, 90, 91, 233, 382
Betta splendens 5, 12, 287, 397, 409
Blennius 240, 422

Blennius cristatus 137, 139, 140
Blennius fluviatilis 102
Blennius gattorugine 397, 409, 419
Blennius pholis 13, 29, 33, 67, 70, 72, 102, 137, 139
Blennius sphinx 102
Blennius vandervekeni 137, 139
Blennius viviparus 50, 416, 422, 461, 465-469
Blicca bjoerkna 110
Boops boops 139
Brachydanio 38, 42, 59, 63, 160, 168, 171, 173, 195, 204, 214
Brachydanio (Danio) rerio 7, 8, 19, 26, 31-33, 35, 44, 49, 59, 60, 65, 67, 68, 81, 84, 85, 109, 115, 117, 118, 134, 136, 157, 160, 161, 165, 168, 171, 178, 179, 185, 194, 202, 205, 211, 221, 226, 240, 245, 246, 276, 294, 298, 337, 392, 393, 394, 409, 411-413, 415, 419, 426, 505

C

Caenorhynchus australis 78
Cairnsichthys rhombosmoides 142
Calanx 5
Callichthys 14
Callionymus 24
Callionymus lyra 78
Caranx trachurus 388
Carassius 178, 252, 260, 438
Carassius auratus 60, 68, 69, 134, 135, 142, 145, 149, 150, 170, 187, 260, 272, 318, 432
Cataetyx 491
Cataetyx memoriabilis 491
Cataetyx messieri 491
Catostomus 173, 214
Catostomus commersoni 255
Cenolebias belotti 65, 93
Ce(y)nolebias ladigesi 65, 68, 71, 93, 99

Cenolebias melanotaenia 65, 93, 99
Centrolabrus exoletus 67
Centrolabrus rupestris 78
Centronotus gunnellus 15, 102
Cephalacanthus volitans 134, 136, 138
Chaenogobius urotaenia 280
Chanodraco myersi 115
Chapalichthys 485
Characodon 486
Chilomycturus antennatus 136, 138, 142
Chionobathiscus dewitti 109, 115
Chionodraco myers 109
Chirostoma 153
Chirostoma notata 95
Chondrostoma nasus 89
Chromis weberi 99
Cichlasoma 63, 65, 66, 68
Cichlasoma bimaculata 444
Cichlasoma meeki 99, 109
Cichlasoma nigrofasciata 69, 98, 99, 109, 246
Cirrhinus mrigala 187
Clarius senegalensis 138, 142
Clinocottus analis 260
Clinus dorsalis 470
Clinus nuchipinnis 137, 140
Clinus superciliosus 469, 470, 503
Clupea 12, 13, 38, 62
Clupea harengus 10, 24, 57, 87, 132, 149, 150, 153, 154, 158, 172, 177, 259, 340, 384, 415, 418
Clupea pallasi 21, 87, 128, 154-159, 161, 162, 171
Clupea sapidissima 150, 172, 384
Clupea sprattus 87, 388
Clupea vernalis 87
Clupeonella delicatula 11, 12
Cobitis 111, 131
Cobitis anableps 477
Cobitis barbatula 90, 111, 114
Cobitis fossilis 111, 114

Cololabis saira 83
Conger muraena 136
Coregonus 5, 59, 84, 201, 262, 263, 424, 450
Coregonus albula 60, 64, 65
Coregonus albus 273
Coregonus alpinus 26, 424, 426, 427, 447, 449
Coregonus clupeaformis 187, 244
Coregonus fera 84, 109
Coregonus lavaretus 13, 60, 72, 77, 84, 109, 115 149, 151, 175, 177, 201
Coregonus macrophthalmus 84, 109, 114
Coregonus nasus 84, 109
Coregonus oxyrhynchus 109
Coregonus palea 20, 83, 149, 151, 240, 259, 382, 422
Coregonus pidschian 109
Coregonus wartmanni 84, 109, 114
Coris julis 78
Cottomorphus 424
Cottus 10, 12, 423, 424, 441, 444, 448
Cottus bubalis 422, 423
Cottus gobio 102, 114, 151
Cottus groenlandicus 13
Cottus scorpius 13, 20
Craterocephalus marjoriae 142
Craterocephalus helenae 142
Craterocephalus stercusmuscarum 142
Crenilabrus 19, 57, 99, 114, 149
Crenilabrus ocellatus 150
Crenilabrus pavo 150, 152, 153, 162, 180, 185, 418
Crenilabrus quinquemaculatus 99, 150
Crenilabrus rostratus 150, 185
Crenilabrus tinca 63, 78, 114, 149, 150, 153
Creratacanthus 153
Cristiceps 63, 187, 260, 424

Cristiceps argentatus 57, 178
Ctenochromis horei 16
Ctenogobius stigmaticus 101
Ctenolabrus 114, 180, 187, 199, 211, 214, 418, 433
Ctenolabrus (Tautogolabrus) adspersus 433
Ctenolabrus rupestris 67
Ctenolabrus tinca 153
Ctenopharyngodon idella 62, 81
Ctenopharyngodon idellus 11
Culaea inconstans 15
Cyclogaster lineatus 62, 124
Cyclopterus 10
Cyclopterus lumpus 12, 50, 102, 176
Cymatogaster 491, 492
Cymatogaster aggregata 134, 136
Cymatogaster aggregatus 19, 492
Cynolebias 6, 63, 93
Cynolebias belotti 65, 93
Cynolebias melanotaenia 68, 71, 78, 93
Cynopoecilus 93
Cyprinodon variegatus 32, 93
Cyprinus 163, 168
Cyprinus alburnus 110
Cyprinus auratus 57
Cyprinus blicca 20, 51, 89, 110, 112, 148, 152, 240
Cyprinus brama 259
Cyprinus carassius 109
Cyprinus carpio 5, 10, 18, 46, 57, 109, 134, 158, 159, 161, 168, 170, 173, 185, 187, 211, 221, 411, 437, 441
Cyprinus erythrophthalmus 89, 148, 152
Cyprinus gobio 114, 149
Cyprinus lobio 151
Cyprinus rufus 89
Cyprinus tinca 110, 186

D

Dactylopterus 136
Dactylopterus (Cephalocanthus) volitans 134, 136
Danio rerio (see Brachydanio rerio)
Dendrochirus brachypterus 106
Dermogenys montanus 492
Dermogenys pusillus 50, 63, 68, 72, 136, 137, 142, 458, 493
Dermogenys 144, 493
Dicentrarchus 68
Dicentrarchus labrax 68, 69, 124, 282, 435
Dinematichthys 491
Diplacanthopoma 491
Discoboli 20
Ditrema temmincki 492
Dorosoma petenense 87

E

Elassichthys 82
Elopichthys bambusa 11
Engraulis 17
Engraulis guineensis 139
Engraulis japonica 17, 109, 121
Engraulis japonicus 81
Entelerus 50
Epiplatys 94
Epiplatys ansorgei 94
Epiplatys bifasciatus 94
Epiplatys chaperi 94
Epiplatys fasciolatus 94
Epiplatys hildegardae 94
Epiplatys lamottei 94
Epiplatys spilargyreius 94
Eptatretus stoutii 142
Erythroculter erythropterus 11
Escheneida 121
Esox 20, 172, 282, 416, 424, 449
Esox bellone 81, 82
Esox lucius 5, 13, 57, 64, 88, 109, 111, 113, 127, 129, 132, 134,

144, 145, 151, 155, 244, 259, 272, 285, 298, 300, 301, 382, 384, 390, 391, 446, 498
Esox masquinongy 88, 204, 205
Esox masquinongy ohioensis 202
Esox reticulatus 20, 52, 57, 62, 114, 129
Ethmalosa fimbriata 136, 137
Eurystole eriarcha 97, 98
Exocoetus 388, 409, 424
Exocoetus exiliens 91
Exocoetus heterurus 92
Exocoetus rondoletti 92
Exocoetus volitans 388, 422

F

Felichthys 106
Felichthys felis 16, 106
Fierasfer 38
Fierasfer acus 11, 17
Fiaresfer dentatus 11
Fistularia tabacaria 138
Fodiator acutus 139
Fundulus 9, 31, 44, 60, 92, 147, 150, 159, 164, 167, 168, 181, 196, 202, 203, 211-214, 217, 221, 222, 230-233, 244, 252-254, 257, 258, 321, 400-402, 409, 419, 435
Fundulus diaphanus 19, 92, 124
Fundulus heteroclitus 7, 8, 19, 31, 32, 35, 38, 44, 45, 50, 57, 62, 63, 65, 66, 68, 69, 72, 92, 93, 120, 124, 136, 142, 161, 162, 167, 178, 181, 182, 197, 203, 265, 294, 300, 398-400, 406, 433, 435, 472, 484, 496
Fundulus majalis 472, 496
Fundulus notatus 92
Fundulus ocellaris 93
Fundulus reticulatus 178

G

Gadus 282
Gadus aeglefinus 11
Gadus callarias 441, 446, 448, 449
Gadus lota 198
Gadus morrhua 11, 17, 31, 62, 67, 74, 114, 150, 158, 166, 172, 174, 177, 185, 340, 382, 411, 418, 438, 439, 441
Galeichthys felis 16, 19
Galeodes decadactylus 139, 140
Gallus domesticus 229, 244
Gambusia 67, 494
Gambusia affinis 136, 494
Gambusia patruelis 496
Gasterosteus 15, 18, 20, 41-43, 57, 85, 86, 111-113, 132, 149, 154, 173, 241, 259, 260, 273, 423
Gasterosteus aculeatus 84, 85, 132, 259, 392, 394, 409, 417
Gasterosteus gymnurus 422, 423
Gasterosteus leiurus 52, 57, 84, 85, 111, 155, 158, 177
Gasterosteus pungitius 52, 57, 84-86, 111, 155, 158, 177, 185, 187
Gasterosteus spinachia 84
Gastrophysus hamiltoni 142
Geophagus jurupari 107
Girardinichthys 485, 486
Girardinichthys innominatus 482, 483
Glossogobius 17
Gnathochromis pfefferi 16
Gnathopogon elongatus 149, 150
Gobio 53, 54, 70, 72, 89, 444, 448
Gobio fluviatilis 53, 89, 101, 111, 114, 151
Gobio gobio 6, 26, 29, 36, 37, 69, 72, 89, 109, 124, 126
Gobiobotia pappenheimi 11
Gobionellus boleosoma 101
Gobionina 6
Gobiosoma bosci 101

Gobius 14, 17, 81, 214, 241, 252, 259, 260, 423, 432, 438
Gobius capito 390, 422, 423
Gobius flavescens 101
Gobius fluviatilis 101
Gobius minutus (see Pomatoschistus minutus)
Gobius niger 13, 17, 100, 101, 153, 255, 259, 422
Gobius ruthensparri 15, 100, 101
Goodea 484-486
Goodea atripinnis 482, 486, 487
Goodea luitpoldi 485-487
Gymnarchus niloticus 144, 155
Gymnarchus 19, 142, 143, 244, 424

H

Haplochilus 424
Haplochromis 231
Haplochromis burtoni 368
Heliasis 36, 99
Heliasis chromis 99, 114, 150, 152
Helicolenus 492
Helicolenus papillosus 493
Hemichromis fasciatus 138, 142
Hemiculter cultratus 11
Hemirhamphodon 144, 493
Hemirhamphodon pogonognathus 136, 137, 142
Hemirhamphus 90
Hemirhamphus marginatus 91
Herotilapia multispinosa 3, 281, 288, 290, 294, 295, 298, 299, 302
Heterandria 472, 475, 476
Heterandria formosa 33, 59, 458, 471, 473-476, 494, 496, 503
Heteroclinus 470
Heteroclinus heptaeolus 470
Heteroclinus perspicillatus 470, 471
Heteropneustes fossilis 35
Hexagrammos octogrammus 74, 101

Hippocampus 16, 50
Hippocampus erectus 65, 68, 71
Hippoglossus hippoglossus 31, 74, 439
Histrio histrio 106
Homo sapiens 244
Hozukius 492
Hubbsina turneri 68, 486
Hyperprosopon argenteum 245, 385
Hypoatherina bleekeri 96
Hypoatherina tsurugae 96
Hypodytes rubripinnis 121
Hypomesus olidus 149, 150
Hypomesus transpacificus nipponendis 128
Hypophthalmichthys molitrix 11, 62, 81
Hyporhampus ihi 91
Hypostoma 305
Hypseleotris 144
Hypseleotris galii 137, 145

I

Ictalurus 314
Ictalurus albidus 90
Ictalurus nebulosus 14
Ictalurus punctatus 144
Ilyodon 485
Inimicus japonicus 121
Italurus nebulosus 90

J

Jenynsia 137
Jenynsia lineata 136, 137, 487, 489
Julis 149, 158, 180, 201
Julis turcia 114
Julis vulgaris 38, 114, 150

K

Kareius bicoloratus 121

L

Labrax lupus 21
Labrus berggylta 99
Labrus festinus 99
Labrus turdus 99
Lactoria diaphana 81
Lactoria fornasini, 121
Lampanyctus 143, 144
Lateolabrax japonicus 121
Lates calcarifer 135, 140
Latimeria chalumnae 3
Lebistes reticulatus (see Poecilia reticulata)
Lepadogaster 17
Lepadogaster bimaculatus 15
Lepadogaster candollii 397, 409
Lepadogaster decandolii 15
Lepadogaster gouani 103
Lepadogaster lepadogaster 102, 135, 136, 144
Lepidogalaxias salamandroides 135, 136
Lepidosteus 124, 125, 185, 242, 378
Lepidosteus osseus 186
Lermichthys 482, 483, 485
Lermichthys multiradiatus 483-484
Leuciscus 153, 388, 409
Leuciscus cephalus 89, 109, 114, 149, 151, 388
Leuciscus erythrophthalmus 57, 89
Leuciscus hakuensis 296-298
Leuciscus leuciscus 142
Leuciscus phoxinus 114, 149, 151
Leuciscus rutilus 89, 114, 132, 259
Leucopsarion petersii 7, 187
Limanda limanda 50
Limanda shrenki 162
Liparis 12
Liparis lineatus 124
Liparis liparis 102
Liparis monagni 102
Liz 140

Lophius 389
Lophius piscatorius 11, 106
Lota vulgaris 13, 57, 153
Loweina 3
Loweina rara 5
Lucifuga 490
Lucifuga subterraneus 458
Luciocephalus 16, 24, 107, 121, 122
Luciocephalus pyriform 17
Luciperca 283
Lutjanus synagris 44, 62
Lycodontis afer 139

M

Maccullochella 137
Maccullochella macquariensis 144
Maccullochella Peeli 137
Macquaria ambigua 137
Macropodus cupanus 124
Macrozoarces americanus 6
Macrurus 81
Macrurus coelorhynchus 78
Maenidia maenidia 96
Mauroclinus mülleri 78, 80
Melanogrammus aeglefinus 11
Melanotaenia 135
Melanotaenia duboulayi 142
Melanotaenia maccullochi 142
Melanotaenia nigricans 44
Menidia beryllina 96
Menidia beryllina cerea 96
Menidia menidia 95
Menidia menidia notata 96
Merlangius merlangus 11, 340
Microbrotula randalli 490
Micrometrus minimus 492
Mimagoniates barberi 134, 136, 140-143
Misgurnus anguillicaudatus 126, 134, 135, 142, 157
Misgurnus fossilis 5, 46, 179, 197, 203, 221

Mola mola 10, 18
Molliensia sphenops 67
Morone 153
Morone americana 57, 62
Motella 17
Mugil capito 52
Mugil cephalus 52, 81
Mullus barbatus ponticus 5
Mullus surmuletus 57
Muraena 432, 433
Muraenidae 136
Mustela vivipera 460
Mylopharyngodon piceus 62, 81

N

Nanichthys 82
Nemachilini 6
Neoceratias 135
Neoceratias spinifer 136, 137
Neoceratodus 378
Neoditrema ransonneti 492
Neophis ophidion 111
Neoophorus 485
Neophorus diazi 487
Neoteca bilineata 486, 488
Neotoca 485
Nerophis 16, 50
Nerophis ophidion 16
Noemacheilus 124, 126
Noemacheilus barbatulus 26, 29, 36, 37, 59, 69, 70, 72, 90, 109, 124
Nomorhamphus 144, 493
Nomorhamphus celebensis 136, 137, 142
Nomorhamphus hageni 492
Notemigonus chrysoleucus 58
Nothobranchius 222
Nothobranchius guentheri 187
Nothobranchius korthausae 65, 298
Novodon modestus 52, 60, 71
Novorhamphus hageni 458

O

Ogilbia cayorum 490, 491
Oligocottus maculosus 136, 142
Oligopus longhursti 490
Ollentodon 485
Oncorhynchus 42, 63, 161, 168, 271
Oncorhynchus gorbuscha 60, 63, 64, 84, 115, 156, 157
Oncorhynchus keta 19, 32, 58-60, 63, 84, 109, 115, 126, 128, 156, 159, 162, 163, 170, 176, 177, 293, 294, 300, 388, 390, 410, 419, 440
Oncorhynchus kisutch 62, 63, 84, 115, 156, 158, 162
Oncorhynchus masu 13
Oncorhynchus mykiss 32, 42, 69, 74, 114, 156, 158, 161, 294, 332, 340
Oncorhynchus nerka 60, 63, 115
Oncorhynchus tshawytscha 63, 115
Ophidion 136
Ophioblennius atlanticus 137, 139, 140
Ophiocephalus 5
Opisthoprochtus 305
Oplegnathus fasciatus 121
Opsanus tau 144
Oreochromis 16, 281
Oreochromis mossambicus 19, 77, 106, 275, 277, 281, 294-296, 298, 299, 301, 337, 437
Oreochromis niloticus 16, 34, 77, 106, 134, 142, 417
Oryzias 44, 94, 95, 120, 147, 161, 162, 167, 169, 170, 171, 181, 203, 211, 222, 232, 276, 471
Oryzias latipes 7, 8, 15, 33, 44, 63, 71, 73, 94-96, 117, 119, 120, 126-128, 150, 156, 157, 159, 160, 162, 166, 168, 170, 172, 175, 178-180, 182, 197, 198,

Species index

204, 205, 245, 275, 294-301, 471
Osmerus 24
Osmerus eperla 87
Osmerus eperlanus 13, 57, 88, 114, 259, 282
Osmerus mordax 87
Ostracion meleagris 121
Oxyporhamphus micropterus 81

P

Pagrus major 109, 121, 341
Pantodon 140
Pantodon buchholzi 134, 136, 137, 142
Papyrocranus afer 139, 142
Parabrotula dentiens 489
Parabrotula plagiophthalmus 490
Paracallionymus costatus 81
Paracheirodon innesi 142
Paraglyphidodon nigroris 99
Paraleucogobio soldatovi 11
Paramonacanthus japonicus 14
Parapercis 140
Parapristipoma octalineatum 139, 140
Parapristipoma trilineatum 121
Parephippus 153
Parophrys vetulus 78, 187, 227, 245
Pecten operculatus 15
Pegusa triophthalmus 136, 142
Perca 2, 11, 52-54, 59, 62, 63, 72, 75, 103, 104, 106, 133, 151, 262, 386, 409, 416, 423, 424, 446
Perca americana 105
Perca flavescens 103, 446
Perca fluviatilis 4, 11, 20, 42, 52, 53, 59, 60, 62, 72, 103-105, 111, 114, 124, 132, 149, 187, 259, 294, 326, 334, 385, 422, 424, 426, 443, 444, 447, 448
Periophthalmus papirio 139

Petromyzon 244
Pholis 423
Phoxinus 44, 58, 62, 89, 111
Phoxinus lacvis 111, 114
Phoxinus phoxinus 44, 45, 59, 60, 62, 64, 67, 89, 109, 169, 271
Pimephales promelas 89
Platichthys flesus 50, 341
Platichthys stellatus 62, 156, 157
Platypoecilus 496
Platypoecilius maculatus 67
Plecoglossus 149
Plecoglossus altivelis 59, 134, 135, 150, 159, 273
Plectropomus lepidorus 142
Pleuronectes americanus 31
Pleuronectes microcephalus 388
Pleuronectes platessa 11, 31, 49, 65, 72, 175, 176, 340, 341
Pleuronectidae 50, 121
Pleuronectinae 50
Pleuronichthys 79
Pleuronichthys coenosus 59, 60, 65, 70, 71, 78-80
Pleuronichthys cornutus 78
Poecilia 2, 321, 424, 428, 451, 494, 496, 502, 503
Poecilia latipinna 136, 471
Poecilia reticulata 2, 4, 7, 9, 24-29, 32, 35, 67, 72, 124, 134, 136, 142, 143, 278, 280, 285, 287, 304, 310, 318, 320, 334, 337, 339, 372, 373, 375, 403, 405, 409, 410, 424, 425, 428, 447, 449, 476, 494-498, 500-503, 507
Poecilia surinamensis 496
Poeciliidae 481
Poeciliopsis 458, 471, 474, 475
Poeciliopsis B 471-474
Poeciliopsis C 474-476
Poeciliopsis turneri 471
Poecilistes D 471, 472, 474
Polyodon spatula 314
Polypterus 282, 378

Pomacentrus 17
Pomasidae 140
Pomatoschistus minutus 7, 15, 17, 57, 60, 65, 71, 100, 101, 114, 153, 259
Porichthys notatus 92, 144
Pseudobalistes fuscus 136, 142
Pseudogobio rivularis 6
Pseudomugil 142
Pseudomugil gertrudae 97
Pseudomugil mellis 97
Pseudomugil signifer 97
Pseudomugil tenellus 97
Pseudopleuronectes americanus 31, 441, 446, 448, 449, 452, 455
Pterophyllum 2, 24, 203, 428, 451
Pterophyllum scalare 3-5, 7, 25, 26, 288, 294, 394, 397, 398, 405, 409, 410, 419, 420, 424, 427, 447-450
Pungitius 109
Pungitius pungitius 392
Pungitius tymensis 86, 112
Pygosteus pungitius 57, 85, 86, 112, 124, 125

Q

Querichthys stramineus 142

R

Raniceps 62
Rhacochilus 492
Rhodeus 181
Rhodeus amarus 13, 17, 111, 114, 149, 152, 155, 185, 214
Rhodeus ocellatus 134, 150, 156, 157, 159-161, 163, 165, 182
Rhombus (spp.) 388
Rhombus maximus 13
Rhynchorhamphus marginatus 90
Rivulus marmoratus 471
Roccus americanus 114

Rudarius ercodes 14
Rutilus rutilus 89

S

Salarias flavoumbricus 46, 102, 124
Salmo 2, 178, 179, 217, 226, 233, 240, 252-254, 257, 385, 418, 422-424, 436, 441, 445
Salmo alpinus 84
Salmo alsaticus 388, 419
Salmo fario 20, 42, 63, 64, 111, 112, 134, 149, 151, 152, 158, 178, 185, 187, 188, 199, 210, 216, 240, 385, 432, 438
Salmo fontinalis 63, 64, 66
Salmo gairdneri 32, 60, 62, 68, 73, 109, 115, 116, 134, 135, 176, 255, 272, 280, 294, 296-298, 300, 339
Salmo iridea 161
Salmo irideus 60, 64, 65, 72, 150, 153-155, 178, 190, 194, 213, 234, 251, 254, 294
Salmo salar 4, 19, 20, 43, 60, 64, 72, 109, 111, 112, 132, 149, 151, 152, 158, 163, 178, 185, 210, 214, 273, 325, 332, 333, 336, 340, 372, 375, 378, 386, 388, 409, 415, 419, 422, 432, 435, 438, 443
Salmo salvelinus 65, 261
Salmo trutta 19, 57, 74, 84, 109, 115, 116, 194, 278, 281, 294, 296-298, 314, 332-334, 337, 372, 375
Salmo trutta fario 161, 190, 298, 300
Salmo trutta morpha fario 109, 115, 155
Salmo trutta morpha lacustris 148
Salmo trutta trutta 60, 64, 65, 72
Salvelinus 60, 65, 251, 252, 262,

264
Salvelinus alpinus 109
Salvelinus fontinalis 60, 62, 63, 72, 84, 109, 115, 149, 150, 176, 255, 273, 294, 296-298, 433
Salvelinus pluvius 294, 296, 297, 298
Sarcocheilichthys variegatus 156
Sardina pilchardus 340
Sardinella aurita 136, 139
Sardinops 19
Sardinops caerulea 280, 340
Sardinops melanostictus 121
Sarotherodon (see Oreochromis)
Sarotherodon mossambicus (see Oreochromis mossambicus)
Sarotherodon niloticus (see Oreochromis niloticus)
Saurida elongata 121
Saurus lacerta 78
Scardinius erythrophthalmus 259, 314
Schizostedion vitreum 117
Sciaenops ocellatus 340
Scomber 62
Scomber scombrus 340
Scomberesox 82, 83
Scomberesox saurus 82
Scopelosaurus lepidus 326
Scophthalmus maximus 13, 74, 341
Scorpaena 149, 158, 162, 180, 384
Scorpaena angolensis 136, 138, 142
Scorpaena porcus 17, 106, 150
Scorpaena rufa 17
Scorpaena scrifa 152
Scorpaena scrofa 38, 106, 114, 150
Sebastes 492, 493, 503
Sebastes diplopora 335, 341
Sebastes paucispinis 493
Sebastes schlegeli 202, 493
Sebasticus 492
Sebastodes 492
Sebastodes rubrovinctus 458
Sebastolobus alascanus 106

Serranus 38, 187, 217
Serranus atrarius 177, 187, 188, 190, 194, 199, 210, 214, 218, 219, 242, 262, 386, 418, 433, 439-444, 448, 451, 498
Serranus heptatus 106
Serrasalmus natteri 89, 120
Sfizostedion vitreum 162
Sillago japonica 121
Silurus glanis 103, 111
Siniperca chua-tsi 11
Siphonostoma typhle 16
Skiffia 485
Solea solea 19, 340
Solea vulgaris 388, 389
Sparus aurata 32
Spinachia 15, 394, 409
Spinachia vulgaris 259
Sprattus sprattus 87
Stenarchus albifrons 135, 140
Stizostedion vitreum 12
Stolephourus 121
Strongylura 90
Strongylura strongylura 95
Sturisoma aureum 117, 122, 123, 157
Stygocola dentata 458
Syngnathus 16, 50, 111, 240, 422
Syngnathus acus 111, 422
Syngnathus fuscus 32, 35, 44, 65, 68, 71
Syngnathus nigrolineatus 16
Syngnathus ophidium 111
Syngnathus scovelli 9, 17, 44, 68, 72, 74

T

Tautoga onitis 11
Tautogolabrus adspersus 21, 433
Thymallus thymallus 390, 409, 410
Thymallus vulgaris 151
Tilapia 16, 77, 99, 106, 294
Tilapia galilea 16, 19, 106, 107

Tilapia guinensis   77
Tilapia leucosticta   345
Tilapia macrocephala   16, 19, 77, 106
Tilapia mossambica (see Oreochromis mossambicus)
Tilapia nilotica   (see Oreochromis niloticus)
Tilapia tholloni   19, 77
Tilapia zillii   77
Tinca   214
Tinca chrysitis   132
Tinca tinca   109, 318
Tinca vulgaris   114, 149, 153, 154, 185
Tomeurus gracilis   471, 493
Trachinocephalus   137
Trachinocephalus myops   136, 138, 142
Trachinus   177
Trachinus vipera   20, 114, 216, 313, 318, 319, 329, 330, 348, 388, 418
Trachypterus   5
Trematomus eulepidus   81
Triacanthus brevirostris   58, 70
Tribolodon hakonensis   69, 89, 149, 150, 275
Trichiurus savala   58, 70
Trichopodus trichoperus   444, 449
Tridentiger trigonocephalus   275
Trigla   20, 214
Trigla gurnardus   20
Tropheus morii   16, 19
Trutta fario   178, 180-182, 187
Trutta iridea   178, 180
Trutta lacustris   161
Trutta morpha lacustris   109
Tylosurus   90
Tylosurus acus   91

U

Upeneus prayensis   136, 138, 140, 142
Uranoscopus   423
Uranoscopus scaber   57, 78, 422

V

Verasper moseri   31, 32
Vibrio anguillarum   176
Vinciguerria   19
Vomer setapinnis   142

X

Xenoophorus   485
Xenoophorus captivus   68
Xenophorus, 486
Xenophorus captivus   68, 481
Xenotoca   485, 486
Xenotoca eiseni   68
Xenotoca   485
Xiphophorus   424
Xiphophorus helleri   32, 67, 136, 477, 496, 498

Z

Zenarchopterus   144, 493
Zenarchopterus dispar   136, 137, 142
Zeus faber   136, 139
Zoarces   461, 462, 465, 466, 468
Zoarces vivipara   462, 466, 502
Zoarces viviparus   9, 50, 62, 67, 132, 382, 415, 457, 460, 461-465, 467
Zoogeneticus   485
Zoogeneticus Hubbsina   485
Zoogeneticus quitzeoensis   488

Subject index*

A

Absorption of embryotrophe by epidermis 470
Absorptive trophotaenial epithelium 486
Accessory cell to chloride cell 281, 284
Accessory outer segment of photoreceptors (AOS) 315-318, 343, 344, 346, 349, 353
Accessory structures of egg envelope 77-108, 507
    Accordion-folded ribbons 103
    Adhesive disc 89, 101, 102, 107
    Adhesive pedestal 99
    Adhesive threads 93, 96
    Anchoring stalk 88
    Annual fishes 77
    Antarctic icefishes 77
    Appendages 86, 87, 89, 101, 107
    Areolae 102
    Blebs 87, 107, 230
    Bristles 82
    Buttons 85
    Cones 84, 87, 89, 90, 99, 100, 102, 105, 107
    Cylindrical processes 89
    Disc 89, 101, 102, 107
    Dome-like projections 84, 85, 107
    Facetted hexagons 89, 105
    Fibres 91, 101
    Fibrils 87, 89, 90-93, 101, 102, 107
    Filaments 82, 83, 87, 90, 92-101, 107, 507
    Formation by follicular cells (granulosa) 107
    Hexagonal pattern 78, 81, 102, 107
    Hooks 98, 107
    Jelly capsule 105
    Jelly plugs 107
    Microplicae 84
    Micropyle 78, 90, 93, 95, 99, 100, 102
    Microvilli 89
    Mouthbrooders 77, 106, 108
    Mushroom type appendages 85
    Netlike string 104
    Nodules 107
    Papillae 107

* Note that in this index only trivial names of fishes are included

Pedestals 99
Pedicles 107
Peduncles 101
Reticula 102
Ribbon-shaped veils 106
Spikes 87, 107
Stalk 87, 102
Structures for floating and attachment 77, 103, 108
Suspensory ligament 87, 107
Tendrils 83, 91
Threads 93, 96, 101, 102
Tufts 83, 90, 91, 94, 95, 97
Accommodation of eye 308, 331
Acetylcholine 197
Acidophilic cells 282
Acrosome 129, 135, 146, 507
Actin 42, 43, 160-162, 166, 172, 179, 213
Actinotrichia 279
Activation 7, 24, 38, 45, 46, 61, 148, 150, 156, 168, 507
Adelphophagy 460, 491
Adenohypophysis 431
Adhesion molecule (occludin) 278
Adhesive 'cement' 288, 290
Adhesive glands 3, 278, 288, 290
Adhesive organs 4, 285
Adhesive phase 288
Adrenal tissue 410
Adult (mature) type (A-erythrocytes) 411
A-granules in cortex 170
Allantois 259, 260, 496, 498, 499
Altricial embryo 1, 7, 12, 508
Alveolar membrane 167
Alveoli 41-48
Amitotical division of periblast nuclei 221, 507
Amnion 472, 474
Amniotic folds 472, 503
Ampulla 271
Anacrosomal sperms 135
Anal fin 120, 278, 288, 424, 425, 427, 468-470, 497
Anal fins modified into gonopodium 458, 469, 481, 482, 487, 493, 507
Anal or postanal vesicle 260
Anchoring stalk 88
Anchovy 278, 280
Animal pole 12, 13, 36
Animal-vegetal axis 190
Annual fishes 6, 77, 108, 222
Annular margin (ora serrata) 304, 339, 364
Annulate lamellae 36, 37
Annulus 185
Anoxia, cause of hatching 300
Antarctic icefishes 77
Anterior epiphyseal vesicle 273
Aorta 226, 227
AOS (accessory outer segment of photoreceptors)
Apical body 135
Apoptosis (normal cell death) 278, 298, 299, 335
Appendices of pylorus 439
Appositions of plasmamembranes 211, 213, 231
Aquasperm 131, 134
Archenteron 234, 260, 261
Archiblast 198, 216
Archinephric duct 378-380
Area (site of acute vision) in retina 326, 368, 369
Argentea enclosing eye 304, 306, 331
Arterial bulb 419
Artificial fertilization 150, 154
Asteriscus 268, 271
Atrium 415, 416, 419, 420
Atrophy of the postanal gut 440
Auditory anlage 239, 241, 270
Auditory organ 268
Auricle 415, 416, 419, 420
Avascular RPE and retina 307
Axial plate 242, 243

Axial (median) vein 226, 385, 386, 388, 422, 423, 429
Axoneme 136, 137, 143, 145, 146, 357, 366

B

Baltic cod 23
Barnacle 15
Basal ganglion of brain 269
Bathypelagic 10
Behaviour 8
Behaviour patterns 6
Belly sac 472-474, 478, 479
Benthic 5, 12
Benthonic 12
Beta components, derivatives of vitellogenin 31, 32
B-granules in cortex 170
Biflagellarity of sperm 144
Biological clock 272
Bipolar cells of retina 272, 329, 330-332, 353, 357-362, 370, 371,
Bipolar differentiation 9, 148, 151, 153, 154, 176-178, 180
Bipolar segregation 38
Birth 496, 501
Bitterling 86, 135, 156, 260
Bivalves 6
Black patch on sclera 307
Black pigment in the RPE 339
Blasteme 232
Blastoderm 208, 215, 218, 220, 508
Blastodisc 12, 13, 110, 151-153, 174, 177, 185, 188-190, 192-194, 196, 208, 212, 214, 215, 218, 254, 508
Blastodisc floor 187
Blastomeres 185, 188-190, 193, 194, 220
Blastopore 209, 210, 216, 218
 Dorsal lip 210, 216, 218, 220, 225, 231, 236, 252, 254, 255
 Lateral lips 225

Blastopore closure 225
Blastoporus (see blastopore)
Blastula 185, 190, 194-197, 202-204, 210, 212, 215, 220, 221, 230, 234, 235, 251, 252, 257, 419, 435
Blebbing 232
Blebs 84, 107, 196
Block of polyspermy 162
Blocking of involution 220
Blood 31-33, 39, 73, 74, 198, 204, 220, 226, 285, 288, 308, 331, 377, 429, 478
Blood anlage (see also haematopoietic sites)
 ICM 252, 377, 380, 381, 385, 389, 393, 395, 396, 398, 403-406
 Islands on yolksac 381, 394, 397, 398, 403, 405, 407, 409, 509
 Posterior islands 404, 405, 407
 Posterior islands + ICM 382, 398, 403, 405, 408
 Praecranial mesoblast 403-405
 Praecranial mesoblast, ICM and posterior islands 403, 404, 409
Blood cells absorbed by rectal wall 465
Blood circulation 76, 205, 282-284, 308, 378, 381, 415, 418, 421-424, 426, 428, 437, 452, 461, 494-496
Blood formation (see haematopoietic sites)
Bloodmesoderm 403-405
Blood vessels 25, 27, 220, 226-228, 278, 286, 307, 337, 377, 420, 422, 428, 442, 445, 449, 452, 453, 483, 494, 495
 Vascular endothelial cells 400
BM (see Bruch's membrane)
Bony fin rays 280
Brain 220, 240, 241, 245, 253, 267-269, 508

Anlage 24
Basal ganglion 269
Bulbus olfactorius 269
C4-isozyme of LDH 374
Central nervous system (CNS)
Cerebellum 268
Cerebral primordium 235
CNS 220, 508
Diencephalic roof 272, 274
Diencephalon 248, 269, 270
Encephalon 268
Epiphysis 269, 271, 273
Epiphysis in embryo 272
Evagination of diencephalon 337, 338
Forebrain 242, 244, 245, 268, 269
Hemisphere 269
Hindbrain 244, 245, 268, 269
Hypophysis (pituitary)
Infundibulum 269
Lobus impar 269
Medulla oblongata 245, 268
Mesencephalon 268-270
Metencephalon 268-270
Midbrain 240, 242, 244, 245, 248, 250, 268, 269
Myelencephalon 268-270
Neuromeres 268-270
Optic tectum 268, 335
Optical part of the brain 372
Pallium 269
Paraphysis 272
Parapineal 272-274
Pinealbody/gland/organ 271-276, 291, 312
Pituitary (hypophysis) 34, 39, 269, 431, 437, 468, 469
Plica ventralis 269
Presumptive telencephalon 268
Primary brain 269
Prosencephalon 244, 268, 269
Prospective telencephalon 268
Rhombencephalon 268, 269

Saccus dorsalis 273
Saccus vasculosus 269
Tectum opticum 268, 335
Telencephalon 269, 270
Third eye (epiphysis) 269, 271, 273
Valvula cerebelli 268, 269.
Velum 272
Velum transversum 273
Branched synapses 357
Branchial circulation 283
Branchial epithelium 281
Branchial placenta 488, 492, 502, 503
Breakdown of cortical alveoli 176
Breakdown of hatching glands 294
Breaking of envelope 293
Bream 5
Bristles 82
Brook trout 251, 294
Brown trout 335, 340-342, 344, 345, 350, 351, 359
Bruch's membrane (BM) 310, 311, 331
Buccal epithelium 502
Buccal placenta 490, 491, 503
Bulbus aortae 415
Bulbus arteriosus 415, 416
Bulbus cordis 415, 419
Bulbus olfactorius 269
Buoyant 10, 11, 62, 81, 103
Buttons 77, 85, 132

C

C4-isozyme of LDH in retinal photo-receptors (and optical parts of the brain) 374
CA (cortical alveoli)
Calcium 61, 75, 148, 156, 170, 171, 176, 197
Calcium wave 179
Calycal processes 344
Calyx nutritius simplex 461

Canalis centralis of neural tube 241
Capeline 61
Carbohydrate yolk 39, 46
Cardiogenic precursors migrate medially 419
Carina 239
Carotenoid pigment 5, 10, 12, 13, 16, 23
Carotenoids, respiratory importance 494
Carp 5, 46, 165, 388, 419
Caspian shad 11
Catfish 16, 433
Cathepsin D 33
Caudal fin 258, 278, 425, 470
Caudal vein 398, 422, 423, 426, 497
Cave fishes 303
Cavitation in nerve rod 239, 249
Cells
   Arrangement into a reticulum 441
   Divisions 185
   Intercalations 247
   Junctions 341
   Lineage analyses 226, 246
   Motility 230
   Movement 230
Cellular envelope 187
Cement 3, 5
Central canal of neural tube 241
Central nervous system (CNS)
Centrioles 140, 143, 146
Centrolecithal 39, 148, 178
Cephalic ectoderm 252, 253
Cephalic flexure 268
Cerebellum 268
Cerebral primordium 235
Cessation of mitoses in periblast 203
Chloride cells 277, 278, 280, 282, 284, 288, 475, 476, 500
   Accessory cell 281, 284
Chondrocytes 276
Chondrostei 185, 186

Chorda dorsalis (see notochord)
Chordamesoderm 207, 209, 225, 226, 377
Choriocapillaris 284, 308, 331
Chorion 52
Chorionase 297, 300
Choroid 304, 331, 337, 338
Choroid fissure 339, 360-362, 371
Choroid gland 308, 331
Chromatophores 276, 278, 279, 338, 508
Chum salmon 42, 44, 63, 68, 126, 128, 163
Cilia 245
Ciliary body 337
Cilium of photoreceptor 265, 315
Circadian oscillator 273, 274
Circadian rhythm of RMM 331
Circadian shedding pattern of photoreceptor outer segments 359
Circulation of serum and finally blood cells 418
Cleavage 185-198, 204
   Animal-vegetal axis 190
   Filipodia 194
   Furrows 188
   Horizontal cleavage 186, 187, 190
   Pattern 204
   Putative regulators of cleavage 197
   Secondary cleavage 199, 202
   Transition from holo- to meroblastic cleavage 185
   Vesicle cleavage 195
Clinid blennies 469
Closed neural fold 240-242, 249
Closure of blastopore 210, 211, 222, 234
Clusters of small melanin granules in RPE 342, 343
CNS (central nervous system) 220, 508
   Canalis centralis 241

Cod 10, 23, 38, 43, 57, 61, 78, 177, 439, 441
Coelom 474
Colchicine 230
Colloid in thyroid 437
Colour cone pattern polymorphism (retina) 360
Colour of the egg 5, 12
Colour vision 508, 509
Common carp 23, 170
Comparison of fate maps 256
Concrescence theory 209
Cone (see retinal cones or cones in accessory structures of envelope)
Cone outer segment (COS) (see under retinal cones)
Conjoined twins (blastoderm) 195
Contraction of the E-YSL 222
Conus arteriosus of heart 416, 420
Conventional synapses 356, 357, 371
Convergence 209, 210
Convergent extension 209, 217, 226, 230, 236, 508
Copulatory apparatus 491
Copulatory organ 461
Cornea 304, 330
Cortex 32, 39, 41-48, 74, 92, 149, 167-169, 171, 507
Cortical alveoli 24, 32, 35, 36, 39, 41-48, 61, 149, 167, 169, 172, 173, 176, 209, 210, 216
Cortical granules 42, 44, 46, 171
Cortical reaction 45, 46, 148, 150, 154, 162, 164-167, 172, 173, 175, 176
Cortical vesicles 45, 46
COS (cone outer segment)
Crucian carp 5
Crystalline pattern of yolk 27, 35, 39
Cupula 274, 276
Cutis 279
Cyprinodont type of ZR 61, 65
Cystovarian 20, 458
Cytochalasin 172, 179

Cytokinesis 186
Cytoplasm mixed with yolk 36, 194
Cytoplasmic streaming towards the animal pole 178
Cytoskeleton 42
Cytotomy 186

D

Dark-adapted retina 328
Dark-adapted retina with bent cone outer segments 355
DC (deep cells)
Death of cells by apoptosis 278, 298, 299, 335
Deep cells (DC) 187, 194, 209, 210, 215-217, 226, 255, 508
Deep sea fish 305
Degenerating spinules in retina 356
Degeneration of embryos as nutrients for the survivors 487
Degree days 1
Dehiscence of CA 168
Delamination 214, 216, 220, 255
Demersal eggs 8, 10, 12, 13, 19, 21, 24, 38, 39, 49, 50, 62, 64, 67, 75, 78, 81, 91, 176, 178, 180, 185, 186, 492, 506, 507
Demersal-adhesive eggs 13, 77, 84, 86, 99, 107
Demersal-non-adhesive eggs 49, 81, 83, 107
Demersal/pelagic eggs 13
Dense yolk 25, 26
Deposition of eggs 20
Dermatome 377, 379
Dermis 279
Dermotrophic transfer sites 502
Dermotrophic uptake 460
Dermotrophy 503
Desmosomes 212, 245, 278
Developing teleost eye 337
Development of Bruch's membrane 339

Development of eye 303
Development of heart 415
Development of heart pulsation pattern 419
Development of photoreceptors 361
Development of retina 339
Development of vitelline circulation 423, 425
Development to term in the follicle 471
Developmental arrest 154
Developmental changes from square to row mosaic of cones 349
Developmental gradient in the retina from fundus to periphery 339
Developmental pattern of phagosome count 350
Developmental sequence in the retina 341
Developmental switching of haemoglobins 415
Diapause 6, 93
Diblastodermic eggs 195
Dichromatic vision 318
Diel RMM 345, 348
Diencephalic roof 272, 274
Diencephalon 248, 269, 270
Differentiating amacrine cells 358, 370
Differentiating bipolar cells 358, 370
Differentiation of entoderm seems to be species-specific 257
Differentiation of horizontal cells 370
Differentiation of retinal rods is ahead of that of cones 345
Dip in dorsal neurochord 243, 245
Direct development 4, 8
Disappearance of UV-cones 332, 334
Discontinuous entoderm 434
Discs of the COS 315
Discs of the ROS, separated into packages of 2 344
   packages of ~9 352

Discus proligerus 110
Displaced amacrine cells 362, 364, 370, 371
Distant touch receptors 276
Distribution of hatching glands 295
Division of photoreceptor inner segment into myoid and ellipsoid 345
Dome stage 210
Dome-like projections 84, 85
Dorsal aorta 226, 228, 249
Dorsal blastopore lip 220, 254
Dorsal dip in nervechord of embryo 242
Dorsal fin 258, 278, 468-470, 487
Dorsal fin fold 5, 278
Dorsal groove in neurula 242
Dorsal light response 322
Dorsal mesocardium 415, 417
Dorsal organizer 207
Dorsal sac 272-274
Dorso-lateral lips of blastopore 220
Double cones (twin cones) 314, 316, 319, 344
Double cones and single cones appear at the same time 359
Double cones subsurface cisternae 344
Ductus Cuvieri (blood vessel) 420
Duplex retina 311, 340, 344
Dusk receptor 272

E

Ear 239, 270
Eccrine secretion 298
Ecomorphology 4, 8
Ectoderm 207, 208, 217, 220, 221, 235, 236, 252
Ectodermal derivatives 267-292
Ectodermal floor plate 248, 250
Ectodermal headfold 406, 495, 503
Ectomesoderm 267
Eelpouts 460
Eggs 9-22

Accesssory structures 77-108
Adhesive 21, 77, 94
Care 14
Colour 5, 12
Demersal 8, 10, 12, 13, 19, 21, 24, 38, 39, 49, 50, 62, 64, 67, 75, 78, 81, 91, 176, 178, 180, 185, 186, 492, 506, 507
Demersal-adhesive 13, 77, 84, 86, 99, 107, 507
Demersal-non-adhesive 49, 77, 107, 507
Demersal/pelagic 13
Deposition 20
Digestion 293
Envelope 49-76
Eyed egg 339, 341
Facetted hexagons on surface 89, 105
Gymnovarian 20
Influx of water into eggs 11
Meroblastic 186
Mosaic egg surface 168
Non-adhesive 13, 21, 87, 95, 197, 231
Number 18, 21
Pelagic 10-12, 14, 19, 21, 24, 26, 38, 39, 49, 50, 52, 62, 65, 67, 75, 77, 78, 82, 87, 106, 121, 147, 176, 177, 180, 185, 201, 211, 216, 261, 262, 388, 389, 428, 438, 507
Pelagic-nonadhesive 10, 50, 77, 99, 107
Pelagic ribbon of eggs 11
planktonic (pelagic)
Plasmamembrane 51
Shape 17, 21
Size 19, 21
Sperm-egg fusion 165
Transparency of live eggs of pelagic species 26
Type 10
Unattended 12

Eggsac (funiculus) 461
Elasmobranchs 452
Electrical coupling between DC 231
Ellipsoid 316
Ellipsosome 318, 320-324, 345, 347, 348, 352, 359, 360
Elliptical, though still immature, erythrocytes (erythrocytes-ima) 410
ELM (external limiting membrane)
Elongated and coiled midgut 478
Emboly 207
Embryo, definition 7, 381
Embryonic and extraembryonic integument 236
Embryonic and mature haemoglobin 415
Embryonic axis 236
Embryonic (choroid) fissure 339, 360-362, 371
Embryonic ectoderm 235
Embryonic erythrocytes (erythrocytes-E) 397
Embryonic nutrients 460
Embryonic shield 207, 209, 217, 218, 226, 229, 239, 240, 267
Embryonic shield attracts cells 230
Embryonic skeleton 276
Embryos versus postembryos 1
Embryotrophe (nutrients) 460, 462, 463, 470
Embryotrophe /histotrophe 462, 492
Encephalon 268
Endocardial blood cells 419
Endocardial cells 412
Endocardial tube 494
Endocardium 382, 384, 390, 410, 411, 415, 416, 418, 419
Endocrine cells of the intestine and rectum 435
Endocrine gland 272
Endocytosis 31, 477
Endoderm (see entoderm)

Endothelium 415, 418, 419
Endothelium of blood vessels 400
Endothelium of heart 390
Endothelium of heart tube 417
Enlarged pericardium 503, 507
Enteroblast 433
Enteroderm (see entoderm)
Enterotrophic transfer sites 502
Enterotrophic uptake 460
Enterotrophy 503
Entoblast 432, 433
Entoderm 207, 208, 217, 220, 231, 235, 236, 242, 243, 251, 252, 254, 257, 260, 262, 508
　Absorption of yolk 452
　Acinar cells of pancreas 431
　Adenohypophysis 431
　Cavity 229
　Cells 217, 229, 509
　Cells, after involution 441
　Derivatives of entoderm 431-456, 508
　Formation of intestine 389
　Hindgut 3, 465, 467-469, 481, 482, 485, 492, 493, 502, 503
　Hypoblast 206, 207, 214, 216, 217, 220, 221, 225, 226, 229, 233, 236, 254, 255, 264, 265, 267, 380, 382, 394, 431-433, 435, 438
　Hypochord 226-229, 236, 248, 249, 392, 431
　Involution 435, 438
　Kupffer's vesicle 259-266, 431
　Lining of digestive system 431
　Origin from the periblast 433
　Parenchymal cells of liver 431
　Part of the gills 281, 431
　Precursory endodermic cells 296
　Pseudobranch 81, 308, 431
　Rectum 466
　Secondary entoderm 431
　Sheet 252
　Swimbladder 431

Entoderm arises as a differentiation from the inflected germ ring 432
Entoderm originates from periblast 433
Entodermal and mesodermal precursors 435
Entodermal cells once involuted grow centripetally to form the intestine 441
Entomesoderm 277
Envelope (periderm, EVL) 9, 49, 55, 56, 66, 507
　Cells 508
　Enveloping layer 210
　Formation of the envelope 74
　Number of layers in the envelope 50
Epaulettes 502
Ependyma 240, 241
Ependymal canal 241
Epiblast 207, 214, 217, 220, 225, 226, 229, 236, 239, 241, 267
Epiboly 207, 208, 210-213, 215, 218-221, 234, 508
Epicardium 415, 420
Epidermal cells 242
Epidermal necrosis 334
Epidermal scales 6
Epidermic stratum 210, 239, 240, 241, 242
Epidermis 210, 220, 236, 243, 267, 278-280, 288
Epimere 379
Epimyocardium 416, 419
Epiphysis cerebri 269, 271
Epiphysis in the embryo 272
Epithelio-chorial pseudoplacenta 498
Erythroblasts (precursors of the red blood cells) 382
Erythrocytes 381
　Adult (mature) type (A-erythrocytes) 411
　Elliptical, though still immature,

erythrocytes (erythrocytes-ima) 410
Embryonic erythrocytes (e-erythrocytes) 397, 410
Ima (immature-adult)-erythrocytes 403, 411
Mature type of erythrocytes (erythrocyte-A) 410
Microspectrophotometry of erythrocytes 321
Primary mature erythrocytes 414
Shift from early/larval to late type erythroblasts 415
Erythropoietic sites in blood islands 403
Erythropoietic sites in ICM 409
Erythropoietic stages in the developing pike, Esox lucius 414
Esterase activity 197
Estrogen (see oestrogen)
European eel 272
Euryhaline 494
Eurythermic 494
EVL (see periderm and envelope)
EVL generating DC 212
Exocytosis 167, 171
Exogenous vitellogenin 32
Exo-rhodopsin 272
Expanded coelom 472
Extended blood mesoderm 403, 405
External bearers 15
External limiting membrane (ELM) 310, 366, 368
External yolk syncytial layer (E-YSL) 213, 214, 221, 222
Extra-embryonic ectoderm 235, 253
Extraretinal photoreceptors 271, 272, 312
Extrusion of supernumerary sperms 162
Eye of sexually mature fish 268, 303-337
  Argentea 304, 306, 331

Blood supply, doubly oxygenated 308
Bruch's membrane 310, 311, 331
Cave fishes 303
Choriocapillaris 284, 308, 331
Choroid 304, 331, 337, 338
Choroid fissure (embryonic fissure) 339, 360-362, 371
Choroid gland 308, 331
Chromatophores 338
Circadian rhythm of RMM 331
Circadian shedding pattern of photoreceptor outer segments 359
Cone outer segments (COS) 315
Cone square mosaic changes to row mosaic 333
Cones arranged in row mosaic 333, 334, 335
Cones, different types 311
Cornea 304, 330
Dichromatic eye 318
Embryonic fissure (see choroid fissure)
Falciform (sickle shaped) process 284, 308, 331, 337
Fovea of twin-cones 326
Ganglion cell layer 310, 330, 371
Glial (Müller) cells 310, 331
Growth zones 333-335, 509
Hyaloid artery 308, 331
Inner nuclear layer 310, 329, 331
Inner plexiform layer 310
Internal limiting membrane (ILM) 310
Iris 305, 331
Lens 304
Lentiform body 308
Light-adapted retina 327
Light- and dark adaptation 331
Open ventral fissure 335
Ophthalmic artery 308
Optic chiasma 269
  Nerve 331

Tectum   335
Vesicle   245
Outer nuclear layer   310, 326, 331
Outer plexiform layer   310, 331
Phagosomes   311, 326, 331
Photoreceptors   311
Pigment layer (RPE)   310
Pseudobranch, provides blood   308
Pupil   331
Putative UV-Cones   314
Regresssion of eye   303
Rete mirabile   308
Retinal pigment epithelium (RPE)   311, 313, 331
Retinomotor movements (RMM)   305, 331
Retinotectal synapses   335
Retractor lentis muscle   308, 331
Rim of the eye   336
Rod outer segment (ROS)   315
Rods   311, 312, 315-317
Rods arranged in tiers   314
Sclera   304, 306, 330
Shedding and renewal of photoreceptor outer segments   326, 331, 345, 348-350
Simplex retina (rods only)   312
Square cone-mosaic   319, 325, 333-335
Telescopic eye   305
Tiered cones   305, 326
Vascular system of the eye   308
Vascular system of the pseudobranch to the eye   308
Ventral embryonic fissure   333
Visual cell layer (VCL)
Eye development   303, 337-376
  Annular margin (ora serrata)   304, 339
  Area (high density and tiering of cones)   371
  Bruch's membrane   339
  Ciliary body   337
  Circadian rhythm of RMM   331
  Circadian shedding pattern of photoreceptor outer segments   359
  Cone outer segments   333, 346, 348, 349, 355, 366, 368, 372
  Cones in row mosaic   359, 367
  Evagination of diencephalon   337, 338
  Eye cup   239, 270
  Eyed ova   339, 341
  Falciform process   337
  Fissure (embryonic/choroid)   337, 338
  Ganglion cells   362
  Growth by cell addition and stretching   337
  Horizontal cells with gap junctions   354
  Hyaloid artery   337
  Inner limiting membrane   357
  Inner nuclear layer   357, 360, 361, 370
  Inner plexiform layer   357
  Iris   337
  Lens anlage   270
  Lens formation   337, 338
  Migration of eye at metamorphosis   6, 8, 340
  Müller (glia) cell   357
  Myeloid body in RPE   311, 342-343, 347, 359
  Ontogeny of RMM   351
  Ophthalmic artery   338
  Optic anlage   246
    Cup   337, 338
    Fissure   338
    Lobe   337
    Nerve   337
    Nerve exit   341
    Primordia   360
    Stalk   337
    Tectum   335
    Vesicle   245, 337

Optical part of the brain 372
Outer nuclear layer 344, 359, 361, 365
Outer plexiform layer 353
Pedicles 197, 353, 359
Phagosomes 311, 326, 331, 342, 359
Photoreceptors 305, 311, 312, 314-316, 318, 326, 329-331, 335, 339, 342, 344, 345, 347, 354, 361, 362, 365, 372, 374, 375
Pinocytotic vesicles in RPE 341
Primordium to optic cup 239, 244, 245, 261, 268, 270, 337, 508
Pseudostratified retina 339, 361
Pure cone retina 340
Reciprocal synapses 359
Retinal pigment epithelium 338, 341-343, 363
Retinomotor movements 341, 342, 344, 348, 351, 359
Retinotectal projection 341
Ribbon synapses 370
Rod outer segment discs 344-349, 352, 356, 365, 366, 368, 372, 375
Rod precursors migrate from from inner to outer nuclear layer 340
Rod terminals (spherules) 360
Rods form part of regular mosaic 335
Rotation of the eye 338, 361
Sclera 337, 339
Spherules 353, 359
Spinules 354, 360
Square mosaic 354, 359, 360
Synaptic contacts between dendrites of photoreceptors, horizontal and bipolar cells 354
Tapetum lucidum 339
Telodendrial contacts between rod and cone terminals 354, 360

Tiered cones in black patch 353
Triad of photoreceptor synapses 366
Two eye rotations 361, 371
Ventral embryonic fissure 337, 338, 339
Visual cell layer (VCL) 359, 365
E-YSL (external yolk syncytial layer)
E-YSL contraction 221

F

Facultative internal bearers 471
Facultative matrotrophy 471
Facultative viviparity 493
Falciform body 284
Falciform (sickle shaped) process of the eye 284, 308, 331, 337
False gill (see pseudobranch)
Fatemaps 220, 231, 236, 251-258, 435, 441
 Comparison of fate maps 256
Fecundation 147, 153
Female pronucleus 180
Fertilization 7, 24, 45, 46, 147-184, 492, 507
 Artificial fertilization 150, 154
 Capacity of sperm 157
 Cone 42, 159-163, 165, 166
 Elevation and hardening of the envelope 150, 167, 169, 172
 Fusion of sperm and egg membranes 159
 Gynogamones 156
 Hardening of the envelope 148, 169, 175
 Internal fertilization (vivparous fish) 136, 458, 502
 Internalization of sperm 161
 In ovarian lumen 461, 493
 In the follicle (intrafollicular) 460, 481, 493
 Ovulation preceding fertilization 460, 502

Perivitelline space after fertilization 172
Fibres 91, 101
Fibrils 7, 89-93, 101, 102, 107
Fibronectin 211
Filaments 42, 82, 83, 87, 90, 92-101, 107, 507
Filolamellipodia 232
Filopodia 194, 230, 236
Fin of sperm tail 144-146
Finfold 278, 288, 502
Fingerprint pattern 277, 288
Fins of cone ellipsoid 144, 316, 319
First erythropoietic site appears in blood islands 403
First gill cleft 436
First haematopoietic sites 382
First polar body 147, 182, 183
First pulsations of the heart 416
Fishes grow throughout life 337
Fissure in (embryonic) eye 337, 338
Flagellum of sperm 134, 137, 140, 141, 143-146
Flat bitterling 135
Flatfish 314, 340, 341
Floating raft 11
Floor plate 226, 229, 248, 249
Flounder 11, 61, 441, 446
Folds of middle intestine 439
Follicle 39, 476, 477, 507
  Cells 34, 53, 93, 95, 507
  Epithelium 35, 58, 60, 82, 107
  Wall 474, 478, 480
Follicular gestation 458, 494
  Granulosa 92, 108
  Placenta 471, 476-478, 494
  Processes 108
  Protuberance 462
  Pseudoplacenta 470, 471, 478, 494, 500, 501, 503
  Theca 82
  Villi 475, 480
Forebrain 242, 244, 245, 268, 269
Formation and regression of pericardial hood (see pericardial hood)
Formation of endocardium 420
Formation of envelope 74
Formation of eye 338
Formation of intermediary cell mass (ICM) 383
Formation of intestine 438
Formation of lens 337
Formation of liver 441
Formation of micropyle 124
Formation of perivitelline space 148, 173, 176
Formation of pharynx 435
Formation of polar bodies 181
Formation of pronuclei 180
Formation of red blood cells 509
Formation of tail bud 231
Formation of thyroid 437
Formation of yolksac (epiboly and involution) 231
Fossa 136, 146
Foureyed Fishes 477
Fovea 326
Free floating discs of ROS separated into
  packages of 2 344
  packages of ~9 352
Freeswimming phase 288
Frog embryo, macrolelcithal 452
Function of solid intestine is osmoregulatory 439
Functional significance of KV 265
Functional significance of the EVL 210
Funiculus (egg sac) 461
Furrow 193
Furrow (groove) 186
Fusion cone 159, 163
Fusion of sperm and egg membranes 159
Future EVL 189
Future germ layers 251

G

Gall bladder 432, 439, 441, 442
Ganglion cell layer 310, 330, 331, 336, 339, 363, 364, 370, 371
Ganglion cells 272, 312, 330, 331, 339, 357, 358, 361, 362, 368, 370, 371
Ganglion of nervus octavus 270
Gap junctional communication 221
Gap junctions 196, 197, 203, 212, 220, 251, 354, 356, 358-360
Gastrulation 207-237, 506-508
    Blastopore 185, 209-211, 216, 218, 222, 225, 234
    Blebs 230
    Blocking of involution 220
    Concrescence theory 209
    Dorso-lateral lips of blastopore 220
    Filolamellipodia 232
    Filopodia 194, 230, 236
    Inflection 214
    Inflection of hypoblast 435
    Invagination 207, 208, 216, 233, 508
    Involution 205, 207, 209, 214, 216, 220, 230, 236, 252, 254, 255, 435, 438, 508
    Lamellipodia 194, 236
    Lobopodia 196, 230, 236
    Ventral lip 216
GATA-1, first haematopoietic and vascular marker 396
GATA-1 and GATA-2 transcription factors 394
Geneses of the periblast nuclei 199
Genesis of lipid (fatty) yolk 29
Genetic studies 7
Germ ring 209, 210, 214, 217-219, 230, 257, 508
Germinal vesicle 24, 147
Gill 278, 280, 432
    Artery 420
    Cleft 435, 436
    Filaments 502
    Formation from part of entoderm 431
Glands 275, 278
Glandula pinealis 271
Glandular organs 5
Glial Müller cells 310, 331, 357
Goblet cells 279
Goldfish 13, 31, 35, 42, 46, 73, 135, 168, 170, 178, 214, 293, 322
Gonadosomatic index (GSI) 31
Gonadotrophic hormones 34, 39
Gonoduct 458, 459, 487
Gonopodium 481, 482, 507
Gonopore 459
GR (germ ring)
Granulosa 58, 107
Greatly enlarged hindgut 469
Growth cone 164, 165
Growth zones 333-335, 509
Guarding 12, 14
Guppy 29, 31, 58, 134, 278, 285, 286, 288, 304-308, 315-318, 320-324, 326, 329, 331, 337, 339-349, 352-354, 357, 359, 360, 370, 374, 404-409, 411, 494, 496, 499-501
Gut enlarged, indicative of nutrient uptake 470.
Gymnovarian 20
Gynogamones 156

H

Haematopoiesis (see blood anlage)
Haematopoiesis of red blood cells and vascular endothelia 380-415
Haematopoietic and vascular marker GATA-1 394, 396
Haematopoietic sites 381, 382, 509
    Blood islands 409
    Head mesenchyme 401
    Heart 414

ICM 409
Kidney 414
  Pro- and mesonephros 382
  Wandering cells from ICM to yolk-sac 401
Haematopoiesis in mature fish takes place in spleen, liver, pro- and mesonephros, intestine, pancreas and gonads 381
Haemoglobin 397, 415
  Developmental switching of haemoglobin. 403, 413, 415
  Fishes with no haemoglobin 415
  Larval haemoglobin 403
  Multiple haemoglobins 415
Halfbeaks 492, 493
Halibut 61
Hardening of the envelope 148, 169, 175
Hatching 293-302, 508
  Anoxia 300
  Apoptosis 298
  Breaking of the envelope 293
  Hatching enzyme 210, 293, 300, 508
  Movements by the embryo 302
Hatching glands 66, 281, 285, 293
  Breakdown of the hatching glands 294
Hatching gland cells 278, 288
  Chorionase 300
  Distribution of hatching glands 294, 295
  Effect of temperature 300
  Goldfish 293
  Histological structure 295
  Holocrine secretion 298
  Mastication by embryo 300
  Origin, ectodermal, some endodermal 302
  Oxygen concentration 300
  Pike 298, 300
  Precursory endodermic cells 296
  Prolactin 298, 573

Proteinase 300
Rainbow trout 300
Secondary necrosis 29
Secretions 298, 302
Secretory granules 295
Triggering of secretion 300
Trout 293, 294, 298
UHG (unicellular hatching gland) 296
UHG in buccal cavity 297
Ultrastructure of UHG 296
Ultrastructure of gland opening 302
Ultrastructural development of UHG, 296
Zebrafish 298
Head 131-134, 146
Headfold 248, 250, 278, 288, 406, 494, 495, 503.
Head mesenchyme 401
Head-mesoderm 225
Heart - Adult 385, 412, 415, 416, 474
  Anlage (solid and paired) 416-420
  Arterial bulb 419
  Atrium 415, 419
  Auricle 415, 416, 419, 420
  Bulbus aortae 415
  Bulbus arteriosus 415, 416
  Bulbus cordis 415, 419
  Cells 417
  Contractions 420
  Conus arteriosus 416, 420
  Ventricle 415, 416, 419, 420
Heart - Development 415-420, 422
  Acquisition of lumen 416
  Differentiation of blood cells 414
  Endothelium (heart tube) 417
  Endothelium 390
  Epicardium 415
  Epimyocardium 416, 419
  Heart cells move to yolksac to form endothelia of blood vessels 418

Mesenchyme of the heart 379
Mesocardium, dorsal 415, 417
Mesocardium, ventral 415, 417
Myocardial cells 419
Myocardium 415, 418
Onset of contractions 417, 418
Positional changes of its parts during development 421
Pulsations are rhythmical 419
Rostral mesentoblast 406
Sinus venosus 415, 416, 419, 420, 422
Hematopoiesis (see haematopoiesis)
Heme pigment 320, 321
Hemibranch 282
Hemoglobin (see haemoglobin)
Hepatic proteins 34
Hepatic vein 420
Hepatocytes 34, 74
Hermaphroditic 471
Herring 11, 24, 26, 36, 50, 51, 61, 63, 87, 114, 132, 147, 153, 156, 158, 159, 178, 187, 227, 259, 261-263, 278, 280, 439
Heterotopic (ectopic) thyroid tissue 437
High and low choriolytic enzyme 301
High Molecular Weight Spawning Femalespecific Substances (H-SF substances) 73
Hindbrain 244, 245, 268, 269
Hindgut of embryo 3, 465, 467-469, 482, 485, 492, 502, 503
Derivatives of the hindgut 481
Major site of nutrient absorption 493, 503
Histiotrophe, embryonic nutrient 460, 503
Histocytic (embryonic nutrients) 460
Histogenic (embryonic nutrients) 460
Holoblastic cleavage 185, 506
Holoblastic cleavage (frog) 453

Holoblastic to meroblastic type 186
Holobranch 282
Holocrine hatching gland 298
Holonephros 378
Holostei 185, 186
Honey-comb structured envelope 107
Hood (see pericardium)
Hooks 98, 107
Horny fibres (actinotrichia) 279
H-SF Substance (High Molecular Weight Spawning Femalespecific Substances) 73
Hyaloid artery 308, 331, 337
Hydration of oocyte during maturation 38
Hyoidean artery 282, 283
Hypertrophied hindgut 481, 492, 502
Hypertrophied midgut 481
Hypertrophied vertical finfold 491
Hypoblast 207, 214, 216, 217, 220, 221, 226, 255, 265, 267, 382, 438
Hypoblast arises by delamination 214, 220, 255
Hypoblast (entoderm and mesoderm) 236
Hypoblast (future entoderm) 380
Hypoblast inflection 435
Hypochord 226, 227, 229, 236, 248, 249, 431
Hypomere (Somite) 378, 379
Hypophysis (pituitary) 34, 39, 269, 431, 437, 468, 469

I

ICM (intermediary cell mass)
Becomes reactive 403
Begins to descend 386
Differentiates into endothelium of the future dorsal aorta 403
Formation 383

Furnishment of endothelia for blood vessels 394
Fusion with putative blood cells of germ ring 405
Incorporation into connective tissue coat of alimentary canal 397
Separation from the primary lateral plates 390
ICM- and bloodisland-embryos 382, 398, 405, 414
Ima (immature-adult)-erythrocytes 403, 411, 413
Imitation gill (pseudobranch)
Immature and mature erythrocytes (e. ima and e.A) 413
Immune system 411
Immunoglobulin producing cells 411
IMP (intramembraneous particles) on head of sperms 134, 135
Impregnation 147, 477
Indifferent cell mass of the tailbud 440
Indirect development 4, 8
Induction of cephalic structures 225
Induction of vitellogenesis by androgens 34
Influx of water into the eggs 11, 75
Infundibulum 269
Ingression of haematopoietic cells 394
Inhibitor of polyspermy 166
INL (inner nuclear layer)
Inner ear 268
Inner endocardial tube. 419
Inner limiting membrane 357
Inner nuclear layer 331
  Inner nuclear layer, adult 310, 329-331
  Inner nuclear layer, embryo 336, 340, 357, 358, 360, 361, 364, 370
Inner plexiform layer 310, 330, 331, 357, 360, 370

Insemination 147
Integument 236, 277
Intercellular junctions 246
Interdigitations 278
Intermediary cell mass (ICM) 252, 377, 380, 381, 385, 389, 393-398, 403-406, 509
Intermediate development 8
Internal bearers 16
Internal fertilization 136, 458, 502
Internal limiting membrane 310
Internal live bearers 457
Internal trophotaeniae 492
Internalization of sperm 161
Interrenal tissue 411
Intestinal epithelium organized into villi 467
Intestinal nutrient transfer 471
Intestinal tube remains distinct from yolk mass 432
Intestine 20, 243, 432, 442 508, 509
  Adult teleost 439
  Anlage suggested to be a solid or a folded cord 438, 441
  Intestine, lumen appears late 435, 441
  Median thickening of the ventral intestinal wall 441
  Origin of the intestine as a closed entodermal fold 438
  Osmoregulatory function 439
  Solid, not yet functional 438
  Solid outgrowth of the dorsal wall 441
Intimate contact between liver cells and periblast of yolk 442-455
Intrafollicular gestation 458, 470, 492, 502
Intralumenal/intraluminal gestation 458, 502
Intramembraneous particles (IMP) on head of sperms 134, 135
Intraovarian cannibalism 488

Intrasperm 131, 134
Intrauterine embryonic cannibalism 460
Intravesicular yolk 24, 35, 46
Intromittent organ for copulation 469
Inturning (involution) 255
Invagination 207, 208, 216, 233, 508
Involution 207, 209, 214, 216, 220, 230, 236, 252, 254, 255, 435, 438, 508
   Involution and convergence are not affected by UV 230
   Involution of entoderm 254, 255
   Involution of mesodermal cells 508
   Inwheeling movements 255
Iris 305, 331, 337

J

Jelly capsule 105
Junctional complex with alternating zonulae adhaerentes and zonulae occludentes between cells of the RPE 343

K

Kidney (see also nephron, pro- and mesonephros) 377, 378, 385, 413, 414
   Aglomerular condition 378
   Archinephric duct 378-380
   Bloodforming tissue 414
   Degenerated glomera 411
   Erythropoiesis 410
   Eythropoiesis-stimulating factor 410
   Glomerular condition 378
   Glomeruli of pro- and mesonephros 410
   Glomerulus 378, 410
   Glomus 378, 412
   Holonephros 378
   Interrenal tissue 411
   Lymphnode 411
   Lymphoid tissue 414
   Mesonephros 378, 379, 410, 411, 414
   Metanephros 378
   Nephron 378
      Connection with coelom, artery, vein and kidney duct 381
   Opisthonephros 378
   Primary nephric ducts 377
   Pronephric tubules developed 411
   Pronephros 379
Kidney of freshwater types 378
Kidney of marine teleosts 378
Killifishes 471
Kupffer's vesicle 227, 234, 259-266, 431, 435, 438-440, 498, 508
KV (Kupffer's vesicle)

L

Labyrinth 268, 271, 341
   Ampulla 271
   Anlage 270
   Asteriscus 268, 271
   Auditory organ 268
   Biological clock 272
   Cupula 274, 276
   Lagena 271
   Lagenolith 268, 271
   Lapillus 268, 271
   Otoliths 268
   Recessus utriculi 271
   Sacculolith 268, 271
   Sacculus 271
   Sagitta 268, 271
   Semicircular canals 271
   Static organ 268
   Stato-acoustic system 274
   Statoliths 268, 271
   Utricolith 268, 271

Labyrinthfish 46
Lactate dehydrogenase (LDH)
Lake trout 42, 61
Lamellary bodies 36
Lamellibranch 15
Lamellipodia 194, 236
Larvae with pure cone retinae 340
Larval fin fold 486
Larval haemoglobin 403
Lateral line 253, 341
Lateral line system 274, 276, 278
Lateral plate 220, 252, 377-80, 387
Lateral plate mesoderm 235, 419, 420
LDH (lactate dehydrogenase)
LDH isozyme pattern (A, B, C)
  Adult Poecilia 373
  Adult salmonids 372
  During development of salmonids 372
  Effect of inhibitor (3M urea) 373
LDH-C in liver and retina 372
LDH-C in newly differentiated photoreceptors 375
LDH-C role in visual pigment regeneration 375
LDH C4-isozyme expressed in retinal photoreceptors (and optical parts of the brain) 374
Lecithotrophy 458, 471, 492-494, 502
Lens (eye) 304
Lens anlage 270
Lentiform body 308
Lifting of the envelope 166, 173
Light- and dark adaptation 305, 326, 327, 331, 341, 349, 353, 356, 359
Lining epithelia of the digestive system (pharynx, intestine) 431
Lipid droplets in RPE 342, 366
Lipid droplets in yolk 24, 29
Lipid yolk 24, 28, 39, 507
Lipid yolk formation 29

Lipids 23
Lipovitellin 31, 32, 39
Lithophilic 5, 8, 12
Livebearers 457-504
Liver 33, 39, 432, 474, 509
  Anlage (primordium) 4, 7
  Development 442, 446
  Formation 441
  Gall bladder 432, 439, 441, 442
  Hepatic proteins 34
  Hepatic vein 420
  Hepatocytes 34, 74
  Yolk uptake 442, 446, 452, 509
Liver-yolksac contact 25-27, 205, 428, 441-446
  Intimate contact between liver cells and periblast of yolk 442-444
  Secondary contact 447, 451
  Secondary direct contact 443, 449, 454
Loaches 13, 135, 203
Lobopodia 196, 230, 236
Lobus impar 269
Location of the prospective entoderm 441
Low Molecular Weight Spawning Femalespecific Substances (L-SF) substances) 73
L-SF (Low Molecular Weight Spawning Femalespecific Substance)
Lumpsucker 61, 62
Lymph nodes 411, 414
Lymphoid haemoblasts 392, 393, 403, 414
Lymphoid organs first appear in pronephros and spleen 394, 411

M

Macrolecithal 36
Macrovilli of follicular cells 59, 60, 64, 66, 75, 127, 507
Main sense organs 220, 270

Male copulatory organ 502
Male pronucleus 166, 183
Margin/rim and ventral fissure of eye 335
Marginal cells during cleavage 187, 188, 190
Massed yolk 27
Mastication by the embryo 300
Maternal embryonic nutrient transfer 480
Maternal immune system 457
Maternal-embryonic relationship 458
Matrotrophic embryonic development 458, 471, 472, 486, 492, 502, 503
Matrotrophic species 460
Medaka 73, 128, 158, 162, 163, 166, 168, 169, 171, 179-181, 245, 249, 272
Median (axial) vein 226, 385, 386, 388, 422, 423, 429
Medulla oblongata 245, 268, 291
Medullary part of neuroectoderm 235
Medullary plate 241
Meiosis 147
Melanosomes 311, 313, 341
Melatonin 272
Meroblastic cleavage 185, 204, 506
Meroblastic eggs 186
Meroblastic taxa 509
Merocrine hatching gland 298
Mesencephalon 268-270
Mesenchymal cells 229, 276, 379, 419, 432, 442
Mesenchyme 377, 392, 401, 404, 483, 495
Mesentoderm 220, 276
Mesoderm and its derivatives 198, 204, 207, 220, 231, 253-255, 257, 258, 377-430
Mesoderm involutes over lateral lips 231

Mesodermal cells disperse 232
Mesomere 378, 379
Mesonephros 378, 379, 410, 411, 414
  Haematopoietic 381, 382
Metachronous waves of YSL mitoses 203
Metamorphosis 4, 6, 8, 12, 340
  Eye migration 8
Metanephros 378
Metaphase of second miotic division of egg 147
Metencephalon 268-270
Mexican topminnows 481
Microfilaments 42, 58, 59, 213, 245
Micropinocytosis of vitellogenin 33
Micropinocytotic vesicles in the follicle cells and the oocytes 59
Microplicae 42, 46, 51, 84, 475
Micropylar apparatus 118
Micropylar cell 124, 126, 127, 129
Micropyle 78, 90, 93, 95, 99, 100, 102, 109-130, 147, 149, 150, 155-157, 162, 164, 166, 168, 507
  Formation of micropyle 124
  Sperm entry through micropyle 158, 164
  Types of micropyle 109
Microridges 277, 278, 288, 290, 471, 500
Microspectrophotometry of ellipsosomes, erythrocytes and photoreceptor outer segments 318-321
Microtubules 93, 214, 245
Microvilli 42, 46, 51, 52, 58-60, 64, 66, 75, 89, 127, 159, 167, 194, 213, 222, 265, 278, 471, 474, 475, 477, 507
Mid-blastula 196, 211, 212, 215, 257
Midbrain 240, 242, 244, 245, 248, 250, 268, 269
Middle intestine 439

Mid-gastrula 211, 220
Midline structures 236, 248, 249
Migration of one eye 340
Mixed feeding (yolk and embryo-trophe) 466
Model tissue for the CNS 303
Modifications of anal fin 458, 469, 481, 482, 487, 493, 507
Modified hindgut indicating oral uptake of food 465, 467, 469, 492, 493, 502, 503
Moribund eggs nutrient source for embryos 493
Morphogenetic gradients 251
Morphogenetic movements 207, 211, 212, 220, 246, 255, 508
Mosaic 312, 348
  Egg surface 168
  Membrane 166
  Patterns 509
Motility-enhancing factor for sperms 166
Motility of blastomeres 194
Motility of sperms 146, 155
Motoneurons 226
Mouthbrooders 21, 77, 106, 108, 506
Movements of the embryo 302
Mucopolysaccharides 69, 76
Mucus cells 58, 277, 282, 288, 295
Mucus gland 279
Müller (glia) cells 310, 331, 357
Multiple haemoglobins 415
Multivesicular bodies 32, 36, 37
Muscles 377
Mushroom-like appendages 85, 86
Myelencephalon 268-270
Myeloid bodies 311, 342-344, 347, 359
Myocardial cells 419
Myocardium 415, 418, 420
Myoid 316, 347
Myosin 42, 43
Myotome 377, 379

N

Na+K+ATPase activity 59
Neckstrap 472, 496, 497, 503
Nephric ducts, primary 377
Nephron (see also kidney) 378
  Aglomerulus 381
  Connection with coelom, artery, vein and kidney duct 381
Nephrotome 378-380
Nerve chord 246
Nerve fibre layer 310
Nervous system 253, 508
Nestfleer 2, 501
Nestsitter 3
Nests 15
Neural crest 239, 245, 249, 276, 339, 379, 508
Neural ectoderm 229, 235, 236, 267
Neurenteric canal 240, 260
Neurochord (Neurocord) 239, 508
Neurocoele 245, 246
Neuromasts 274, 276, 278
Neuromeres 268-270
Neurula 239
Neurulation 239-250, 508
  Auditory anlage 239, 241, 270
  Carina 239
  Cavitation in nerve rod 239, 249
  Closed neural fold 240-242, 249
  Dorsal groove 242
  Ear anlage 239, 241, 270
  Embryonic shield 239, 240
  Eyecup 239, 244, 270
  Neural cavity 246
  Neural chord 239, 240, 242, 243, 246, 267
  Neural fold 240
  Neural keel 239, 242, 243, 245, 246, 248, 267
  Neural plate 218, 220, 239, 243, 245, 246, 248
  Neural ridge (see neural keel)

Neural rod 220
Neural tube 226, 239, 243, 246-250
Neurenteric canal 240
Neurochord 239
Neurocoele 245, 246
Nidicolous 1, 4, 5, 7
Nidifugous 1, 4, 7, 496
Nodules 107
Non-guarding type 12
Notochord 220, 225-229, 231, 235, 236, 249, 252, 253, 377
Nucleus (germinal vesicle) 36
Number of eggs deposited 18, 21
Number of layers in envelope 50

O

Obligate internal bearers 460
Occludin (adhesion molecule) 278
Occlusion or disappearance of radial canals 60
Ocular blood pressure 281
Oesophagus 432, 435, 439, 440
Oestrogen 33, 34, 39, 74, 76
Oil drop, globule, droplet 5, 7, 9-12, 13, 24, 255, 318, 507,
Olfactory anlage 239, 241, 270
Olfactory organ 268
ONL (outer nuclear layer)
Onset of epiboly 196
Ontogeny of radial canals 58
Ontogeny of rod and double cone RMM 351
Oocytes 74, 462
Oolemma 46, 58, 64, 51, 75
Oophagy 460
Ooplasmic segregation 38, 177
Open blastomeres 203
Open ventral fissure of eye 335
Opening of hatching gland cells 301, 302
Opercula opened 488
Opercular gill (see pseudo-branch)

Opercular membranes 281
Opercular movements 300, 484
Opercular respiratory movements 497
Operculum 282
Ophthalmic artery 282, 283, 308, 338
Opisthonephros 378
Optic
  Anlage 239, 241, 270
  Chiasma 269
  Cup 337, 338
  Fissure 338
  Lobe 337
  Nerve 331, 337
  Nerve exit 341
  Primordia 360
  Stalk 337
  Tectum 335
  Vesicle 245, 337
  Part of the brain 372
Ora serrata 304, 339, 364
Oral gestation 16
Oral incubators 106
Oral placenta 490, 491, 503
Organizer 229
Origin of blood 226
Origin of the zona radiata 70
Ostracophilic 5, 8, 12, 13
Otoliths 268
Outer myocardial tube 419
Outer nuclear layer 310, 326, 331, 344, 359, 365
Outer plexiform layer 310, 331, 353
Outer segments of rods (ROS)
Ovarian fluid is absorbed 482
Ovarian liquid (embryotrophe or histiotrophe) 460, 461-463, 465-467, 470, 492, 504
Ovarian lumen 507
Ovarian septum 459, 486
Ovary 9, 35, 39
  Cystovarian 458
  Dermotrophic uptake 460, 503

Double (paired) 458, 459, 493
　Partially fused 75, 458, 459
Oviduct 20, 85, 86, 102, 150, 459
Ovigerous bulbs 490
Oviparous 9
Oviposition 20, 189
Ovoviviparous 16, 21, 77, 107, 108
Ovulation 460, 502
Oxygen concentration 300

P

Pacific salmon 64
Pacific sardine 278
Paired ovary 458, 459, 480
Pallium 269
Pancreas 431, 432
Papillae 84, 89, 107
Parablast 198, 382
Paracrystalline bodies 30, 36, 37
Paraphysis 272
Parapineal organ 272-274
Parasynchronous mitosis 188
Parental care 14
Parturition 496, 507
Pavement cells 278
Pectoral fins 6
Pedicles 97, 107, 353, 354, 359, 360, 370
Peduncles 101
Pelagic eggs 10, 21, 24, 38, 49, 75, 507
Pelagic-nonadhesive eggs 77, 78, 107
Pelagic ribbon of eggs 11
Pelagophilic 5, 8
Perch 36, 54-56, 125, 132, 214, 240, 264, 443, 451, 507
Periblast (YSL) 185, 187, 190, 192, 198, 205, 209, 211, 213, 214, 216, 217, 222, 262, 265, 382, 507, 508
　Central periblast 188, 199
　Destiny of periblast 198

External periblast 209, 213
Functional significance of periblast 204
Internal periblast 209
Origin of periblast cells 199
Role of periblast in gastrulation 204
Periblast/liver contact (see liver/yolksac contact)
Peripheral periblast 188, 189, 193
Periblast nuclei 188, 189, 198, 202, 221, 448
　Amitotic division 188, 202, 205, 448, 507
　DNA 204
　Increase in size 188
　Fate 450
　Gigantic size 203-205
Pericardium 415-418, 472-474, 484, 503
　Amniochorion 502
　Cavity 419
　Headfold 278, 288
　Hood 278, 288, 424, 428, 471, 472, 474, 484, 494-498, 500, 503
　Membrane 280
　Neckstrap 472, 496, 497, 503
　Sac 472, 474, 481, 484, 485, 488, 500, 502
　Trophoderm 502
Periderm (EVL) 187, 209, 211-213, 216, 267, 277, 278, 288, 290, 508
Periorator (acrosome) 135
Peripheral corner cells 189
Peripheral mesoblast 404
Perivitelline fluid 294
Perivitelline fluid, oral uptake 496
Perivitelline space 11, 75, 148, 149, 152, 163, 164 166, 167, 174, 190
Permanent interphase 213
Peroxidase activity, superficial egg layer 89

Phagosomes (shed tips of photoreceptor outer segments) 311, 326, 331, 342, 359
  Rhythmic fluctuations 342
Pharyngeal pouch 435
Pharynx 437, 440
Phosvitin 31, 32, 39
Photomechanical changes (see retinomotor movements)
Photoneuroendocrine transducers 273
Photoreceptor outer segments 273, 275, 366
Photoreceptor ribbon synapses form triads 331, 370
Photoreceptors 272, 273, 274, 311
  Accessory outer segment 315-318, 343, 344, 346, 349, 353
  Visual pigments 318, 321, 322, 332
Phytophilic fish 5, 8, 12
Pigment cells 277
Pike 13, 23, 36, 58, 88, 110, 111, 113, 129, 132, 149, 151, 152, 155, 158, 175, 185, 188, 214, 240, 242, 244, 259, 288, 294, 298, 300, 326, 382, 386, 388, 390, 409, 414, 416, 418, 419, 422, 440, 443, 451, 506
Pikeperch 23
Pineal body/gland/organ 271-276, 291, 312
  Anterior epiphyseal vesicle 273
  Biological clock 272
  Photoreceptors 272
  Pineal window 271
Pinocytotic (coated) vesicles in RPE 341
Pipe-fish 33
Pituitary (see hypophysis)
Placenta (oral, buccal) 476, 498
Plaice 61, 175, 176, 341, 439
Planktonic 10
Plasma cells 413

Plasma vitellogenin 32
Plasmamembrane of the egg 51
Plasmolecithal 36
Platelets 25
Plica ventralis 269
Plugs 60, 62, 75, 89, 107, 158, 163
Polar body 42
Polylecithal 36
Polyploid 204, 508
Polyploidy 148, 205
Polyspermy 163
Pore canals 64, 66, 107
Porphyropsin 318
Portal blood system 473, 475
Portal capillary 487
Portal network 474, 485
Postanal gut 439, 440, 441, 498
Postanal vesicle 260
Postembryo, definition 1, 7
Posterior blood anlage 405
Posterior blood anlage and ICM 408
Posterior blood anlage in germ ring 407
Posterior blood islands 404
Posterior portion of the intestine 479
Posterior trunk 232
Praecranial blood islands 404, 405
Praecranial mesoblast, ICM and posterior blood islands 409
Prechordal plate 225, 231, 236, 248, 252, 253, 255
Precocial embryo 1, 4, 7, 508
Precursor of protein yolk 30, 39
Precursory endodermic cells 296
Pregastrula twinning 195
Presumptive areas 235, 251, 257, 258
Presumptive rod cells migrate from INL to ONL 340
Prevention of polyspermy 176
Previtellogenesis 27
Primary and secondary direct liver/yolksac contact 451
Primary blood circulation of a fish

421
Primary brain 269
Primary entoderm 406, 431
Primary hypoblast 217, 254, 431, 433
Primary lateral plate 377, 387
Primary mature erythrocytes 414
Primary nephric ducts 377
Primary neurulation 239, 240, 244, 247
Prolactin 298, 573
Pronephric anlage 380
Pronephric chamber 378, 380
Pronephric tubules have degenerated 411
Pronephros (see also kidney and nephron) 252, 378-380, 410, 412-414
Pronephros and spleen first lymphoid organs 411
Pronephros is haematopoietic 381, 382
Pronephros remains haematopoietic to the mature stage 410
Pronucleus 42, 180, 183
Prosencephalon 244, 268, 269
Prospective area 235, 251, 257, 258
Prospective (presumptive) fate 251
Prospective (presumptive) potency 251
Prostomal thickening 433
Proteid yolk 28, 507
Proteid yolk platelets 39
Protein portion of the yolk 36
Protein yolk 39
Protuberances 463, 464
Provisional blood 204
Psammophilic 5, 6, 8, 12
Pseudoacrosome 135
Pseudoamnion 472
Pseudobranch (false gill; opercular gill, imitation gill) 218, 281, 288, 308, 431, 432
  Accessory cells 281, 284
  Attached type 281, 283
  Covered and glandular type 282
  Cells 277, 282
  Doubly oxygenated blood supply 282, 283, 285, 308
  Embryonic pseudobranch cell 287
  Non-covered/free 282
  Tubules 286, 287
  Vascular system to the eye 308
Pseudochorion 472, 473
Pseudoflagellum 142
Pseudoplacenta 470, 471, 478, 494, 500
Pseudoplacental barrier 501
Pseudostratified retina 361
Pupil 331
Pure cone retina 340
Putative UV-cones 314
Pyloric caeca or appendices 435, 439

Q

Quadruple retinal cones 314

R

Radial canals 75
Rainbow trout 31, 33, 69, 211, 281, 282, 300, 339, 341, 342, 344, 345, 361, 401, 409, 413, 414
Receptor-mediated endocytosis 31, 477
Recessus utriculi 271
Reciprocal synapses 359
Rectum 439, 465, 466
Redd 50, 83
Reduced envelope (ZR) 460
Reissner's fibres 249
Reissner's substance 248
Relation between yolk and cytoplasm 36
Release of embryotrophe by apocrine secretion 470

Reproductive styles 10
Respiration shifted to the gills 484
Respiratory importance of carotenoids 494
Rete mirabile 308
Retia mirabilia 308
Reticula 102
Retina 508
  Accommodation 308, 331
  Amacrine cells 358
  Avascular 307
  Bipolar cells 358, 360
  Connecting cilia 344, 363
  Darkadaption 328, 355
  Development 337-376
  Duplex 311, 340, 344
  Electrondense zone between telodendria 358
  Elevated oxygen pressure 307
  Fovea 326
  Light- and dark adaptation 305, 331, 341, 349, 356, 361
  Müller (glia) cells 310, 331, 357
  Respiration 282
  Stage 361, 368
Retinal cones 87, 89, 90, 99, 100, 102, 105, 107, 272, 311, 312, 316, 317, 331, 344
  Absorption spectra of cone types 320
  Accesssory outer segment 315, 316, 344
  Age-related loss of UV-cones 335
  Area (high density of cones) 326, 348, 368, 369, 371
  Arrangement in row mosaic 333-335, 359
  Colour pattern polymorphism 360
  Cone outer segments (COS) 327-329, 346-349, 353, 355, 365, 368, 372
  Cone to rod ratio 345
  Connecting cilia 344, 363
  COS (cone outer segment)
  Different types of cones 311
  Differentiation completed 372
  Double cones 314, 316, 344
  Ellipsoid 316
  Ellipsoidal fins 316, 319, 330, 348
  Ellipsosome 318, 320, 323, 345, 347, 348, 352, 359, 360
  Fenestrated cisternae between cones 345, 348
  Mitochondria display vitreo-scleral size gradient 316
  Mosaic 312, 314, 319, 320, 325, 331-336, 348, 349, 354, 359, 360, 366, 367, 509
  Mosaic changes 336, 348, 349, 359
  Myoids show fenestrated subsurface cisternae in all contact zones with cones 345
  Pedicles 360
  Row mosaic 333-336, 348, 349, 359, 367
  Square mosaic 319, 320, 325, 333-336, 348, 354, 359, 360
  Subsurface cisternae (membranes) 316, 344, 345, 347, 348, 354
  Subsurface membranes 'disappear' in the dark 360
  Terminals (pedicles) 360
  Trichromatic vision 318
  Tetrachromatic vision 318
  Tiered arrangement 348, 368
  Triple cone 319
  UV-cones 325, 332, 508
    Age related loss 335
    Disappearance 332
    Radiation 213
    Receptors 318
    'Regresssion' 334
Retinal pigment epithelium (RPE) 310, 311, 313, 331, 338, 341-343, 359, 361
  Aggregates of melanosomes bound-

Subject index 629

ed by membrane 311, 313, 342
  Avascular 307
  Black pigment 339
  Circadian rhythm of RMM 331
  Circadian shedding pattern of photoreceptor outer segments 359
  Eyed egg 339, 341
  First pigmentation of the eye 4
  Rod-shaped granules 342, 344, 359, 491
  Spherical granules 311, 342, 343, 359
  Zonulae adherentes 189, 311, 341, 343, 344, 363, 365, 366, 368
  Zonulae occludentes 278, 311, 341, 343, 500
Retinal rods 311, 312, 316, 344 315, 317
  Accessory outer segment 315-317, 344, 349
  Addition of basal discs in outer segments 329
  Arrangement in tiers 314
  Differentiation of rods ahead of cones 345
  Formation of a regular mosaic 335
  Free floating discs in outer segment 344
  Terminals (spherules) 360
Retinomotor movements (RMM) 305, 311, 326, 331, 341, 342, 344, 348, 359
Retino-tectal projections 341
Retinotectal synapses 335
Retractor lentis muscle 308, 331
Rhodopsin 272
Rhodopsin and porphyropsin 318
Rhombencephalon 268, 269
Ribbon-shaped veils 106
Rim of the eye 304, 336, 339
RMM (retinomotor movements)
Roach 5, 33

Rockfish 492, 507
Rod outer segments (ROS)
Rods (see retinal rods)
Rohon-Beard cells 249
Role of periblast in gastrulation 204
ROS 315, 329, 333, 348, 349, 352, 356, 368, 372, 375
Rose bitterling 182
Rostral mesentoblast 406
Rosy barb 220, 226, 435
Rotation of the eye 361
Row mosaic 333-336, 338, 349, 359
RPE (retinal pigment epithelium)
Ruffles 194, 197, 265
Rupture of the cortical alveoli 166

S

Saccules (cone outer segments) 315
Sacculolith 268, 271
Sacculus 271
Saccus dorsalis 273
Saccus vasculosus 269
Sagitta 268, 271
Salmon 26, 42, 44, 52, 53, 56, 60-64, 68, 112, 126, 128, 132, 152, 158, 163, 175, 214, 219, 229, 258, 262, 269, 279, 285, 325, 332, 335, 340, 344, 345, 372, 386, 418, 436, 439, 440, 443, 445, 451, 506, 509
Salmonid type of ZR 61, 63, 75
Schwann cells 276
Sclera 304, 330, 337, 339
Sclerotome 377, 379
Scorpionfishes 492
Sea bass 33, 227, 229, 434
Sea horse 33
Sea trout 31
Second polar body 148, 179, 182, 183
Secondary embryonic induction by the notochord 248

Secondary entoderm or hypoblast 382, 434
Secondary entoderm or intestinal epithelium 431
Secondary haematopoietic sites in the embryo (endocardium and kidney) 410
Secondary hypoblast 217, 225, 229, 431
Secondary lateral plate 377, 387
Secondary necrosis 298
Secondary neurulation 239, 240, 244, 247, 249
Secondary proteolysis 31, 39
Secretory granules 295
Secretory pores 290
Segmentation cavity 188, 190, 196
Semicircular canals 271
Sense organs 268
Sensory neurons 276, 277
Sensory plates 243
Separation of intestine and yolk 443
Sequential appearance of red blood cells 391
Serial synapses involving three consecutive conventional synapses 359
Serum 422
SES (sperm entry site)
Sexually mature fish 415, 421
SF (spawning female specific protein) 33, 68, 73, 74, 76
Shape of the eggs 17, 21
Shark 442, 454
Shedding and renewal of photoreceptor outer segments 326, 331, 345, 348-350
Shoal formation 341
Simplex retina (rods only) 312
Single ovary 459
Sinus cephalicus 5
Sinus venosus 415, 416, 419, 420, 422
Size of eggs 19, 21

Skin 279, 280, 508
Small cones 84
Small cylindrical processes on egg surface 89
Sockeye 68
Sole 78, 80, 136, 142, 187, 227, 245, 314, 340, 341, 389
Solid CNS 241
Solid eye primordium 343
Solid intestine 260
Solid intestinal rod 438
Solid intestinal rod or folded cord 441
Solid lens 338
Solid medullary plate 242
Solid neural keel 244
Solid neural rod 239
Solid paired primordium of heart 420
Solid postanal gut 259, 261, 265, 440
Somatopleura 220, 377, 417, 472, 494
Somite 220, 226, 235, 250, 253, 377-380
Source for ZR outside ovary 72
Spawning Female-specific Substance (see SF)
Sperm 131-146, 507
  Anacrosomal 135
  Aquasperm 131, 134
  Biflagellarity of sperm 144
  Capacity of sperm 157
  Duct 379
  Entry site (SES) 159, 160, 161, 163, 182
  Entry through micropyle 158, 164
  Extrusion of supernumerary sperms 162
  Fin of sperm tail 144-146
  Flagellum 134, 141, 143-146
  Guidance role 157
  Guidance system 107
  IMP (intramembraneous particles)

Inhibitor of polyspermy 166
Intramembraneous particles (IMP) on head of sperms 134, 135
Intrasperm 131, 134
Length 131, 144
Motility of sperms 146, 155
Polyspermy 163
Prevention of polyspermy 176
Spermatozoon 146, 491
Sperm-egg fusion 165
Spermiogenesis 140
Sperm-stimulating factor 156
Storage of sperm in ovarial folds 458, 492, 502
Supernumerary sperms 163, 164, 166
Tubular anal fin transfers sperm 481
Spermatophore 134, 146, 481, 491, 492
Spermatozeugma 134, 146, 481, 492, 494
Spermatozoon 146, 491
Spherical granules in RPE 311, 342
Spherules (rod terminals) 326, 353, 354, 359, 360, 370
Spikes 87, 107
Spinules 354, 356, 360, 631
Spiracle 282, 283
Splanchnopleura 220, 377, 384, 417, 420
Spleen 386, 403, 411, 413, 414, 437, 498, 499
Square mosaic of cones 319, 320, 325, 333-336, 348, 354, 359, 360
Stageing index 210
Stalk 87, 102
Static organ 268
Stato-acoustic system 274
Statoliths 268, 271
Steelhead trout 115
Stickleback 12, 33, 85, 107, 199, 273

Stomach 435, 439, 440
Storage of sperm in the ovarial folds 502
Stratum germinativum 279
Structures for floating and attachment 77, 103 108
Subgerminal cavity 190, 210
Subnotochordal rod 226, 249
Subretinal space 337
Subsurface cisternae (membranes) 316, 344, 345, 347, 348, 354
Superembryonation 458, 469
Superfoetation 469, 470, 472, 486, 494
Supernumerary sperms 163, 164, 166
Surface appendices of envelope 82
Surface projections of blastomeres 196
Surfperches 491
Suspensory ligament 87, 107
Sweetfish 135
Swim bladder 431, 432, 439
Swim up stage 288, 496
Switch from larval to mature haemoglobins 413
Sympathetic neurons 276, 277
Synapse 347
Synaptic contacts between photoreceptors and horizontal and bipolar cell dendrites 354
Syncytial layer (see periblastl)
Synthesis of vitellogenin 33
Synthesis of ZR by follicle 72
Synthesis of ZR by oocyte 70

T

T4 and T3 (thyroglobulins) 437
T4 or T3 delay secretion of hatching enzymes 437
Tail 131-134, 146, 232, 258
Tailbud 232, 234, 236, 247, 265, 472

Tailbud region 239
Tailbud-blastema 240
Tail filament 143
Tapetum lucidum. 339
Tectum opticum 268
Telencephalon 269, 270
Telescopic eye 305
Telodendrial contacts between rod and cone terminals 354, 360
Telolecithal 39, 148, 178, 507
TEM (transmission electron microscopy) of yolk 29
Tench 318
Tendrils 83, 91
Tertiary egg membranes 51
Tetrachromatic vision 318
Perivitelline space following fertilization 172
Theca folliculi 92
Third eye (see pineal body)
Threads 101, 102
Thyroid
   Heterotopic tissue 437
   Scattered follicles 437
   Stimulation by the pituitary 437
   Thyroglobulin (T3 and T4) 437
   Thyroid hormones 437
   Thyroxine 332
   Walls of follicles composed of a secretory epithelium 437
Tiered cones 305, 326, 353
Time-lapse filming of involution 230
Transcription factors GATA-1 and GATA-2 394
Transparency of live eggs of pelagic species 26
Triad of photoreceptor synapse 366
Trichromatic vision 318
Triggering of vitellogenin 33
Triglyceride fat 24
Triple cone 319
Trophoderm 478
Trophodermal matrotrophy 471
Trophodermic skin 470

Trophodermy 503
Trophotaeneal placenta 482, 503
Trophotaeniae 460, 481-490, 502, 503
Trout 20, 23, 31, 33, 41, 52, 56, 61, 69, 111-113, 150-152, 155, 158, 160, 175, 189, 192, 193, 198, 204, 214, 226, 229, 233, 235, 240-243, 249, 252, 253, 255, 259, 261, 264, 272, 273, 281, 282, 293, 294, 298, 305, 316, 332, 335, 340-346, 348-353 355-360, 367, 372, 380, 382-388, 390, 405, 414, 419, 433, 434, 438, 441, 506
Tubular anal fin is used to transfer sperm 481
Tubulin 161
Tuft of microvilli 160, 165, 166
Tufts on egg surface 83, 90, 91, 94, 95, 97
Twin cones 314, 316, 319, 344
Two eye rotations 338, 361, 371
Two types of pigment granules in the REP 363
Types of eggs 10

U

UHG (unicellular hatching gland)
Ultrastructure of the oocytes 462
Ultraviolet light (see UV)
Unattended eggs 12
Unicellular hatching glands (see Hatching gland cells)
Uptake of embryotrophe by the gut 467
   Ingested blood cells 468
   Iodide 437
   Nutrients by vertical fins 503
   Oxygen 10
   Water 20, 21, 61, 150
Urinary bladder 259, 499
   Bilobed 496, 498, 503

Collapse 496, 497, 499
Development 497
Enlarged 467, 498
Emptied 496
Restored 501
Urogenital organ 481
Uterine milk 462
Utricolith 268, 271
UV (ultraviolet)
UV-cones 325, 332, 508
   Regression 334
UV-radiation 213
   Affects epiboly 213
   Damages microtubules 213
   Epidermal necrosis 334
   No effect on involution and convergence 230
   Receptors 318

V

Vacuolar yolk 46
Vacuomes 46
Valvula cerebelli 268, 269
Variation in number of ZR layers 67
Vascular bulbs 478
Vascular endothelial cells 400
Vascularized ovarian folds 503
VCL (visual cell layer)
Velum 272
Velum transversum 273
Venae cardinales posteriores 422
Vena caudalis 4, 398, 421-427, 497
Venae hepaticae 422
Vena subintestinalis 422
Ventral embryonic fissure 333, 337-339
Ventral fin rays 502
Ventral lip of blastopore 216
Ventral mesocardium 415, 417
Ventral mesoderm 253
Vertical fins highly vascularized 492
Very high density lipid (VHDL) 31
Very high density lipoprotein II (VHDLII) 31
Vesicula germinativa 147
Vestige of the archenteron 260
VHDL (very high density lipid)
VHDLII (very high density lipoprotein II)
Villi 94, 474, 476
Villous follicular wall 474
Viscous layer 51, 75
Visual cells (see photoreceptors)
Visual cell layer (VCL) 310, 359, 365
Visual pigments 318
Vitamin A 24
Vitelline circulation 423, 428, 471
Vitellogenesis 24, 26-28 34, 39, 60
Vitellogenin 23, 24, 31-34, 39, 74
   Beta components 31, 32
   Liver vitellogenin 35
   Plasma vitellogenin 32
   Triggering of vitellogenin 33
   Vitellogenin receptors 32
   Vitellogenin synthesis 35
   Vitellogenin uptake 33
Vitellus (yolk)
Viviparity 16, 21, 77, 108, 457-504, 506, 507
   Absorbed blood cells by rectum 465
   Absorption of embryotrophe by epidermis of ventral yolksac and pericardium) 470
   Absorptive trophotaenial epithelium 486
   Adelphophagy 460, 491
   Allantois 496, 498, 499
   Amnion 474
   Anal fins modified into gonopodium 507
   Atrophy of postanal gut 440
   Belly sac 472-474, 478, 479
   Birth 496, 501
   Buccal placenta 491, 503
   Chloride cells 475, 500

Copulatory organ 461, 491, 507
Embryotrophe 463, 470
Embryotrophe/histiotrophe 462, 492
Endocardial tube 494
Endocytosis of follicle cells 477
Enlarged pericardium 503
Enterotrophic uptake 460, 503
Epithelio-chorial pseudoplacenta 498
Follicular development 458
Follicular gestation 458, 494
Follicular placenta 471, 476-478, 494
Follicular protuberance 462
Follicular villi 474-476, 480
Formation and regression of pericardial hood 495
Foureyed fishes 477
Funiculus 461
Gonoduct 458, 459, 487
Gonopodium 481, 482,
Gonopore 459
Guppy 29, 58, 286, 306, 308, 315-318, 320-324, 326, 337, 339-349, 352-354, 357, 359, 360, 370, 374, 404, 406, 408, 409, 411, 494, 496, 499, 501
Gut 474
Halfbeaks 492, 493
Hermaphrodites 471
Hindgut food uptake 469, 479, 485, 493, 503,
Hood 503
Hypertrophied hindgut 468, 481, 492, 502
Hypertrophied midgut 481
Hypertrophied vertical finfold 491
Intestinal epithelium organized into villi 467
Intestinal nutrient transfer 471
Intrafollicular development 458, 470, 492, 502
Intralumenal development 458, 502
Intrauterine embryonic cannibalism 460
Intromittent organ for copulation 469
Male anal fin remodelled 458
Male copulatory organ 502
Maternal embryonic nutrient transfer 480
Maternal immune system 457
Maternal-embryonic relationship 458
Matrotrophic embryonic development 458, 471, 472, 486, 492, 502, 503
Matrotrophic species 460
Mexican topminnows 481
Microplicae of embryonic epidermis 475
Microridges of embryonic epidermis 471, 500
Midgut, elongated and coiled 478
Mixed feeding (yolk and embryotrophe) 466
Modifications of anal fin 493, 507
Modified hindgut indicates oral uptake of food 465
Moribund eggs as nutrient source 493
Neckstrap 472, 496, 497, 503
Nestfleer 501
Nidifugous 496
Number of embryos dependent on age of mother 461
Nutrient uptake by gut 470
Nutrients uptake from ovarian fluid 489
Obligate internal bearers 460
Oocytes, ultrastructure 462
Parturition 496,
Pericardial hood 471, 472, 474, 484, 494-496
Pericardial sac 472, 474, 481, 484, 485, 488, 500, 502

Pericardial trophoderm 502
Pericardium 472-474, 484, 503
Perivitelline fluid, oral uptake 496
Placenta 476, 498
Pseudoplacenta 470, 471, 478, 494, 500, 503
Rectum takes up food 470
Rectum with deep folds 465, 466
Respiration shifted to gills 484
Respiratory function of carotenoids 494
Scorpionfishes 492
Somatopleura 472, 494
Spermatophores 481, 491, 492
Spermatozeugmata 481, 492
Sperms stored in ovary 458
Sperm transfer by tubular anal fin 481
Superembryonation 458, 469
Superfoetation 469, 470, 472, 486, 494
Theca 82
Trophoderm 478
Trophodermal matrotrophy 471
Trophodermic skin 470
Trophodermy 503
Trophotaeniae 460, 481-490, 502, 503
Trophotaenial placenta 482, 503
Urinary bladder
  Collapsed 496, 497
  Development 497
  Greatly enlarged 467, 498
  Restored 501
Urogenital organ 481
Uterine milk 462
Vertical fins, highly vascularized 492
Vertical fins take up nutrients 503
Yolksac 502
Yolksac bulbs 480
Yolkvessels, devoid of cellular walls 422

W

Wandering cells from the ICM unto the yolksac 401
Warts 102
Water entry into unfertilized egg 149
Water uptake 20, 21, 62
Waves of contraction after fertilization 177
Weberian ossicles 268
Wolffian duct 252

Y

YCL (yolk cytoplasmic layer)
Yelk-hypoblast 384
Yolk (vitellus) 23-40, 507
  Absorption 506
  Absorption by liver 444
  Absorption by vitelline circulation 444
  Blood islands and ICM 382
  Carbohydrate yolk 39, 46
  Cell 24, 211
  Cellular layer 189
  Centrolecithal 39, 148, 178
  Cytoplasmic layer 187, 188, 209, 211
  Dense 25, 26
  Distinct from intestinal tube 432
  Endogenous formation (intraovarian, autosynthetic) of proteid yolk 30, 35, 36
  Entodermal absorption 452
  Food, mixed with embryotrophe) 466
  Formation of yolk 27-37
  Genesis of lipid (fatty) yolk 29
  Globules 26, 32
  Intravesicular 24, 35, 46
  Lipid yolk 24, 28, 39, 507
  Lipid yolk formation 29
  Macrolecithal 36

Massed (proteid) yolk 27
Mixed with cytoplasm 36, 194
Nuclei 198
Platelets 26, 37
Precursor of protein yolk 30, 39
Precursors 37
Proteid yolk 28, 36, 39, 507
Proteid yolk dual origin (exo- and endogenous) 35
Relation between yolk and cytoplasm 36
Resorbed by vitelline circulation 444
Separation from intestine 443
Source of the mesoderm-inducing signal 377
Surface 382, 406
Telolecithal 39, 148, 178, 507
TEM of yolk 29
Transfer 204
Uptake of yolk 442, 509
Uptake by the liver 452
Vacuolar yolk 46
Vesicles 35, 36, 46
Whole yolk surface 390
Yolk syncytial layer (YSL)
Yolksac 236, 280, 502
  Bulbs 480
  Blood islands 381, 394, 397, 398, 403, 405, 409 509
  Capillary network 403
  Chloride cells with accessory cells 281, 284
  Circulation 232, 421, 428
  Envelope 280
  Formation of the yolksac (epiboly and involution) 231
  Portal system. 428
  Vessels 418, 422
  Wandering cells on the yolksac 401
  Yolksac-liver contact 25-27, 205, 428, 442-446
YSL (yolk syncytial layer, periblast)

Z

Zebrafish 1, 3, 39, 41- 44, 46, 51, 65, 70, 75, 107, 160, 168, 171, 179, 182, 194, 195, 198, 202-204, 207-210, 213-217, 220, 226, 228, 229, 246-249, 257, 258, 265, 267, 268, 272, 273, 276, 298, 312, 326, 335, 339, 340, 360-369, 371, 375, 377, 392-396, 409, 412, 415, 419, 426, 428, 431, 435, 442
Zona pellucida 28, 58, 71 75
Zona perforata 52
Zona radiata (ZR) 51-54, 57, 58, 61, 62, 66, 74, 75, 82, 92, 93, 104, 105, 108, 127, 507
  Chemical composition 69
  Hardening 61, 75
  Proteins 74
  Cyprinodont type 61, 63, 65, 75
Salmonid type 61, 63, 75
Zona radiata externa 37, 62, 69, 75, 82, 86, 101, 110, 115, 119, 126, 128, 129, 149, 164, 165, 298
Zona radiata interna 37, 63, 69, 75, 82, 86, 110, 115, 120, 124, 126, 129, 149, 163-165, 175, 298
ZR (Zona radiata)
ZRE (Zona radiata externa)
ZRI (Zona radiata interna)

Fish and Fisheries Series

1. R.J. Wootton: *Ecology of Teleost Fishes*. ISBN 0-412-31730-3
2. M.A. Keenleyside: *Cichlid Fishes Behaviour.* Ecology and Evolution.
   ISBN 0-412-32200-5
3. I.J. Winfield and J.S. Nelson (eds.): *Cyprinid Fishes Systematics*. Biology and Exploitation. ISBN 0-412-34920-5
4. E. Kamler: *Early Life History of Fish*. An Energetics Approach. ISBN 0-412-33710-X
5. D.N. MacLennan and E.J. Simmonds (eds.): *Fisheries Acoustics*.
   ISBN 0-412-33060-1
6. T.J. Hara: *Fish Chemoreception*. ISBN 0-412-35140-4
7. T.J. Pitcher: *Behaviour of Teleost Fishes*. Second Edition.
   PB; ISBN 0-412-42940; HB; ISBN 0-412-42930-3
8. C.R. Purdom: *Genetics and Fish Breeding*. ISBN 0-412-33040-7
9. J.C. Rankin and F.B. Jensen (eds.): *Fish Ecophysiology*. ISBN 0-412-45920-5
10. J.J. Videler: *Fish Swimming*. ISBN 0-412-40860-0
11. R.J.H. Beverton and S.J. Holt (eds.): *On the Dynamics of Exploited Fish Populations*.
    ISBN 0-412-54960-3
12. G.D. Pickett and M.G. Pawson (eds.): *Sea Bass*. ISBN 0-412-40090-1
13. M. Jobling: *Fish Bioenergetics*. ISBN 0-412-58090-X
14. D. Pauly: *On the Sex of Fish and the Gender of Scientists*. A Collection of essays in fisheries sciences. ISBN 0-412-59540-0
15. J. Alheit and T. Pitcher (eds.): *Hake*. Fisheries, Products and Markets.
    ISBN 0-412-57350-4
16. M. Jobling: *Environmental Biology of Fishes*. ISBN 0-412-58080-2
17. P. Moller: *Electric Fishes*. History and Behavior. ISBN 0-412-37380-7
18. T.J. Pitcher and P.J.B. Hart (eds.): *The Impact of Species Changes in African Lakes*.
    ISBN 0-412-55050-4
19. J.F. Craig: *Pike*. Biology and Exploitation. ISBN 0-412-42960-8
20. N.V.C. Polunin and C.M. Roberts (eds.): *Reef Fisheries*. ISBN 0-412-60110-9
21. R.C. Chambers and E.A. Trippel (eds.): *Early Life History and Recruitment in Fish Populations*. ISBN 0-412-64190-9
22. S.J.M. Blaber: *Fish and Fisheries in Tropical Estuaries*. ISBN 0-412-78500-5
23. T.J. Pitcher, P.J.B. Hart and D. Pauly (eds.): *Reinventing Fisheries Management*. 1999
    ISBN 0-7923-5777-9

Fish and Fisheries Series

24. R.J. Wootton: *Ecology of Teleost Fishes.* Second Edition. 1999
    PB; ISBN 0-412-64200-X; HB; ISBN 0-412-84590-3
25. M.C.M. Beveridge and B.J. McAndrew (eds.): *Tilapias: Biology and Exploitation.* 2000
    PB; ISBN 0-7923-6391-4 HB; ISBN 0-412-80090-X

KLUWER ACADEMIC PUBLISHERS – BOSTON / DORDRECHT / LONDON